# Data Structures

# Little, Brown Computer Systems Series

## Gerald M. Weinberg, *Editor*

# Data Structures

Edward M. Reingold
University of Illinois at Urbana-Champaign

Wilfred J. Hansen
Carnegie-Mellon University

 Little, Brown and Company
Boston   Toronto

Library of Congress Cataloging in Publication Data

Reingold, Edward M., 1945-
    Data structures.

    Includes bibliographies.
    1. Data structures (Computer science)
I. Hansen, Wilfred J.   II. Title.
QA76.9.D35R44   1983        001.64'2        82-12697
ISBN 0-316-73951-0

Library of Congress Catalog Card No. 82-12697

ISBN 0-316-73951-0

9  8  7  6  5  4

MV

Published simultaneously in Canada by Little, Brown & Company (Canada) Limited

Printed in the United States of America

The cover is a reproduction of the first panel of the triptych *Three Trees* by Karel Appel, reproduced by permission of Karel Appel.


# Acknowledgment

The authors gratefully acknowledge permission to use material from Reingold/Nievergelt/Deo, *Combinatorial Algorithms*: *Theory and Practice*, © 1977, pp. 280–315. Reprinted by permission of Prentice-Hall, Inc., Englewood Cliffs, N.J.

To our children, without whom
this book would not have been necessary

# Preface

I saw, when at his word the formless mass,
This world's material mould, came to a heap:
Confusion heard his voice, and wild uproar
Stood ruled, stood vast infinitude confin'd;
Till at his second bidding darkness fled,
Light shone, and order from disorder sprung.

*Paradise Lost,* John Milton

**A**ll information processed by a computer is ultimately encoded as a sequence of bits; the specialized field of data structures considers how to impose order and structure on those bits so that the encoded information is readily available and easy to manipulate. This field thus includes the design, implementation, and analysis of structures and techniques for information processing at all levels of complexity—from individual bits, characters, and words to aggregates such as records and files, and from abstract structures such as stacks, trees, and graphs to algorithms for searching, sorting, and storage management.

Data structures are central to computer science in general and to the discipline of programming in particular. In the more analytic areas of computer science, appropriate data structures have often been the key to significant advances in the design of algorithms. Their role in programming is no less profound: in most cases, once the appropriate data structures are carefully defined, all that remains to be done is routine coding. A comprehensive understanding of data structure techniques is thus essential in the design of algorithms and programs for all but the simplest applications.

Where there is such practical importance, college courses and textbooks are sure to follow. Since the publication of *Curriculum 68* by the Association for Computing Machinery in 1968, a course in data structures has become a core requirement in virtually every undergraduate and graduate program in computer science. A number of texts have appeared, which by now seem outdated or inadequate. Moreover, in our teaching we have adopted a number of approaches as preferable to those in most texts. In the present text, we have assembled a core of material that is unlikely to be supplanted or revised by further research.

**Organization.** The chapters are organized in increasing degree of complexity and abstraction, so each can be based on earlier ones. Throughout the book, for each abstract structure we emphasize its conceptual identity as a set of operations and its possible implementations in terms of the lower level structures already discussed.

Chapter 1 introduces the algorithmic and mathematical notations we employ throughout the book by discussing a sample table search problem. This discussion also serves to show the reader the scope of the techniques presented in later chapters. Chapter 2 discusses elementary data objects at the machine level—integers, characters, and so on—and how they are represented in bits (in some curricula this material is covered in a different course; if so, it can be freely omitted without interfering with later chapters). Chapter 3 then considers primitive data structures composed of aggregates of primitive objects. It shows how structures such as arrays, records, and pointers are represented in machines and in typical high-level programming languages.

Building on this basis, Chapter 4 presents material on lists, their various implementations, and the applications to, for example, stacks and queues, sparse matrix representation, and graph representation. Chapter 5 discusses trees in similar fashion: implementations and applications. These two chapters form the core of the course, presenting between them the most important tools in the design of data structures.

The next three chapters cover various more specialized problems that have wide applicability. Chapter 6 examines the techniques used to allocate and deallocate storage, Chapter 7 examines the organization of data for efficient search, and Chapter 8 examines techniques for sorting.

The material in this book is more than sufficient for a one-semester course in data structures; we have provided enough to fill a two quarter course. By choosing only the first five chapters and selected material from the rest, the instructor could cover most important topics in a single quarter, while a semester would allow the inclusion of important additional topics. A two-semester course might include discussions of the more important exercises as well as outside reading to supplement their exposition.

**Presentation.** Our presentation is unique in several ways. We present only a carefully chosen fraction of the material available but supplement it with a wide variety of exercises, many of which lead the student to discover interesting alternatives. The more complex exercises and those requiring advanced techniques are marked with a ★.

No single book or course can successfully discuss all known data structures and algorithms; far too many minor variants have been devised for special purposes. Rather than an encyclopedic catalog, we present the *art* of designing data structures to prepare the student to devise his own special-purpose structures for problems he will encounter.

Examples illustrating the techniques presented have been selected from many different application areas, in order to indicate the importance and ubiquity of data structures. In selecting examples and applications for presentation, we have taken care to keep the presentations self-contained and to avoid undue digressions.

Our presentation is machine and programming language independent. We use **if-then-else, while-do, repeat-until**, and **case** for flow of control, and we have chosen a functional notation for reference to the fields of a record. In this way our notation is readily understood and unambiguous, but unencumbered by the syntax of any specific language.

The presentation has been organized to be clear and interesting to both undergraduate and graduate students. The material covered is accessible to students who have completed an introductory programming course.

We recognize that the student must eventually be able to choose among implementations on the basis of the analyses of the behavior of the corresponding algorithms, but it is beyond the scope of this book to teach any but a few of the basic mathematical techniques of algorithm analysis. We have skirted this issue in part by giving brief sketches of the methods of analysis for certain key algorithms, but mainly by just summarizing the results of analysis for most algorithms. Where mathematical arguments have been unavoidable, we have emphasized the intuition behind the argument.

We have not cluttered the presentation with involved discussions of the origins of the various techniques, except where such discussions are necessary to put the material in proper perspective. The annotated bibliography that concludes each chapter provides sources for students interested in deeper treatments of the topics.

## ACKNOWLEDGMENTS

> Nothing corrupts a man as deeply as writing a book,
> the myriad of temptations are overpowering.
>
> Nero Wolfe,    in John McAleer, *Rex Stout: A Biography*

No good textbook can be written in a vacuum. Without the comments and suggestions of critical readers, the authors' impatience would introduce many errors and careless presentations.

We are fortunate to have had the benefit of comments from a number of critical readers, some voluntary (colleagues and reviewers) and some involuntary (students). All added immeasurably to this book, but it is with special gratitude that we thank Amitava Bagchi, Marcia Brown, and Nachum Dershowitz for the time they spent in looking at the manuscript and making comments about it.

A very special acknowledgment is due to John S. Tilford who, in writing the very complete solution manual available for this text, suggested innumerable improvements in the presentation.

E.M.R.
W.J.H.

## TO THE READER

Si qua videbuntur chartis tibi, lector, in istis
    sive obscura nimis sive Latina parum,
non meus est error: nocuit librarius illis
    dum properat versus adnumerare tibi.
quod si non illum sed me peccasse putabis,
    tunc ego te credam cordis habere nihil.
"Ista tamen mala sunt." quasi nos manifesta negemus!
    haec mala sunt, sed tu non meliora facis.

Martial, *Epigrams, II, viii*

# Contents

# Chapter 1

# Concepts
# and Examples

So she went on, wondering more and more at
every step, as everything turned into a tree the moment
she came up to it.

*Through the Looking Glass*,   Lewis Carroll

**D**ata structures represent *data* and relationships among data for manipulation by computer systems; *data bases* are an important class of computer system. To distinguish these three terms and to establish the context for our study of data structures, this section briefly explores the topics of data and data bases.

*Data* are considerably more than just a collection of numbers; each number must be *interpreted* as part of a *model*. For example, the values 153 and 169 are meaningless until described as the 1970 and 1980 counts of the number of United States cities with population over 100,000. To a company that sells to large cities, these values help define the market and its possible growth, but only when they are interpreted as part of a model of cities. In general, a model has *representations* of *objects*. In typical models the objects might be physical objects, persons, events, or abstract concepts. For the purpose of the model, certain *attributes* of each object are selected and the value for each is recorded. The collection of attribute values is then the representation of the object.

The usual purpose of a model is to analyze the behavior of the objects represented. For instance, models of wing sections of aircraft are evaluated in wind tunnels to determine whether the corresponding real wing will fly without turbulence; models of the economy are evaluated to determine the effect of possible changes in the money supply. Because models do not reflect all properties of the real object, care must be taken in evaluating the results of computations with the model. The models of the wings or the economy may predict good behavior, but disaster can occur in reality if crucial parameters have been omitted or misrepresented.

A computer data base is one common form of model. The simplest data bases have one *record* for each object, and each record contains values corresponding to the object's various attributes. The data base itself is a model of the entire collection of objects, while each record is a model of a specific object. The term "record" derives from manual filing systems where, say, a personnel file will have one folder (record) for each employee (object). Each captioned space of a form in the folder is an attribute of the employee.

Whether manual or mechanized, the personnel data base solves both *periodic* and *one-time* problems. One periodic problem is that of paying all employees each week; this requires a sequential process that goes through the entire data base writing checks. A possible one-time problem is that of finding an employee with the skill for some special task—for example, guiding a visitor who speaks Mandarin Chinese. Such a request for information from a data base is called a *query*. It may be met by scanning the entire file, but if an appropriate index exists, only a fraction of the records need be examined.

Large corporations maintain data bases to model many aspects of their business. Human relations are covered by records on employees, departments, suppliers, customers, and stockholders. The work is represented by files on plans, designs, warehouse contents (inventory), and projects in progress. The financial picture is given by budget, accounts payable, accounts receivable, general ledger, assets, and liabilities. As with all data bases, there are many relationships among these files. An employee is in a department, is working on projects, is accounted for by a budget line, may be a customer, and may be a stockholder. A project is performed by employees, may have budget items, requires material from suppliers, and may be slated for delivery to some customer. Typical contents of corporate files are shown in Table 1.1, and possible relationships are illustrated in Figure 1.1.

Table 1.1 shows two visits on April 1. In the first, Willie Clout proposed purchase of a software package for sorting. In the second, Al Bennet, software salesman, presented the benefits and costs of enhanced software for the visitor's log. From the budget data it seems the log has higher priority than sorting.

With on-line data base systems a manager can pose *queries* and get answers quickly. For example, suppose the president is impressed by Clout's sort presentation. The decision-making process might involve several queries, perhaps starting with

> Find *budget_item* with name containing "sort."
> Show *amount_budgeted*.

In response to this query the system searches the data base, or part of it, to find the appropriate record(s).

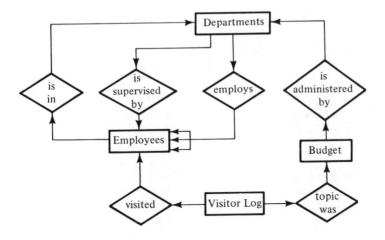

**Figure 1.1**
Relationships among the data bases of a corporation. Rectangles are data bases; diamonds are relationships between them. Items in the data base at the tail of an arrow refer to items in the data base at the head of the arrow. For example, the upper right diamond represents the fact that each budget item lists the department responsible for administering that item. The multiheaded arrow indicates that departments employ more than one person.

Such search problems are quintessential to computer science; they are the principal focus of the study of data structures. In the remainder of this chapter we will use a simple search problem together with two elementary solutions to illustrate our notation and important aspects of algorithmic analysis. More difficult search problems and intricate solutions are presented in Chapter 7. It is often necessary to sort data into order (for example, alphabetical or numerical) to accomodate searching it; sorting algorithms are discussed in Chapter 8. To facilitate searching and sorting, data are stored in lists and trees, presented in Chapters 4 and 5, respectively. In turn, the manipulation of lists and trees requires that memory be organized for rapid allocation. Such memory management techniques are treated in Chapter 6. Chapters 2 and 3 present the basic components from which lists, trees, and other structures are composed.

**Exercises**

Complex exercises and those requiring advanced techniques are keyed ★.

1. Typical automated weather forecast systems subdivide a region and represent weather conditions in each sector. What data attributes might be chosen for a sector? What is a reasonable size for a sector? With your above choices, how many data values must be stored for the $10^8$ square miles of the earth?

**Departments**

| No. | Name | Supervisor | Personnel |
|-----|------|------------|-----------|
| 37 | Information Systems | (R. S. Teague) | (P. Larson,...) |
| 52 | Executive Suite | (H. R. Ahner) | (R. Stocks,...) |

**Budget**

| No. | Name | Dept. | Amount | Increase (%) |
|-----|------|-------|--------|--------------|
| 37–291.5 | Sort Package | (37) | 4750 | 4 |
| 52–153 | Visitor Log | (52) | 7000 | 15 |

**Employees**

| No. | Name | Dept. | Salary | Hire Date |
|-----|------|-------|--------|-----------|
| 1728 | H. R. Ahner | (52) | 65,200 | 5/28/79 |
| 1967 | R. S. Teague | (37) | 54,360 | 4/11/67 |
| 2053 | P. Larson | (37) | 13,200 | 10/30/81 |
| 2271 | R. Stocks | (52) | 25,600 | 1/1/82 |

**Visitor Log**

| Date | Time | Min. | Visited | Visitor | Budget Item | Topic |
|------|------|------|---------|---------|-------------|-------|
| 4/1/84 | 11:30 | 13 | (Ahner) | W. Clout | (37-291.5) | Better sort |
| 4/1/84 | 11:55 | 86 | (Ahner) | A. Bennet | (52-153) | Improved log |

**Table 1.1**

Data bases for a corporation. The attributes shown are only a selection of the much larger number of attributes in the file. A value in parentheses is a reference to a record in another file; it may be just the value shown, or it may be an encoding of the location of the appropriate record.

2. Sometimes it is difficult to distinguish an object from its representations. For instance, a computer program may have many different physical representations. List at least five and for each, argue that it, rather than the others, is the "program" itself.

3. Each record in a payroll file represents a person as its object. What attributes might objects in this file have?

4. What steps might a system take to respond to a query from the president for all budget changes that were above average in value?

5. The visitor log data base contains a reference to the budget item discussed during the visit. This might be implemented as the number of the budget item or as the *address* that describes where the budget item is stored. Each solution has an advantage over the other; give these advantages. Consider speed of operation and the difficulty of changing the data base.

6. What data bases might a college or university implement? What are some queries that might be generated to work on the following problems?

   (a) How can we give raises when total income has not risen?

(b) Should we limit the number of faculty given tenure?

(c) Have academic standards changed? Is there grade inflation or deflation or has the quality of the students changed?

(d) Shall we build a new building for the Computer Science Department?

## 1.1   DATA STRUCTURES

A data structure can be defined informally as an organized collection of values and a set of operations on them. For the integer data structure, the operations are just addition, multiplication, and so on; for a data base, the operations may include complex queries. But most data structures fall between these extremes: an array has storage and retrieval via subscripts; a table of names has the operations of insertion and search.

Somewhat more formally, we define a data structure to have three components:

A set of *function definitions*: each function is an operation available to the rest of the program.

A *storage structure* specifies classes of values, collections of variables, and relations between variables as necessary to implement the functions.

A set of algorithms, one for each function: each algorithm examines and modifies the storage structure to achieve the result defined for the corresponding function.

The function definitions separate the implementation of the data structure from the construction of the rest of the program. They define the externally observable behavior of the data structure, while the storage structures and algorithms are the internal details. The latter can be changed without modifying routines that use these functions.

One simple example of a data structure is an array. As an introduction, we will define an array of twenty **real** *values* with **integer** *subscripts*. The set of functions is

*store*: Must be given a value and a subscript. Associates the value with that particular subscript.

*retrieve*: Must be given a subscript. Reports back the value most recently associated with that subscript.

As these descriptions show, functions must sometimes be given values. We can also say the values are required by the function or are its *arguments*. A function may report back or *return* a value, as does *retrieve*. A function may also communicate back a value by assigning it to a variable.

Chapter 3 will present a reasonable implementation for arrays; for this example, however, we unreasonably choose a storage structure consisting of twenty **real** variables:

$$A1, A2, A3, A4, A5, A6, A7, A8, A9, A10, A11,$$

$$A12, A13, A14, A15, A16, A17, A18, A19, A20 : \textbf{real}$$

⟦locations for array; $Ai$ will contain the
value associated with subscript value $i$⟧

With this storage structure, the algorithms $A$ for $F$ are somewhat tedious:

⟦store *value* with *subscript*⟧
    **if** *subscript* $= 1$ **then** $A1 \leftarrow value$
    **else if** *subscript* $= 2$ **then** $A2 \leftarrow value$
    **else if** *subscript* $= 3$ **then** $A3 \leftarrow value$
    $\cdots$
    **else if** *subscript* $= 20$ **then** $A20 \leftarrow value$
    **else** *ERROR*

⟦retrieve *value* for *subscript*⟧
    **if** *subscript* $= 1$ **then** *value* $\leftarrow A1$
    **else if** *subscript* $= 2$ **then** *value* $\leftarrow A2$
    $\cdots$
    **else if** *subscript* $= 20$ **then** *value* $\leftarrow A20$
    **else** *ERROR*

where *ERROR* performs some appropriate action.

The class of values a variable may take on is called its *domain*. Programming languages commonly provide primitive domains such as **integer**, **real**, **character**, and **boolean** (see Chapter 2). They also provide nonprimitive values such as arrays, records, and pointers (see Chapter 3). In the array above, *value* can take on any **real** value, but the domain of *subscript* is only the **integer** values 1 through 20. In defining a storage structure, it is common to specify such specialized domains.

Examples in the remainder of this chapter will explore a simple data structure for a *table*. Such a structure stores some number of *elements* by storing a set of attribute values for each. The most common operations are insertion of a new element and search to find an element with a given value for one of the attributes. Usually one attribute, such as name or part number, is the most common search target. This attribute is called the *key* and is specially treated.

Tables are an important class of data structures because they have many diverse applications. We will cover them in depth in Chapter 7, discussing various alternative storage structures. For the moment, however, we will implement a table as an array with the subscript as the key. In a budget data base, each budget line might be an element, the budget item number might be the key, and other attributes might include the name of the item, the dollar amount allocated, and the responsible department. For other applications see Exercises 4 and 5.

A specification for the table data structure must include the domain of each attribute. The table could then be defined as having a retrieve function for each attribute. This function would be given the value of a key, it would search for the element with that key and return the corresponding variable for the specified attribute. Thus in a table of budget items the operation to add 1000 to the *amount_allocated* attribute of the item whose number is in *budg_item* would be

$$amount\_allocated(budg\_item) \leftarrow amount\_allocated(budg\_item) + 1000$$

The drawback to this approach is that the search for *amount_allocated(budg_item)* is slow and this simple code fragment would require this same search twice. If several attributes of *budg_item* were accessed, the search would be repeated for each.

To reduce the number of searches, we introduce a domain *element_location* whose values designate the location of elements. When an element is first introduced to the system, a function call will enter it into the table and assign it an *element_location* value. Then whenever a particular element is required, another function is called to search for that element's key and return its *element_location*. This is then supplied as an argument to the attribute functions, which can use it to rapidly access the appropriate attribute.

The set of functions for our table can be described thus:

*initialize_table*: Sets a table to have no entries. Must be called before any other table function.

*enter_element*: Given a key value, *enter_element* creates a table element with that value for the key attribute; returns the *element_location* of the new element. Signals an error if an existing element has the same key.

*find_element*: Given a key value, *find_element* returns the *element_location* of the corresponding element. Returns an error value if there is no such element.

*attribute_name*: Given an *element_location*, returns the variable corresponding to the attribute of the element whose location is given.

There is one such function for each attribute. A value can be assigned to the returned variable, or its current value can be accessed.

At least three errors can be committed by calls to the corresponding algorithms. For the array example, errors were treated by calling *ERROR*, which may do as little as just halting the program. To provide graceful error recovery with our table functions, our algorithms should return special values for errors; these values will be constants, but they will be given the following names:

*table_full*: This value is returned by *enter_element* when there is no room to insert the new identifier in the table.

*duplicate_key*: This value indicates that an attempt was made to enter a key that was already in the table. This error can occur even if the attempt would otherwise produce a *table_full* error.

*key_absent*: This value indicates that *find_element* did not find an element with *key_value*.

Note that the algorithms will not create two entries for the same *key_value*; this is reasonable because a subsequent *find_element* operation can return only one value.

### Exercises

1. Specify a set of functions that might be provided by a data structure for calendar dates. What storage structure would you propose?

2. Define a domain for dollar amounts and a suitable set of functions. The domain should not contain values with a fraction of a cent. Include a function that divides one dollar amount by another. What is the domain of its result?

3. For subscript value 20, the array *store* algorithm performs twenty **if** tests. Rewrite the algorithm so that no subscript value requires more than five tests. [*Hint*: Start with "**if** *subscript* ≤ 10 **then**".] What changes, if any, would be needed in the rest of the program after this revision?

4. Suppose a table contains descriptions of university courses. What attribute might be the key? List some other attributes that each table element might include.

5. Compilers and assemblers store the identifiers they encounter in a *symbol table* as they process a program. List some attributes a symbol might have. What functions should this data structure support? Can you suggest a storage structure?

6. For this problem you are to define further functions for the table data structure. (Specify domains, but do not write algorithms.)

   (a) Define a function to delete an element from the table.

   (b) Define two functions, one to take note of the current contents of the table and another to restore the contents to the situation at one of the times when note was taken of the contents. You will need a domain of values to represent the contents.

   (c) Define a function that will apply some procedure to all elements in the table.

   (d) Define a pair of functions that can be used to *traverse* the table. One is to produce the "first" element in the table and the other is to produce the "next" element following any given element. The order can be chosen arbitrarily, but successive calls of next must ultimately visit all elements.

   (e) Explain why the function *find_element* is not needed if the functions in (d) are implemented. Why is it preferable to have *find_element* anyway?

7. In the text, error conditions arising in table functions are signaled by special values in the *element_location* domain. Alternatively, each function could return an additional value in a separate domain. In addition to the already defined error conditions, this domain would also include a value to indicate successful completion. Produce a complete description of the domains and functions of a table data structure that employs this alternate approach.

8. In most higher-level languages, **integer** addition returns an **integer** result. However, it is possible to define an integer data structure where the value of one of the operands is modified to contain the result. Define this data structure.

9. In the table functions, the data structure "owns" the key attribute storage space in that it does not give this space to the calling routine but only reports the location of elements within it. Suppose that integer operations were similarly defined: an arithmetic operation would return not an integer value but an indication of where that value was stored. How does this data structure permit two interpretations of an assignment statement? For each, show the final result of executing the assignments $b \leftarrow 6$;  $a \leftarrow b$;  $b \leftarrow 5$;  $c \leftarrow a$.

## 1.2 NOTATION

Programming languages provide notations for both declarations and algorithms. Since our major concern in this text will be with algorithms, we will only sketch here some notation for declarations. Later in this section we will introduce our notation for algorithms. Examples will be in terms of the table data structure whose functions were defined above.

To describe a storage structure we must define how each domain will be represented, declare variables to store values, and specify relationships that are to hold between values during execution. For instance, the several storage structures implied by the twenty-element-array example must include specification of the domain of subscript values, declaration of the $Ai$ variables, and recognition of the relationship that the value in $Ai$ is associated with subscript value $i$.

The table data structure needs only a definition of the *element_location* domain. For our purposes, prose descriptions suffice to define how values in a domain are represented. If we are storing elements of the table in an array, the location of an element will be its subscript: An *element_location* value is an integer; it is a subscript into an array of elements.

After specifying domain representations, we next define variables for the data structure. It is important to distinguish between a variable and an identifier. A *variable* is a location in which to store a value during execution of a program; an *identifier* is a means to refer to this variable while writing the program. The relationship can be established by a declaration:

$$\text{variable-name}, \ldots, \text{variable-name: domain}$$

where each variable-name is an identifier.

Using the above form, if $x$ is to represent a variable capable of holding **real** values, we write

$$x: \textbf{real} \quad \llbracket \text{store unknown value} \rrbracket$$

and if $i, j$, and $k$ are to represent subscript variables, we write

$$i, j, k: \textbf{integer} \quad \llbracket \text{subscripts for array } A \rrbracket$$

In diagrams of memory, variables will be shown as rectangles preceded by the identifier and a colon. Thus the above declarations establish the contents of memory as

$$x: \boxed{\phantom{xxxxxxxxxxx}} \qquad i: \boxed{\phantom{xxxxxxxx}}$$
$$j: \boxed{\phantom{xxxxxxxxxxx}} \qquad k: \boxed{\phantom{xxxxxxxx}}$$

We are able to limit the storage structure for our table data structure to just an array containing the keys:

*table*: **array** [1 : *table_size*] **of** *key_value*
    ⟦stores keys that have been entered in the table⟧

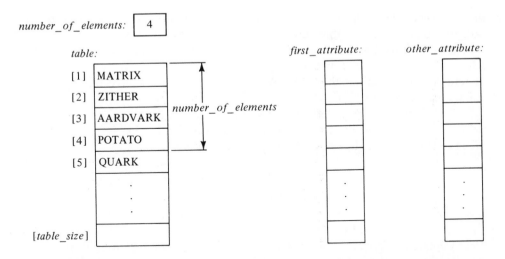

**Figure 1.2**
One possible storage structure for a table. Elements have been entered with the four keys MATRIX, ZITHER, AARDVARK, and POTATO. The value QUARK is in the array but is not in the table because it is outside the range 1, 2, ..., *number_of_elements*.

where *table_size* is some constant value. The domain specification after the colon produces an array of *key_values* with subscripts 1 through *table_size*. In addition to this array, we need a variable to keep track of how many locations in the array contain elements:

> *number_of_elements*: **integer**     ⟦valid *key_values* occupy
> the first *number_of_elements* positions of table⟧

To illustrate these definitions, Figure 1.2 shows the table after *enter_element* has been called for four different keys. Note that *number_of_elements* has the value four.

Also shown in Figure 1.2 is the implementation of the attribute functions. Attribute values are stored in arrays parallel to the table array. For the element whose key is in *table*[*k*], the attribute value with name *attr* is in location *attr*[*k*]. The arrays are parallel in the sense that when their diagrams are side-by-side, all attributes of any given element are in adjacent locations. Each attribute array would be declared something like this:

> *attr*: **array** [1 : *table_size*] **of integer**

The specification of the attribute functions in the previous section indicated that they returned a variable as their value. In many programming languages it is not possible to return a variable; however, with the declaration here, the attribute functions need no algorithms at all. The variable that is returned can just be the selected array location. (Of course, this allows an external routine to perform undefined operations on the table. Such operations are a source of bugs when storage structures are changed.)

The final problem in defining the storage structure for tables is to define the three *element_location* values for errors. One way to do this is to give them **integer** values that cannot possibly be subscripts of *table*:

$$table\_full = -1,$$
$$duplicate\_key = -2,$$
$$key\_absent = -3.$$

Next we extend *table* and the attribute arrays so they have locations corresponding to the error values. Then user routines can use these attribute values for some steps and defer the check for an error until perhaps a higher level of the software.

While defining variables and values, we must also define relationships among them. For example, one important relationship for our table is that *number_of_elements* cannot be greater than *table_size*. This relationship can be indicated by the boolean expression

$$0 \leqslant number\_of\_elements \leqslant table\_size$$

which also specifies that *number_of_elements* is not negative. Another relationship is that every valid *element_location* must refer to a table position that contains an element. To be precise, however, we must also accept the error values:

$$1 \leqslant element\_location \leqslant number\_of\_elements$$
$$\textbf{or } element\_location = table\_full$$
$$\textbf{or } element\_location = duplicate\_key$$
$$\textbf{or } element\_location = key\_absent$$

Such relationships specify *invariant conditions*—conditions that are to hold at all times during execution (except temporarily when the variables involved are being changed). As such, the set of relationships constitutes a "checklist"; when the value of one variable in a relationship is changed, then the values of all others related to it should probably change, as well.

Having discussed the definition of the function and storage structure components of data structures, we turn now to a notation for *algorithms*.

The term "algorithm" usually refers to the abstract technique used to deal with the storage structure, rather than to any specific implementation of that technique. To be part of a program, an algorithm must be expressed in some *notation* (programming language) and incorporated into the program. The algorithm may be a separate *subroutine*, or it may be a section of code to be written into the program wherever the data structure function is required. If it is a subroutine, it may be either a *procedure* or a *function*; the former does not return a value, but the latter does. The use of the term "function" to refer both to the abstract operations provided by a data structure and to a type of subroutine presents no difficulty because the two do not occur in the same context and, moreover, a function subroutine is usually an implementation of a data-structure function.

In this text, algorithms are expressed in a high-level notation similar to many programming languages. The statement forms used are summarized in Table 1.2. Statements are written on separate lines and generally have no punctuation between them. *When a statement is indented, it is controlled by the unindented statement above it.* Thus all statements within a subroutine are indented from the subroutine heading. Similarly the bodies are indented in **for**, **while**, and **repeat** loops and under **if** and **case**. ⟦Comments are written inside special comment brackets like this.⟧ Examples will appear below.

The algorithm for *initialize_table* need do no more than indicate there are no elements in the table:

$$number\_of\_elements \leftarrow 0 \qquad \llbracket\text{start with no elements}\rrbracket$$

This is an instance of an assignment statement. The expression on the right of the arrow is evaluated and the resulting value is placed in the variable named on the left. Expressions can have operators and operands and the latter can be variables, constants, function calls, and parenthesized expressions. The most common operators are shown in Table 1.3, but we will introduce others when appropriate.

The various operators have precedence relations, so that if an expression is written without parentheses, the order of evaluation is as in algebra. For instance, the expression $x + a * 3$ is understood to mean $x + (a * 3)$ rather than $(x + a) * 3$. In Table 1.3, operators with higher precedence levels are evaluated before those at lower levels. However, expressions inside parentheses are evaluated before the operators outside.

Unlike other operators, **and** and **or** evaluate only as many of their arguments as necessary. Thus the expression

$$i \leqslant number\_of\_elements \textbf{ and } table[i] \neq z$$

evaluates without error; the second condition is not tested when $i$ might be

| Category | Construct | Example |
|---|---|---|
| Routine | procedure | **procedure** $GREETINGS(x)$<br>    print ("Hello", $x$) |
| | function | **procedure** $ADD1(x)$ **returns integer**<br>    $x$: **integer**<br>    **return** $x + 1$ |
| Control statements | if-then-else | **if** $x > 0$ **then** $x \leftarrow x - 1$<br>        **else** $x \leftarrow x + 1$ |
| | while-do | **while** $x > 0$ **do**<br>    $x \leftarrow x - d$<br>    $q \leftarrow q + 1$ |
| | repeat-until | **repeat**<br>    $x \leftarrow x - 1$<br>    $y \leftarrow y - 1$<br>    **until** $x \leqslant 0$ |
| | for | **for** $i = 5$ **to** 30 **by** 6 **do**<br>    print($i$, " is prime") |
| | case | **case**<br>    $x < 0$: $x \leftarrow x + 1$<br>    $x = 0$: $[\![$do nothing$]\!]$<br>    $x > 0$: $x \leftarrow x - 1$ |
| Simple statements | assignment | $a \leftarrow ADD1(a)$ |
| | swap values | $x \leftrightarrow y$ |
| | procedure call | $GREETINGS$ ("Ed") |
| | sentence | Add one to $x$. |

**Table 1.2**
Statements used in the notation. Indentation subordinates the indented statement to the overhanging subroutine definition or control statement.

greater than *table_size*. (Note that in many languages **and** and **or** always evaluate both arguments.)

The above expression might appear in a loop to search an array:

$$\textbf{while } i \leqslant number\_of\_elements \textbf{ and } table[i] \neq z \textbf{ do}$$
$$i \leftarrow i + 1$$

The **while** construction evaluates the expression and executes the statement if it is **true**. After executing the statements, control returns to the **while** and the process repeats. When the expression yields **false**, control proceeds to the unindented statement after **while**. In this case, since $i$ is continually getting larger inside the loop, the loop will eventually terminate, even if it does not find a table entry

| Name | Symbol | Type | Example |
|---|---|---|---|
| functions | | | |
| logarithm base 2 | lg | **integer, real** | lg 8 $\Rightarrow$ 2.0 |
| floor | $\lfloor \ \rfloor$ | **real** | $\lfloor 1.732 \rfloor \Rightarrow 1$ |
| ceiling | $\lceil \ \rceil$ | **real** | $\lceil 1.732 \rceil \Rightarrow 2$ |
| length | length | **string** | length("emr") $\Rightarrow$ 3 |
| absolute value | $\| \ \|$ | **integer, real** | $\| -3 \| \Rightarrow 3$ |
| square root | $\sqrt{\ }$ | **integer, real** | $\sqrt{3} \Rightarrow 1.732$ |
| unary operations | | | |
| minus | $-$ | **integer, real** | $-(1 - 2) \Rightarrow 1$ |
| negation | **not** | **boolean** | **not true** $\Rightarrow$ **false** |
| level 6 | | | |
| exponent | (raised) | **integer, real** | $1.732^2 \Rightarrow 3.00$ |
| level 5 | | | |
| multiply | $*$ | **integer, real** | $2 * 1.5 \Rightarrow 3.0$ |
| divide | $/$ | **integer, real** | $3/2 \Rightarrow 1.5$ |
| modulus | mod | **integer** | 11 mod 4 $\Rightarrow$ 3 |
| concatenation | $\|\|$ | **string** | "wj"$\|\|$"h" $\Rightarrow$ "wjh" |
| level 4 | | | |
| add, subtract | $+, -$ | **integer, real** | $1 + 2 \Rightarrow 3$ |
| level 3 | | | |
| comparison | $<, >, \leqslant, \geqslant$ | **integer, real, string** | "wjh" > "emr" $\Rightarrow$ **true** |
| equality | $=, \neq$ | **integer, real, string, boolean** | **true** $\neq$ **false** $\Rightarrow$ **true** |
| level 2 | | | |
| and | **and** | **boolean** | **true and** $3 < 5 \Rightarrow$ **true** |
| level 1 | | | |
| or | **or** | **boolean** | **false or** $3 < 5 \Rightarrow$ **true** |

**Table 1.3**
Operators often used in the notation. Unless otherwise indicated by parentheses, operators at higher levels are performed first.

equal to $z$. (The **repeat-until** construction executes the indented statements and *then* tests the final expression. The loop repeats if the expression is **false**.)

In addition to constants, variables, and expressions, another way to compute a value is to call a function. One function required for the table data structure is *search_table*, as defined in the next section. This function requires two arguments, so a call might look like this:

$$search\_table(z, index)$$

There is one argument for each formal parameter specified in the first line of the procedure definition. We will see that $z$ may be any expression that computes a key, while *index* must be a variable in the *element_location* domain. The procedure *search_table* searches *table* for an element with key *z* and sets *index* to where it is or should be. The procedure returns a **boolean** value: **true** if it finds *z* and **false** otherwise.

Since it returns a **boolean** value, *search_table* can be called in the control expression between **while** and **do** or between **if** and **then**. Although **while-do** creates a loop, **if-then-else** merely chooses between two sets of statements. The statements under **then** are executed if the control expression is **true** and those under **else** otherwise. For example, consider the following implementation of *find_element*. The algorithm is expected to find the key *z* in the table and store its *element_location* in *loc*. It uses a temporary variable *index*, which will be set by the call on *search_table*.

> **if** *search_table*(*z*, *index*) **then**                    ⟦is *z* in the table?⟧
>               *loc* ← *index*                    ⟦yes, say where⟧
>       **else**
>               *loc* ← *key_absent*           ⟦no, signal an error⟧

When *search_table* finds the key, the first assignment to *loc* is executed, using the value placed in index by *search_table*. But if *search_table* returns **false**, the assignment under **else** is executed, setting *loc* to an error value.

Sometimes, instead of writing the implementation of a function as in-line code, we will write it as a procedure. Algorithm 1.1 shows a procedure definition for *enter_element*. The routine calls *search_table* to be certain the key is not already in the table; it checks for possible errors; and finally it stores the new key in the array and returns the location where it is stored.

> **procedure** *enter_element*(*z*) **returns** *element_location*
>       ⟦store *z* in table after checking for errors⟧
>    *z*: *key_value*                              ⟦key to be stored⟧
>    *index*: *element_location*                 ⟦temp for search result⟧
>    **if** *search_table*(*z*, *index*) **then return** *duplicate_key_error*
>    **if** *number_of_elements* ⩾ *table_size* **then return** *table_full_error*
>    *number_of_elements* ← *number_of_elements* + 1
>                        ⟦count new element⟧
>    *table* [*number_of_elements*] ← *z*        ⟦store key⟧
>    **return** *number_of_elements*              ⟦return its location⟧

**Algorithm 1.1**
Implementation of the *enter_element* function for the table data structure.

Procedures may have *local* variables such as *index*. Space is allocated for an instance of each such variable every time execution of the procedure begins, but the space is released when the procedure execution ends. Variable $z$ is a *formal parameter* of the procedure. When the procedure is called, as it would be by

$$enter\_element(\text{``FLAG''}),$$

there must be one *actual* parameter for each formal parameter. As execution of the procedure begins, the formal parameters are initialized with the value of the corresponding expression in the list of actual parameters. Then during execution, except when otherwise noted, the formal parameter variable behaves exactly like any other local variable of the procedure during its execution. (In other words, we assume "call by value" unless otherwise stated.)

**Exercises**

1. Suppose the keys for the table are strings. Show the contents of the table after execution of the following calls:

    *initialize_table*
    *enter_element*("MOTHER")
    *enter_element*("APPLE")

    What value will be stored in *loc* by *find_element* if $z$ is "MOTHER"? If it is "PIE"?

2. Show a complete set of declarations for the table data structure. Be sure to modify those in the text to account for the *element_location* values that represent error conditions. Also, revise Figure 1.2 appropriately.

3. Revise the definition of the table data structure so that instead of the variable *number_of_elements* there is a variable that has the subscript of the location in which to store the next entered element. What should its initial value be? Show the complete revised declarations, comments, and relationships. Show a diagram to correspond to Figure 1.2.

4. Show a storage structure for the functions developed for Exercise 7 of Section 1.1 where the error conditions are signaled by values in a separate domain rather than in the domain of *element_locations*.

5. Write a set of table functions to implement the scheme of Exercise 4.

6. Show where parentheses are assumed in the evaluation of the expression

$$i \leqslant number\_of\_elements \textbf{ and } table[i] \neq z$$

according to Table 1.3. Show also where the parentheses would be if **and** were assumed at a higher level than $\leqslant$ and $\neq$ (as it is in PASCAL).

7. Use the priorities of Table 1.3 to show where parentheses are assumed in the evaluation of the expression

$$a + b * c > d \text{ or not } p \text{ and } q > 5.$$

8. The treatment of **and** and **or** in this text means that evaluation of a boolean expression is equivalent to evaluation of a procedure containing **if-then-else**'s and **return**'s of **true** and **false**. For example, when $p$ and $q$ are boolean variables, the expression $p$ **or** $q$ is equivalent to calling

> **procedure** $p\_or\_q$ **returns boolean**
> **if** $p$ **then return true**
> **else return** $q$

Notice that $p$ and $q$ are *not* parameters to the procedure because, if so, both would be evaluated before entering the procedure, and it is precisely this evaluation that is avoided by our definition of logical expressions.

(a) Show the procedure corresponding to $p$ **and** $q$.

(b) Show a procedure corresponding to

$$p \text{ and } q \text{ or } t \text{ and } \mathbf{not}(u \text{ or } v).$$

(c) State a condition under which a compiler could improve efficiency by reversing the order of a boolean expression. Be sure that the reversals allowed by your condition do not allow errors that would not occur with the original expression.

9. The square root of a value $x$ can be computed with a **while** loop and a few arithmetic expressions using a technique called Newton-Raphson approximation. The algorithm starts by making a guess $r$ as to the square root and then computing a better approximation $r'$ by the formula

$$r' = \tfrac{1}{2}(r + x/r).$$

This process is repeated until the difference between $r$ and $x/r$ is less than a millionth of the value of $r$. Write a function to find square roots using this method.

10. Repeat the previous exercise but use a **repeat-until** and a temporary variable to avoid dividing twice for each execution of the loop.

11. It is possible to multiply positive integers using operations no more complex than addition and division by two (done with a shift):

```
procedure MULTIPLY(a, b) returns integer
    a, b, p: integer              [operands and result]
    p ← 0                         [start with result zero]
    while a > 0 do
        if odd(a) then            [test low-order bit]
            p ← p + b             [and maybe increase result]
        b ← b + b                 [double one operand...]
        a ← shift_right(a, 1)     [and halve the other]
    return p                      [return the product]
```

Prove that this procedure computes the product correctly. Do this by showing that the value of $a*b + p$ is the same every time $a > 0$ is tested.

12. As with *MULTIPLY* in the previous exercise, it is usually possible to express an operator in terms of more primitive operators. Express the following as procedures that use only the operators addition, subtraction, and integer comparison.

   (a) odd($i$)—returns **true** if the integer $i$ is odd, **false** if it is even.

   (b) div($i$, $j$)—returns the integer quotient resulting from dividing the integer $i$ by the integer $j$.

   (c) $\lfloor r \rfloor$—returns the largest integer that is not greater than the real value $r$. Be careful that your solution works when $r$ is negative.

## 1.3  SEARCH

Seek and ye shall find.
   *Matthew 7, 7*

Although the search problem will be considered at length in Chapter 7, it is appropriate to define the problem here and give a first solution. The most general form of the problem is that one is given a collection of elements and is required to find those meeting some criterion. In simple cases, such as our table data structure, the criterion is just a match on the value in one attribute. More complex cases include all the possibilities for data base queries.

   The search in the table data structure has already been specified as the routine *search_table*, which is to find the key and return **true** if it is indeed present. One possible implementation is that shown as Algorithm 1.2.

**procedure** *search_table*$(z, i)$ **returns boolean**

⟦find where $z$ should be in the table and set $i$ to that value;
return **true** if $z$ is in table; $i$ is passed by reference⟧

$i \leftarrow 1$

| | |
|---|---|
| **while** $i \leqslant$ *number_of_elements* | ⟦check occupied elements...⟧ |
| **and** *table*$[i] \neq z$ **do** | ⟦...looking for $z$⟧ |
| $i \leftarrow i + 1$ | ⟦not $z$, go to next element⟧ |
| **return** $(i \leqslant$ *number_of_elements*$)$ | ⟦if $z$ is not there, $i$ will |
| | be $1 +$ *number_of_elements*⟧ |

**Algorithm 1.2**

Procedure to locate a key $z$ in the table. This algorithm returns a boolean value indicating whether the key was in the table. It also stores the location of the key in the variable that is the second argument supplied by the calling routine.

The fact that the parameter $i$ is passed by reference means that the final value of that variable will be copied back to the actual parameter. This is how *index* gets set in *find_element* and *enter_element*. Variable $i$ is used in Algorithm 1.2 as the point that is currently being searched. It is initialized at the beginning of the procedure, tested in the **while** condition, and incremented every time the **while** condition is met. The final step of *search_table* is to return a boolean value; observe that this value is computed from the value of $i$ and determines which of the two conditions caused termination of the **while** loop.

An important characteristic of algorithms is that we can prove statements about them. For example, we can prove that *search_table* finds the desired value $z$ if it is in the table. To prove this assertion, we begin with a *loop invariant*, a boolean condition that is true whenever execution begins for the body of the loop. For the **while** loop in *search_table*, the appropriate invariant is

$$i \leqslant 1 + \textit{number\_of\_elements} \textbf{ and } z \text{ is not in } table[1 \ldots i - 1] \qquad (1.1)$$

where the "..." indicates the sequence of elements *table*[1], *table*[2],..., *table*$[i - 1]$. Note that this invariant is true prior to initial execution of the **while** loop because $i$ is 1. (The range $1 \ldots 0$ is taken to contain no elements.) If the expression between **while** and **do** is false, the loop body will not be executed. In this case the invariant is still true because no values have been changed. Upon exiting from the loop, if $i \leqslant$ *number_of_elements*, then $z$ must have been found, because the loop could have exited only via the $z \neq table[i]$ clause. Otherwise, $i = 1 +$ *number_of_elements*, so the second part of the invariant states that $z$ is not in *table*$[1 \ldots (1 + \textit{number\_of\_elements}) - 1]$ and thus is not present at all.

The crucial step of the proof is to show that the loop invariant is still true if the body of the loop is executed. To do so, we begin by assuming that the invariant (1.1) is true at the start of the body and then show that it must still be true afterward. If the loop body executes, the expression between **while** and **do** was tested, so the following is true at the start of the body:

$$i \leqslant number\_of\_elements \textbf{ and } z \neq table[i]. \tag{1.2}$$

Combining the second terms of (1.1) and (1.2), we find that

$$z \text{ is not in } table[1 \ldots i]. \tag{1.3}$$

The body of the loop makes $i$ larger by one, so the first term of (1.2) becomes

$$i \leqslant 1 + number\_of\_elements$$

and (1.3) becomes

$$z \text{ is not in } table[1 \ldots i - 1],$$

which together are exactly the invariant (1.1). Since the body of the loop preserves the truth of the invariant, our assumption is justified that it was correct upon entry to the body. Consequently, *search_table* does find $z$ if it is present. Much of the work of proving program properties involves boolean expressions, which are covered in more detail in Section 2.1.

Because search is a frequent operation on a data structure, its efficiency is important. The *search_table* routine of Algorithm 1.2 is inefficient because it doggedly checks keys one after another. Fortunately, there are more efficient search techniques.

To avoid examining every element of the array, *search_table* may be able to use the binary search technique. The central assumption of this method is that the keys to be searched are ordered. Then, by examining the middle element of the array, we can quickly eliminate all those elements above or below the middle. One way to express this algorithm is to write English phrases as part of the routine:

> Consider entire array as the "current section."
> **while** current section has elements **do**
>     Set *xm* to key in middle of current section.
>     **case**
>         $xm < z$: Reduce current section to
>             the half above the middle.
>         $xm > z$: Reduce current section to
>             the half below the middle.
>         $xm = z$: **return** middle    ⟦found $z$⟧
>   **return** current section   ⟦$z$ belongs here⟧

The **case** statement evaluates the logical expressions following it and executes the first statement indented under the expression having value **true**. Note that each of the first two cases eliminates half the current section. This is considerably better than Algorithm 1.2, which eliminates only one value with each iteration. The half eliminated cannot contain any key equal to $z$ because of the ascending order. When $xm < z$, all keys below the middle are also less than $z$; and similarly for " $>$ ".

Using binary search, the body of Algorithm 1.2 can be replaced with Algorithm 1.3. In either case the external description of the function does not change; it states $i$ is set to the location that should contain $z$ and a boolean value is returned indicating whether $z$ was found. For binary search, however, the values in the table must be arranged in ascending order. To preserve this relationship when new keys are entered, some of the keys already in the table may have to be moved. If *element_location* values are implemented just as before, values in all attribute arrays will have to be moved too. But the table routines may not know what attributes exist. To avoid telling them, we extend the storage

⟦find where $z$ should be in the table and set $i$ to that value:
return **true** if $z$ is in table, **false** otherwise⟧
$l \leftarrow 1$                                              ⟦start with portion...⟧
$h \leftarrow number\_of\_elements$                           ⟦...that has all the elements⟧
$found \leftarrow$ **false**
**while** $l \leqslant h$ **and not** $found$ **do**

$$m \leftarrow \left\lfloor \frac{l + h}{2} \right\rfloor$$           ⟦find middle element⟧

$\quad\quad xm \leftarrow table[m]$                           ⟦and its value⟧
$\quad\quad$ **case**
$\quad\quad\quad xm < z: l \leftarrow m + 1$                  ⟦reduce to upper half⟧
$\quad\quad\quad xm > z: h \leftarrow m - 1$                  ⟦reduce to lower half⟧
$\quad\quad\quad xm = z:$
$\quad\quad\quad\quad\quad i \leftarrow m$                    ⟦found; note where...⟧
$\quad\quad\quad\quad\quad found \leftarrow$ **true**         ⟦...and signal success⟧
**if not** $found$ **then** $i \leftarrow l$                  ⟦note where $z$ belongs...⟧
**return** $found$                                           ⟦...and signal absence⟧

**Algorithm 1.3**
Binary search to locate a key in *table*. (Replaces body of sequential version in Algorithm 1.2.) Binary search proceeds by considering a section of the array and successively dividing it in half. The variables $l$ and $h$ point to the ends of the current section; $m$ points to the middle element. Note that none of the elements between $l$ and $h$ (inclusive) have been compared to $z$, so the loop repeats even when $l = h$.

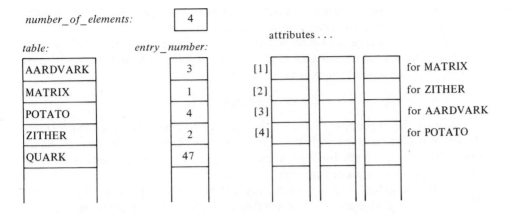

**Figure 1.3**
Storage structure for a table with ordered key values and an *entry_number* array to contain *element_location* values. The elements are presumed to have arrived in the order MATRIX, ZITHER, AARDVARK, and POTATO, so they have *element_location* values that are increasing integers in that order. The *table* and *entry_number* arrays are parallel to each other but not to the attribute arrays. The location of the attributes for a given element is in the parallel position of the *entry_number* array.

**if** *search_table* (*z*, *index*) **then return** *duplicate_key*
**if** *number_of_elements* ≥ *table_size* **then return** *table_full*
**for** *i* = *number_of_elements* **to** *index* **by** −1 **do**
        *table*[*i* + 1] ← *table*[*i*]        ⟦move entries for...⟧
        *entry_number*[*i* + 1] ← *entry_number*[*i*]  ⟦...keys above *z*⟧
*number_of_elements* ← *number_of_elements* + 1
*table*[*index*] ← *z*
*entry_number*[*index*] ← *number_of_elements*    ⟦use count...⟧
**return** *entry_number*[*index*]        ⟦... as element location⟧

**Algorithm 1.4**
Enter a key in table organized for binary search. This algorithm corresponds to Algorithm 1.1 for the sequential search case. The crucial step is to move all entries up one position with the **for** loop. The *entry_number* value assigned to an element is identical to its *element_location* value as used in Algorithm 1.1. The value never changes, even though the table location of the key itself may change when new elements are entered.

structure to include an array to keep the *element_location* value for each table element. This is the array *entry_number* illustrated in Figure 1.3.

With this storage structure, *initialize_table* and the attribute scheme given for Algorithm 1.2 are unchanged. Only one line of *find_element* must be changed; instead of returning the index it must return

$$entry\_number\,[\,index\,],$$

thus getting the appropriate *element_location* for the key. The procedure *enter_element* must be replaced with the version in Algorithm 1.4. The major change from Algorithm 1.1 is the **for** loop to move keys so the new one can be inserted at its correct position in ascending order.

A **for** statement like that in the algorithm consists of a *control line* and a *body* of indented statements. The control line declares an integer variable—in this case *i*—and defines a sequence of integer values. For each member of the sequence the body is executed with the current member as the value of the control variable. The sequence is the arithmetic sequence that starts with the value after "=", increments by the value after "**by**", and ends with the largest value that is not greater than that after "**to**". If the **by** value is omitted, the value one is assumed. In this case the **by** value is negative, so the sequence descends from *number_of_elements* to *index*.

Recursion sometimes provides a clear description of an algorithm. In binary search, for example, the search examines the current section of the file. This can be expressed recursively by introducing the procedure of Algorithm 1.5 among the declarations of *search_table*. The body of the latter would then reduce to

$$BINSRCH(1,\ number\_of\_elements\,)$$
$$\textbf{return}\ table[\,index\,] = z$$

**procedure** *BINSRCH(l, h)*
    *l, h*: **integer**　　〚bounds of current section〛
    **if** $l > h$ **then** *index* ← *l*
        **else**
$$m \leftarrow \left\lfloor \frac{l+h}{2} \right\rfloor$$
           *xm* ← *table*[*m*]
        **case**
           $xm < z$: *BINSRCH(m + 1, h)*
           $xm > z$: *BINSRCH(l, m − 1)*
           $xm = z$: *index* ← *m*

**Algorithm 1.5**
Recursive binary search procedure. The arguments to *BINSRCH* delimit the current interval; they are set either by the initial call or by one of the calls within the **case**.

*BINSRCH* always examines a current interval defined by its two parameters, *l* and *h*. The first line checks that the interval has at least one element and exits if not. If there are elements, the succeeding lines split them in two just as the loop in Algorithm 1.3 did. Recursive routines such as *BINSRCH* create separate local variables every time the routine is called. Thus if *BINSRCH* has been called from *BINSRCH*, there are at least two different sets of variables in existence, both of which have values for *l* and *h*.

In one version of the table search problem all *enter_element* operations are done before any *find_element* operations. In this case, *enter_element* need not move keys, it can store them sequentially; then the table is sorted before the first *find_element* operation. Indeed sorting, which is discussed at length in Chapter 8, is an important tool in making search routines run faster. One of the simpler sorting methods is the selection sort, given in Algorithm 1.6.

⟦Sort the elements of table in ascending order
and rearrange the values in *entry_number*⟧
**for** $i = 1$ **to** *number_of_elements* $-$ 1 **do**
      ⟦process successively smaller subarrays⟧
      Set *j* to subscript value of smallest value
      in *table*[*i*...*number_of_elements*]
      ⟦swap current entry with smallest⟧
      *table*[*i*] ↔ *table*[*j*]
      *entry_number*[*i*] ↔ *entry_number*[*j*]

**Algorithm 1.6**
Simple selection sort. In the *i*th step, the *i*th largest element is found and put into place.

The first statement in the loop of Algorithm 1.6 suggests a search operation but does not say how to do it (see Exercise 9). The next two statements use the double-arrow notation to indicate an interchange of the values in the named variables. That is, the statement *table* [*i*] ↔ *table* [*j*] is equivalent to the three statements:

$$t \leftarrow table[i]$$
$$table[i] \leftarrow table[j]$$
$$table[j] \leftarrow t$$

A final example of search techniques is the use of indexes to permit a space-time trade-off for data base design. Suppose we wish to be able to find names when given phone numbers. If we already have an alphabetical phone directory file, we have three choices: scan the existing file with sequential search, copy and sort the file for a binary search, or create an index to the existing file. The sequential search would require a considerable amount of computing but

Phone Directory File

| file location | name | address | number |
|---|---|---|---|
| [1] | Anderson | 13 Main | 977–4231 |
| [2] | Black | 27 Second Ave | 964–3210 |
| [3] | Carson | 42 Elm | 967–8901 |
| [4] | Green | 135 Hill | 321–4567 |
| [5] | van der Mer | 3 Easy | 455–1000 |

Index File

| number | entry |
|---|---|
| 321–4567 | 4 |
| 455–1000 | 5 |
| 964–3210 | 2 |
| 967–8901 | 3 |
| 977–4231 | 1 |

**Figure 1.4**
Example of an index for a data base. To find a number given a name, the phone directory file can be searched by an efficient technique like binary search. For the inverse problem of finding a name from a number the index file can be searched. The entry attribute in the index file gives the location of the complete element in the main file.

little additional space. The copy of the file would double the space requirements but would provide fast answers. A good compromise is an index like that shown in Figure 1.4; each entry in the index contains only the phone number and the location of the corresponding element in the main file. To find a name, the search routine first finds the entry in the number index and then follows the location indicator to find the element. This scheme is slightly slower than using a copy of the file, but it can require considerably less space. Indeed, indexes are frequently used in data bases where elements have so many attributes that multiple copies would be impossible; in such cases it is not uncommon for a file to have indexes for numerous attributes. With this approach, however, changes to the main file must be carefully reflected in the indexes.

**Exercises**

1. Prove that the **while** loop in Algorithm 1.2 will always exit after a finite number of executions of the body.

2. Revise the table algorithms to eliminate the double test in the **while** loop of *search_table* (Algorithm 1.2). Do this by first storing the value $z$ as a dummy key at the end of the table. Note that one location of the table must never be filled. Describe a simplification of *enter_element* that becomes possible.

3. If the **case** statement in Algorithm 1.3 were replaced by an **if-then-else**, it could test the various possibilities in six different orders. Specifically, the first test could be equality, less than, or greater than. Which one of these three choices is less efficient and results in performing almost twice as many tests?

4. Suppose Algorithm 1.3 has proceeded to the point where $l = h - 1$; that is, there are just two possible elements that might be equal to $z$. When we consider that either of these elements might be equal to, less than, or greater than $z$, we have a maximum of nine possible continuations for the algorithm.

   (a) Show that four of the nine possibilities are excluded by the storage structure definition.

   (b) For each of the other five cases show the outcome of further execution. How many more times is the loop executed? To what values are $l$ and $h$ set in each case? What value is returned? Is $i$ set to indicate the location where $z$ should be?

5. Suppose the assignment to $m$ in Algorithm 1.3 used $\lceil \ \rceil$ instead of $\lfloor \ \rfloor$. What changes to the remainder of the algorithm would be needed?

6. Prove that if $l > 1$ in Algorithm 1.3, then all elements with subscripts 1 through $l - 1$ are less than $z$.

7. It is not necessary to move the keys in Algorithm 1.4. They can be treated as any other attribute and accessed via *entry_number*. In this scheme the smallest key is in *table*[*entry_number*[1]] rather than in *table*[1]. Revise Algorithms 1.3 and 1.4 for this scheme.

8. Rewrite Algorithm 1.2 to use recursion.

9. Rewrite Algorithm 1.6 so the statement "Set $j$..." is replaced by a function call. (The returned value should be assigned to variable $j$.) Write the function.

10. Prove that these two assertions are true whenever Algorithm 1.6 "executes" the English sentence.
    (a) All keys in *table*$[1 \ldots i - 1]$ are sorted in ascending order.
    (b) All keys in *table*$[i \ldots number\_of\_elements]$ are at least as large as *table*$[i - 1]$.

## 1.4  EFFICIENCY

Had we but world enough and time.
>      A. Marvell, *To His Coy Mistress*

The efficiency of a process is its relative total cost. It cannot be said to be efficient in itself; it can only be less expensive than some alternative. Thus to say that running is an efficient way to build heart muscles, is only to say that as an activity it builds more muscle per hour than alternatives such as walking, watching television, or talking on the telephone. The major costs considered in this text are those for computer memory and program execution time. In this section we discuss techniques for evaluating these costs.

### 1.4.1  Expected Value and Worst Case

A friend offers to sell you a lottery ticket that will pay you \$600 if the three-digit number on it is equal to the one drawn at random next Thursday. If neither you nor your friend is to make a profit on the transaction, what price should you pay for the ticket? This sort of question launched the development of probability theory and can be answered by computing the *expected value* of the ticket. In this case, there is one chance in a thousand the ticket will pay off. This probability multiplied by the size of the payoff gives the expected value: 60 cents. (If you can't in fact get tickets at that price, *someone* is making a profit!)

The general formula for computing an expected value $E$ of a situation depends on dissecting the situation into a number of cases—say $n$ of them—each with a known probability $p_i$ and an individual value $v_i$. The expected value is then

$$E = p_1 v_1 + p_2 v_2 + p_3 v_3 + \cdots + p_n v_n$$
$$= \sum_{i=1}^{n} p_i v_i. \tag{1.4}$$

Exercise 1 shows that this formula can be derived from the formula for an average. For the lottery-ticket problem there are two cases: payoff = 600 and payoff = zilch. The probability of any particular three-digit number is $1/1000$, so the expected-value formula gives

$$\text{ticket value} = 0.999 \times 0 + 0.001 \times 600 = 0.60.$$

Expected-value computations can help describe the behavior of programs; that is, they can describe how the utilization of space and time changes as conditions change. Some programs behave in a constant fashion, always using the same amount of time or memory. Others have behavior that varies in proportion to some parameter such as the number of data values, the size of the largest data value, the number of pairs of values that are equal, and so on. For the table algorithms, the parameter affecting the cost of execution is the number of elements in the table. The number of distinct values the parameter may have is then the value $n$ for the expected-value computation.

To illustrate, let us compute the expected cost of a successful application of the sequential search in Algorithm 1.2. The major cost of this algorithm is the execution time for the **while** loop. (When the number of elements in the table is large enough so that the total cost is significant, the relative cost of the statements outside the loop is negligible.) The **while** loop consumes time in proportion to the number of evaluations of its boolean expression: there is one evaluation if the desired key is first, two if it is second, and so on up to *number_of_elements* tests for the last key in the table. Thus the cost $c_i$ to find the $i$th key is proportional to $i$. In the absence of better information, we assume that each key in the table is equally likely, so all have probability $p_i$ equal to $1/number\_of\_elements$. Thus the expected cost to find a key in the table is proportional to

expected number of boolean evaluations

$$= \sum_{i=1}^{n} p_i c_i = \sum_{i=1}^{n} \frac{1}{n} i = \frac{1}{n} \sum_{i=1}^{n} i$$

$$= \frac{n+1}{2},$$

where $n$ represents *number_of_elements*. (See Exercise 2.) Note that the expected cost of $(n+1)/2$ for linear search is proportional to $n$ itself.

The expected cost of an algorithm is not the only measure of its efficiency, nor is it always the best. In many situations it is necessary to use the *worst-case* cost for an algorithm. In parallel with our discussion of the expected cost above, we have

$$W = \max_{1 \leq i \leq n} \{v_i\}.$$

For the sequential search of Algorithm 1.2, the worst-case cost is proportional to

worst-case number of boolean evaluations.

$$= \max_{1 \leq i \leq n} \{c_i\} = \max_{1 \leq i \leq n} \{i\}$$
$$= n.$$

For the binary search of Algorithm 1.3, the cost depends on the expected number of loop executions. This cost can be evaluated in terms of Equation (1.4), as shown in Section 5.3. To demonstrate the efficiency of binary search, however, it is enough to show an upper bound for the worst-case number of loop executions, because this bound is itself considerably less than the cost for sequential search. The upper bound can be derived by considering that each execution of the loop cuts the size of the current section of the array in half. Thus we have

number of loop executions
    = number of times *number_of_elements* can be divided
        in half with a result greater than one
    = number of times to double one to reach a value
        greater than *number_of_elements*
    = $l$, where $l$ is such that
        $2^{l-1} < number\_of\_elements \leqslant 2^l$.

But by the definition of logarithms, when $2^x = n$ we have $x = \lg n$ (where $\lg n$ is the logarithm to the base 2 of $n$). Thus

$$\text{number of loop executions} = \lceil \lg n \rceil$$

and the cost of binary search is at worst proportional to $\lg n$.

The above discussions have shown these proportional costs:*

Sequential search        $n$
Binary search            $\lg n$

Although $\lg n$ is smaller than $n$ for all positive values of $n$, as suggested by Table 1.4, the cost of binary search could exceed sequential search on small tables if the cost for each loop iteration were large. Evaluation of this factor requires the operator-count techniques below (see Exercise 11 of Section 1.4.3). In practice, however, binary search has been found preferable for twenty or more elements and not too far off for fewer elements. For under twenty elements, though, neither approach costs enough to affect the total cost.

Analysis of computer algorithms usually finds that their cost is proportional to one of the few expressions shown in Figure 1.5. For large $n$, these expressions have the relationship

$$1 < \lg \lg n < \lg n < n < n \lg n < n^2 < n^3 < 2^n.$$

----

*As a shorthand expression the proportional cost of an algorithm is sometimes called its "order," as in "Binary search is an order $\lg n$ algorithm." We discuss this in more depth at the beginning of Chapter 7.

| number_of_elements | $n$ | 10 | 20 | 100 | 200 | 1,000 | 1,000,000 |
|---|---|---|---|---|---|---|---|
| Sequential search | $n/2$ | 5 | 10 | 50 | 100 | 500 | 500,000 |
| Binary search | $\lg n$ | 4 | 5 | 7 | 8 | 10 | 20 |

**Table 1.4**
Comparison of number of loop executions for each of two search techniques
when the table size is *number_of_elements*

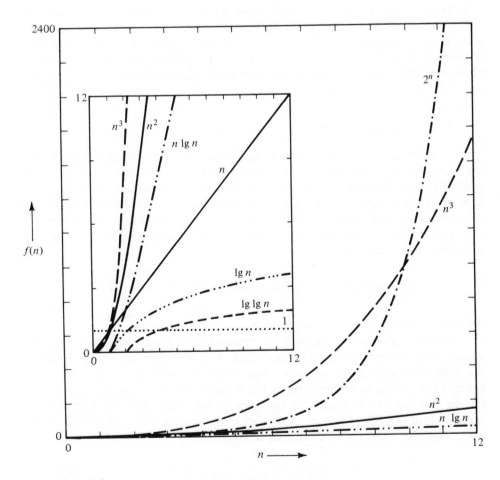

**Figure 1.5**
Common patterns of growth. Each line is labeled with the function that
produces it.

Algorithms are generally practical if their cost is proportional to any but the last of these expressions. Costs proportional to $2^n$, however, are disastrously expensive for even moderate $n$.

Among the possible cost expressions listed above is the constant 1. Algorithms with cost proportional to a constant exhibit no growth in cost no matter how the parameter changes. One curious example of such an algorithm is a table search that can find an alphabetic key in a fixed amount of time. Suppose keys are one letter in length. Then the table can be an array of 26 integer values, each giving the *element_location* value for one key. When keys are allowed to have two letters, the table must have $26^2 = 676$ entries, and so on for larger maximum lengths. A table permitting keys of $m$ letters occupies space proportional to $2^m$, even though the time is still constant. Since this table consumes extra space and binary search consumes extra time, these methods can be compared only by the space-time method presented below (see Exercise 15 of Section 1.4.3). In practice, however, both methods are usually inferior to the hashing methods presented in Chapter 7, which have reasonable costs for both space and time.

### 1.4.2   Operator Counts

In some situations two alternate algorithms both have execution times with the same proportional costs, so we must analyze their details to judge their relative merits. One suitable measure for the cost of execution is the expected number of "operators" each executes in performing its task. However, the count must include a variety of items beyond the ordinary arithmetic operations, as shown in Table 1.5. On this basis, there are seven operators in the following statement from Algorithm 1.2:

$$\textbf{while } i \leqslant \textit{number\_of\_elements} \textbf{ and } \textit{table}[i] \neq z \textbf{ do } i \leftarrow i + 1 \qquad (1.5)$$

The operators are: **while**, $\leqslant$ , **and**, [   ], $\neq$ , $\leftarrow$ , $+$.

Computation of the number of operator executions must consider the number of times each expression is evaluated. In a **while** loop or other loop where the body is usually executed more than once, the execution count can be a variable introduced as a parameter of the analysis. Thus $n$ could be a parameter for the number of executions of the loop above. Note that the loop control expression of a **while** is actually executed one more time than the body of the loop. Thus the total cost of the fragment is

$$5(n + 1) + 2n = 7n + 5.$$

---

data-manipulation operators
+, < , **and**, $|x|$,... (See Table 1.3)
(Some functions such as $\sqrt{x}$ and lg are more expensive than one operator)
flow-of-control operators
**if, case**
**while, for, until**    (Count once each time around loop)
miscellaneous
←
array subscript, field selection
procedure call, function call, **return**

---

**Table 1.5**
Categories of operators to be counted. Each of these items counts as one operator each time it is encountered during execution of a program. "Field selection" will be discussed in Section 3.2.

The alternate version using a dummy key in Exercise 2 of Section 1.3 eliminates the ≤ and the **and**, so it would have a cost of execution of $5n + 3$, which is somewhat better than (1.5). This analysis, however, does not present strongly enough the case for using dummy values. As will appear often in Chapters 4 through 8, dummy values can eliminate many special-case tests.

When an algorithm includes an **if**, the statements under **then** are not executed every time the **if** is. Instead, they are executed with probability $p$, where $p$ is the probability that the **if** test is **true**. Similarly the **else** statements, if any, are executed with probability $1 - p$. For example, suppose we rewrite the **case** statement of Algorithm 1.3 as an **if-then-else**:

$$\textbf{if } xm < z \textbf{ then } l \leftarrow m + 1$$
$$\textbf{else if } xm > z \textbf{ then } h \leftarrow m - 1$$
$$\textbf{else}$$
$$i \leftarrow m$$
$$\textbf{return true} \qquad (1.6)$$

There are ten operators in this fragment: **if**, < , ← , +, **if**, > , ← , − , ← , **return**. If we denote by $p$ and $q$ the respective probabilities of < and > , the expected operator execution count is

$$2 + 2p + (1 - p)(2 + 2q + 2(1 - q)) = 6 - 2p.$$

(Why is it reasonable that there is no $q$ term in the final result?) Note that the probability of execution of all operators after the first **else** is multiplied by $1 - p$, the probability of **not**($xm < z$). The last line then has probability $(1 - p)(1 - q)$.

This general principle of multiplying probabilities applies whenever control constructions are nested (see Exercise 10 of Section 1.4.3).

In a similar vein, we can analyze the worst-case operator count of the fragment (1.6): when $xm < z$, four operators are required. When $xm \geq z$, six are required. The worst-case is thus six operators.

We can compare the cost of fragment (1.6) with an alternative that is often used unthinkingly (Exercise 3 of Section 1.3):

$$
\begin{aligned}
&\textbf{if } xm = z \textbf{ then} \\
&\qquad i \leftarrow m \\
&\qquad \textbf{return true} \\
&\textbf{else if } xm > z \textbf{ then } h \leftarrow m - 1 \\
&\qquad \textbf{else } l \leftarrow m + 1
\end{aligned}
$$

The primary difference is that equality is tested first. This alternative has the same worst-case cost as (1.6), but an expected cost of

$$2 + 2(1 - p - q) + (p + q)(2 + 2p + 2(1 - p)) = 4 + 2(p + q).$$

Since $p$ and $q$ are close to one-half when there are a large number of elements, the cost of (1.6) will be about five while the cost of the alternative will be near six. The difference of 20 percent is an excess cost, since there is no additional physical effort to write form (1.6) in the first place.

### 1.4.3  Space-Time Analysis

Although previous sections have compared algorithms by studying their execution time, the choice is often a "space-time" trade-off: one algorithm will use more space and the other will use more time. Such algorithms are compared by a space-time "integral." This value is the product of the program size and its execution time. If the program changes size during execution, the product must be computed for each size. A program that occupies 70 space units for its first 10 seconds and 100 units for another 20 seconds will have a space-time integral of

$$70 \times 10 + 100 \times 20 = 2700 \text{ unit-seconds.}$$

It is convenient to view this integral as a diagram with time on the $x$ axis and space on the $y$ axis. In these terms the example program uses resources in this manner:

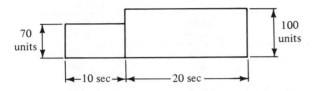

At many computer installations a charge $c_t$ is made for the total execution time of the program in addition to a charge $c_s$ for the space-time integral. The time charge is incorporated in the analysis by converting it to a space overhead of $c_t/c_s$ space units. (Derive this expression for yourself.) This charge is shown as additional space at the bottom of a diagram:

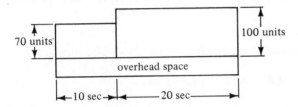

(Typical university computer centers charge a space overhead in the neighborhood of 10,000 to 200,000 words.)

Space-time analysis permits a complete comparison of the search methods in Algorithms 1.2 and 1.3. Our version of sequential search is slower but avoids the use of the *entry_number* array, so the situation can be shown by these two diagrams:

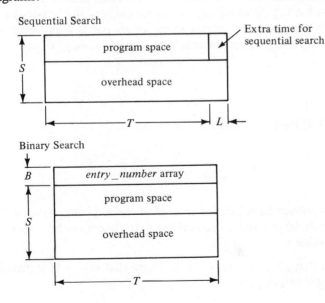

| Number of Elements | Linear Search | Binary Search |
|---|---|---|
| 10 | 98 | 98 |
| 100 | 103 | 100 |
| 1,000 | 158 | 116 |
| 10,000 | 1,088 | 273 |

**Table 1.6**
Relative cost of sequential and binary search versions of a typical program. The costs are based on treating binary search with 100 elements as having a cost of exactly one hundred. Because there is considerable processing other than searching, the cost is not much lower when there are only ten elements. Note that with more elements the cost rises much more rapidly for sequential search.

From the areas of the rectangles, it is clear binary search will be superior if

$$(B + S)T < (T + L)S.$$

For a reasonable set of assumptions about an imaginary program, Table 1.6 shows the relative space-time costs for these search methods when different numbers of items are stored in the table (see Exercise 13). The two schemes are close when there are not many items, but binary search is increasingly superior for larger table sizes.

Space-time analysis is also the best method to compare the constant-time search method suggested at the end of Section 1.4.1 with other search techniques. When the space of possible keys is small, say one or two letters, the constant-time approach is feasible. When there are four or more letters per key, however, the table space is considerably more expensive than any possible savings in execution time (see Exercise 14).

**Exercises**

1. Derive Equation (1.4) from the equation for averages:

$$\overline{X} = \frac{1}{m} \sum_{j=1}^{m} x_j,$$

   where the $x_j$ are the observed values for each of $m$ cases. [*Hint*: Assume that $m$ is some sufficiently large fixed value and use $p_i$ to determine how many $x_j$ should have each value $v_i$.]

2. Show by induction that $\sum_{i=1}^{n} i = n(n + 1)/2$. Prove that $n(n + 1)$ is divisible by two for all integer values of $n$.

3. The analysis of sequential search in Section 1.4.1 assumes that every search finds a key in the table. Extend the analysis by assuming that the search is successful only for fraction $s$ of all attempts.

4. Suppose the entries in *table* are sorted as for binary search, but a sequential search is chosen instead.
   (a) Show a revised version of Algorithm 1.2 that takes advantage of the array order to reduce the cost of searches when the element is not found.
   (b) Compute the expected number of times the **while** condition is tested in your solution to (a).
   (c) Rewrite your solution to (a) to also use a dummy key as in Exercise 2 of Section 1.3.

5. The computation of the number of loop executions for binary search given in the text was only an upper bound. Suppose the table has 100 elements; compute exactly the worst-case number of loop executions for a search that succeeds.

6. (a) Evaluate the eight expressions in Figure 1.5 for $n = 10$.
   (b) For what value of $n$ is $2^n = n^2$? $2^n = n^3$?
   (c) What is the smallest value attained by $n \lg n$?

7. "As I was going to St. Ives, I met a man with seven wives. Each wife had seven sacks, each sack had seven cats, and each cat had seven kits. Kits, cats, sacks, and wives, how many were going to St. Ives?" Rather than how many were going to St. Ives, let us consider how many I met. Suppose we have $m$ levels of objects where each possesses seven of the next higher level. (In the riddle, $m$ is 5 and the objects are man, wife, sack, cat, and kit.) Show that the total number of objects is proportional to $7^m$.

In the next two exercises the phrase "What is the order of" means "To what expression is the cost proportional?"

8. (a) What is the order of the usual algorithm to add two $n \times n$ matrices?
   (b) What is the order of the usual algorithm to multiply two $n \times n$ matrices?

9. One way to sort $n$ keys is to insert each in turn in a table using Algorithm 1.4.
   (a) What is the order of the number of binary search loop executions required for this sort?
   (b) What is the order of the number of times a key must be moved from one position in the array to another?

10. Suppose the statement **if** $B$ **then** $S$ is executed with the probability $p$ and suppose that $B$ has probability $q$ of being **true**; then $S$ will be executed with

probability $pq$. In general, the probability of execution of a statement is the product of the probabilities along the path to the statement. Compute the expected number of operators executed for the binary search in Algorithm 1.3. Use $w$ to denote the number of executions of the statements within the **while** and $t$, $u$, and $v$ to denote the probabilities of each of the alternatives of the **case**.

11. Compare sequential and binary search.
    (a) Compute the expected number of operators executed for the sequential search in Algorithm 1.2.
    (b) Compare the results of (a) with those of Exercise 10. (Assume the number of binary search loop executions is $\lg n$.) For what values of *number_of_elements* is binary search preferable?

12. The text compares two **if-then-else** implementations of the binary search loop body by assuming that each comparison requires an operator execution. Suppose that the result of the first comparison can be saved, so the second comparison is never made. What then are the relative costs of the two implementations?

13. Develop the formulas used to generate the data of Table 1.6. The following assumptions were used:

    There are $n$ elements in the table.
    Each element occupies 0.02 space units.
    The program other than the table occupies 30 space units and the overhead charge $(c_t/c_s)$ is 200 units.
    Each execution of the loop in sequential search takes one time unit.
    Each execution of the binary search loop takes two time units.
    After each search, the program spends 1000 time units on other processing.

    Reproduce the values found in Table 1.6. What would the results have been if the program spent only 100 time units on other processing after each search?

14. Extend the analysis of the previous exercise to the constant-time search sketched at the end of section 1.4.1. The appropriate parameter for analysis is the number of letters per key rather than the number of keys in the table.

# Chapter 2

# Elementary
# Data Objects

*Parvis e glandibus quercus.*
*[Tall oaks from little acorns grow.]*

Latin motto

The smallest, most primitive storage structure is the single bit—a unit so elementary that a bit variable can have only two values. These are usually called zero and one or **true** and **false**. Computer hardware does not usually operate on individual bit variables but instead deals with groups of bits called *words*. This chapter discusses how the various forms of values that can be stored in words are represented in terms of the bits that comprise the word. Later chapters show how higher storage structures are based on words.

A *word* is a sequence of bits—that is, a number of bits organized in a row. We can picture a word like this:

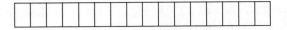

where each little box contains one of the two possible values for a bit. Bits at the left are called *high-order*, those at the right *low-order*. Usually a word value is written without the boxes, simply as a string of values, such as

0101011101001000

Since it is inconvenient to express a bit string as a sequence of ones and zeros, values are usually expressed in *hexadecimal notation* (base 16). Bits are grouped into blocks of four, and a single symbol is associated with each block to express

| Bit String Block | Hexadecimal Representation |
|:---:|:---:|
| 0000 | 0 |
| 0001 | 1 |
| 0010 | 2 |
| 0011 | 3 |
| 0100 | 4 |
| 0101 | 5 |
| 0110 | 6 |
| 0111 | 7 |
| 1000 | 8 |
| 1001 | 9 |
| 1010 | A |
| 1011 | B |
| 1100 | C |
| 1101 | D |
| 1110 | E |
| 1111 | F |

**Table 2.1**
Hexadecimal representation. In a bit string each block of four bits is represented by the corresponding symbol from the right-hand column.

the value in the block. There are sixteen possible configurations in a block of four bits, so sixteen symbols are used, as shown in Table 2.1. Using this notation, we would write the bit string above as

$$5748_{16}$$

where the subscript 16 indicates that the value is in hexadecimal.

The words above have sixteen bits, so the *word size* or *word length* is 16. Typical computers have only one or two word sizes, and all values are constrained to fit into one of those lengths. Early computers used six bits for each character and had word lengths such as 12, 18, 36, and 60. The IBM/360, introduced in 1964, stored characters in eight-bit *bytes* and the word length was 32. Since that time, most new computers have used eight-bit bytes for characters and two or four bytes for words. (A few computers have had words of twenty-four or forty-eight bits, sizes which are multiples of both six and eight.) In this chapter we will use word lengths of sixteen and thirty-two, but the rest of the book is independent of the specific word length.

Despite its elementary nature, a word is indeed a data structure:

- There are several sets of functions on words, as described below.
- Words are defined in terms of a more primitive storage structure, the single bit.
- The functions on words are implemented by algorithms that operate on bits.

Logical

| 0 1 0 1 | 0 1 1 1 | 0 1 0 0 | 1 0 0 0 |     $5748_{16}$

Integer

| 0 0 0 1 0 1 1 0 0 1 1 1 0 1 0 0 |     5748

| 0 1 0 1 0 1 1 1 0 1 0 0 1 0 0 0 |     22344

Floating point

| 0 | 0 1 1 0 1 | 1 0 1 1 0 0 1 1 1 0 |     $0.5744 \times 10^4 = 0.701172 \times 2^{13}$

| 0 | 1 0 1 0 1 | 1 1 0 1 0 0 1 0 0 0 |     $0.4004 \times 10^{-3} = 0.820313 \times 2^{-11}$

Character string

| 0 0 1 1 0 1 0 1 | 0 0 1 1 0 1 1 1 |     "57"

| 0 1 0 1 0 1 1 1 | 0 1 0 0 1 0 0 0 |     "WH"

**Figure 2.1**
Alternate interpretations of words. The second word in each pair contains the same bits as in the logical value (top line); the differing interpretations are shown at the right. The first word in each pair shows the bits that correspond to the version of "5748" shown at the right.

In the case of words, the algorithms are implemented by the hardware; for most operations—for example, adding two integers—there is a single hardware instruction. Four types of interpretation of words are provided:

*Logical.* The value is a string of bits, each of which may be interpreted independently of the others. See Section 2.1.

*Integer.* The values are treated according to the rules of the common arithmetic operations—addition, subtraction, multiplication, and division.

*Floating Point.* As with integers, the operations are those of arithmetic. However, a portion of each word is used to indicate the location of the decimal point.

*Character.* There is a translation between the bit value of each byte and a printed graphic character. This translation is implemented in the hardware of peripheral devices, such as terminals and printers.

For examples of words with each of these interpretations, see Figure 2.1.

This chapter will first explore these four interpretations of words. Logical values are discussed at length to show some of the ramifications of the operations available on bits. Then integers and conversion to decimal are shown to be related to the representation of "enumerated" types and characters. A section on "packed" words demonstrates that values in words are sometimes composed of several integerlike fields. All of these are discussions of the general concept of "representation": the use of one value to record another. The final section of the chapter discusses two important mathematical aspects of representations: reliability and efficiency.

**Exercises**

1. One modern technique for storing bits is in an integrated-circuit chip. For example, a square 0.2 inches on a side may contain 65,536 bits. How many eight-bit bytes can be stored on such a chip? What is the surface area devoted to each bit? If this chip was created by 100 : 1 photoreduction, what surface area was covered by the bit in the original design of the chip?

2. Complete the following chart of equivalences between logical and hexadecimal values.

| Logical | Hexadecimal |
|---|---|
| 1100000011011110 | $\text{C0DE}_{16}$ |
| 10110 | $16_{16}$ |
| 1011101010111110 | |
| 11111101101 | |
| 1111000111010000 | |
| 1110011101101 | |
| 110000001101 | |
| | $\text{B4}_{16}$ |
| | $\text{C0FFEE}_{16}$ |

3. In computer programming languages there are many notations for distinguishing hexadecimal from decimal values. Find three and invent one of your own. What are their advantages and disadvantages with respect to ease of interpretation by both human and computer?

4. Get a list of the instruction mnemonics and their meanings for some computer. Classify each instruction as to whether it deals with bits, integers, floating-point numbers, characters, or none of these.

# 2.1 BOOLEAN OPERATIONS AND LOGICAL VALUES

The operators defined for bit values are rather simple, but they have many important applications, including program implementation, proof of correctness, and even design. Although we cannot consider all applications, we will consider bit operations in some depth.

The operations on single bit values are the operators of boolean algebra: **and**, **or**, **not**, **xor**, and twelve others not named here. From the name of the algebra, bit variables and expressions are called *boolean variables* and *boolean expressions*. Examples in this chapter will use $p$, $q$, and $r$ as boolean variables.

Although boolean variables are conceptually single bits, their values are referred to as **true** and **false** instead of one and zero. This is especially appropriate when the boolean expression is part of the condition in an **if** or **while** statement. Since boolean variables have two possible values, there are exactly four possibilities for the input of a boolean function of two arguments:

| State | First argument | Second argument |
|:-----:|:--------------:|:---------------:|
| 1 | true | true |
| 2 | true | false |
| 3 | false | true |
| 4 | false | false |

To define a boolean operator, then, it is sufficient to specify its output for each of the four possible input states. This kind of table is called a *truth table*.

In Table 2.2 a truth table defines the four most common boolean operators: **and**, **or**, **not**, and **xor**. When expressed in terms of **true** and **false**, these definitions are more or less intuitively obvious: **and** returns **true** when its first *and* second arguments are **true**. **or** returns **true** when either its first *or* second argument or both are **true**. **xor** is an abbreviation for "exclusive or"; it is **true** only when *exactly* one of the arguments is **true**. **not** is a function of one argument; in the table its output depends on the value of $p$, but not $q$. The function **not** is true when its argument is **false** and vice versa.

Truth tables can be used to evaluate boolean expressions. The variables and the expression are written across the top of the table, and then all possible combinations of values for the variables are written below them. If there are $m$ variables, there will be $2^m$ combinations, and the table will have that many lines.

| $p$ | $q$ | $p$ and $q$ | $p$ or $q$ | $p$ xor $q$ | not $p$ |
|-----|-----|-------------|------------|-------------|---------|
| true | true | true | true | false | false |
| true | false | false | true | true | false |
| false | true | false | true | true | true |
| false | false | false | false | false | true |

**Table 2.2**
Truth-table definitions of the four most common boolean operators. The columns labeled $p$ and $q$ specify the values used for each of the other four columns. For example, the third line says that when $p$ is **false** and $q$ is **true** then $p$ **and** $q$ is **false** but the other three operators produce **true**.

| $p$ | $q$ | $r$ | ($p$ **and** $q$) | **not** $r$ | ($p$ **and** $q$) **or not** $r$ |
|------|------|------|------|------|------|
| true | true | true | true | false | true |
| true | true | false | true | true | true |
| true | false | true | false | false | false |
| true | false | false | false | true | true |
| false | true | true | false | false | false |
| false | true | false | false | true | true |
| false | false | true | false | false | false |
| false | false | false | false | true | true |

**Table 2.3**
Truth-table evaluation of the expression ($p$ **and** $q$) **or not** $r$. The left three columns list all possible values for $p$, $q$, and $r$.

The result value for each combination of inputs is written under each operator of the expression. Beginning with subexpressions inside parentheses, **not**'s are done first, then **and**'s, and finally **or**'s and **xor**'s. Table 2.3 shows the process for the expression

$$( p \text{ and } q ) \text{ or not } r.$$

Truth tables require enumeration of all possible input states; if there are many variables, it is easier to manipulate the boolean expression according to the rules of boolean algebra. These rules are similar to—and in some cases simpler than—the rules of arithmetic; see Table 2.4. The most complex and counterintuitive—DeMorgan's laws—are also among the most important, so we use them in Tables 2.5 and 2.6 to illustrate proof by truth table and proof by algebraic manipulation.

**Logical Variables.** Each boolean operator produces a one-bit result from one-bit arguments. These operators can be extended to operate on entire words by specifying that each bit position is treated separately; the result in position $i$ depends only on the values in position $i$ of the operands. Values and expressions of this type are called *logical*. Some examples are given in Table 2.7.

In addition to the boolean operators **and**, **or**, **not**, and **xor**, several other functions are useful for logical variables:

shift_left(*logical_word*, *shift_count*). The *logical_word* is shifted left the number of places indicated by *shift_count*. The leftmost bits of *logical_word* are lost and the rightmost bits are set to zero. For example:

$$\text{shift\_left}(1011011000101101, 4) \Rightarrow 0110001011010000.$$

shift_right(*logical_word*, *shift_count*). Same as shift_left, but the shift is to
the right.

integer(*logical_word*). This function does nothing to *logical_word*. Its purpose is to allow use of *logical_word* in contexts where an integer is
necessary. The integer value derived is the one with the same hexadecimal representation as the value of *logical_word*:

$$\text{integer}(1011011) \Rightarrow 91.$$

logical(*integer_value*). This function is the inverse of the integer function
above. It converts its argument so it can be used in contexts demand-
ing logical values:

$$\text{logical}(91) \Rightarrow 1011011.$$

| | |
|---|---|
| Commutativity | (1) $p$ **and** $q$ = $q$ **and** $p$ |
| | (2) $p$ **or** $q$ = $q$ **or** $p$ |
| Associativity | (3) $p$ **and** ($q$ **and** $r$) = ($p$ **and** $q$) **and** $r$ |
| | (4) $p$ **or** ($q$ **or** $r$) = ($p$ **or** $q$) **or** $r$ |
| Distribution | (5) $p$ **and** ($q$ **or** $r$) = ($p$ **and** $q$) **or** ($p$ **and** $r$) |
| | (6) $p$ **or** ($q$ **and** $r$) = ($p$ **or** $q$) **and** ($p$ **or** $r$) |
| Absorption | (7) $p$ **and** ($p$ **or** $q$) = $p$ |
| | (8) $p$ **or** ($p$ **and** $q$) = $p$ |
| | (9) $p$ **and** $p$ = $p$ |
| | (10) $p$ **or** $p$ = $p$ |
| Zero and identity | (11) $p$ **and true** = $p$ |
| | (12) $p$ **and false** = **false** |
| | (13) $p$ **or true** = **true** |
| | (14) $p$ **or false** = $p$ |
| | (15) $p$ **and not** $p$ = **false** |
| | (16) $p$ **or not** $p$ = **true** |
| Double negative | (17) **not not** $p$ = $p$ |
| DeMorgan's laws | (18) **not** ($p$ **and** $q$) = **not** $p$ **or not** $q$ |
| | (19) **not** ($p$ **or** $q$) = **not** $p$ **and not** $q$ |

**Table 2.4**
Rules for boolean algebra. These rules are similar to those of arithmetic and
can be similarly used. When these rules are applied, the letters $p$, $q$, and $r$ can
be replaced with constant values, other variables, or entire expressions.

| *Input Values* | | *Expression 1* | | *Expression 2* | | |
|---|---|---|---|---|---|---|
| *p* | *q* | **not** | (*p* **and** *q*) | **not** *p* | **or** | **not** *q* |
| true | true | false | true | false | false | false |
| true | false | true | false | false | true | true |
| false | true | true | false | true | true | false |
| false | false | true | false | true | true | true |

**Table 2.5**

Proof of equivalence by truth tables. By writing truth tables for two expressions we can determine if they are equivalent. In this example the two expressions are the two sides of DeMorgan's law (rule 18). Since the values of both expressions (boxed) are the same, the two expressions are equivalent.

| **not**( *p* **or** *q*) | |
|---|---|
| = **not**(**not not** *p* **or not not** *q*) | rule 17 (twice) |
| = **not**(**not**(**not** *p* **and not** *q*)) | rule 18 |
| = **not** *p* **and not** *q* | rule 17 |

**Table 2.6**

Proof by transformation. Rules (and other theorems) can be proved by transformations applying other rules. In this example, one DeMorgan's law (rule 19) is proved by application of the other DeMorgan's law (rule 18).

| a) | **and** | first operand | 0110100011101000 |
|---|---|---|---|
| | | second operand | 1000110111001101 |
| | result = | first **and** second | 0000100011001000 |
| (b) | **or** | first operand | 0110100011101000 |
| | | second operand | 1000110111001101 |
| | result = | first **or** second | 1110110111101101 |
| (c) | **xor** | first operand | 0110100011101000 |
| | | second operand | 1000110111001101 |
| | result = | first **xor** second | 1110010100100101 |
| (d) | **not** | operand | 0110100011101000 |
| | result = | **not** operand | 1001011100010111 |

**Table 2.7**

The extension of boolean operations to entire words. The bit values 1 and 0 are used for the boolean values **true** and **false**, respectively. Note that the result in any given output position depends only on the operand bits in that same position. Thus for **and** the result has ones only in positions where both operands have ones.

As an example of these functions, consider multiplication by a power of two. Since each bit position corresponds to one higher multiple of two, a shift of one position to the left corresponds to multiplication by two. For instance, six is 0110, and when this value is shifted left one it becomes 1100, which is twelve, exactly the double of six. To multiply $k$ by 8, it would be shifted left three times, because 8 equals $2^3$:

$$\text{integer}(\text{shift\_left}(\text{logical}(k), 3))$$

In most computers, shifts are much faster than multiplication, so this technique can be a useful optimization.

**Exercises**

1. Find an expression for $p$ **xor** $q$ in terms of **and, or,** and **not.** Demonstrate the equivalence with a truth table.

2. What is the value of $p$ **xor** $q$ **xor** $p$? Prove it with a truth table. Suppose your computer has an instruction "$t \leftarrow t$ **xor** $v$" which forms the **xor** of $t$ and $v$ and stores the result in $t$. Show how to interchange the values of two variables using only this instruction.

3. Find rules for **xor** like those for **and** and **or** in Table 2.4. Prove the commutativity rule by application of the equivalence between **xor** and the other operators. Prove distributivity by a truth table.

4. What is the analog for **xor** of DeMorgan's laws? Prove it with a truth table. Prove it again by application of the equivalence with **and, or,** and **not.**

5. An important technique for devising boolean expressions is to work backward from the truth table. For example, we may desire an expression for the truth table shown in Table 2.8. One possible such expression is the *disjunctive normal form*; the expression is written as the **or** of a number of terms, each of which corresponds to a row of the table that produces a true output. In the example, the term corresponding to the third row would be

$$p \text{ and } (\text{not } q) \text{ and } r;$$

that is, it contains "variable" for **true** inputs and "**not** variable" for **false** inputs.

(a) Give the disjunctive normal form for Table 2.8.
(b) Find a much simpler expression for the table.
(c) Prove by truth tables that your answers to (a) and (b) are correct.
(d) Prove by transformations that your answers to (a) and (b) are equivalent expressions.

| Inputs | | | Output |
|---|---|---|---|
| $p$ | $q$ | $r$ | |
| true | true | true | false |
| true | true | false | false |
| true | false | true | true |
| true | false | false | true |
| false | true | true | false |
| false | true | false | false |
| false | false | true | true |
| false | false | false | false |

**Table 2.8**

Exercise 5

6. There are sixteen boolean functions of two boolean variables—one corresponding to each of the sixteen possible combinations of values in the result column of a truth table. However, all these can be represented by expressions involving only one of **not** or **xor** and one of **and** or **or**. For some entries in Table 2.9 we have given the result column; for others we have given an expression equivalent to the result column in terms of **not** and **and**. Fill in the other entries of the table.

| Function name | $p$: $q$: | Results when $p$ and $q$ have these values: true true | true false | false true | false false | Written with only **not** and **and** |
|---|---|---|---|---|---|---|
| **false** | | false | false | false | false | $p$ and not $p$ |
| **not**( $p$ **or** $q$ ) | | false | false | false | true | |
| **not**( $q$ **implies** $p$ ) | | false | false | true | false | $q$ and not $p$ |
| **not** $p$ | | | | | | not $p$ |
| **not**( $p$ **implies** $q$ ) | | | | | | |
| **not** $q$ | | | | | | not $q$ |
| $p$ **xor** $q$ | | false | true | true | false | (see Exercise 1) |
| **not**( $p$ **and** $q$ ) | | | | | | not( $p$ and $q$ ) |
| $p$ **and** $q$ | | | | | | $p$ and $q$ |
| $p$ **equiv** $q$ | | true | false | false | true | |
| $q$ | | | | | | $q$ |
| $p$ **implies** $q$ | | true | false | true | true | |
| $p$ | | | | | | $p$ |
| $q$ **implies** $p$ | | | | | | |
| $p$ **or** $q$ | | | | | | |
| **true** | | | | | | |

**Table 2.9**

Exercise 6

**procedure** *RECOGNIZE_IDENTIFIER* **returns boolean**
  *count*: **integer**  ⟦the number of valid identifier characters so far⟧
  **if** *char* < "A" **or** *char* > "Z" **then return false**
  *STORE*(*char*)
  *GET_CHARACTER*
  *count* ← 1
  **while** *count* _____ **do**  ⟦keep identifier short⟧
    **if** _____  ⟦check for valid characters⟧
       **then return true**
   *STORE*(*char*)
   *GET_CHARACTER*
   *count* ← *count* + 1
 **return false**

**Algorithm 2.1**
Exercise 7

7. A FORTRAN identifier must begin with a letter and contain no more than six letters and digits. Algorithm 2.1 is part of a compiler and has the function of recognizing an identifier. That is, it returns **true** if the next characters in the input stream constitute an identifier. It also extracts the identifier for further processing by calling *STORE* for each of its characters. The input stream is represented by successive calls of *GET_CHARACTER*; the assumption is that *RECOGNIZE_IDENTIFIER* and all other recognizer procedures expect an initial value in *char* and leave the character following the recognized item in *char* when they finish.

 (a) Fill in the blanks so that Algorithm 2.1 fulfills its intention.

 (b) Under what circumstances will *char* contain a letter or digit after Algorithm 2.1 exits?

8. Algorithm 2.2 is outrageously bloated with unnecessary code. Its purpose is to read an integer from an input stream and leave its value in *val*. Its assumptions about the input stream are the same as those for the previous exercise. Rewrite this program to make it shorter; the original version only took four lines. Try to describe some of the transformations you use to reduce this procedure; avoid the approach of simply writing your own version to accomplish the same function.

9. Write boolean expressions for *c* and *s* in terms of *p*, *q*, and *r* as specified by Table 2.10.

 (*Note*: If **true** is one and **false** is zero, *c* and *s* are the carry and sum from adding the binary values of *p*, *q*, and *r*. Consequently your expressions for *c* and *s* specify circuits for calculating binary sums.)

*val*: **integer**                          ⟦value of the integer found⟧
*ininteger*: **boolean**                    ⟦**true** until *char* is a nondigit⟧
*toolarge, toosmall*: **boolean**   ⟦**true** if *char* > "9" or < "0", respectively⟧
*convchar*: **integer**                     ⟦the decimal value of the digit in *char*⟧

*val* ← 0
**if** *char* ≤ "9" **then**
    **if** *char* ≥ "0" **then**
        *ininteger* ← **true**
        **while** *ininteger* **do**
            *convchar* ← (integer value corresponding to
                        digit in *char*   ⟦see Section 2.2.3⟧)
            **if** *convchar* < 0 **then** *convchar* ← 0
            **if** *convchar* > 9 **then** *convchar* ← 9
            *val* ← 10 * *val* + *convchar*   ⟦see Equation (2.2)⟧
            *char* ← *GET_CHARACTER*
            **if** *char* > "9" **then** *toolarge* ← **true**
                        **else** *toolarge* ← **false**
            **if** *char* < "0" **then** *toosmall* ← **true**
                        **else** *toosmall* ← **false**
            *ininteger* ← *not*(*toolarge* **or** *toosmall*)

**Algorithm 2.2**
Exercise 8

| *p* | *q* | *r* | *c* | *s* |
|------|------|------|------|------|
| true | true | true | true | true |
| true | true | false | true | false |
| true | false | true | true | false |
| true | false | false | false | true |
| false | true | true | true | false |
| false | true | false | false | true |
| false | false | true | false | true |
| false | false | false | false | false |

**Table 2.10**
Exercise 9

10. Assume that your computer has operations for **and, or,** and **xor,** and that every word is sixteen bits. Write a single statement with one operator that will set to 1 the third bit from the left in the logical variable *ioflag*. Write another statement to set it to 0 and a third to invert the value of the bit.

11. *char* is a character variable. For what values of *char* will this program print "digit"?

$$\textbf{if } char \leq \text{``9''} \textbf{ or } char \geq \text{``0''} \textbf{ then } \text{print (``digit'')}$$
$$\textbf{else } \text{print (``not digit'')}$$

12. Write a function called *count_ones* that has a logical value as its argument and returns an integer giving the number of one bits in the argument. (This is an excellent problem for assembly language experimentation. What is the smallest number of instruction executions required in your machine? The least amount of memory?)

## 2.2 INTEGERS

What is "00010011"?

It might be a representation of the string of bits in one byte of computer memory. Equally well, it might be a representation of a binary number having the decimal value 19, as we will see below. The fact that a string of bit values can be interpreted as a binary number is called the *natural interpretation* of a bit string as a binary number. In this section we will consider two applications of this natural interpretation to represent values. First, however, we must consider how to translate values between bases—for instance, between binary and decimal.

### 2.2.1 Conversion of Integers From One Base to Another

A *positional* number system expresses a number in terms of a *base* or *radix* and a set of *digit symbols*. The base is an integer assigned, for example, the value $b$. The digit symbols are an ordered set and represent the values $0, 1, 2, 3, \ldots,$ $b - 1$. In the familiar decimal system the base is ten and the digit symbols are 0, 1, 2, 3, 4, 5, 6, 7, 8, and 9, which have their respective values. A number $n$ is represented by writing a string of digit symbols, $a_{k-1}a_{k-2} \cdots a_2a_1a_0$, and is interpreted by multiplying the value of each digit by a corresponding power of the base:

$$n = a_{k-1}b^{k-1} + a_{k-2}b^{k-2} + \cdots + a_1b^1 + a_0b^0$$
$$= \sum_{i=0}^{k-1} a_i b^i. \tag{2.1}$$

In computer science, four bases are common: binary (base 2), octal (base 8), decimal (base 10), and hexadecimal (base 16). The digit symbols are the same as those used for the decimal system with the addition of six more for base 16, as shown in Table 2.1. In this book the base of a number will generally be clear from context; binary numbers will have only zeros and ones and will seldom be used; octal numbers will not be used; decimal values are not marked; and hexadecimal numbers are followed by subscript 16.

In Equation (2.1) the coefficients $a_i$ and the powers of $b$ can be expressed in any base, as long as the arithmetic is carried out using the rules for that base. This fact leads directly to a technique for transformation from one base to another, as illustrated by these three examples:

**Example 1.**
What is 10011 if it is assumed to be a number in the binary number system? If we do the calculations and representation in decimal, the result will be the required value:

$$10011 \quad \text{(base 2)}$$

$$
\begin{aligned}
&= 1 \qquad 0 \qquad\quad 0 \qquad\quad 1 \qquad\quad 1 \qquad\qquad \text{(base 2)}\\
&= 1 \times 2^4 + 0 \times 2^3 + \ 0 \times 2^2 + 1 \times 2^1 + 1 \times 2^0 \\
&= 1 \times 16 + 0 \qquad + 0 \qquad + 1 \times 2 + 1 \times 1 \\
&= 16 + \ 2 \ + 1 \quad = 19
\end{aligned}
\left.\rule{0pt}{40pt}\right\} \text{(base 10)}
$$

**Example 2.**
What is 10011 if it is assumed to be a number in the decimal system? We will do the calculation in the decimal system to derive a decimal result.

$$10011 \quad \text{(base 10)}$$

$$
\begin{aligned}
&= 1 \qquad\qquad 0 \qquad\quad 0 \qquad\quad 1 \qquad\quad 1 \\
&= 1 \times 10^4 \quad + 0 \times 10^3 + 0 \times 10^2 + 1 \times 10^1 + 1 \times 10^0 \\
&= 1 \times 10{,}000 + 0 \qquad + 0 \qquad + 1 \times 10 + 1 \times 1 \\
&= 10{,}000 \qquad + 10 \qquad + 1 \\
&= 10{,}011 \qquad \text{(no surprise)}
\end{aligned}
\left.\rule{0pt}{50pt}\right\} \text{(base 10)}
$$

**Example 3.**
What is 10011 if it is assumed to be a number in the decimal system? This time we will do the calculation in binary: the base and all multiplications

and additions will be binary, and so will the result. Note that the decimal value 10 is expressed in binary notation as 1010.

10011   (base 10)

$$
\begin{aligned}
&=1 & 0 \qquad\quad & 0 \qquad\quad 1 \qquad\quad 1 \\
&=1 \times 1010^{100} & + 0 \times 1010^{11} + {}& 0 \times 1010^{10} + 1 \times 1010^{1} + 1 \times 1010^{0} \\
&=1 \times 10011100010000 + 0 & + 0 \qquad\quad & + 1 \times 1010 \ + 1 \times 1 \\
&=10011100010000 & + 1010 \qquad\quad & + 1 \\
&=10011100011011
\end{aligned}
\right\} \quad \text{(base 2)}
$$

Equation (2.1) is somewhat inconvenient for computer calculations because it requires computing large powers of the base. To avoid this, the equation can be rewritten as

$$
\begin{aligned}
n = (((\cdots \, ((a_{k-1}b + a_{k-2})b + a_{k-3})b + \cdots \\
+ a_3)b + a_2)b + a_1)b + a_0.
\end{aligned}
\tag{2.2}
$$

At each step of this computation, called *Horner's method*, the present value of the result is multiplied by the base and the next digit is added. Like Equation (2.1), Equation (2.2) lends itself to conversions in both directions between binary and decimal. We will leave most of these to the exercises but will illustrate Equation (2.2) at the end of section 2.2.3.

Many normal arithmetic operations are implemented for integers in computer hardware. These include addition, subtraction, multiplication, and division. All these can generate one error or another: division by zero is not possible, just as it is not possible in ordinary arithmetic. Moreover, the fact that integers must be represented in finite words means that an operation can generate a value that is larger than can fit in a word. For addition and subtraction, hardware usually treats this possibility by signaling an "arithmetic overflow" error. For multiplication, hardware usually simply produces a two-word result as the product of two one-word operands. In higher-level languages the leftmost of the two words is often ignored.

**Exercises**

1. What is the magnitude of the largest integer that can be represented in a 32-bit word?

2. Prove that

$$2^k - 1 = \sum_{i=0}^{k-1} 2^i.$$

(In other words, prove that the binary representation consisting of $k$ ones has the value $2^k - 1$ and is thus one less than the smallest value in $k + 1$ bits.)

3. How many distinct values can be represented by $k$ bits? Is it possible to partition these values into positive, negative, and zero while having exactly as many positive as negative? How? Or why not?

4. Show how to use successive decimal divisions by the base to convert a number from decimal to some other base. Illustrate this process by converting 497 to base 7.

5. Suppose your computer has 32-bit words and that eight four-bit quantities have been read into word $v$. Each quantity represents a decimal digit by the natural interpretation. Write a *loop-free* program using only one temporary location to convert $v$ to binary. [*Hint:* Since the binary for 10 is 1010, you can multiply by ten by shifting and adding.]

The next three exercises explore computer techniques for conversion from binary to decimal.

6. Assume your computer permits representation of decimal values as an array of bytes, with each byte containing a value between 0 and 9. Under this assumption, suppose there is an instruction to add one such array to another. Using this instruction and a table of arrays giving the values of the powers of two, write a program to implement Equation (2.1) for conversion from binary to decimal.

7. What will be the remainder if the quantity $n$ in Equation (2.2) is divided by the base $b$? What will be the quotient? Use these facts as the basis for a program to convert from binary to decimal by successive divisions.

8. Since multiplication is usually faster than division, write a version of the program in the previous exercise that uses multiplication by $1/b$ instead of division by $b$.

9. It is not necessary that the positions in an integer representation have the values of powers of some base. For example, in the Fibonacci number system the successive positions have the values: $1, 2, 3, 5, 8, 13, 21, 34, 55, 89, 144, \ldots$ (each is the sum of the preceding two). In this system 19 can be represented as 101001 ($13 + 5 + 1$). Derive rules so that the representation of every

integer is unique, using only zeroes and ones as digits. Show how to add two integers represented in the Fibonacci number system.

10. Write a sequence of statements to multiply the values in $a$ and $b$ and store the result in $rhi$ and $rlo$, respectively. Assume that $a$ and $b$ are less than $10^8$ and that the true result is given by $rhi \times 10^8 + rlo$. Use only the operations of addition, subtraction, multiplication, and integer division. Assume that multiplication yields only the low-order end of the product. [*Hint:* Split $a$ and $b$ into two parts and perform four multiplications.]

11. There are four missing terms in the sequence below; what are they?

$$10, 11, 12, 13, 14, 15, 16, 17, 20, 22, \underline{\phantom{0}}, \underline{\phantom{0}}, \underline{\phantom{0}}, \underline{\phantom{0}}, 10000.$$

[*Hint:* The terms preceding 10 may all be written as G. There is only one possible term after 10000.]

### 2.2.2 Enumerated Types

In many programming situations a variable is needed that will take on only a small number of values, and the programmer wishes to give a name to each such value. There may be as few as two values, as in the case of variables that represent sex and take on the values *MALE* and *FEMALE*. Or there may be many values, as in the case of a variable to represent the type of account for a bank's data base. The latter might take on values such as *SAVINGS, REGULAR_CHECKING, LOW_VOLUME_CHECKING, AUTOLOAN_ CHECKING, PHONE_PAYMENT, TRUST,* and so on.

Variables of this variety are said to be of *enumerated type*. Declaration of such a variable proceeds by first defining a type and enumerating the possible values for variables of that type. Then variables are declared to have that type. The account type could be declared in PASCAL, for example, as

TYPE *ACCOUNT_TYPE* = (*SAVINGS, REGULAR_CHECKING,*
      *LOW_VOLUME_CHECKING, AUTOLOAN_CHECKING,*
      *PHONE_PAYMENT, TRUST*)

Later a variable of this type would be declared by writing

*CUST_ACCT*: *ACCOUNT_TYPE*

The result of these declarations is that six names for values have been created. These names can be used to assign the corresponding value:

*CUST_ACCT* ← *LOW_VOLUME_CHECKING*

or they can be used for comparison:

$$\textbf{if } CUST\_ACCT = TRUST \textbf{ then} \ldots$$

Variables of enumerated type are implemented directly as integers. At the simplest, the names in the list are assigned successively higher integers starting with one. Statements referring to these named constants are translated into operations referring to the appropriate integers. Thus the assignment above would assign the integer value 3 to $CUST\_ACCT$ and the **if** predicate would test for value 7.

## Exercises

1. Write a declaration for a variable $DAY$, the values of which are to be the days of the week. Write a **case** statement to advance $DAY$ to the next day of the week. (PASCAL provides an operator called SUCC that performs most of this computation by doing an integer addition. The change from the last day to the first would still require a separate test.)

2. As part of a program to create new bank accounts, there is a routine to read in information about the new account. Some implementations permit the programmer to write

$$\text{READ}(CUST\_ACCT)$$

to retrieve an $ACCOUNT\_TYPE$ value from the input. Input values would be names such as $REGULAR\_CHECKING$, $PHONE\_PAYMENT$, and the rest; these would be converted to integers by the input routines. This conversion requires a symbol table during program execution to find the appropriate values for names. What attributes should be defined for each entry in this symbol table?

### 2.2.3   Character Representations

One of the most common instances of enumerated types is the representation of characters: letters, digits, punctuation, and others. The desired characters are listed in a sequence, and the code value for each is its location in the sequence. For historical reasons the sequence is called the *collating sequence* of the characters; examples are shown in Tables 2.11 and 2.12. In these tables the column and row indexes taken together form the hexadecimal value corresponding to the location of the character in the sequence. Thus the top left character has the code 00 and the eighth down has code 07. In USASCII the latter corresponds to the control character BEL, which has the effect of sounding the bell if transmitted to

| | 0_ | 1_ | 2_ | 3_ | 4_ | 5_ | 6_ | 7_ |
|---|---|---|---|---|---|---|---|---|
| _0 | NUL | DLE | space | 0 | @ | P | ` | p |
| _1 | SOH | DC1 | ! | 1 | A | Q | a | q |
| _2 | STX | DC2 | " | 2 | B | R | b | r |
| _3 | ETX | DC3 | # | 3 | C | S | c | s |
| _4 | EOT | DC4 | $ | 4 | D | T | d | t |
| _5 | ENQ | NAK | % | 5 | E | U | e | u |
| _6 | ACK | SYN | & | 6 | F | V | f | v |
| _7 | BEL | ETB | ' | 7 | G | W | g | w |
| _8 | BS | CAN | ( | 8 | H | X | h | x |
| _9 | HT | EM | ) | 9 | I | Y | i | y |
| _A | LF | SUB | * | : | J | Z | j | z |
| _B | VT | ESC | + | ; | K | [ | k | { |
| _C | FF | FS | , | < | L | \ | l | \| |
| _D | CR | GS | - | = | M | ] | m | } |
| _E | SO | RS | . | > | N | ^ | n | ~ |
| _F | SI | US | / | ? | O | _ | o | DEL |

**Table 2.11**

USASCII character code. The code for each entry in the table is the two-digit hexadecimal value formed by taking the digits at the top of the column and the left of the row. The code for "A" is $41_{16}$ (that is, 65 decimal). The two- and three-character names are names for control codes used in communication. For example, BEL at $07_{16}$ rings the bell at typewriterlike terminals.

| | 0_ | 1_ | 2_ | 3_ | 4_ | 5_ | 6_ | 7_ | 8_ | 9_ | A_ | B_ | C_ | D_ | E_ | F_ |
|---|---|---|---|---|---|---|---|---|---|---|---|---|---|---|---|---|
| _0 | NUL | DLE | DS | | space | & | - | | | | | | { | } | \ | 0 |
| _1 | SOH | DC1 | SOS | | | | / | | a | j | ~ | | A | J | | 1 |
| _2 | STX | DC2 | FS | SYN | | | | | b | k | s | | B | K | S | 2 |
| _3 | ETX | TM | | | | | | | c | l | t | | C | L | T | 3 |
| _4 | PF | RES | BYP | PN | | | | | d | m | u | | D | M | U | 4 |
| _5 | HT | NL | LF | RS | | | | | e | n | v | | E | N | V | 5 |
| _6 | LC | BS | ETB | UC | | | | | f | o | w | | F | O | W | 6 |
| _7 | DEL | IL | ESC | EOT | | | | | g | p | x | | G | P | X | 7 |
| _8 | GE | CAN | | | | | | | h | q | y | | H | Q | Y | 8 |
| _9 | RLF | EM | | | | | | ` | i | r | z | | I | R | Z | 9 |
| _A | SMM | CC | SM | | ¢ | ! | | : | | | | | | | | LVM |
| _B | VT | CU1 | CU2 | CU3 | . | $ | , | # | | | | | | | | |
| _C | FF | IFS | | DC4 | < | * | % | @ | | | | | ⌠ | ⌐ | | |
| _D | CR | IGS | ENQ | NAK | ( | ) | _ | ' | | | | | | | | |
| _E | SO | IRS | ACK | | + | ; | > | = | | | | | | ⌐ | | |
| _F | SI | IUS | BEL | SUB | \| | ¬ | ? | " | | | | | | | | ED |

**Table 2.12**

EBCDIC character code. This chart is read the same way as Table 2.11. The code for "A" is $C1_{16}$ (that is, 193 decimal).

a device that recognizes this code. In EBCDIC this same control character has code 2F, while code 07 corresponds to DEL, which is intended to be ignored.

Other than assignment and input/output, the principal operation on characters is to compare two characters to test for equality or to see which comes first in the collating sequence. When writing programs with such comparisons, it is useful to have some notion of where the characters come in the sequence. For USASCII, this information can be summarized by

controls < space < special characters < digits < capitals < lower case,

while for EBCDIC the sequence is summarized by

controls < space < special characters < lower case < capitals < digits.

In both cases these relationships omit a few special characters which are inserted at other points in the sequence. Note especially that in EBCDIC some special characters appear between the letters.*

It is also possible to do arithmetic on the integer values that encode the characters. To permit this, we have a function "integer" that returns the integer giving the location of the character in the collating sequence. Note that this function has no effect on the actual value; it just causes the value to be treated as an integer rather than a character. Thus in USASCII integer("A") yields $41_{16} = 65$ as an **integer** value.

To illustrate the concept of dealing with the integer values equivalent to character strings, consider writing an algorithm to read a string of digit characters and convert the value to an integer. Although such an algorithm can easily be written as a while loop (Exercise 2), it is written below recursively to illustrate recursion. The algorithm *READ_INTEGER* requires as its arguments an integer value and a character that may be a digit. If it is, the routine includes the digit at the right end of the integer argument and calls itself to process digits further to the right. The value gathered from the input is returned as the final value. A compiler would call this routine to recognize an integer from the input stream; the initial call would be

$$READ\_INTEGER(0,\ last\_input\_char\,)$$

where *last_input_char* is expected to be the first digit of an integer. (What happens if it is not a digit?)

---

*History: In years before EBCDIC, capital letters were punched on cards with two holes for each letter. One of these holes was assigned a digit value, and this value became the low-order hexadecimal digit of the EBCDIC code. The reason "S" is assigned a code ending in two is allegedly that the engineers were uncertain the code ending in one could be reliably interpreted by the card readers of the time.

> **procedure** *READ_INTEGER*(*val*, *ch*) **returns integer**
>         *val*: **integer**          ⟦part of integer processed so far⟧
>         *ch*: **character**     ⟦next character in the input⟧
> **if** "0" > *ch* **or** *ch* > "9" **then return** *val*     ⟦end of integer reached⟧
> **return** *READ_INTEGER* 10 * *val* + (integer(*ch*) − integer("0")),
>   *GET_NEXT_CHARACTER*)
>                                ⟦see Equation (2.2)⟧

Note that integer("0") is both more meaningful to the reader and more portable than explicit integer values such as 48 and 240 as would be used for USASCII and EBCDIC.

## Exercises

1. If USASCII is used, what are the values of "A" < "a", "3" < "8", "9" < "a", " " < "2", "," < "a", "," < "#", integer("3") − integer("0"), integer("h") − integer("a"), and integer("z") − integer("a")? Contrast the values of all the preceding expressions with their values if the character code were EBCDIC.

2. Write an iterative version of *READ_INTEGER*. (See also the solution to Exercise 8 of Section 2.1.)

3. If the character code is USASCII, we can determine if variable *ch* contains a capital letter by testing

$$\text{"A"} \le ch \text{ and } ch \le \text{"Z"}.$$

   (a) Is this correct for EBCDIC?
   (b) Show how to use an array of boolean values to determine if the value in *ch* is a capital letter.
   (c) Describe how to extend the solution of (b) to determine whether *ch* contains a letter and whether it is capital or lower case.
   (d) Write a routine that correctly initializes the array needed for (c).

4. Write an expression such that if *ch* contains a lower-case letter, the value of the expression will be the corresponding capital letter. Your expression should work for both USACII and EBCDIC and must not reference an array. To convert from an integer to a character value, use the function called "character" that performs the operation inverse to "integer."

5. Write a program that has no input and has as its only output an exact copy of itself. This trick can be performed in all languages but is easier for some. In those that use two quotation marks to represent a single quotation in a constant, the program contains one such constant and prints it or portions twice to get the double quotation mark in the output.

## 2.3  PACKED WORDS

In many situations data items do not need to occupy an entire word. For example, sex can be encoded in a single bit, and age can usually be encoded in fewer than eight bits (why?). When information is encoded with several data items in a single word, that word is said to be *packed*.

The operation of storing and accessing packed data is called *packing* and *unpacking*. A simple way to do this is with ordinary arithmetic operations; for example age and sex can be stored in a single word using the low-order bit for sex:

$$agesex \leftarrow 2 * age + sex.$$

This assignment results in the packed word:

| age | sex |
|-----|-----|

Division can unpack a word to access its components:

$$age \leftarrow \lfloor agesex/2 \rfloor$$

$$sex \leftarrow agesex \textbf{ mod } 2$$

However, integer arithmetic is a poor packing technique because multiplication and division are usually rather slow operations.

A more efficient technique for packing and unpacking data is to use logical and shift operations. A given field is extracted from a word by **and**'ing the word with a *mask* containing one bits in the appropriate positions and then shifting the result so the desired field is in the low-order end of a word. For example, suppose *age*, *sex*, *number_of_siblings*, and *number_of_children* are packed in a word called *person_data* with fields of seven, one, four, and four bits each:

| age | sex | # sibs | # kids |
|-----|-----|--------|--------|

Then *number_of_siblings* can be extracted by the expression:

$$\text{shift\_right}( \textit{person\_data} \textbf{ and } 00F0_{16}, 4)$$

The inverse operation of inserting a value for *number_of_siblings* is more complex, because the values in other fields must be preserved. The trick is to first **and** *person_data* with a mask to delete the old value for *number_of_siblings* and then **or** the new value into the result:

$$person\_data \leftarrow (\, person\_data \text{ and } \text{FF0F}_{16})$$
$$\text{or } (\text{shift\_left}(\, number\_of\_siblings, 4) \text{ and } \text{00F0}_{16}) \qquad (2.3)$$

The final **and** is to ensure that the new value of *number_of_siblings* does not interfere with the values for age and sex.

Another approach to creating a value for *person_data* is to construct the entire value at once:

$$person\_data \leftarrow$$
$$\text{shift\_left}(age, 9)$$
$$\text{or } \text{shift\_left}(sex \text{ and } 1_{16}, 8)$$
$$\text{or } \text{shift\_left}(number\_of\_siblings \text{ and } \text{F}_{16}, 4)$$
$$\text{or } (number\_of\_children \text{ and } \text{F}_{16}) \qquad (2.4)$$

These two approaches illustrate an often-made choice in the design of data structures. Functions can either modify existing values or create entirely new values. The latter approach sometimes makes clearer programs, because in a newly created value there is less uncertainty as to what the other fields are.

. The use of packed data is an excellent example of a space-time trade-off problem. Unpacked data may occupy a large amount of space, but packing and unpacking consume processing time. (See Exercise 3.) Often, hardware provides special conversion operations for a few data representations that are processed by other operations. As discussed in succeeding subsections, these include signed integers, floating-point values, character strings, and decimal digit strings.

### Exercises

1. There are two ways to increment the number of siblings for the value in *person_data*. The first uses the pack and unpack operations shown above, and the second adds a value to *person_data* directly.

   (a) Show instructions for both methods.

   (b) Show how to extend the second method so the person will not undergo a sex change when there are too many siblings.

2. What error might occur undetected on execution of the second **and** in statement (2.3)? Why is there no need for an **and** with age in (2.4)?

3. Data values may be stored in either packed or unpacked form. The value 250,000,000 (roughly the population of the United States) can be stored in

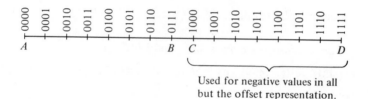

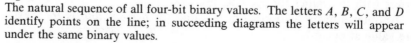

Used for negative values in all
but the offset representation.

**Figure 2.2**
The natural sequence of all four-bit binary values. The letters $A$, $B$, $C$, and $D$
identify points on the line; in succeeding diagrams the letters will appear
under the same binary values.

many ways, including one character per word for nine words or as an integer
in a 32-bit word.

(a) Estimate how many operations would be required to convert a nine-digit
value to binary using the *READ_INTEGER* routine of Section 2.2.3. (See
Section 1.4.2.)

(b) The cost of writing a record to disk can be written as $c_1 + nc_2$, where $c_1$
and $c_2$ are cost constants and $n$ is the number of words in the record.
Roughly speaking, $c_1$ is at least a thousand operations and $c_2$ is about one.
Given these factors, discuss whether conversion to binary of nine-digit
values would reduce costs.

### 2.3.1   Signed Integers

The correspondence between bit strings and binary numbers generates a
straightforward representation for positive integers, as discussed in Section 2.2.
Unfortunately, there is no equally direct representation for negative integers; all
choices have computational difficulties either in the calculation of sums or in
overflow. The four most common representations are presented below: two's
complement, one's complement, signed magnitude, and offset. To give an intuitive notion of the meaning of these representations, they will be expressed in
terms of the number line—the natural progression of $k$-bit values shown in
Figure 2.2. The first three representations use the second half of the number line,
from $C$ to $D$, to represent the negative values, so in these cases the high-order bit
of a negative value is one and of a positive value is zero. Where the discussions
include arithmetic expressions, any overflow from the $k$th to the $(k + 1)$st bit is
ignored.

The *two's-complement* representation is formed by placing the last half of
the number line in front of zero. Thus $-1$ is represented by the bit string that
was last in the natural order; see Figure 2.3. The name of the representation
derives from the fact that if $n$ is a positive integer of $k - 1$ bits, the $k$-bit
two's-complement representation of $-n$ is given by $2^k - n$. One simple way to

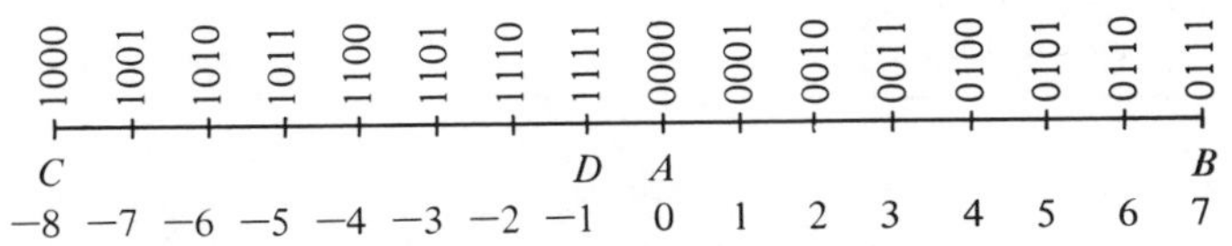

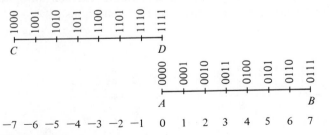

**Figure 2.3**
The two's complement representation. Note that there are more negative values than positive.

compute the two's-complement negative of a binary value is to first convert all ones to zeros and vice versa and then add one to the result. For example, with four bits the representation of $-2$ is given by $0010 \to 1101 \to 1110$, which is indeed the next-to-last value in the natural progression (and is also $2^4 - 2$).

The *one's-complement* representation also places the last half of the line in front of the positive integers, but the largest number and zero are made to overlap; there are two representations for zero, as shown in Figure 2.4. The $k$-bit one's-complement representation of $-n$ is given by $2^k - 1 - n$. (This is exactly one smaller than the two's complement, as could be expected from the diagram.) In one's complement representation the negative of a value is computed by simply changing all ones to zeros and vice versa. Thus the four-bit representation of $-2$ is $0010 \to 1101$. (Verify that this is $2^4 - 1 - 2$.)

The *signed-magnitude* representation is a packed value; the $k - 1$ bits on the right are used to represent the absolute value of the integer, and the high-order bit is set to one if the value is negative. This corresponds to moving the second half of the number line to the front of zero, but reversing it end for end and overlapping the previously middle value with zero, as depicted in Figure 2.5. As with one's complement, there are two representations for zero. To manually convert a value to negative, the first bit is simply reversed in value. This corresponds to the expression $2^{k-1} + n$ for negative $n$.

The *offset* representation leaves the number line in its order and the integers in their order as well; however, it abandons the natural representation. To

**Figure 2.4**
One's complement overlaps two values, so there are two representations of zero.

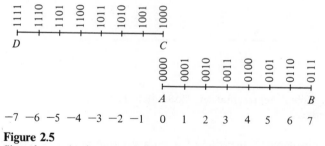

**Figure 2.5**
Signed magnitude reverses the order of the negative values.

achieve this, the positive integers are represented by the right half of the number line and the negative integers by the left; observe Figure 2.6. The representation of any $n$ is given by $2^{k-1} + n$; for positive values of $n$ this merely means setting the high-order bit to one. Negative values, however, are represented with the two's complement on $k - 1$ bits; that is, when $n$ is positive, the representation of $-n$ is $2^{k-1} - n$. Offset representations are often used for the exponent of floating-point values.

**Exercises**

1. When two $k$-digit positive numbers are added, they are said to *overflow* if the sum requires $k + 1$ bits. For example, the sum of 14 and 2 overflows in four-bit arithmetic because the sum, 16, requires five binary digits. For each of the four notations discussed in the text derive rules for adding two $(k - 1)$-bit values. Your rules must properly set the sign of the result and must determine if an overflow has occurred. Consider adding positive values, negative values, and values of opposite sign. For what representation are the rules simplest?

2. Multiplication and division are frequently done by making both operands positive and adjusting the sign of the result. Determine whether multiplication can be done more directly for any of the four representations.

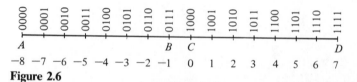

**Figure 2.6**
Offset retains both the order of the integers and the order of the binary values, but the representation of the positive values has the high-order bit set to one.

3. Explore the "signed-minitude" representation. In this representation the high-order bit is changed as in signed magnitude, but one is also subtracted from the value. Thus, with four bits, the representation of $-1$ is 1000. Moreover, there is only one representation of zero, and there are more negative than positive values. What is the addition algorithm? Is there a simple rule for deriving the representation of $-n$ from the representation of $n$? What is the mathematical expression for $-n$?

4. Show the representation of the following numbers in six-bit binary, using each of the four representation schemes: $0, 1, -1, -6, -12, -31, -32$.

5. Explain why the circuitry for comparing integers must usually be different from that for comparing characters. Under which representation could they be the same?

6. In one's complement and signed magnitude there are two representations for zero. Computers that use one or the other of these representations usually convert negative zero to positive zero in one (or more) of the following situations: after performing an arithmetic operation, before comparing two values, or before storing a value. Suppose we have three otherwise identical computers, each of which converts negative zero in a different one of the three cases. For each computer describe a program fragment that will perform as expected on that computer but fail on one of the others.

7. Note that there are more negative integers than positive integers with two's complement. What does this fact imply for the correctness of the statement $x \leftarrow -x$? Describe precisely what this statement does on your local computer.

8. Show that two's complement is "natural" in the following sense: when two integers of whatever sign are added, the result has the proper sign and representation if no overflow has occurred. Argue first with the number line and then with expressions involving $2^k$ and computations modulo $2^k$.

## 2.3.2  Floating-Point

In the notation introduced in Chapter 1, a variable may be declared to be in the **real** domain, a domain that is meant to include fractional values as well as integers. Conceptually, the values in this domain are all those that can be expressed in *scientific notation*—that is, with a mantissa multiplied by ten raised to some *exponent*. Thus the following values in scientific notation are also **real** values:

$6.02 \times 10^{23}$ (Avogadro's number, the number of molecules in a mole).

$3.34 \times 10^{-9}$ seconds (the time for light to travel one meter).

6.96 years $(= 6.96 \times 10^0)$ (the time for capital to double at 10 percent interest).

One advantage of scientific notation is that the decimal point can be adjusted to avoid leading or trailing zeros. The first of the examples is considerably more compact than 60200000000000000000000. The rule for moving the decimal point is to add one to the exponent for each place left the decimal point is moved. This rule amounts to multiplication by 10/10; the coefficient is divided by the ten in the denominator, and the ten in the numerator is absorbed by adding 1 to the exponent.

Hardware representations of real values are called *floating point* because each value includes the exponent that shows where the decimal point belongs in the final value. Also encoded in a floating-point value are the mantissa and sign fields; the format used for Figure 2.1 was:

| sign | exponent | mantissa |
|------|----------|----------|

where the exponent is a five-bit integer in two's-complement form and the sign is a one-bit field that gives the sign of the mantissa in signed-magnitude form. Since the values are represented in binary, it is more convenient to use the exponent as the position of the *binary point* of the value rather than the decimal point. The binary point of the mantissa is assumed to be at its left, so the value

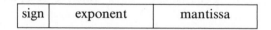

is interpreted as $0.101_2 \times 2^{10_2}$ or, in decimal, $0.625 \times 2^2$, which is 2.5. The exponent shift rule applies to binary values, so the above value could also be interpreted by shifting the binary point right two places and reducing the exponent to zero. This gives the binary value $10.1_2$, which is indeed the binary for 2.5.

Almost every hardware design has utilized a different format for floating-point values; most of the differences are in the representation of negative exponents and negative mantissa. A major difference is that a few designs (notably those of IBM) shift the mantissa four bits for every change of the exponent by one. This scheme provides a wider range of exponent, but the accuracy of the mantissa is reduced because the mantissa must often begin with insignificant zeros.

Not all values in scientific notation are values in floating point. In practice, the range of values is limited by the number of exponent bits, and the accuracy is limited by the number of mantissa bits. In addition, some hardware operations on floating values introduce additional problems. These will be illustrated in subsequent paragraphs, assuming a computer with decimal digits in the mantissa and exponent, to avoid inconvenient conversions between decimal and binary.

Consider the addition of two floating-point values. This is done in three steps, starting with lining up the decimal points by increasing the smaller exponent. For example, if the values to be added are $0.5 \times 10^{-7}$ and $0.3 \times 10^{-4}$, the former is converted to $0.0005 \times 10^{-4}$. The second step is to add the values, producing in this case the value $0.3005 \times 10^{-4}$. The third step is to adjust the result so the decimal point is just before the leftmost nonzero digit. This would occur in adding 0.6 to 0.4 or in adding values of opposite sign. For instance, the sum of $0.3005 \times 10^{-4}$ and $-0.2991 \times 10^{-4}$ is $0.0014 \times 10^{-4}$, which is adjusted by shifting the mantissa left to become $0.14 \times 10^{-6}$.

This process of shifting the result is called *normalization* because it is considered "normal" to have a nonzero digit after the decimal point. It has the advantage of increasing the accuracy of multiplication (Exercise 3). Note that when a value is normalized by shifting left, zeros are inserted at the right, which may not be correct for the situation modeled with the floating value. In the last example above, both operands have four valid digits of information, but the result has only two.

The fact that only finite accuracy is available can lead to incorrect results and results that depend on the order of operations. To illustrate, consider the sum of $0.3003$, $0.1 \times 10^1$, and $0.3007$ with four-digit precision. Added in this order, the first partial sum is $0.1300 \times 10^1$ and the final result is $0.1600 \times 10^1$. However, if the two smaller values are added first, the result is $0.1601 \times 10^1$. The discrepancy is small for this example, but it can accumulate to a large value for a long chain of calculations, or lead to strange results when testing for equality.

It is a curious fact that floating-point values are "closer together" near zero. On our decimal computer, one is given by $0.1 \times 10^1$ and the closest value is $0.9999 \times 10^0$, which is at a distance of $0.1 \times 10^{-3}$. (Note that 1.001 is ten times farther away.) Consider, however, $0.1 \times 10^{-50}$, which is nearer to zero than one is; the closest value is at the distance of only $0.1 \times 10^{-54}$. We can illustrate these facts by examining the number line in Figure 2.7 and labeling some of the points.

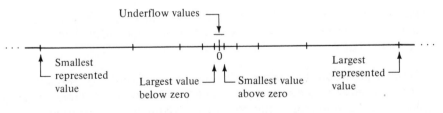

**Figure 2.7**
Number line.

Just as overflow can occur by adding values that produce an exponent larger than can be represented, a floating-point operation can also *underflow* by producing a value with an exponent too small to be represented, that is, in the gap between zero and the smallest non-zero value representable. Hardware often signals this error and automatically chooses the value zero if the program does not otherwise respond to the signal.

### Exercises

1. Convert the following values to the binary floating-point representation described in the beginning of this section, showing the result in hexadecimal: 1, $-1$, 0.5, 1024, 0.1, $-0.2$, 91.

2. Find the smallest and largest nonzero positive values that can be represented with the floating-point scheme above. Express them in hexadecimal and decimal. Under what condition is it not possible to normalize a value?

3. Show in binary the result of multiplying 0.50390625 by 0.65625 with and without normalization. Which form of the answer is closer to the correct product of these values?

4. Suppose a floating-point value in the format given in the text is generated with zeros in the first two binary positions and random ones and zeros in the remaining fourteen positions. What is the expected value of the result?

5. Prove that a greater number of distinct values can be represented with a one-word **integer** variable than a one-word **real** variable. [*Hint:* Consider the representation of zero.]

6. The Newton-Raphson approximation to the square root of $x$ is given by iteratively finding a new approximation $r'$ from the old approximation $r$, using

$$r' = \tfrac{1}{2}(r + x/r)$$

(see also Exercise 12 of Section 1.2.3). On some calculators, this approximation always produces a value that is exact or is too small. Why is the result never too large?

7. Write a procedure to perform a floating-point addition of two values represented as given in the second paragraph of this section. Assume the hardware has integer and logical operations but no floating-point operations.

### 2.3.3 Strings

One way to store a sequence of character values is to store one character in each word and use a sequence of words as long as the sequence of characters. This scheme has advantages if the hardware does not otherwise allow direct access to individual characters of the sequence, but it is wasteful of space. In a computer with 32-bit words and eight-bit characters the scheme would utilize only one quarter of the memory devoted to strings. Instead, a form of packed data can be used, storing more than one character in each word. In the example computer, four characters could be stored in each word, so a sequence $c$ characters long would need only $\lceil c/4 \rceil$ words. Thus the string of characters "CURLY MOE" would take nine words with the first scheme:

| "C" | "U" | "R" | "L" | "Y" | " " | "M" | "O" | "E" |
|---|---|---|---|---|---|---|---|---|

and would take only three words with the packed approach:

| "CURL" | "Y MO" | "E    " |
|---|---|---|

In the latter approach, the last word may be filled out with some special character that cannot be part of any string value, or left blank, with total length of the string indicated elsewhere.

Some representations of character strings include the length of the string as part of the first word of the packed value. Usually two character positions are enough for the length, since this would give sixteen bits and a maximum string length of 65,535 characters. In this case the string would look like

| 9 "CU" | "RLY  " | "MOE  " |
|---|---|---|

where the first word would appear as (assuming USASCII)

$$00000000\ 00001001\ 01000011\ 01010101$$

A character string can be used as an alternative to the binary form of Section 2.2 for storing integers. The value is represented as the sequence of characters that make up the printed version of the value. Thus the decimal value of the approximate population of the United States—250,000,000—could be stored in binary as the value $0EE6B280_{16}$, but it could equally well be stored as the string of characters "2" "5" "0" "0" "0" "0" "0" "0" "0". In some situations this might be appropriate to avoid the expense of conversion to the binary form (Exercise 3 of Section 2.3). There is a representation of integers intermediate between binary and character strings, called *packed decimal*, which stores only the four low-order bits for each digit. In both USASCII and EBCDIC these bits are the natural representation of the digit value. For example, the low-order four bits of the character "3" are the binary for that digit value: 0011. With this scheme, the hexadecimal representations of the packed-decimal version of 250,000,000 is

$$2500000_{16}\, 0xxxxxxx_{16}$$

where the "$x$" digit means that the value is unimportant since it is not part of the information. As with packed-character strings, the length of a packed-decimal value must be indicated somehow—either as a first digit or elsewhere.

With packed-decimal data two additional items may be encoded: the sign of the value and a decimal point. Convenient encodings for these are possible with four-bit values because the ten decimal digits require only ten of the sixteen values that can be represented in four bits. It is an easy matter to assign three of the six extra codes to represent "$+$", "$-$", and "$x$". (In USASCII a natural choice would be B, D, and E. Why?) Unfortunately, most hardware does not make specific assignments to these codes, so any arithmetic with such values would have to be performed with a subroutine rather than a hardware instruction.

One common application of strings of characters is the storage of non-numeric data values such as names and dates. Both of these raise problems. In principle, a person's name is entered in a string variable, but in practice the problem is hard because names must be intelligible to humans and yet the computer must be able to put them in alphabetical order. The problem is complicated by the fact that there are many common forms for any single name; for example, Edward Bernard Grogan, Junior, may be referred to as E. B. Grogan, Jr., Ed Grogan, Mr. Grogan, Grogan, E. B., and many others. He may even receive a "personalized" letter from a fund raiser addressing him as "Mr. Junior." If Grogan had a doctoral degree, he could also be referred to as Dr. Grogan or E. B. Grogan, Jr., Ph.D. Names with various heritages offer additional challenges: the surname is in the middle of Spanish names and at the beginning of many oriental names. Names sometimes have embedded blanks and capital letters, as in McDonald and da Vinci.

Any form of a name can be stored in a file, but the appropriate record can be recovered later only if the searcher has some idea of the form in which the name was originally entered. One suitable standard format might be:

- All titles and honorary suffixes are removed.
- The name is divided into a "last name," a "first name and initials," and a "suffix" (Jr., Sr., III, ...). Suffixes are abbreviated.

- None of these contains a comma. They may contain letters, capitalization, spaces, hyphens, and periods.
- The name is written with last name, comma, and first name. If there is a suffix, it is placed at the end after a second comma.

For sorting names in this format, the names must first be transformed to remove spaces and convert lower case to upper case. (Why?) Then when the names are sorted according to the collating sequence in use, they will be ordered as would normally be expected. A user could then easily find any name in a printed listing of the file. When the file is interrogated with an on-line query, the system converts the name to the above format and searches the file, perhaps using one of the search techniques of Chapter 7.

Strings are also employed to record dates, but the task of finding an appropriate encoding is harder than that for names because the encoding must allow human reading, comparison of dates, and arithmetic to determine the elapsed time between two dates. Among the many date forms are

| | | |
|---|---|---|
| May 28, 1941 | month-name dd, yyyy | Old-time standard. Sometimes the name is abbreviated |
| 5/28/41 | mm/dd/yy | North American |
| 28/5/41 | dd/mm/yy | European |
| 41/5/28 | yy/mm/dd | Data processing standard |
| 41.148 | yy.ddd | "Julian"; the ddd is the number of days past December 31. |

The yy/mm/dd form provides for human readability and comparison, but it must be converted to perform arithmetic. The yy.ddd form is suitable for arithmetic but is not very readable by humans. Indeed, to determine precisely what day a Julian date represents, one must first determine if the given year is a leap year. Some systems have adopted a standard form for date and time and convert all other forms to it for internal processing. One possibility is to count the number of milliseconds since the birth of Christ (assumed to have occurred at the midnight after December 31 in the year that would be A.D. 0 by the Julian calendar). A six-byte field is sufficient to distinguish any millisecond between then and A.D. 8000. This form is excellent for comparison and arithmetic but makes no concession to human readability.

For many varieties of data items other than names and dates there are existing standards that can be employed without the effort of developing a code. For example, the National Bureau of Standards has produced standard two-character codes for states of the United States. For more detailed geographic encoding, the Bureau of the Census has divided the country into county-sized units and has defined the extent of the large metropolitan areas. Additional codes have been developed under the auspices of the American National Standards Institute.


**Exercises**

1. Show the hexadecimal representations of the words containing "Y MO" and 9 "CU". Use both USASCII and EBCDIC.

2. Write a procedure to select the $k$th character from a string packed with four characters per word. Arguments to the procedure should be the length of the string, an array containing the words, and $k$. Assume that the length is not stored in the first word of the string. The character selected should be returned in the leftmost bits of the result word.

3. Write a procedure to add two packed-decimal values and replace the second with the sum. The arguments to the procedure should be arrays containing the two values and the lengths of the values. The procedure should return a result code indicating whether any error occurred during execution. Be sure to carefully consider what errors can occur. Do not expect decimal points and signs in the values.

4. It is not necessary to assign the values 8, 4, 2, 1 to the bits representing a digit; another possibility is $-1, 2, 3, 6$. Show the code for each digit using the latter values. Find another set of values that includes neither 1 nor $-1$.

5. Early RCA computers used a code where zero was 0011 and succeeding decimal digits had successively higher binary values with 1100 for nine. Why was this code called XS3? (Please do not groan too loudly.) Show the more-or-less simple rule for adding two decimal digits encoded in XS3 and determining whether to carry a one into the next higher digit position. In your judgment, is the simplicity of this rule a sufficient justification for choosing XS3 representation?

6. Write a procedure to check the form of names stored with the standard form sketched in the text. The input should be a name in an array of characters. The output should be a value indicating the nature of the first error detected, if any. The hardest part of this exercise is to develop the list of errors.

7. In some situations, such as airline reservations systems, it is valuable to encode names so they can be located even if the exact spelling is not known. One such code is used, for example, in the driver-licensing system of many states. The code for a name starts with the first letter of the name. The

remainder of the code is the first three digits derived from letters in this table:

| | |
|---|---|
| BFPV | 1 |
| CGJKQSXZ | 2 |
| DT | 3 |
| L | 4 |
| MN | 5 |
| R | 6 |
| other | ignore |

Double letters such as LL are reduced to a single letter, and the combination SC at the beginning of a name is absorbed by the initial S. Thus SHELL and SCHULL are both encoded as S400. Hansen is H525 and Reingold is R524.

(a) Encode your name in this code.

(b) Find three names that have at least four variant spellings each and show the encoding for all twelve names.

(c) Try to find a name with a variant spelling such that the encoding maps the variants into different codes.

8. Design a procedure to convert dates to the yy/mm/dd format. The input should be a date in an array, one character per array element; the output should be the date in another array in the prescribed format. Your routine should accept at least three different input formats, but it should assume that xx/xx/xx dates are in the order MMDDYY if the first two digits are not a year value (if they are, no conversion is needed). Complications to consider include dashes, spaces, commas, or slashes between date elements; alphabetic month specifications; year value absent (use current year) or specified with two or four digits; Julian dates.

9. The most common abbreviations for the names of the months are three and four characters long, but a shorter representation could save space. A one-character code would be too terse and artificial because more than one month starts with J and M. Devise a two-character alphabetic code for the month names. An overriding criterion should be recognizability by humans; a secondary concern is that humans ought to be able to encode month values quickly and reliably with little training.

10. In many situations a word typed to a computer is supposed to match one of a set of words. If the typed word is misspelled, a procedure can check to see if it is a misspelling of a correct word according to one of these possibilities:

- A letter has been omitted.
- A letter has been added.

- A letter has been changed to another.
- Two letters have been interchanged.

Write a procedure to check whether one word is a misspelling of another. The input for each word should be a length and an array with one character in each element. The output should be signals indicating whether there is a potential match, and if so, what and where the error is. [*Hint:* Use the lengths to determine which errors are possible.]

## 2.4  RELIABILITY AND EFFICIENCY

Previous sections have covered representations of values of various kinds. Two areas of mathematics deal with properties that such representations can have. One of these areas, *coding theory*, deals with the question of how message transmission can be made reliable; that is, how the sender of a message can ensure that the recipient gets the intended message instead of some garbled version. The other area, *information theory*, deals with the question of how efficiently information can be transmitted—how quickly and at what minimum cost. Reliability can be illustrated by computer memories, which often use *nine* bits to encode each 8-bit byte of memory. The ninth bit is called the *parity* bit and is set so that the total number of one bits is odd. As a result, if any bit of memory is accidentally changed, the change will be detected as a byte with an even number of one bits. The resulting memory is more reliable to the extent that, while it may not reconstruct correct values, it will at least indicate that it does not have the correct value.

Efficiency is illustrated by the Morse code shown in Table 2.13, where different numbers of dots and dashes are sent for each character. For example, "e" and "t" are high-frequency letters and are represented by only one dot or dash. At the low-frequency end of the alphabet are letters such as "j" and "q" with four dots or dashes each. If all codes were of the same length, each would have to be five dots or dashes just to distinguish among the letters and six of the ten digits. There would then be no provision for the other four digits or the punctuation marks.

In data base applications, many values are nonnumerical and must be encoded in strings. These include sex, language skills, warehouse locations, product colors and styles, and so on. All such encodings raise the issues of reliability and efficiency. A one-bit code for sex is an efficient use of storage, but it may not be reliably interpreted by humans. (Which sex is **true** and which is **false**?) A more reliably interpreted code might be "M" and "F", even though these require an entire byte if stored in this form. A trade-off between reliability and efficiency can be seen in the choice of "MALE" and "FEMALE" for sex codes; these can be reliably interpreted, but they make inefficient use of space.

| | | | |
|---|---|---|---|
| A | • — | N | — • |
| B | — • • • | O | — — — |
| C | — • — • | P | • — — • |
| D | — • • | Q | — — • — |
| E | • | R | • — • |
| F | • • — • | S | • • • |
| G | — — • | T | — |
| H | • • • • | U | • • — |
| I | • • | V | • • • — |
| J | • — — — | W | • — — |
| K | — • — | X | — • • — |
| L | • — • • | Y | — • — — |
| M | — — | Z | — — • • |

**Table 2.13**
International Morse code. The "—" (pronounced "dah") symbols are three times the length of the "•" ("dit") symbols when the code is transmitted as sound signals. Codewords for letters are separated by "letter space" gaps as long as a dit; words are separated by "word spaces" as long as a dah.

Sex can be encoded with a long code for data entry and then translated to a single bit for storage. Other types of data may raise problems if fixed-length fields are stored. For example, in 1964 there was a one-digit expansion in the size of serial numbers for officers in the United States Army. The Army had to revise all programs that processed files containing these numbers.

Because of the problems of getting names correct, most accounting systems assign account numbers to every individual in the system. Such numbers are also used to identify projects, parts, and other entities. One of the principal problems with account numbers is ensuring that a number has been correctly entered by a data entry clerk or other system user. Entry of a wrong number might mean that Mrs. Birdwhistle got charged for Mr. Blacktree's andirons. The most important protection in interactive systems is that the operator must verify the name and address from the display after entry of the account number. In addition, most systems employ *check digits* in every account number; the account number is made longer with an extra digit computed from the other digits of the number. Thus, after reading an account number, the system examines the check digit to verify the typing of the account number.

Account numbers are usually all the same length, so errors of insertion or deletion of digits are readily detected. The most common other errors are substitution of digits and transposition of digits, either adjacent or separated by one or two positions. For instance, 4152 might be rendered as 4512, 4125, 4251, 9152, or one of many other possibilities. The most common failure is that at most one such error has occurred, so it is not necessary to protect against multiple errors with as much care as against single errors. Let us consider three typical check-digit schemes.

**Scheme 1—Casting Out Nines.** In this method the account number is divided by nine and the remainder is the check digit. (Exercise 1 shows the common method to achieve this division.) This method is about 98 percent effective against substitution of one digit for another (Exercise 2) but is worthless against transpositions. (The result of Exercise 1 explains why.)

**Scheme 2—Double Even.** The check digit is chosen so the following process yields zero when the check digit is included in calculation:

1. Double the digits in the even positions (that is, the ones position, the hundreds position, and so on).

2. Add all digits in the results to all digits in odd positions.

3. Examine the low-order digit of the sum.

Thus account number 36415209 yields 12, 2, 4, 18, and the sum $3 + 4 + 5 + 0 + 1 + 2 + 2 + 4 + 1 + 8$, which is 30, so it is a valid account number. The doubling protects against adjacent transpositions; the inclusion of the check digit in the calculation protects against transpositions with it; and the addition of the tens digit from the doubling avoids the difficulty that the low-order digit from doubling can be only 0, 2, 4, 6, or 8. Note, for example, that 4 and 9 are often confused in handwriting and yet when doubled both produce a low-order digit of 8.

**Scheme 3—Divisible by Thirteen.** In some cases it is possible to assign entirely new account numbers. If so, they can be chosen so they are all divisible by thirteen. Such numbers are protection against all single-digit replacements and transpositions across zero, one, two, and three positions (Exercise 4). Though there is no extra check digit with this protection scheme, account numbers will be somewhat longer than without it. The efficiency of the scheme can be measured by considering that it uses only one-thirteenth of all possible values with a given number of digits. This is about 0.0769, where the double-even scheme uses 0.1000 of all possible values. (Why?)

**Exercises**

1. The process of "casting out nines" is as follows: add together all of the digits of a decimal integer getting another integer. Repeat the process until the result is a single integer. Prove that the result is the remainder when the original number is divided by 9.

2. Show that a casting-out-nines check digit will detect almost 98 percent of all instances where one digit is replaced by another.

3. (a) Suppose these are account numbers with the check digit represented by $x$: $472215x, 982327x, 111x$. Compute the double-even check digit for each.

   (b) Using the same method, verify that the last digit is a correct check digit for these numbers: $123459, 00000, 99994$.

4. (a) Show that divisible-by-thirteen account numbers protect against all possible single-digit replacements.

   (b) Show that divisible-by-thirteen account numbers provide protection against all transpositions between adjacent digits and digits separated by one, two, or three others.

5. Four familiar numbers in the United States are zip codes, social security numbers, driver's license numbers, and ten-digit phone numbers. All these codes have interesting features: zip codes have low error protection but gain some when combined with the two-letter state code; social security numbers are not unique (in fact, one particular number at one time served for twenty-nine different people); driver's license numbers often use the code of Exercise 7 of Section 2.3.3; and the second digit of area codes is either zero or one. For one or more of the above codes, do the following:

   (a) In many cases a number has components; a geographic code may use two digits for state, three for county, and two for town. For the code you are studying, describe the breakdown into components.

   (b) Find out what restrictions there are on the values in each component.

   (c) (Efficiency) Compute the number of possible codes when the restrictions are followed and how large a fraction that is of all codes if there were no restrictions.

   (d) (Reliability) Compute the probability of detecting an error caused by changing a single digit.

   (e) What is the cost of an error in interpretation of a codeword? Would this cost justify addition of a check digit?

### 2.4.1 Reliable Codes

The mathematical theories of reliability and efficiency consider a model where a *sender* is sending *messages* to a *receiver* over a *channel*. Each message is chosen from a set (usually finite) of possible messages and transmitted over the channel by the sending of a *codeword*. A *code* is a table that specifies the relationship between the messages and the codewords; for example, Table 2.11 and 2.13 are codes for USASCII and Morse code, respectively. For our purposes, a codeword can be thought of as a sequence of binary digits, but the theory is general enough to encompass codewords composed of arbitrary symbols. Binary digits are sufficient here because they are the only means of storing or transmitting information in computers.

Coding theory deals with reliability as one aspect of *block codes*. These are codes like USASCII where all codewords have the same length. The basis for analyzing such a code is the Hamming distance: The *Hamming distance* $d(c_1, c_2)$ between two codewords in a block code is the number of bit positions in which the two codewords differ. For example, in USASCII, $d(\text{"H"}, \text{"R"}) = d(01001000, 01010010) = 3$ because the codewords differ in the fourth, fifth, and seventh positions. Exercise 1 shows that the Hamming distance is a "distance" as mathematicians define that term. One way to think of it as a distance is to imagine the map of a country where the city names are binary numbers and roads exist between cities only if they differ in one digit of their name. The Hamming distance between two cities is then the minimum number of roads that have to be traversed to get from one to the other.

The simplest reliability property of a code is its detection of errors; this can be done by a parity bit as described above. In terms of the imaginary map, parity partitions all cities into two classes such that it is impossible to get from a city to another in the same class without going through a city of the other class. That is, if a bit is changed in a codeword with an odd number of one bits, the value produced must have an even number of one bits. Thus to change one code to another requires a transitions to an error code and then another into valid code. Because a change in one bit position has a rather low probability, the change of two separate bits is far more improbable, so this code is sufficiently reliable for ordinary computer applications.

It is possible for codes to *detect* errors in more than one position:

**Theorem 1**

*Error detection.* Errors in $k$ or fewer positions can be detected if all codewords of a code differ by Hamming distances at least $k + 1$.

Thus, in the parity example, all codewords differ by at least two, so any error in no more than one bit position can be detected. In general, Theorem 1 divides the cities of a map into three classes: the original codewords, those within $k$ steps of each original codeword, and those that may be further than $k$ steps from any codeword. The codewords in the first class are accepted as being correct, while those in the latter two classes are taken as errors.

This same division into classes is the basis for construction of codes that *correct* errors in transmission:

**Theorem 2**

*Error correction.* Errors in $k$ or fewer positions can be corrected by the receiver if all codewords of a code differ by Hamming distances at least $2k + 1$.

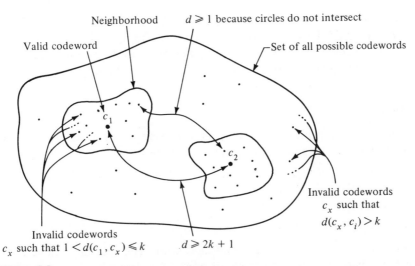

**Figure 2.8**
Neighborhoods within Hamming distance $k$ of codewords. Since $d(c_1, c_2) \geq 2k + 1$, the circles with "radius" $k$ do not intersect. Any invalid codeword in a circle can be corrected to the valid codeword at the center of the circle.

In an error-correcting code, the erroneous codewords are partitioned in groups so that each is interpreted as an error variation of some particular codeword. The situation is sketched in Figure 2.8, where each valid codeword is shown surrounded by a "circle" of invalid codewords that are closer to it than to any other codeword.

One important question in coding theory is to determine how big the codewords have to be to encode information with a given degree of reliability. This problem can be approached by starting with Figure 2.8 and computing the number of codewords that can be encoded with a given length and a given minimum Hamming distance between codewords. With codewords $n$ bits in length, the available number of codewords is $2^n$. For $k$ bit error correction, the neighborhood of each codeword must include the codeword itself plus all codewords reachable in $k$ or fewer errors. In exactly $j$ errors the number reachable is $\binom{n}{j} \dfrac{n!}{j!(n-j)!}$, the number of ways to choose $j$ items out of $n$. Combining the above, the number of codewords $N$ is given by

$$N \leq \frac{\text{number of potential codewords}}{\text{number reachable in } k \text{ or fewer steps}}$$

$$= \frac{2^n}{\displaystyle\sum_{j=0}^{k} \binom{n}{j}}. \tag{2.5}$$

Thus with messages of length sixteen, a code for at most 478 messages could be created that would detect and correct all errors in no more than two bit positions. Note that as shown by Exercise 3 it may not be possible to construct a code for the maximum number of codewords given by the equation.

### Exercises

1. Verify that the Hamming distance is a measure of distance. In other words, show that if $c_1$, $c_2$, and $c_3$ are codewords, then
   (a) $d(c_1, c_2) \geqslant 0$.
   (b) $d(c_1, c_2) = d(c_2, c_1)$.
   (c) $d(c_1, c_2) = 0$ if and only if $c_1 = c_2$.
   (d) $d(c_1, c_2) + d(c_2, c_3) \geqslant d(c_1, c_3)$.

2. Find two block codes for five messages each. The first must have Hamming distance two or greater between all pairs of codewords; the other must have distances of three or greater. (Codewords should be as short as possible.) What degree of error detection is afforded by each of these codes? Error correction?

3. Using Equation (2.5), determine the minimum number of bits required for codewords in a code that can send four messages in such a way that all errors of two or fewer bits can be detected and corrected. Show that no such code is possible with the minimum bit length. Find a suitable code using more bits.

4. The IBM 650 computer encoded information in a "two-out-of-five" code: five bits were used for each decimal digit; they had the values 0, 1, 2, 3, 6; and exactly two bits had to be set to represent a digit correctly. Since there are two possible representations for three with the above bit values, one of them was used to represent zero.
   (a) What is the reliability of this code? Prove your assertion from the viewpoints of the Hamming distance and of the design of the code itself.
   (b) What is the efficiency of this code? That is, of the possible values with the given number of bits, what fraction are actually available as codewords?

## 2.4.2  Efficient Codes

A codeword from a message source consumes resources, space in a storage medium or time on a transmission channel. One goal of information theory is to reduce this consumption by construction of more efficient codes. With block codes, every codeword has the same length, so the only way to reduce the resource requirement is to encode more than one message with some of the codewords.

For example, English text could be represented more efficiently with eight-bit bytes if some of the unassigned codes were assigned to common letter pairs such as "TH", "AS", and "ED". More often, efficient codes are constructed by using fewer bits for the more frequent messages, just as the Morse code of Table 2.13 uses fewer positions for the more frequent letters.

For the present discussion we require that codes be *instantaneously decodable*: no codeword is allowed to begin with a bit sequence that constitutes another codeword. Note that Morse code fails this test because (for example) many codewords begin with " • ", the codeword for "E". With an instantaneously decodable code, a receiver can extract each codeword as soon as it is completely received, without waiting to see if subsequent bits transform it into a longer codeword.

The resource consumption of an encoding is measured by the expected number of bits to encode a message. (See the discussion of expected values in Section 1.3.2.) Thus if $m$ messages have codewords with lengths $t_1, t_2, \ldots, t_m$ and probabilities $p_1, p_2, \ldots, p_m$, then the expected length of the representation of a message is

$$\text{expected length} = \sum_{i=1}^{m} p_i t_i. \tag{2.6}$$

As an example, consider a source of messages for the outcome of hockey games. The possible messages might be: H—home team wins, A—away team wins, T—tie, or S—stoppage (riot, tornado, avalanche, or whatever). A block code for these messages might be

$$H = 10, \quad A = 01, \quad T = 11, \quad S = 00,$$

and the expected length for a message would be

$$2p_H + 2p_A + 2p_T + 2p_S = 2(p_H + p_A + p_T + p_S) = 2.$$

Since we assume the message must be one of the four, the probabilities must sum to one, so the expected message length is two, as would be expected.

An instantaneously decodable variable-length code for the hockey outcomes is

$$H = 0, \quad A = 10, \quad T = 110, \quad S = 111$$

with expected length

$$1p_H + 2p_A + 3p_T + 3p_S.$$

If the messages are equally likely, the expected length is $(1 + 2 + 3 + 3)/4 = 2.25$, worse than the block code. In practice, however, wins are more common than ties, while stoppages are rare, so the probabilities might be something like

$$p_H = 0.47, \quad p_A = 0.44, \quad p_T = 0.07, \quad p_S = 0.02.$$

With these probabilities the average length with the variable-length code is 1.62, an improvement over the block code.

Since a change in the code has reduced the expected length, it is reasonable to ask whether some other encoding might do still better. There must be some lower limit, however, because clearly some number of bits must be used. Information theory shows that this lower limit is given by

$$\text{minimum expected length} = \sum_{i=1}^{m} p_i(-\lg p_i) \tag{2.7}$$

which is called the *entropy* of the message source. The entropy of the hockey-outcomes code is

$$0.47(-\lg 0.47) + 0.44(-\lg 0.44) + 0.07(-\lg 0.07) + 0.02(-\lg 0.02) = 1.41.$$

Thus the variable-length code above is 19 percent longer than the minimum, a level that is quite reasonable. To get closer to the entropy of a message source generally requires a considerably more complex encoding.

Suppose we have a message source that produces exactly two possible messages. Since these can be encoded in a single bit, the entropy of this source describes the capacity of a single bit. If the probability of one of the messages is $p$, then the probability of the other is $(1 - p)$, so the entropy is

$$-p \lg p - (1 - p)\lg(1 - p). \tag{2.8}$$

This expression approaches a value of zero when $p$ approaches zero or one. This corresponds to the idea that a message source does not give as much information if the probability of one message is small. (There is little surprise in receiving the high-probability message.) The maximum value for the expression is one, its value when $p = \frac{1}{2}$. *Thus a bit encodes at most one "bit's worth" of information and often less.*

A process called *Huffman's algorithm* computes the most efficient possible variable-length code for a set of messages with given probabilities. This process, has two parts, as illustrated in Figure 2.9. The first part builds a diagram linking the messages; the second assigns a codeword to each message. Initially the diagram has an isolated element for each message with the probability $p_i$ in the element. At each step of building the diagram, the two elements of smallest probability are connected into a new element containing the sum of their

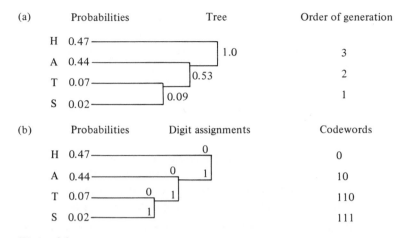

**Figure 2.9**
Illustration of Huffman algorithm. (a) The diagram is constructed in the order shown. (b) After 0's and 1's have been assigned, the code for each message is read off by tracing the line from right to left and recording the bit values encountered. For this example, the expected number of bits per codeword is 1.62.

probabilities. This construction ends when there is only a single element, which by definition has probability one. Part (a) of Figure 2.9 shows the outcome of this construction. The second part of the algorithm is to label one connection from each node with "1" and the other with "0". The codeword for any given node is then read off as the sequence of labels encountered in traveling to the message from the right. This process ends as shown in part (b) of Figure 2.9. One interesting property of the algorithm is that it maximizes the entropy of each bit, because it tends to make each bit choose between options that are as nearly as possible equal in probability.

The algorithm can be proved correct by a recursive (inductive) argument starting with consideration of the two elements with the smallest probabilities. These elements must differ by a single bit in at least one code that produces the minimum expected length. This is so because the two smallest-probability codewords can be given messages of the same length, and these should be the longest codewords. After combining the two smallest elements, the same argument applies recursively to a set with one element less.

The foregoing discussions suggest that a code can be either reliable or efficient but not both. Actually, messages can be encoded first in a variable-length code, and then the stream of variable-length segments can be split into fixed-length blocks. Finally, each block can be encoded with enough extra bits to achieve the desired degree of error protection. As an illustration, the hockey outcomes H, A, A, H, H, T, A, H, S, T, A, H, A would be first encoded as 0, 10, 10, 0, 0,

110, 10, 0, 111, 110, 10, 0, 10 and then grouped together to form 01010001, 10100111, 11010010. For single error detection, a parity bit could then be appended to each block: 010100011, 101001111, 110100100. Note that in this case the encoding takes only one more bit than the two-bit block code for the messages, even though the latter offered no error protection at all. (See Exercise 10.)

**Exercises**

1. Note that Equation (2.7) for entropy computes the expected value for the expression $-\lg p_i$. This quantity, $-\lg p_i$, is called the *information content* of message $i$; it depends only on the probability of that message. In a sense, this quantity gives the number of bits to encode a given message.

   (a) Show that the information content in two independent messages is equal to the sum of the information contents for the two messages.

   (b) Suppose a source is capable of generating $2^m$ possible messages and all are equally likely. Show that messages can be encoded in $m$ bits. Show that the information content of a message is $m$.

2. Consider Equation (2.8) for the entropy of a message from a source with only two possible messages. Show that the entropy cannot exceed one. This means that a bit cannot encode more than one "bit's worth" of information.

3. The nonnumeric component of a bridge bid has eight possibilities: Pass, Club, Diamond, Heart, Spade, No Trump, Double, Redouble.

   (a) Suppose these are equally probable; what is the expected length of a codeword to send one of these messages?

   (b) Suppose the probabilities are the following (more reasonable) set: $0.43, 0.10, 0.08, 0.10, 0.12, 0.12, 0.04, 0.01$, respectively. What is the entropy of this message source? Use the Huffman algorithm to derive a code with one codeword for each message. What is the expected length of a message in your code?

4. If a code encodes four messages, essentially two different sets of codewords can be derived by Huffman's algorithm. What are they? Under which conditions is each the result of the algorithm?

5. Show that the process of Huffman's algorithm gives the most accurate method to sum floating-point numbers. (See the example of summing three such numbers in Section 2.3.2.)

6. One way to reduce the bit requirement for messages is to send more than one message with each codeword. The $k$th extension of a message source is the set of all messages consisting of $k$ messages chosen from the original set. The

| Information | Probability | Code |
|---|---|---|
| HH | 0.2209 | 00 |
| HA | 0.2068 | 01 |
| AH | 0.2068 | 10 |
| AA | 0.1936 | 110 |
| TH | 0.0329 | 11100 |
| HT | 0.0329 | 11101 |
| TA | 0.0308 | 111100 |
| AT | 0.0308 | 111101 |
| SH | 0.0094 | 1111100 |
| HS | 0.0094 | 1111101 |
| SA | 0.0088 | 1111110 |
| AS | 0.0088 | 111111100 |
| TT | 0.0049 | 111111101 |
| TS | 0.0014 | 111111110 |
| ST | 0.0014 | 1111111110 |
| SS | 0.0004 | 1111111111 |

**Table 2.14**
Second extension of the hockey-outcome code. The code shown here is instantaneously decodable but is not optimum for the probabilities given.

second extension for the hockey-outcome code is shown in Table 2.14 along with one possible code for the extended messages.

(a) What is the expected number of bits per original message (H, A, T, or S) with the code shown in Table 2.14?
(b) Use the Huffman algorithm to derive a more efficient code for the messages.
(c) With your new code, what is the expected number of bits for an original message?

7. The second extension in Table 2.14 uses sixteen codewords. One way to use fewer codewords is to represent only the high-probability pairs with special codewords. For example, we could have codewords for the messages HH, HA, AH, AA, H, A, T, and S, where H and A would be used only as the last codewords encoding a string of an odd number of H's and A's. When the probabilities of H, A, T, and S are 0.47, 0.44, 0.07, and 0.02, respectively, the probabilities of the eight codewords are

HH  0.2041778     HA  0.1911452   AH  0.1911452   AA  0.1789444

H   0.0390979      A   0.0366023    T   0.1235789    S   0.0353083

(T and S have increased probabilities because relatively fewer total codewords will be used for strings of H and A messages.)

(a) Use the Huffman algorithm to generate a code for the eight codewords above.

(b) Compute the expected number of bits required by the eight-codeword code to represent one of the original four messages (in a sequence).

★ (c) Derive the probabilities given above for the eight codewords. [*Hint*: Use Markov chains.]

8. Zipf's law states that the probability of using a word in a human vocabulary is inversely proportional to its position in a list ordered according to frequency of use. Thus if there are $n$ words, the probability of the $i$th is proportional to $1/i$ and is given by

$$ p_i = \frac{1}{iH_n} \quad \text{where} \quad H_n = \sum_{i=1}^{n} \frac{1}{i} \approx \ln n. $$

$H_n$ is the sum of the first $n$ terms of the harmonic sequence $1, \frac{1}{2}, \frac{1}{3}, \ldots$; it appears in the divisor of $p_i$ to normalize the probabilities so they sum to one.

(a) With a vocabulary of $n$ words, how many bits would be required to express each if they were equally probable?

★ (b) Show that the entropy of the probabilities in Zipf's law is about $\frac{1}{2}\lg n$.

(c) How many bits per human word are implied by the relation in (b)? How does this compare with the number derived in (a)?

9. Computer memories typically contain many bytes with the value zero; one study showed that as many as forty percent were zero. Given this fact, the contents of a memory can be encoded by using a single zero bit for zero bytes and a one followed by the contents for the other bytes. Determine the saving in bits if memory is in fact forty percent zeros. What is the major disadvantage of this scheme?

10. In the example in the final paragraph of this section the thirteen messages averaged 2.077 bits each.

(a) Compute the expected number of bits per message with this encoding. (It is less than 2.077.)

(b) Write an expression that gives the expected number of bits when messages with expected length $m$ are encoded in blocks of size $b$ having $p$ additional bits for protection.

11. One interesting property of a code is its "resynchronization distance": if a stream of bits encoding messages is interpreted starting at an arbitrary point, how many bits will be examined before the bits are being correctly interpreted? For example, if the stream in the final paragraph of this section is interpreted starting with the third bit, only one bit will be incorrectly

interpreted. However, if the start is at the fifteenth bit (ignoring parity bits), five bits will be misinterpreted. What is the expected number of bits that will be misinterpreted for the entire sequence, assuming the interpretation is equally likely to start at any of the 24 bits (ignoring parity bits)?

## 2.5 REMARKS AND REFERENCES

The focus of this chapter has been on the programming side of bits and their interpretations as boolean, integer, real, and character values. A discussion of the machine architect's view of the same material can be found in

Mano, M. M., *Digital Logic and Computer Design*. Englewood Cliffs, N.J.: Prentice-Hall, Inc., 1979.

The representation of numbers, and especially real numbers (as opposed to integers), is fraught with computational difficulty. See the October 1979 issue of the *ACM SIGNUM Newsletter*, which is a special issue devoted to the various difficulties in the choice of a representation.

The representation of characters or messages by bit sequences is an area of considerable depth in mathematics. See

MacWilliams, F. J., and N. J. A. Sloane, *The Theory of Error-Correcting Codes*, Vol. I (1977) and Vol. II (1980). Amsterdam: North-Holland Publishing Co.

This work has come to be known as the bible of coding theory.

# Chapter 3

# Elementary
# Data Structures

A teacher who can arouse a feeling for one single good action, for one single good poem, accomplishes more than he who fills our memories with rows and rows of natural objects, classified with name and form.

*Elective Affinities*,   Goethe

The next storage structures beyond words are arrays and records, the topics of this chapter. Just as a sequence of bits is the storage structure used to implement the word data structure, a sequence of words is used to implement these higher structures. An array or a record allows us to deal with a group of related variables as a unit; in later chapters these two structures will serve in turn as storage structures for even higher data structures.

## 3.1  ARRAYS

If a programming language offers any data structure at all, it most likely offers one-dimensional arrays. Such an array is just a sequence of elements, any one of which can be selected by specifying its position in the sequence with an integer *subscript*. Each element is itself a data structure—an integer, logical, or other word value, or even an array or record; the only restriction is that all elements be of the same type and occupy the same amount of space. The specification of an array normally requires the name of the array, the type of its elements, and the range of permissible values for its subscripts. For example, a typical specification might be

$$signs: \textbf{array } [37:239] \textbf{ of integer}$$

which specifies that *signs* is an array of 203 integer elements with permissible

|  | [37] | [38] | [39] | [40] |  | [238] | [239] |
|---|---|---|---|---|---|---|---|
| signs: | 12 | 13 | 10 | 6 | $\cdots$ | 22 | 35 |

**Figure 3.1**
An example of an array. As declared in the text, the array *signs* has 203
integer elements with subscript values 37, 38,..., 239. In this diagram, 12 is
the value in the array element with subscript value 37.

subscripts ranging from 37 to 239, as shown in Figure 3.1. The limits for the
subscript values are called the *bounds* of the array.

As a data structure, one-dimensional arrays include only the function of
*subscripting*: this function gives the array element corresponding to a given
subscript. The array element thus specified can be used either for its value or as
the target for an assignment. We will indicate the operation of subscripting by
placing the subscript expression in square brackets after the array name. Thus the
array of Figure 3.1 has elements *signs*[37], *signs*[38],..., *signs*[239].

The most common application of arrays is to store a value for each of a
number of cases. For example, the array of Figure 3.1 might be part of a system
for studying highway beautification, the values in it representing the number of
billboards on each mile of Interstate 94 between Ann Arbor and Chicago. In this
and similar applications, the array elements are usually referenced in sequential
order. For example, the total number of signs could be computed by

$$n \leftarrow 0$$
$$\textbf{for } i = 37 \textbf{ to } 239 \textbf{ do}$$
$$n \leftarrow n + signs[i]$$

The elements of *signs* are accessed in order of increasing subscript.

The elements of an array can be bit strings of any length; it is not necessary
that each element take one word or more. If small enough, they can be stored
with one value per word or they can be *packed* with several values per word (see
Exercise 10). One important form of an array of packed information is a *string*, a
sequence of character values. If each character takes only six or eight bits, several
can be stored in each word. A 32-bit word, for example, can hold four characters
of eight bits each. Another important instance of a packed array is a *boolean
array* or *bit table*; in this structure a single bit is set aside for each of a number of
cases. For example, we might have a bit to indicate whether there is an exit ramp
in each mile of the Interstate studied above:

$$exit\_ramp: \textbf{array } [37:239] \textbf{ of boolean}$$

In many applications of arrays the elements of one array are used as indices
into other arrays. Suppose, for example, that we have an array $A$ whose elements
are many words long, making it time-consuming to rearrange the elements of $A$.

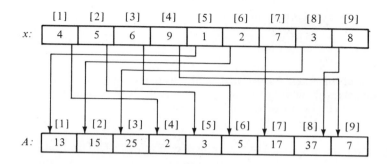

**Figure 3.2**
Example of an array used to index another array. The elements of $A$ are data values and the elements of $x$ are subscript values indicating elements in $A$. The values in $x$ are such that they give the elements of $A$ in ascending order. Thus $x[1]$ contains the value 4 and $A[4]$ is the smallest data value. The arrows from $x$ to $A$ indicate the element of $A$ specified by each element of $x$.

To reduce the cost of sorting $A$, we can use an index array $x$ as follows: $x[1]$ will contain the subscript of the smallest element of $A$, $x[2]$ will contain the subscript of the second smallest element of $A$, and so on as illustrated by Figure 3.2. We can construct the array $x$ given the array $A$ by initially setting $x[i] \leftarrow i$ and sorting with comparisons between $A[x[i]]$ and $A[x[t]]$ but with interchanges between $x[i]$ and $x[t]$, as in Algorithm 3.1. This particular sorting algorithm is *not* recommended in general, since, as we will see in Chapter 8, it is inefficient. However, the technique of making changes in an array other than the one to be sorted can be used in conjunction with just about any sorting technique.

$$\textbf{for } t = n \textbf{ to } 2 \textbf{ by } -1 \textbf{ do}$$
$$\quad i \leftarrow 1$$
$$\quad \textbf{for } k = 2 \textbf{ to } t \textbf{ do if } A[x[i]] < A[x[k]] \textbf{ then } i \leftarrow k$$
$$\quad x[i] \leftrightarrow x[t]$$

**Algorithm 3.1**
Simple indirect selection sort. (Compare this with Algorithm 8.5)

A similar use of arrays occurs when the elements to be in an array are of varying lengths. For example, names are generally less than eight letters long, but names of fifteen or more letters occasionally occur. If we allocated the maximum possible length for all names, the table would take up a great deal of space, most of it unused and unusable. Instead, we can store the characters of the names in a character array *chars* and store the subscript of the end of $i$th name in *name*$[i]$. In this technique the $i$th name consists of the characters

$$chars[name[i-1]+1], \ldots, chars[name[i]].$$

Figure 3.3 shows an example. Notice that the 0th element of *name* has been given the value 0, so that the above is correct even for $i = 1$.

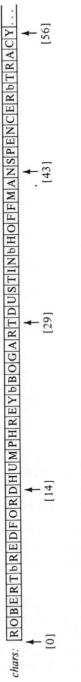

**Figure 3.3**
Names of varying lengths packed into a single long array with a second array
used as an index array to indicate element boundaries in the first array

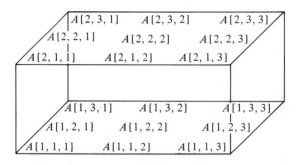

**Figure 3.4**
The elements of a three-dimensional array specified as
$A$: **array** [1 : 2, 1 : 3, 1 : 3] **of integer**

In addition to one-dimensional arrays, most programming languages allow multidimensional arrays. These are specified by giving more than one pair of dimension limits in the description of the array. For example, the three-dimensional array shown in Figure 3.4 could be declared by

$$A: \textbf{array} \; [1 : 2, 1 : 3, 1 : 3] \; \textbf{of integer}$$

In general, the ranges of the subscripts can be arbitrary integers.

We will discuss representations of multidimensional arrays in Section 3.3.1.

**Exercises**

1. Write a loop to find the cities along Interstate 94 from the information in the array *signs*, assuming that a city will have a cluster of at least five consecutive miles where the number of signs is above average. Store the location of each city in an array *cities* by storing the number of the middle mile in the cluster.

2. Design a procedure to generate all $n!$ permutations of the $n$ elements $A[1], A[2], \ldots, A[n]$. [*Hint:* Make your procedure recursive, choosing each element to be the first in turn and calling itself to generate all permutations of the remaining elements.]

3. Design a program to build and print histograms. The input is a sequence of integers with values between zero and 100 (say, grades on an exam). The output should be 101 lines, each containing an integer at the left and a row of asterisks proportional to the number of times that that integer occurred in the input. The lines should be in order by the numerical values of the integers.

[This exercise is intended to show the use of an array for both random access (counting the occurrences) and sequential access (printing the lines).]

4. Array subscript values can sometimes be calculated directly from data values, especially if the data values are from a small range. For example, a compiler must scan the input looking for identifiers, integers, and special characters; this is called *token scanning*. As part of this process it is useful to have a "type" value associated with each character to indicate whether it is a letter, a digit, or some other character. This is most simply accomplished by establishing an array *type* with one entry for each of the 256 possible characters:

$$type: \mathbf{array}\ [0:255]\ \mathbf{of\ integer}$$

Values in the array *type* are used by the token scanner to determine the type of each input character. A location in *type* is selected by converting the character value to an integer by the integer function, which returns the integer corresponding to the code for the character (see Section 2.2.3). Thus the type of character $c$ is given by *type*[integer[$c$]]. The array *type* could be initialized with constants, but the program can be machine independent if character constants in the program are used to generate the locations to store values. For example, to set the type for all letters we could use this program fragment:

> *alphabet* ← "ABCDEFGHIJKLMNOPQRSTUVWXYZ"
> **for** $i = 1$ **to** 26 **do**
>     *char* ← *alphabet*[$i$]     ⟦select the $i$th letter of *alphabet*⟧
>     *type*[integer(*char*)] ← 1     ⟦set type to 1 for that letter⟧

(a) Write a program to initialize *type* so that the type of all digits is 3.

(b) Write a program to initialize *type* so the following special characters have the indicated value:

| *value*: | 2 | 3 | 4 | 5 | 6 | 7 | 8 | 9 | 10 | 11 | 12 | 13 | 14 | 15 |
|---|---|---|---|---|---|---|---|---|---|---|---|---|---|---|
| *character*: | E | ↑ | . | + | − | * | / | ( | ) | = | , | b | $ | ' |

(c) Assume that *buffer* is a string and that *buffer*[*cursor*] contains the first digit of a sequence defining a nonnegative integer, terminated by the first succeeding nondigit character. Write a procedure to find the value of the integer.

5. Rewrite Algorithm 8.1, the simple insertion sort, in a manner parallel to Algorithm 3.1.

6. Write a procedure to make an entry in a table represented as in Figure 3.3. Assume that the characters of the name to be added are in *new*[1], *new*[2],...,*new*[*n*] and that the first unused location in *name* is given by *free*.

7. (a) Write a procedure to compare two names represented as in Figure 3.3, indicating which of the names would be first in a telephone directory. (Assume the names are in last-name-first form.)

   (b) Use the result of (a) along with Algorithm 3.1 to get a procedure to produce a sorted index array for a table of names represented as in Figure 3.3.

   (c) Write a version of binary search for index arrays as in (b).

   (d) Suppose that a search as in (c) discovered that a name was not in the table. Give an algorithm to insert it, keeping the index array up-to-date.

8. A simple alternative to the representation of names in Figure 3.3 is to have an array of fixed-length strings, each as long as the longest possible string. If the longest string has $l$ characters and the average length of the $n$ strings is $\bar{l}$, which approach is more economical?

9. Another alternative to the representation of Figure 3.3 is to use an array of characters, an array of subscripts *start*, and an array of lengths. If we have

$$\text{\textit{chars}:} \quad \text{SELLSHELLSEASHOREBYTHE}$$
$$\begin{array}{cccccc} \uparrow & \uparrow & \uparrow & & \uparrow & \uparrow \\ 1 & 5 & 10 & & 18 & 20 \end{array}$$

then we can represent the popular tongue twister by

$$\begin{array}{ll} \textit{start}: & 5\ 1\ 10\ 5\ 18\ 20\ 10 \\ \textit{length}: & 3\ 5\ \ 3\ 6\ \ 2\ \ 3\ \ 8 \end{array}$$

Note that values can overlap, as do "SELLS" and "SHELL".

   (a) The array *chars* contains 22 characters but represents a phrase with 30 nonblank characters, giving a compression of $22/30 \approx 73$ percent, neglecting the extra space for the arrays *start* and *length*. Encode some other phrases in the same manner and compute the compression ratios.

   (b) It would be more realistic to compute the compression ratio including the space required for the arrays *start* and *length*. Do it for the example above and your examples in part (a).

   (c) How would you sort a set of strings represented as above?

10. Packed arrays of characters and bits are two examples of the general case consisting of an array of $k$-bit values, where $k$ is less than the word size $w$. For such a structure the two functions needed are *STORE* and *RETRIEVE*. The function *STORE* puts a given-$k$-bit value in the array at the location specified. The function *RETRIEVE* gets the value of the $k$-bit element stored in the array at the specified location. There are at least three possible storage structures for this data structure: one value per word, continuous without straddled words, and a continuous stream of bits, as shown below:

one value per word:

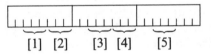

[1]     [2]     [3]     [4]     [5]

no straddled words:

[1]  [2]    [3] [4]    [5]

continuous bit stream:

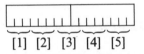

[1]  [2]  [3]  [4]  [5]

In this example, $k = 3$ and $w = 8$ with five values being stored. (Often $k$ divides $w$ evenly, but this need not be the case.)

  (a) Write *STORE* and *RETRIEVE* procedures for each of the three cases above, using **and** and shift as needed to extract the proper bits (see Section 2.1).

  (b) For each of the three cases above, how many words are needed to store $n$ elements for given values of $k$ and $w$?

  (c) Describe circumstances under which each of the above cases would be the best choice to use.

11. Suppose that three edges of a square sheet of copper are maintained at 0°C and the fourth is kept at 100°C. We want to compute the temperatures at points on the interior of the sheet. Approximate temperatures can be computed by representing the square as a collection of smaller squares, each with a uniform temperature. The temperature of each such smaller square can

then be kept in an element of a two-dimensional array. The computation begins by setting the temperatures of outside squares to either 0°C or 100°C (at the corners between the 0°C and 100°C edges there is a discontinuity and the temperature may be set, perhaps, as 50°C), and the temperatures of the squares not on the boundary are initialized to some value, say 0°C. Then, each square not on the boundary has its temperature set to the average of the temperatures of the eight surrounding squares; this is done repeatedly until it yields only insignificant changes in the interior temperatures. Perform such a calculation using a $10 \times 10$ array. How much longer would it take to use a $100 \times 100$ array?

12. Write a program to print an organization chart. The input is a file of no more than 100 lines, each containing the name of an executive and the names of his subordinates; the lines are in alphabetical order by executive. Each name is limited to ten characters, and each executive is limited to a maximum of seven subordinates. The final line specifies which of the executives is the president —it contains blanks in the first field and the name of the president in the second field. The output is to list one name per line, indented to indicate the number of executives above this person. The president's name is to be unindented on the first line. Subordinates' names are to appear in the listing below the name of their boss and before any other executive at the same level as their boss.

## 3.2   RECORDS AND POINTERS

Just as an array is a sequence of elements all having the same type and size, a *record* or *structure* is a sequence of elements having diverse types and sizes. The elements of a record are called *fields*. For example, a record representing an entry in a phone directory might contain fields for the name, address, and phone number of a listing, as shown in Figure 3.5. As in Figure 3.5, it is customary to draw records as rectangles with labeled subrectangles for the fields. The differing sizes of the rectangles typically reflect the different space requirements of the fields.

The most frequently applied operation on records is that of *field selection*. This operation accesses a specified field of a record in order to find the field's value or to assign it a new value. To represent field selection we use the field name as a function operating on the record; thus the phone number in the record *dad* is given by

$$phone\_number(dad)$$

and it can be assigned a value by

$$phone\_number(dad) \leftarrow 7521129$$

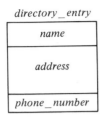

**Figure 3.5**
An example of a record containing three fields—*name*, *address*, and *phone_number*. The subrectangles are shown in different sizes to reflect the different space requirements of each.

or used for its value as in the statement

$$\text{print}(\textit{phone\_number}(\textit{dad}))$$

The same operations apply to a field as to any value of the same type. For example, since integers can be added, subtracted, and so on, so can integer fields.

To specify the fields that records of some type will have, most programming languages that allow records require the declaration of a *record type*. Such a declaration might look something like

> *directory_entry*: **record type**
>     *name*: **string length** = 20
>     *address*: **string length** = 20
>     *phone_number*: **integer**

This specifies a record type called *directory_entry* which has three fields—*name*, *address*, and *phone_number*. The type of each of these fields is also specified.

Once a record type has been defined as above, variables can be declared to be of that type, just as variables can be declared as **real**, **integer**, and so on. Thus the record variable *callee* might be declared as

> *callee*: *directory_entry*

It is important to note that the declaration of the record type name does not set aside any storage; only when a record variable of that type is declared is space reserved for it.

The types of the fields within a record are restricted only in that they must have known lengths; indeed, a field may even be an array or a record. For example, we could define a record type for addresses and use it within the declaration of directory entries:

> *address_type*: **record type**
>     *street_number*: **integer**
>     *street_name*: **string length** = 16
> *directory_entry2*: **record type**
>     *name*: **string length** = 20
>     *address*: *address_type*
>     *phone_number*: **integer**

Now we can refer to either an entire address, as in

> *address*(*customer*)

or to the street-name portion of the address:

> *street_name*(*address*(*customer*))

This hierarchical structure helps to make programs clearer, for it permits us to move entire records with a single assignment; for example, if *payer* is a record of type *directory_entry2*, we can assign its address from *customer* by

> *address*(*payer*) ← *address*(*customer*)

In traditional data processing, a *record* is the unit of communication between external files and a program. A *file* is like an array of records, except that it takes a relatively long time to access individual elements. In the file of Figure 3.6, the records are the phone company's list of customers. An input operation from this file makes a copy of one record into some record variable such as *customer*. It is then processed and perhaps written out again to the same file or another.

The picture of records we have given so far is inadequate in one respect. We will frequently have record types in which one or more of the fields are themselves records of the same type. For example, the phone company might wish to allow each entry in the directory to have a subordinate entry, such as when a physician lists a residence number immediately after his office number, or a family lists the children's number after the family number. This subordinate entry could have a name, an address, and a phone number, so it has the same structure as a directory entry. However, if we make the obvious declaration

> *directory_entry*: **record type**
>     *name*: **string length** = 20
>     *address*: *address_type*
>     *phone_number*: **integer**
>     *subsidiary_listing*: *directory_entry*

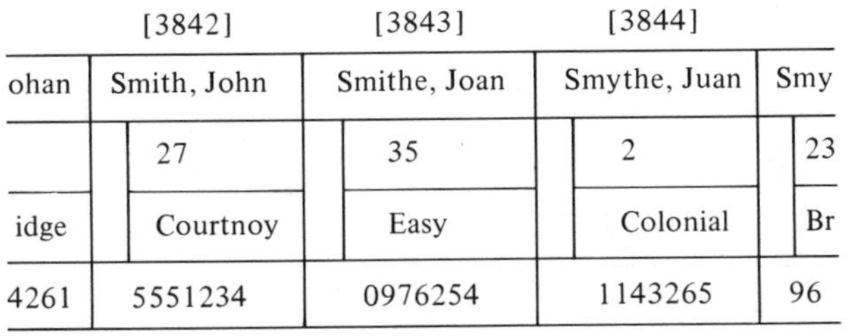

| [3842] | | [3843] | [3844] | |
|---|---|---|---|---|
| ohan | Smith, John | Smithe, Joan | Smythe, Juan | Smy |
| | 27 | 35 | 2 | 23 |
| idge | Courtnoy | Easy | Colonial | Br |
| 4261 | 5551234 | 0976254 | 1143265 | 96 |

(a) External file

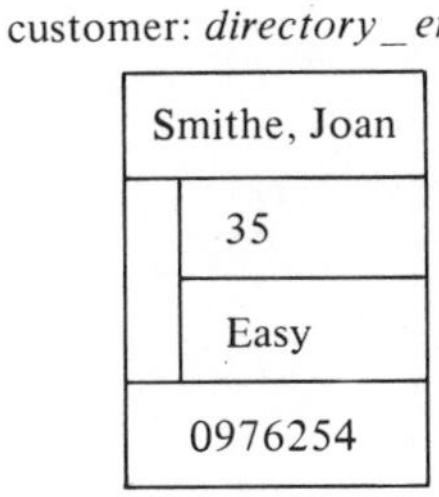

customer: *directory_entry2*

| Smithe, Joan | |
|---|---|
| | 35 |
| | Easy |
| 0976254 | |

(b) In memory

**Figure 3.6**
In data processing an external file contains many different records of the
same type (*directory_entry2*, in this case). The record *customer*—a variable in
memory—contains a copy of one record from the file.

we find that a variable declared to be of this type appears to require infinite
storage! For instance,

*listing*: *directory_entry*

would need the storage layout shown in Figure 3.7.

To avoid the need for infinite storage, we will not have record variables per
se, but rather *pointer variables*. Such a variable does not itself contain the fields
of a record; instead, it specifies the location of the actual record value. This can
be thought of as though each pointer variable were a rope and each record value
had a handle to which the rope might be attached. Assignment of a record value
to a pointer would cause the rope to be tied to the handle. Declarations of the
form

*variable_name*: *record_type*

where *record_type* has been declared to be a **record type**, are interpreted as
meaning that the given variable will be a pointer to a record of the type specified.

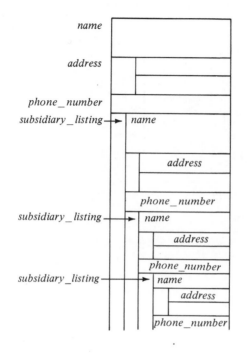

**Figure 3.7**
The storage layout for a *directory_entry* requiring infinite storage

Such a declaration causes space to be allotted only for a pointer, not for a record. To create the record values themselves we will use the primitive **getcell**. This primitive is used in assignments such as

$$listing \leftarrow \textbf{getcell}$$

where it finds space for a *directory_entry* and returns the "handle" of the space, that is, a pointer to it. **getcell** must determine from the context the type and size of the record requested. (In some situations the type and size information would be supplied as parameters to **getcell**.) The *directory_entry* record thus created will have space for a name and a phone number but not for an address; it will have space only for a pointer to an address record. We must create one by

$$address(\,listing\,) \leftarrow \textbf{getcell}$$

Similarly, we can create a subsidiary entry by

$$subsidiary\_listing(\,listing\,) \leftarrow \textbf{getcell}$$

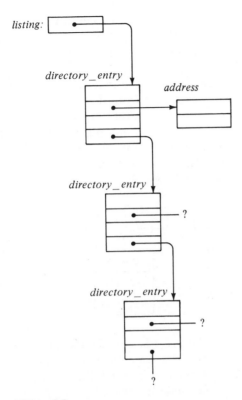

**Figure 3.8**
Example of record variables implemented as pointers to records. A question
mark indicates a pointer that has not been assigned a value; unpredictable
results follow when one of these pointers is used as a value in a computation.

or even, in the case of a physician's residence with a child's phone,

$$subsidiary\_listing(subsidiary\_listing(listing)) \leftarrow \textbf{getcell}$$

The results of the above calls of **getcell** are illustrated in Figure 3.8.

The only constant value for a pointer variable is **nil**, the *empty* or *null pointer*
that does not point to anything. The null pointer is usually indicated in drawings
by a diagonal line through the pointer field. Thus

$$subsidiary\_listing(subsidiary\_listing(listing)) \leftarrow \textbf{nil}$$

will modify Figure 3.8 so that listing has only one subsidiary listing and no
subsubsidiary. This will cause the bottom record to no longer be attached and it
must be returned to the storage pool from which it came. We will ignore the

question of how it is returned to the storage pool until Chapter 6; meanwhile we simply assume that records are automatically returned when they can no longer be accessed from program variables, either directly or through a sequence of pointers.

As data structures themselves, pointers can be operated on by the operation of field selection in which a data value is extracted from the record pointed to. This operation is just as described previously for records. In addition, pointers can be compared for equality and inequality, just as any other value can. It is important to note, however, that the comparison of two pointers does *not* compare the records to which they point! Thus if *A*, *B*, *C*, and *D* are all pointers to directory listing records and they point as follows:

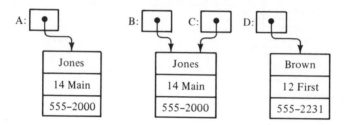

then *A* does not equal *B* (even though the records contain the same values); however, *B* does equal *C* (they both point to the same record). To compare the contents of two records, the record type name can be used like a field name to select the entire record. Thus the following would be true:

$$directory\_entry(A) = directory\_entry(B)$$

The same limitation applies to assignment of values: the result of $C \leftarrow D$ is to make *C* point to the record already pointed to by *D*; it has no effect on *B*. However, the assignment

$$directory\_entry(C) \leftarrow directory\_entry(D)$$

changes the contents of the record pointed to by *C*. Thus it does not cause *C* to point to a different record but alters memory so that the record pointed to by *C* *and B* contains new values. "Back-door" assignments like that to *B* can introduce very subtle bugs into programs.

## Exercises

1. Define a record type for playing cards; it should include fields for suit, rank, and whether the card is face up. Declare two variables for card records, one to represent the top card of a draw pile and the other to represent the top

card of a discard pile. Imagine that cards are face down in the draw pile and face up in the discards, and assume that there is initially no card in the discard pile. Write a fragment of code to move the card from the draw pile to the discard pile; be sure to "turn it over" so it is face up.

2. A data base management system must keep a dictionary of the variable names for each data record the system stores. For example, a payroll record may have identifiers such as *name, date_of_birth, sex, pay_grade, date_hired, address,...* . This dictionary is usually implemented as a set of records, with one for each identifier.

   (a) Define a record type for dictionary entries. Each record should include at least the identifier and the field type.

   (b) The *address* variable has subordinate variables such as *street_number, street_name, city*, and *zip*. One way to represent this relationship is to store the name of the superordinate variable in the record for the subordinate name. Why is this simpler than storing the names of the subordinates?

   (c) Instead of storing the name, we can store a pointer to the superordinate record. In what way would this be more efficient?

   (d) Suppose we have an array of pointers to dictionary records. Write a procedure to find and print the names of all the subordinates of a given dictionary entry.

3. Write a procedure to read information from an input file for a *directory_entry2*, create one, and set its fields from the information read. Assume there are procedures *READ_INT* and *READ_STRING* that return the next integer or string from the input. Procedure *READ_STRING* has one argument that gives the length of the string to be read.

4. Consider the declarations

$$\text{*testcell*: } \textbf{record type}$$
$$\text{value: } \textbf{integer}$$
$$\text{*child, sib*: *testcell*}$$
$$A, B, C, D: \text{*testcell*}$$

At the two points indicated in the program below, show the configuration of the records. In particular, which point to which, which pointers are **nil**, and which records are no longer accessible?

$$A \leftarrow \textbf{getcell}$$
$$sib(A) \leftarrow \textbf{nil}$$
$$B \leftarrow \textbf{getcell}$$
$$child(B) \leftarrow A$$
$$C \leftarrow \textbf{getcell}$$

$$sib(C) \leftarrow A$$
$$child(C) \leftarrow B$$
$$D \leftarrow child(child(C))$$
$$child(D) \leftarrow \textbf{nil}$$
[show the structure here]
$$A \leftarrow \textbf{getcell}$$
$$child(A) \leftarrow A$$
$$B \leftarrow \textbf{getcell}$$
$$sib(A) \leftarrow B$$
$$sib(B) \leftarrow B$$
$$child(B) \leftarrow child(C)$$
$$C \leftarrow \textbf{getcell}$$
$$sib(C) \leftarrow C$$
$$child(C) \leftarrow \textbf{nil}$$
$$D \leftarrow child(sib(B))$$
[show the structure here]

5. Suppose we have an array of pointers to *testcell* records, as declared in the previous exercise. Write a procedure to sort the array of pointers so that the first points to the *testcell* record with the smallest value, and succeeding pointers point to *testcell* records with successively higher values. Do not examine or modify the *child* or *sib* fields.

6. Write a procedure to count the depth of a *directory_entry* record—that is, to determine how "deep" the list of subsidiaries goes. [*Hint:* If the subsidiary listing field is **nil**, the depth is zero; otherwise it is one more than the depth of the record that comprises its subsidiary listing.] Write two versions of the procedure, one iterative and the other recursive.

7. At one point in a program for phone-directory maintenance it has been determined that the *directory_entry* L must have a new subsidiary listing even though one may be there already. The new record is to have the existing subsidiary listing, if any, as its own subsidiary. Suppose that N, A, and P are, respectively, the name, address pointer, and phone number for the new subsidiary listing. Write a sequence of assignment statements to make the insertion.

8. (a) Define a record type for representing a person. The information should be such that a program can determine for each person the name, date of birth, sex, and the location of records for parents and children.

   (b) Write a procedure to find a person's oldest sister. [*Hint:* You may want to redefine your structure for part (a) so that age ordering determines the order of children on lists.]

(c) Write a procedure to find and print the names of a person's mother's brothers.

9. Most programming language implementations store real values directly in **real** variables. However, it is possible to implement them so a **real** variable is just a pointer to the real value. This would make sense if real values were multiword values, as is often the case in high precision calculations or in machines with small word lengths.

   (a) Unfortunately, a pointer implementation of **real** values must invoke **getcell** frequently. Show this by outlining a procedure to add two real values, assuming there is a subroutine to perform the actual addition. Explain why the result cannot be stored in one of the argument records.

   (b) Suppose assignment is implemented by just copying one pointer into another. Can the system then have a "normalize" operation that does not call **getcell**? How about a procedure to increment a real value?

   (c) Can we implement a system such that comparison of real values for equality is simply a process of comparing the pointers to the values? Explain the difficulties in this approach.

10. In many respects an implementation of records as variables is the same as an implementation by pointers; however, there are important differences. Suppose that we have two *directory_entry* records $P$ and $Q$ with the same values for the fields. Using assignment and field selection, show how you *might* detect that the implementation of records was using pointers. Under what condition would you be able to detect the difference?

### 3.2.1 Examples of Records and Pointers

Records and pointers occur in many guises; this section considers four.

**Random Access in Arrays.** Any time a variable holds a subscript value, that variable is serving as a pointer to an element of an array. For example, the variables $l$ and $h$ in binary search (Algorithm 1.3) point to the low and high ends of the array segment where the desired element may be found; the variable $m$ in that algorithm points to the element that will be tested next. Similarly, in the array of subscripts illustrated in Figure 3.2, each element of the array is a pointer to an element in another array.

**Call-by-Reference.** When a formal parameter $F$ is specified to be call-by-reference (also known as call-by-variable), the corresponding actual parameter $A$ must be a variable. Any request for the value of $F$ is met by the current value of $A$; any changes to $F$ are completed by modifying $A$. The typical implementation of call-by-reference is to pass a pointer to $A$ as the value for $F$; operations on the formal parameter are then carried out by following the pointer to find the actual parameter variable. In this case, the record pointed to may have just a single field—the actual parameter variable. Note that the space allocated for $F$ is always only one pointer, no matter how big the actual parameter is, even if it is a large array. We can illustrate the effect of call-by-reference with this program:

$$A \leftarrow 3$$
print($A$)     ⟦3 is printed⟧
$P(A)$     ⟦call with parameter passed by reference⟧
print($A$)     ⟦4 is printed⟧

where $P$ is

**procedure $P(F)$**
$$F \leftarrow F + 1$$

During the execution of the procedure $P$, the value of $F$ is a pointer to the location used by the calling procedure to store $A$. Except as otherwise noted, we will assume that parameters are passed to procedures by value, so the procedure cannot modify them.

**Variable-Length Strings.** In many applications strings need to be allowed to vary dynamically in length, rather than have their exact length specified at the time of declaration. We can do this with records and pointers by implementing the string variable as a pointer to a record for the string value. This record can contain a length field followed by a string field containing the characters of the string. If the length of the string value is changed, **getcell** must be called to allocate a new string value.

For example, if $A$ and $B$ were string variables of variable length with values "ABC" and "MICHIGAN", respectively, the situation would be this:

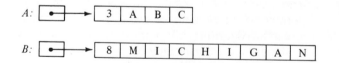

If $A$ were assigned the value "ILLINOIS", the result would be:

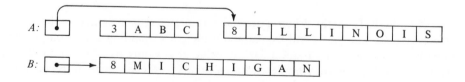

If there were no other pointer to the record for | 3 | A | B | C | it would
be returned to the pool of available storage.

**Secondary Storage.** The concept and term "record" evolved from the
storage of data records on external (secondary) storage, as illustrated in Figure
3.6(a). When needed, such records are transferred from the secondary storage
device to record variables in memory. In such cases there are not pointers like
those we have been using to implement record variables and there are not
pointers between records, since such pointers work by giving the location in
memory where the other record is to be found—the other may or may not be in
memory.

In secondary storage applications the term "record" has two meanings. It
refers not only to actual data records but also to the block of data transferred to
or from the secondary device in a single unit. These are not the same, because the
high overhead in initiating and completing a transfer makes it more economical to
move large numbers of data records together. This is called *blocking* the records,
and the number of records in each block is the *blocking factor*. To distinguish the
two kinds of records, the actual data records are called *logical records* while the
groups transferred are called *physical records*.

When a physical record is read, it is transferred in its entirety from the
storage device. Since only one logical record is generally needed at a time, some
strategy must be used to pass on the particular logical record of interest. There
are two common strategies: *copy mode* and *locate mode*. In both methods the
physical record is copied from the secondary storage device into a large
input/output area called a *buffer*. In copy mode there is also a separate data
record area into which the desired logical record is transferred. In locate mode
there is only a pointer variable that refers to the proper place in the buffer. This
avoids the costly operation of copying the data.

**Exercises**

1. Parameter-passing mechanisms are a considerable source of difficulty, both in
   implementation and in understanding by programmers. Consider this terrible
   program:

$$temp, var: \textbf{integer}$$
$$\textbf{procedure } SWAP(u, v)$$
$$u, v: \textbf{integer}$$
$$temp \leftarrow u$$
$$u \leftarrow v$$
$$v \leftarrow temp$$

The terrible part of this program is the use of a global variable (*temp*) for a purely local purpose in *SWAP*. Trouble arises if we have the call *SWAP*(*temp*, *var*) in a call-by-reference implementation. Explain the problem, showing the status of the variables and the sequence of events as the procedure is executed.

2. One alternative to call-by-reference is call-by-value-result. In this method a location is set aside for the parameter in the procedure itself. The actual parameter value is copied into the formal parameter space when the procedure is entered and is copied back when the procedure terminates. Do the previous exercise assuming that *u* and *v* are passed with call-by-value-result.

3. Write a procedure to determine whether parameters are passed by reference or by value-result. Do not use any global variables (as in Exercises 1 and 2). [*Hint:* Pass the same actual parameter to two different formal parameters.]

4. There are other problems in parameter passing, besides those of the previous exercises. For example:

   (a) Suppose we call *SWAP*(*i*, *A*[*i*]). Show the possible outcomes for suitable values of *i* and *A*. Among the alternatives to be considered is whether to evaluate the subscript expression once on entry, or to reevaluate it every time an array reference is made.

   (b) Suppose we make the call *STORE*(*j*, *j*), where *STORE* is

   > **procedure** *STORE*(*x*, *y*)
   >  [[*x* and *y* are passed by value-result]]
   >  $x \leftarrow 1$
   >  $y \leftarrow 2$

   What value will *j* have after the call? Could it have another value?

5. The principal operations on strings are concatenation and extraction of substrings. Assume that strings of undeclared length are implemented as in the text with a pointer to a record containing first the length and then the characters of the string. Show how to implement concatenation and extraction of substrings, given a version of **getcell** that takes an integer parameter specifying how long a string to create.

6. An alternate implementation of strings is to keep the actual characters in a separate area and implement each string value as a record containing the length of the string and a pointer to the first of the string characters. In this representation extraction of the substrings is especially simple; a new record is created describing the length and location of the substring. In some cases, however, a substring can also be used as the target for an assignment. Show how to implement the operation of replacing the third, fourth, and fifth characters of string *C* with another three characters. What happens if string *C*

points to characters that are also part of some other string $D$? Suppose the replacement is four characters; can the operation be done?

7. (a) Assume that *projectfile* contains records of type *project*, blocked ten to a physical record. Outline a procedure *READ* that will copy the next project record into an area specified as a parameter to *READ*. *READ* should return **false** if an end of file is reached—that is, if the previous record returned was the last—and **true** otherwise. Assume that you have a procedure *READFILE(buffer)* to transfer a physical record to buffer and that *READFILE* will return **false** when an end of file is reached. Specify the initial values to be stored in the variables of your procedure; these values will be set when the *OPEN* operation is performed on the file.

   (b) Outline a *WRITE* procedure for the same file as in (a), assuming a procedure *WRITEFILE(buffer)* is given.

   (c) Do (a) and (b) using locate mode instead of copy mode.

### 3.2.2  Typical Programming Language Notations

Most modern programming languages provide facilities for records and pointers. In this section we compare these facilities in PASCAL, PL/I, ALGOL 68, ALGOLW, and ADA. The problem chosen for the illustration is a portion of the code for implementing a symbol table for a compiler. Each record has fields giving the identifier for the variable, the number of occurrences of that identifier in the program, and a pointer to the next record of the same type in alphabetical order by identifier. After declaring a record type and a variable $R$ to point to records of that type, the fragment gives a procedure to print all identifiers. The fragment shows how to create a record, set $R$ to point to it, and initialize the number of occurrences to zero, increment the number of occurrences by one, and finally print the list of identifier records starting with the record pointed at by $R$. This fragment is presented in our notation in Algorithm 3.2(a). Algorithms 3.2(b), (c), (d), (e), and (f) give comparable fragments in PASCAL, PL/I, ALGOL 68, ALGOLW, and ADA, respectively.

```
PROGRAM example;
TYPE varpointer = ↑ variable;
     variable = RECORD
         identifier: array [1..31] of char;
         freq: integer;
         next: varpointer
     END;
VAR R: varpointer;
     ...
PROCEDURE listing(P: varpointer);
BEGIN
     WHILE P <> nil DO BEGIN
         writeln(P ↑.identifier);
         P := P ↑.next
     END;
     ...
END;
     ...

BEGIN
     new(R)
     R ↑. freq := 0;
     ...
     R ↑. freq := R ↑. freq + 1;
     ...
     listing(R);
     ...
END.
```

**Algorithm 3.2(b)**
The fragment in PASCAL

```
variable: record type
    identifier: string length = 31     [identifier name]
    freq: integer                      [number of occurrences]
    next: variable                     [next identifier]
R: variable
    ...
procedure LISTING(P)
P: variable
while P ≠ nil do
    print(identifier(P))
    P ← next(P)                        [go on to next variable]
    ...

R ← getcell
freq(R) ← 0
    ...
freq(R) ← freq(R) + 1
    ...
LISTING(R)
    ...
```

**Algorithm 3.2(a)**
The fragment used to illustrate records and pointers in several
programming languages

```
example: PROCEDURE;
    DECLARE 1 variable based,
        2 identifier char(31),
        2 freq fixed bin(31),
        2 next pointer;
    DECLARE R pointer;
    ...
listing: PROCEDURE(P);
    DECLARE P pointer;
    DO WHILE(P ¬ = null);
        PUT LIST(P − > identifier);
        P = P − > next;
    END;
END listing;
    ...
ALLOCATE variable set(R);
R − > freq = 0;
    ...
R − > freq = R − > freq + 1;
    ...
CALL listing(R);
    ...
END example;
```

**Algorithm 3.2(c)**
The fragment in PL/I

```
begin
record variable
    (string(31)identifier;
    integer freq;
    reference(variable) next);
reference(variable) R;
    ...
procedure listing(reference(variable) value P);
    while p ¬ = null do begin
        print(identifier(P));
        P := next(P) end listing;
    ...
R := variable( , 0, null);
    ...
freq(R) := freq(R) + 1;
    ...
listing(R);
    ...
end
```

**Algorithm 3.2(d)**
The fragment in ALGOLW

```
begin
  mode variable = struct
    (string identifier,
     integer freq,
     ref variable next);
  ref variable R;
    ...
  proc listing (ref variable P); begin
    ref variable T := P;
    while T :/= :ref variable(nil) do begin
      print(identifier of T);
      T := next of T
    end
  end;
    ...
  R := heap variable := (skip, 0, nil);
    ...
  freq of R := freq of R + 1;
    ...
  listing(R)
    ...
end
```

**Algorithm 3.2(e)**
The fragment in ALGOL68

```
procedure example is
  type var_record;
  type variable is access var_record;
  type var_record is record
    identifier: string(1..31);
    freq: integer;
    next: variable; end record;
  R: variable;
    ...
  procedure listing(P: variable) is
    T: variable := P;
  begin
    while T/= null loop
      print(T.identifier);
      T := T.next;
    end loop
  end;
    ...
begin
  R := new variable(identifier = >"" ,
                    freq = > 0,
                    next = > null);
    ...
  R.freq := R.freq + 1;
    ...
  listing(R);
    ...
end example;
```

**Algorithm 3.2(f)**
The fragment in ADA

**Exercises**

1. Work various exercises from Section 3.2 with the five sample languages used in this section.

2. Explore how the five sample languages treat record variables as opposed to records implemented with pointers. In particular, determine how the languages express the copying and comparing of entire records. Show also how the languages distinguish declaration of pointer variables from declaration of record variables for records of the same class.

## 3.3   IMPLEMENTATIONS IN MEMORY

In this section we present the details of storage structures for implementing arrays and records. We begin with a brief discussion of the overall arrangement of computer memory.

The memory of a computer is composed of a sequence of words as shown in Figure 3.9. Each word is referred to by its unique *address*—that is, its position in the sequence of words. The address of the first word is normally zero, since this allows a memory of $2^k$ words to have addresses that are $k$ bits long (see Exercise 1).

A computer program is a sequence of instructions, each of which generally refers to only one or a few words of memory. The reference is made by specifying the address of the desired word (the address value can be thought of as a pointer to the word). Figure 3.10 shows a snapshot of the contents of memory as it might appear during execution of a program, although it is typical for more than one program to share the memory at a time. The first portion of memory contains the instructions and data that constitute the operating system—the monitor program responsible for starting and stopping programs and giving them access to system resources. Next come the instructions of the program being run and the data areas for variables, arrays, records, and input/output buffers. Programs usually occupy a contiguous group of words because instructions are executed sequentially; execution of the instruction in location $w$ is followed by execution of the one in $w + 1$ unless the one in location $w$ modifies the sequence of execution of instructions.

We use the notation loc(*variable*) to represent the address to be used in an instruction referring to the variable. When a compiler encounters a declaration for a variable, say $x$, the compiler implements $x$ by setting aside one of the words of memory as the location to store the value of $x$. The address of the word set aside is loc($x$).

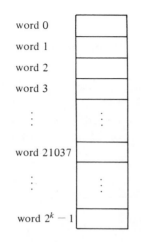

**Figure 3.9**
The sequence of locations in a computer memory

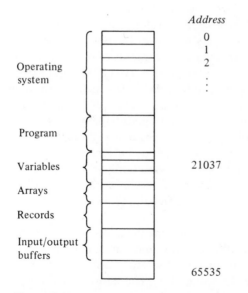

**Figure 3.10**
Diagram of a computer memory with $2^{16}$ words

## Exercises

1. Suppose a memory has $2^k$ words and addresses start with 1. How many bits are needed to express the address of the penultimate word of memory? How many bits are needed to express the address of the last word of memory?

2. If we consider the size of a program as a parameter, its space requirement will not grow as a linear function of the number of pointers in the program; as the program grows, each pointer requires more bits. Let $n$ be the number of $k$-bit words required by a program; find an expression for the number of bits required by the program, assuming that a fraction $f$ of the words are pointers.

3. Suppose you have a computer with $2^{16}$ words of sixteen bits each. Under what conditions will storing two variables $A$ and $B$ in successive words with $A$ in the first and $B$ in the second result in having the address of $B$ numerically less than the address of $A$? This might be important if a program were doing address arithmetic to sequence through an array. For example, a loop might test for an address greater than the last in the array, but that address might be less than the beginning of the array. What approaches can be taken to avoid the problem?

4. If a program is moved from one area of memory to another (nonoverlapping) area, all the pointers in the program must be adjusted. Write a routine to move a program. It has four arguments: the old location, the new location, the number of words in the program, and a bit map. The bit map has one bit for every word in the program; it is zero if the word does not contain a pointer, and one if it does. By what amount will pointers have to be adjusted? How many words does the bit table occupy?

### 3.3.1  Arrays

An array

$$A : \textbf{array}[\,l : u\,] \textbf{ of } item \qquad [\![\text{where } item \text{ is } w \text{ words long}]\!]$$

is easily arranged in order in adjacent memory locations. Thus $A[l]$ is stored, say, beginning in location $L$; $A[l + 1]$ is stored beginning in location $L + w$; $A[l + 2]$ is stored beginning in location $L + 2w$, and so on. Figure 3.11 shows this schematically. A total of $(u - l + 1)w$ words of memory are used.

In this implementation there is a simple, fixed relationship between an index $i$ and $\text{loc}(A[i])$, $l \leqslant i \leqslant u$:

$$\text{loc}(A[i]) = \text{loc}(A[l]) + (i - l)w.$$

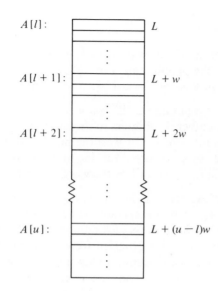

**Figure 3.11**
Diagram of an array $A[l:u]$ as arranged in memory, each element of the array requiring $w$ words of storage

This relationship permits direct, immediate access to any element of the array, as needed for example in the binary search procedure of Algorithm 1.3.

The implementation of multidimensional arrays can be derived from that of one-dimensional arrays. We do this first for a two-dimensional array, and then extend it to higher dimensions. To implement the $n \times m$ array

$$
\begin{pmatrix}
A[1,1] & A[1,2] & \cdots & A[1,m] \\
A[2,1] & A[2,2] & \cdots & A[2,m] \\
& \vdots\ \vdots & \vdots & \\
A[n,1] & A[n,2] & \cdots & A[n,m]
\end{pmatrix}
\tag{3.1}
$$

consider it as an array $B[1],\ldots,B[n]$ in which each $B[i]$ is in turn an array of $m$ elements consisting of the $i$th row of the matrix. Thus $\hat{w}$, the number of words needed for an element $B[i]$, is $mw$, where $w$ is the number of words needed for each element $A[i,j]$ of (3.1). Since the array $B[i]$ begins at location

$$
\begin{aligned}
\mathrm{loc}(B[i]) &= \mathrm{loc}(B[1]) + (i-1)\hat{w} \\
&= \mathrm{loc}(B[1]) + (i-1)mw,
\end{aligned}
$$

the location of $A[i, j]$ is therefore

$$
\begin{aligned}
\mathrm{loc}(A[i, j]) &= \mathrm{loc}(B[i]) + (j - 1)w \\
&= \mathrm{loc}(B[1]) + [(i - 1)m + (j - 1)]w \\
&= \mathrm{loc}(A[1, 1]) + [(i - 1)m + (j - 1)]w.
\end{aligned}
$$

This representation is known as *rowwise* storage of the matrix; the *columnwise* storage is obtained by considering the array (3.1) to be an array $C[1], \ldots, C[m]$ in which each $C[i]$ is an array of $n$ elements consisting of the $i$th column of the matrix.

More generally, suppose we have an $n$-dimensional array $A[l_1 : u_1, l_2 : u_2, \ldots, l_n : u_n]$. If $n = 1$, the representation is as already discussed:

$$
\mathrm{loc}(A[i]) = \mathrm{loc}(A[l_1]) + (i - l_1)w. \tag{3.2}
$$

This is the basis of a recursive description of the rowwise representation of $A$. We consider $A$ to be an $(n - 1)$-dimensional array $\hat{A}[l_1 : u_1, l_2 : u_2, \ldots, l_{n-1} : u_{n-1}]$, each of whose entries $\hat{A}[i_1, i_2, \ldots, i_{n-1}]$ is the one-dimensional array $A[i_1, i_2, \ldots, i_{n-1}, j]$, $l_n \leqslant j \leqslant u_n$. Then, using (3.2),

$$
\mathrm{loc}(A[i_1, i_2, \ldots, i_n]) = \mathrm{loc}(\hat{A}[i_1, i_2, \ldots, i_{n-1}]) + (i_n - l_n)w. \tag{3.3}
$$

Continuing, using (3.3),

$$
\mathrm{loc}(\hat{A}[i_1, i_2, \ldots, i_{n-1}]) = \mathrm{loc}(\hat{\hat{A}}[i_1, i_2, \ldots, i_{n-2}]) + (i_{n-1} - l_{n-1})\hat{w},
$$

where $\hat{w}$ is the number of words required for an array element of $\hat{A}$—that is, $\hat{w} = (u_n - l_n + 1)w$—so

$$
\begin{aligned}
\mathrm{loc}(A[i_1, i_2, \ldots, i_n]) = {} &\mathrm{loc}(\hat{\hat{A}}[i_1, i_2, \ldots, i_{n-2}]) \\
&+ (i_{n-1} - l_{n-1})(u_n - l_n + 1)w + (i_n - l_n)w.
\end{aligned}
$$

Applying (3.3) again and again eventually yields

$$
\mathrm{loc}(A[i_1, i_2, \ldots, i_n]) = \mathrm{loc}(\hat{\hat{A}}[i_1]) + w \sum_{j=2}^{n} \left[ (i_j - l_j) \prod_{k=j+1}^{n} (u_k - l_k + 1) \right].
$$

A final application of (3.2) gives

$$\text{loc}\left(\overset{\uparrow}{A}[i_1]\right) = \text{loc}\left(\overset{\uparrow}{A}[l_1]\right) + (i_1 - l_1)\overset{\uparrow}{w},$$

where $\text{loc}(\overset{\uparrow}{A}[l_1])$ is the first location of the entire array, that is, $\text{loc}(A[l_1, l_2, \ldots, l_n])$, and $\overset{\uparrow}{w}$ is the number of words required for an array element of $\overset{\uparrow}{A}$, that is, $\overset{\uparrow}{w} = w \prod_{k=2}^{n}(u_k - l_k + 1)$. Thus

$$\text{loc}(A[i_1, i_2, \ldots, i_n]) = \text{loc}(A[l_1, l_2, \ldots, l_n])$$

$$+ w \sum_{j=1}^{n}\left[(i_j - l_j) \prod_{k=j+1}^{n} (u_k - l_k + 1)\right]. \quad (3.4)$$

**Exercises**

1. Simplify (3.4) for the following relatively common cases:
   (a) $n = 1$.
   (b) $n = 2$.
   (c) $l_j = 0$ for all $j$.
   (d) $l_j = 1$ for all $j$.
   (e) $l_j = -u_j$ for all $j$.

2. How should the formula in (3.4) be evaluated? Write an *efficient* procedure to do the evaluation. [*Hint:* Model your procedure after Horner's method, Equation (2.2).]

In Exercises 4 through 7 you are to design storage structures that allocate space only to non-zero elements.

3. Develop the columnwise analog of (3.4).

4. A *lower triangular matrix* is a matrix $A = (a_{ij})$ in which $a_{ij} = 0$ for $i < j$, and thus it is written

$$A = \begin{pmatrix} a_{11} & & & & \\ a_{21} & a_{22} & & & \\ a_{31} & a_{32} & a_{33} & & \\ \vdots & \vdots & \vdots & \ddots & \\ a_{n1} & a_{n2} & a_{n3} & \cdots & a_{nn} \end{pmatrix}$$

Design a sequential allocation for such matrices. The location of $a_{ij}$ should be a simple function of $i$, $j$, and the location of $a_{11}$. Generalize your result for *tetrahedral arrays*: a $k$-dimensional tetrahedral array has the property that $a_{i_1 i_2 \cdots i_k}$ is nonzero only for $0 < i_k \leqslant i_{k-1} \leqslant \cdots \leqslant i_1 \leqslant n$.

5. A *tridiagonal matrix* is a matrix $A = [a_{ij}]$ in which $a_{ij} = 0$ if $|i - j| > 1$, and thus it is written

$$
A = \begin{pmatrix}
a_{11} & a_{12} & & & & \\
a_{21} & a_{22} & a_{23} & & & \\
& a_{32} & a_{33} & a_{34} & & \\
& & & \ddots & \ddots & \ddots \\
& & & & & a_{n-1, n} \\
& & & & a_{n, n-1} & a_{n, n}
\end{pmatrix}
$$

Design a sequential allocation for such matrices. The location of $A[i, j]$ should be a simple function of $i$, $j$, and the location of $A[1, 1]$.

6. Use your result from Exercise 5 to obtain a sequential allocation for

$$
\begin{pmatrix}
& & & & a_{1, n-1} & a_{1, n} \\
& & & & & a_{2, n} \\
& & & \ddots & & \\
& & \ddots & \ddots & \ddots & \\
& & & \ddots & & \\
a_{n-1, 1} & & & & & \\
a_{n, 1} & a_{n, 2} & & & &
\end{pmatrix}
$$

7. We want to store values $A[1]$, $A[2]$,...,$A[7]$ corresponding to the vertices of a triangular grid, as in Figure 3.12. There are three axes $x$, $y$, and $z$ as shown. Notice, however, that a vertex is determined by specifying any two of its three coordinates; thus, for example, $A[5]$ could be specified by $x = -1, z = 0$ or by $x = -1, y = 1$ or by $y = 1, z = 0$. Find three addressing functions $f(x, y)$, $g(x, z)$, and $h(y, z)$ that determine the array entry specified by given coordinates. In the example we would have $f(-1, 1) = g(-1, 0) = h(1, 0) = 5$.

8. In some FORTRAN implementations, arrays are stored in consecutive segments of memory. That is, if $A$ and $B$ are declared with

INTEGER $A(100)$, $B(50)$

then 150 words are allocated with the first 100 devoted to $A$ and the rest to $B$.

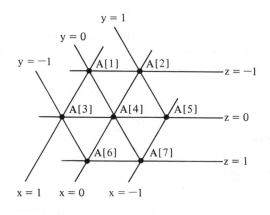

**Figure 3.12**
Exercise 7

Since many of these implementations do not check the size of subscript values, it is possible to refer to elements of $B$ with subscripts on $A$. Thus, for example, the following fragment initializes both $A$ and $B$ to zero:

$$\text{DO } 10 \text{ I} = 1,150$$
$$10 \quad \text{A(I)} = 0$$

What are the benefits and disadvantages of this trick? Consider space to store code, execution time, the probability of bugs, and the ability to move the code to another implementation of FORTRAN.

### 3.3.2   Records and Pointers

The implementation of records and pointers is simpler than that for arrays. First of all, note that a pointer value can be implemented simply as the address of the corresponding record. Second, a record is just a consecutive block of words. All we need discuss is the implementation of field selection. Suppose our record type is declared as

$$\textit{type\_name}: \textbf{record type}$$
$$\textit{field\_name}_1: \text{item}_1,$$
$$\textit{field\_name}_2: \text{item}_2,$$
$$\cdots$$
$$\textit{field\_name}_n: \textit{item}_n$$

where the $i$th item has length $len_i$ words. Whenever $P$ points to a record of type

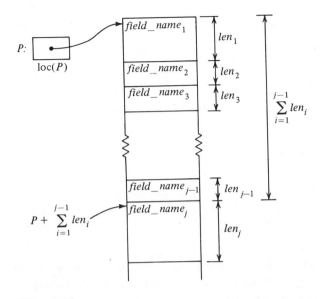

**Figure 3.13**
Layout of a record. Note that loc($P$) is irrelevant to the location of the fields.

*type_name*, then the address of *field_name*$_j$($P$) can be calculated by computing the length of the preceding $j - 1$ fields. Thus the address of the $j$th field is given by

$$\text{loc}\big(\text{field\_name}_j(P)\big) = P + \sum_{i=1}^{j-1} len_i \tag{3.5}$$

This is illustrated in Figure 3.13. Note that Equation (3.5) uses the value of $P$ rather than loc($P$) [compare this with Equation (3.4)], since $P$ is a pointer to the physical location of the record.

In the preceding discussion we have assumed that each of the fields consists of an integral number of words. This need not be the case, and we can have *packed records*, but field selection is more complex than for unpacked records. As noted in Chapter 2, masking and shifting must be done to get the value from the field. More generally, fields may extend from the middle of one word through another word and on into a third or beyond. To permit this we simply consider the rightmost bit of one word to immediately precede the leftmost bit of the next word. In order to write a general procedure for field selection in packed records, we need a way to specify the location of the field. For unpacked records this is done by an offset to the word; for packed records we need a *descriptor* consisting of an offset giving the number of bits from the start of the record and the number of bits in the field (see Figure 3.14). Since these values will be fairly small

| Field | Descriptors | |
|---|---|---|
| Ⓐ | 0 | 21 |
| Ⓑ | 21 | 16 |
| Ⓒ | 37 | 62 |
| Ⓓ | 99 | 29 |
| Ⓔ | 128 | 32 |

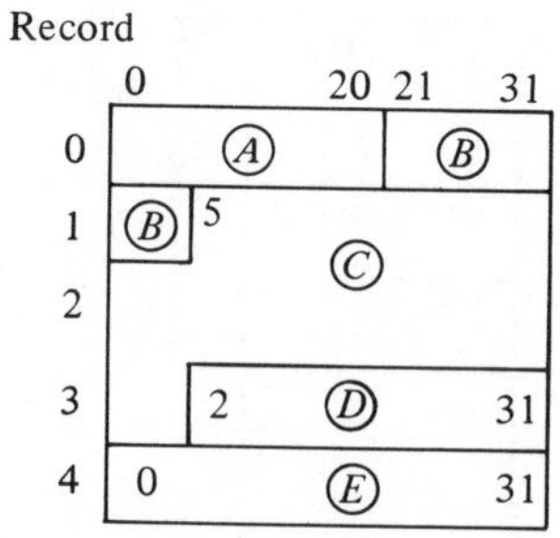


**Figure 3.14**
Descriptors for fields in a packed record, assuming 32-bit words. The record has five fields, and each has a descriptor containing the start bit and length of the field. For example, the field *C* starts in the thirty-eighth bit of the record and is 62 bits long.

compared to the maximum value for a word, they can themselves be packed into a single word. The descriptor might also contain an indication of the type of the field: **integer**, **real**, **character**, or whatever.

### Exercise

Write out the details of the procedure for field selection in packed records. The arguments of the procedure are a pointer to the record, a pointer to the descriptor, and a pointer to an area in which to store the extracted field. Assume that there are *w* bits per word.

## 3.4 REMARKS AND REFERENCES

The problems involved in implementing arrays, records, and pointers are relatively simple in the static case—that is, when all size information is available at compile time. For declarations and allocations at run time the difficulties and possible trade-offs are many; some will be discussed in Chapter 6. Arrays with bounds known only at run time can be handled with *dope vectors*, a simple instance of records and pointers. See, for example,

Gries, D., *Compiler Construction for Digital Computers*. New York: John Wiley & Sons, Inc., 1971.

This book also contains a discussion of the various parameter-passing mechanisms. Similar discussions can also be found in

Aho, A. V., and J. D. Ullman, *Principles of Compiler Design*. Reading, Mass.: Addison-Wesley Publishing Co., Inc., 1977.

# Chapter 4

# Lists

I've got a little list, I've got a little list.

*The Mikado,*   Gilbert and Sullivan

**A** *list* data structure is a finite sequence of *elements*

$$s_1, s_2, \ldots, s_n. \tag{4.1}$$

For example, a shopping list is a list of items to purchase, a library catalog is a list of the contents of the library, and a telephone book is a list of names, addresses, and telephone numbers. The *length* of the list is the number of elements on it—$n$ in (4.1). Among the functions that can be defined for a list data structure are *insertion* and *deletion* of elements and an operation to find the *next* element after a given element. We will use the constant **nil** to represent the empty list—that is, the list with no elements. When *next* is applied to the last element of a list, **nil** is the value returned.

To illustrate these functions, consider writing a program to add an element *new* to a phone list $L$ stored in order by phone number. The first portion of the program must find the proper position in the list for *new*:

$$x \leftarrow L$$
$$\textbf{while } x \neq \textbf{nil and } number(x) < number(new) \textbf{ do } x \leftarrow next(x) \tag{4.2}$$

If we neglect the ends of the list, this loop will exit with the condition

$$number(\text{element before } x) < number(new) \leqslant number(x) \tag{4.3}$$

At this point the *insert* function is invoked to put *new* into place before $x$ in $L$.

The simplest storage structure for a list such as (4.1) is *sequential allocation* in a one-dimensional array of records (see Chapter 3). If the elements of the list are in an array $A$, and $s_1$ is in $A[1]$, then $s_i$ is in $A[i]$. Or, in general, if the first element is in $A[k]$, then $s_i$ is in $A[k + i - 1]$. With sequential allocation, the algorithm for *next* is trivial: the subscript value is increased by one. In (4.2), $x$ can be a subscript value, so the *next* operation is

$$x \leftarrow x + 1. \tag{4.4}$$

For this simple storage structure, the value of **nil** is one plus the number of elements in the list. With sequential allocation, however, the search of (4.2) could be replaced by the more efficient binary search of Section 1.3.

Unlike *next*, *insert* is cumbersome with sequentially allocated lists. For example, if the phone list of (4.2) is in an array $L$ of $n$ elements, the following loop is required to insert a record just before $x$:

$$\textbf{for } j = n \textbf{ to } x \textbf{ by } -1 \textbf{ do } L[j + 1] \leftarrow L[j]$$
$$L[x] \leftarrow \textit{new}$$

The time for this operation is proportional to $n$, the length of the list. There are also two additional costs of this *insert*. First, enough space must be allocated so that the array is large enough for the maximum number of elements that will ever be needed; in many cases programs are run with only a fraction of the maximum possible data, so this overhead may be large. Second, a graceful recovery is rarely possible if the number of elements allocated for the array turns out to be insufficient—the run must be restarted with a larger array. Since runs with more data generally take more time, the cost of restarting can be large. Fortunately, the problems of space and time can often be solved by the technique of linked lists introduced next. Linked lists are so important that the term "list" almost always refers to a linked-list implementation.

**Exercises**

1. Explain how the program fragment (4.2) can be used to sort a list into order.

2. Using a sequential allocation, write a program to delete from the phone list the entry for a phone that has been discontinued. Be sure to account for all special cases.

3. Expand condition (4.3) to include the situation at the ends of the list. Under what conditions is it possible to insert *new* by placing it after the element before $x$?

4. Show that a dummy list element with phone number 999-9999 would simplify the program fragment (4.2). [*Hint:* Where in the list would this dummy element be?]

5. Write a program using the operations of *next*, *insert*, and *delete* to decompose one list and produce another in reverse order.

6. An integer can be represented as a list of decimal digits. Use the list operations to calculate $F_{85}$, where $F_0 = 0$, $F_1 = 1$, and for $i \geq 2$, $F_i = F_{i-1} + F_{i-2}$. The required value has more than sixteen decimal digits.

7. Show that if all the $n + 1$ possible locations for an insertion are equally likely, then the expected number of elements that must be moved for an insertion into a sequentially allocated list is $n/2$.

8. Show in detail how to implement the insertion into a phone list when the list is sequentially allocated and maintained in order by phone number. Include declarations and considerations of insertion at the ends of the list.

9. Explain why an assembly language version of (4.4) would be

$$x \leftarrow x + a$$

where $a$ is the number of words in each record.

★10. Suppose the number of elements actually occupied in an array $R$ is a normally distributed random variable with mean $\mu$ and standard deviation $\sigma$.
    (a) Show that 97.7 percent of all runs will succeed if the space allocated for $R$ is $\mu + 2\sigma$.
    (b) With the space allocated in (a), what is the expected percentage of the allocated space that is occupied?

## 4.1 LINKED LISTS

The inefficiency of inserting and deleting elements from a list that is sequentially allocated in an array occurs because the order of the list of elements is recorded *implicitly*. Adjacent records of the list must be in adjacent memory locations, so a number of records may have to be moved for an insertion or deletion. This movement can be avoided if the order of the elements is recorded explicitly instead of implicitly. In particular, a *linked-list* implementation associates with each list element $s_i$ a pointer $LINK_i$ to indicate the address at which $s_{i+1}$ is stored. There is also a pointer $P$ to the first element—the *head*—of the list; that is, $P$ contains the address of $s_1$. These relationships are illustrated in Figure 4.1.

Each node (record) in Figure 4.1 consists of two fields, the list element itself in the INFO field and the pointer to its successor in the LINK field. Since there is no successor to $s_n = \text{INFO}(a_n)$, we use the notation $LINK_n = \text{LINK}(a_n) = \textbf{nil}$, where **nil** is the *empty* or *null* pointer. The exact locations of $a_1, a_2, \ldots, a_n$ are unimportant in this context, so it is more common to depict the list of Figure 4.1 as shown in Figure 4.2.

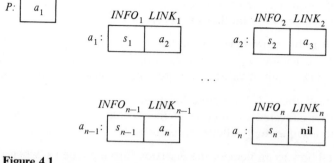

**Figure 4.1**
Representation of the list (4.1) as a linked list. Each record has an INFO field containing an element of the list and a LINK field containing the address of the successor record. The end of the list is indicated by the value **nil** in the LINK field.

This linked representation facilitates the operations of insertion and deletion of an element after $s_i$. All that is necessary is to change the values of some pointers. For example, to delete the element $s_2$ from the list in Figure 4.2, it is only necessary to set $\text{LINK}(a_1) \leftarrow \text{LINK}(a_2)$, and the element $s_2$ is no longer in the sequence (see Figure 4.3). To insert a new element $s_{1.5}$ into the sequence in Figure 4.2, it is only necessary to create a new element at some location $a_{1.5}$ with $\text{INFO}(a_{1.5}) = s_{1.5}$ and $\text{LINK}(a_{1.5}) = \text{LINK}(a_1)$ and to set $\text{LINK}(a_1) \leftarrow a_{1.5}$ (see Figure 4.4). (What is required to add an element $s_{0.5}$ to the list?) Concatenation of lists and splitting a list into lists are also easy.

Use of linked allocation implies the existence of some mechanism for allocating new records (list elements) as they are needed and collecting old records when they are no longer part of a list. The latter situation arises when a list element is deleted, as was $s_2$ in Figure 4.3. We will assume that some specialized software is written to keep track of unused records and to make them available as requests arise. Requests will be assumed to be met by **getcell**; as discussed in Chapter 3, **getcell** is used as the sole right-hand side of an assignment statement and returns the address (location) of a new record of whatever size is required for the variable on the left-hand side of the assignment. The record returned has no particular contents, so it must be initialized before it is attached

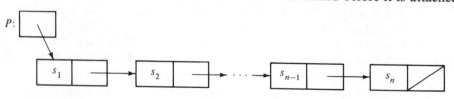

**Figure 4.2**
An alternate, more common way of depicting the list in Figure 4.1

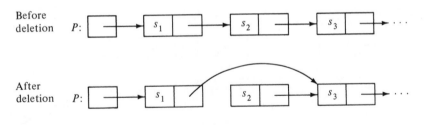

Before deletion

After deletion

**Figure 4.3**
Deletion of an element from a linked list

to a list. Thus to add the element $s_{1.5}$ as in Figure 4.4 we would actually have

$$t \leftarrow \textbf{getcell} \qquad \llbracket \text{allocate a new record} \rrbracket$$
$$\text{INFO}(t) \leftarrow s_{1.5} \qquad \llbracket \text{initialize it} \rrbracket$$
$$\text{LINK}(t) \leftarrow \text{LINK}(a_1) \qquad \llbracket \text{put it in the list} \rrbracket$$
$$\text{LINK}(a_1) \leftarrow t$$

where $a_{1.5}$ is the value returned by **getcell** and stored in $t$.

Details of various mechanisms for allocation and collection of records are presented in Chapter 6. For the time being, the reader can imagine that all records are the same size and each request is met by taking a record from a *free list* of unallocated records. A record that becomes unused is automatically returned to this list.

It is instructive to consider the general case of moving an element from the middle of one list to the middle of another. Suppose $x$ and $y$ point to elements of lists and we want the element after $x$ to be removed and inserted after $y$. A

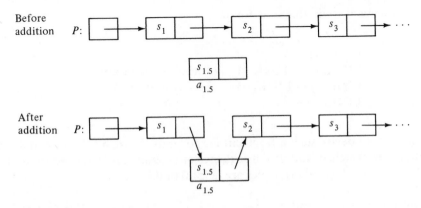

Before addition

After addition

**Figure 4.4**
Insertion of an element in a linked list

(a) Before

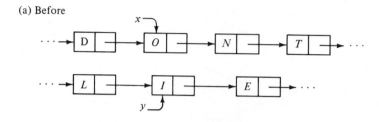

(b) Sequence of assignments

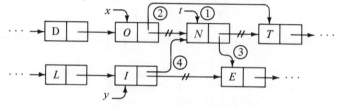

(c) After

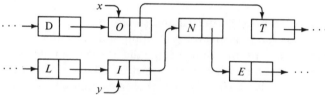

**Figure 4.5**

Moving an element from one list to another. The element $N$ is moved from after $x$ to be after $y$. Circled digits in (b) show the order of the assignments, and the slashes show the pointers that are destroyed by those assignments.

program fragment to do this is as follows:

$$t \leftarrow \text{LINK}(x)$$
$$\text{LINK}(x) \leftarrow \text{LINK}(t) \qquad [\![\text{delete element after } x]\!]$$
$$\text{LINK}(t) \leftarrow \text{LINK}(y) \qquad [\![\text{insert it after } y]\!]$$
$$\text{LINK}(y) \leftarrow t \qquad\qquad\qquad\qquad\qquad\qquad (4.5)$$

The best way to devise such a fragment is shown in Figure 4.5. Before and after conditions are sketched, and then a sequence of assignments is devised to produce the desired result. Several principles are helpful in this regard:

1. Copy a pointer value and then replace it [ $\cdots \leftarrow \text{LINK}(\sim)$; $\text{LINK}(\sim) \leftarrow \cdots$ ]. Note that each $\text{LINK}(\sim)$ in (4.5) appears first on the right side of an

assignment to make a copy, and then on the left side of an assignment where its value is changed. The entire sequence of statements can be thought of as a circular "daisy chain," since the $t$ at the beginning appears also at the end. Neglect of this principle can lead to "losing" a record because there is no pointer to it. Note, for example, the problems that arise if the first step of (4.5) is omitted or the steps are performed out of order.

2. Disconnect a record before modifying it [LINK($x$) ← LINK($t$)].

3. Initialize a record before attaching it [LINK($t$) ← LINK($y$)].

Although correct programs can be written that violate these rules, these rules do give an order for assignments and help ensure that all necessary actions are taken: disconnection, modification, and attachment. The suggested order follows from considering a list as an entity in itself. If the rules are followed, the lists are always intact; the record moved is modified only while it is not part of any list. Such considerations are essential for correct programs in cases where more than one process is accessing the list simultaneously.

The ease of insertion and deletion in linked lists is not without its cost. The fixed relationship between $i$ and the location of $s_i$ in a sequential allocation allows us, for example, to have immediate and direct access to any element of the list. In a linked allocation there is no such relationship, and access to list elements other than the first is indirect and inefficient. For instance, given the length of a list, it is easy to find the middle element if the list is sequentially allocated, but relatively difficult if the list is linked. Furthermore, a price is paid in terms of storage overhead for the LINK fields.

In choosing between sequential and linked allocation for a particular application, we must examine the types of operations that will be performed on the list and their relative frequencies in order to make an intelligent decision. If the operations are largely accessing random elements, searching for specific elements (see Chapter 7), or sorting the elements into order (see Chapter 8), then sequential allocation is usually better. On the other hand, linked allocation is preferable if the operations are largely inserting and/or deleting elements and/or splitting lists.

As an example of the manipulation of linked lists, let us return to the problem of insertion of a new phone record in a list $L$ ordered by phone number. Our first problem is to find the proper position in the list, which means finding the first element $p$ with a phone number larger than the new number. The new record is to be inserted just after the predecessor of $p$—that is, the record pointing to $p$. Accessing the predecessor of a selected element is a task that is often needed; the standard technique is to scan the list, keeping track of both the element currently being scanned *and* its predecessor in the list. Thus if $p$ is the element currently being scanned and *pred* is its predecessor on the list, the loop would be

$$\textbf{while } number(\,p\,) < number(\,new\,) \textbf{ do}$$
$$pred \leftarrow p$$
$$p \leftarrow \text{LINK}(\,p\,)$$

Notice that the relationship between *p* and *pred* is preserved with each iteration of the loop. When the loop ends, we insert the new record after *pred* by the assignments

$$\text{LINK}(new) \leftarrow \text{LINK}(\,pred\,)$$
$$\text{LINK}(\,pred\,) \leftarrow new$$

This, of course, is just the bare bones; we have left three important points unresolved: What if the new record belongs at the end of the list? What if it belongs before the first element? What if the list is empty?

For the first point, it is sufficient to change the test in the **while** loop to "*p* ≠ **nil** and *number*(*p*) < *number*(*new*)," allowing the loop to end gracefully when the end of the list is encountered. Although a conditional test is needed in many algorithms to determine why the loop ended, it is sufficient in this case to always insert the new element after *pred*, no matter how the loop ended. The algorithm is thus as given in Algorithm 4.1.

$$pred \leftarrow L$$
$$p \leftarrow \text{LINK}(L)$$
$$\textbf{while } p \neq \textbf{nil and } number(\,p\,) < number(\,new\,) \textbf{ do}$$
$$pred \leftarrow p$$
$$p \leftarrow \text{LINK}(\,p\,)$$
$$\text{LINK}(new) \leftarrow \text{LINK}(\,pred\,)$$
$$\text{LINK}(\,pred\,) \leftarrow new$$

**Algorithm 4.1**
Insertion of a new record in a sorted list

The crux of the second point is that *pred* is necessarily undefined for the first element of the list: This possibility is handled by not executing the **while** loop of Algorithm 4.1 if the number in *new* should precede that in the first element of the list. None of this will work, however, if the list is empty, so the third point is handled by a special test for *L* = **nil**. The complete algorithm is then

$$\textbf{if } L = \textbf{nil or } number(new) < number(L)$$
$$\textbf{then}$$
$$\text{LINK}(new) \leftarrow L$$
$$L \leftarrow new$$
$$\textbf{else } [\![\text{continue with Algorithm 4.1}]\!]$$

(a) Empty list

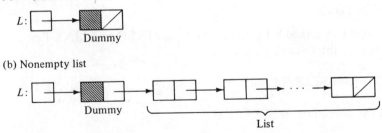

(b) Nonempty list

**Figure 4.6**
A linked list $L$ with a dummy element. The shaded INFO field contains no meaningful information.

Notice that the first assignment statement assigns the value **nil** to LINK(*new*) if the list is empty and otherwise assigns the pointer to the first element.

This algorithm can be written more cleanly, however, if we introduce a *dummy element* as a perpetual first element in the list. The INFO field of the dummy will have no particular value, but the LINK field will point to the first actual element of the list. Figure 4.6 shows both empty and nonempty lists with this dummy element. With the addition of the dummy element, the special tests for the second and third points can be omitted, and the entire algorithm is just as given in Algorithm 4.1. Observe for yourself what happens if the list is empty or the new record belongs before the first element. Note the extra space required for the dummy element can be considered repaid by the reduction in the space required for the algorithm.

The technique of designing (or modifying the design of) a data structure to simplify an accompanying algorithm is one that occurs frequently, and the reader should review this example carefully.

## Exercises

1. Follow the three principles given in the text to devise a program that moves the first element of a list $P$ to become the first element of a list $Q$.

2. Draw three versions of Figure 4.5(a) showing the pointers after each assignment of the program fragment (4.5).

3. Show how the **while** loop of Algorithm 4.1 can be simplified by the use of the dummy element 999-9999 of Exercise 4, page 124.

4. Write two versions of a program to delete an element from a phone list. One version should assume no dummy elements and the other should use dummy

elements. If the element to be deleted is not found in the list, no action should be taken.

5. If $P$ points to a linked list of elements having INFO and LINK fields, what is the effect of the following statement?

$$P \leftarrow MYSTERY(P, \textbf{nil})$$

where the procedure $MYSTERY$ is

> **procedure** $MYSTERY(x, y)$ **returns pointer**
>    **if** $x = $ **nil then return** $y$
>        **else**
>              $t \leftarrow MYSTERY(\text{LINK}(x), x)$
>              $\text{LINK}(x) \leftarrow y$
>        **return** $t$

6. Let $x$ and $y$ each point to linked lists of the form described in the previous exercise. What value is returned by the following procedure?

> **procedure** $MYSTERY(x, y)$ **returns pointer**
>    **if** $x = $ **nil then return** $y$
>        **else**
>              $l \leftarrow$ **getcell**
>              $\text{INFO}(l) \leftarrow \text{INFO}(x)$
>              $\text{LINK}(l) \leftarrow MYSTERY(\text{LINK}(x), y)$
>        **return** $l$

7. Rewrite the mysterious procedures in the previous two exercises as more straightforward, nonrecursive procedures.

8. Design a procedure that returns as its value a copy of the list given to it as an argument. Write your procedure in two different ways, one straightforward, nonrecursive, and the other in the style of the mysterious procedures in Exercises 5 and 6, above.

9. State reasonable conditions under which the efficiency of storage utilization is
   (a) Better with sequential allocation than with linked allocation.
   (b) Better with linked allocation than with sequential allocation.

10. A variation of the technique of introducing a dummy header record into a list $L$ is to make that dummy record the *location of* $L$. Comment on the advantages and disadvantages of this trick, contrasting the cases of assembly language programming and high-level language programming.

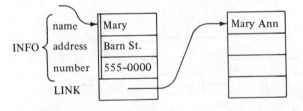

**Figure 4.7**
Example of a typical list element implemented as a record. A link to a second
record is also shown.

### 4.1.1  Implementations of List Elements

Although we have so far implied that a list element is implemented as a
record, an element is in fact a data structure with its own functions and several
possible storage structures. The principal functions are extraction of the contents
of the element and creation of new elements. When records are the storage
structure for elements, the algorithms for these functions are field selection and
**getcell**, respectively. A diagram of a typical record is in Figure 4.7.

Another possible implementation is with *parallel arrays*, as were used in
Chapter 1 to implement the name table. There is a separate array for each
component of the list element, as shown in Figure 4.8. A pointer to an element is
a subscript value that selects a "row," one entry from each of the arrays.
Selection of a component of an element is done by using the pointer as a
subscript in the array containing the desired component. Creation of a new
element can only occur if there is an unused row across the arrays. Often the

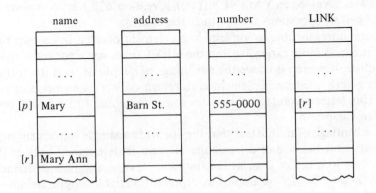

**Figure 4.8**
Implementation of elements in parallel arrays. The value $p$ is a pointer to the
record in the row with subscript value $p$.

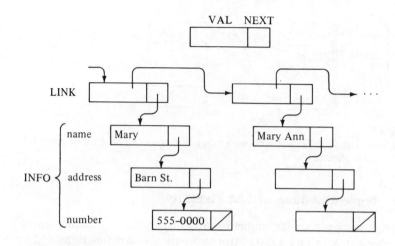

**Figure 4.9**
Implementation of list elements with a list of mini-elements, each having VAL
and NEXT fields

initialization phase links together all rows of the array as a *free list*. Each request
for a new element is met by taking the first element off this list.

A third possibility is to implement each element as a list itself! To avoid
confusion, we can temporarily call this a "minilist" composed of "mini-elements"
with fields VAL and NEXT. The most extreme approach is to put each component in a separate mini-element. Done this way, elements in our phone-listing
example would each require four mini-elements, as depicted in Figure 4.9. Field
selection is implemented by the *next* operation on the minilist. Thus the name
field of record $P$ is accessed as VAL(NEXT($P$)). Creation of a record is done by
a succession of **getcell** operations to construct the minilist.

It may seem unreasonable to implement an element of a list as its own tiny
list. Not only is access more expensive, but the NEXT fields can double the space
required. Minilists, however, do have the advantage of simplicity, and the further
advantage that **getcell** need be implemented to return only mini-elements of one
specific size. This latter advantage eliminates some waste due to fragmentation,
as we will see in Chapter 6.

Another advantage of minilists is that they provide a simple implementation
with the property that the size of a particular element is determined only at the
moment of its creation, rather than at the time the class of elements is defined.
This flexibility allows us, for example, to represent variable-length strings of
characters, without having to specify a length large enough for the longest
possible string. In the case of phone records, for example, we can allow names to
be 40 characters long without wasting space, even when the average length of a

name is only eight characters. For similar reasons, the minilist implementation allows for optional fields: a phone listing may have subsidiary numbers or it may have references to advertisements, but it is wasteful to allow space for these in every record; they can be linked into the minilist as required.

**Exercises**

1. Show how to do the insertion of Figure 4.4 using minilists.

2. Design a representation for variable-length strings using the minilist idea. Assume that the mini-elements consist of enough space to store the pointer and ten alphanumeric characters.

3. In one application, the observed record sizes were 5, 17, 20, 9, and 12, with the probabilities of occurrence being 0.3, 0.1, 0.1, 0.3, and 0.2, respectively. Suppose these records are implemented with a minilist scheme with fixed-size mini-elements. Show that the optimum mini-element size is five and that nine is nearly optimum.

### 4.1.2  Sublists and Recursive Lists

To derive the rowwise storage of a matrix in Section 3.3.1, we considered the elements in an array to be arrays themselves. We now return to that notion, examining some of the ramifications when we consider linked lists of linked lists.

With this in mind, we extend the definition of a list given at the beginning of this chapter: a *list* is a finite sequence of zero or more elements $(s_1, s_2, \ldots, s_n)$, each of which either is *atomic* or is a list itself, called a *sublist* of the list. "Atomic" means only that the element is not itself a list, not that it is otherwise indivisible; in practice it could be a record, a string, an array, and so on. In order to display the structure of lists and sublists properly, we will enclose the elements of a list in parentheses. Thus the list of zero elements, called the *empty* or *null* list, is written ( ). The list

$$((0, 1, 2), (\ ), (\text{blue}, \text{red})) \tag{4.6}$$

is a list of three elements; the first is the list $(0, 1, 2)$, the second is the empty list, and the third is the list (blue, red). Notice that ( ) and (( )) are not the same! The first is the empty list, while the second is a list of one element and that element is the empty list.

Linked representation of a list is just as before, only now the INFO field can be either an atomic element (as it has been previously) or a pointer to a sublist. Thus the list (4.6) would be represented as

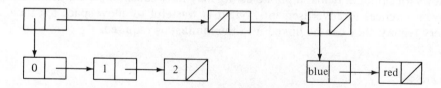

Notice that the second element of the list, the empty list, is represented by a **nil** pointer. This is consistent with our use of the **nil** pointer in the LINK field of the last element of a list or sublist, because we consider the LINK field of an element to point to the remainder of the list, which, for the last element, is the empty list.

Because of the increased flexibility of such general lists (as compared to the linked structures so far discussed), they have very wide applicability and, in fact, have even been the basis for some programming languages. They are especially useful in representing structural information, as we will see in Section 4.3. In the present section, however, we will focus on one small yet important example in which they are used.

We want to describe patterns of alphanumeric characters. For instance, the compass directions would be described by

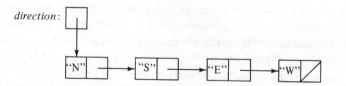

where *direction* is a pointer to the list. The list itself is viewed as a sequence of alternatives, an "N", an "S", an "E", or a "W". In general, each of the alternatives is a concatenation of strings. Thus in Figure 4.10, written also as

$$(("n", "o", "r", "t", "h"), ("s", "o", "u", "t", "h"),$$
$$("e", "a", "s", "t"), ("w", "e", "s", "t")),$$

is a pattern describing the four alternatives "north", "south", "east", "west". More generally still, each of the elements being concatenated is an alternation, and we have the following recursive definition of a pattern: a *pattern* is either

1. A single alphanumeric character or ε, the empty string.

2. A linked list of alternatives. An alternative is a single alphanumeric character, ε, or a linked list of patterns that are to be concatenated together.

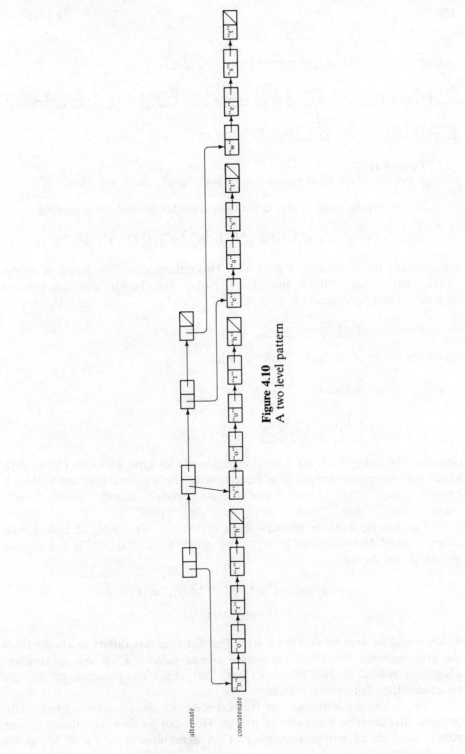

alternate

concatenate

**Figure 4.10**
A two level pattern

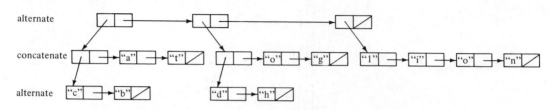

**Figure 4.11**
A pattern for the set of strings "cat", "bat", "dog", "hog", and "lion"

To see the meaning of this definition, consider the following pattern:

$$(((\text{"c"}, \text{"b"}), \text{"a"}, \text{"t"}), ((\text{"d"}, \text{"h"}), \text{"o"}, \text{"g"}), (\text{"l"}, \text{"i"}, \text{"o"}, \text{"n"})),$$

which would be drawn as in Figure 4.11. This pattern describes the set of strings "cat", "bat", "dog", "hog", and "lion". Notice that the lists alternate between alternation and concatenation, so that

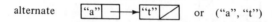

describes the set of strings "a" and "t", while

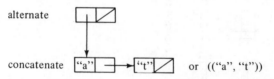

describes the string "at". As a final example of this type, consider Figure 4.12 which uses the empty string $\varepsilon$ in its description of the set of strings "bed", "best", "bead", "beast", "led", "lest", "lead", "least", "char", "chan", "chain", "star", "stan", "stair", "stain", "par", "pan", "pair", and "pain".

Suppose we want to describe pairs of vowels. We could, of course, just create a list of 25 possibilities $((\text{"a"}, \text{"a"}), (\text{"a"}, \text{"e"}), \ldots, (\text{"u"}, \text{"u"}))$, but instead we can define the lists

$$vowel = (\text{"a"}, \text{"e"}, \text{"i"}, \text{"o"}, \text{"u"})$$

$$pair = ((vowel, vowel))$$

which would be drawn like Figure 4.13. This list structure differs markedly from the previous ones because it contains a *shared sublist*—a list that is simultaneously a sublist of two different lists. This slight complication of the list structure offers self-evident economy.

The machinery introduced so far in this section allows us to construct only patterns that describe finite sets of strings. How can we describe infinite sets of strings such as all strings consisting of an even number of "a"s? We build

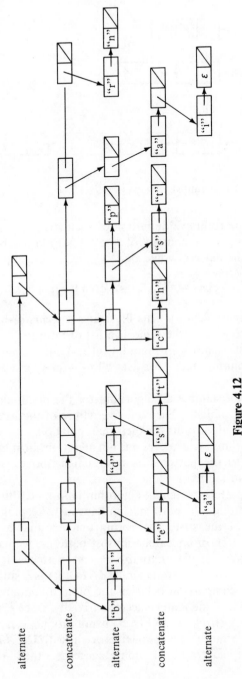

**Figure 4.12**
Use of the empty string ε in a pattern

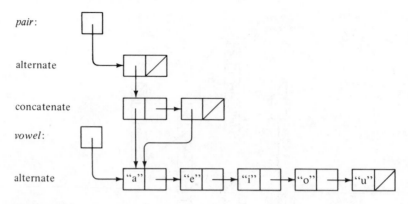

*pair*:

alternate

concatenate

*vowel*:

alternate

**Figure 4.13**
A pattern with a shared sublist

such a pattern from the recursive definition of such strings: a string of "a"s has even length if it is empty, or consists of two "a"s followed by an even-length string of "a"s. Thus the pattern would be

$$even = (("a", "a", even), \varepsilon),$$

drawn as shown in Figure 4.14(a). This illustrates a *recursive* list—one that is a sublist of itself. It is effectively the same as a list of infinite depth; in the case of (4.7) the equivalent list is shown in Figure 4.14(b). Unlike shared sublists, which are only a simple economy, recursive lists allow a new, more powerful use of linked lists.

It is instructive to examine an algorithm for the manipulation of recursive linked lists with shared sublists. The recursive nature of the lists usually imposes a recursive structure on algorithms for their manipulation, and you must learn to *think* about the algorithms with recursion, not iteration, in mind. An algorithm for matching strings to the patterns as represented above will exemplify this recursive algorithmic structure.

The idea of the pattern-matching algorithm is to match the pattern against the string by tracing through the pattern, simultaneously keeping track of the first unmatched character of the string. Mirroring the fact that a pattern either is atomic or is an alternation of concatenations of patterns, we have two recursive procedures. $OR(P)$ takes a list (pattern) $P$ and matches it starting at the character of the string indicated by *cursor*. If $INFO(P)$ is a single character, the match involves only a comparison between the pattern character and the string character; if $INFO(P)$ is $\varepsilon$, the match succeeds trivially, since $\varepsilon$, the empty string, matches anyplace in any string. If $INFO(P)$ is not atomic, it is an alternation of concatenations, and we apply the parallel procedure $CONCAT$ to each of the alternatives until one succeeds; if none succeeds, then the match fails.

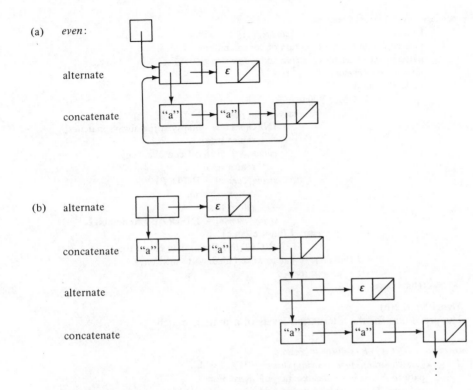

**Figure 4.14**
A recursive pattern (a) and its infinite equivalent (b)

$CONCAT(P, i)$ works identically, except that it applies $OR$ to each of the patterns being concatenated, and all must succeed for the match to succeed. The procedure $MATCH$ initiates the entire process. The details are given in Algorithms 4.2(a), (b) and (c).

> **procedure** $MATCH(pat, string)$ **returns boolean**
>     $cursor \leftarrow 1$    ⟦index position in $string$ currently being matched⟧
>     $l \leftarrow \text{length}(string)$
>     **if** $OR(pat)$ **then**
>             **if** $cursor = l + 1$ **then**
>                          ⟦pattern matched entire string⟧
>                          **return true**
>     **return false**

**Algorithm 4.2(a)**
The driver for the procedures $OR$ and $CONCAT$ of Algorithms 4.2(b) and 4.2(c), respectively

**procedure** $OR(P)$ **returns boolean**

⟦match *string*[*cursor*], *string*[*cursor* + 1], ... with
pattern $P$ that is an alternation of concatenations of
patterns. The variables *l*, *string* and *cursor* are global.⟧
*savecursor* ← *cursor*
**while** $P \neq$ **nil do**
    **if** ATOM($P$) **then**
        **case**
            INFO($P$) = $\varepsilon$: ⟦empty string always matches⟧
                **return true**
            *cursor* > *l*: ⟦ran off end of *string*⟧
                **return false**
            *string*[*cursor*] = INFO($P$):
                *cursor* ← *cursor* + 1
                **return true**
            *string*[*cursor*] $\neq$ INFO($P$): ⟦do nothing⟧
        **else** ⟦ $P$ not atomic⟧
            **if** $CONCAT$(INFO($P$)) **then return true**
    $P$ ← LINK($P$)    ⟦alternative failed, go to next one⟧
    *cursor* ← *savecursor*
**return false**

### Algorithm 4.2(b)
The procedure for the alternation levels of a pattern match

**procedure** $CONCAT(P)$ **returns boolean**

⟦match *string*[*cursor*], *string*[*cursor* + 1], ... with
pattern $P$ that is a concatenation of alternations of
patterns. The variables *l*, *string*, and *cursor* are global.⟧
**while** $P \neq$ **nil do**
    **if** ATOM($P$) **then**
        **case**
            INFO($P$) = $\varepsilon$: ⟦empty string always matches–do
                            nothing⟧
            *cursor* > *l*: ⟦ran off end of *string*⟧
                **return false**
            *string*[*cursor*] = INFO($P$):
                *cursor* ← *cursor* + 1
            *string*[*cursor*] $\neq$ INFO($P$): **return false**
        **else** ⟦ $P$ not atomic⟧
        **if not** $OR$(INFO($P$)) **then return false**
    $P$ ← LINK($P$)    ⟦pattern matched, go on to next one⟧
**return true**

### Algorithm 4.2(c)
The procedure for the concatenation levels of a pattern match

Notice that in Algorithm 4.2 we have used a boolean field ATOM($P$) to
determine whether INFO($P$) is an atomic element or a pointer to a pattern. We
did not include such a field in the drawings above in order to make them more
legible.

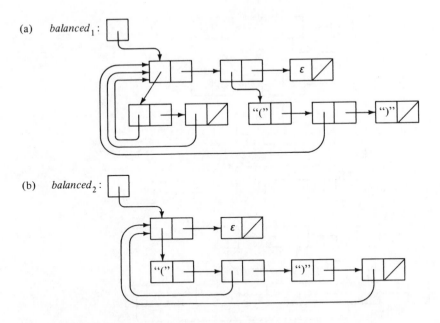

**Figure 4.15**
Patterns for balanced strings of parentheses. The pattern in (a) causes Algorithm 4.2 to go into an infinite loop; the pattern in (b) works correctly

Our interest in Algorithm 4.2 is solely as an example of list manipulation,
but we would be remiss if we did not point out that it has two serious
shortcomings as a pattern-matching technique. First, many proper patterns and
strings will cause the algorithm to go into an infinite loop. Consider, for example,
the pattern in Figure 4.15(a). This pattern should match any string of balanced
parentheses, but the use of recursion in the pattern is such that an infinite loop
results on any attempted match. (This type of pattern is sometimes called *left
recursive* because the first alternative begins with the pattern itself.) The pattern
in Figure 4.15(b) avoids the difficulty, and Algorithm 4.2 will correctly match it
with balanced strings of parentheses.

A second difficulty with Algorithm 4.2 is that the strings matched will
depend on the order of the alternatives in a list. This is quite a flaw, since we
normally think of the operation "or" as commutative. In *balanced*$_2$ above, for
instance, interchanging the order of the alternatives makes the pattern useless as
far as Algorithm 4.2 is concerned. It is always possible to redesign a pattern so
that it will work correctly with Algorithm 4.2, and we give an important example
of this in Exercise 4.

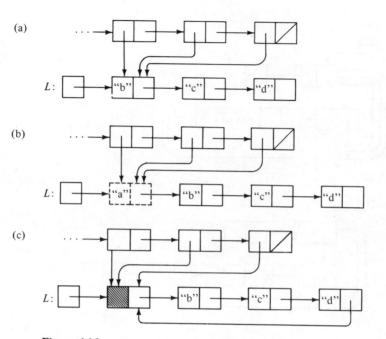

**Figure 4.16**
The difficulty of modifying lists with shared or recursive sublists. Inserting a new element into (a) to obtain (b) is difficult; the same insertion in (c) is relatively simple

Algorithms that manipulate lists with shared or recursive sublists generally involve the obvious sort of operations such as insertion, deletion, concatenation, splitting, and traversal. Since Algorithm 4.2 is not concerned with where the list representing the pattern came from, it illustrates only the traversal operation. In algorithms that construct such lists (see, for example, Exercise 6), however, the other operations are important and present a new difficulty that we must cope with. That difficulty is that a shared or recursive sublist will be referred to from several places in the overall structure; if we modify such a sublist, we must take care that all references to it are properly updated.

Consider the example in Figure 4.16(a). We have a sublist $L$ that is shared and we need to modify $L$ by, say, adding "a" to the beginning. Since there are three references to $L$, all three must be corrected to obtain the list shown in Figure 4.16(b), where $\boxed{\text{"a"}}$ is a new node that has been added to the structure.

In general it is difficult or impossible to locate all references to a sublist, so we must find a different approach. We need to centralize the references to a sublist, and we can do this by having a *header node* for each nonempty sublist. In the example above, we would have instead the structure shown in Figure 4.16(c).

To add [ "a" ] to *L* now, we need only insert it after the header node of *L*. There is a slight overhead in memory locations for the header nodes, but this is often mitigated by various uses that can be made of their unused fields. For example, we will find in Chapter 6 that header nodes can be useful in keeping track of how many references there are to a sublist (see also Exercise 9).

**Exercises**

1.  Design a list for a pattern to match the set of strings "bed", "beds", "bead", "beads", "beard", "beards", "rot", "rots", "root", "roots", "roost", "roosts".

2.  Design a list for a pattern to match any string of an odd number of "a"s.

3.  Design a list for a pattern matching any string of "a"s and "b"s in which the number of "a"s is equal to the number of "b"s.

4.  The pattern in Figure 4.17, correctly describes syntactically correct arithmetic expressions constructed from "x", "+", "*", "(", and ")".

    (a) Explain why Algorithm 4.2 will not work correctly with it.

    (b) Redesign it so that Algorithm 4.2 *does* work correctly with it.

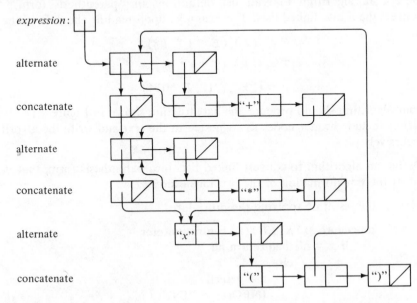

**Figure 4.17**
A pattern for arithmetic expressions

5. A pattern could have been defined somewhat differently, by the following sequence of definitions: A *pattern* is either

      (i) a single alphanumeric character, or $\varepsilon$, the empty string, or

      (ii) an alternative list.

An *alternative list* is either

      (i) empty, or

      (ii) a node $N$, where INFO($N$) is a concatenative list and LINK($N$) is an alternative list.

A *concatenative list* is either

      (i) empty, or

      (ii) a node $N$, where INFO($N$) is a pattern and LINK($N$) is a concatenative list.

  (a) Redraw some of the patterns of this section according to this new definition (including some of the recursive patterns).

  (b) Rewrite the matching procedures of Algorithm 4.2 for this new definition.

6. Design an algorithm to read list definitions in "parenthesis form" and convert them into linked lists. For example, upon reading the definitions

$$E = ((T),(E,``+",T))$$
$$T = ((F),(T,``*",F))$$
$$F = (``x",(``(",E,``)"))$$

your algorithm should produce the list structure shown in Figure 4.17 above. [*Hint*: Include header nodes as suggested in the text and write the algorithm recursively.]

7. Design an algorithm to convert linked lists into parenthesis form, that is, to invert the transformation of the previous exercise.

8. What is the result of applying the procedure

      **procedure** *MYSTERY*($x$) **returns pointer**
          **if** $x =$ **nil then return nil**
               **else**
                        $t \leftarrow$ **getcell**
                        INFO($t$) $\leftarrow$ LINK($x$)
                        LINK($x$) $\leftarrow$ *MYSTERY*(INFO($x$))
                        **return** $t$

to the following list $p$?

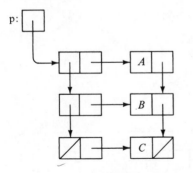

(*A*, *B*, and *C* are unspecified pointers.)

9. Design a recursive algorithm to copy a list structure that can have shared sublists and/or recursive sublists. Assume that sublists are always referenced through a header node that is distinguishable from ordinary nodes and that has sufficient room to store a pointer field and a few extra bits, if needed. [*Hint:* Use the pointer field of the header node to point to the copy of the sublist, after the copy has been made.]

10. Assuming the structure of lists described in the previous exercise, design an algorithm to test two lists for equality. First you must decide exactly what equality of lists should mean.

11. Modify Algorithm 4.2 so that it does the pattern match on strings represented by "minilists" as discussed in Section 4.1.1.

### 4.1.3  Common Variants of Linked Lists

A trivial modification of the linked list in Figure 4.2 gives a slightly more flexible linked list: if we set $LINK_n$ to point to $s_1$, as shown in Figure 4.18, we obtain a *circular list*. This results in a simple type of recursive list in which it is possible to reach (albeit indirectly) any element of the list from any other element.

To familiarize ourselves with some of the ramifications of this modification, let us try to modify Algorithm 4.1 so it works correctly on a circular list. Notice

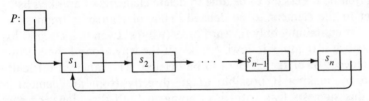

**Figure 4.18**
A circular list

(a) Empty circular list

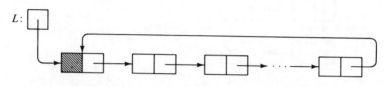

(b) Nonempty circular list

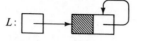

**Figure 4.19**
The circular analogs of the lists in Figure 4.6

that the test in the **while** loop of Algorithm 4.1 needs to be changed, since in a nonempty circular list we would always have $p \neq$ **nil**. The corresponding test for circular lists is $p \neq L$, for we want the loop to stop when we have gone full circle around the list. As in the case of a noncircular list, everything works smoothly if we add a dummy element at the beginning of the circular list, as we did in Figure 4.6. Figure 4.19 shows the circular analogs of the lists in Figure 4.6. We leave it to the reader to satisfy himself that, after changing the test $p \neq$ **nil** to $p \neq L$, Algorithm 4.1 works properly for the circular lists of Figure 4.19.

As a second example of an algorithm that operates on circular lists, let us consider the problem of creating a copy of such a list. That is, we want to take records provided by **getcell** and form a list identical to the original list. The difficulty here is that as we create a copy of a record of the original list, we do not know where the LINK field should point, because we have not yet copied its successor! We solve this problem by leaving the new LINK field empty until the successor is copied; at that time we can fill in the LINK field. Put another way, at the same time we copy a record we also fill in the LINK field in the copy of its predecessor. This technique is embodied in Algorithm 4.3, which copies a list at $P$ to create a list at $Q$. Notice that this algorithm correctly copies an empty list as well as a nonempty one.

It will often be necessary to be able to delete elements of a linked list, given *only* a pointer to the element to be deleted. This, of course, is impossible in a simple linked list containing only forward links (why?). Even in a circular list it is inconvenient, since it requires following links all the way around the list in order to find the predecessor of the element to be deleted. The inherent difficulty can be eliminated by making it possible to go directly from an element to its predecessor; this suggests lists with links going in both directions, or a *doubly*

$Q \leftarrow$ **getcell**                          ⟦create a dummy element⟧
$y \leftarrow$ LINK($P$)
$prevx \leftarrow Q$
**while** $y \neq P$ **do**
       $x \leftarrow$ **getcell**                    ⟦$x$ is the copy of $y$⟧
       INFO($x$) $\leftarrow$ INFO($y$)
       LINK($prevx$) $\leftarrow x$        ⟦fill in LINK field of previous $x$⟧
       $y \leftarrow$ LINK($y$)            ⟦move on to next record⟧
       $prevx \leftarrow x$
LINK($prevx$) $\leftarrow Q$                   ⟦fill in final LINK field⟧

**Algorithm 4.3**
Copy a circular list of the type shown in Figure 4.18

*linked list*, as illustrated in Figure 4.20. Each node contains two pointers, FLINK and BLINK, used as forward and backward links, respectively.

In the simplest case, given a doubly linked list and a pointer $p$ to an element in it, we can delete the element as follows:

$$\text{FLINK}(\text{BLINK}(p)) \leftarrow \text{FLINK}(p) \tag{4.7}$$
$$\text{BLINK}(\text{FLINK}(p)) \leftarrow \text{BLINK}(p)$$

These statements are insufficient if the element is at one of the two ends of the list (the list elements containing "A" or "Z" in Figure 4.20, for example), since one of the links referred to does not exist. In general, we need

**if** BLINK($p$) $\neq$ **nil then** FLINK(BLINK($p$)) $\leftarrow$ FLINK($p$)
**if** FLINK($p$) $\neq$ **nil then** BLINK(FLINK($p$)) $\leftarrow$ BLINK($p$)

Just as it is sometimes convenient to make a singly linked list circular, it also is useful to have circular doubly linked lists. The doubly linked version of the circular list of Figure 4.18 is shown in Figure 4.21. Notice that statements in (4.7) above also suffice for deletion of an element in a circular doubly linked list, *provided that we are not deleting the element pointed to by the pointer to the list* ($L$ in Figure 4.21). In fact, this could be a serious difficulty, and we are led to the

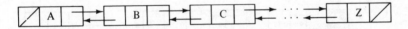

**Figure 4.20**
A doubly linked list

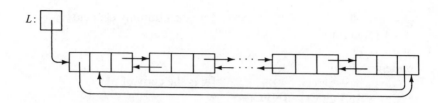

**Figure 4.21**
A doubly linked circular list

doubly linked analog of the circular list shown in Figure 4.22. The dummy element $D$ again has a form that is consistent with that of other list elements, but in this case that means it contains *two* pointers $F$ and $B$ such that $F$ is stored in FLINK($D$) and $B$ is stored in BLINK($D$).

Deleting an element $x$ from the doubly linked circular list of Figure 4.22 is done exactly by the statements (4.7) above, even when the last element is being deleted to result in the empty list. Insertion of a new element after an element $y$ in the list is done by

$$\text{FLINK}(new) \leftarrow \text{FLINK}(y)$$

$$\text{BLINK}(new) \leftarrow y$$

$$\text{FLINK}(y) \leftarrow new$$

$$\text{BLINK}(\text{FLINK}(new)) \leftarrow new$$

We leave it to the reader to verify that this works properly when the element is to be added to an empty doubly linked circular list.

(a) Empty doubly linked circular list

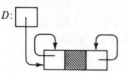

(b) Nonempty doubly linked circular list

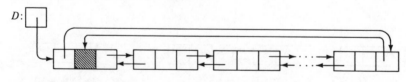

**Figure 4.22**
The doubly linked analog of the lists in Figure 4.19

**Exercises**

1. Design an algorithm to insert a new element $s_0$ as the first element of a circular list like that in Figure 4.18.

2. Write a simple, nonrecursive algorithm to *reverse* a circular list of the type in Figure 4.19.

3. Suppose we want to maintain a circular list to be searched occasionally and in which the outside reference to the list (for example, $P$ in Figure 4.18) is to be a "roving" pointer—after the list is searched, the pointer is left wherever the search ended. Furthermore, we want to be able to insert and delete elements of the list. Discuss the applicability of the various forms of circular lists we have presented. Choose the most suitable and design the needed algorithms for insertion and deletion.

4. It is possible to achieve the effect of a doubly linked list with only one pointer field per record. Develop the following idea: The LINK field of a node will contain a value $L$ **xor** $R$, where $L$ is the address in binary of the preceding node and $R$ is the address in binary of the succeeding node. [*Hint:* $(L$ **xor** $R)$ **xor** $L = R$ and $(L$ **xor** $R)$ **xor** $R = L$.]

### 4.1.4  Orthogonal Lists

Suppose that we need to organize the employee records of a company in which each employee falls into a job category $1, 2, \ldots, j$ and into a salary category $1, 2, \ldots, s$. The records need to be examined with regard to both the job categories and the salary categories: it must be convenient, for example, to scan the records of all employees in job category $i$, for any $i$, or similarly to scan all records of employees in salary category $k$, for any $k$. Data of this type can be organized into *orthogonal lists* in which there are linked lists for each job category $1, 2, \ldots, j$ and for each salary category $1, 2, \ldots, s$; each employee record is simultaneously an element of one of the job-category lists and one of the salary-category lists. Each record thus contains, in addition to whatever data fields there are, two pointer fields JOBLINK and SALLINK by which the linked lists are formed. Figure 4.23 shows such a configuration, its structure illustrating why the lists are called orthogonal. Structures like these in which the elements are members of lists in accordance with some property or category are sometimes called *inverted lists*.

The example of an orthogonal list in Figure 4.23 is greatly oversimplified for purposes of illustration. A more typical such list would be a jumble of links that would be impossible to portray legibly. So, in order to illustrate a fairly typical algorithm for orthogonal lists, we will restrict our attention to a more specialized type of orthogonal list having all the salient features, but easier to visualize.

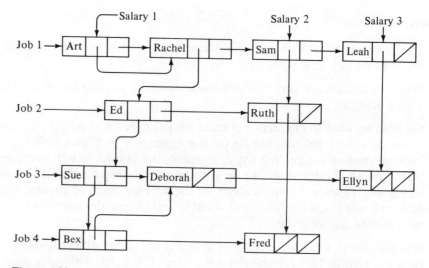

**Figure 4.23**
Employee records formed into lists by job and salary category. Each record
has the form

| NAME | SALLINK | JOBLINK |
|------|---------|---------|

Suppose we want to represent *sparse matrices*—that is, matrices in which many or
most of the entries are zero. In such a matrix it would be wasteful to allocate a
position for each entry, and instead we will allocate positions only for the nonzero
entries, the others being considered zero by omission. Of course we must also
explicitly record the subscripts of the entries (they were implicit in the matrix
representation discussed in Section 3.3.1; the subscripts could be deduced from
the location). Our representation here will use a record containing five fields for
each nonzero entry:

| ROW | COL | VALUE |
|-----|-----|-------|
| DOWN | | RIGHT |

RIGHT and DOWN are pointers, used to form circular linked lists for each row
and column, respectively, in which the nonzero entry appears. ROW and COL
are integers, the row and column subscripts of the entry, and VALUE is the value
of the matrix entry.

The RIGHT and DOWN pointers are used in the natural way. RIGHT
points to the next nonzero entry to the right of the given entry; similarly, DOWN

points to the next nonzero entry below the given entry. There are dummy elements for every row and column in which a nonzero entry appears. The dummy elements in the RIGHT lists have the value ∞ for COL; those in the DOWN lists have the value ∞ for ROW. The RIGHT pointer in a RIGHT dummy element points to the first (leftmost) element in the row; similarly, the DOWN pointer in a DOWN dummy element points to the first (lowest row number) element in the column.

The RIGHT dummy elements themselves are formed into a circular list by their DOWN pointers: for a given RIGHT dummy element, the DOWN pointer points to the RIGHT dummy element of the next row down that has a nonzero entry. Similarly, the DOWN dummy elements are formed into a circular list by their RIGHT pointers. The dummy element of both these lists is a "super dummy element" that has ROW = COL = ∞, RIGHT pointing to the leftmost DOWN dummy element, and DOWN pointing to the highest RIGHT dummy

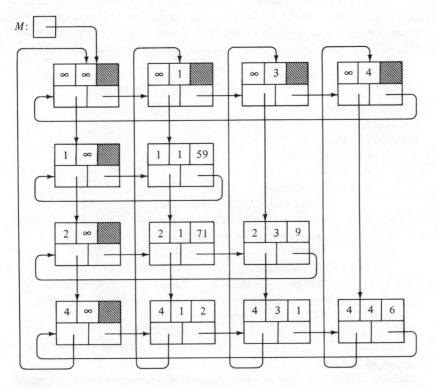

**Figure 4.24**
Representation of the matrix given in (4.8)

element. The variable $M$ points to this super dummy element. For example, the matrix

$$\begin{pmatrix} 59 & 0 & 0 & 0 \\ 71 & 0 & 9 & 0 \\ 0 & 0 & 0 & 0 \\ 2 & 0 & 1 & 6 \end{pmatrix} \tag{4.8}$$

is represented as shown in Figure 4.24.

We have used the value $\infty$ for the ROW and COL fields of the dummy elements and chosen to have the circular lists go in ascending order (by row and column numbers). In some cases, however, it might be more convenient or natural to use 0 instead of $\infty$; this would probably make it more reasonable to have the circular lists go in descending order (why?).

As a simple example of an algorithm for manipulating orthogonal lists, we will present a procedure to delete the $(i, j)$ entry from the orthogonal-list representation of a sparse matrix as outlined above. Although simplified by the straightforward nature of the underlying orthogonal list, this procedure [given as Algorithm 4.4(a)] displays some of the typical intricacies of one that modifies a structure in which the elements are simultaneously members of more than one list.

$RW \leftarrow predDOWN(M,i)$
$P \leftarrow DOWN(RW)$
**if** $ROW(P) = i$ **then**
       $CL \leftarrow predRIGHT(P, j)$
       $Q \leftarrow RIGHT(CL)$
       **if** $COL(Q) = j$ **then**
               ⟦at this point $CL$ is the entry in row $i$ before the $j$th⟧
               $RIGHT(CL) \leftarrow RIGHT(Q)$    ⟦delete $(i, j)$ entry from its row⟧
               ⟦delete row $i$ dummy element if row $i$ is now empty⟧
               **if** $RIGHT(CL) = CL$ **then** $DOWN(RW) \leftarrow DOWN(P)$
               $x \leftarrow predDOWN(Q, i)$
               $DOWN(x) \leftarrow DOWN(Q)$    ⟦delete $(i, j)$ entry from its column⟧
               ⟦delete column $j$ dummy element if column $j$ is now empty⟧
               **if** $DOWN(x) = x$ **then**
                       $y \leftarrow predRIGHT(x, j)$
                       $RIGHT(y) \leftarrow RIGHT(x)$

**Algorithm 4.4(a)**
Deletion of the $(i, j)$ entry in a sparse matrix M represented by an orthogonal list as shown in Figure 4.24. The procedure *predDOWN* is given in Algorithm 4.4(b), and the analogous procedure *predRIGHT* is left to Exercise 1.

**procedure** *predDOWN*(*T*, *k*) **returns pointer**
   *P* ← *T*
  **repeat**
    *pred* ← *P*
    *P* ← DOWN(*P*)
    **until** ROW(*P*) = *k* **or** *P* = *T*
  **return** pred

**Algorithm 4.4(b)**
Procedure to find the predecessor of row *k*, starting at *T*. The analogous procedure *predRIGHT* is left to Exercise 1.

The algorithm begins by using the procedure *predDOWN* [Algorithm 4.4(b)] to scan down the column of ROW dummy elements to find the dummy element for the *i*th row and the row preceding it. The algorithm then scans across the *i*th row to find the (*i*, *j*) entry (using a procedure *predRIGHT*, which we leave for Exercise 1). This entry is then deleted from its row and column lists. In both deletions we must be careful to delete the dummy elements if the row and/or column becomes empty.

**Exercises**

1. Write out the details of the procedure *predRIGHT* omitted in the text.

2. Design an algorithm based on Algorithm 4.3, for copying a sparse matrix represented as an orthogonal list as illustrated in Figure 4.24.

3. Design an algorithm to multiply two sparse matrices represented as orthogonal lists; the product should also be represented as an orthogonal list.

4. Suppose we represent matrices as orthogonal lists in which the lists are circular and doubly linked. Rewrite Algorithm 4.4 for this case.

5. Design an algorithm to insert an (*i*, *j*) entry in an orthogonal-list representation of a matrix in which that entry is missing.

## 4.2 STACKS AND QUEUES

In this section we discuss two extremely important data structures that can be implemented with lists as the storage structure. The principal functions for each of them are the insertion of items and their later retrieval, with the order of the retrieval being dictated by the order of their insertion. A *push-down stack*, or *stack* for short, is a list in which all insertions and deletions occur at only one end, in this case called the *top* of the stack—the other end is the *bottom*. The list

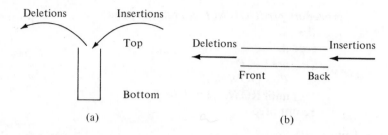

**Figure 4.25**
Schematic diagrams of (a) a stack and (b) a queue

elements enter and leave the stack in a last-in, first-out order. A *queue* is a list in which all insertions are made at one end of the list (the *rear* or *back*), and all deletions are made at the other end (the *front*). In contrast to a stack, a queue operates in a first-in, first-out order. Figure 4.25 shows schematic diagrams of the operation of a stack and a queue.

The importance of stacks and queues lies in their use as bookkeeping devices. In order to perform some task, we may first have to perform a number of subtasks. Each subtask may, in turn, lead to other subtasks that must be performed. Both stacks and queues provide a mechanism for keeping track of the subtasks that are yet to be performed, and most importantly, the order in which they must be performed. In some cases that order is last-in, first-out, and so a stack is appropriate; in other cases the order is first-in, first-out and a queue is appropriate.

In the next two subsections we present two different strategies for implementing stacks and queues, one based on sequentially allocated lists and the other on linked lists. In the final subsection we show some typical examples of the use of stacks and queues, but many other important uses occur later in the book.

Since the operations of insertion and deletion from stacks and queues occur frequently, we use the following notation:

> $D \Leftarrow x$: Add $x$ to $D$. If $D$ is a stack, $x$ is added at the top. If $D$ is a queue, $x$ is added at the rear.
>
> $x \Leftarrow D$: Set the value of $x$ to be the element at the top of $D$ if $D$ is a stack, or the element at the front of $D$ if $D$ is a queue. *This element is deleted from $D$.*

The initialization of a stack or queue will be indicated by an assignment statement such as "$S \leftarrow$ empty stack" or "$Q \leftarrow$ empty queue". The actions required by such an initialization will become clear as we discuss the particular implementations. In algorithms using stacks it is occasionally necessary to know the top stack entry, without needing to remove it. For this purpose we use

> $top(S)$: The element at the top of stack $S$. *Such a reference does not change the contents of S.*

All the above notations will be used without regard for the technique used to implement the stack or queue.

Both stacks and queues are special cases of a data structure called a *priority queue*. In a priority queue each element has an associated number, its priority, and the structure operates on a highest-priority-out-first basis. Thus the order in which the elements are removed is determined solely by the associated priorities, not by the order of insertion. If the priority of each element inserted is higher than that of the preceding element we have a stack; if it is lower we have a queue. In general, priority queues cannot be efficiently implemented by the techniques discussed in this chapter; efficient implementation will be discussed in Chapter 7.

**Exercises**

1. Suppose letters arrive, one at a time in alphabetical order, from an input string. We can take an arriving letter and add it to the end of the output string, or we can put the arriving letter onto a stack. At any time we can remove the top stack entry and add it to the output string. Thus, for example, if the input string were "aepr", it could be permuted into the output string "pear" by stacking the "a", stacking the "e", adding the "p" to the output string, taking "e" off the stack and adding it onto the output string, taking "a" off the stack and adding it onto the output string, and finally adding "r" onto the output string.

   (a) Could "aepr" be permuted into "pare"? Into "reap"? Into "aper"?

   (b) Into what English words can "aelst" be permuted?

2. Work Exercise 1 with "queue" substituted for "stack."

3. Formulate and prove a rule for determining which permutations $\Pi = (\pi_1, \pi_2, \ldots, \pi_n)$ of $1, 2, \ldots, n$ can be obtained with a stack used as in Exercise 1.

4. In a "series" of stacks the top element of each serves as the input to the next. Show that if there are $\lceil \lg n \rceil$ stacks in a series, then all $n!$ permutations can be obtained (see Exercise 3).

### 4.2.1 Sequential Implementation

The natural storage structure for a stack is a list, and especially simple algorithms are possible for the stack operations when the list is implemented as

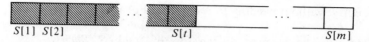

**Figure 4.26**
Sequential implementation of a stack. The shaded elements are the stack contents, with $S[1]$ at the bottom and $S[t]$ at the top. For the empty stack, $t = 0$.

an array. We need only the array $S$ of $m$ elements and an integer $t \leqslant m$ to keep track of the top of the stack. At any time, the current stack contents are $S[1]$, $S[2],\ldots,S[t]$, and $t = 0$ means that the stack is empty (see Figure 4.26). The algorithms for insertion and deletion are simply

$$S \Leftarrow x: \textbf{if } t \geqslant m \textbf{ then overflow}$$
$$\textbf{else}$$
$$t \leftarrow t + 1$$
$$S[t] \leftarrow x$$

$$x \Leftarrow S: \textbf{if } t = 0 \textbf{ then underflow}$$
$$\textbf{else}$$
$$x \leftarrow S[t]$$
$$t \leftarrow t - 1$$

An **underflow** means that an attempt has been made to delete an element from an empty stack; it is generally a meaningful end condition in an algorithm. Conversely, an **overflow** means that there is no more room to add $x$ to the stack, which usually means trouble.

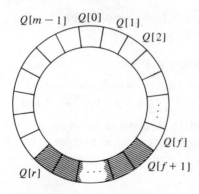

**Figure 4.27**
Sequential implementation of a queue. The shaded elements are the queue contents, with $Q[f + 1]$ at the front and $Q[r]$ at the rear. For the empty queue, $f = r$.

The sequential allocation of a queue is more complicated because it grows at one end and shrinks at the other; if we are not careful, it can inch its way along and try to overrun the locations set aside for it. Thus we use the $m$ locations $Q[0], Q[1], \ldots, Q[m-1]$ allocated for a queue in a circular fashion, and we consider $Q[0]$ to follow $Q[m-1]$. Using $f$ as a pointer to the location *just before* the front of the queue and $r$ as the pointer to the rear of the queue, the queue consists of the elements $Q[f+1], Q[f+2], \ldots, Q[r]$. (See Figure 4.27.) With this definition, the empty queue corresponds to $r = f$. We have

$$Q \Leftarrow x: s \leftarrow (r+1) \bmod m$$
$$\textbf{if } s = f \textbf{ then overflow}$$
$$\textbf{else}$$
$$r \leftarrow s$$
$$Q[r] \leftarrow x$$

$$x \Leftarrow Q: \textbf{ if } r = f \textbf{ then underflow}$$
$$\textbf{else}$$
$$f \leftarrow (f+1) \bmod m$$
$$x \leftarrow Q[f]$$

As in the case of a stack, **underflow** is generally a meaningful end condition and **overflow** is generally trouble.

Notice that **overflow** occurs while attempting to add an $m$th element to a queue containing $m-1$ elements; this means that one of the allotted $m$ locations is wasted (see Exercise 1). The insertion algorithm is carefully designed to modify $r$ only after it is certain there will be no **overflow**.

Generally speaking, sequentially allocated stacks and queues are simpler than the linked versions, less flexible, and more economical in memory usage. This situation exactly mirrors that of lists in general: when more than one stack or queue is required, it is usually more economical to use the linked implementations discussed in the next section. Two stacks can, however, coexist nicely in a single array. One stack grows from left to right, as did the example in Figure 4.26, while the other grows from right to left from the other end of the array. An **overflow** occurs when the tops of the two stacks collide—in other words, only when all the locations are full (Exercise 4). Other combinations of sequentially implemented stacks and/or queues are not as fortuitous, and **overflow** can occur even when there are still unused locations; this means that the stacks and queues need to be shifted around to make the free locations available where they are needed—a very time-consuming operation. Techniques for simultaneous sequential implementation of stacks and queues that do not overflow until all locations are used are described in Exercise 6.

**Exercises**

1. Modify the algorithms for the sequential implementation of queues so **overflow** does not occur until the attempted insertion of the $(m + 1)$st element. [*Hint*: Use an extra flag bit.]

2. Rewrite the queue insertion/deletion algorithms without using the mod function. Under what conditions is the modified version faster than the original?

3. A *deque* (*d*ouble *e*nded *que*ue, pronounced "deck") is a list in which all insertions and deletions are made at the ends of the list. The queue operations given in this section suffice for "insert at rear" and "delete at front," so all that is needed to implement a deque are the operations "insert at front" and "delete at rear." Design algorithms for these operations.

4. Design insertion and deletion algorithms for two stacks coexisting in a single array as described in the text.

5. Suppose that in addition to the insertion and deletion operations we want to be able to reverse the order of the elements in a queue. Suggest modifications to the sequentially allocated queue and the algorithms given in the text to facilitate the reversal operation.

6. We want to have $n$ sequentially allocated stacks coexisting in a single array. "Local" **overflow** occurs if one stack runs out of space while the others still have space remaining; at that time the stacks are shifted around to give more space to the one that ran out. "Global" **overflow** occurs when all the space is used and hence reorganization is not possible.

   (a) Suppose the stacks are in the array as follows:

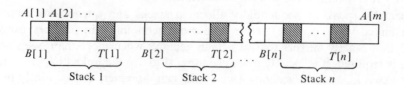

   with $B[i]$ pointing one position to the left of the base of the $i$th stack and $T[i]$ pointing to its top element, each stack growing from left to right. Design algorithms for insertion into and deletion from the $i$th stack, and design an algorithm to relocate the stacks when a local **overflow** occurs. The relocation process consists of assigning new locations for each stack and then actually moving them. Suggest various heuristics for the assignment phase and give algorithms for them. Are your algorithms sufficient if there is a mixture of stacks and queues instead of just stacks?

   (b) Do part (a) so that the stacks grow toward each other in pairs.

7. Describe algorithms for the sequential implementation of a priority queue. How efficient are your algorithms?

8. Suppose it is known that the elements that will be put on the stack are all the same. Suggest a good way to implement the stack. What if it is a queue instead of a stack?

### 4.2.2  Linked Implementation

The linked implementation of a stack is as easy as the sequential implementation. We maintain a pointer $t$ to the top stack element and use the LINK field of a stack element to point to the element below it on the stack. The bottom stack element has the pointer **nil** in its LINK field, and $t =$ **nil** corresponds to the empty stack. (See Figure 4.28.) We have

$$S \Leftarrow x: l \leftarrow \textbf{getcell}$$
$$\text{INFO}(l) \leftarrow x$$
$$\text{LINK}(l) \leftarrow t$$
$$t \leftarrow l$$

$$x \Leftarrow S: \textbf{if } t = \textbf{nil then underflow}$$
$$\textbf{else}$$
$$x \leftarrow \text{INFO}(t)$$
$$t \leftarrow \text{LINK}(t)$$

In this case, the **overflow** condition occurs in the operation **getcell** when no more records are available.

For queues, the linked implementation is essentially the same as for stacks [that is, LINK($x$) points to the element behind $x$ in the queue], except that we use $f$ instead of $t$ as a pointer to the front, and we use $r$ to point to the rear, as shown

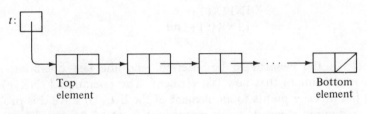

Top
element

Bottom
element

**Figure 4.28**
Linked implementation of a stack

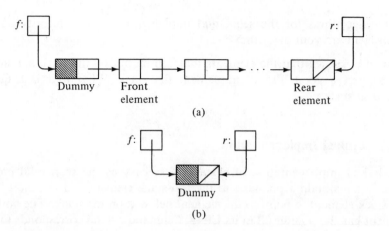

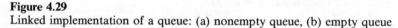

**Figure 4.29**
Linked implementation of a queue: (a) nonempty queue, (b) empty queue

in Figure 4.29(a). To add an element to the queue we use

$$Q \Leftarrow x: l \leftarrow \textbf{getcell}$$
$$\text{INFO}(l) \leftarrow x$$
$$\text{LINK}(l) \leftarrow \textbf{nil}$$
$$\text{LINK}(r) \leftarrow l$$
$$r \leftarrow l$$

As with the stack, the **overflow** condition is hidden in the call to **getcell**.

It would be convenient to have the empty queue represented by $f = \textbf{nil}$, but we must insure that the value of $r$ is such that the insertion algorithm above works properly when the first element is inserted. The first three operations of the insertion algorithm

$$l \leftarrow \textbf{getcell}$$
$$\text{INFO}(l) \leftarrow x$$
$$\text{LINK}(l) \leftarrow \textbf{nil}$$

correctly construct the element to be inserted, and the last operation $r \leftarrow l$ correctly sets $r$ to point to that new last element. The operation $\text{LINK}(r) \leftarrow l$, however, will fail unless $r$ points to an element of the list. To solve this problem we introduce a dummy element as the permanent front of the list. The empty queue is thus represented as shown in Figure 4.29(b), and deletion is accomplished by

$$x \Leftarrow Q: t \leftarrow \text{LINK}(f)$$
$$\quad \textbf{if } t = \textbf{nil then underflow}$$
$$\qquad \textbf{else}$$
$$\qquad\quad x \leftarrow \text{INFO}(t)$$
$$\qquad\quad \text{LINK}(f) \leftarrow \text{LINK}(t)$$
$$\qquad\quad \textbf{if } \text{LINK}(f) = \textbf{nil then } r \leftarrow f$$

**Exercises**

1. Design a linked implementation and algorithms for a deque (see Exercise 3 of Section 4.2.1).

2. Design a linked implementation and algorithms for a priority queue and compare its efficiency to that of the sequential implementation of Exercise 7 of Section 4.2.1.

3. Suppose we need to keep track of the middle of a linked queue with a pointer $M$. $M$ should point to the middle element if there are an odd number of elements on the queue, or just in front of the middle if there are an even number. Modify the insertion and deletion algorithms to keep track of $M$ in such a way that the time required to do so is independent of the length of the queue. Be sure that your algorithms work for the empty queue.

### 4.2.3  Applications of Stacks and Queues

The applications of stacks and queues are legion, and we will see many throughout this book. In this section we present three examples of the use of stacks that are more or less unrelated to the material of the other chapters. Some applications of queues are described briefly, but since queues are generally needed in situations that do not lend themselves to the extraction of simple examples, we do not give any specific algorithms. Specific algorithms are found in Algorithms 5.8, 5.9, and 8.8.

**Stacks and Recursion.** Throughout this book we develop and present recursive algorithms, and it is necessary to know how recursion can be transformed into iteration. The necessity is obvious if we are called upon to work in nonrecursive programming languages, but it is just as important to know how to economize by taking advantage of the idiosyncrasies of a particular situation. For these reasons, our first application of stacks is an example of the implementation of recursive procedures.

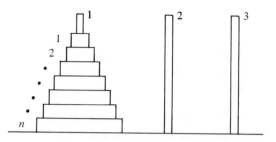

**Figure 4.30**
The Towers of Hanoi problem

A recursive procedure is one that, in the process of doing its computations, invokes itself as a subprocess. By the nature of the computation, this subprocess must be completed before the computation can continue; in other words, the last invoked procedure must finish before any other procedure. This corresponds precisely to the last-in, first-out nature of a stack, and so a stack is the appropriate mechanism to save (and later restore) the values of parameters and local variables on a recursive call.

Consider the *Towers of Hanoi problem*. There are three poles numbered 1, 2, and 3. On pole 1 there are $n$ disks with diameters $1, 2, \ldots, n$ arranged in a pile in order by size; the other two poles are empty (see Figure 4.30). The problem is to move all $n$ disks from pole 1 to pole 3 so that, as originally on pole 1, the disks are in order by size. The disks may be moved only one at a time, and no disk may ever be placed on one of smaller diameter.

In what order should the disks be moved? Actually, we should first ask *if* it is possible to solve the problem, but the fact that it is possible and the method of solution are entwined, and we will prove that it can be done by giving a way of doing it. If we had only one disk, the problem would be trivial, since we could simply move the disk from the source to the destination. With $m \geqslant 2$ disks we proceed as follows: recursively move $m - 1$ disks to the intermediate pole, move the $m$th disk to the final pole, and then recursively move the $m - 1$ disks from the intermediate pole to the final pole using the original pole as an intermediate. (Prove to yourself that, at lower levels of the recursion, any disks still on the original pole are larger than those to be moved there while using it as an intermediate.) This technique tells us how to move the $m$ disks, assuming we already know how to move $m - 1$ disks, and we have the trivial case of moving one disk as the basis for the process. The recursive procedure that uses this method is given in Algorithm 4.5. That procedure has four parameters, one giving the number of disks to be moved, and the other three specifying the initial, intermediate, and final poles.

**procedure** $HANOI(n, i, j, k)$
⟦move $n$ disks from pole $i$ to pole $k$ using pole $j$⟧
**if** $n = 1$ **then** move the top disk from pole $i$ to pole $k$
   **else**
      $HANOI(n - 1, i, k, j)$ ⟦move the top $n - 1$ disks to the
                                       intermediate pole⟧
      move the top disk from pole $i$ to pole $k$
      $HANOI(n - 1, j, i, k)$ ⟦move top $n - 1$ disks from the
                                       intermediate to the final pole ⟧

**Algorithm 4.5**
Recursive algorithm for the Towers of Hanoi problem

As an example, you are urged to verify that a call $HANOI(3, 1, 2, 3)$
produces the following sequences of moves:

$$1 \rightarrow 3$$
$$1 \rightarrow 2$$
$$3 \rightarrow 2$$
$$1 \rightarrow 3$$
$$2 \rightarrow 1$$
$$2 \rightarrow 3$$
$$1 \rightarrow 3$$

where "$x \rightarrow y$" means "move the top disk from pole $x$ to pole $y$." You should
also verify that this sequence of moves correctly solves the Towers of Hanoi
problem with three disks.

To solve the problem iteratively, we use a stack $S$ to store the values of the
parameters on a recursive call and to restore them afterward. Each stack element
will be a quadruple $(n, i, j, k)$; such an element means that the top $n$ disks on
pole $i$ are to be moved to pole $k$ using pole $j$. The key idea is that *a call to
HANOI will be replaced by the insertion of the corresponding quadruple into the
stack*. Thus the heart of Algorithm 4.5 becomes

**if** $n = 1$ **then** move the top disk from pole $i$ to pole $k$
   **else**
      $S \Leftarrow (n - 1, j, i, k)$
      $S \Leftarrow (1, i, j, k)$
      $S \Leftarrow (n - 1, i, k, j)$

Notice that the **else** clause here contains two essential differences from that in Algorithm 4.5. First, instead of saying "move the top disk from pole $i$ to pole $k$" it says "$S \Leftarrow (1, i, j, k)$". As stated above, this stack element means exactly that the top disk from pole $i$ is to be moved to pole $k$; it is put on the stack because this operation must be done *in between* the two operations of moving the remaining $n - 1$ disks. The second difference is that the order in which the elements are added to the stack is the opposite of the order in which the recursive calls are made in Algorithm 4.5. This is necessary since the stack is last-in, first-out and we want $(n - 1, i, k, j)$ to be finished *before* $(n - 1, j, i, k)$.

This code fragment must be repeated until there are no stack elements left, and thus the complete nonrecursive algorithm is as given in Algorithm 4.6. The process is started by putting an initial quadruple $(n, 1, 2, 3)$ on the stack.

$S \leftarrow$ empty stack
$S \Leftarrow (n, 1, 2, 3)$ 〚the initial quadruple to move $n$ disks from pole 1 to pole 3 using pole 2〛
**while** $S \neq$ empty **do**
    $(n, i, j, k) \Leftarrow S$
    **if** $n = 1$ **then** move the top disk from pole $i$ to pole $k$
        **else**
            $S \Leftarrow (n - 1, j, i, k)$
            $S \Leftarrow (1, i, j, k)$
            $S \Leftarrow (n - 1, i, k, j)$

**Algorithm 4.6**
Stack algorithm for the Towers of Hanoi problem

Suppose, for example, that $(3, 1, 2, 3)$ is the initial quadruple used in Algorithm 4.6. When the while loop begins, the stack looks like

$$\lfloor (3, 1, 2, 3) \rfloor$$

The quadruple is removed and three quadruples replace it, leaving

$$\begin{array}{|l}
(2, 1, 3, 2) \\
(1, 1, 2, 3) \\
(2, 2, 1, 3) \\
\hline
\end{array}$$

The top quadruple is removed and again three quadruples replace it, leaving

$$
\begin{vmatrix}
(1, 1, 2, 3) \\
(1, 1, 3, 2) \\
(1, 3, 1, 2) \\
(1, 1, 2, 3) \\
(2, 2, 1, 3)
\end{vmatrix}
$$

As each of the top four elements is removed from the stack, it causes a disk to be moved in the following sequence: pole 1 to pole 3, pole 1 to pole 2, pole 3 to pole 2, pole 1 to pole 3. The stack now contains

$$
\begin{vmatrix}
(2, 2, 1, 3)
\end{vmatrix}
$$

The quadruple is removed and three quadruples replace it, leaving

$$
\begin{vmatrix}
(1, 2, 3, 1) \\
(1, 2, 1, 3) \\
(1, 1, 2, 3)
\end{vmatrix}
$$

Each of these elements is removed from the stack, causing disks to be moved as follows: pole 2 to pole 1, pole 2 to pole 3, pole 1 to pole 3. All three disks are now in order on pole 3 as desired. The stack has precisely captured the essence of the recursive algorithm.

**Stacks and Arithmetic Expressions.** The problem of evaluating an arithmetic expression arises at all levels of computation, from hand calculations to using a pocket calculator to programming large computers. It is a nontrivial task to transform the expression into a sequence of simple arithmetic operations. In this section we examine two important algorithms for such manipulations of arithmetic expressions; both algorithms center around the use of stacks. In a sense, however, this material could fall under the previous heading of "stacks and recursion," since it is possible to view these algorithms as iterative implementations of their recursive counterparts (see Exercises 6 and 8).

Before constructing algorithms for any task, we must have a clear understanding of what the task is. For the evaluation of arithmetic expressions, what we want to do is fairly obvious, except for one thing: given the arithmetic expression $A + B * C$, we want to compute $A + (B * C)$ and *not* $(A + B) * C$. Without parentheses, how do we know which one we want? The answer lies in the notion of *precedence*. It is the convention that the multiplicative operations of multiplication and division take precedence over the additive operations of addition and subtraction; in other words, when there are no parentheses to make it otherwise, the multiplicative operations are done first, before the additive operations. All right, but how do we evaluate $A - B + C$? Should it be $A - (B + C)$ or $(A - B) + C$? Of course we want $(A - B) + C$, because it is the convention that operations of equal precedence are performed from left to right. As a further example, $A/B*C$ would be evaluated as $(A/B)*C$ and *not* $A/(B*C)$.

Our problem is thus to evaluate arithmetic expressions that contain variables, $+$, $-$, $*$, $/$, and parentheses, with the conventions that $*$ and $/$ will be done before $+$ and $-$ (in the absence of parentheses, of course!) and that sequences of $+$s and $-$s or sequences of $*$s and $/$s will be done from left to right. This evaluation is tricky and can be done by a very clever recursive algorithm (see Exercise 6), but here we will do it by a simpler two-stage process that emphasizes the use of stacks. We will first show how to evaluate the expression, assuming it has been converted into an intermediate form, and then show how to do the conversion to that form.

The intermediate form is *Polish postfix* notation, in which the operator follows its two operands, rather than separating them as in conventional notation. For example, instead of $A + B$ we would write $AB +$. (To eliminate the ambiguities possible when multicharacter variable names are adjacent to one another, we will insist that all variable names be single characters; this restriction is easy to overcome by using separator characters, but at the expense of some clarity of presentation.) For $A + B*C$ we would have $ABC* +$, which is interpreted as follows: The two operands for the $*$ are the $B$ and $C$ that precede it; the two operands of the $+$ are $A$ and the expression $BC*$. If instead we want $(A + B)*C$, we write $AB + C*$, so that the two operands for the $+$ are the $A$ and $B$ that precede it, while the two operands for the $*$ are the expression $AB +$ and $C$. The examples $ABC* +$ and $AB + C*$ illustrate the most important characteristic of Polish postfix notation: it does not need parentheses or precedence conventions to indicate the order of the computation; the order is defined completely by the relative order of the operands and the operators.

We can define the class of Polish postfix expressions recursively as follows: such an expression is either a simple variable, or consists of two Polish postfix expressions followed by an operator. The recursive definition gives us the key to the evaluation of postfix expressions. Consider, for example, the expression

$$AB + CD - E*F + *,$$

which corresponds to the expression

$$(A + B)*((C - D)*E + F).$$

The order of evaluation is as follows:

$$A \ \ B \ + \ \ C \ \ D \ - \ \ E \ * \ \ F \ + \ \ *$$

$$A + B$$

$$C - D$$

$$(C - D) * E$$

$$(C - D) * E + F$$

$$(A + B) * ((C - D) * E + F)$$

The general rule used in this example is that whenever we find two operands followed by an operator, that operator is applied to those operands and the result replaces the substring consisting of the operands and operator. Thus, in the above example, we replaced the substring "$AB +$ " by the value of $A + B$; then we replaced the substring "$CD -$ " by the value of $C - D$; then we replaced the substring consisting of the value of $C - D$ followed by "$E *$" with the value of $(C - D) * E$, and so on, finally replacing the entire string with the value of $(A + B) * ((C - D) * E + F)$.

More precisely, the algorithm to evaluate postfix expressions operates by scanning the expression one character at a time from left to right. Operands are placed on a stack and operators are applied to the top two stack entries, which are deleted and replaced by the result of the operation. Algorithm 4.7 is a straightforward implementation of this process, assuming that the postfix expression is stored in an array $P[1], P[2], \ldots, P[n]$.

We are left with the problem of converting the usual infix expression into its equivalent Polish postfix form. To convert an expression into postfix form we must repeatedly replace an operand-operator-operand sequence by operand-oper-

$S \leftarrow$ empty stack
**for** $i = 1$ **to** $n$ **do**
    **if** $P[i]$ is an operand **then** $S \Leftarrow P[i]$
                **else**
                    $y \Leftarrow S$
                    $x \Leftarrow S$
                    $S \Leftarrow$ value of the operator $P[i]$
                    applied to $x$ and $y$

**Algorithm 4.7**
Evaluation of Polish postfix expression $P[1], P[2], \ldots, P[n]$.

and-operator. For example, the following illustrates the transformation of

$$(A + B) * ((C - D) * E + F)$$

into postfix form:

$$(A\ +\ B)\ *\ (\ (C\ -\ D)\ *\ E\ +\ F\ )$$

$$\underbrace{A\ B\ +}$$

$$\underbrace{C\ D\ -}$$

$$\underbrace{C\ D\ -\ E\ *}$$

$$\underbrace{C\ D\ -\ E\ *\ F\ +}$$

$$\overline{A\ B\ +\ C\ D\ -\ E\ *\ F\ +\ *}$$

To do this as we scan the expression from left to right, we use a stack as follows. When we scan an operator, then we know its left operand has already been converted to postfix and is in the output string. So, we store the operator on the stack and process its right operand. After finishing with its right operand, the operator will conveniently be at the top of the stack; we remove it and add it to the output string.

It is clear from this description of the process that operands must go directly into the output string and operators go into the stack. However, if we have just finished the second of the two operands of an operator, then that operator will be on top of the stack and we must recognize that it is time to put it into the output string. The end of the second of the operands occurs for some operator at a closing parenthesis, at another operator for which the preceding was the first operand, or at the end of the input string. The case of another operator is handled by observing that, if this incoming operator has lower or equal prece-dence to the one on top of the stack, then we must have completed the second operand of the operator on top of the stack; that operator can now be popped off the stack and added to the output string. This process must be repeated for the new top stack element, and so on. We now have only finished the first operand for this incoming operator, and it is added to the stack. We handle the case of the end of the input string by adding a special end-of-string-marker, a "$" that is treated as a very low precedence operator, causing the above outlined loop to dump out the stack when the "$" is encountered.

Algorithm 4.8 embodies the procedure just outlined; it converts an expression in an array $E$ to Polish form in array $P$. In the algorithm the bottom of the stack is marked by the symbol "#" that is treated as an operator of even lower precedence than the "$" that marks the end of the input string. This causes the "$" to be put onto the stack, the end condition of the algorithm. The algorithm uses the precedence function *PREC* as given in Table 4.1 to determine the relative precedences of two operators. Notice that the "(" has precedence lower than the arithmetic operations in order to keep it in the stack appropriately.

$S \leftarrow$ empty stack
$S \Leftarrow$ "#"     〚bottom of stack marker〛
$j \leftarrow 0$     〚output string cursor〛
$i \leftarrow 0$     〚input string cursor〛
**while** $top(S) \neq$ "$" **do**
         〚"$" is the end of input string marker; its precedence is set so that it will cause the stack to be emptied before it is pushed on〛
         $i \leftarrow i + 1$     〚process next character of input string〛

    **case**

         $E[i]$ is an operand:
                 〚transfer it to the output string〛
                 $j \leftarrow j + 1$
                 $P[j] \leftarrow E[i]$
         $E[i] =$ "(": $S \Leftarrow E[i]$     〚stack left parenthesis〛
         $E[i] =$ ")":
                 〚empty stack contents down to the matching parenthesis〛

                 $x \Leftarrow S$
                 **while** $x \neq$ "(" **do**
                         $j \leftarrow j + 1$
                         $P[j] \leftarrow x$
                         $x \Leftarrow S$
         $E[i]$ is an operator:
                 〚empty stack contents down to first operator
                 with lower precedence, then stack the operator)〛
                 **while** $PREC(E[i]) \leq PREC(top(S))$ **do**

                         $j \leftarrow j + 1$
                         $P[j] \Leftarrow S$

                 $S \Leftarrow E[i]$

**Algorithm 4.8**
Conversion of an infix expression $E[1], E[2],\ldots,E[n]$ into its equivalent Polish postfix form $P[1], P[2],\ldots,P[m]$. The *PREC* function used is defined in Table 4.1.

Algorithm 4.8 is really only the bare bones. To be useful, such an algorithm must check for syntactic errors (how does Algorithm 4.8 react to invalid expressions?) and allow for other operations such as exponentiation and unary minus. These issues are the subject of Exercises 9 and 10.

| Character | Precedence |
|:---:|:---:|
| # | 0 |
| $ | 1 |
| ( | 2 |
| +, − | 3 |
| *, / | 4 |

**Table 4.1**

The precedence function *PREC* used in Algorithm 4.8 for the conversion of infix expressions to their equivalent postfix form. The character "#" marks the bottom of the stack and "$" marks the end of the input string.

**Queues.** Applications of queues tend to be too intricate to allow the extraction of a concise example. For example, queues are needed in the simulation of various business systems requiring processing customers, orders, jobs, or requests in the order that they arrive. The operation of some computer systems requires that jobs be executed in their order of submission; in this case, again, a queue is mandated. Within the computer itself, queues are needed to keep track of input/output requests—since they are so time consuming relative to internal operations, they can accumulate; if their order is not carefully adhered to, a program might end up, for instance, trying to read a record that has not yet been written.

**Exercises**

1. In the Towers of Hanoi problem, how many moves are required to move the pile of $n$ disks from the original pole to the final pole?

2. Modify Algorithms 4.5 and 4.6 to keep enough information so that after each move a picture can be printed of the current arrangement of the disks on the poles.

3. Modify Algorithm 4.5 so that the "basic" case is $n = 0$, not $n = 1$. Compare the efficiency of this modification to the original algorithm.

4. Suppose there are $n$ disks and *four* poles. Design an algorithm to do the moving in this case.

5. Consider the following recursively defined sequences of integers: $T_1 = (1)$, $T_{n+1} = (T_n, n + 1, T_n)$. Thus, for example, $T_2 = (1, 2, 1)$ and $T_3 = (1, 2, 1, 3, 1, 2, 1)$. Design a nonrecursive algorithm based on a stack to generate $T_n$. Find a different algorithm based on divisibility by 2. What is the relation of this sequence $T_n$ to the Towers of Hanoi problem?

6. Suppose we have arithmetic expressions over $+$, $*$, parentheses, and variable names, with each expression terminated by a "$". Such an expression (if

syntactically correct) will be properly evaluated by a call to $EXP(1, value)$; its value will be left in the parameter *value*:

**procedure** $EXP(i, val)$     ⟦evaluate a sum of terms⟧
    ⟦$i$ and *val* are passed by reference⟧
    $TERM(i, val)$
    **while** $E[i] = $ "$+$" **do**
                $i \leftarrow i + 1$
                $TERM(i, v)$
                $val \leftarrow val + v$

**procedure** $TERM(i, val)$     ⟦evaluate a product of factors⟧
    ⟦$i$ and *val* are passed by reference⟧
    $FACTOR(i, val)$
    **while** $E[i] = $ "$*$" **do**
                $i \leftarrow i + 1$
                $FACTOR(i, v)$
                $val \leftarrow val * v$

**procedure** $FACTOR(i, val)$     ⟦evaluate a factor⟧
    ⟦$i$ and *val* are passed by reference⟧
    **if** $E[i]$ is an operand **then**
                $val \leftarrow$ value of the operand $E[i]$
                $i \leftarrow i + 1$
            **else** ⟦must be a parenthesized expression⟧
                $i \leftarrow i + 1$     ⟦skip opening parenthesis⟧
                $EXP(i, val)$ ⟦evaluate expression inside⟧
                $i \leftarrow i + 1$     ⟦skip closing parenthesis⟧

(a) Explain the relationship between these procedures and the pattern given in Exercise 5 of Section 4.1.2.

(b) Modify the above procedures to include the operations $-$ and $/$.

(c) Modify them further to include the operation $\uparrow$ (exponentiation) so that $a \uparrow b \uparrow c$ is interpreted as $a \uparrow (b \uparrow c)$—that is, so that it associates from right to left, unlike the other operations.

7. Design an algorithm analogous to that of the previous exercise to evaluate Polish postfix expressions.

8. Design an algorithm analogous to those in the previous two exercises to convert an infix expression to postfix.

9. Algorithm 4.8 assumes that the input expression is syntactically correct. Modify it to recognize when the input expression is not syntactically correct.

10. Modify Algorithm 4.8 so that it will properly convert expressions with the ↑ operator [see Exercise 6(c) above for a description of how this operator associates]. The precedence of ↑ should be 5 in Table 4.1.

11. What would be the effect of changing "$PREC(E[i]) \leqslant PREC(top(S))$" in Algorithm 4.8 to "$PREC(E[i]) < PREC(top(S))$"?

12. Show that by assigning the proper value as the precedence of ")", it is possible to combine the last two cases in the **case** statement of Algorithm 4.8.

13. Peano's *dot notation* is used to define parenthesis-free expressions of a single operator " * " as follows. The largest number of consecutive dots in any expression divides the expression into its two principal subexpressions, and each subexpression is evaluated using the same rule recursively; ties are broken by going from left to right (that is, by doing the leftmost operator first). For example, the following are some expressions and dot notation equivalents:

| Expression | Dot Form |
|---|---|
| $a*(b*c)$ | $a.bc$ |
| $a*b*c$ | $abc$ |
| $(a*b)*c*(d*(e*f))$ | $a.b.c...d..e.f$ |

[Notice that the notation is not unique: $a..b..c$ and $abc$ both represent $(a*b)*c$.] Give an algorithm to convert a dot notation expression into its Polish postfix form. For example, $a.b.c...d..e.f$ should yield $ab*c*def***$.

14. Let $s_1 s_2 \ldots s_n$ be a sequence of operands and operators $+$, $-$, $*$, and $/$. Let $f$ be a function defined by

$$f(0) = 0,$$

$$f(i) = \begin{cases} f(i-1) + 1 & \text{if } s_i \text{ is an operand,} \\ f(i-1) - 1 & \text{if } s_i \text{ is an operator.} \end{cases}$$

Prove that $s_1 s_2 \ldots s_n$ is a syntactically correct Polish postfix expression if and only if $f(i) \geqslant 1$ for $1 \leqslant i < n$ and $f(n) = 1$. What does $f$ correspond to in Algorithm 4.7?

15. The Polish prefix form of an arithmetic expression is defined analogously to the postfix form, except that the operator precedes (rather than follows) the

operands. Find a relationship between the prefix and postfix forms, and use this relationship to design an algorithm for the conversion of infix to prefix.

16. State and prove an analog of Exercise 14 for Polish prefix expressions.

## 4.3   GRAPHS

We end this chapter on lists with a discussion of an important application of linked lists. A *graph* $G = (V, E)$ consists of finite set of *vertices* $V = \{v_1, v_2, \ldots\}$ and a finite set of *edges* $E = \{e_1, e_2, \ldots\}$. To each edge there corresponds a pair of vertices; if the pair is *ordered* the graph is called *directed*; if the pair is *unordered* the graph is called *undirected*. The vertices corresponding to an edge are said to be *incident* on the edge. To draw a picture of a graph, we use dots for vertices and line segments for edges. If the graph is directed, the line segments have arrowheads showing the direction. Figure 4.31 shows examples of directed and undirected graphs.

Graphs are an extremely versatile mathematical structure. They can be used to represent diverse types of physical structures such as networks of roads between cities, connections among the components of an electrical circuit, or bonds in an organic compound. They can also be used to represent abstract relationships such as social connections, priority of tasks, or flow of control. Their versatility makes the representation of graphs and the algorithms for their manipulation a most important application of the list structure technique presented in previous sections of this chapter. We begin by describing how graphs can be conveniently represented by linked lists, and we then present several basic algorithms on graphs so represented.

We will represent a graph as an *adjacency structure* in which all the "adjacencies" are explicitly recorded as pointers in a linked list. A vertex $y$ in a directed graph is called a *successor* of another vertex $x$ if there is an edge from $x$ to $y$; in an undirected graph, two vertices are *neighbors* if there is an edge between them. The adjacency structure of a graph is a linked list for every vertex $v$ of the successors (neighbors) of $v$; the relative order of the elements of the list is unimportant. The linked lists of successors are themselves formed into a linked list, as Figure 4.32 shows for the graphs of Figure 4.31.

Thus, each vertex is represented by a record

with possibly other information as needed for the vertex (coordinates, size, and so on), while in a list of edges each edge is a record of the form

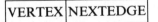

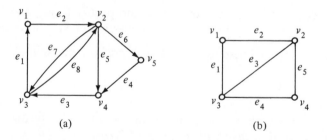

(a)                                              (b)

**Figure 4.31**
Examples of graphs: (a) a directed graph of five vertices and eight edges;
(b) an undirected graph of four vertices and five edges

with possibly other information as needed for the edge (a name, length, capacity, and the like). Notice that the number of records required for an adjacency structure is $|V| + |E|$ for a directed graph and $|V| + 2|E|$ for an undirected graph (why?) where $|V|$ and $|E|$ are the number of vertices and edges, respectively.

### Exercises

1. (a) Explain how an electrical circuit could be represented as a graph. Is the graph directed or undirected? What kinds of additional information might be stored in the records for the vertices and edges?

   (b) Repeat part (a) but for a highway system instead of an electrical network.

2. (a) An edge of a graph is a *self-loop* if it begins and ends at the same vertex. Do adjacency structures allow the representation of graphs with self-loops?

   (b) Two edges are called *parallel* if they start and end at the same vertices (and have the same direction in a directed graph). Do adjacency structures allow the representation of graphs with parallel edges?

3. Show how a graph can be represented by a $|V| \times |V|$ bit matrix. Compare such representations to adjacency structures in terms of the storage required. Can parallel edges or self-loops be represented? When will the matrix be symmetric?

4. A graph is *weighted* if with every edge there is a number called the *weight* of the edge. Show how a weighted graph can be represented by a $|V| \times |V|$ matrix.

5. The *incidence matrix* of an undirected graph $G = (V, E)$ is the $|V| \times |E|$ bit matrix $A$ defined by $A[i, j] = 1$ if and only if the $i$th vertex is incident on the

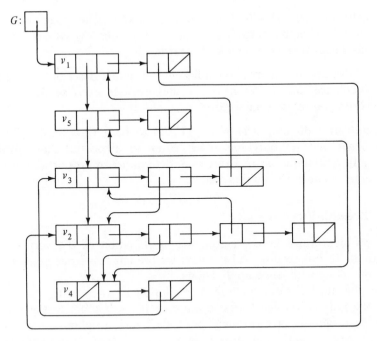

(a)

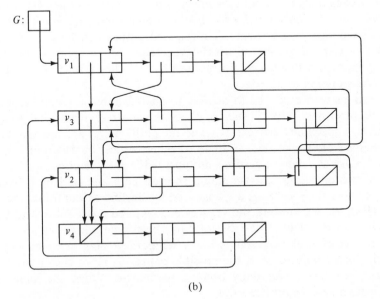

(b)

**Figure 4.32**
Adjacency structures: (a) the adjacency structure for the directed graph of
Figure 4.31(a); (b) the adjacency structure for the undirected graph of Figure
4.31(b)

*j*th edge; otherwise $A[i, j] = 0$. Can self-loops and parallel edges be represented? Can this idea be extended to directed graphs? What are the advantages and disadvantages of incidence matrices?

6. Design an algorithm to construct the adjacency structure described in the text from the incidence matrix described in the previous exercise. How many operations does your algorithm require?

7. Define a *sink* in a directed graph $G = (V, E)$ to be a vertex with $|V| - 1$ incoming edges and no outgoing edges. Find an algorithm that, given a directed graph represented as in Exercise 3, determines in proportional to $|V|$ bit inspections whether $G$ contains a sink.

### 4.3.1   Breadth-First Search

All graph algorithms require systematic examination of the vertices and edges of a graph. In this section and the next we present two strategies for such an examination, along with some applications.

Perhaps the most natural method of exploring a graph is to start at an arbitrary vertex and fan outward, first examining the neighbors of the starting vertex, then the neighbors of those neighbors, then their neighbors, and so on. This technique, called *breadth-first search*, is extremely useful in designing algorithms for answering questions about distances between vertices in graphs. In this section we will consider only *weighted graphs* in which each edge has a positive number associated with it, its weight. Unweighted graphs can be considered a special case of weighted graphs in which every edge has weight 1. Weighted graphs are represented with adjacency structures by including a field WEIGHT in the record corresponding to an edge.

The questions we would like to be able to answer about a graph (given its adjacency structure) are: What is the length of the shortest (least total weight) path between two specified vertices? What is the path? What are the lengths of the shortest paths from a specified vertex to *all* other vertices in the graph? What are the paths? Various related questions are pursued in the exercises.

The first two questions, determining a shortest path and its length from a starting vertex *s* to a final vertex *f*, are answered by starting at *s* and fanning out until *f* is reached. As we fan out, the vertices encountered are labeled with their distance from *s*, so that when the label of *f* has been determined we are done. Actually, each vertex will start out with a temporary label representing its distance from *s* using only some of the possible paths. As more and more paths are considered, the labels eventually become permanent. When the label of *f* becomes permanent, the algorithm stops.

We begin by considering *none* of the paths, so that *s* is labeled by 0 and every vertex except *s* is labeled by ∞. We then iterate as follows, with each iteration making one of the temporary labels permanent. Let *last* be the vertex whose label was just made permanent. Every vertex $v$ with a temporary label is relabeled with the smaller of

1. The current label of $v$.
2. The sum of the label of *last* and the distance from *last* to $v$.

Then the smallest of the temporary labels is found and made permanent; in the case of a tie, any of the candidates is chosen. When the label of $f$ becomes permanent, the process ends—$f$ having been labeled with its distance from $s$.

Why does this algorithm work? We can understand it by understanding the meaning of the labels. The label of a vertex $v$ is the length of the shortest path from $s$ to $v$ that consists only of permanently labeled vertices other than *last*. Inductive reasoning verifies this assertion, for it is true initially when $last = s$ and all labels except that of $s$ are infinite and temporary and the label of $s$ is 0 and permanent. Furthermore, if this assertion is true on one iteration, it will be true on the next—the smallest of the temporary labels cannot get smaller by considering paths using other vertices, since all the as-yet-unconsidered vertices have labels at least as large, and all edges have positive length. We leave it to Exercise 1 to demonstrate that the labels of the vertices are made permanent in nondecreasing order of their distances from $s$.

A record of the path itself can be kept as the algorithm proceeds: when the label on a vertex $v$ becomes permanent, a pointer PRE($v$) is set to the closest vertex to $v$ that has a permanent label. In other words, PRE($v$) is the vertex preceding $v$ along the shortest path from $s$ to $f$. The shortest path from $s$ to $f$ is thus

$$s,\ldots,\text{PRE}(\text{PRE}(\text{PRE}(f))),\text{PRE}(\text{PRE}(f)),\text{PRE}(f),f.$$

We leave it to Exercise 3 to modify this idea to keep a record of *all* shortest paths from $s$ to $f$.

Algorithm 4.9 gives the details of the above-outlined shortest-path algorithm. In addition to the fields described earlier and shown in Figure 4.32, each vertex record contains a numerical field LABEL, a pointer field PRE, and a boolean field FINAL (indicating whether or not the LABEL is permanent). Each edge record additionally contains a numerical field LENGTH. Algorithm 4.9 executes the body of the outer **while** loop at most $|V| - 1$ times, and the number of operations required for each of those times is proportional to $|V|$ just to determine $w$, the vertex whose label will become permanent. The total time required is thus proportional to $|V|^2$ operations.

⟦initially, all labels are ∞ and temporary, except that
of $s$ which is 0 and permanent⟧
$v \leftarrow G$
**while** $v \neq$ **nil do**
      LABEL$(v) \leftarrow \infty$
      FINAL$(v) \leftarrow$ **false**
      $v \leftarrow$ NEXTVERTEX$(v)$
LABEL$(s) \leftarrow 0$
FINAL$(s) \leftarrow$ **true**
$last \leftarrow s$
**while not** FINAL$(f)$ **do**
      ⟦until the vertex $f$ gets its final label, continue with
      the breadth-first search⟧
      $x \leftarrow$ EDGELIST$(last)$
      **while** $x \neq$ **nil do**
            ⟦update the label of every vertex with a temporary
            label to which there is a shorter path via $last$⟧
            $v \leftarrow$ VERTEX$(x)$
            **if not** FINAL$(v)$ **and** LABEL$(v) >$ LABEL$(last) +$ LENGTH$(x)$
                **then**
                        LABEL$(v) \leftarrow$ LABEL$(last) +$ LENGTH$(x)$
                        PRE$(v) \leftarrow last$
            $x \leftarrow$ NEXTEDGE$(x)$
      ⟦make the smallest temporary label permanent because there
      can be no shorter path from $s$ to it⟧
      Let $w$ be any vertex with the minimum temporary LABEL
      FINAL$(w) \leftarrow$ **true**
      $last \leftarrow w$

**Algorithm 4.9**
Finds a shortest path from $s$ to $f$ in a graph $G = (V, E)$ represented as in
Figure 4.32 with additional fields as described in the text

## Exercises

1. Show that Algorithm 4.9 labels the vertices permanently in nondecreasing
   order of their distance from $s$.

2. Would Algorithm 4.9 work properly if the lengths of the edges could be
   negative?

3. How should Algorithm 4.9 be modified to compute all shortest paths from $s$ to
   $f$? [*Hint*: The difficult part of this exercise is to choose the proper data
   structure to represent the paths.]

4. Devise an algorithm comparable to Algorithm 4.9, but designed to work on a graph represented as a weight matrix $W$, where $W[i, j]$ is the length of the edge from vertex $i$ to vertex $j$.

5. Design an algorithm to compute all shortest paths—that is, for all possible starting vertices and all possible final vertices. Assume the graph is represented as weight matrix as in the previous exercise. As in Exercise 3, you must first decide how the paths are to be represented.

6. A *spanning tree* of a connected, undirected graph $G = (V, E)$ is a graph $T = (V, E')$, where $E' \subseteq E$ and $T$ is *connected* and *acyclic*; that is, there is exactly one path in $T$ between any pair of vertices.

   (a) How many edges does $T$ have?

   (b) Give an algorithm to find a spanning tree of a graph $G$.

   ★(c) A *minimum spanning tree* of a weighted graph is a spanning tree with the additional property that the sum of the lengths of its edges is minimal. Give an algorithm to find a minimum spanning tree of a weighted graph $G$.

## 4.3.2 Depth-First Search

Instead of the breadth-first approach of the previous section, we could examine a graph by a *depth-first* strategy in which we attempt to go deeper and deeper into the graph before examining neighboring vertices. When we are examining a vertex $v$, we follow one of the edges $(v, w)$ out of $v$. If all edges out of $v$ have been considered, we go back along the edge $(u, v)$ that led to $v$ and continue the exploration from $u$. The process ends when we attempt to back up from the vertex at which the whole exploration began. In this section we will illustrate this important technique with two examples, one on undirected graphs and one on directed graphs.

**Connected Components.** An undirected graph is called *connected* if there is at least one path between every pair of vertices in the graph. A *connected component* of a graph is a maximal connected subgraph; that is, every vertex in a connected component is "reachable" from every other vertex in the component, and any vertex not in the component is not "reachable" from vertices in the component. In the undirected graph of Figure 4.31(b) there is only one connected component, the graph itself. Deleting edges $e_4$ and $e_5$ from that graph would leave a graph with two connected components: $(\{v_1, v_2, v_3\}, \{e_1, e_2, e_3\})$ is one and $(\{v_4\}, \emptyset)$ is the other.

We want to use a depth-first search to label each vertex of the graph with a *component number* in such a way that all vertices in the same connected component have the same component number, and vertices in different connected

⟦give all vertices an initial component number of 0⟧
$v \leftarrow G$
**while** $v \neq$ **nil do**
            COMPNUM$(v) \leftarrow 0$
            $v \leftarrow$ NEXTVERTEX$(v)$
$c \leftarrow 0$        ⟦current component number⟧
⟦apply depth-first search to each vertex of $G$⟧
$v \leftarrow G$
**while** $v \neq$ **nil do**
            **if** COMPNUM$(v) = 0$ **then**
                                                ⟦start a new component⟧
                                                $c \leftarrow c + 1$
                                                $COMP(v)$
            $v \leftarrow$ NEXTVERTEX$(v)$
**procedure** $COMP(x)$
            COMPNUM$(x) \leftarrow c$
            ⟦explore unexplored vertices adjacent to $x$⟧
            $w \leftarrow$ EDGELIST$(x)$
            **while** $w \neq$ **nil do**
                    $u \leftarrow$ VERTEX$(w)$
                    **if** COMPNUM$(u) = 0$ **then** $COMP(u)$
                    $w \leftarrow$ NEXTEDGE$(w)$

**Algorithm 4.10**
Numbering the vertices of $G$ according to connected components—each
connected component has a unique number

components have different component numbers. This can be done by first
assigning all vertices a component number of zero and then exploring the graph
in a depth-first manner starting at every vertex in turn. If a vertex already has a
nonzero component number, the search goes no deeper. If a vertex has a
component number of zero, it is assigned the current component number, and the
search continues deeper into the graph. Algorithm 4.10 gives the details. Each
vertex record is assumed to have a numerical field COMPNUM in which the
component number is stored.

The time required by Algorithm 4.10 is proportional to $|V| + |E|$. This can
be understood by observing that, aside from the initialization which requires time
proportional to $|V|$, each pointer in the adjacency structure is examined (that is,
followed) exactly once. Moreover, the amount of work done before the next
pointer is followed is bounded by a small constant.

**Topological Numbering.** We introduce this problem by an example. Suppose we need to schedule the individual tasks involved in a construction project.

The various tasks are related by the fact that some tasks cannot be started until others have been completed; this precedence among tasks can be represented by a directed graph in which the vertices are the tasks and there is an edge from $v$ to $w$ if task $v$ must be completed before task $w$ is started. Figure 4.33 shows ten construction tasks and their relationships to one another. For simplicity, we assume that for some reason (such as inadequate manpower) it is possible to work on only one task at a time and that the tasks cannot be further subdivided. We would like to know the order in which the tasks should be performed by the work crew. In the case of the example in Figure 4.33, we can order the tasks as follows:

1. Fence site
2. Erect site workshops
3. Dig foundations
4. Install concrete plant
5. Bend reinforcement

6. Fabricate steelwork
7. Paint steelwork
8. Place reinforcements
9. Pour foundations
10. Erect steelwork

This arrangement has the property that no task requires a higher-numbered task to precede it. Such a numbering is called a *topological numbering* or a *topological*

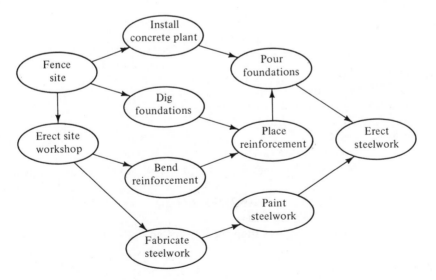

**Figure 4.33**
Tasks in a construction project, with their precedences shown by arrows

*sort* of the vertices of a graph. The general problem is to label the vertices of a directed graph $G = (V, E)$ with the integers $1, 2, \ldots, |V|$ so that if there is an edge from vertex $v$ to vertex $w$, then $\text{LABEL}(v) < \text{LABEL}(w)$. If the directed graph contains a *cycle* of edges $v_0 \to v_1 \to \cdots \to v_0$, such a labeling is, of course, not possible, because it would require that $\text{LABEL}(v_0) < \text{LABEL}(v_0)$, a contradiction.

We can perform a topological numbering of the vertices of a directed, acyclic graph by a depth-first search strategy: recursively label all vertices "descended from" the current vertex in decreasing order, then label the current vertex with a number less than any number already used. Algorithm 4.11 embodies this idea by initially marking each vertex as yet unvisited and unlabeled; the depth-first labeling procedure is then applied to each unvisited vertex in turn. This procedure first marks the vertex $x$ as visited and then recursively

⟦initially, all vertices are labeled 0 and are unvisited⟧
$v \leftarrow G$
**while** $v \neq$ **nil do**
      $\text{LABEL}(v) \leftarrow 0$
      $\text{VISITED}(v) \leftarrow$ **false**
      $v \leftarrow \text{NEXTVERTEX}(v)$
$j \leftarrow |V| + 1$
⟦process each unvisited vertex⟧
$v \leftarrow G$
**while** $v \neq$ **nil do**
      **if not** $\text{VISITED}(v)$ **then** $TOPSORT(v)$
      $v \leftarrow \text{NEXTVERTEX}(v)$
**procedure** $TOPSORT(x)$
    $\text{VISITED}(x) \leftarrow$ **true**
    ⟦process descendants of $x$⟧
    $w \leftarrow \text{EDGELIST}(x)$
    **while** $w \neq$ **nil do**
        $u \leftarrow \text{VERTEX}(w)$
        **if** $\text{VISITED}(u)$ **then if** $\text{LABEL}(u) = 0$ **then** ⟦$G$ has a cycle and
                                          cannot be labeled⟧
                **else** $TOPSORT(u)$
        $w \leftarrow \text{NEXTEDGE}(w)$
    ⟦label $x$ with a number less than the label of any of its descendants⟧
    $j \leftarrow j - 1$
    $\text{LABEL}(x) \leftarrow j$

**Algorithm 4.11**
Topological numbering of the vertices in a directed graph

processes each vertex to which there is an edge from $x$. If one of those vertices has already been visited but is yet unlabeled, we have found a cycle in the graph —that is, we have determined that no topological numbering is possible. When we have finished all the descendants of $x$, we label $x$ with the next lower number and return. This algorithm is subtle, and you should trace through it carefully on the graph of Figure 4.33.

As in the case of Algorithm 4.10, Algorithm 4.11 requires only time proportional to $|V| + |E|$ because each edge and each vertex is "examined" only once.

### Exercises

1. Simplify Algorithm 4.10 so that it only counts the number of connected components, without labeling the vertices with the component number.

2. Rewrite Algorithm 4.10 in an iterative fashion.

3. Why is $j$ initialized to $|V| + 1$ instead of $|V|$ in Algorithm 4.11?

4. Rewrite Algorithm 4.11 in an iterative fashion.

5. Simplify Algorithm 4.11 so that all it does is to determine whether the graph has a cycle.

★6. A *bridge* in an undirected, connected graph is an edge whose removal disconnects the graph. Develop a depth-first search algorithm to determine all the bridges of a graph. [*Hint*: Assign the direction followed to an edge of the graph as it is "traversed" by the depth-first search method. Call an edge a *back edge* if it leads to an already visited vertex, and a *tree edge* otherwise. Number the vertices $1, 2, \ldots, |V|$ in the order in which they are visited by the depth-first search, and define LOWPT($v$) to be the lowest-numbered vertex that can be reached from $v$ by a sequence of zero or more (directed) tree edges followed by at most one (directed) back edge. Show that these LOWPT labels can be computed on a depth-first search and that they can be used to determine the bridges.]

★7. An *Eulerian path* in an undirected, connected graph is a path through the graph that traverses every edge of the graph exactly once.

(a) Prove that a graph has such a path if and only if it has at most two vertices of odd degree. (The *degree* of a vertex is the number of edges going out of it.)

(b) Design an algorithm to find an Eulerian path in a graph or determine that none exists.

## 4.4 REMARKS AND REFERENCES

Much of the material in this chapter was known for years only in the folklore of programming. It was in

> Knuth, D. E., *The Art of Computer Programming*, Vol. 1, *Fundamental Algorithms*. Reading, Mass.: Addison-Wesley Publishing Co., 1st ed. 1968, 2d ed. 1973.

that the various data structures and algorithms first coalesced; Knuth's book remains an encyclopedic source for material on lists, stacks, and queues, as well as trees and memory management, which we cover in the next two chapters.

The pattern-matching example of Section 4.1.2 was inspired by SNOBOL4, although it is enormously simpler than the pattern matching of facility SNOBOL4. See

> Griswold, R. E., *The Macro Implementation of SNOBOL4*. San Francisco: W. H. Freeman and Co., 1972.

The difficulties of the pattern-matching strategy of Algorithm 4.4 are actually an instance of the problems that occur with parsing by recursive descent. For a general presentation of the method and a discussion of the possible solutions to its difficulties, see

> Aho, A. V., and J. D. Ullman, *Principles of Compiler Design*. Reading, Mass.: Addison-Wesley Publishing Co., 1977.

This book is also recommended in regard to the evaluation of arithmetic expressions (given as an example in Section 4.2.3), which is a simple case of operator-precedence parsing.

In Section 4.3 we barely scratched the surface of the very broad and deep area of graph algorithms. For a more comprehensive treatment, see

> Reingold, E. M., J. Nievergelt, and N. Deo, *Combinatorial Algorithms*: *Theory and Practice*. Englewood Cliffs, N.J.: Prentice-Hall, Inc., 1977.

# Chapter 5

# Trees

"I wonder about trees."

*The Sound of Trees*,   Robert Frost (1916)

In this chapter we will examine trees, a simple form of the general lists introduced in Section 4.1.2, discussing them in terms of the nonlinear, hierarchical organization they epitomize. Hierarchical organizations are so common and so useful that trees are one of the most important data structures covered in this book.

The best-known nontechnical example of a tree structure used to organize information is a *family tree*. In such a tree we show an individual and as many of his ancestors as known or needed. Consider the family tree in Figure 5.1. The principal of the tree, Gerhard Nothmann, is in the middle on the right. Just to his left are his parents, Rudolf Nothman and Margarete Caro; to their left are their parents, and so on; in each case a line is drawn to show the parent-child relationship.

A related nontechnical example is a *lineal chart*, in which we display a person's descendants, rather than his ancestors. Figure 5.2 gives a (partial) lineal chart for Johanna Caro; as in the family tree of Figure 5.1, the parent-child relationship is indicated by a line. In Figure 5.2, however, we have drawn the principal of the tree on the left, as is customary with lineal charts. This is consistent with Figure 5.1 in that the child appears to the right of the parent, and it is indicative of the generative character of the parent-child relationship since (in English) we read from left to right—that is, from parent to child.

Informally, a tree is a collection of elements, of which one is the *root* and the rest are partitioned into trees, called the *subtrees* of the root. For example, the root in the lineal chart of Figure 5.2 is Johanna Caro and the subtrees of the root are lineal charts for the children (Berta Speyer, Friederike Fränkel, and so on). In terms of this recursive definition, a lineal chart for a person would be defined as

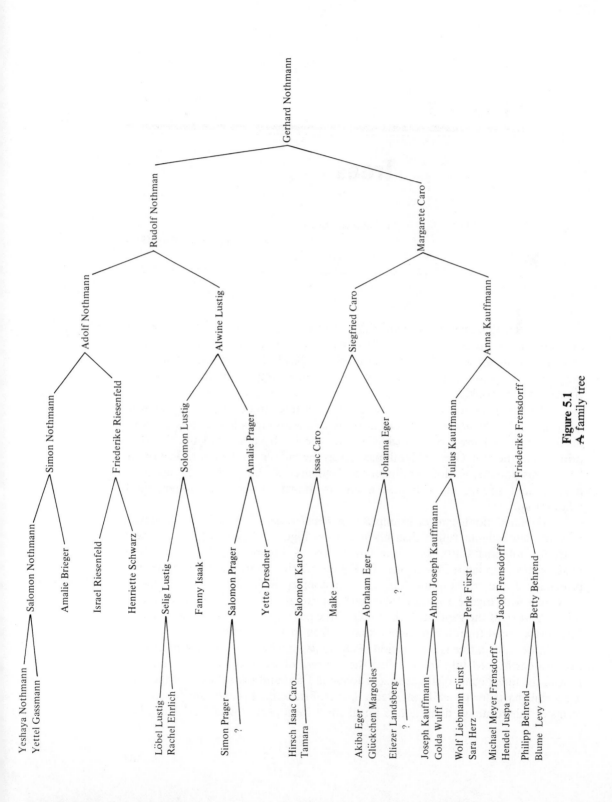

**Figure 5.1**
A family tree

the person's name (the root) and lineal charts for each of his children (the subtrees of the root).

Figure 5.3 shows two ways to depict another example of a tree structure, the various procedures that make up a modest compiler written in a block-structured programming language. Figure 5.3(a) shows the compiler's structure much as it would appear in actual code, in a representation by the indentation of the procedure names. Figure 5.3(b) shows the same tree, only drawn in a more conventional manner.

There are so many common examples of tree structures used to present information that we could go on and on giving them: the Dewey decimal system, outlines, biological classification of organisms, organization charts, and so on. Whenever a hierarchy of elements is involved, there is a natural representation as a tree. In this and later chapters we shall see many ways in which trees can be used to organize information. Moreover, we will have other important uses of trees; in particular, we will see that trees are used in organizing hierarchical processes (such as knock-out tournaments, for example) and are an important device for the analysis or understanding of various algorithms.

Formally, a tree $T$ is defined as a nonempty finite set of labeled nodes such that there is one distinguished node, called the *root* of the tree, and the remaining nodes are partitioned into $m \geqslant 0$ disjoint subtrees $T_1, T_2, \ldots, T_m$. Nodes that have no subtrees are called *leaves* or *external nodes*; the remaining nodes are called *internal nodes*. These concepts are illustrated in Figure 5.4, which shows a tree with eleven nodes labeled $A$ through $K$. The nodes labeled $D, E, F, H, J,$ and $K$ are leaves; the other nodes are internal nodes. The node labeled $A$ is the root.

In describing the relationships between nodes in a tree it has become customary to use the terminology of lineal charts. Thus all the nodes in a tree are said to be *descendants* of its root; conversely, the root is an ancestor of all of its descendants. Furthermore, we refer to the root as the *father* of the roots in its subtrees; these nodes are, in turn, the *sons* of the root. Sons of the same node are called *brothers*. For example, in Figure 5.4; node $A$ is the father of nodes $B, G,$ and $I$; $J$ and $K$ are the sons of $I$; and $C, E,$ and $F$ are brothers.

Just as there are many species of natural trees, there are many possible variations on the tree structures we will be using. For instance, all the trees considered in this book will be *ordered*; that is, the relative order of the subtrees of each node is important. Thus we consider

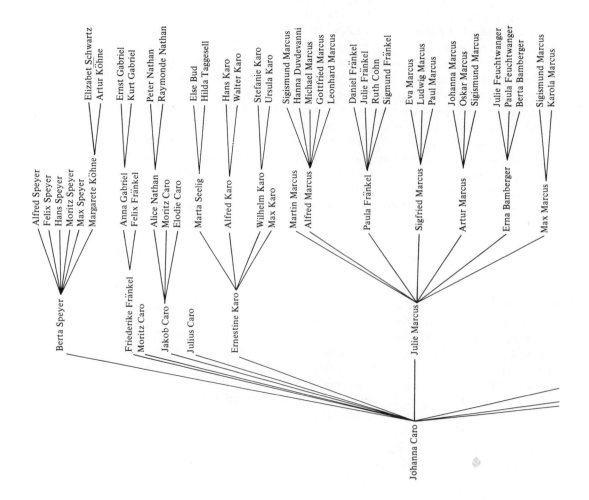

Alfred Speyer
Felix Speyer
Hans Speyer
Moritz Speyer
Max Speyer
Margarete Köhne

Elizabet Schwartz
Artur Köhne

Ernst Gabriel
Kurt Gabriel

Anna Gabriel
Felix Fränkel

Peter Nathan
Raymonde Nathan

Alice Nathan
Moritz Caro
Elodie Caro

Else Bud
Hilda Taggesell

Marta Seelig

Hans Karo
Walter Karo

Alfred Karo

Stefanie Karo
Ursula Karo

Wilhelm Karo
Max Karo

Sigismund Marcus
Hanna Duvdevanni
Michael Marcus
Gottfried Marcus
Leonhard Marcus

Martin Marcus
Alfred Marcus

Daniel Fränkel
Julie Fränkel
Ruth Cohn
Sigmund Fränkel

Paula Fränkel

Eva Marcus
Ludwig Marcus
Paul Marcus

Sigfried Marcus

Johanna Marcus
Oskar Marcus
Sigismund Marcus

Artur Marcus

Julie Feuchtwanger
Paula Feuchtwanger
Berta Bamberger

Erna Bamberger

Sigismund Marcus
Karola Marcus

Max Marcus

Berta Speyer

Friederike Fränkel
Moritz Caro

Jakob Caro

Julius Caro

Ernestine Karo

Julie Marcus

Johanna Caro

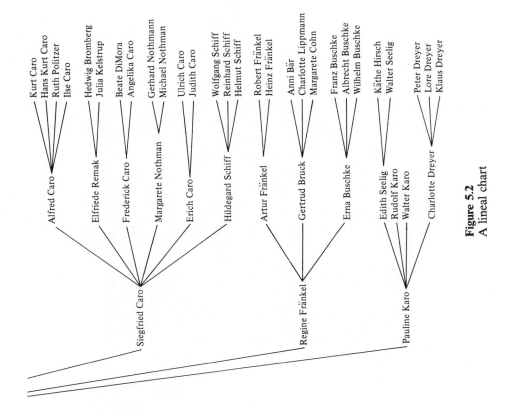

**Figure 5.2**
A lineal chart

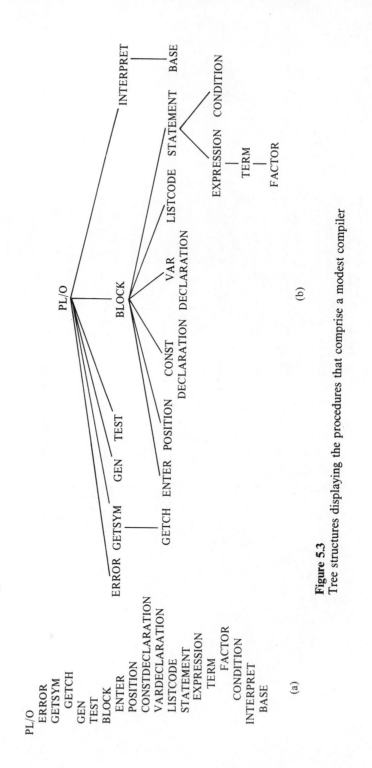

**Figure 5.3**
Tree structures displaying the procedures that comprise a modest compiler

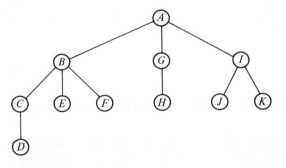

**Figure 5.4**
A tree with eleven nodes labeled $A$ through $K$. The nodes labeled $D$, $E$, $F$, $H$, $J$, and $K$ are leaves; the other nodes are internal nodes. The node labeled $A$ is the root.

to be different trees, although this is not apparent from the definition. We can define a *forest* as an ordered set of trees and so rephrase the definition of a tree: a tree is a nonempty set of nodes such that there is one distinguished node, called the root of the tree; and the remaining nodes are partitioned into a forest of $m \geq 0$ subtrees of the root. A forest, of course, may also be considered a tree in which the root, although implicit, does not explicitly appear.

Perhaps the most important tree variant is the binary tree. A *binary tree T* either is *empty* or consists of a distinguished node called the root and two binary subtrees $T_l$ and $T_r$, the left and right subtrees, respectively. Binary trees differ from the trees so far considered in two important ways: a binary tree may be empty while a tree cannot be, and, more importantly, the distinction of left and right causes

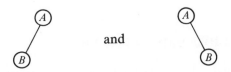

to be different binary trees; yet as forests they are both indistinguishable from

In this chapter we will concentrate primarily on binary trees, since, as we shall see, trees and forests can be neatly encoded into binary trees. In later chapters, moreover, binary trees will prove more useful as a structure and an analytical tool.

### Exercises

1. Define the relationships "cousin," "uncle," and "second cousin once removed" in terms of trees.

2. The Strahler numbering of nodes in a binary tree (useful in hydrology and botany) is defined as follows. The empty tree has Strahler number 0. If the binary tree $T$ has subtrees $T_l$ and $T_r$, the Strahler number $S(T)$ of $T$ is defined as

$$S(T) = \begin{cases} \max[S(T_l), S(T_r)], & \text{if } S(T_l) \neq S(T_r), \\ S(T_l) + 1, & \text{otherwise.} \end{cases}$$

   Compute the Strahler numbers of some binary trees. What is the smallest binary tree $T$ with $S(T) = 3$?

3. Write down all the different trees that can be made from three nodes $A$, $B$, and $C$. Write down all the different binary trees that can be made from these nodes.

4. Show that if a binary tree has $n \geq 1$ nodes and each node has either 0 or 2 sons (that is, no node has only 1 son), then $n$ is an odd number. Show that such a tree has $(n - 1)/2$ internal nodes (nodes with sons) and $(n + 1)/2$ external nodes (nodes without sons).

## 5.1  LINKED REPRESENTATIONS

In Figures 5.1, 5.2, and 5.3 we have seen several ways of drawing trees. We have had the root at the right, the left, and the top, although not at the bottom as with trees in nature. The orientations of the trees in these figures has, in each case, arisen from the attempt to have the orientation depend on the nature of the information represented. In the same way, different computer representations of trees are convenient, depending on the application.

   In this section we will examine the fundamental ways of representing trees by nodes and pointers. Important variations will be introduced in later sections and chapters. Also, because sequential representations must, by their nature, impose a linear ordering on the nodes of the tree, we discuss such techniques in a later section.

### 5.1.1 LEFT and RIGHT Pointers

Almost all computer representations of trees are based on pointers that explicitly convey the hierarchical relationships. Thus each node of the tree consists of some information and some pointers; for simplicity we will assume that there is a single information field, INFO, and pointer fields as required by the particular linking techniques.

Most computer uses of trees require easy movement down a tree from ancestors to descendants. Such movement generally requires that the tree be represented with pointers that go from fathers to sons; this is complicated because, although a node has at most one father, it can have arbitrarily many sons. In other words, the nodes in the representation will need to vary in size—a definite inconvenience. For the moment we will ignore this problem by concentrating on binary trees, since they are easily represented with nodes of fixed size. Each node has three fields: LEFT (pointing to the root of the left subtree), INFO (the contents of the node), and RIGHT (pointing to the root of the right subtree). This is illustrated in Figure 5.5.

An example of the use of such a structure is the decoding of characters represented in the International Morse Code (see Table 2.13, p. 75). The tree in Figure 5.6 represents this code in the following way. If we start at the root and follow any valid sequence of dots (left branches) and dashes (right branches) down the tree, the INFO field of the node at which we stop is the character represented by that sequence of dots and dashes. Thus, for example, the sequence — • — • causes us to go first right, then left, then right again, and finally left; the character thus found in the INFO field is "C", and indeed, "C" is represented by — • — • in the International Morse Code. If a sequence of dots and dashes does not correspond to a character, then the path down the tree thus followed will end

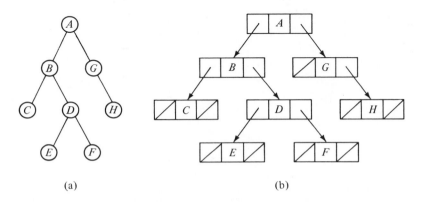

(a)                    (b)

**Figure 5.5**
A binary tree (a) and its representation (b) by nodes with the three fields LEFT, INFO, and RIGHT

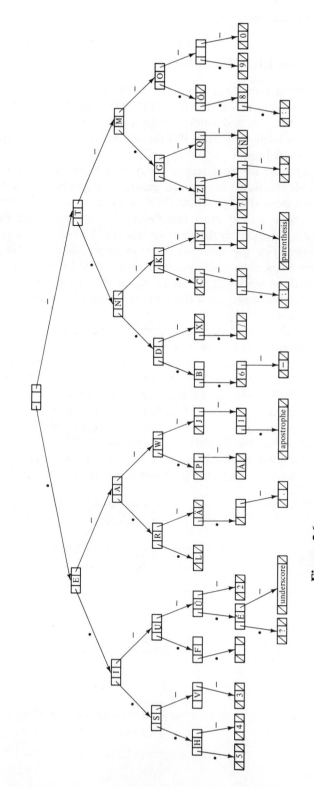

**Figure 5.6**
Tree used by Algorithm 5.1 for decoding Morse code (see Table 2.13)

[[$i$ is an index into the string $d_1d_2...d_n$ of dots and dashes;
$p$ is a pointer to a node in the Morse code tree]]
$i \leftarrow 0$
$p \leftarrow$ root of the Morse code tree
**while** $i < n$ **and** $p \neq$ **nil do**
    $i \leftarrow i + 1$
    **if** $d_i =$ " • " **then** $p \leftarrow$ LEFT($p$)
                 **else** $p \leftarrow$ RIGHT($p$)
**if** $p \neq$ **nil and** INFO($p$) $\neq$ " "**then** the character is in INFO($p$)
                                   **else** the sequence is invalid

**Algorithm 5.1**
Decoding a sequence $d_1d_2...d_n$ of dots and dashes into the corresponding
character in the international Morse code, using the tree in Figure 5.6

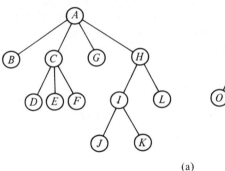
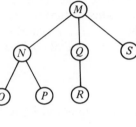

(a)

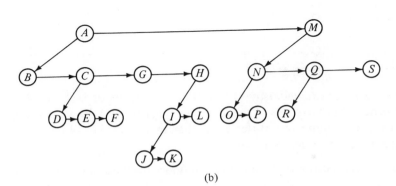

(b)

**Figure 5.7**
A forest (a) and its binary tree representation (b) under the natural correspondence

at a node with a blank INFO field or will "fall off" the bottom of the tree. For instance, following $\cdot\cdot - - \cdot$ leads to a blank node, while $- \cdot\cdot - -$ leads off the tree; these two sequences do not, therefore, correspond to characters.

We can use the tree of Figure 5.6 to write a simple decoding algorithm. Suppose we are given the sequence of dots and dashes $d_1 d_2 \ldots d_n$. To decode it, we use it to follow links down the tree from the root, left and right accordingly, as indicated by the $d_i$. If we arrive at a blank node or fall off the tree, then $d_1 d_2 \ldots d_n$ is not a valid character; otherwise, it is the character found in the INFO field of the last node. The algorithm is given formally by Algorithm 5.1.

Returning to the problem of representing trees in general, we observe that we can represent trees as binary trees (using nodes of fixed size) by representing every node in a forest as a node consisting of LEFT, INFO, and RIGHT fields and by using the LEFT field of a node to point to the leftmost son of that node and the RIGHT field of a node to point to the next brother of that node. For example, the forest shown in Figure 5.7(a) is transformed into the binary tree shown in Figure 5.7(b). Thus we are using the LEFT field of a node to point to a linked list of the sons of that node; that list is linked together by RIGHT fields. We call this the *natural correspondence* between forests and binary trees; it will be useful and natural in several contexts.

**Exercises**

1. Draw a tree analogous to Figure 5.6 for the following code:

| | | |
|---|---|---|
| A $- - \cdot -$ | J $- - - - - \cdot\cdot$ | S $\cdot - \cdot -$ |
| B $\cdot\cdot - - \cdot -$ | K $- - - - - - \cdot -$ | T $\cdot\cdot\cdot$ |
| C $\cdot - - \cdot\cdot$ | L $\cdot - - - -$ | U $\cdot - - - \cdot$ |
| D $\cdot\cdot - \cdot$ | M $\cdot - - \cdot -$ | V $\cdot\cdot - - \cdot\cdot$ |
| E $- \cdot -$ | N $- - \cdot\cdot$ | W $\cdot\cdot - - - -$ |
| F $- - - - \cdot\cdot$ | O $- - - \cdot$ | X $- - - - - - - - \cdot$ |
| G $\cdot\cdot - - - \cdot$ | P $- - - - \cdot -$ | Y $- - - - - \cdot$ |
| H $\cdot - \cdot\cdot$ | Q $- - - - - - - \cdot$ | Z $- - - - - - - -$ |
| I $- \cdot\cdot\cdot$ | R $- \cdot\cdot -$ | |

(This code was constructed using the Huffman algorithm of Section 2.4.2 with the frequency of occurrences of the letters in English text.) In what important way does the resulting tree differ from that in Figure 5.6? What does this difference mean in terms of the code?

2. What binary tree corresponds to Figure 5.8 under the natural correspondence?

3. To what forest does the binary tree in Figure 5.9 correspond, via the natural correspondence?

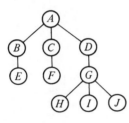

**Figure 5.8**
Exercise 2

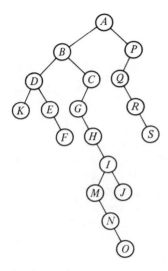

**Figure 5.9**
Exercise 3

4. To what binary tree does the forest in Figure 5.10 correspond, via the natural correspondence? To what forest does it correspond (considering it as a binary tree)?

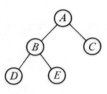

**Figure 5.10**
Exercise 4

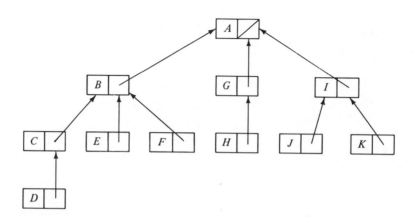

**Figure 5.11**
The tree in Figure 5.4 represented by nodes with an INFO field and a
FATHER pointer

## 5.1.2 Father Pointers

There are occasions when trees will be used that will require easy movement
*up* the tree from descendants to ancestors. In Section 5.2.2 we will see an indirect
way of doing this within the context of LEFT/RIGHT pointers. In this section
we will examine the direct way and an important application of it.

Figure 5.11 shows a linked representation of the tree in Figure 5.4 based on
pointers from son to father: each node consists of the INFO field and a single
pointer, FATHER. This representation is useful if, as is occasionally the case, we
need to move up a tree, from descendants to ancestors.

An important example of the usefulness of this representation is found in a
set-manipulation problem that occurs frequently in combinatorial algorithms.
Suppose that we want to manipulate disjoint subsets of a set $S = \{s_1, s_2, \ldots, s_n\}$.
The operations to be performed are merging two of the disjoint subsets and, given
an element $s_i$, finding which of the subsets contains $s_i$. At any given time, we thus
have a partition of $S$ into nonempty disjoint subsets.

For identification, each of the disjoint subsets of $S$ will have a name. The
name is simply one of the elements of the subset and can be thought of as a
subset representative. When we refer to the name of a subset, we are referring to
its subset representative. Consider, for example,

$$S = \{A, B, C, D, E, F, G, H, I, J, K\}$$

partitioned into four disjoint subsets

$$\{A, F, \textcircled{G}, H, K\}, \quad \{\textcircled{B}\}, \quad \{\textcircled{C}, D, E\}, \quad \{I, \textcircled{J}\}; \qquad (5.1)$$

in each case the circled element is the name of the subset. If we ask to find the subset in which $H$ is contained, the answer we expect is $G$, the name of the subset containing $H$. If we ask to take the union of the subsets named $B$ and $J$, we want the resulting partition of $S$ to be

$$\{A, F, \text{\textcircled{G}}, H, K\}, \quad \{\text{\textcircled{C}}, D, E\}, \quad \{\text{\textcircled{B}}\} \cup \{I, \text{\textcircled{J}}\},$$

in which the name of the set $\{B\} \cup \{I, J\}$ can be chosen as either $B$ or $J$.

We assume that initially we have the partition of $S = \{s_1, s_2, \ldots, s_n\}$ into $n$ singleton sets

$$\{\text{\textcircled{$s_1$}}\}, \quad \{\text{\textcircled{$s_2$}}\}, \quad \ldots, \quad \{\text{\textcircled{$s_n$}}\}, \tag{5.2}$$

in which each set is named after its only element. This partition is modified by a sequence of union operations in which find operations are intermixed. This seemingly contrived problem is quite useful in certain combinatorial algorithms; it was first motivated by the processing of EQUIVALENCE statements in FORTRAN.

We will give procedures $UNION(x, y)$ and $FIND(x)$ to implement the union and find operations. $UNION(x, y)$ takes the names of two different subsets $x$ and $y$ and creates a new subset containing all the elements of $x$ and $y$. $FIND(x)$ returns as its value the name of the subset containing $x$. For example, if we want to cause the set containing $a$ to be merged with the set containing $b$, we use the sequence of instructions

$$x \leftarrow FIND(a)$$
$$y \leftarrow FIND(b)$$
$$\textbf{if } x \neq y \textbf{ then } UNION(x, y)$$

Suppose that we have a sequence of $u$ union operations intermixed with $f$ find operations and we start with $S = \{s_1, s_2, \ldots, s_n\}$ partitioned into the singleton sets of (5.2). We want a data structure to represent the disjoint subsets of $S$ so that such a sequence of operations can be efficiently performed. The data structure that we will use is a forest representation based on father pointers, as illustrated in Figure 5.11. Each set element $s_i$ will be a node in the forest, and the father of set element $s_i$ will be another element in the same subset as $s_i$. If an element has no father (that is, is a root), then it is the name of its subset. Thus the partition (5.1) might be represented as shown in Figure 5.12.

With this representation, the operation $FIND(x)$ consists of following father pointers up from $x$ to the root (that is, name) of its subset. The operation $UNION(x, y)$ consists of somehow hooking together the trees rooted at $x$ and $y$; for example, this could be done by making $y$ the father of $x$.

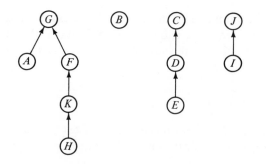

**Figure 5.12**
A forest representation of the partition (5.1)

After $u$ union operations, the largest subset possible in the resulting partition of $S$ contains $u + 1$ elements. Furthermore, since each union reduces the number of subsets by one, the sequence of operations can contain at most $n - 1$ unions; thus $u \leqslant n - 1$. Since each union operation changes the name of the subset containing some of the elements, we can assume that each union is preceded by at least one find, and hence we assume that $f \geqslant u$. The problem then is to efficiently perform a sequence of $u \leqslant n - 1$ union operations intermixed with $f \geqslant u$ find operations. The time required by the union operations is clearly proportional to $u$, because only the small constant amount of work needed to rearrange some pointers is necessary for each union operation. We can therefore concentrate on the time required by the $f$ find operations.

If the operation $UNION(x, y)$ is performed by making $x$ the father of $y$, then it is possible, after a sequence of $u$ union operations, to produce the forest shown in Figure 5.13. In this case if the $f$ find operations are done after all the union operations, and each find starts at the bottom of the chain of $u + 1$ set elements, it is clear that the time required by the find operations will be

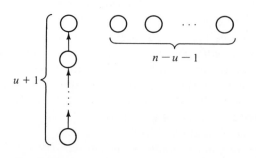
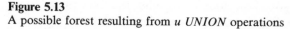

**Figure 5.13**
A possible forest resulting from $u$ *UNION* operations

proportional to $f \times (u + 1)$. Obviously, it could not be worse than proportional to $f \times (u + 1)$.

By being more clever we can reduce this worst case considerably. If the operation $UNION(x, y)$ keeps the trees in the forest "balanced" by making the root of the larger subset the father of the root of the smaller subset (ties can be broken arbitrarily), then we pay a slight premium in storage, since each node in the forest must contain information about the size of the subtree beneath it. However, we will see that the time required by the find operations will then be at most proportional to $f\lg(u + 1)$. First, we need some definitions.

The *level* of a node $p$ in a tree $T$ is defined recursively as 0 if $p$ is the root of $T$, otherwise the level of $p$ is $1 + \text{level}(\text{FATHER}(p))$. The *height* $h(T)$ of a tree $T$ is defined by

$$h(T) = \max_{\substack{\text{nodes} \\ p \text{ in } T}} \text{level}(p)$$

The level of a node is thus its distance from the root, while the height of a tree is the distance from the root to the furthest leaf.

Going back to the $UNION/FIND$ problem, for any $s_i$ in $S$ let $h(s_i)$ be the height of the subtree rooted at $s_i$ and let $w(s_i)$ be the number of nodes in that subtree. Observe that we always have $w(s_i) \geq 2^{h(s_i)}$. This is clearly true for $u = 0$, since then each element $s_i$ in $S$ is the root of a tree consisting of a single node, and so we have $h(s_i) = 0$ and $w(s_i) = 1$. Suppose that $w(s_i) \geq 2^{h(s_i)}$ for $u = k$ and consider what happens on the $(k + 1)$st union operation, $UNION(x, y)$. By induction, we have (before the union operation) $w(x) \geq 2^{h(x)}$ and $w(y) \geq 2^{h(y)}$. Without loss of generality, let $w(x) \geq w(y)$. Then $UNION(x, y)$ causes $x$ to become the father of $y$, and the height of $x$ after the union is $\max[h(x), h(y) + 1]$. After the union we will have $w(x) \geq 2w(y) \geq 2^{h(y)+1}$; and both before and after the union we have $w(x) \geq 2^{h(x)}$. We conclude that after $u$ union operations each of the elements $s_i$ satisfies

$$0 \leq h(s_i) \leq \lg(u + 1),$$

and thus $f$ find operations will require time at most proportional to $f\lg(u + 1)$. Furthermore, since $f \geq u$, the *total* time for unions *and* finds is at most proportional to $f\lg(u + 1)$. It is easy to see that a tree of height $\lg(u + 1)$ can result from the $u$ unions (how?), and therefore there is an example in which the finds can actually achieve the bound of $f\lg(u + 1)$.

We can improve the efficiency of the find operations by using *path compression*: after the operation $FIND(x)$, $x$ and all the vertices on the path between $x$ and the root are made sons of the root. For example, if we did $FIND(H)$ on the forest in Figure 5.12, the value returned would be $G$ as before, but in the meantime the forest would have changed to the one shown in Figure 5.14. Path

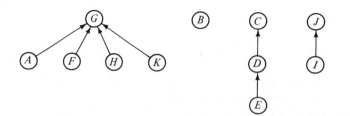

**Figure 5.14**
The result of *FIND(H)* on the forest in Figure 5.12 when path compression is used

**procedure** *UNION*(*x*, *y*)
    ⟦*x* and *y* are assumed to be roots of trees in the forest stored at locations $l_x$ and $l_y$, respectively⟧
    **if** $\text{SIZE}(l_x) < \text{SIZE}(l_y)$ **then**
                               $\text{FATHER}(l_x) \leftarrow l_y$
                               $\text{SIZE}(l_y) \leftarrow \text{SIZE}(l_x) + \text{SIZE}(l_y)$
        **else**
                               $\text{FATHER}(l_y) \leftarrow l_x$
                               $\text{SIZE}(l_x) \leftarrow \text{SIZE}(l_y) + \text{SIZE}(l_x)$

**Algorithm 5.2**
The union operation with balancing

**procedure** *FIND*(*x*) **returns pointer**
    ⟦*x* is a node in the forest stored at location $l_x$⟧
    $Q \leftarrow$ empty queue    ⟦*Q* is used to store the path to be compressed⟧
    $t \leftarrow l_x$
    **while** $\text{FATHER}(t) \neq$ **nil do**
        $Q \Leftarrow t$
        $t \leftarrow \text{FATHER}(t)$
    $\Delta \leftarrow 0$
    **while** $Q \neq$ empty **do**
        $v \Leftarrow Q$
        $\text{FATHER}(v) \leftarrow t$
        $temp \leftarrow \text{SIZE}(v)$
        $\text{SIZE}(v) \leftarrow \text{SIZE}(v) - \Delta$
        $\Delta \leftarrow temp$
    **return** *t*

**Algorithm 5.3**
The find operation with path compression. See Exercises 3 and 4.

compression causes only a minor increase in the cost of a find operation, and, as we will see, its use significantly affects the time required by a sufficiently large number of find operations.

The union operation with balancing is given in Algorithm 5.2, and the find operation with path compression is given in Algorithm 5.3. In each algorithm we assume that the set element $s_i$ is represented by a forest node at location $l_i$ that consists of three fields: $ELT(l_i)$ is $s_i$; $FATHER(l_i)$ is the location of the node corresponding to the father of $s_i$; and $SIZE(l_i)$ is $w(s_i)$, the number of elements in the subtree rooted at $s_i$.

The performance of Algorithms 5.2 and 5.3 is startlingly good. It can be shown, although the proof is beyond the level of this book, that the time required for $f \geqslant n$ finds and $n - 1$ unions is at most proportional to $n\alpha(f, n)$ where $\alpha(f, n)$ is a function that grows so slowly that $\alpha(f, n) \leqslant 3$ for

$$
n \leqslant 2^{2^{\cdot^{\cdot^{\cdot^{2}}}}} \left. \right\} \begin{array}{l} 65,537 \\ \text{twos} \end{array}
$$

**Exercises**

1. What structure results if we start with the sets $\{A\}$, $\{B\}$, $\{C\}$, $\{D\}$, $\{E\}$, $\{F\}$, $\{G\}$, $\{H\}$ and perform the following sequence of operations with Algorithms 5.2 and 5.3?

   > $UNION(A, B)$
   > $UNION(FIND(A), C)$
   > $UNION(D, E)$
   > $UNION(F, G)$
   > $UNION(H, FIND(F))$
   > $UNION(FIND(H), FIND(D))$
   > $FIND(G)$
   > $FIND(B)$
   > $FIND(E)$

2. Let $f(k)$ be the maximum height of a tree produced by a sequence of $k$ union/find operations with initially given sets $\{1\}, \{2\}, \ldots, \{2k\}$, when balancing (but not path compression) is used. Thus $f(0) = 0$, $f(1) = f(2) = 1$, $f(3) = 2$, and so on. Prove that $f(2n + 1) \geqslant f(n) + 1$ for all $n \geqslant 0$.

3. Why are the statements *"temp* ← SIZE($v$), SIZE($v$) ← SIZE($v$) − Δ, Δ ← *temp"* not really needed in Algorithm 5.3?

4. Rewrite Algorithm 5.3 without the queue.

## 5.2  TRAVERSALS

In many applications of trees and forests it is necessary to traverse them—that is, to visit systematically each of the nodes and process each of them in some manner. The visit at each node might be as simple as printing its contents or as complicated as a major computation. The only assumption that we make about the visit is that is does not change the structure of the forest. The four basic traversal orders that we will find useful are *preorder*, *postorder*, *level order*, and, for binary trees, *inorder*.

*Preorder* visits the nodes of a forest as described by the following recursive procedure.

1. Visit the root of the first tree.

2. Traverse the subtrees, if any, of the first tree in preorder.

3. Traverse the remaining trees, if any, in preorder.

For instance, in the forest shown in Figure 5.7(a), the nodes would be visited in the order $A, B, C, D, E, F, G, H, I, J, K, L, M, N, O, P, Q, R, S$. The name "preorder" refers to the fact that we visit the root before traversing the remainder of a tree.

For a binary tree, the recursive procedure simplifies to

1. Visit the root of the binary tree.

2. Traverse the left subtree in preorder.

3. Traverse the right subtree in preorder.

In this case the empty tree is traversed by doing nothing. Notice that traversing a forest in preorder is exactly the same as the preorder traversal of the binary tree arising by the natural correspondence. This fact makes the correspondence "natural."

*Postorder* visits the nodes of the forest according to the following recursive procedure.

1. Traverse the subtrees of the first tree, if any, in postorder.

2. Visit the root of the first tree.

3. Traverse the remaining trees, if any, in postorder.

The name "postorder" refers to the fact that at the time a node is visited, all its descendants have already been visited. In the forest in Figure 5.7(a) this order visits the nodes in the order $B, D, E, F, C, G, J, K, I, L, H, A, O, P, N, R, Q, S, M$. The recursive procedure for the postorder traversal applied to binary trees simplifies to

1. Traverse the left subtree in postorder.

2. Traverse the right subtree in postorder.

3. Visit the root.

*Inorder* or *symmetric order* is defined recursively for binary trees as

1. Traverse the left subtree in inorder.

2. Visit the root.

3. Traverse the right subtree in inorder.

This is also known as *lexicographic order* (for reasons that will become clear in Chapter 7). Notice that traversing a forest in postorder is equivalent to traversing the binary tree corresponding to the forest (by the natural correspondence) in inorder.

Comparing the recursive procedures for the preorder, inorder, and post-order binary tree traversals, we find considerable similarity. In all cases the left subtree is visited before the right subtree; only the time when the root is visited differentiates between the orders:

| | |
|---|---|
| Preorder: | before the subtrees |
| Inorder: | between the subtrees |
| Postorder: | after the subtrees |

This similarity allows us to construct a general nonrecursive algorithm that can be adapted to each of these orders for binary tree traversal. We use a stack $S$ to store pairs consisting of a node in the binary tree and an integer $i$ whose value tells which of the three operations of Table 5.1 (the first, second, or third) is to be performed when the pair reaches the top of the stack. This general algorithm is shown in Algorithm 5.4.

So, for example, the straightforward specialization of Algorithm 5.4 to the preorder binary tree traversal yields Algorithm 5.5(a). We can simplify Algorithm 5.5(a) by noticing that when $(p, 2)$ or $(p, 3)$ comes to the top of the stack, the

| | | Order | |
|---|---|---|---|
| *Operation* | *Preorder* | *Inorder* | *Postorder* |
| First | visit node $p$ | $S \Leftarrow (\text{LEFT}(p), 1)$ | $S \Leftarrow (\text{LEFT}(p), 1)$ |
| Second | $S \Leftarrow (\text{LEFT}(p), 1)$ | visit node $p$ | $S \Leftarrow (\text{RIGHT}(p), 1)$ |
| Third | $S \Leftarrow (\text{RIGHT}(p), 1)$ | $S \Leftarrow (\text{RIGHT}(p), 1)$ | visit node $p$ |

**Table 5.1**
The operations for the general binary tree traversal of Algorithm 5.4

$$S \leftarrow \text{empty stack}$$
$$S \Leftarrow (\text{root}, 1)$$
**while** $S \neq$ empty **do**
$\quad\quad (p, i) \Leftarrow S$
$\quad\quad$ **if** $p \neq$ **nil then**
$\quad\quad\quad\quad\quad\quad$ **case**
$\quad\quad\quad\quad\quad\quad i = 1$:
$\quad\quad\quad\quad\quad\quad\quad\quad S \Leftarrow (p, 2)$
$\quad\quad\quad\quad\quad\quad\quad\quad$ first operation
$\quad\quad\quad\quad\quad\quad i = 2$:
$\quad\quad\quad\quad\quad\quad\quad\quad S \Leftarrow (p, 3)$
$\quad\quad\quad\quad\quad\quad\quad\quad$ second operation
$\quad\quad\quad\quad\quad\quad i = 3$:
$\quad\quad\quad\quad\quad\quad\quad\quad$ third operation

**Algorithm 5.4**
General binary tree traversal, with operations as indicated in Table 5.1

$$S \leftarrow \text{empty stack}$$
$$S \Leftarrow (\text{root}, 1)$$
**while** $S \neq$ empty **do**
$\quad\quad (p, i) \Leftarrow S$
$\quad\quad$ **if** $p \neq$ **nil then**
$\quad\quad\quad\quad\quad\quad$ **case**
$\quad\quad\quad\quad\quad\quad i = 1$:
$\quad\quad\quad\quad\quad\quad\quad\quad S \Leftarrow (p, 2)$
$\quad\quad\quad\quad\quad\quad\quad\quad$ visit node $p$
$\quad\quad\quad\quad\quad\quad i = 2$:
$\quad\quad\quad\quad\quad\quad\quad\quad S \Leftarrow (p, 3)$
$\quad\quad\quad\quad\quad\quad\quad\quad S \Leftarrow (\text{LEFT}(p), 1)$
$\quad\quad\quad\quad\quad\quad i = 3$:
$\quad\quad\quad\quad\quad\quad\quad\quad S \Leftarrow (\text{RIGHT}(p), 1)$

**Algorithm 5.5(a)**
Straightforward specialization of Algorithm 5.4 to the preorder traversal of a binary tree

$$S \leftarrow \text{empty stack}$$
$$S \Leftarrow \text{root}$$
**while** $S \neq$ empty **do**
$$\quad p \Leftarrow S$$
$$\quad \textbf{if } p \neq \textbf{nil then}$$
$$\qquad\qquad \text{visit node } p$$
$$\qquad\qquad S \Leftarrow \text{RIGHT}(p)$$
$$\qquad\qquad S \Leftarrow \text{LEFT}(p)$$

**Algorithm 5.5(b)**
Preorder binary tree traversal, a simplified version of Algorithm 5.5(a)

only thing that happens is that $(\text{LEFT}(p), 1)$, or $(\text{RIGHT}(p), 1)$ is put on the stack. This step can be done earlier, when we first visit node $p$, so we can simplify Algorithm 5.5(a) to Algorithm 5.5(b).

Specializing Algorithm 5.4 for the inorder traversal of binary trees and simplifying the result as above, we get Algorithm 5.6. Specializing Algorithm 5.4 for the postorder results in Algorithm 5.7. (See Exercises 5 through 8 for further comments on these traversal algorithms.)

The final traversal order is the *level order*. This traversal visits the nodes of the forest from the left to right, level by level from the roots down. Thus the nodes in the forest in Figure 5.7(a) would be visited in the order $A, M, B,$ $C, G, H, N, Q, S, D, E, F, I, L, O, P, R, J, K$. In a binary tree this traversal is accomplished by having $S$ be a queue instead of a stack in Algorithm 5.5(b) and by interchanging the last two statements of the **while** loop (why?), yielding Algorithm 5.8.

For forests represented by binary trees via the natural correspondence, the level-order traversal is a bit more complicated. At each node we must follow

$$S \leftarrow \text{empty stack}$$
$$S \Leftarrow (\text{root}, 1)$$
**while** $S \neq$ empty **do**
$$\quad (p, i) \Leftarrow S$$
$$\quad \textbf{if } p \neq \textbf{nil then}$$
$$\qquad\qquad \textbf{if } i = 1 \textbf{ then}$$
$$\qquad\qquad\qquad S \Leftarrow (p, 2)$$
$$\qquad\qquad\qquad S \Leftarrow (\text{LEFT}(p), 1)$$
$$\qquad\qquad \textbf{else}$$
$$\qquad\qquad\qquad \text{visit node } p$$
$$\qquad\qquad\qquad S \Leftarrow (\text{RIGHT}(p), 1)$$

**Algorithm 5.6**
Inorder binary tree traversal, a simplified specialization of Algorithm 5.4

$$S \leftarrow \text{empty stack}$$
$$S \Leftarrow (\text{root}, 1)$$
**while** $S \neq \text{empty}$ **do**
$$\quad (p, i) \Leftarrow S$$
$$\quad \textbf{if } p \neq \textbf{nil then}$$
$$\qquad \textbf{case}$$
$$\qquad\quad i = 1:$$
$$\qquad\qquad S \Leftarrow (p, 2)$$
$$\qquad\qquad S \Leftarrow (\text{LEFT}(p), 1)$$
$$\qquad\quad i = 2:$$
$$\qquad\qquad S \Leftarrow (p, 3)$$
$$\qquad\qquad S \Leftarrow (\text{RIGHT}(p), 1)$$
$$\qquad\quad i = 3:$$
$$\qquad\qquad \text{visit node } p$$

**Algorithm 5.7**
Postorder binary tree traversal, a specialization of Algorithm 5.4

$$Q \leftarrow \text{empty queue}$$
$$Q \Leftarrow \text{root}$$
**while** $Q \neq \text{empty}$ **do**
$$\quad p \Leftarrow Q$$
$$\quad \textbf{if } p \neq \textbf{nil then}$$
$$\qquad \text{visit node } p$$
$$\qquad Q \Leftarrow \text{LEFT}(p)$$
$$\qquad Q \Leftarrow \text{RIGHT}(p)$$

**Algorithm 5.8**
Level order traversal of a binary tree

$$Q \leftarrow \text{empty queue}$$
$$Q \Leftarrow \text{root}$$
**while** $Q \neq \text{empty}$ **do**
$$\quad p \Leftarrow Q$$
$$\quad \textbf{while } p \neq \textbf{nil do}$$
$$\qquad \text{visit node } p$$
$$\qquad \textbf{if } \text{LEFT}(p) \neq \textbf{nil then } Q \Leftarrow \text{LEFT}(p)$$
$$\qquad p \leftarrow \text{RIGHT}(p)$$

**Algorithm 5.9**
Level-order traversal of a forest represented as a binary tree via the natural correspondence

RIGHT links to visit the brothers and simultaneously keep track of the LEFT links ("eldest" sons in the underlying forest). Thus we arrive at Algorithm 5.9.

**Exercises**

1. Which of the traversal orders described is appropriate for computing the Strahler numbers of Exercise 2, page 194?

2. Find a simple relationship between the preorder traversal of a binary tree and the postorder traversal of its mirror image.

3. Is each of the following true or false? Prove your answers.
   (a) In a binary tree the node $x$ is a descendant of the node $y$ if and only if $x$ follows $y$ in preorder and precedes $y$ in postorder.
   (b) In a binary tree the node $x$ is a descendant of the node $y$ if and only if $x$ follows $y$ in preorder and precedes $y$ in inorder.
   (c) The leaves of a binary tree occur in the same relative order in preorder, inorder, and postorder. What about level order?
   (d) The leaves of a forest occur in the same relative order in preorder and postorder.

4. Research on Rhesus monkeys has shown that females take their place in the dominance hierarchy directly below their mother, and above *all* of the females that their mother dominates, including any of their older sisters. Assuming that the daughters of a node are in decreasing order by age from left to right, to what forest traversal order does this dominance relation correspond?

5. In Algorithm 5.4, its specializations, and their simplifications much energy is expended moving the empty tree [that is, pairs (**nil**, $i$)] on and off the stack. Modify these algorithms so that only nonempty trees are put on the stack.

6. Why is the right son put on the stack before the left son in Algorithm 5.5(b)?

7. Design an inorder traversal algorithm that uses a stack to store tree nodes so that a node is put on and removed from the stack only once (as compared to Algorithm 5.6, in which each node is put on and removed from the stack twice). [*Hint*: Arrange it so that at the time a node is removed from the stack its left subtree has already been traversed; it is then sufficient to visit the node and then traverse its right subtree.]

8. Design a postorder traversal algorithm that uses a stack to store tree nodes so that a node is put on and removed from the stack only twice (as compared to Algorithm 5.7, in which each node is put on and removed from the stack three times).

9. Find all binary trees with the property that the preorder traversal and inorder traversal visit the nodes in the same order.

10. Describe the set of nonempty binary trees with the property that the preorder traversal and the level-order traversal visit the nodes in the same order.

11. Design an algorithm to reconstruct a binary tree from the preorder and inorder lists of nodes. Do corresponding algorithms exist for the preorder and postorder lists or the inorder and postorder lists?

12. Suppose the nodes of a tree are linked together in preorder with the LEFT pointers and in inorder in the RIGHT pointers. In other words, there is a linked structure of nodes like

$$\text{preorder} \quad \overset{\displaystyle N}{\underset{}{\boxed{\phantom{x}\,N\,\phantom{x}}}} \quad \text{inorder}$$
$$\text{successor of } N \qquad\qquad\qquad \text{successor of } N$$

Devise an algorithm to construct the tree *in the given nodes* (see Exercise 11).

13. Design an algorithm for deleting a node in a tree in such a way that the inorder of the remaining nodes is unchanged. Design similar algorithms for preorder and postorder.

14. Consider binary trees represented with LEFT and RIGHT pointers in which each node also contains an additional one-bit field BIT. A *mixed-order traversal* of such a tree is defined recursively as

    (a) If BIT of the root is 0, then
        (1) Visit the root.
        (2) Traverse the left subtree in mixed order.
        (3) Traverse the right subtree in mixed order.
    (b) If BIT of the root is 1, then
        (1) Traverse the left subtree in mixed order.
        (2) Traverse the right subtree in mixed order.
        (3) Visit the root.

    For example, the mixed-order traversal of the tree in Figure 5.15 visits the nodes in the order $B, D, E, F, G, I, J, H, C, A$. Design a nonrecursive algorithm to traverse such a tree in mixed order.

15. Give a nonrecursive algorithm to traverse a binary tree in *dual order* defined recursively as

    (a) Visit the root for the first time.
    (b) Traverse the left subtree in dual order.

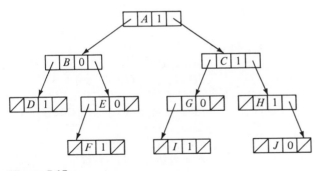

**Figure 5.15**
Exercise 14

(c) Visit the root for the second time.

(d) Traverse the right subtree in dual order.

16. Suppose we want to traverse a binary tree in preorder, inorder, and postorder *simultaneously* in the following recursively defined "triple-order" traversal:

(a) Visit the root for the preorder traversal.

(b) Traverse the left subtree in triple order.

(c) Visit the root (again) for the inorder traversal.

(d) Traverse the right subtree in triple order.

(e) Visit the root (again!) for the postorder traversal.

Design an algorithm analogous to Algorithm 5.4 for the triple-order traversal of a binary tree.

17. The following seven operations can be rearranged in $7! = 5040$ ways to yield recursive binary tree-traversal algorithms:

   (i) Visit the root.

   (ii) Visit the left son of the root, if it exists.

   (iii) Traverse the left subtree of the left son, if it exists.

   (iv) Traverse the right subtree of the left son, if it exists.

   (v) Visit the right son of the root, if it exists.

   (vi) Traverse the left subtree of the right son, if it exists.

   (vii) Traverse the right subtree of the right son, if it exists.

(a) Express preorder, inorder, and postorder in terms of the above steps.

(b) Can the following traversal be expressed in terms of the above steps? If so, do it; if not, explain why not.

**procedure** *TRAVERSE*(*t*)
    **if** *t* ≠ **nil then**
        **if** RIGHT(*t*) = **nil then**
                        visit node *t*
                        *TRAVERSE*(LEFT(*t*))
            **else**
                        *TRAVERSE*(LEFT(RIGHT(*t*)))
                        visit node *t*
                        *TRAVERSE*(LEFT (*t*))
                        visit node *t*   RIGHT(*t*)
                        *TRAVERSE*(RIGHT(RIGHT(*t*)))

(c) Design a general traversal analogous to Algorithm 5.4 but based on the above operations.

18. In what order does the following algorithm traverse a binary tree?

    *t* ← root
    **while** *t* ≠ **nil do**
        **if** LEFT(*t*) = **nil then**
                        visit node *t*
                        *t* ← RIGHT(*t*)
            **else**
                      *p* ← LEFT(*t*)
                      **while** RIGHT(*p*) ≠ **nil and** RIGHT(*p*) ≠ *t* **do**
                                      *p* ← RIGHT(*p*)
                      **if** RIGHT(*p*) = **nil then**
                            RIGHT(*p*) ← *t*
                            *t* ← LEFT(*t*)
                      **else**
                          visit node *t*
                          RIGHT(*p*) ← **nil**
                          *t* ← RIGHT(*t*)

Explain how the algorithm works. Design similar algorithms for other traversal orders.

19. Let *T* be a binary tree in which associated with each edge (FATHER(*v*), *v*) there is a nonnegative *capacity* *c*(FATHER(*v*), *v*), the amount of water that can flow from FATHER(*v*) to *v*. We assume that an infinite supply of water is available at the root of *T*. A *flow* in *T* is a nonnegative function *f* on the set

of edges such that

$$f(\text{FATHER}(v), v) \leq c(\text{FATHER}(v), v)$$

and

$$f(\text{FATHER}(v), v) = f(v, \text{LEFT}(v)) + f(v, \text{RIGHT}(v))$$

The *value* of a flow is defined as the total flow through the root. A flow $f$ is a *maximum flow* if its value is the largest possible.

(a) Explain why the value of $f$ is $\sum_{v \text{ a leaf}} f(\text{FATHER}(v), v)$.

(b) Devise an algorithm to compute the *value* of a maximum flow $f$ for a binary tree $T$ with its associated capacity function. [*Hint:* Base your algorithm on a traversal.]

(c) Devise an algorithm to compute the values $f(\text{FATHER}(v), v)$ for a maximum flow.

20. Define a *priority forest* as a forest of numerical values that either is empty or has the shape

where $m \geq 1$, $x_1, \ldots, x_m$ are values, and $F_1, \ldots, F_m$ are priority forests; furthermore, the preorder traversal of the forest yields the values in the nodes in *increasing order*. Since $F_m$ will always be empty, we will write this forest as $(x_1 F_1 x_2 \ldots F_{m-1} x_m)$. The insertion of a new value into a priority forest is done by the following recursive strategy. Insertion of $x$ into the empty forest yields $(x)$. Insertion of $x$ into $(x_1 F_1 \ldots F_{m-1} x_m)$ yields $(x x_1 F_1 \ldots F_{m-1} x_m)$ if $x \leq x_1$; or $(x_1 \ldots x_i \hat{F}_i x_{i+1} \ldots x_m)$ if $x_i < x \leq x_{i+1}$, where $\hat{F}_i$ is the result of recursively inserting $x$ into $F_i$; or $(x_1 F_1 \ldots F_{m-1} x_m x)$ if $x > x_m$. This recursive description of the algorithm insures that the values in the priority forest are always increasing in preorder; in particular, $x_1 < \cdots < x_m$, and the value of $i$ is thus uniquely defined. It also insures that $F_m$ is always empty.

(a) Show the priority forest that results from inserting $16, 11, 1, 3, 6, 2, 4, 5, 12, 15$ into the originally empty forest.

(b) Suppose the priority forests are represented as binary trees according to the natural correspondence so that each node looks like

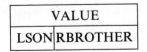

where LSON points to the leftmost son, if any, and RBROTHER points to the next brother to right, if any. Assume further that the forest is pointed to by a header node HEAD of the same form where LSON(HEAD) points to the root of the binary tree representing the forest. Give an algorithm to delete the element of a *nonempty* forest having the lowest VALUE. Your algorithm must leave the remaining values in a proper priority forest.

(c) Give both recursive and nonrecursive versions of the insertion algorithm described above.

21. A binary tree can be laid out compactly in the plane by the following technique:

Design traversal algorithms to assign the $(x, y)$ coordinates to a node, as it will appear in the above examples.

### 5.2.1 Applications

The systematic traversal of the nodes of a tree is required in many applications. In this section we will examine some typical problems in which such traversals play a central role.

**Copying a Binary Tree.** Suppose we are given the root of a binary tree and we wish to create a copy of that tree. We can do so by traversing the tree in preorder, with the visit at a node copying its contents. It is easiest to understand the copying process when it is specified recursively: to copy a tree we create a new root, store in its INFO field the contents of the INFO field of the original tree, and then recursively copy the left and right subtrees of the original tree. Algorithm 5.10 embodies this process with a recursive procedure *COPY*. *COPY* has as its single argument the root of the tree to be copied and it returns as its value the root of the copy produced.

The use of recursion in Algorithm 5.10 makes it crystal clear that this procedure is just a preorder traversal. This observation is obscured slightly in the iterative version given in Algorithm 5.11. In this version we have chosen to create a new node at the time its father is being copied. This allows us to set the values of LEFT and RIGHT pointers for the father at that time. The stack is used to store pairs $(p, q)$, where $p$ is a node in the original tree and $q$ is the corresponding node in the copy being produced.

```
procedure COPY(T) returns pointer
   if T = nil then COPY ← nil
         else
               t ← getcell
               INFO (t) ← INFO(T)
               LEFT(t) ← COPY(LEFT(T))
               RIGHT(t) ← COPY(RIGHT(T))
               return t
```

**Algorithm 5.10**
A recursive procedure (based on a preorder traversal) to produce a copy of
binary tree $T$

```
if root = nil then newroot ← nil
         else
               S ← empty stack
               newroot ← getcell
               S ⇐ (root, newroot)
               while S ≠ empty do
                    (old, new) ⇐ S
                    INFO(new) ← INFO(old)
                    if RIGHT(old) = nil then RIGHT(new) ← nil
                                        else
                                              x ← getcell
                                              S ⇐ (RIGHT(old), x)
                                              RIGHT(new) ← x
                    if LEFT(old) = nil then LEFT(new) ← nil
                                       else
                                             x ← getcell
                                             S ⇐ (LEFT(old), x)
                                             LEFT(new) ← x
```

**Algorithm 5.11**
An iterative procedure (based on a preorder traversal) to produce a copy of a
binary tree $T$

**Printing a Binary Tree.** Suppose we want to print a binary tree so that its
"shape" is properly displayed. For example, we might want to print the tree of
Figure 5.5(a) as shown in Figure 5.16. The lines of the output must be printed
from top to bottom; this requires that the tree be traversed in "reverse inorder."

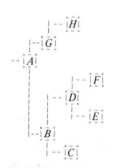

**Figure 5.16**
A "computer printed" version of Figure 5.5(a)

In other words, we need to traverse the tree according to the recursively defined order

1.  Right subtree

2.  Visit root

3.  Left subtree

   We will simplify the problem slightly by eliminating the connective lines between the nodes of the tree and also the characters that form the box around the nodes. Since this "window dressing" is not too difficult to add, once the basic idea is understood, we leave it to Exercise 3. Thus, consider the problem of printing the tree of Figure 5.5(a) as shown in Figure 5.17.

   The basic idea is to write a recursive procedure, essentially a traversal in reverse inorder, that has an additional parameter, say *pre*, whose value is the string of characters to be printed as a prefix at the left end of every line of the tree. Initially *pre* is $\varepsilon$, the empty string, but as we go deeper in the tree, blanks are

|   |   |   |   |
|---|---|---|---|
|   |   | H |   |
|   | G |   |   |
| A |   |   |   |
|   |   |   | F |
|   |   | D |   |
|   |   |   | E |
|   | B |   |   |
|   |   | C |   |

**Figure 5.17**
A simplified version of Figure 5.16

added to it; these blanks are removed as we go back up toward the root. Algorithm 5.12 embodies this idea.

> $PRINTTREE(\varepsilon, T)$
> **procedure** $PRINTTREE(pre, t)$
> > **if** $t \neq$ **nil then**
> > > $PRINTTREE(pre \parallel$ " ", $RIGHT(t))$
> > > print $pre \parallel INFO(T)$
> > > $PRINTTREE(pre \parallel$ " ", $LEFT(t))$

**Algorithm 5.12**
Printing a binary tree; $\parallel$ is the operation of concatenation of strings

**Arithmetic Expressions.** In Section 4.2.3 we examined the conversion of an arithmetic expression from infix notation to postfix notation. If we now examine the process from the perspective of binary trees, we see that if we take an arithmetic expression and write it in the obvious way as a binary tree, then the postorder traversal of that tree produces the postfix form of the arithmetic expression. For example, consider the expression

$$(B - (B \uparrow 2 - 4 * A * C) \uparrow 0.5) / (2 * A).$$

As a binary tree we would write it as shown in Figure 5.18. A postorder traversal of this tree would visit the nodes in the order

$$BB2 \uparrow 4A * C * -0.5 \uparrow -2A * /,$$

exactly the postfix of the expression considered. Similarly, the preorder traversal of such a tree corresponds to prefix form of the expression, while the inorder traversal corresponds to the normal infix form with the parentheses deleted.

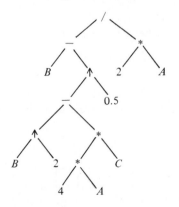

**Figure 5.18**
An arithmetic expression represented by a binary tree

The connection between the postfix form of an arithmetic expression and the postorder traversal of the corresponding binary tree allows us to evaluate an arithmetic expression from its corresponding binary tree. We use a simple combination of the postorder traversal of Algorithm 5.7 and Algorithm 4.7, which computes the value of an arithmetic expression from its postfix form; the result is Algorithm 5.13.

$$S \leftarrow \text{empty stack}$$
$$S \Leftarrow (root, 1)$$
$$V \leftarrow \text{empty stack}$$
**while** $S \neq \text{empty}$ **do**
$$\quad (p, i) \Leftarrow S$$
$$\quad \textbf{case}$$
$$\qquad i = 1:$$
$$\qquad\quad S \Leftarrow (p, 2)$$
$$\qquad\quad \textbf{if } \text{LEFT}(p) \neq \textbf{nil then } S \Leftarrow (\text{LEFT}(p), 1)$$
$$\qquad i = 2:$$
$$\qquad\quad S \Leftarrow (p, 3)$$
$$\qquad\quad \textbf{if } \text{RIGHT}(p) \neq \textbf{nil then } S \Leftarrow (\text{RIGHT}(p), 1)$$
$$\qquad i = 3:$$
$$\qquad\quad \textbf{if } \text{INFO}(p) \text{ is an operand}$$
$$\qquad\qquad\quad \textbf{then } V \Leftarrow \text{INFO}(p)$$
$$\qquad\qquad\quad \textbf{else}$$
$$\qquad\qquad\qquad y \Leftarrow V$$
$$\qquad\qquad\qquad x \Leftarrow V$$
$$\qquad\qquad\qquad z \leftarrow \text{value resulting from applying}$$
$$\qquad\qquad\qquad\qquad \text{the operation INFO}(p) \text{ to } x \text{ and } y$$
$$\qquad\qquad\qquad V \Leftarrow z$$

**Algorithm 5.13**
Evaluation of an arithmetic expression represented as a nonempty binary tree

## Exercises

1. Devise an algorithm that takes a binary tree represented with LEFT and RIGHT pointers and produces a copy of the binary tree with FATHER pointers. In other words, the tree

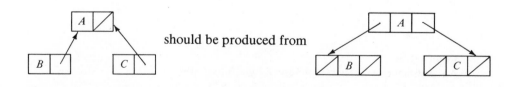

should be produced from

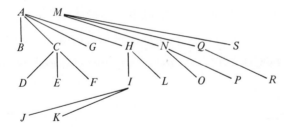

**Figure 5.19**
Exercise 4

2. Devise an algorithm to compute the height of a binary tree (see page 203).

3. Modify Algorithm 5.12 so it prints trees with all the window dressing of Figure 5.16.

4. Design an algorithm that prints forests by placing all nodes at a given level as far to the left as possible. For example, assuming that the tree was drawn on a plotter or that connecting lines were added by hand to printed output, the forest of Figure 5.7(a) would appear as in Figure 5.19.

5. How would Algorithm 5.12 print the tree in Figure 5.20? Design a tree-printing algorithm that produces a more aesthetic output! [*Hint*: Combine the idea of Algorithm 5.12 with that in the previous exercise.]

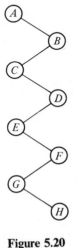

**Figure 5.20**
Exercise 5

6. Rewrite Algorithm 5.12 using iteration and a stack in place of recursion.

7. Rewrite Algorithm 5.13 using recursion for the traversal instead of the stack $S$.

8. Devise an algorithm that changes a tree into its mirror image (so that left and right are reversed).

### 5.2.2 Threaded Trees

In our conventional representation of trees with pointers from fathers to sons we find it generally inconvenient to move up from sons to fathers and to move laterally between brothers. Of course we might choose to use both the FATHER pointers and the LEFT/RIGHT pointers, giving us three pointers for each node in a tree. This would be overkill, however, because in most instances we need to move up or laterally in a tree in only restricted ways; in any case, the storage requirements for the pointers of the tree would then be increased by 50 percent. In this section we discuss a more economical means of facilitating upward and lateral movement in trees.

First of all, notice that any binary tree represented with LEFT/RIGHT pointers contains many unused pointers. In a binary tree with $i$ nodes there are (obviously) $2i$ pointers. However, each of the nodes except the root has only one father and hence only one pointer pointing at it. Thus only $i - 1$ of the pointers are nonnull, so that $2i - (i - 1) = i + 1$ of the LEFT/RIGHT pointers are unused in a binary tree of $i$ nodes. We can utilize this wasted space.

For motivation, consider Figure 5.7(b), a forest being represented by a binary tree in the natural way. Observe that we could use the brother pointer of the youngest son to point to its father, as shown in Figure 5.21. This pointer is otherwise unused, and our use of it necessitates only that we be able to distinguish

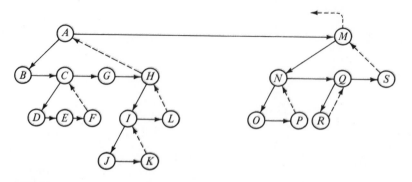

**Figure 5.21**
Figure 5.14 shown with extra pointers added from youngest son to father

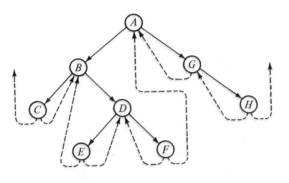

**Figure 5.22**
A threaded tree

it from other brother pointers. The distinction can be easily made by using an extra bit in each tree node to tell us whether the node is a youngest son—that is, whether the brother pointer is being used to indicate the father. Looking at Figure 5.21 more carefully and considering it as a binary tree, we notice that the newly added dashed right pointers indicate the successors of the respective nodes in inorder (ignore for the moment the dashed arrow from $M$, which leads nowhere). Furthermore, this suggests that any unused LEFT pointers might well be used to point to the inorder predecessors of the respective nodes. Figure 5.22 shows a binary tree in which the unused LEFT/RIGHT pointers are so utilized; such pointers are called *threads* and such trees are called *threaded binary trees*.

In order to distinguish the threads from the regular pointers we need to add two additional bits to each node: LTH and RTH are **true** or **false** according as LEFT and RIGHT are threads or pointers, respectively [see Figure 5.23(a)]. Thus the threaded tree of Figure 5.22 would be represented as in Figure 5.23(b). The addition of these two bits per node is an excellent investment, as we shall see.

We must decide what to do with the leftmost and rightmost threads—that is, the left thread from the node that has no inorder predecessor (the first node in inorder) and the right thread from the node that has no inorder successor (the last node in inorder). These are the threads that we left dangling in Figures 5.22 and 5.23(b). However, the decision will hinge on the various algorithms that use the threads, and so we will postpone the issue temporarily.

Given the nature of the threads, the most obvious algorithms to construct are those for determining the inorder successor and predecessor of a given node. These algorithms are, in a sense, mirror images of each other, so we arbitrarily choose to design the successor algorithm first. Given a node in a binary tree, where is its inorder successor to be found? Since inorder traverses the (unthreaded) tree in the order left subtree, root, and right subtree, the successor of a node $n$ with RIGHT($n$) $\neq$ **nil** is the first node in the inorder traversal of its right

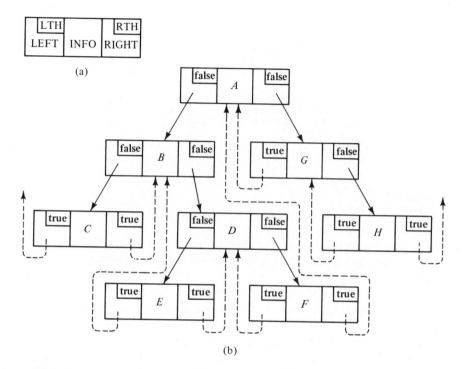

(a)

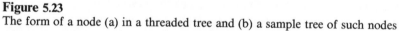

(b)

**Figure 5.23**
The form of a node (a) in a threaded tree and (b) a sample tree of such nodes

subtree; thus if RIGHT($n$) $\neq$ **nil** the inorder successor of $n$ is found by repeatedly following LEFT pointers beginning at RIGHT($n$). If, on the other hand, the unthreaded tree had RIGHT($n$) = **nil**, the threaded-tree representation would have RIGHT($n$) pointing to the inorder successor. Recalling that in a threaded tree the test for an empty LEFT or RIGHT pointer becomes a test on the LTH or RTH fields, respectively, the procedure in Algorithm 5.14 for inorder successor becomes apparent.

> **procedure** *INORDERSUC*($n$) **returns pointer**
> $\quad s \leftarrow$ RIGHT($n$)
> $\quad$ **if not** RTH($n$) **then while not** LTH($s$) **do** $s \leftarrow$ LEFT($s$)
> $\quad$ **return** $s$

**Algorithm 5.14**
A procedure to compute the inorder successor of a node $n$ in a threaded binary tree

For an algorithm to compute the inorder predecessor of a node $n$ we observe that we are computing the inorder successor in the mirror-image tree (the tree obtained when LEFT and RIGHT are consistently interchanged. Thus the desired algorithm is obtained by reversing the roles of left and right in Algorithm 5.14, as shown in Algorithm 5.15.

> **procedure** *INORDERPRED*($n$) **returns pointer**
>     $s \leftarrow$ LEFT($n$)
>     **if not** LTH($n$) **then while not** RTH($n$) **do** $s \leftarrow$ RIGHT($s$)
>     **return** $s$

**Algorithm 5.15**
A procedure to compute the inorder predecessor of a node $n$ in a threaded binary tree. The procedure is obtained from Algorithm 5.14 by interchanging the roles of left and right.

With these two simple algorithms in mind, we can now resolve the question of what to do with the dangling threads of Figures 5.22 and 5.23(b). The resolution is based on how we might use Algorithms 5.14 and 5.15 to find the first and last nodes (respectively) in inorder, given, as usual, a pointer to the root of the tree. Assume that the pointer to the root of the tree is stored in a location $l$; we would like, for example, *INORDERSUC*($l$) to yield the first node in inorder. This means immediately that $l$ must itself have the form of the tree node of Figure 5.23(a) with the fields LEFT, RIGHT, and so on, for these fields are referred to by the algorithms.

Now, the first node in inorder is found by starting at the root and following LEFT pointers down as far as possible. This is exactly what Algorithm 5.14 would do *if* the root of the tree were the RIGHT son of $l$ and RTH($l$) were **false**. The last node in inorder is found by starting at the root and following RIGHT pointers down as far as possible. This is exactly what Algorithm 5.15 would do *if* the root of the tree were the LEFT son of $l$ and LTH($l$) were **false**. We settle this conflict of which pointer will point to the root of the tree by observing that if the root of the tree were given by the LEFT pointer, then Algorithm 5.15 would work correctly as noted, and we can trick Algorithm 5.14 into working correctly by having RIGHT($l$) = $l$. Both LTH($l$) and RTH($l$) must be **false**.

We have effectively considered establishing the header node $l$ of a tree as the inorder predecessor of the first node in inorder and as the inorder successor to the last node in inorder. Thus we set the dangling threads accordingly, and Figure 5.24 shows the completed threaded tree of Figure 5.23(b).

How is the empty tree to be represented? The obvious choice is simply to set the LEFT pointer of the header node, which points to the root of the tree, to be **nil**. However, a threaded tree contains no **nil** pointers, since a pointer that would be **nil** is used to point to the inorder successor or predecessor. For an

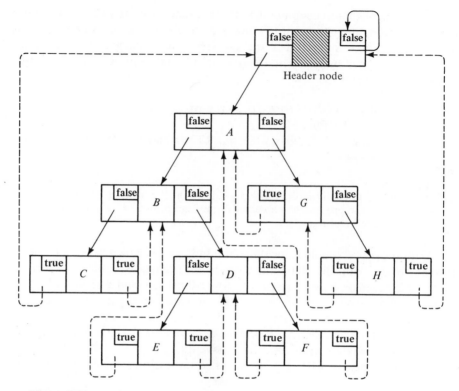

**Figure 5.24**
The completed threaded tree of Figure 5.23 in which the dangling threads
have been resolved

empty tree both the inorder successor and predecessor of the header node should
be the header node itself, and so to be consistent we choose to represent the
empty tree by the header node

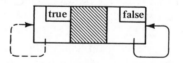

This choice insures that Algorithms 5.14 and 5.15 work correctly when applied to
the header node of an empty tree.

We can apply Algorithm 5.14 repeatedly to obtain an iterative algorithm for
inorder traversal that does not require a stack. This gives us Algorithm 5.16,
whose input is the header node of a threaded binary tree. Notice that this
algorithm also behaves properly on the empty tree.

$$s \leftarrow INORDERSUC(head)$$
**while** $s \neq head$ **do**
       visit $s$
       $s \leftarrow INORDERSUC(s)$

**Algorithm 5.16**
Inorder traversal of a threaded binary tree, given its header node *head*

It is hardly surprising that the inorder successor, predecessor, and traversal algorithms worked out so easily and neatly, since the threads are based on inorder. It *is* surprising that threaded trees also have similar algorithms for finding the father, the preorder successor, and the postorder predecessor of a node, and slightly more complicated algorithms for the preorder predecessor and the postorder successor of a node. We will present algorithms for finding the father of a node and the preorder successor of a node, leaving the remaining algorithms to the exercises.

To find the father of a node $n$, consider the inorder sequence of nodes, and the subsequence consisting of $n$ and all of its descendants:

$$\ldots, x, \ldots, n, \ldots, y, \ldots$$

the inorder sequence of
the subtree rooted
at node $n$

If $n$ is the right son of its father, then by the definition of inorder the father of $n$ must be $x$ (why?). In this case we find $x$ by first following LEFT pointers from $n$ until we get to the "leftmost" descendant of $n$ and then following the left thread from that node. Of course, we do not know if the node thus found is the father of $n$, since $n$ may have been the left son of its father. So, we must check to see if $n$ is the right son of the node thus found:

$q \leftarrow n$
**while not** LTH($q$) **do** $q \leftarrow$ LEFT($q$)
〖at this point $q$ is the leftmost descendant of $n$〗
$q \leftarrow$ LEFT($q$)
**if** RIGHT($q$) $= n$ **then** $q$ is the father

If instead $n$ is the left son of its father, the final test above will fail and we know by the definition of inorder that the father must be $y$. This node is then found in like manner, by following RIGHT pointers from $n$ until we get to the "rightmost" descendant of $n$ and then following the right thread from that node. The complete procedure is given in Algorithm 5.17.

**procedure** *FATHER*(*n*) **returns pointer**
   $q \leftarrow n$
   **while not** LTH($q$) **do** $q \leftarrow$ LEFT($q$)
   ⟦at this point $q$ is the leftmost descendant of $n$⟧
   $q \leftarrow$ LEFT($q$)
   **if** RIGHT($q$) = $n$ **then return** $q$
               **else**
                   ⟦$n$ is the left son of its father⟧
                   $q \leftarrow n$
                   **while not** RTH($q$) **do** $q \leftarrow$ RIGHT($q$)
                   ⟦at this point $q$ is the rightmost descendant of $n$⟧
                   **return** RIGHT($q$)

**Algorithm 5.17**
Finding the father of a node $n$ in a threaded tree

Observe that in an unthreaded tree it would be necessary to traverse the entire tree to find the father of a node (assuming, of course, that we knew the root of the tree; without that knowledge the problem would be impossible to solve). This would require time proportional to the number of nodes in the tree, while Algorithm 5.17 will require at most time proportional to the height of the tree. As will become clear in Section 5.3, this represents a considerable gain in efficiency. Moreover, unlike an algorithm based on one of the traversal algorithms for unthreaded trees, Algorithm 5.17 does not use a stack.

We leave it to the reader to verify that Algorithm 5.17 works correctly on the root of the tree, giving the header node as the result, and on the header node itself, giving the same result.

In order to derive an algorithm for the preorder successor of a node in a threaded tree, we need to characterize the node for which we are searching: The preorder successor of a node $n$ is either

1. its left son, if it exists,

2. its right son, if it exists and the left son does not, or

3. if $n$ is a leaf, the right son of its lowest ancestor $q$ such that $n$ is in the left subtree of $q$ and $q$ has a right son.

The first two cases are obvious from the definition of preorder. To understand the third case, let $x$ be the right son of the specified $q$. Now, observe that by the nature of preorder $x$ must certainly appear somewhere after $n$ in preorder; furthermore suppose there is a node $y$ between them in preorder. The preorder is then

$$\ldots, q, \ldots, n, \ldots, y, \ldots, x, \ldots$$

and $y$ must be a descendant of $q$ (since both $x$ and $n$ are descendants and $y$ comes between them in preorder), but since it precedes $x$, the right son of $q$, $y$ must be in the left subtree of $q$. However, since it appears after $n$ in preorder it cannot be an ancestor of $n$ and it cannot be a descendant of $n$ because we have assumed in case 3 that $n$ is a leaf. Let $a$ be the lowest common ancestor of $n$ and $y$; we have just shown that such an $a$ exists and is in the left subtree of $q$. Since $a$ is the lowest such ancestor and since $n$ precedes $y$ in preorder, $n$ must be in the left subtree of $a$ and $y$ in the right. This contradicts our assumption that $q$ was the lowest ancestor of $n$ such that $n$ was in the left subtree of $q$ and $q$ had a right son. We conclude that no such $y$ can exist and so, indeed, case 3 gives the correct preorder successor for a leaf.

Now we need to relate case 3 to the threads of a threaded tree: the inorder successor of a leaf $n$ (that is, the node pointed to by the RIGHT thread) is the lowest ancestor $r$ of $n$ such that $n$ is in the left subtree of $r$. This follows directly from the definition of inorder and an argument similar to one in the previous paragraph.

The desired algorithm is now apparent: the preorder successor of a node $n$ that is not a leaf is the left son if it exists, and if not, the right son. If $n$ is a leaf, the preorder successor is found by following RIGHT pointers as long as they are threads and then following one more RIGHT pointer. Algorithm 5.18 gives this procedure, simplified slightly by combining cases 2 and 3 in an obvious manner.

**procedure** *PREORDERSUC*( $n$ ) **returns pointer**
**if not** LTH( $n$ ) **then return** LEFT( $n$ )
        **else**
                $q \leftarrow n$
                **while** RTH( $q$ ) **do** $q \leftarrow$ RIGHT( $q$ )
                **return** RIGHT( $q$ )

**Algorithm 5.18**
Preorder successor of a node $n$ in a threaded binary tree

Like the other threaded tree algorithms presented in this section, Algorithm 5.18 works appropriately for the header node, the last node in preorder, and the empty tree. Like Algorithms 5.14, 5.15, and 5.17 it requires at worst time proportional to the height of the tree (instead of the number of nodes in the tree) and no stack.

We conclude this section with some assorted remarks about threaded trees. First, it is easy to insert new nodes at the leaves of a threaded tree or to delete leaves, keeping the tree properly threaded after the insertion or deletion. For example, to insert the node $x$ as the left son of a node $a$ (assuming, of course, that $a$ has no left subtree) we need only do the following:

$$\text{LEFT}(x) \leftarrow \text{LEFT}(a)$$
$$\text{RIGHT}(x) \leftarrow a$$
$$\text{LTH}(x) \leftarrow \text{RTH}(x) \leftarrow \textbf{true}$$
$$\text{LEFT}(a) \leftarrow x$$
$$\text{LTH}(a) \leftarrow \textbf{false}$$

The steps needed for deletion or insertion at other points in the tree are comparable (see Exercises 8 and 9).

Our second remark is that threaded trees do *not* make all traversals easy. For example, it does not appear to be much simpler to determine the preorder predecessor or the postorder successor of nodes in a threaded tree than it is in an unthreaded tree. The only advantages of threaded trees in these cases are that a stack is not needed and we do not need to know the root of the tree.

Our final remark is that it may be useful in certain circumstances to use trees that are only partially threaded. We might, for instance, have an application in which only LEFT threads were needed; in such a case we would not need the RTH field of a tree node, since we would let null RIGHT pointers remain null. We leave it as Exercise 10 to determine which of the various algorithms presented in this section would work (with minor modifications) for some form of partially threaded trees.

### Exercises

1. Verify that Algorithm 5.17 works correctly for the root, header node, and empty tree.

2. Suppose that in writing Algorithm 5.17 we had chosen to test first if $n$ was a left son of its father (by reversing LEFT and RIGHT and LTH and RTH, respectively). How would the resulting algorithm behave when applied to the header node of a tree?

3. Prove that the right thread of a node points to its lowest ancestor $a$ such that the node is in the left subtree of $a$. Prove a parallel result for left threads.

4. Devise an algorithm to thread an unthreaded tree whose nodes have the LTH and RTH fields available (but unused).

5. Prove that a node has a left thread pointing to it if and only if that node has a nonempty right subtree. Prove a parallel result for right threads.

6. Design an algorithm for determining the postorder predecessor of a node in a threaded tree. Your algorithm should not use a stack, and it should require at worst time proportional to the height of the tree.

7. Design an algorithm to copy a threaded binary tree. [*Hint*: Use Algorithm 5.18.]

8. Modify the program fragment given for inserting a node $x$ as the left son of $a$ so that if $a$ already has a left son then $x$ is added "between" $a$ and its left son. That is,

would become

9. (a) Give the code fragment necessary to delete a leaf in a threaded tree so that the tree remains properly threaded.

   (b) Use your answer to (a) to design an algorithm that deletes a given node $x$ (not necessarily a leaf) from a threaded tree so that the inorder of the remaining nodes is unchanged and the tree remains properly threaded.

10. Determine which of the algorithms given in this section can be easily modified to work for partially threaded trees—that is, trees threaded only in either the LEFT or RIGHT pointers.

11. Examine the advantages and disadvantages of threads based on preorder, postorder, or some combination of the orders (for example, left threads pointing to the preorder successor and right threads pointing to the inorder successor).

12. Let $N_1$, $N_2$,..., $N_n$ be the nodes of a threaded binary tree in preorder. Suppose **true** is represented as 1 and **false** is represented as 0 in the LTH and RTH fields. Prove that

$$\sum_{i=1}^{k} (\text{LTH}(N_i) + \text{RTH}(N_i) - 1)$$

is 1 for $k = n$ and is at most 0 for $1 \leqslant k < n$.

13. (a) Let the nodes of two threaded binary trees $T$ and $T'$ be $N_1, N_2, \ldots, N_n$ and $N'_1, N'_2, \ldots, N'_{n'}$, respectively, in preorder. Prove that $T$ and $T'$ are identical if and only if $n = n'$ and $\mathrm{INFO}(N_i) = \mathrm{INFO}(N'_i)$, $\mathrm{LTH}(N_i) = \mathrm{LTH}(N'_i)$, and $\mathrm{RTH}(N_i) = \mathrm{RTH}(N'_i)$ for $1 \leqslant i \leqslant n$. [*Hint*: Use the result of Exercise 12.]

(b) Use the result in (a) to design an algorithm for testing whether two threaded binary trees are identical.

### 5.2.3  Sequential Representations of Trees

The representations of trees considered in Section 5.1 are all based on explicitly storing pointers to the ancestors or descendants (or both) for each node of the tree. In this section, by contrast, we will examine three representations in which we have only a sequential list of the nodes (in one of the traversal orders)

|  |  |
|---|---|
| 1. Gerhard Nothmann | 25. Malke |
| 2. Rudolf Nothman | 26. Abraham Eger |
| 3. Margarete Caro | 28. Ahron Joseph Kauffmann |
| 4. Adolf Nothmann | 29. Perle Fürst |
| 5. Alwine Lustig | 30. Jacob Frensdorff |
| 6. Siegfried Caro | 31. Betty Behrend |
| 7. Anna Kauffmann | 32. Yeshaya Nothmann |
| 8. Simon Nothmann | 33. Yettel Gassmann |
| 9. Friederike Riesenfeld | 40. Löbel Lustig |
| 10. Salomon Lustig | 41. Rachel Ehrlich |
| 11. Amalie Prager | 44. Simon Prager |
| 12. Isaac Caro | 48. Hirsch Isaac Caro |
| 13. Johanna Eger | 49. Tamara |
| 14. Julius Kauffmann | 52. Akiba Eger |
| 15. Friederike Frensdorff | 53. Glückchen Margolies |
| 16. Salomon Nothmann | 54. Eliezer Landsberg |
| 17. Amalie Brieger | 56. Joseph Kauffmann |
| 18. Israel Riesenfeld | 57. Golda Wulff |
| 19. Henriette Schwarz | 58. Wolf Liebmann Fürst |
| 20. Selig Lustig | 59. Sara Herz |
| 21. Fanny Isaak | 60. Michael Meyer Frensdorff |
| 22. Salomon Prager | 61. Hendel Juspa |
| 23. Yette Dresdner | 62. Philipp Behrend |
| 24. Salomon Karo | 63. Blume Levy |

**Figure 5.25**
A "genealogical numbering" for the family tree in Figure 5.1

together with some limited additional information from which the structure of the tree can be deduced. These representations are not as useful as the linked representations already considered.

Genealogists long ago discovered an interesting sequential representation of family trees. The principal of the tree (the one whose ancestors are being presented) is numbered 1; the others in the tree are numbered according to the rule that if a person is numbered $i$, his father is numbered $2i$ and his mother is numbered $2i + 1$. Thus we would encode the family tree of Figure 5.1 as the list of Figure 5.25. Notice that this numbering is according to the level-order traversal of the tree, with the missing numbers corresponding to missing nodes of the tree.

We can use this "genealogical numbering" to get a sequential representation of binary trees. We have a one-dimensional array, say $A$, and we store the root in $A[1]$; in general, the left son of $A[i]$ is stored in $A[2i]$ and the right son in $A[2i + 1]$. It is easy to move down, up, or laterally in a tree thus represented. For example, $A[\lfloor \frac{1}{2}i \rfloor]$ is the father of $A[i]$, $A[i + 1]$ is the brother of $A[i]$ if $i$ is even, and so on. We will, in fact, have an important use for just this representation of a tree in Chapter 8, but for now let us only mention its major disadvantage: storing a tree of height $h$ requires an array of about $2^{h+1}$ locations, regardless of the actual number of nodes in the tree. Thus to represent the tree

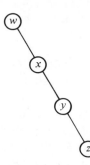

we would have to have an array $A[1]$, $A[2], \ldots, A[15]$ in which $A[1] = w$, $A[3] = x$, $A[7] = y$, $A[15] = z$, and the rest of the array is empty. It is clear that this representation is suitable only for short, full binary trees and not tall, scrawny ones.

A general sequential representation of (not necessarily binary) trees can be deduced from Polish notation for arithmetic expressions. In Section 4.2.3 we remarked that arithmetic expressions could be written uniquely in either prefix or postfix form, and in Section 5.2.1 we observed the connection between these forms and binary tree traversals. In effect, we can reconstruct the tree corresponding to an arithmetic expression from a list of its nodes in either preorder or postorder. Perhaps we can do that in general.

How do we know that, say, the prefix expression $* A + B C$ corresponds to
the tree

while we cannot tell whether the preorder traversal *uvwxy* corresponds to

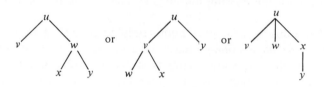

or some other tree? The answer, of course, is that in $* A + B C$ we know by
inspection which nodes are operands and which are operators, while the preorder
traversal *uvwxy* gives us no comparable information about which nodes are leaves
and which are not. The key observation here is that in $* A + B C$ we know the
*degree* of each node—that is, how many sons it has in the tree. If we add that
information to the preorder list of nodes, we can reconstruct the tree.

Suppose we are given the lists

> preorder:  *A B C D E F G H I J K L*
> degree:    4 0 3 0 0 0 0 2 2 0 0 0

We can retrieve the structure of the tree by reasoning as follows. Examining the
lists from right to left and always remembering that the list is in preorder, we see
that I must be the father of *J* and *K*. With the same reasoning *H* must be the
father of *I* and *L*, *C* must be the father of *D*, *E*, and *F*, and finally *A* must be the
father of *B*, *C*, *G*, and *H*. The tree is thus the one in the left of Figure 5.7(a).

The tree could also have been reconstructed from the lists

> postorder:  *B D E F C G J K I L H A*
> degree:     0 0 0 0 3 0 0 0 2 0 2 4

by a similar process. Examining the lists from left to right this time, we see that
*D*, *E*, and *F* must be sons of *C*; *J* and *K* must be sons of *I*; *I* and *L* must be sons
of *H*; and finally *B*, *C*, *G*, and *H* must be sons of *A*.

These two sequential representations of trees are useful only rarely, the
main example being the application to arithmetic expressions already discussed.

Their usefulness is limited because it is impossible to identify the son or father of a particular node except by reconstructing the entire tree.

**Exercises**

1. Why is it possible to thread trees in which the sons of $A[i]$ are $A[2i]$ and $A[2i + 1]$ without the use of LTH and RTH fields?

2. How much additional information is needed per node to reconstruct a binary tree from a preorder list of nodes together with the LTH and RTH values for each node?

3. Prove formally that the postorder list of nodes and their associated degrees uniquely determine the forest from which they were derived.

4. Design an algorithm to construct the binary tree corresponding (via the natural correspondence) to the forest represented by the arrays INFO[$i$] and DEGREE[$i$] given in preorder.

5. Given a (not necessarily binary) tree, we will walk around it as shown in Figure 5.26 by dashed lines, writing the sequence of up and down moves as we walk:

$$\text{DUDDUDUDUUDUDDDUDUUDUU.}$$

We add an extra U at the end of this sequence and consider the sequence to be the preorder list of nodes of a binary tree in which each D is an internal node (with two sons) and each U is a leaf. Using the method discussed in this section, we can reconstruct the binary tree of D's and U's. What is the relationship between the binary tree thus constructed and the original tree? Prove your answer.

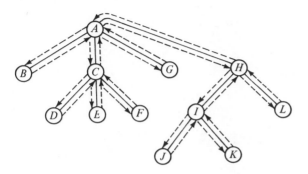

**Figure 5.26**
Exercise 5

## 5.3   QUANTITATIVE ASPECTS OF BINARY TREES

In addition to finding trees invaluable as a data structure when dealing with hierarchical information, we will also find them useful as a tool in the analysis of various searching and sorting algorithms. In this context we will need quantitative measures of certain aspects of binary trees.

Two important quantitative aspects of trees have already been introduced and used in Section 5.1.2. There we defined the *level* of a node $N$ to be 0 if $N$ is the root of the tree and $1 + \text{level}(\text{FATHER}(N))$ otherwise. We also defined the *height* $h(T)$ of a tree $T$ to be the maximum level of any node in $T$.

To understand the usefulness of these and the other quantitative measures that will be introduced, we will make a brief digression to the binary search procedure (Algorithm 1.3) presented in Section 1.3.1. Binary trees and the results discussed in this section will provide the techniques needed for the analysis of this important procedure.

It is useful to describe as a binary tree the sequence of comparisons "$z : x_i$" made by binary search in looking for $z$ in $x_1, x_2, \ldots, x_n$. The first comparison is the root of the tree

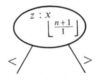

and the left and right subtrees are the tree structures of the succeeding comparisons when $z < x_{\lfloor (n+1)/2 \rfloor}$ and $z > x_{\lfloor (n+1)/2 \rfloor}$, respectively. The tree corresponding to a binary search of the range $x_l, x_{l+1}, \ldots, x_h$ is defined recursively by Equation (5.3).

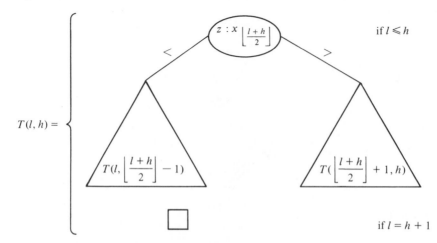

Thus $T(1, 10)$, the tree corresponding to a binary search for $z$ in a table containing ten names, is shown in Figure 5.27.

We can consider such a tree to be a flowchart of the binary search algorithm. The algorithm stops when it finds $z = x_i$ or when it reaches a leaf. The leaves correspond to the "gaps" between $x_i$ and $x_{i+1}$ in which $z$ might lie, with the obvious interpretation for the leftmost and rightmost leaves. For example, in Figure 5.27 the left and right sons of the node $z : x_4$ are leaves corresponding to the conditions $x_3 < z < x_4$ and $x_4 < z < x_5$, respectively.

The height of the tree just derived from binary search has an important significance: it is the number of comparisons that will be made by the algorithm in the worst case on a search that ends unsuccessfully (that is, without finding $z$ among $x_1, x_2, \ldots, x_n$). This leads us to ask what the height of the tree $T(1, n)$ is, or, more generally, what the height of the tree $T(l, h)$ is. We leave it as Exercise 1 to prove by induction that $T(l, h)$ has height $\lceil \lg(h - l + 2) \rceil$. Thus $T(1, n)$ has height $\lceil \lg(n + 1) \rceil$, and this then is the number of comparisons made by binary search in the worst-case unsuccessful search. The worst-case successful search (in which $z$ is found among $x_1, x_2, \ldots, x_n$) will require the same number of comparisons (why?).

To analyze the average behavior of binary search we need some more machinery—some additional quantitative measures of binary trees. For $U_n$, the average number of comparisons $z : x_i$ in an unsuccessful search of $x_1, x_2, \ldots, x_n$, we need to compute the average distance from the root to one of the square nodes at the bottom of the corresponding tree $T(1, n)$. In computing this average we will assume that each of the possible unsuccessful outcomes is equally probable, so that the probability of having the search end at any one of the $n + 1$ square nodes is $1/(n + 1)$. Thus the average number of comparisons for an unsuccessful search is

$$U_n = \frac{\text{sum of levels of the } n + 1 \text{ square nodes}}{n + 1}.$$

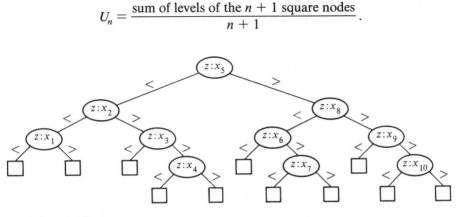

**Figure 5.27**
$T(1, 10)$, the tree corresponding to binary search in a table with ten names

For example, for $n = 10$, Figure 5.27 tells us that this value is

$$U_{10} = \frac{3 + 3 + 3 + 4 + 4 + 3 + 4 + 4 + 3 + 4 + 4}{11}$$

$$\approx 3.55,$$

and so the average unsuccessful search will require about 3.55 comparisons $z : x_i$.

For $S_n$, the average number of comparisons $z : x_i$ in a successful search of $x_1, x_2, \ldots, x_n$, we need the average distance from the root to one of the circular nodes in the tree $T(1, n)$. In parallel to the case of $U_n$, we will assume that each of the possible successful outcomes is equally probable, so that the probability of having the search find that $z = x_i$ is $1/n$ for $i = 1, 2, \ldots, n$. Thus the average number of comparisons for a successful search is

$$S_n = \frac{\text{sum of } (1 + \text{level}) \text{ of each of the } n \text{ circular nodes}}{n}.$$

In our example $n = 10$ we find that

$$S_{10} = [(1 + 0) + (1 + 1) + (1 + 1) + (1 + 2) + (1 + 2)$$
$$\qquad + (1 + 2) + (1 + 2) + (1 + 3) + (1 + 3) + (1 + 3)]/10$$

$$= 2.9,$$

and so the average successful search will require 2.9 comparisons $z : x_i$.

Before continuing our analysis, it is important to understand exactly what the analysis describes. We have made an assumption that is not generally true: only rarely will each of the $x_i$ be the search object with equal likelihood or will the "gaps" (square nodes) be equally likely. Why bother, then, with an analysis whose cornerstone is improbable? The answer is that in writing the binary search procedure we have assumed implicitly that we knew nothing about the probabilities involved. In fact, as we will see in Chapter 7, if something is known about the probabilities we will be able to construct much more efficient search techniques. In the absence of any such knowledge we have made the most reasonable assumption, and we expect it to provide us with the right order of magnitude in the general case.

With the above introduction as motivation, we now backtrack slightly to introduce some general concepts and results that will be applied to binary search at the end of this section. Suppose we are given a binary tree (of circular nodes) as in Figure 5.28(a). We extend the tree by adding a square node, usually called

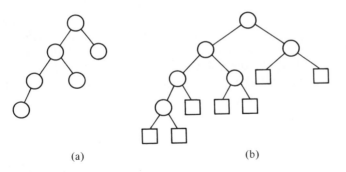

(a)                              (b)

**Figure 5.28**
(a) A binary tree and (b) the same tree extended by the addition of external
nodes

an *external node*, in place of every empty subtree to obtain an *extended binary tree*
as in Figure 5.28(b). The circular nodes are then called *internal nodes*. An
extended binary tree of $n$ internal nodes always contains $n + 1$ external nodes
(Exercise 2).

The *external path length* $E(T)$ of an extended binary tree $T$ with $n$ internal
nodes is the sum of levels of all the external nodes; the *internal path length* $I(T)$ is
the sum of the levels of all the internal nodes. Therefore in Figure 5.28(b) the
external path length is 21 and the internal path length is 9. These sums divided
by the number of external nodes or the number of internal nodes, respectively,
give the averages needed for the analysis of binary search:

$$\text{average distance to an external node} = \frac{E(T)}{n + 1},$$

$$\text{average distance to an internal node} = \frac{I(T)}{n}.$$

It is easy to define both $E(T)$ and $I(T)$ recursively for an extended binary
tree $T$ with $n$ internal nodes:

$I(\square) = 0,$

$$I\left(T = \overset{\circ}{\underset{T_l \qquad T_r}{\triangle \qquad \triangle}}\right) = I(T_l) + I(T_r) + n - 1, \qquad\qquad (5.4)$$

$E(\square) = 0,$

$$E\left(T = \overset{\circ}{\underset{T_l \qquad T_r}{\triangle \qquad \triangle}}\right) = E(T_l) + E(T_r) + n + 1. \qquad\qquad (5.5)$$

To understand the recursive part of the definition (5.4) notice that $T_l$ and $T_r$ contain between them $n - 1$ internal nodes and that adding an internal node above them as the root increases the level of each by 1. Similar reasoning explains the recursive part of the definition (5.5). We can relate $E(T)$ and $I(T)$ by considering the difference $D(T) = E(T) - I(T)$. Using (5.4) and (5.5), we find the results,

$$D(\square) = 0$$

$$D\left( T = \overset{\circ}{\underset{T_l \quad T_r}{\triangle \quad \triangle}} \right) = D(T_l) + D(T_r) + 2, \qquad (5.6)$$

which tell us (Exercise 3) that $D(T) = 2n$ for an extended binary tree $T$ with $n$ internal nodes. Thus

$$E(T) = I(T) + 2n, \qquad (5.7)$$

and hence we need only study one of $E(T)$ or $I(T)$ to determine the properties of both.

We are particularly interested in the range of values of $E(T)$ and $I(T)$. For instance, over all extended binary trees with $n$ internal nodes, one with the maximum value of $I(T)$ is the tree shown in Figure 5.29. In this case

$$I(T) = \sum_{i=0}^{n-1} i = \tfrac{1}{2}n(n - 1)$$

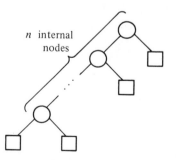

$n$ internal nodes

**Figure 5.29**
An extended binary tree with the largest possible internal and external path lengths

and

$$E(T) = \tfrac{1}{2}n(n + 3).$$

(See Exercise 4.)

Deriving the minimum values of $I(T)$ and $E(T)$ is more complicated, but we will need these results for the analysis of binary search in the average case, as well as for a benchmark against which to compare the binary search trees of Chapter 7. The derivation is based on two observations and some elementary algebra:

**Observation 1**

An extended binary tree of $n$ internal nodes with minimum external path length has all of its external nodes on levels $l$ and $l + 1$, for some $l$. Such a tree is called a *completely balanced binary tree* of $n$ internal nodes.

**Proof**

Let $T$ be an extended binary tree with minimum external path length and let $L$ and $l$ be the maximum and minimum, respectively, of all levels on which external nodes appear, $L \geqslant l$. Suppose $L \geqslant l + 2$. Remove two external nodes that are brothers from level $L$ and make their father an external node. Then put them below an external node on level $l$, which thus becomes an internal node. This process is shown in Figure 5.30. Such a modification of the tree preserves the number of internal and external nodes, but it decreases the external path length by $L - (l + 1)$, which is positive since $L \geqslant l + 2$. This contradicts the minimality of the external path length of $T$.

**Observation 2**

If $l_1, l_2, \ldots, l_{n+1}$ are the levels of the $n + 1$ external nodes in an extended binary tree with $n$ internal nodes, then $\sum_{i=1}^{n+1} 2^{-l_i} = 1$.

**Proof**

The proof is by mathematical induction on the height of an extended binary tree $T$. If the height of $T$ is 0, then $T = \square$, $n = 0$, $l_1 = 0$, and indeed

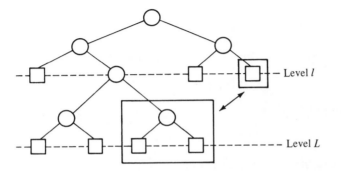

**Figure 5.30**
An example of how the external path length can be decreased if $L \geqslant l + 2$ in Observation 1

$2^0 = 1$. If the height of $T$ is at least 1, we can write

$$T = $$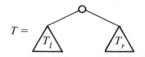

Since the level of each external node of $T_l$ and $T_r$ is one greater with respect to $T$ than with respect to $T_l$ and $T_r$, we see that if $l_1, l_2, \ldots, l_{n_1+1}$ are the levels in $T$ of the $n_1 + 1$ external nodes in $T_l$ and similarly $l_{n_1+2}, l_{n_1+3}, \ldots, l_{n_1+n_2+2}$ are the levels in $T$ of the $n_2 + 1$ external nodes in $T_r$ $(n_1 + n_2 + 2 = n + 1)$, then

$$\sum_{i=1}^{n+1} 2^{-l_i} = \sum_{i=1}^{n_1+1} 2^{-l_i} + \sum_{i=1}^{n_1+n_2+2} 2^{-l_{n_1+1+i}}$$

$$= \frac{1}{2}\left( \sum_{i=1}^{n_1+1} 2^{-(l_i-1)} + \sum_{i=1}^{n_1+n_2+2} 2^{-(l_{n_1+1+i}-1)} \right)$$

$$= \tfrac{1}{2}(1 + 1) \qquad \text{(by induction)}$$

$$= 1$$

as desired.

Now we are ready to compute the minimum external path length of an extended binary tree with $n$ internal nodes and $n + 1$ external nodes. Using the first observation, let there be $k$ external nodes on level $l$ and $n + 1 - k$ on level $l + 1, 1 \leq k \leq n + 1$ (that is, all the external nodes may be on level $l$). The second observation tells us that

$$k2^{-l} + (n + 1 - k)2^{-l-1} = 1$$

and hence

$$k = 2^{l+1} - n - 1. \tag{5.8}$$

Since $k \geq 1$, $2^{l+1} > n + 1$ and, since $k \leq n + 1$, we have $2^l \leq n + 1$; that is,

$$l = \lfloor \lg(n + 1) \rfloor. \tag{5.9}$$

Combining (5.8) and (5.9) gives

$$k = 2^{\lfloor \lg(n+1) \rfloor + 1} - n - 1,$$

and the minimum external path length is thus

$$lk + (l + 1)(n + 1 - k) = (n + 1)\lfloor \lg(n + 1) \rfloor + 2(n + 1) - 2^{\lfloor \lg(n+1) \rfloor + 1}.$$

Letting $\theta = \lg(n + 1) - \lfloor\lg(n + 1)\rfloor$, $0 \leqslant \theta < 1$, the minimum external path length becomes

$$(n + 1)\lg(n + 1) + (n + 1)(2 - \theta - 2^{1-\theta}). \qquad (5.10)$$

The function $f(\theta) = 2 - \theta - 2^{1-\theta}$ is small on the interval $0 \leqslant \theta < 1$; more precisely, $0 \leqslant f(\theta) \leqslant 0.0861$ on this interval.

In light of the above discussion, we can conclude our analysis of binary search by observing that

$$S_n = 1 + \frac{1}{n}[\text{internal path length of } T(1, n)]$$

and similarly

$$U_n = \frac{1}{n + 1}[\text{external path length of } T(1, n)].$$

Combining these equations, Equations (5.6) and (5.10), and the fact that $T(1, n)$ is a completely balanced binary tree with $n$ internal nodes (why?), we derive

$$U_n = \lg(n + 1) + 2 - \theta - 2^{1-\theta}$$
$$S_n = \left(1 + \frac{1}{n}\right)[\lg(n + 1) + 2 - \theta - 2^{1-\theta}] - 1,$$

where $\theta = \lg(n + 1) - \lfloor\lg(n + 1)\rfloor$, as before.

### Exercises

1. Prove that the height of $T(l, h)$ is $\lceil\lg(h - l + 2)\rceil$. [Hint: Use mathematical induction on $h - l$.]

2. Explain why an extended binary tree with $n$ internal nodes contains $n + 1$ external nodes.

3. Prove that $D(T) = 2n$, where $T$ is an extended binary tree with $n$ internal nodes and $D(T) = E(T) - I(T)$.

4. Prove that the extended binary tree of Figure 5.29 has the largest possible internal and external path lengths of all binary trees of $n$ internal nodes.

★5. Consider the sequence of extended binary trees $T_0, T_1, T_2, \ldots$ defined by

$$T_0 = \square, \quad T_1 = \text{(a)}, \quad \text{and} \quad T_n = \text{(b)} \quad \text{for} \quad n > 1.$$

Compute the height, the number of internal and external nodes, and the internal and external path lengths of $T_n$, in terms of $n$.

6. (a) Find a formula analogous to Equation (5.7) for *extended t-ary trees* with $n$ internal nodes.

(b) Find the formula for the minimum and maximum internal and external path lengths for extended $t$-ary trees.

7. Let $l_1, l_2, \ldots, l_{n+1}$ be the levels of the $n + 1$ external nodes in an extended binary tree, and let $s_1, s_2, \ldots, s_n$ be the numbers of external nodes descended from the $n$ internal nodes. Prove that

$$\sum_{i=1}^{n} s_i = \sum_{i=1}^{n+1} l_i.$$

8. What does Observation 1 imply about binary search?

9. Define $F(T)$ as

$$F(\Box) = 0,$$

$$F(T = \text{}) = F(T_l) + F(T_r) + \min(h(T_l), h(T_r)),$$

where $h(T)$ is the height of $T$. Find a simple closed formula for $F(T)$. Generalize your result for $m$-ary trees and for forests, replacing "min" by "sum-of-all-but-the-largest."

## 5.4 REMARKS AND REFERENCES

This chapter has only scratched the surface of the large volume of material on trees. We will go into the application of trees to the organization of data in Chapter 7. In these remarks, therefore, we concentrate on their structural and mathematical aspects.

In Chapter 2 of

Knuth, D. E., *The Art of Computer Programming*, Vol. 1, *Fundamental Algorithms*. Reading, Mass.: Addison-Wesley Publishing Co., 1st ed. 1968, 2d ed. 1973.

the reader will find a great wealth of material on trees, their representations, mathematical properties, and some applications. By and large, the only topics lacking in this work are some recent innovations in algorithms for tree traversals (for example, see Exercise 18 of Section 5.2), which are not of fundamental importance.

For more detailed presentations of the union/find application of Section 5.2 the reader is referred to

> Reingold, E. M., J. Nievergelt, and N. Deo, *Combinatorial Algorithms*: *Theory and Practice*. Englewood Cliffs, N.J.: Prentice-Hall, Inc., 1977.

and to

> Aho, A. V., J. E. Hopcroft, and J. D. Ullman, *The Design and Analysis of Computer Algorithms*. Reading, Mass.: Addison-Wesley Publishing Co., Inc., 1974.

Very deep mathematical analyses can be found in

> Tarjan, R. E., "Efficiency of a Good but Not Linear Set Union Algorithm," *J. ACM*, **22** (1975), 215–225.
>
> Tarjan, R. E., "Applications of Path Compression on Balanced Trees," *J. ACM*, **26** (1979), 690–715.
>
> Knuth, D. E., and A. Schönhage, "The Expected Linearity of a Simple Equivalence Algorithm," *Theoretical Comp. Sci.*, **6** (1978), 281–315.

# Chapter 6

# Memory Management

Your borrowers of books—those mutilators of collections, spoilers of the symmetry of shelves, and creators of odd volumes.

*"The Two Races of Men," Essays of Elia,* Charles Lamb

In this chapter we present the basic techniques used to manage the pool of available storage. In previous chapters we have used the notation "$x \leftarrow$ **getcell**" to indicate that somehow a program can obtain an unused record. The possible implementations of such a feature are quite varied, depending on the characteristics of the use of the storage that will be requested. Different mechanisms of implementation have different strengths and weaknesses, and we must understand them to select an appropriate method.

Memory management strategies are distinguished by several characteristics: the ability to allocate records of varying as opposed to fixed sizes, the ability to recognize which records are currently in use and which are not, and the ability to relocate records that are currently in use. As we discuss the various techniques, we will see how these characteristics interact and interrelate with each other.

All the techniques use the same basic idea and terminology: when a record is in use it is called *active*. A linked list is kept of all *inactive* records—that is, those available for use. This list, called the *free list* or the *list of available records*, is used to dispense records as requested and to keep track of records that have been released.

## 6.1 UNIFORM SIZE RECORDS

When we used "$x \leftarrow$ **getcell**" in previous chapters we intended that the record provided by the statement would conform to whatever size was being used.

246

Although in many cases that assumption is really too strong, we will continue to make it in this section; the next section discusses how records of varying sizes are handled. For the purposes of this section we shall assume that the records being requested and released are of uniform size. Inactive records contain a field called LINK which is used by the memory management routines to connect inactive records in the list of available records. In active records the space of the link field can be used for other purposes.

Since the records to be allocated are uniform in size, they are considered interchangeable in that it makes no difference which particular record is allocated for any particular request. Thus it suffices to maintain the list of available records as a *stack*, popping off records as they are requested and pushing them on again as they are released. Initially, we place onto the stack all the records available for allocation; if these records are *record*[1], *record*[2],...,*record*[*m*] and *free* is to point to the list of available records (that is, to the top of the stack), then this is done by Algorithm 6.1. For our present purpose, the integer values $1, 2, \ldots, m$ also serve as pointer values (see Section 4.1.1 on the implementation of lists in parallel arrays).

> *free* ← 1
> **for** $i = 1$ **to** $m - 1$ **do** LINK(*record*[*i*]) ← $i + 1$
> LINK(*record*[*m*]) ← **nil**

**Algorithm 6.1**
Initialization of the list of available records

A request for a record "$x$ ← **getcell**" is thus handled by Algorithm 6.2. Notice that when all the records have been allocated, the allocation procedure indicates an overflow condition. If records are never made inactive, or *released*, by the requesting program, **overflow** will occur on the $(m + 1)$st call to **getcell**. Usually, however, programs release records explicitly—by a call to the memory management routines—or implicitly—by ignoring inactive records and expecting the memory management routines to find them. These two possibilities are discussed in the subsections that follow.

> **procedure** *GETCELL* **returns pointer**
>     **if** *free* = **nil then overflow**     〚no space available〛
>                **else**
>                     $x$ ← *free*
>                     *free* ← LINK(*free*)
>                     **return** $x$

**Algorithm 6.2**
Implementation of **getcell**

**Exercise**

Modify Algorithm 6.2 so that Algorithm 6.1 is no longer needed. In other words, rewrite Algorithm 6.2 so that it initially allocates records *sequentially* from $record[1]$, $record[2]$, ..., $record[m]$, switching to the linked allocation only later.

### 6.1.1   Explicit Release

In the simplest memory management routines, records are explicitly released when they are no longer needed. Algorithm 6.3 shows how a record thus released would be added to the list of available records.

$$\textbf{procedure } RELEASE(x)$$
$$\text{LINK}(x) \leftarrow free$$
$$free \leftarrow x$$

**Algorithm 6.3**
Procedure to return record $x$ to the list of available records

While this method is the easiest for the memory management routines, it can be inconvenient for programs using the records. For example, to release an entire list of records it is necessary that they be released one by one unless they are (fortuitously) already linked together by the LINK field (see the Exercise). Furthermore, since each record may conceivably be part of several structures simultaneously, it is not always clear when a record should be released.

One obvious solution to this problem is to keep track of how many pointers currently point to a given record. This information could be stored in the record itself in a new field called REF (for "reference counter"), which is initialized upon allocation. In this strategy we know that a record has become inactive when its REF field becomes zero. Of course, this technique will require that every time a pointer variable has its value changed, the corresponding record must have its REF field updated. Thus the simple assignment of pointer variables $p \leftarrow q$ would require

$$\text{REF}(q) \leftarrow \text{REF}(q) + 1$$
$$\text{REF}(p) \leftarrow \text{REF}(p) - 1$$
$$\textbf{if } \text{REF}(p) = 0 \textbf{ then } \text{RELEASE}(p)$$
$$p \leftarrow q$$

This is inconvenient, to say the least.

Furthermore, considerable extra storage is occupied by the REF fields. This could be alleviated somewhat by using the REF fields only in the header nodes of lists or trees, but then our use of pointers into the list or tree would be restricted

—we would not be able to have shared sublists or subtrees. Actually, the most significant objection to the use of reference counters is that circular (self-referencing) structures are effectively prohibited: they will always have a nonzero REF field, even when they are indeed inactive. For these reasons reference counters are useful only in limited circumstances.

**Exercise**

Design an algorithm to return the records in a circular linked list to the list of available records. Assume that the list records are connected by the LINK field. Your algorithm should require only a constant number of operations, no matter the length of the list.

### 6.1.2 Garbage Collection

Instead of having programs explicitly release records they no longer need, it is typically more convenient simply to request records as needed and to ignore those no longer needed. Of course, sooner or later the list of available records will become empty, and a request will lead to **overflow** in Algorithm 6.1. When this happens, the memory management routines will begin a process called *garbage collection*, in which all the inactive records are determined and gathered together to form a new list of available records. Garbage collection is accomplished by *marking* all the active records, that is, all those that may be reached (via a sequence of pointers) from pointer variables within the using programs. The records that remain unmarked after this process are considered to be inactive ("garbage").

Algorithms 6.4 and 6.5(a) give procedures to mark all the active records and collect all inactive records into a new list of available records. The marking is done by a recursive traversal algorithm reminiscent of a recursive preorder tree traversal (see Section 5.2). As each record $r$ is visited it is marked by setting a bit ACTIVE($r$) to **true**; initially, of course, all these bits are set to **false**. The bits used for marking constitute a one-bit-per-record overhead necessary in all garbage collection schemes. These bits may be stored in the records themselves or in a separate bit table. The collection phase links all records marked **false** into the new list of available records. Notice that if we still have *free* = **nil** after the collection procedure, there is an **overflow**.

Notice that we have implicitly assumed that all the pointers in a given record can be recognized. This is a strong assumption that may cause difficulties in practice; unfortunately, there is no other way to mark all the active records.

The marking method of Algorithm 6.5(a) has a very serious shortcoming: because it is recursive it will need (implicitly) a push down stack at the very moment that memory is at a premium (otherwise, why would we be collecting the

⟦mark all records inactive⟧
**for** $i = 1$ **to** $m$ **do** ACTIVE(*record*[*i*]) ← **false**
⟦mark all accessible records active⟧
**for** program pointer variables $p$ **do** MARK($p$)
⟦form list of all inaccessible records⟧
*free* ← **nil**
**for** $i = 1$ **to** $m$ **do**
        **if not** ACTIVE(*record*[*i*])
            **then**
                    LINK(*record*[*i*]) ← *free*
                *free* ← *i*
    **if** *free* = **nil then overflow**

**Algorithm 6.4**
Garbage collection

    **procedure** *MARK*($p$)
        **if** $p \neq$ **nil and not** ACTIVE($p$) **then**
            ACTIVE($p$) ← **true**
                **for** every pointer $q$ in record $p$ **do** MARK($q$)

**Algorithm 6.5(a)**
Marking for garbage collection

inactive records?). However, we can eliminate the stack by making the algorithm nonrecursive and reserving a few extra bits per record for a clever programming trick.

The trick is to store the stack in the pointer fields of the very records being marked; the few extra bits per record allow us to do this without loss of information. We will assume that each record $r$ contains an extra field COUNT($r$) which is large enough to contain a count from 1 to $n$, the number of pointers per record; the purpose of this field will be clear in a moment. The heart of the trick is the use of the pointer fields of the records being traversed as the stack: each pointer is "inverted" as we follow it from one record to the next; the pointers are restored to their original values as we retrace our way back. At any moment a record will have at most one pointer whose value has been modified—the value of the COUNT field indicates which. Algorithm 6.5(b) shows the details of this marking technique. The notation $v(1)$ used there means the first pointer in the record pointed to by $v$; similarly, $u(\text{COUNT}(u))$ means the COUNT($u$)th pointer in the record pointed to by $u$. Figure 6.1 shows a typical list structure with a stack stored in it by this trick.

It is best to think of the algorithm as a traversal of the records in which the pointer $v$ indicates the current record and the pointer $u$ indicates the top record on the "stack" that keeps track of the path from the origin to the current record. The path back to $p$, the starting record, consists of $v, u_0, u_1, u_2, \ldots, u_j, p$, where

**procedure** *MARK*( *p* )
$u \leftarrow$ **nil**      ⟦first record of the "stack"⟧
$v \leftarrow p$      ⟦current record⟧
*down* $\leftarrow$ **true**
**repeat**
    **if** *down* **then**
        **if** $v =$ **nil or** ACTIVE($v$) **then**
            ⟦prepare to go up⟧
            *down* $\leftarrow$ **false**
        **else**
            ⟦mark and go down⟧
            ACTIVE($v$) $\leftarrow$ **true**
            COUNT($v$) $\leftarrow$ 1
            $T \leftarrow v(1)$
            $v(1) \leftarrow u$
            $\left. \begin{matrix} u \leftarrow v \\ v \leftarrow T \end{matrix} \right\}$      ⟦add $v$ to top of "stack"⟧
    **else if** COUNT($u$) $< n$  **then**
            ⟦next pointer⟧
            $T \leftarrow u(\text{COUNT}(u))$
            $u(\text{COUNT}(u)) \leftarrow v$      ⟦restore original value⟧
            COUNT($u$) $\leftarrow$ COUNT($u$) + 1
            $v \leftarrow u(\text{COUNT}(u))$
            $u(\text{COUNT}(u)) \leftarrow T$
            *down* $\leftarrow$ **true**      ⟦prepare to go down⟧
        **else**
            ⟦go up⟧
            $T \leftarrow u(\text{COUNT}(u))$
            $u(\text{COUNT}(u)) \leftarrow v$      ⟦restore original value⟧
            $\left. \begin{matrix} v \leftarrow u \\ u \leftarrow T \end{matrix} \right\}$   ⟦pop up "stack"⟧
  **until** $u =$ **nil**

**Algorithm 6.5(b)**
Marking for garbage collection. The notation $v(1)$ means the first pointer in the record pointed to by $v$, while $u(\text{COUNT}(u))$ means the $\text{COUNT}(u)$th pointer in the record pointed to by $u$.

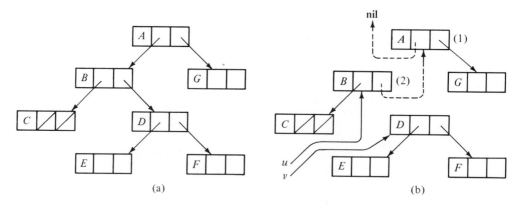

**Figure 6.1**
Storing the stack within the structure being traversed, as in Algorithm 6.5(b):
(a) the structure; (b) the structure while processing node $D$. The stack is
shown by dashed pointers; nonzero COUNT fields are in parentheses.

$u_0 = u$ and $u_{i+1} = u_i(\text{COUNT}(u_i))$; notice that $p(\text{COUNT}(p)) = $ **nil**, thus when
$u = $ **nil** the stack is empty.

Since it temporarily modifies the structure in memory, Algorithm 6.5(b)
must be used with great care in situations where multiple users share the same
data.

**Exercises**

1. Rewrite Algorithm 6.5(a) so that it uses a stack instead of recursion.

2. Design a marking algorithm based on "backtracking." First, mark all records
   directly accessible by program pointer variables. Then scan $record[1]$,
   $record[2],\ldots,record[m]$ as follows: maintain a counter $i$ which is initially 1
   and iterates by (a) adding 1 to $i$ if $record[i]$ is unmarked, (b) marking all
   records pointed to by $record[i]$ if $record[i]$ is marked and then setting $i$ to the
   minimum of $i + 1$ and the index positions of all records accessible from
   $record[i]$ and just marked in this step. Is this algorithm more or less efficient
   than Algorithms 6.5(a) and 6.5(b) with respect to storage? With respect to
   time?

3. Combine the ideas in the algorithms of the previous two exercises to obtain an
   algorithm in which a stack of *fixed size* is used until it is full, at which point
   the strategy of Exercise 2 is employed.

4. In what order are the nodes in Figure 6.2 marked by Algorithm 6.5(b),
   assuming the algorithm starts at $A$?

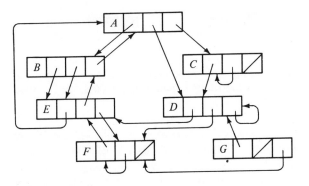

**Figure 6.2**
Exercise 4

5. Specialize Algorithm 6.5(b) for the case of one pointer per record (called LINK).

6. Specialize Algorithm 6.5(b) for the case of two pointers per record (called ALINK and BLINK).

7. Design an inorder binary tree-traversal algorithm based on the idea of Algorithm 6.5(b)—that is, one that temporarily modifies the tree during the traversal, instead of using recursion or a stack.

8. Design a recursive algorithm to copy an arbitrary list structure. Assume that each record has an unused pointer field NEWLOC that can be used during the copying. Assume further that all the NEWLOC fields are initially **nil**. [*Hint*: Look at Algorithm 6.5(a).]

## 6.2 DIVERSE SIZE RECORDS

In most applications the records needed at different times or by different programs have diverse sizes, hence the uniform size techniques discussed in the previous section are limited in usefulness. We can get around this problem by making the fixed size of our records large enough to satisfy any possible demand that could occur, but this could waste a great deal of memory due to *internal fragmentation*: the unavailability of memory because it is scattered around unused and unusable in active records throughout memory. The amount of internal fragmentation could be reduced by maintaining several lists of available records, one for each of a few possible record sizes. This strategy is worthwhile only when there are a limited number of possible record sizes. In this section we discuss the problem of allocating diverse size records when there will be too many sizes for the multiple-list approach to be feasible. In Section 6.3 we discuss a limited form of the multiple-list technique.

0          15,000        35,000        55,000        75,000    90,000  100,000

**Figure 6.3**
A possible memory configuration. The dark regions are active; there are
50,000 inactive words, but no request larger than 20,000 words can be filled
due to external fragmentation.

The memory management strategies discussed in this section suffer very
little from internal fragmentation (but see the discussion of $\varepsilon$ in Section 6.2.1).
However, they do suffer from some degree of *external fragmentation*: the inactive
areas of memory may become broken into pieces that are too small to satisfy most
requests. To understand this pervasive problem, consider the following situation.
Suppose 100,000 words are available to our memory management system and
through requests and releases the inactive words are spread throughout memory
as shown in Figure 6.3. If a request for 25,000 words occurs, it cannot be filled
because there is no contiguous block of words large enough, even though in total
only 50,000 words of storage are active.

In Section 6.1 we did not face problems of external fragmentation because
the records were of uniform size and any request could be satisfied by any
available record. In this section the main emphasis is on how to avoid or
minimize external fragmentation. One technique is to keep a list of free blocks
and search it for each request. The alternative is *memory compaction*: all active
records are moved so that there is only one free block. These two approaches are
discussed in the next two subsections.

We will assume here that the memory available for allocation consists of $m$
words $word[1], word[2], \ldots, word[m]$, that requests occur for records or *blocks* of
arbitrary sizes, and that the release of blocks no longer needed is explicit. The
integer values $1, 2, \ldots, m$ serve as the pointer values.

**Exercises**

1. (a) Suppose $word[1], \ldots, word[23]$ are available for allocation and suppose that
       all requests are for records of one or two words. Develop an allocation
       scheme that will not overflow if there are never more than sixteen allocated
       words.

   (b) Suppose there are only 22 words available. Prove that *any* allocation
       scheme can be forced to reach a state in which only fourteen of the words
       are allocated, but a request for a record of two words cannot be honored.

2. As in Exercise 1(a) above, determine, as a function of $n$, the minimum number
   of words needed to guarantee that any request for one- or two-word records
   can be honored, as long as there are never more than $n$ allocated words.

### 6.2.1   Allocation by Search

As in the case of uniform size records, it is convenient to maintain a list of available blocks of words. Initially the list consists of a single record containing all $m$ words. In general, a request for a record is filled by choosing some available record from the list and, if necessary, breaking it into two pieces, one to be allocated and the other to be put back on the list of available records. Although this is the general outline, we need to resolve various questions of implementation such as how to choose a record to be allocated, in what order to maintain the list, and how to return a released record. It is not surprising that these issues are interrelated and must be considered all together rather than individually.

To begin, consider the process of allocating a record of $n$ words to fill a request. In searching through the list of available records, should we be looking for the record closest in size and no smaller than the request (*best fit*) or should we stop at the first record large enough that is encountered (*first fit*)? While the best-fit strategy is plausible, it performs poorly in practice: in order to find the best fit we must search the entire list of available records, or if it is ordered by size, about half of it on the average, to do the allocation (see Section 7.1.1). This can be expensive! Furthermore, by choosing the smallest record large enough to fill the request and making the excess into a new record, we may be causing a proliferation of small, unusable records. The first-fit technique works much better, and we will use it here.

When first fit is used and the search for the record to be allocated is always started at the beginning of the list of available records (that is, from *free*), whatever small records occur through division of large records to satisfy small requests will tend to occur at the front of the list of available records, cluttering it up and increasing the average time required to find a suitable record. It is thus a good idea to keep the list of available records as a circular list and to have *free* point to a "random" record in the list. This can be achieved by just leaving *free* wherever it is after an allocation and beginning the search in the next allocation at that point. We call this method *next fit*.

We can further inhibit the occurrence of small unusable records by not dividing a record when it is only a little larger than the record requested. In other words, we establish a minimum size for records on the list of available space, say $\varepsilon$. Whenever a record of $n$ words is requested and a record of size $k > n$ is found for the allocation, we leave a free record of size $k - n$ only if $k - n \geqslant \varepsilon$; otherwise we fill the request with the entire record of $k$ words.

These ideas are embodied in the allocation procedure given in Algorithm 6.6(a). In this algorithm we have made the list of available records into a circular list using the pointer FLINK; in addition to FLINK, each record on the list contains a field SIZE that gives its size. In an actual implementation these fields might be put into the first several words of the record; for simplicity, however, we will assume that all the fields we will use will fit into the first word. (Notice that the number of words used for these fields will affect the value of $\varepsilon$.) We assume further that the SIZE field of this word is left unchanged while the record is active. In order for the algorithm to work properly we include a sentinel record of SIZE zero on the list of available records; this insures that the list is never empty. Without such a sentinel we would have to complicate the algorithm slightly to compensate for the different things that would need to be done when the final free record was allocated.

**procedure** $ALLOCATE(n)$ **returns pointer**
    ⟦ *free* is a *global* variable; the circular list of available records,
    into which *free* points, is never empty—it always contains a
    sentinel record of SIZE zero⟧
    $p \leftarrow \text{FLINK}(\textit{free})$    ⟦the value of *free* is left from the previous
                                                        allocation⟧
    $q \leftarrow \textit{free}$            ⟦the predecessor of $p$ in the list⟧
    **while** $p \neq \textit{free}$ **and** $\text{SIZE}(p) < n$ **do**
          $q \leftarrow p$
          $p \leftarrow \text{FLINK}(q)$
    **if** $\text{SIZE}(p) < n$ **then overflow**    ⟦no available record is large enough⟧
               **else**
                    **if** $\text{SIZE}(p) - n < \varepsilon$ **then**
                          ⟦allocate entire record: remove it from list⟧
                          $\text{FLINK}(q) \leftarrow \text{FLINK}(p)$
                          $\textit{free} \leftarrow q$
                **else**
                    ⟦allocate right end of record: split it in two⟧
                    $\textit{free} \leftarrow p$
                    $\text{SIZE}(p) \leftarrow \text{SIZE}(p) - n$
                    $p \leftarrow p + \text{SIZE}(p)$
                    $\text{SIZE}(p) \leftarrow n$
                ⟦a record of length $\text{SIZE}(p)$ is reserved,
                beginning at $\textit{word}[p]$⟧
                **return** $p$

**Algorithm 6.6(a)**
Next-fit allocation procedure for diverse size records

We must now devise the procedure to return a released record to the list of available records. The most simplistic technique is to insert the record at an arbitrary position on the list, say at *free*. Unfortunately this will work poorly, since the records of storage available will become smaller and smaller through splittings, until all are too small to satisfy requests of even moderate size. The procedure to return released records must therefore recombine adjacent available records. In order to do the recombination for a newly released record $p$ we must find the records physically adjacent to $p$ in memory; the record $p$ may be combined with one or both, depending on which are currently unallocated.

One approach to the problem of finding the physical neighbors of $p$ is to search the list of available records. We reject this approach immediately, since we want to avoid the time overhead involved in such searches; even if the list of available records were kept in order of physical location we would have to search about half the list, on the average (see Section 7.1.1).

Actually, we can easily determine the record physically following $p$: it begins in $word[p + SIZE(p)]$. The problem is how to know whether it is allocated. By using one extra bit per record we can keep this information in the record itself. We will assume that each record $r$ has a field ACTIVE($r$) that is **true** if the record is currently allocated and **false** if the record is on the list of available records. We are not quite done, however, because just knowing that a record $q$ is on the list does not allow us to delete it easily—we must know its predecessor on the list. For this reason we will maintain the list of available records as a *doubly linked* circular list, recalling that in such a list we can easily delete any element without a search. Thus we will consider the first word of each record to have fields BLINK, SIZE, ACTIVE, and FLINK, where BLINK points to the predecessor of the record on the list of available space, and the other fields are as already described. The BLINK and FLINK fields will be used only for unallocated records, while the SIZE and ACTIVE fields will be in use for all records, allocated or not.

Getting to the record physically preceding record $p$ is only slightly more complicated. We have direct access to the last word in that record, $word[p - 1]$, and so we can store in that location a pointer FRONT to the first word of the record. This allows us to reach the first word of the record physically preceding $p$: it is $word[\text{FRONT}(p - 1)]$. Of course since our goal is the recombination of available records, we will only need to reach this word if the preceding physical record is inactive; furthermore, it is definitely undesirable to have unusable space in the last word of an active record. It is thus better to have the field FRONT defined only for inactive records and to store a copy of the ACTIVE bit of a record in the physically following record. This means that we will consider the first word of each record to have the form

| SIZE | ACTIVE | PREACTIVE | FLINK | BLINK |
|------|--------|-----------|-------|-------|

where PREACTIVE indicates whether the preceding physical record is active.

The structure of a record is thus as shown in Figure 6.4. We have used some additional space per record for these new fields, but only in inactive records, and

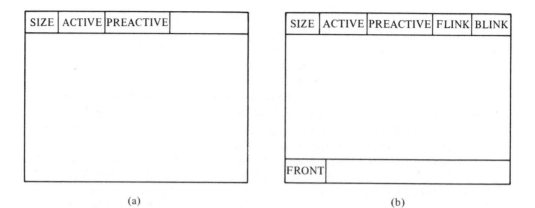

(a)                                    (b)

**Figure 6.4**
The structure of records for memory management by Algorithms 6.6(b) and
6.7: (a) an active record—ACTIVE is **true**; (b) an inactive record—ACTIVE
is **false** and FRONT points to the first word of the record

we gain the important advantage of being able to release records and recombine
them without searching the list of available records.

The modified version of Algorithm 6.6(a) necessary for this new structure is
given as Algorithm 6.6(b). Algorithm 6.7 gives the procedure to release a record,
recombining it with its physical neighbors if necessary. To simplify the release
algorithm we have established a sentinel record in $word[m + 1]$; this record is
considered allocated and thus allows us to use the same algorithm for releasing
records at the boundaries and in the interior. We have also used the record in
$word[m + 1]$ as the sentinel record on the list of available records. Figure 6.5
shows the initial situation. Because of the additional information kept at the
boundaries of each record, this technique has become known as dynamic storage
allocation with *boundary tags*.

We can combine the ideas presented in this section with those presented in
our discussion of garbage collection to eliminate the need for an explicit release
procedure. When overflow occurs on an attempted allocation, we initiate a
marking phase and follow it by a sequential scan of all records to combine
adjacent available records and reform them into a list of available space. Since
the marking phase requires the ability to recognize all the pointers in a record,
and since such an ability would also allow us the flexibility needed to relocate
active records, we will discuss this topic together with compaction in Section
6.2.2.

**Performance.** Almost no analytical results are known about the performance of Algorithms 6.6(b) and 6.7. Instead we must rely on empirical evidence

**procedure** $ALLOCATE(n)$ **returns pointer**

⟦ *free* is a *global* variable; the circular, doubly linked list of
available records, into which *free* points, is never empty—it always
contains a sentinel record of SIZE zero⟧

$p \leftarrow free$     ⟦the value of *free* is left from the previous allocation⟧

⟦find next record that is large enough, if any⟧

**while** $p \neq \text{BLINK}(free)$ **and** $\text{SIZE}(p) < n$ **do** $p \leftarrow \text{FLINK}(p)$

**if** $\text{SIZE}(p) < n$ **then overflow**     ⟦no available record is large enough⟧

   **else**

    ⟦set *free* for next allocation⟧

    $free \leftarrow \text{FLINK}(p)$

    **if** $\text{SIZE}(p) - n < \varepsilon$ **then**

                ⟦allocate entire record: remove
                it from list⟧

                $\text{FLINK}(\text{BLINK}(p)) \leftarrow \text{FLINK}(p)$

                $\text{BLINK}(\text{FLINK}(p)) \leftarrow \text{BLINK}(p)$

    **else**

                ⟦allocate right end of record:
                split it in two⟧

                $\text{SIZE}(p) \leftarrow \text{SIZE}(p) - n$

                $\text{FRONT}(p + \text{SIZE}(p) - 1) \leftarrow p$

                $p \leftarrow p + \text{SIZE}(p)$

                $\text{SIZE}(p) \leftarrow n$

                $\text{PREACTIVE}(p) \leftarrow$ **false**

    ⟦a record of length $\text{SIZE}(p)$ is reserved,
    beginning at $word[p]$⟧

    $\text{ACTIVE}(p) \leftarrow$ **true**

    $\text{PREACTIVE}(p + \text{SIZE}(p)) \leftarrow$ **true**

    **return** $p$

**Algorithm 6.6(b)**
Next-fit allocation procedure with boundary tags

that comes from extensive simulations. For example, it has been observed that, on the average, the **while** loop of Algorithm 6.6(b) goes through less than three iterations. Thus the technique of having *free* be a "roving" pointer pays off. The rest of the allocation procedure clearly requires only a small constant amount of time, as does the release procedure also.

The more important question, however, is the degree to which external fragmentation prevents the memory available from being fully utilized. Of course we can make any allocation seem foolish by examining its behavior on a particularly bizarre sequence of requests and releases; on the average, though, the

**procedure** *RELEASE(p)*
⟦merge *p* with the physically following record, if it is not allocated⟧
*t* ← *p* + SIZE(*p*)    ⟦pointer to physically following record⟧
**if not** ACTIVE(*t*) **then**

⟦delete record *t* from list⟧
FLINK(BLINK(*t*)) ← FLINK(*t*)
BLINK(FLINK(*t*)) ← BLINK(*t*)
**if** *t* = *free* **then** *free* ← BLINK(*t*)
⟦merge records *p* and *t* into *p*⟧
SIZE(*p*) ← SIZE(*p*) + SIZE(*t*)
FRONT(*p* + SIZE(*p*) − 1) ← *p*

**else** PREACTIVE(*t*) ← **false**

⟦merge *p* with physically preceding record, if it is not allocated⟧
**if not** PREACTIVE(*p*) **then**

*t* ← FRONT(*p* − 1)     ⟦pointer to physically
preceding record⟧
⟦merge records *p* and *t* into *t*; *t* is already on the list⟧
SIZE(*t*) ← SIZE(*p*) + SIZE(*t*)
FRONT(*t* + SIZE(*t*) − 1) ← *t*

**else**

⟦put record *p* on the list and mark it inactive⟧
FLINK(*p*) ← *free*
BLINK(*p*) ← BLINK(*free*)
FLINK(BLINK(*free*)) ← *p*
BLINK(*free*) ← *p*
ACTIVE(*p*) ← **false**

**Algorithm 6.7**
Release procedure for boundary-tag method

strategy presented in Algorithms 6.6(b) and 6.7 performs reasonably well. Simulations have shown that if the average storage utilization exceeds $\frac{2}{3}m$ (recall that *m* is the number of words available for allocation), then overflow tends to occur. Also, if the sizes of requests are small compared to *m*, then the memory can be used effectively, even up to 90 percent utilization; a good rule of thumb is to limit the maximum record size to $m/10$.

## Exercises

1. Find sequences of requests and releases with the property that:
   (a) Best fit succeeds where both first fit and worst fit fail.
   (b) First fit succeeds where both best fit and worst fit fail.

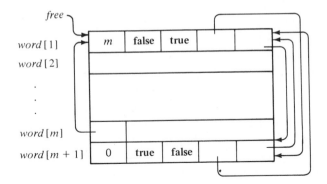

**Figure 6.5**
The initial configuration of memory for the boundary-tag method of dynamic storage allocation

(c) Worst fit succeeds where both best fit and first fit fail.
(Worst fit always uses the largest available block to fill a request.)

2. Suppose that requests and releases are made in a last-in, first-out manner so that no allocated record is released until after all subsequently requested records have been released. Design a memory management strategy for this case. Does the first-fit/best-fit distinction make sense?

3. Rewrite Algorithm 6.6(a) so that it uses best fit instead of next fit.

### 6.2.2 Compaction

We can combine the flexibility of the allocation procedure for diverse size records given in the previous section and the convenience of a garbage-collection technique instead of explicit release. Recall that garbage collection requires the ability to recognize all the pointers in a record; it is not unreasonable to expect this ability in records of fixed size as in Section 6.1.2, but it is not so reasonable for records of varying sizes. If we do have that ability, however, then we can both perform garbage collection and relocate active records. Relocating active records means that we can compact all active records into a contiguous area of memory. This is of little consequence for records of fixed size, since in that case it makes no difference which particular record is used to fill a request, but for records of varying sizes this is of great consequence, since it allows us to combine all available records into a single large record and completely eliminate external fragmentation.

In this section we will assume that all pointers in a record can be recognized. Furthermore, we will use a very simple allocation scheme: there is only one block available at any time, pointed to by *free*. Everything physically before *free*

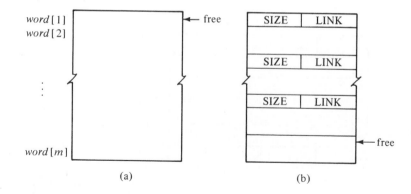

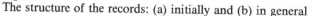

**Figure 6.6**
The structure of the records: (a) initially and (b) in general

in memory is assumed to be active (although parts of it may be inactive), and a request for a record is handled by allocating the front part of the only available record, the remainder of which then becomes the available record. For convenience, each allocated record is assumed to have a SIZE field as before and a LINK field to be used during garbage collection and allocation. Initially we would have the configuration of Figure 6.6(a) with the general case shown in Figure 6.6(b).

Under these conditions the allocation procedure is as given by Algorithm 6.8. The LINK field is set to **nil** for later use—it will always be **nil** except during garbage collection, when it is used to store the stack (see Section 6.1.2), and during relocation, when it is used to store the new location of the record.

> **procedure** $ALLOCATE(n)$ **returns pointer**
>     **if** $free + n - 1 > m$ **then** $COMPACT$
>     **if** $free + n - 1 > m$ **then overflow** ⟦compaction did not help⟧
>                 **else**
>                         ⟦allocate a record of $n$ words⟧
>                         $p \leftarrow free$
>                         $free \leftarrow free + n$
>                         $SIZE(p) \leftarrow n$
>                         $LINK(p) \leftarrow$ **nil**
>                         **return** $p$

**Algorithm 6.8**
Allocation of a record of size $n$

The procedure $COMPACT$, given in Algorithm 6.9, has a multifaceted task. First, it marks all active records by doing a garbage-collection procedure, Algorithm 6.9(a), based on Algorithm 6.5(a) but storing the needed stack in the LINK fields; the LINK field of each active record is thus left nonnull, marking it as active. Next, Algorithm 6.9(b) scans all records (in their physical order), determining a new location for each active record. Then all pointers are changed by Algorithm 6.9(c) to refer to the new (future) locations of the records to which they point. Finally the records themselves are moved to their new locations by Algorithm 6.9(d). Notice that we have implicitly assumed that pointers must point *only* to the first word of a record.

**procedure** *COMPACT*

⟦At this point all LINK fields are null; all available records are marked by leaving the LINK field nonnull. The marking is done using the LINK fields to form the stack.⟧

*top* ← *free*
LINK( *free* ) ← **nil**     ⟦bottom of the stack sentinel⟧
**for** each program variable *P* **do**
    **if** *P* ≠ **nil and** LINK(*P*) = **nil then**
                              ⟦add *P* to stack, since it is not
                              null and not already on the stack⟧
                              LINK(*P*) ← *top*
                              *top* ← *P*
*MARK*     ⟦Apply Algorithm 6.9(a) to mark the active records.⟧

⟦At this point *p* = *free* and *MARK* has left each active record with a nonnull LINK field. The compacting phase begins with assigning new locations for each active record.⟧

*ASSIGN*     ⟦Apply Algorithm 6.9(b) to change all pointers
               to their future values.⟧

⟦The LINK field of each active record now contains its future address; we must change all references accordingly.⟧

*CHANGE*     ⟦Apply Algorithm 6.9(c) to change all pointers
                to their future values.⟧

⟦All pointers are now set. We can now move the active records to their new locations using Algorithm 6.9(d).⟧

*MOVE*
*free* ← *q*     ⟦Reset first available location.⟧

**Algorithm 6.9(a)**
Compaction of unused records

**procedure** *MARK*
    ⟦Mark all active records using the LINK fields to form the stack.⟧
    $p \leftarrow top$
    $top \leftarrow \text{LINK}(top)$
    **while** $top \neq$ **nil do**
        **for** each pointer $P$ in *record*($p$) **do**
            **if** $P \neq$ **nil and** $\text{LINK}(P) =$ **nil then**
                ⟦push $P$ onto stack⟧
                $\text{LINK}(P) \leftarrow top$
                $top \leftarrow P$
      ⟦pop up stack⟧
      $p \leftarrow top$
      $top \leftarrow \text{LINK}(top)$

**Algorithm 6.9(b)**
Marking active records for compaction

**procedure** *ASSIGN*
    ⟦Assign new locations to each active record.⟧
    $q \leftarrow 1$
    $p \leftarrow 1$
    **while** $p \neq free$ **do**
        **if** $\text{LINK}(p) \neq$ **nil then**
            ⟦give $p$ a new location, $q$⟧
            $\text{LINK}(p) \leftarrow q$
            $q \leftarrow q + \text{SIZE}(p)$
        $p \leftarrow p + \text{SIZE}(p)$

**Algorithm 6.9(c)**
Assignment of new locations

**procedure** *CHANGE*
    ⟦Change all pointers to their future values.⟧
    $p \leftarrow 1$
    **while** $p \neq free$ **do**
        **if** $\text{LINK}(p) \neq$ **nil then**
            **for** each pointer $P$ in *record*($p$) **do**
                **if** $P \neq$ **nil then** $P \leftarrow \text{LINK}(P)$
      $p \leftarrow p + \text{SIZE}(p)$

**Algorithm 6.9(d)**
Modifying pointers

**procedure** *MOVE*
⟦Move all active records to their new locations.⟧
$p \leftarrow 1$
**while** $p \neq free$ **do**
     $t \leftarrow \text{LINK}(p)$
     **if** $t \neq$ **nil then**
          ⟦reset LINK for next time and move the record⟧
          $\text{LINK}(p) \leftarrow$ **nil**
          copy contents of *record*($p$) into *record*($t$)
     $p \leftarrow p + \text{SIZE}(t)$

**Algorithm 6.9(e)**
Moving records

**Exercises**

1. In the last line of Algorithm 6.9(e), why do we have "$p \leftarrow p + \text{SIZE}(t)$" instead of "$p \leftarrow p + \text{SIZE}(p)$"?

2. Suppose we have records of fixed size containing two pointers each, ALINK and BLINK. Design an algorithm for compaction. You may not use an explicit stack, and there is no LINK field as used in Algorithm 6.9, but your algorithm is to move and compact the structure into an entirely new area of memory, and you can distinguish between addresses in the new and old areas.

## 6.3   THE BUDDY SYSTEM

In the two previous sections we have presented memory management techniques for fixed record size (Section 6.1) and for diverse record sizes (Section 6.2); in this section we describe a system that compromises between these two extremes. In the *buddy system* there is a sequence $s_0, s_1, s_2, \ldots, s_t$ of possible record sizes. Of course, we could use the techniques of Section 6.1 by maintaining entirely separate areas for each record size, but it would then be difficult to use the words of memory for different size records at different times. The key idea of the buddy system is that records are allowed to be split or recombined in a carefully controlled manner: In general we have $s_n = s_{n-1} + s_{n-j}$, so that a record of $s_n$ words can be split only into two records of $s_{n-1}$ and $s_{n-j}$ words, respectively. The two records are called *buddies*. One of the buddies will be used to satisfy the request (with further splittings as needed) and the other will be free. If, as a result of a release, two buddies are both free, they are combined to restore the original record (which may be further recombined).

   The buddy system, if properly implemented, allows easy access to the buddy of a newly released record. Of particular interest in this regard is the *binary buddy system*, in which $s_n = s_{n-1} + s_{n-1}$, $s_0 = 1$, and hence $s_k = 2^k$. This method is well suited to binary computers because there is a simple relationship between the size of a record, its location (that is, the location of its first word), and the location of its buddy. To derive this relationship, consider a memory of $2^4 = 16$ locations $word[0]$, $word[1], \ldots, word[15]$. This record of $2^4$ locations is split into buddies as shown in Figure 6.7. This small example suggests that the first word of a record of $2^k$ words is at a location divisible by $2^k$. Indeed, this is true in general, for it is true initially and if a block of $2^{k+1}$ words begins at $word[l]$, the buddies of $2^k$ words begin at $l$ and at $l + 2^k$; since $2^{k+1}$ divides $l$, we see that $2^k$ divides $l$ and $l + 2^k$. We thus know that the buddy of a record of $2^k$ words beginning at location $l$ begins at either $l + 2^k$ or $l - 2^k$. Which one? The answer depends on $l$: If $2^{k+1}$ does not divide $l$, then $l$ cannot be the first location of a record of $2^{k+1}$ words; that is, it cannot be the first of two buddies. Hence

$$\begin{array}{l}\text{first word of the} \\ \text{buddy of a record of} \\ 2^k \text{ words beginning} \\ \text{at location } l\end{array} = \begin{cases} l + 2^k, & \text{if } 2^{k+1} \text{ divides } l, \\ l - 2^k, & \text{if not.} \end{cases}$$

The advantage of the binary buddy system on a binary computer now becomes apparent: the first word of the buddy is easily computed from the binary representation of the address of the first word of the record: it is $l$ **xor** $2^k$.

In the remainder of this section we outline the rather straightforward algorithms for allocation and release in the binary buddy system. Suppose we have $m = 2^p$ words. We will maintain $p + 1$ linked lists $free[0]$, $free[1], \ldots, free[p]$ for the available records of sizes $2^0, 2^1, \ldots, 2^p$, respectively. In addition, to simplify the reservation algorithm, we will have a dummy list $free[p + 1]$ which is always non-**nil**. Initially each of the $p + 2$ lists is empty except $free[p]$ which contains the record of $2^p$ words that comprises the available memory. To fill a request for an $n$-word record we first find the smallest $j \leqslant p$ such that $2^j \geqslant n$ (that is, $j \geqslant \lceil \lg n \rceil$) and $free[j]$ is not empty. If no such $j$ exists, the request cannot be filled and an **overflow** condition is indicated. If such a $j$ exists, the first record on $free[j]$ is removed and is used to fill the request. If $n > 2^{j-1}$, the entire record is allocated, otherwise it is split into buddies of $2^{j-1}$ words; one of these buddies is used to fill the request (with further splitting if $2^{j-2} \geqslant n$) and the other is put onto $free[j - 1]$.

When a record of $2^j$ words is released, we see if its buddy is on $free[j]$. If so, we combine the buddies and continue further combinations with buddies as possible. When all such combinations (if any) are done, the record is then added to the appropriate list.

As in other memory allocation schemes, it is convenient to maintain the lists of available records as doubly linked lists, since it is then easy to delete an

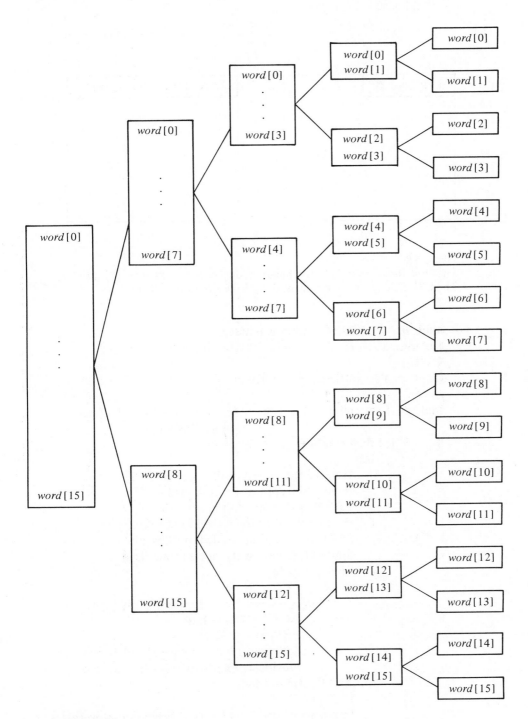

**Figure 6.7**
A tree showing how $2^4 = 16$ words are split into buddies of sizes 8, 4, 2, and 1

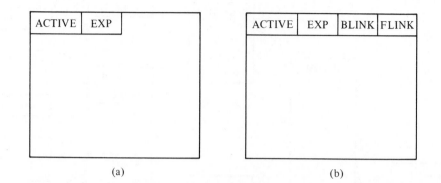

Figure 6.8
The structure of records in the binary buddy system: (a) active, (b) inactive.
The EXP field gives the exponent of 2 of the size of the record

**procedure** *BINBUDRES*($n$) **returns pointer**
    ⟦reserves a record of $2^k$ words, $k = \lceil \lg n \rceil$⟧
    $k \leftarrow \lceil \lg n \rceil$
    ⟦find smallest sufficient nonactive block⟧
    $j \leftarrow k$
    **while** *free*[$j$] = **nil do** $j \leftarrow j + 1$    ⟦at worst, *free*[$p + 1$] $\neq$ **nil**
                                                   stops this loop⟧
    **if** $j = p + 1$ **then overflow**
              **else**
                      ⟦reserve the record, splitting as needed⟧
                      remove a record of $2^j$ words from
                      FREE[$j$] and set $l$ to the subscript of
                      the first word in the record (See Exercise 1)
                      ⟦split the record as needed, putting the
                      unused buddies on the appropriate lists⟧
                      **while** $j > k$ **do**
                            $j \leftarrow j - 1$
                          $b \leftarrow l$ **xor** $2^j$    ⟦address of buddy⟧
                          ACTIVE ($b$) $\leftarrow$ **false**
                          EXP($b$) $\leftarrow j$
                          add record $b$ to list *free*[$j$] which was
                          previously empty (why?) (See Exercise 1)
                  ACTIVE($l$) $\leftarrow$ **true**
                  EXP($l$) $\leftarrow k$
                  ⟦a record of size $2^k$ is reserved beginning at *word*[$l$]⟧
                  **return** $l$

Algorithm 6.10
Reservation in the binary buddy system

element at any point in the list without knowing its predecessor. As in the other schemes also, the size of a record will be stored in all records, but here it is only necessary to store the exponent of 2. The structure of the records in the binary buddy system is shown in Figure 6.8.

Algorithms for reservation and release in the binary buddy system are given as Algorithms 6.10 and 6.11, respectively. They work in a straightforward manner, and the details left to Exercise 1 are not too difficult to work out.

**procedure** $BINBUDREL(l)$      ⟦releases the record beginning at $word(l)$⟧
    $k \leftarrow EXP(l)$
    $b \leftarrow l$ **xor** $2^k$      ⟦address of buddy⟧
    **while** $k \neq p$ **and not** $ACTIVE(b)$ **and** $EXP(b) = k$ **do**
        ⟦combine with buddy⟧
        delete $b$ from list $free[k]$ (see Exercise 1)
        $k \leftarrow k + 1$
        $l \leftarrow \min(l, b)$
        $b \leftarrow l$ **xor** $2^k$
    add $b$ to list $free_k$ (see Exercise 1)

**Algorithm 6.11**
Release in the binary buddy system

In the more general buddy system (that is, when we have $s_n = s_{n-1} + s_{n-j}$), it is necessary to store in a block an indication of whether it is a left or right buddy. This could be done with a single bit, LEFT, but the problem arises of how to set the bit in a block resulting when two buddies are recombined into a larger block. To solve this problem, we introduce a *storage bit* S. When a block is split, its LEFT bit is stored in the S bit of its left son and its S bit is stored in the S bit of its right son. When the buddies are recombined, their S bits are used to set the S and LEFT bits of the resulting block. (See Exercise 5 below.)

## Exercises

1. Fill in the list-manipulation details that have been omitted from Algorithms 6.10 and 6.11. See Section 4.2.1 for two possible implementations of doubly linked lists. What changes (if any) need to be made to the algorithms if dummy elements are used?

2. Can it happen that two records of $2^k$ words each can be adjacent and inactive simultaneously (that is, not buddies)?

3. How can the test in the **while** loop of Algorithm 6.11 be simplified by having a dummy record in $word[2^p + 1]$?

4. By experiments determine whether the binary buddy system is superior to the *Fibonacci buddy system*, which is based on the recurrence $F_1 = 1$, $F_2 = 2$, $F_n = F_{n-1} + F_{n-2}$.

5. Use the LEFT and S bit approach to implement the Fibonacci buddy system of Exercise 5.

## 6.4   REMARKS AND REFERENCES

The particular needs of any system will necessarily make memory management idiosyncratic, so in this chapter we have introduced only the basic and most general techniques. A great many strategies have been proposed, and the reader is referred to the following for detailed discussions of implementations and simulation results:

Knuth, D. E., *The Art of Computer Programming*, Vol. 1, *Fundamental Algorithms*. Reading, Mass.: Addison-Wesley Publishing Co., 1st ed. 1968, 2d ed. 1973.

Standish, T. A., *Data Structure Techniques*. Reading, Mass.: Addison-Wesley Publishing Co., Inc., 1980.

These books contain extensive bibliographies that the reader may find useful.

# Chapter 7

# Searching

Thou preparest a table before me.    *Psalms* 23:5

This chapter examines various techniques for organizing data to be searched efficiently. Rather than introduce new storage structures, we will demonstrate the *use* of structures such as arrays, linked lists, and trees in the organization of tables to be searched. For a given problem, the particular choice of a storage structure depends on the nature of the storage medium (internal memory, magnetic tape, disk, or whatever), on the nature of the data being organized (does it change through insertions or deletions, is it alphabetic or numeric, are some elements more likely to occur as search objects than others, and so on), and on the requirements of the search (must it be as fast as possible on the average or in the worst case, how much information is available, and the like). We will present the most important table organizations and discuss their relative strengths and weaknesses.

Throughout this chapter we will assume that we are searching a *table* of *elements*, where each element has a collection of values associated with it, one value for each of a number of *attributes*. One of these attributes is special and will be called the *key* or *name* that is used to refer to the element. For example, the key might be a person's name and the other attributes might be the address, phone number, age, sex, salary, and so on. A *search* will consist of examining the table to find the element (or elements) containing some particular value for the key. For simplicity of presentation, our storage structures will consist *only* of the value of the keys. If the structure is a record, other attributes can be imagined to be stored in additional fields; if the structure is an array, other attributes can be stored in parallel arrays (see Sections 1.3 and 4.1.1). The value of the key for which we are searching will be assumed to be in the variable $z$. The result of the search routine will be to choose between two pieces of code—one to be executed when $z$ is found in the table, and the other when it is not found.

*O* **Notation.** In order to compare searching and sorting algorithms, we will need to analyze their behavior under various assumptions. Often, such an analysis is difficult or impossible to do precisely and we will need some notation (in this and the following chapter) that allows us to describe an analysis in terms of the *growth rate* of a function. We will write the equation $f(n) = O(g(n))$, read "$f(n)$ is big-oh of $g(n)$," when we mean that $f(n)$ grows no faster than $g(n)$ as $n$ gets large. Specifically, it means that as $n$ gets bigger and bigger, the value of $|f(n)/g(n)|$ does not grow without bound. The similar notation $f(n) = o(g(n))$ ("little-oh") means that as $n$ gets bigger and bigger, the value of $|f(n)/(g(n)|$ gets closer and closer to zero; in other words, $f(n) = o(g(n))$ means that $f(n)$ grows strictly more slowly than $g(n)$.

The equation $f(n) = h(n) + O(g(n))$ is shorthand for $f(n) - h(n) = O(g(n))$ and, similarly, $f(n) = h(n) + o(g(n))$ is shorthand for $f(n) - h(n) = o(g(n))$. Thus, for example, when we state that the average number of "probes" into a table is $\lg(n + 1) + o(1)$, we are saying that the number of probes differs from $\lg(n + 1)$ by an amount that becomes vanishingly small as $n$ gets big. When we state that the external path length of some tree is $E_n = (2 \ln 2) n \lg n + O(n)$, we are saying that $E_n$ differs from $(2 \ln 2)n \lg n$ by an amount that grows no faster than $n$ as $n$ gets large.

These notations, once they are mastered, give us a way to describe the dominant behavior of an algorithm without getting mired in minutiae.

## 7.1  SEARCHING LISTS

In many cases the simplest and most obvious storage structure for a table is a list data structure. The algorithms are short and there is little overhead in wasted space; there is *no* wasted space if an array of the proper size is used to implement the list. The simplicity is deceptive, however, since a number of subtleties need examination.

Given that we are going to organize a table as a linear list, we can vary only two things: the order of the elements in the list and the implementation as either sequential or linked. The elements may be in no particular order, in an order based on their frequencies as search objects, or in their natural order (alphabetic or numeric). The cases in which the elements are in no particular order or in an order based on their frequencies are similar, and we treat them together in Section 7.1.1. The case of a linear list in natural order is the subject of Section 7.1.2.

We will find the same trade-off as in Chapter 4 between sequential and linked lists: the ease with which we can directly access any element in a sequential list makes it an ideal structure under certain conditions, while under other conditions the ease of insertion and deletion makes a linked list more appropriate. Situations also occur in which both efficient access and ease of modification are needed simultaneously. Such situations cannot be properly handled by the techniques presented here, since they require a compromise between the two conflicting properties; techniques for these situations are discussed in Section 7.2.2.

### 7.1.1 Sequential-Search Techniques

The essence of sequential search is obvious: begin at the start of the linear list and examine each element in turn to see if it is the one sought. This process continues either until the element is found or until all the elements in the list have been checked. For a sequential list (array) $x_1, x_2, \ldots, x_n$ and a search object $z$ this amounts to the loop given in Algorithm 7.1(a). When the list is linked rather than sequential and $L$ points to its first element, we have the loop of Algorithm 7.1(b) instead. In either case the order of the elements in the list does not affect the correctness of these algorithms, only the amount of time they require.

$i \leftarrow 1$
**while** $i \leq n$ **and** $z \neq x_i$ **do** $i \leftarrow i + 1$
**if** $i \leq n$ **then**　〚found: $x_i$ is what we want〛
　　　　**else**　〚not found: $z$ is not among $x_1, x_2, \ldots, x_n$〛

(a) Sequential-list search

$p \leftarrow L$
**while** $p \neq$ **nil and** $z \neq \mathrm{KEY}(p)$ **do** $p \leftarrow \mathrm{LINK}(p)$
**if** $p \neq$ **nil then**　〚found: $p$ points to the element〛
　　　　**else**　〚not found: $z$ is not in the list〛

(b) Linked-list search

**Algorithm 7.1**
Sequential search of a list for the element $z$

In order to compare the performance of linear search under various orderings, we must establish some basis of comparison. The basis that we choose is the number of probes into the list: a *probe* is a comparison between the search object $z$ and the key of some element of the table being searched. We will evaluate search strategies by the number of probes required to find an object, both in the worst case and on the average. The amount of work in searching for an element is not entirely in the probes, of course, but the total work done is usually proportional to the number of probes, since only a constant number of operations are done per probe. Thus if one method requires $n$ probes and another $\lg n$ probes, the constants of proportionality are not too important, since $\lg n$ grows so much more slowly than $n$. The constants of proportionality become important only when comparing methods with comparable numbers of probes, or when the constant is so large as to make a method impractical for table sizes occurring in practice. The number of probes will be our measure of efficiency, not only for the linear search algorithms of this section but throughout the chapter.

Let us examine Algorithm 7.1, the simple linear-search algorithm, from the perspective of the number of probes. In this case a probe is a comparison "$z = x_i$" or "$z = \text{KEY}(p)$". In the worst case each of the $n$ elements in the table must be compared to the search object; thus $n$ probes will be required for the worst-case search, no matter whether it ends successfully or unsuccessfully. In the best case a successful search will end after a single probe (this is true for all search strategies), but an unsuccessful search will still require $n$ probes. To consider the behavior on an "average" search, we must have a precise notion of what an "average" instance of a search is. For example, on an unsuccessful search, $n$ probes will always be required, so that the worst case, best case, and average behavior coincide—all are $n$ probes. A successful search will end having found $z$ in the table, and the average such search will depend on the probability that $x_i$ is the search object. If $p_i = \text{Pr}\,(z = x_i)$, that is $p_i$ is the probability that $z = x_i$, then the expected number of probes required is

$$1p_1 + 2p_2 + 3p_3 + \cdots + np_n, \qquad (7.1)$$

since $i$ probes are required to find $z$ if it is $x_i$, an event which happens with probability $p_i$. We can summarize the behavior of linear search with the following table:

|                              | *Worst Case* | *Best Case* | *Average Case* |
|------------------------------|:------------:|:-----------:|:--------------:|
| *Search Ends Successfully*   | $n$ probes   | 1 probe     | $\sum_{i=1}^{n} ip_i$ probes |
| *Search Ends Unsuccessfully* | $n$ probes   | $n$ probes  | $n$ probes     |

Usually we are most concerned with the average behavior of a search strategy, and this means evaluating formula (7.1) for particular probabilities $p_1, p_2, \ldots, p_n$. In the absence of any other information we may as well assume that each $x_i$ is equally likely to be the object of the search—that is, that $p_i = 1/n$ for all $i, 1 \leqslant i \leqslant n$. Formula (7.1) becomes

$$1 \times \frac{1}{n} + 2 \times \frac{1}{n} + \cdots + n \times \frac{1}{n} = \frac{1}{n} \sum_{i=1}^{n} i = \frac{1}{n} \frac{n(n+1)}{2} = \frac{n+1}{2},$$

which tells us that an average successful search will use $(n + 1)/2$ probes—that is, it will examine about half the entries in the table.

From the table above it is clear that the only possibility for improving the linear-search strategy is to arrange the list so that the value of formula (7.1) is made small. The minimum value occurs when

$$p_1 \geqslant p_2 \geqslant \cdots \geqslant p_n.$$

To understand why, suppose that $p_2 > p_1$. Then, since $p_1 + 2p_2 > p_2 + 2p_1$, the value of (7.1) is reduced if we interchange the elements $x_1$ and $x_2$ (and hence $p_1$ and $p_2$). Similarly, if any $p_i > p_j$, for $i > j$, then $ip_i + jp_j > ip_j + jp_i$ so the value of (7.1) is reduced by interchanging $x_i$ and $x_j$. It follows that for (7.1) to be minimized there can be no $p_i > p_j$ for $i > j$. In other words, the best arrangement of the table elements is in nonincreasing order of their probabilities as search objects.

Changing the order of the elements can have an enormous effect on the number of probes in an average successful search. Consider, for example, $p_1 = \frac{1}{2}$, $p_2 = \frac{1}{4}$, $p_3 = \frac{1}{8}, \ldots, p_{n-1} = 1/2^{n-1}$, $p_n = 1/2^{n-1}$. If the elements are in decreasing order of probability, then the expected number of probes for a successful search is

$$\sum_{i=1}^{n} ip_i = \sum_{i=1}^{n-1} \frac{i}{2^i} + \frac{n}{2^{n-1}} = 2 - \frac{1}{2^{n-1}} < 2.$$

If the elements are in increasing order of probability, this value becomes

$$\sum_{i=1}^{n} ip_{n+1-i} = \sum_{i=2}^{n} \frac{i}{2^{n+1-i}} + \frac{1}{2^{n-1}} = n - 1 + \frac{1}{2^{n-1}}.$$

The difference is staggering; in the first case only a small constant number of probes are expected on a successful search, while in the second case the entire table generally needs to be examined. In practice, of course, such extreme probabilities are unlikely, but it is just as unlikely that the probabilities will be uniform (all equal to $1/n$). Thus it is worthwhile, when possible, to arrange in decreasing order of probability a table that will be searched sequentially. (Exercise 4 describes a generalization of this result that is applicable when searching tables stored on tape.)

Just as few tables are governed by uniform access probabilities, it is seldom possible to determine the access probabilities a priori. Even empirical observation may not give an accurate picture of the probabilities if they tend to fluctuate in time. We can still take advantage of nonuniform access probabilities, however, by allowing the order of the elements in the table to change in such a way that those frequently accessed move to the front of the table while those infrequently accessed move toward the rear. Such a table is called *self-organizing*.

The basic idea is that when an element $z$ is accessed, it is moved to a position closer to the beginning of the table. The amount of work to do this movement must be reasonable, and so the possibilities are limited. If the table is a sequential list, we can interchange $z$ with its predecessor, the *move-ahead-one strategy*, or we can interchange $z$ with the first element of the list, the *interchange-to-the-front strategy*. If the table is a linked list we can, in addition, simply move $z$ to the front of the table, the *move-to-front strategy*. (This strategy is too time-consuming in sequential lists.)

The move-ahead-one strategy, applicable to either linked or sequential lists, works very well to keep the table well arranged if the table order is not too far from the desired order. However, it will take quite a while initially for the popular elements to move to the beginning of the list, since they move so slowly. On the other hand, the move-to-front strategy, applicable only to linked lists, works well to order the elements quickly when they are far out of order, but it causes erratic behavior in a table that is nearly in order. The interchange-to-the-front strategy is even worse in this regard. Thus it is most reasonable to apply the move-to-front strategy initially until the table order settles down a bit and continue thereafter with the move-ahead-one strategy.

**Exercises**

1. We can make Algorithm 7.1(a) a bit faster on unsuccessful searches by first adding $z$, the object of the search, at the end of the list and rewriting the loop:

$$x_{n+1} \leftarrow z$$
$$i \leftarrow 1$$
**while** $z \neq x_i$ **do** $i \leftarrow i + 1$
**if** $i \leqslant n$ **then**   $[\![$found: $x_i$ is what we want$]\!]$
              **else**   $[\![$not found: $z$ is not among $x_1, x_2, \ldots, x_n]\!]$

   Compare the efficiency of this algorithm with that of Algorithm 7.1(a). Use the operator count analysis of Section 1.4.2.

2. Suppose the access probabilities for $x_1, x_2, \ldots, x_n$ are $p_i = c/i, 1 \leqslant i \leqslant n$, where $c = 1/\sum_{i=1}^{n} 1/i$. Compare the behavior of sequential search when the table is in decreasing order by access probability to that when the table is in increasing order by access probability. [*Hint:* You may consider that $\sum_{i=1}^{n} 1/i \approx \ln n$.]

3. Design and analyze the behavior of a sequential-search algorithm on a circular linked list in which the search begins wherever it left off on the previous search [see Algorithm 6.6(b)].

4. Assume there is a cost $c_i$ involved in examining an element $x_i$ *after* the position of $x_i$ has been found. Prove that for successful searches the minimum average total cost is achieved when the elements are arranged so that

$$\frac{p_1}{c_1} \geqslant \frac{p_2}{c_2} \geqslant \cdots \geqslant \frac{p_n}{c_n}.$$

5. Perform simulations to determine the behavior of the various strategies for self-organizing lists.

### 7.1.2  Lists in Natural Order

In many instances it is possible to maintain the list in some natural order (such as numeric or alphabetic) and it is almost always advantageous to do so. Algorithm 7.1 can be speeded up somewhat for *unsuccessful* searches because it can now stop when it discovers the first element beyond $z$ in the natural order, rather than go all the way to the end of the list. The algorithm becomes somewhat simpler, too, if we append an element with the key value $\infty$ at the end of the list (compare this with Exercise 1 of Section 7.1.1). Algorithm 7.2 gives this modified version of Algorithm 7.1. The behavior of Algorithm 7.2 is no different from that of Algorithm 7.1 with respect to *successful* searches.

$i \leftarrow 1$
**while** $z > x_i$ **do** $i \leftarrow i + 1$
**if** $z = x_i$ **then**   [found: $x_i$ is what we want]
             **else**   [not found: $z$ is not among $x_1, x_2, \ldots, x_n$;
                    it would have been just before $x_i$]

(a) Sequential-list search in an ordered list

$p \leftarrow L$
**while** $z > \text{KEY}(p)$ **do** $p \leftarrow \text{LINK}(p)$
**if** $z = \text{KEY}(p)$ **then**   [found: $p$ points to the element]
             **else**   [not found: $z$ is not in the list; it
                    would have been just before the
                    element pointed to by $p$]

(b) Linked-list search in an ordered list

**Algorithm 7.2**
Sequential search of an ordered list for the element $z$. *The list is assumed to contain a dummy final element whose value is $\infty$.*

The improvement for unsuccessful search times in tables in the natural order is minor in contrast to the fact that a single probe into the table can now get a good deal more information than when the table is in some other order. If we find that $z > x_i$, then $z$ cannot be one of $x_1, x_2, \ldots, x_i$, or if we find that $z < x_i$, then $z$ cannot be one of $x_i, x_{i+1}, \ldots, x_n$. This observation is behind the binary search procedure, Algorithm 1.3, discussed in Sections 1.3.1 and 5.3. For convenience we repeat binary search here as Algorithm 7.3 and summarize the results of Section 5.3 with the following table:

|  | *Worst Case* | *Best Case* | *Average Case* |
|---|---|---|---|
| *Search Ends Successfully* | $\lceil \lg(n+1) \rceil$ probes | 1 probe | $\left(1 + \dfrac{1}{n}\right)\lg(n+1) + o(1)$ probes |
| *Search Ends Unsuccessfully* | $\lceil \lg(n+1) \rceil$ probes | $\lfloor \lg(n+1) \rfloor$ probes | $\lg(n+1) + o(1)$ probes |

$l \leftarrow 1$
$h \leftarrow n$
*found* $\leftarrow$ **false**
**while** $l \leq h$ **and not** *found* **do**
    ⟦at this point $1 \leq l \leq h \leq n$ and $z$ is not
    among $x_1, \ldots, x_{l-1}, x_{h+1}, \ldots, x_n$⟧
    $m \leftarrow \left\lfloor \dfrac{l+h}{2} \right\rfloor$
    **case**
        $z < x_m$: $h \leftarrow m - 1$
        $z > x_m$: $l \leftarrow m + 1$
        $z = x_m$: *found* $\leftarrow$ **true**
**if** *found* **then**   ⟦$x_m$ is the element sought⟧
      **else**   ⟦not found: at this point $l = h + 1$ and $x_h < z < x_{h+1}$⟧
**Algorithm 7.3**
Binary search

Recall that binary search requires the direct access to all parts of the list that sequential representation allows, but linked representation does not. This means that Algorithm 7.3 cannot be adapted to linked lists and so it may not be applicable in certain circumstances. For example, when order is important, it is easy to allow insertions and deletions if the table is implemented as a linked list; in the case of a sequential list, as required by binary search, the insertion or deletion of an element is a time-consuming operation. This is only the first instance in this chapter of the conflict between fast search times and ease of modification—this conflict is ubiquitous in designing search algorithms.

Recall also that the values given for the number of probes in the average successful and unsuccessful searches were based on the assumption that for successful searches each of the $n$ elements was equally probable as the place for the search to end. As we mentioned in discussing the average behavior of sequential search, this assumption is rarely justified in practice but is the only reasonable one in the absence of any information. When the access probabilities are known, it is possible to use the analogs of binary search discussed in Section 7.2.1.

So far we have considered only the frequency with which various elements will be the object of a search or the frequency with which the search will fail in a specified way. There are, however, other useful statistical properties that the table elements often have in practice. For example, in looking up the name "Smith" in a phone book we would be unlikely to probe first at the half point and then at the three-quarters point, and so on, as in binary search. Instead, we would assume that under normal conditions the name "Smith" would be found near the end of the listings, and we would begin our search nearer to the expected location of the search object. This idea leads to *interpolation search*.

For simplicity let us assume that we are dealing with numeric values $x_1 < x_2 < \cdots < x_n$ that are uniformly distributed in the range $(x_0, x_{n+1})$; extensions to nonnumeric keys and nonuniform distributions are not difficult (see Exercise 11, for example). If we are searching such a table for $z$, where $x_0 < z < x_{n+1}$, the uniform distribution suggests that we interpolate linearly to determine the expected location of $z$. That expected location is $n(z - x_0)/(x_{n+1} - x_0)$, and this is where we should probe first. In general, assume we know that $x_l < z < x_h$; then we should probe at location

$$l + \frac{z - x_l}{x_h - x_l}(h - l - 1).$$

Following a style similar to binary search, Algorithm 7.4 implements interpolation search. Notice, however, that it is more convenient to begin at $l = 0$ and $h = n + 1$ and have the condition for continuing the loop be "$h - l > 1$"; this happens because the range under consideration, $x_l$ to $x_h$, is now exclusive of the endpoints, while in binary search it was inclusive of the endpoints.

The number of probes in the best and worst cases of interpolation search is obvious, and the analysis of the average number involves mathematics that is well beyond the level of this book, so we will simply summarize the results in the following table:

|  | Worst Case | Best Case | Average Case |
|---|---|---|---|
| *Search Ends Successfully* | $n$ probes | 1 probe | lg lg $n$ probes |
| *Search Ends Unsuccessfully* | $n$ probes | 2 probes | lg lg $n$ probes |

$$l \leftarrow 0$$
$$h \leftarrow n + 1$$
*found* $\leftarrow$ **false**
**while** $h - l > 1$ **and not** *found* **do**
$\qquad$ ⟦at this point $0 \leqslant l \leqslant h \leqslant n + 1$ and $x_l < z < x_h$⟧

$$m \leftarrow \left\lceil l + \frac{z - x_l}{x_h - x_l}(h - l - 1) \right\rceil$$

$\qquad$ **case**
$\qquad\qquad z < x_m: h \leftarrow m$
$\qquad\qquad z > x_m: l \leftarrow m$
$\qquad\qquad z = x_m: \textit{found} \leftarrow \textbf{true}$
**if** *found* **then** $\quad$ ⟦$x_m$ is the element sought⟧
$\qquad\quad$ **else** $\quad$ ⟦not found: at this point $h = l + 1$ and $x_l < z < x_{l+1}$⟧

**Algorithm 7.4**
Interpolation search for $z$, $x_0 < z < x_{n+1}$ in a table $x_1 < x_2 < \cdots < x_n$ of elements uniformly distributed over $(x_0, x_{n+1})$

It is crucial to understand that the average behavior of interpolation search is a much different "average" than was considered for either sequential search or binary search. In those cases the table of elements was fixed and the average was over the occurrences of the various elements as search objects; in this case the average is over search objects *and* tables whose elements follow a certain statistical pattern. This means that if a particular table does not follow that pattern, the average search in that table may be much poorer than expected.

Simulation results suggest that interpolation search is inferior to binary search unless the tables are much larger than most tables that occur in practice in internal memory. The difficulty stems from the greatly increased cost per probe in interpolation search as compared to binary search, where a probe can be made in only a few machine instructions. Thus, for example, in a table of 4000 elements binary search will need twelve probes and interpolation search an average of four probes, but since each of the latter probes is at least twice as costly, there will be little savings (in fact, interpolation search will be *slower* than binary search on small tables). However, for larger tables, the fact that lg lg $n$ grows so much more slowly than lg $n$ will outweigh the increased cost per probe, making interpolation search worthwhile.

### Exercises

1. Under whatever probabilistic assumptions you deem reasonable, analyze the average unsuccessful search time for Algorithm 7.2.

2. Not all of the results in the table summarizing the performance of binary search were derived in Section 5.3. Derive the number of probes in the best

case for both successful and unsuccessful searches and the worst case for successful searches.

3. What happens if we change the statement "$m \leftarrow \lfloor (l + h)/2 \rfloor$" to "$m \leftarrow l$" in Algorithm 7.3? What about "$m \leftarrow h$"?

4. In first attempts at writing a binary search procedure novice programmers often use "$h \leftarrow m$" and "$l \leftarrow m$" instead of "$h \leftarrow m - 1$" and "$l \leftarrow m + 1$", respectively. What kind of error does this cause in Algorithm 7.3? Modify the rest of Algorithm 7.3 so that it works correctly with "$h \leftarrow m$" and "$l \leftarrow m$".

5. Give a proof that Algorithm 7.3 works correctly for both successful and unsuccessful searches.

6. Rewrite Algorithm 7.3 so that instead of three pointers, $l$, $h$, and $m$, only two values are kept—the current position $m$ and its rate of change $\delta$. After an unequal comparison the algorithm should do something like $m \leftarrow m \pm \delta$ and $\delta \leftarrow \delta/2$. Analyze this search algorithm.

★ 7. Analyze the behavior of Algorithm 7.3 with the statement "$m \leftarrow \lfloor (l + h)/2 \rfloor$" changed to "$m \leftarrow \text{rand}(l, h)$" where $\text{rand}(l, h)$ is a randomly chosen integer, $l \leqslant \text{rand}(l, h) \leqslant h$.

8. Algorithm 7.3 requires a division by 2 at each iteration to evaluate $\lfloor (l + h)/2 \rfloor$. Use the relation $F_h = F_{h-1} + F_{h-2}$, $F_0 = 0$, $F_1 = 1$ to replace the division by 2 by a subtraction. Analyze the worst- and average-case behavior of the resulting search algorithm. [*Hint:* The search effectively bisects the interval at the $(\sqrt{5} - 1)/2$ point rather than in the middle.]

9. Under what conditions will binary search be less efficient than sequential search, considering the *total search time*, not just the number of probes? (See Exercise 16 of Section 1.4.3.)

10. Let $x_1 < x_2 < \cdots < x_n$ be an array of integers in increasing order. Give an algorithm that requires only time proportional to $\log n$ time to determine if any $x_i = i$, finding such an $i$ if one exists. Can it be done in less than a logarithmic number of probes?

11. Design a search technique requiring at most $2 \lg n$ probes in the worst case and $2 \lg \lg n$ probes on the average. [*Hint:* Use a hybrid of binary search and interpolation search.] When might such a technique be useful?

12. Analyze the average-, worst-, and best-case behavior of the following variant of interpolation search. The first probe is at $x_{\lceil pn \rceil}$, $p = (z - x_0)/(x_{n+1} - x_0)$. If $z > x_{\lceil pn \rceil}$, then $z$ is successively compared with $x_{\lceil pn + i\sqrt{n} \rceil}$, $i = 1, 2, \ldots$ . Similarly, if $z < x_{\lceil pn \rceil}$, $z$ is successively compared

with $x_{\lceil pn-i\sqrt{n}\rceil}$, $i = 1, 2, \ldots$ . In any case, the subtable of size $\sqrt{n}$ thus found is then searched by applying the same technique recursively.

★13. Modify Algorithm 7.4 so that it interpolates correctly for any probability distribution over the table elements, subject to the restriction that the cumulative distribution function $F(x) = \Pr(X \leq x)$ is continuous. [*Hint:* The distribution of $F(x)$ is uniform over $(0, 1)$ and $F$ is order preserving.]

## 7.2 BINARY SEARCH TREES

As we saw in Section 5.3, the order in which the elements are examined by binary search is governed by an implicit binary tree on the table elements (see Figure 5.27, for an example). Similarly, we could view linear search on a table in natural order and interpolation search in terms of a binary tree. For linear search the implicit tree is as shown in Figure 7.1, and for interpolation search the tree varies with the search argument—$x_{\lceil pn\rceil}$ is the root, $p = (z - x_0)/(x_{n+1} - x_0)$, and so on. In this section we will discuss the benefits of making such a binary tree structure explicit instead of implicit. Those benefits are twofold. For tables in which the elements do not change through insertions and deletions (*static tables*), an explicit tree structure can be used to take advantage of a known distribution of the frequency of access of the elements. For *dynamic tables* that change through insertions and deletions, an explicit tree structure gives us the flexibility to search the table in logarithmic time and to make insertions and deletions also in logarithmic time.

A *binary search tree* is a binary tree in which the inorder traversal of the nodes gives the elements stored therein in the natural order. In other words, every node $p$ in the tree has the property that elements in its left subtree are before KEY($p$) in the natural order and those in its right subtree are after KEY($p$) in the natural order. For descriptive purposes we will consider a binary search tree

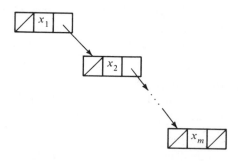

**Figure 7.1**
The implicit tree corresponding to linear search

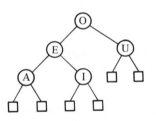

**Figure 7.2**
A binary search tree: the inorder traversal gives the elements in the natural
order

to be an extended binary tree (see Figure 5.28), that is, a tree with explicit nodes
for empty subtrees. In practice, these nodes are not implemented; the pointers to
them are simply **nil**. Figure 7.2 shows a binary search tree for the set of names
{A, E, I, O, U}.

The structure of the binary search tree makes it easy to search for an
element $z$. We compare $z$ to the root; if it is equal, the search ends successfully,
and if it is not, we search the left or right subtree according to whether $z$ is less
than or greater than the root, respectively. Algorithm 7.5 gives this procedure
explicitly. Observe that unsuccessful searches terminate at external nodes of the
extended tree. Also, notice the similarity between this algorithm and binary
search.

$p \leftarrow T$ 　　[start at root of the tree]
*found* $\leftarrow$ **false**
**while** $p \neq$ **nil and not** *found* **do**
　　　　**case**
　　　　　　$z <$ KEY($p$): $p \leftarrow$ LEFT($p$)
　　　　　　$z >$ KEY($p$): $p \leftarrow$ RIGHT($p$)
　　　　　　$z =$ KEY($p$): *found* $\leftarrow$ **true**
　　**if** *found* **then** 　[found: $p$ points to the element sought]
　　　　　　**else** 　[not found: $z$ is not in the tree]

**Algorithm 7.5**
Searching a binary search tree $T$ for the element $z$

In Sections 7.2.1 and 7.2.2 we will examine the two aspects of binary search
trees—their application to static tables and dynamic tables.

## Exercises

1. How would we produce a list of the elements of a binary search tree in their
   natural order?

2. Does Algorithm 7.5 work for empty trees?

★3. Given $n$ elements, how many possible different binary search trees are there?

4. Suppose $x_1 < x_2 < \cdots < x_n$ are formed into a binary search tree and $x_i$, for some $i \geqslant 2$, has LEFT($x_i$) = **nil**. Prove that a search for $x_i$ by Algorithm 7.5 will make a probe at $x_{i-1}$.

5. Suppose $x_1 < x_2 < \cdots < x_n$ are formed into a binary search tree. Prove that a search for $z$, $x_i < z < x_{i+1}$, by Algorithm 7.5 will make probes at both $x_i$ and $x_{i+1}$.

6. Given a binary search tree $T$ and a new element $x$ not in the tree, how many possible ways are there to insert $x$ into the tree so that it remains a binary search tree? Consider two cases: (a) $x$ must be inserted at a leaf, and (b) $x$ can be inserted anywhere. In case (b), consider only those insertions that can be accomplished by changing a single pointer in the tree (along with the two pointers in the new node).

### 7.2.1  Static Trees

The application of binary search trees to static tables is concerned entirely with arranging the tree so as to minimize search time; we assume that the table is constructed once and that its contents will change either never or so infrequently that it will be possible to reconstruct the entire table to make a change. If we want to minimize the worst-case search time, we simply use the tree corresponding to binary search (why?), and we do not need an explicit tree at all. The more difficult problem is to minimize the average search time, given some distribution of how the search will end. If the table consists of elements $x_1 < x_2 < \cdots < x_n$, then the search can end successfully at any of the $x_i$ (internal nodes) and unsuccessfully in any of the $n + 1$ gaps between the $x_i$ and at the endpoints (external nodes $y_i$). Throughout this section we assume that we have values $\beta_1, \beta_2, \ldots, \beta_n$ and $\alpha_0, \alpha_1, \ldots, \alpha_n$, were $\beta_i$ is the relative frequency with which a search will end successfully at $x_i$ and $\alpha_i$ is the relative frequency with which the search for $z$ will end unsuccessfully at $y_i$, that is, with $x_i < z < x_{i+1}$ (defining $x_0 = -\infty$ and $x_{n+1} = \infty$). As a continuing example, we will use the vowels A, E, I, O, U as table elements with their frequencies of occurrence in English text as the $\beta_i$ and the frequencies of occurrence of the intervening letters as the $\alpha_i$. These frequencies are shown in Figure 7.3. Notice that the $\alpha_i$ and $\beta_i$ are *not* probabilities; to compute the equivalent probabilities it is necessary to divide each $\alpha_i$ and $\beta_i$ by sum $\alpha_0 + \beta_1 + \alpha_1 + \cdots + \beta_n + \alpha_n$.

The problem is to choose among the many possible binary trees with $n$ internal nodes for a particular set of values $\alpha_i$ and $\beta_i$. We will measure the desirability of a tree by the cost of an average search; as before, the cost will be

| A | E | I | O | U |
|---|---|---|---|---|
| $\beta_1 = 32$ | $\beta_2 = 42$ | $\beta_3 = 26$ | $\beta_4 = 32$ | $\beta_5 = 12$ |

$\alpha_0 = 0 \qquad \alpha_1 = 34 \qquad \alpha_2 = 38 \qquad \alpha_3 = 58 \qquad \alpha_4 = 95 \qquad \alpha_5 = 21$

**Figure 7.3**
A sample set of frequencies. The $\beta_i$ are the approximate frequencies of occurrence of vowels in English text and the $\alpha_i$ are the approximate frequencies of occurrence for the intervening letters.

the number of probes. In Section 5.3 we introduced such a measure, the internal path length (or the related external path length). That measure is not sufficient for our purpose here because it does not take the varying frequencies into account. However, we can generalize the notion of path length as follows: the *weighted path length* of a binary tree $T$ with internal nodes $x_1, x_2, \ldots, x_n$, external nodes $y_0, y_1, \ldots, y_n$, and $\alpha_i$ and $\beta_i$ as defined above is

$$\sum_{i=1}^{n} \beta_i[1 + \text{level}(x_i)] + \sum_{i=0}^{n} \alpha_i \, \text{level}(y_i). \tag{7.2}$$

Notice that this is $\Sigma\alpha_i + \Sigma\beta_i$ times the average search time, since the search ends successfully at internal node $x_i$ with frequency $\beta_i$ and a cost of $[1 + \text{level}(x_i)]$ probes and unsuccessfully at external node $y_i$ with frequency $\alpha_i$ and a cost of $\text{level}(y_i)$ probes. As in the cases of the external and internal path lengths, it is convenient to define the weighted path length recursively.

$$W(\square) = 0$$

$$W(T = \qquad ) = W(T_l) + W(T_r) + \Sigma\alpha_i + \Sigma\beta_i, \tag{7.3}$$

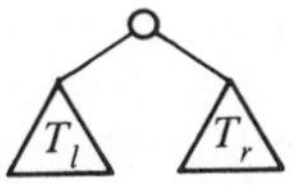

where the summations $\Sigma\alpha_i$ and $\Sigma\beta_i$ are over all $\alpha_i$ and $\beta_i$ in $T$. (Exercise 1 is to prove that these two definitions of weighted path length are equivalent.)

Let us consider the tree of Figure 7.2 as an example. Figure 7.4 shows the same tree, but with the corresponding frequencies from Figure 7.3 given below each node. The weighted path length of this tree is computed as follows:

$$\sum_{i=1}^{n} \beta_i[1 + \text{level}(x_i)] = 32 \times 3 + 42 \times 2 + 26 \times 3 + 32 \times 1 + 12 \times 2 = 314$$

and

$$\sum_{i=0}^{n} \alpha_i \text{level}(y_i) = 0 \times 3 + 34 \times 3 + 38 \times 3 + 58 \times 3 + 95 \times 2 + 21 \times 2 = 622.$$

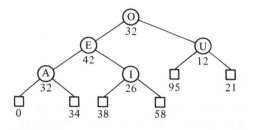

**Figure 7.4**
A weighted version of the binary search tree of Figure 7.2 using the weights
from Figure 7.3. The weighted path length is 936 and the number of probes
on an average search is 936/390 = 2.4. It turns out that this tree is optimal
for the given weights.

The weighted path length is thus $314 + 622 = 936$. If we divide this by $\Sigma\alpha_i + \Sigma\beta_i = 390$, we find that the number of probes in the table on an average search will be $936/390 = 2.4$.

Our problem is to determine the binary search tree that will have an optimal (minimal) weighted path length, given the frequencies $\alpha_i$ and $\beta_i$. Since the number of possible trees is *exponentially* large as a function of $n$ (Exercise 3 of Section 7.2), we cannot do the obvious of examining all possibilities, computing the weighted path length of each, and choosing the smallest. In fact, the large number of possibilities makes it seem doubtful that there is any reasonable way to make the determination. However, a simple but crucial observation about the nature of the weighted path length of a tree will show us the way to proceed.

The observation is that subtrees of an optimal tree must themselves be optimal. More precisely, if $T$ is an optimal binary search tree on weights $\alpha_0, \beta_1, \alpha_1, \ldots, \beta_n, \alpha_n$ and it has weight $\beta_i$ at the root, then the left subtree must be optimal over weights $\alpha_0, \beta_1, \alpha_1, \ldots, \beta_{i-1}, \alpha_{i-1}$ and the right subtree must be optimal over weights $\alpha_i, \beta_{i+1}, \alpha_{i+1}, \ldots, \beta_n, \alpha_n$. To see why this *optimality principle* must hold, suppose that some tree over $\alpha_0, \beta_1, \alpha_1, \ldots, \beta_{i-1}, \alpha_{i-1}$ had lower weighted path length than the one that is the left subtree of $T$. Then by (7.3) we could get a tree $T'$ with lower weighted path length than $T$ by replacing the left subtree of $T$ by the one of lower weighted path length we have supposed to exist. This contradicts the assumed optimality of $T$. We can argue similarly about the right subtree of $T$, and in fact *any* subtree of $T$. This optimality principle is the basis of a technique called *dynamic programming*, which we will use to compute optimal binary search trees and which in general is an extremely powerful technique in combinatorial algorithms.

The optimality principle together with (7.3) allows us to write the following recursive description of optimal binary search trees: Let $C_{ij}, 0 \leq i \leq j \leq n$, be the cost of an optimal tree (although the minimum cost is unique, it may be achieved by more than one tree) over the frequencies $\alpha_i, \beta_{i+1}, \ldots, \beta_j, \alpha_j$. Then

$$C_{ii} = 0$$

and

$$C_{ij} = \min_{i < k \leqslant j} \left( C_{i,k-1} + C_{kj} \right) + \sum_{t=i}^{j} \alpha_t + \sum_{t=i+1}^{j} \beta_t$$

by (7.3), since the optimality principle guarantees that if $x_k$ is the root of the optimal tree, then $C_{i,k-1}$ and $C_{kj}$ are the costs of the left and right subtrees, respectively. Defining

$$W_{ii} = \alpha_i$$
$$W_{ij} = W_{i,j-1} + \beta_j + \alpha_j, \quad i < j, \tag{7.4}$$

so that $W_{ij} = \alpha_i + \beta_{i+1} + \cdots + \beta_j + \alpha_j$, we get

$$C_{ii} = 0,$$
$$C_{ij} = W_{ij} + \min_{i < k \leqslant j} \left( C_{i,k-1} + C_{kj} \right). \tag{7.5}$$

Equations (7.4) and (7.5) form the basis of our computation of the optimal search tree; in evaluating (7.5) to get $C_{0n}$, the cost of the optimal tree over $\alpha_0, \beta_1, \ldots, \beta_n, \alpha_n$, we need only keep track of the choices of $k$ that achieve the minimum in (7.5). We thus define

$$R_{ij} = \text{a value of } k \text{ that minimizes } C_{i,k-1} + C_{kj} \text{ in (7.5).} \tag{7.6}$$

$R_{ij}$ is the root of an optimal tree over $\alpha_i, \beta_{i+1}, \ldots, \beta_j, \alpha_j$.

We are left with the problem of organizing the computation from (7.4), (7.5), and (7.6). Of course, we could simply make (7.4) and (7.5) into recursive procedures as they stand, but that would lead to an exponential time algorithm because many computations would be repeated over and over. For example, the computation of $C_{1,3}$, the optimal subtree over $\alpha_1, \beta_2, \alpha_2, \beta_3, \alpha_3$ of Figure 7.3, would be done in the computation of $C_{0,3}$ and $C_{1,4}$; in larger examples the number of repeated computations becomes staggering. The obvious way to avoid this difficulty is to insure that each $C_{ij}$ is computed only once. We do this by observing that the value of $C_{ij}$ in (7.5) depends only on values below and/or to the left of $C_{ij}$ in the matrix, as illustrated in Figure 7.5. We thus compute the matrix $C$ (and, in parallel, $W$ and $R$) starting from the main diagonal and moving up one diagonal at a time. First, $C_{ii} = 0$, $0 \leqslant i \leqslant n$ by (7.5). Then we compute

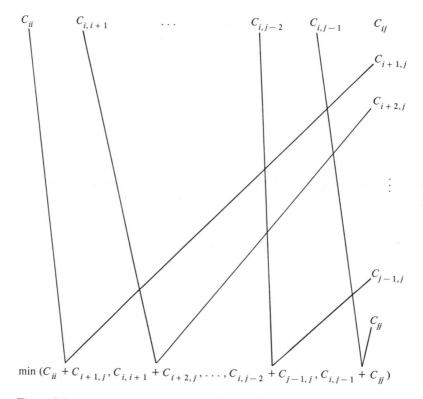

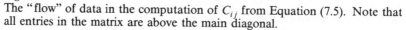

**Figure 7.5**
The "flow" of data in the computation of $C_{ij}$ from Equation (7.5). Note that
all entries in the matrix are above the main diagonal.

$C_{i,i+1}$, $0 \leq i \leq n - 1$, then $C_{i,i+2}$, $0 \leq i \leq n - 2$, and so on. Algorithm 7.6
embodies this idea.

An example of the computations of Algorithm 7.6 is shown in Figure 7.6.
The frequencies used are those given for the sample problem of Figure 7.3. To
recover the tree from the data in Figure 7.6, we look at the $(0, 5)$ matrix entry and
see that $R_{0,5} = 4$; this tells us that $x_4$, corresponding to "O", is the root of the
tree. The left and right subtrees are found similarly, looking at the $(0, 3)$ and $(4, 5)$
matrix entries, respectively. The optimal tree thus computed turns out to be the
one shown in Figure 7.4. Figure 7.7 shows optimal trees constructed for the
frequencies of the example in Figure 7.3 with successful searches only [Figure
7.7(a)] and unsuccessful searches only [Figure 7.7(b)]; notice that the shape of the
optimal tree is different in each case, and each is different from the optimal tree
of Figure 7.4.

⟦initialize the main diagonal⟧
**for** $i = 0$ **to** $n$ **do**

$\quad\quad\quad R_{ii} \leftarrow i \quad$ ⟦$R_{ij}$ = index of the root of the optimal
$\quad\quad\quad\quad\quad\quad\quad\quad\quad$ tree over $\alpha_i, \beta_{i+1}, \ldots, \beta_j, \alpha_j$⟧

$\quad\quad\quad W_{ii} \leftarrow \alpha_i \quad$ ⟦$W_{ij} = \alpha_i + \beta_{i+1} + \cdots + \beta_j + \alpha_j$⟧
$\quad\quad\quad C_{ii} \leftarrow 0 \quad$ ⟦$C_{ij}$ = cost of an optimal tree over $\alpha_i, \beta_{i+1}, \ldots, \beta_j, \alpha_j$⟧

⟦visit each of the $n$ upper diagonals⟧
**for** $l = 1$ **to** $n$ **do**

$\quad\quad\quad$⟦visit each entry in $l$th diagonal⟧
$\quad\quad\quad$**for** $i = 0$ **to** $n - l$ **do**

$\quad\quad\quad\quad\quad\quad j \leftarrow i + l \quad$ ⟦the elements on the $l$th diagonal
$\quad\quad\quad\quad\quad\quad\quad\quad\quad\quad$ have $j - i = l$⟧

$\quad\quad\quad\quad\quad\quad$⟦compute $(i, j)$ entries; $R_{ij}$ is a value
$\quad\quad\quad\quad\quad\quad$ of $k, i < k \leqslant j$ minimizing $C_{i, k-1} + C_{kj}$⟧

$\quad\quad\quad\quad\quad\quad R_{ij} \leftarrow i + 1$
$\quad\quad\quad\quad\quad\quad$**for** $k = i + 2$ **to** $j$ **do**
$\quad\quad\quad\quad\quad\quad\quad\quad\quad$**if** $C_{i, k-1} + C_{kj} < C_{i, R_{ij}-1} + C_{R_{ij}j}$ **then** $R_{ij} \leftarrow k$

$\quad\quad\quad\quad\quad\quad W_{ij} \leftarrow W_{i, j-1} + \beta_j + \alpha_j$
$\quad\quad\quad\quad\quad\quad C_{ij} \leftarrow C_{i, R_{ij}-1} + C_{R_{ij}j} + W_{ij}$

**Algorithm 7.6**
Construction of an optimal binary search tree over the frequencies
$\alpha_0, \beta_1, \alpha_1, \ldots, \beta_n, \alpha_n$

The running time of Algorithm 7.6 will be roughly proportional to the number of comparisons "$C_{i, k-1} + C_{kj} < C_{i, R_{ij}-1} + C_{R_{ij}j}$" made in the innermost loop. That **if** statement is executed

$$\sum_{l=1}^{n} \sum_{i=0}^{n-l} \sum_{k=i+2}^{i+l} 1 = \sum_{l=1}^{n} \sum_{i=0}^{n-l} (l - 1)$$

$$= \sum_{l=1}^{n} (n - l + 1)(l - 1)$$

$$= \tfrac{1}{6} n^3 + O(n^2) \tag{7.7}$$

times. Algorithm 7.6 thus runs in time proportional to $n^3$—not very acceptable in constructing search trees of several thousand elements.

Algorithm 7.6 can be speeded up considerably by an important observation: there is an optimal tree over $\alpha_i, \beta_{i+1}, \ldots, \beta_j, \alpha_j$ whose root $R_{ij}$ satisfies $R_{i, j-1} \leqslant R_{ij} \leqslant R_{i+1, j}$, where $R_{i, j-1}$ and $R_{i+1, j}$ are roots of optimal trees over $\alpha_i, \beta_{i+1}, \ldots, \beta_{j-1}, \alpha_{j-1}$ and $\alpha_{i+1}, \beta_{i+2}, \ldots, \beta_j, \alpha_j$, respectively (see Exercise 7).

|   | 0 | 1 | 2 | 3 | 4 | 5 |
|---|---|---|---|---|---|---|
| 0 | $R_{00} = 0$ $W_{00} = 0$ $C_{00} = 0$ | $R_{01} = 1$ $W_{01} = 66$ $C_{01} = 66$ | $R_{02} = 2$ $W_{02} = 146$ $C_{02} = 212$ | $R_{03} = 2$ $W_{03} = 230$ $C_{03} = 418$ | $R_{04} = 3$ $W_{04} = 357$ $C_{04} = 754$ | $R_{05} = 4$ $W_{05} = 390$ $C_{05} = 936$ |
| 1 |   | $R_{11} = 1$ $W_{11} = 34$ $C_{11} = 0$ | $R_{12} = 2$ $W_{12} = 114$ $C_{12} = 114$ | $R_{13} = 3$ $W_{13} = 198$ $C_{13} = 312$ | $R_{14} = 3$ $W_{14} = 325$ $C_{14} = 624$ | $R_{15} = 4$ $W_{15} = 358$ $C_{15} = 798$ |
| 2 |   |   | $R_{22} = 2$ $W_{22} = 38$ $C_{22} = 0$ | $R_{23} = 3$ $W_{23} = 122$ $C_{23} = 122$ | $R_{24} = 4$ $W_{24} = 249$ $C_{24} = 371$ | $R_{25} = 4$ $W_{25} = 282$ $C_{25} = 532$ |
| 3 |   |   |   | $R_{33} = 3$ $W_{33} = 58$ $C_{33} = 0$ | $R_{34} = 4$ $W_{34} = 185$ $C_{34} = 185$ | $R_{35} = 4$ $W_{35} = 218$ $C_{35} = 346$ |
| 4 |   |   |   |   | $R_{44} = 4$ $W_{44} = 95$ $C_{44} = 0$ | $R_{45} = 5$ $W_{45} = 128$ $C_{45} = 128$ |
| 5 |   |   |   |   |   | $R_{55} = 5$ $W_{55} = 21$ $C_{55} = 0$ |

**Figure 7.6**
The computation of Algorithm 7.6 to obtain the optimal binary search tree for the sample problem of Figure 7.3. The tree obtained is that shown in Figure 7.4.

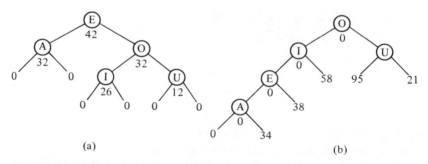

(a)                                                                        (b)

**Figure 7.7**
Optimal binary search trees for the frequencies in the example of Figure 7.3 with (a) successful searches only and (b) unsuccessful searches only. For (a) the average number of probes is $284/144 \approx 1.97$ and for (b) $598/246 \approx 2.43$.

Based on this observation we can replace the statement "$R_{ij} \leftarrow i + 1$" by "$R_{ij} \leftarrow R_{i,j-1}$" and the innermost loop "**for** $k = i + 2$ **to** $j$" by "**for** $k = R_{i,j-1} + 1$ **to** $R_{i+1,j}$". Equation (7.7) becomes

$$\sum_{l=1}^{n} \sum_{i=0}^{n-l} \sum_{k=R_{i,j-1}}^{R_{i+1,j}} 1 \tag{7.8}$$

which, after some calculation (Exercise 8), is at most proportional to $n^2$. In other words, the modified Algorithm 7.6 runs in time proportional to $n^2$.

Even the improved version of Algorithm 7.6, however, may not be efficient enough in certain circumstances. If $n$ is several thousand, it may be quite expensive to construct the optimal tree; furthermore, the frequencies $\alpha_i$ and $\beta_i$ are rarely known with any accuracy and it would be foolish to invest much computation time to get an optimal tree from inaccurate frequencies. In such cases a tree that approximates the optimal tree may be satisfactory and will certainly be less expensive to construct. We now examine heuristics for the construction of "near optimal" binary search trees.

Given frequencies $\alpha_i$ and $\beta_i$, two heuristics immediately suggest themselves; we discuss them in turn. The *monotonic* rule constructs a binary search tree by choosing the root to be $x_i$, where $\beta_i$ is the largest $\beta$ value, and then proceeding recursively on the left and right subtrees. Figure 7.8 shows the tree resulting when the monotonic rule is applied to the sample problem of Figure 7.3. This tree costs more than the optimal tree, but perhaps this is because the large $\alpha_i$ ends up too far from the top of the tree; in fact, the optimal tree when all $\alpha_i = 0$ [Figure 7.7(a)] is exactly the tree of Figure 7.8. This suggests that perhaps the monotonic rule may work very well in the special case of only successful searches occurring. Unfortunately, that is not the case: the monotonic rule produces poor trees in general, even when all the $\alpha_i = 0$: *on the average a tree constructed according to the*

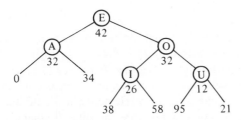

**Figure 7.8**
The tree constructed by the monotonic rule for the example in Figure 7.3. The weighted path length is 988, so the expected number of probes is $988/390 \approx 2.53$.

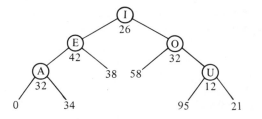

**Figure 7.9**
The tree constructed by the balancing rule for the example in Figure 7.3. The weighted path length is 948, so the expected number of probes is $948/390 \approx 2.43$, only slightly worse than for the optimal tree of Figure 7.4.

*monotonic rule is no better than a tree constructed at random*! (See Exercises 9 and 10.)

The second heuristic is the *balancing rule*: choose the root so as to equalize (as much as possible) the sum of the frequencies in the left and right subtrees, breaking ties arbitrarily. Figure 7.9 shows the tree constructed by the balancing rule for the example of Figure 7.3. The cost of the resulting tree is almost exactly that of the optimal tree; is this coincidental? No, the cost of the tree resulting from the balancing rule is always extremely close to the cost of the optimal tree. In particular, suppose the frequencies are normalized so that $\Sigma\alpha_i + \Sigma\beta_i = 1$ and let*

$$H = \sum_{i=0}^{n} \alpha_i \lg \frac{1}{\alpha_i} + \sum_{i=1}^{n} \beta_i \lg \frac{1}{\beta_i} \ ;$$

$H$ is called the *entropy* of the frequency distribution. Then it can be shown that

$H - \lg H - \lg e + 1 \leqslant$ weighted path length of the optimal tree

$\leqslant$ weighted path length of the tree derived by the balancing rule

$\leqslant H + 2.$

Since the maximum value of $H$ is $\lg(2n + 1)$, which occurs for $\alpha_i = \beta_i = 1/(2n + 1)$, the balancing rule always comes within $\lg H + 2.443 \approx \lg \lg (2n + 1) + 2.443$ of giving the optimal weighted path length. (See also Exercises 11 and 12.)

That the balancing rule does so well is even more remarkable since it can be implemented in time proportional to $n$, the number of elements in the table! We want to choose the root to equalize as much as possible the total frequencies of the left and right subtrees. In other words, we need to find an $i$ such that

---

*$0 \lg \frac{1}{0}$ is taken as 0.

$|(\alpha_0 + \beta_1 + \cdots + \beta_{i-1} + \alpha_{i-1}) - (\alpha_i + \beta_{i+1} + \cdots + \beta_n + \alpha_n)|$ is minimized, and we must repeat the computation recursively for $\alpha_0, \beta_1, \ldots, \beta_{i-1}, \alpha_{i-1}$ and $\alpha_i, \beta_{i+1}, \ldots, \beta_n, \alpha_n$ to find the left and right subtrees, respectively. The computation is organized as follows. We first compute the $W_{0i}$ of Algorithm 7.6

$$W_{0i} = \alpha_0 + \beta_1 + \cdots + \beta_i + \alpha_i \qquad 0 \leqslant i \leqslant n$$

by the recurrence relation

$$W_{00} = \alpha_0,$$
$$W_{0,i+1} = W_{0i} + \beta_{i+1} + \alpha_{i+1}.$$

The computation of the $W_{0i}$ thus requires only time proportional to $n$. Given the $W_{0i}$, we can immediately get any needed $W_{ij}$ with two subtractions, since

$$W_{ij} = \alpha_i + \beta_{i+1} + \cdots + \beta_j + \alpha_j$$
$$= W_{0j} - W_{0,i-1} - \beta_i$$

We will describe the computation for the root of the entire tree; the process is identical when applied to subtrees. To find where $|W_{0,i-1} - W_{in}|$ is minimized, we need to find where $W_{0,i-1} - W_{in}$ changes sign; the $i$ we want will be on either side of the sign change. More exactly, if $W_{0,k-1} - W_{kn} \leqslant 0 < W_{0k} - W_{k+1,n}$, then we want either $i = k$ or $i = k + 1$, depending on whether or not $|W_{0,k-1} - W_{kn}|$ is less than $|W_{0k} - W_{k+1,n}|$. We can find $k$ by a binary search type of process. Initially we know that $1 \leqslant k \leqslant n$; in general, if $l \leqslant k \leqslant h$, check the sign of $W_{0,m-1} - W_{mn}$, where $m = \lfloor (l+h)/2 \rfloor$. If it is positive, set $h \leftarrow m$, or if negative set $l \leftarrow m$ and continue; if it is zero we are done. (By using $h \leftarrow m$ and $l \leftarrow m$ instead of $h \leftarrow m - 1$ and $l \leftarrow m + 1$, respectively, we ensure that the differences of W's have opposite signs at the end points.) Finding $i$ in this way will require work proportional to $\lg n$ (why?), and the total amount of work will be given by the recurrence relation:

$$T(n) \leqslant \max_{1 \leqslant i \leqslant n} [T(i-1) + T(n-i) + c \lg n],$$

where the $T(i-1)$ and $T(n-i)$ are the work to find the left and right subtrees, respectively, once the root is found. The $c \lg n$ term is the time required to find the root $i$. $T(0)$ is some constant. The solution to this recurrence relation gives $T(n)$ proportional to $n \lg n$ and *not* to $n$ as was promised.

We can reduce the computation time needed by searching for $i$ in a slightly different way. We find the spot where $W_{0,k-1} - W_{kn}$ changes sign by checking $k = 1, k = n, k = 2, k = n - 1, k = 4, k = n - 3, k = 8, k = n - 7, \ldots$. In other words, we check from the left and right simultaneously, doubling the interval at each step. In this way we spend time proportional to $\min[\lg i, \lg(n - i)]$ to find an interval containing $i$. This interval has length proportional to $\min[i, n - i]$ and $i$ can be located by binary search in time proportional to $\min[1 + \min[\lceil\lg i\rceil, \lceil\lg(n - i + 1)\rceil]]$ The recurrence relation for $T(n)$ becomes

$$T(n) \leqslant \max_{1 \leqslant i \leqslant n} \{T(i - 1) + T(n - i) + d(1 + \min[\lceil\lg i\rceil, \lceil\lg(n - i + 1)\rceil])\},$$

and the solution of this gives $T(n)$ proportional to $n$, as desired. We leave the details to Exercises 14 and 15.

Exercise 16 describes a variant of the balancing heuristic whose behavior is slightly better.

### Exercises

1. Show the equivalence of the two definitions (7.2) and (7.3) for weighted path length.

2. Assuming successful searches only, to what does the weighted path length of the Morse code tree of Figure 5.6 correspond?

3. In what way is the Morse code tree of Figure 5.6 a binary search tree?

4. By hand calculation determine the optimal binary search tree for the frequencies $\alpha_0 = 17$, $\beta_1 = 6$, $\alpha_1 = 5$, $\beta_2 = 10$, $\alpha_2 = 14$, $\beta_3 = 1$, $\alpha_3 = 0$, $\beta_4 = 11$, $\alpha_4 = 3$, $\beta_5 = 7$, $\alpha_5 = 3$. Determine the trees that result from the monotonic and balancing rules.

5. Modify Algorithm 7.6 so that it computes *all* optimal binary search trees instead of just one of them.

6. Find a set of frequencies that have the tree shown in Figure 7.10 as their optimal search tree.

7. (a) Prove that the $W_{ij}$ defined by Equation (7.4) satisfy

$$W_{ij} + W_{i'j'} = W_{i'j} + W_{ij'} \qquad \text{for } i \leqslant i' \leqslant j \leqslant j'.$$

★ (b) Prove that the $C_{ij}$ defined by Equation (7.5) satisfy

$$C_{ij} + C_{i'j'} \leqslant C_{i'j} + C_{ij'} \qquad \text{for } i \leqslant i' \leqslant j \leqslant j'.$$

[*Hint*: Use induction on $j' - i$ and the result from part (a), considering the four cases $i = i'$, $j = j'$, $i < i' = j < j'$, and $i < i' < j < j'$.]

(c) Modify Equation (7.6) so that $R_{ij}$ is the *largest* value of $k$ that minimizes $C_{i,k-1} + C_{kj}$ in Equation (7.5). Using part (b), show that $R_{i,j-1} \leqslant R_{ij} \leqslant R_{i+1,j}$ for $i < j$.

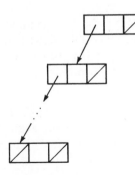

**Figure 7.10**
Exercise 6

8. Evaluate the triple summation to verify that Formula (7.8) is $O(n^2)$.

9. Find a set of frequencies for which the monotonic rule produces a tree with weighted path length proportional to $n$, but the balancing rule produces a tree with weighted path length proportional to $\log n$. [*Hint:* Let the $\alpha_i = 0$ and the $\beta_i$ be almost all equal, but differing from each other by enough so that the monotonic rule produces a tree that is essentially a linear list.]

10. Suppose the $\alpha_i = 0$ and the $\beta_i$ are drawn at random from some distribution over the nonnegative real numbers with average $\bar{\beta}$. Let $R_n$ be the expected weighted path length of a random binary search tree on $n$ nodes, defined by the property that each of the $n$ elements has equal probability $1/n$ of being the root, and if $\beta_i$ is the root, then it has a random binary search tree of elements $\beta_1, \ldots, \beta_{i-1}$ as its left subtree and a random binary search tree of elements $\beta_{i+1}, \ldots, \beta_n$ as its right subtree.

   (a) Verify the recurrence relation

   $$R_n = \bar{\beta}n + \frac{1}{n} \sum_{i=1}^{n} (R_{i-1} + R_{n-i}).$$

   (b) What is $R_0$?

   (c) Show how to transform the recurrence relation to

   $$R_n = \bar{\beta}n + \frac{2}{n} \sum_{i=0}^{n-1} R_i.$$

   (d) Show that the solution to this recurrence relation is

   $$2\bar{\beta}(n+1)H_n - 3\bar{\beta}n, \qquad \text{where } H_n = \sum_{i=1}^{n} \frac{1}{i} \approx \ln n.$$

This shows that $R_n \approx (2 \ln 2) \bar{\beta} n \lg n$. Defining $M_n$ to be the weighted path of the tree resulting from applying the monotonic rule to $n$ weights drawn at random from the same distribution as above, it can be shown that $M_n \approx (2 \lg 2) \bar{\beta} n \lg n$, so that the monotonic rule is (in a probabilistic sense) no better than constructing a random tree!

11. Compute $B_n$, the expected cost of the tree that results from applying the balancing rule to $n$ weights drawn from a distribution with average $\bar{\beta}$. [*Hint*: $B_n = n\bar{\beta} + B_{\lfloor (n-1)/2 \rfloor} + B_{\lceil (n-1)/2 \rceil}$. Approximate this with $B_n = \bar{\beta} n + 2 B_{n/2}$; your answer should be that $B_n \approx \bar{\beta} n \lg n$.]

★12. Compute $O_n$, the expected cost of the tree that results from applying Algorithm 7.6 to $n$ weights drawn from a distribution with average $\bar{\beta}$.

13. Compute the entropy $H$ for the frequencies of the example in Figure 7.3. [*Hint*: Remember to convert them to probabilities first.]

14. Write out the details of the algorithm to compute a binary search tree by the balancing rule.

15. Prove that the second recurrence relation given for $T(n)$, the time required to implement the balancing heuristic, is proportional to $n$.

16. Consider the following variation on the balancing heuristic. Suppose $\Sigma \alpha_i + \Sigma \beta_i = 1$. Define

$$ s_i = \alpha_0 + \beta_1 + \cdots + \alpha_{i-1} + \beta_i + \frac{\alpha_i}{2}, \qquad 0 \leqslant i \leqslant n. $$

The root of the search tree is chosen as $x_k$, where $s_{k-1} \leqslant \frac{1}{2}$ and $s_k \geqslant \frac{1}{2}$, and the heuristic continues on $s_0, \ldots, s_{k-1}$ and $s_k, \ldots, s_n$ using $\frac{1}{4}$ and $\frac{3}{4}$, respectively, in place of $\frac{1}{2}$, and so on for smaller intervals. Show that this heuristic can be implemented in time proportional to $n$. It can be shown that the weighted path length of the tree produced by this heuristic is at most $H + 1 + \Sigma \alpha_i$.

17. Study the properties of the *min-max* heuristic: choose as the root of $x_i, x_{i+1}, \ldots x_j$ the element $x_k$ for which $\max(W_{i,k-1}, W_{kj})$ is minimized.

### 7.2.2  Dynamic Trees

In the previous section we used binary search trees to improve the binary search technique in the case of a static table with nonuniform access frequencies. In this section we will show how binary search trees can be used to get a binary search technique in dynamic tables—that is, tables whose contents change because of insertions and deletions. There is a conflict between fast-search algorithms and fast-modification algorithms: Fast search requires a rigid structure, while fast modification obviously needs a flexible structure; binary search trees provide a compromise between the two requirements.

Inserting a new element $z$ into an existing binary search tree $T$ is not difficult if we do not care what the effect is on the shape of the tree. If the elements in the tree are $x_1 < x_2 < \cdots < x_n$ and $x_i < z < x_{i+1}, 0 \le i \le n$ (with $x_0$ and $x_{n+1}$ considered as $-\infty$ and $\infty$, respectively), then the $i$th external node can simply be replaced with

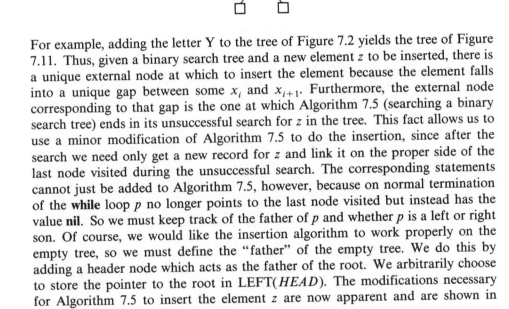

For example, adding the letter Y to the tree of Figure 7.2 yields the tree of Figure 7.11. Thus, given a binary search tree and a new element $z$ to be inserted, there is a unique external node at which to insert the element because the element falls into a unique gap between some $x_i$ and $x_{i+1}$. Furthermore, the external node corresponding to that gap is the one at which Algorithm 7.5 (searching a binary search tree) ends in its unsuccessful search for $z$ in the tree. This fact allows us to use a minor modification of Algorithm 7.5 to do the insertion, since after the search we need only get a new record for $z$ and link it on the proper side of the last node visited during the unsuccessful search. The corresponding statements cannot just be added to Algorithm 7.5, however, because on normal termination of the **while** loop $p$ no longer points to the last node visited but instead has the value **nil**. So we must keep track of the father of $p$ and whether $p$ is a left or right son. Of course, we would like the insertion algorithm to work properly on the empty tree, so we must define the "father" of the empty tree. We do this by adding a header node which acts as the father of the root. We arbitrarily choose to store the pointer to the root in LEFT($HEAD$). The modifications necessary for Algorithm 7.5 to insert the element $z$ are now apparent and are shown in Algorithm 7.7.

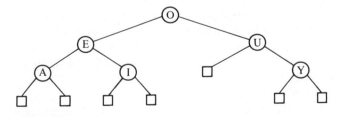

**Figure 7.11**
The tree of Figure 7.2 with the letter Y added at its proper place

$p \leftarrow \text{LEFT}(HEAD)$        ⟦the root of the tree⟧
$direction \leftarrow$ "left"       ⟦direction of $p$ from its father⟧
$father \leftarrow HEAD$          ⟦father of $p$⟧
$found \leftarrow$ **false**
**while** $p \neq$ **nil and not** $found$ **do**
            **case**
                    $z < \text{KEY}(p)$:
                            $direction \leftarrow$ "left"
                            $father \leftarrow p$
                            $p \leftarrow \text{LEFT}(p)$
                    $z > \text{KEY}(p)$:
                            $direction \leftarrow$ "right"
                            $father \leftarrow p$
                            $p \leftarrow \text{RIGHT}(p)$
                    $z = \text{KEY}(p)$: $found \leftarrow$ **true**
    **if** $found$ **then**    ⟦already in tree, no need to add it⟧
            **else**
                    ⟦add $z$ to the tree⟧
                    $p \leftarrow$ **getcell**
                    $\text{LEFT}(p) \leftarrow \text{RIGHT}(p) \leftarrow$ **nil**
                    $\text{KEY}(p) \leftarrow z$
                    **if** $direction =$ "left" **then** $\text{LEFT}(father) \leftarrow p$
                                       **else** $\text{RIGHT}(father) \leftarrow p$

**Algorithm 7.7**
Insertion of a new element $z$ into a binary search tree

What happens if we use this algorithm to construct search trees? In the worst case, of course, the tree can degenerate into a linear list; this happens, for example, if the order of insertion is A, E, I, O, U. (Even more innocent looking insertion orders can be problematic; consider A, U, E, O, I.) Are things really that bad on the average? If we have a random insertion order, what will the average search time be in the tree constructed? To answer, we recall that the external path length is the measure of the average search time (since it is the sum of the number of probes over all possible unsuccessful searches and differs only slightly from the internal path length, which is the sum of the number of probes over all possible successful searches). We want to compute the expected external path length in a tree constructed by Algorithm 7.7 for random insertion order. Without further information, we may as well assume that each of the $n!$ permutations of the $n$ elements is equally likely as the insertion order. Let $E_n$ be the expected external path length in a tree constructed from $n$ elements taken in a random order. [The nonmathematical reader should skip to Equation (7.16).] To develop a recurrence relation for $E_n$ we observe that if the $n$ elements are in random order, then the probability that any particular one is first is $1/n$, and that the remaining $n-1$ are again in random order. Furthermore, if the first one happens to be $x_i$, the $i$th element of the $n$ elements (in the natural order), then those elements less than $x_i$ (that is, $x_1, x_2,\ldots,x_{i-1}$) are in a random order, as are those larger than $x_i$ (that is, $x_{i+1}, x_{i+2},\ldots,x_n$). Thus if $x_i$ happens to be the first element inserted into the tree (as the root), it will have, by the nature of the insertion process of Algorithm 7.7, a random tree made up of $x_1, x_2,\ldots,x_{i-1}$ as its left subtree and a random tree made up of $x_{i+1}, x_{i+2},\ldots,x_n$ as its right subtree. Together with Equation (5.5) this gives us

$$E_0 = 0,$$

$$E_n = \sum_{i=1}^{n} (n + 1 + E_{i-1} + E_{n-i})\Pr(i \text{ will be the root}).$$

Since the probability that $i$ will be the root is equal for all $i$, it is $1/n$, and the above equation becomes

$$E_n = \sum_{i=1}^{n} \frac{1}{n}(n + 1 + E_{i-1} + E_{n-i}),$$

which by some elementary algebra can be rewritten as

$$E_n = n + 1 + \frac{2}{n}\sum_{i=0}^{n-1} E_i.$$

Recurrence relations of this type are not difficult to solve if one knows the trick, and since they occur not only here but also in Chapter 8, we now make a short detour to solve

$$t_n = an + b + \frac{2}{n}\sum_{i=0}^{n-1} t_i, \qquad n \geqslant n_0 \tag{7.9}$$

for $t_n$ in terms of $n, a, b, n_0, t_0, t_1,\ldots,t_{n_0-1}$.

To eliminate the summation from (7.9) we first multiply each side by $n$ to obtain

$$nt_n = an^2 + bn + 2\sum_{i=0}^{n-1} t_i, \qquad n \geqslant n_0. \tag{7.10}$$

Replacing $n$ by $n-1$, we get

$$(n-1)t_{n-1} = a(n-1)^2 + b(n-1) + 2\sum_{i=0}^{n-2} t_i, \qquad n \geqslant n_0 + 1. \tag{7.11}$$

Subtracting (7.11) from (7.10) gives

$$nt_n - (n - 1)t_{n-1} = 2t_{n-1} + 2an + b - a, \qquad n \geqslant n_0 + 1,$$

or

$$nt_n - (n + 1)t_{n-1} = 2an + b - a, \qquad n \geqslant n_0 + 1.$$

Divide this by $n(n + 1)$ and we have

$$\frac{t_n}{n + 1} - \frac{t_{n-1}}{n} = \frac{3a - b}{n + 1} + \frac{b - a}{n}, \qquad n \geqslant n_0 + 1.$$

Replacing $n$ by $i$ and summing gives

$$\sum_{i=n_0+1}^{n} \left( \frac{t_i}{i + 1} - \frac{t_{i-1}}{i} \right) = \sum_{i=n_0+1}^{n} \left( \frac{3a - b}{i + 1} + \frac{b - a}{i} \right). \tag{7.12}$$

The left-hand side is the telescoping sum

$$\frac{t_n}{n + 1} - \frac{t_{n-1}}{n} + \frac{t_{n-1}}{n} - \frac{t_{n-2}}{n - 1} + \cdots - \frac{t_{n_0+1}}{n_0 + 2} + \frac{t_{n_0+1}}{n_0 + 2} - \frac{t_{n_0}}{n_0 + 1}$$

$$= \frac{t_n}{n + 1} - \frac{t_{n_0}}{n_0 + 1}$$

and the right-hand side can be rewritten as

$$2aH_n - 2aH_{n_0} - (3a - b)\left( \frac{1}{n_0 + 1} - \frac{1}{n} \right),$$

where

$$H_n = \sum_{i=1}^{n} \frac{1}{i} \tag{7.13}$$

are called the *harmonic numbers*. Thus (7.12) yields

$$t_n = 2anH_n + n\left( \frac{t_{n_0} - 3a + b}{n_0 + 1} - 2aH_{n_0} \right) + 2aH_n$$

$$+ \frac{t_{n_0} - 3a + b}{n_0 + 1} + 3a - b - 2aH_{n_0}, \tag{7.14}$$

where

$$t_{n_0} = an_0 + b + \frac{2}{n_0} \sum_{i=0}^{n_0-1} t_i$$

from (7.9). Since $H_n = \ln n + O(1)$ (see Exercise 1), (7.14) tells us that

$$\begin{aligned} t_n &= 2an \ln n + O(n) \\ &= (2a \ln 2)n \lg n + O(n). \end{aligned} \tag{7.15}$$

For $E_n$, this yields

$$\begin{aligned} E_n &= (2 \ln 2)n \lg n + O(n) \\ &\approx 1.38n \lg n. \end{aligned} \tag{7.16}$$

Equation (7.16) tells us that in binary search trees built at random, the average search time will be about 1.38 lg $n$, or about 38 percent longer than in an optimal (unweighted) tree. The simplicity of Algorithm 7.7 would make it acceptable in spite of the increase over the minimum search time, except for an important fact: our analysis assumed that the insertion order was random, and this is almost never true in practice, since there are often sequences of elements arriving in their natural order. Thus, despite Equation (7.16), Algorithm 7.7 must be considered unreliable except in truly random circumstances.

So far we have considered only insertions. What about deletions? Deletions are somewhat more complex than insertions, because insertions cause changes only in the external nodes, but a deletion affects the internal nodes as well. There is no problem if the element to be deleted has two **nil** sons; we just replace the pointer to it by **nil**. Also, if the element to be deleted has only one **nil** son, we replace the pointer to it with a pointer to its single son. These two easy cases are illustrated in Figure 7.12. What if the element to be deleted has two non**nil** sons? Then the element has an inorder predecessor which has a **nil** right son and it has an inorder successor which has a **nil** left son (Exercise 3). Thus we can replace the element to be deleted by either its predecessor or its successor, deleting *that* element's node as shown in Figure 7.12. The details of the entire process are left to Exercise 4.

There is no known analysis of the expected search time in binary search trees constructed through random insertions and deletions, as we gave above for the case of random insertions only. But it is unlikely that results more favorable than (7.16) are true, especially since there is an inherent asymmetry in the deletion algorithm (do we replace an element with two sons by its predecessor or successor?). Furthermore, the possibility of biased sequences of insertions and

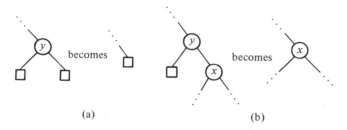

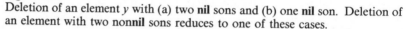

      (a)                                             (b)

**Figure 7.12**
Deletion of an element $y$ with (a) two **nil** sons and (b) one **nil** son. Deletion of
an element with two **nonnil** sons reduces to one of these cases.

deletions is very real in practice and makes these algorithms unreliable. Suitably
adapted by the techniques described below, however, these algorithms form the
basis of very flexible table organizations that have logarithmic search and
insertion/deletion times even in the worst case.

    Logarithmic search times can be achieved by keeping the tree perfectly
balanced at all times (as is implicit in binary search). Unfortunately, when the
tree is thus constrained, it is more costly to insert or delete an element than to
rearrange elements in the sequentially allocated arrays required by binary search.
Instead, our goal is to allow more flexibility in the shape of the tree so that
insertions and deletions will not be so expensive, yet search times will remain
logarithmic. In the remainder of this section we present two techniques that
achieve this goal. Both techniques keep the trees "balanced" so that they cannot
become too skewed (and hence degenerate to linear search times). The height of
such trees of $n$ elements will be $O(\log n)$, so that search times are logarithmic and
insertions and deletions will require only local changes along a single path from
the root to a leaf, requiring only time proportional to the height of the tree—that
is, $O(\log n)$.

    **Height-Balanced Trees.** An extended binary tree is *height-balanced* if and
only if it consists of a single external node, or the two subtrees $T_l$ and $T_r$ of the
root satisfy

1. $|h(T_l) - h(T_r)| \leqslant 1$.

2. $T_l$ and $T_r$ are height-balanced.

In other words, at any node in a height-balanced tree the two subtrees of that
node differ in height by at most one. Figure 7.13 shows two trees, one height-
balanced and the other not.

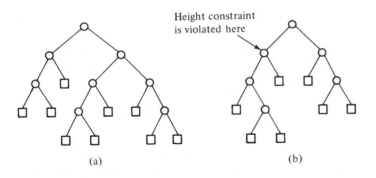

**Figure 7.13**
Two extended binary trees: (a) height-balanced and (b) not height-balanced

We want to use height-balanced trees as a storage structure for dynamic tables, but in order for height-balanced trees to be useful we must demonstrate that search times are at worst $O(\log n)$ and that insertions and deletions are easily and efficiently accommodated. Once we have shown that a height-balanced tree of $n$ nodes has height $O(\log n)$, then the worst-case search time is $O(\log n)$ and, since the insertion/deletion time will also be proportional to the height, we will be done.

What is the height of the tallest height-balanced tree containing $n$ internal nodes and $n + 1$ external nodes? To answer this question we will turn it around and ask what is the least number of internal nodes necessary to achieve height $h$ in a height-balanced tree.

Let $T_h$ be a height-balanced tree of height $h$ with $n_h$ internal nodes, the fewest possible. Obviously,

$$T_0 = \square, \qquad n_0 = 0,$$

and

$$T_1 = \triangle, \qquad n_1 = 1, \qquad (7.17)$$

since these are the only extended binary trees with heights 0 and 1, respectively. Now consider $T_h$, $h \geqslant 2$. Since $T_h$ is height-balanced and has height $h$, it must have a tree of height $h - 1$ as its left or right subtree and a tree of height $h - 1$ or $h - 2$ as its other subtree. For any $k$, a tree of height $k$ has a subtree of height $k - 1$ and thus $n_k > n_{k-1}$; this tells us that $T_h$ has one subtree of height $h - 1$ and the other of height $h - 2$, for if $T_h$ had two subtrees of height $h - 1$, we would replace one of them by $T_{h-2}$ and, since $n_{h-1} > n_{h-2}$, this would contradict the assumption that $T_h$ had as few nodes as possible for a height-balanced tree of height $h$. Similarly, the two subtrees of $T_h$ must be height-balanced and have the fewest nodes possible, for otherwise we could replace one or both subtrees with same height subtrees of fewer nodes, again contradicting the assumption that $T_h$ has as few nodes as possible. Thus $T_h$ has $T_{h-1}$ as one subtree and $T_{h-2}$ as the other,

$$T_h = \overset{\bigcirc}{\underset{T_{h-1} \quad T_{h-2}}{\bigwedge}}, \qquad n_h = n_{h-1} + n_{h-2} + 1. \tag{7.18}$$

Readers familiar with Fibonacci numbers will notice the resemblance of the construction of $T_h$ and the recurrence relation for $n_h$ to those for the Fibonacci numbers.

The solution of $n_h$ in terms of $h$ can be shown to be

$$n_h = \frac{1}{\sqrt{5}} \left( \frac{1 + \sqrt{5}}{2} \right)^{h+2} - \frac{1}{\sqrt{5}} \left( \frac{1 - \sqrt{5}}{2} \right)^{h+2} - 1$$

or, one less than the $(h + 2)$nd Fibonacci number. Since $|(1 - \sqrt{5})/2| < 1$, the term $((1 - \sqrt{5})/2)^{h+2} / \sqrt{5}$ is always quite small, so that

$$n_h + 1 = \frac{1}{\sqrt{5}} \left( \frac{1 + \sqrt{5}}{2} \right)^{h+2} + O(1)$$

Since the tree of height $h$ with the fewest nodes has $n_h$ nodes, it follows that any tree with fewer than $n_h$ nodes has height less than $h$. Therefore, if a height-balanced tree of $n$ nodes has height $h$, then

$$n + 1 \geqslant n_h + 1 = \frac{1}{\sqrt{5}} \left( \frac{1 + \sqrt{5}}{2} \right)^{h+2} + O(1),$$

implying that

$$h \leqslant \frac{1}{\lg \dfrac{1 + \sqrt{5}}{2}} \lg(n + 1) + O(1)$$

$$\approx 1.44 \lg(n + 1)$$

(we have written $n + 1$ and not $n$ so the function will be properly defined for $n = 0$, that is, a tree with only a single external node).

Thus in the worst case the number of probes by Algorithm 7.5 into a height-balanced tree of $n$ internal nodes will be about $1.44 \lg(n + 1)$ and the total search time will be logarithmic, as desired. It remains to be seen, however, whether insertions and deletions can be done efficiently so that a height-balanced tree remains height-balanced afterward.

To make an insertion or deletion we will use the method outlined above for the case when the tree can change in an unconstrained manner, but we will follow it with a rebalancing pass that verifies or restores the height-balanced state of the tree. In order to verify/restore the tree we need to be able to test whether the element inserted or deleted has changed the relationship between the heights of the subtrees of a node so as to violate the height constraints. For this purpose, we will store a *condition code* in each node of a height-balanced tree. The condition code is one of the following:

/    Means the left subtree of this node is taller (by one) than its right subtree.

=    Means the two subtrees of this node have equal height.

\    Means the right subtree of this node is taller (by one) than its left subtree.

Storing condition codes requires an extra two-bit field per node in the tree. This very modest additional storage requirement can be made even more modest by the techniques of either Exercise 15 or Exercise 16, which reduce the condition code to a single bit, but at the expense of the efficiency and (relative) simplicity of the algorithms for rebalancing a tree that has become unbalanced through an insertion or deletion.

Roughly speaking, the rebalancing pass consists of retracing the path upward from the newly inserted node (or from the site of the deletion) to the root. If FATHER pointers are available, they are the most efficient way to accomplish this. If they are not available—and they usually are not—then we have two choices: either store the path, node by node, on a stack as we go down the tree from the root to the site of the modification, or use the trick of Algorithm 6.5(b) (see Exercise 6 of Section 6.1.2) to change the pointers as we go down the tree; this latter choice will require an additional two bits per node. In the case of an insertion there is a third possibility, that of simply retracing a portion of the path downward—see Exercise 11, page 324.

As the path is followed upward, we check for instances of the taller subtree growing taller (on an insertion) or the shorter subtree becoming shorter (on a deletion). When we find such an occurrence, we apply a local transformation to the tree at that point. In the case of an insertion it will turn out that applying the transformation at the first such occurrence will completely rebalance the tree. In the case of a deletion the transformations may need to be applied at many points

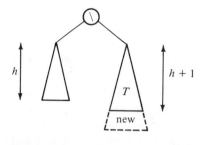

**Figure 7.14**
Insertion making a node unbalanced

along the way up to the root. We must be careful that any transformation made does not affect the inorder of the nodes in the tree (why?).

Since the rebalancing after an insertion is a bit simpler than after a deletion, we consider it first. What could have happened using Algorithm 7.7 to add an element to a height-balanced tree? The only thing is that the new element may have been added to the bottom of the taller of two subtrees of some node. Without loss of generality, suppose the right subtree was the taller before the insertion, as in Figure 7.14.

The way to repair the newly created imbalance depends on where within the taller subtree $T$ the insertion was made. Suppose it was in the right subtree of $T$; we then have a situation that can be repaired as shown in Figure 7.15(a). The transformation shown there is called a *rotation*, and it is considered to be applied to the element $A$. Obviously, if the left subtree in Figure 7.14 had been taller and the insertion made it even taller, we would have to rotate in the other direction, using the mirror image of the transformation shown in Figure 7.15(a).

The transformation of Figure 7.15(a) would not have helped if the insertion had been to the left subtree of $T$ in Figure 7.14—that is, to $T_2$ in Figure 7.15(a). In this case the repair is made as shown in Figure 7.15(b). This transformation, called a *double rotation*, is considered to be applied at $A$. The new element can be at the bottom of either $T_2$ or $T_3$. Both $T_2$ and $T_3$ can be empty, in which case $B$ is the new element (see Exercise 9). Again, a mirror-image transformation would be needed in the comparable case where the left subtree in Figure 7.14 was the taller.

Before continuing, the reader should try enough examples to be convinced that Figures 7.15(a) and (b), along with their mirror images, are the only possible cases.

The transformations of Figure 7.15 have two critical properties: the inorder of the elements of the tree remains the same after the transformation as it was before, and the overall height of the tree is the same after the transformation as it was before the insertion. The first property is necessary if we are to be able to search the tree properly by Algorithm 7.5. The second property means that after

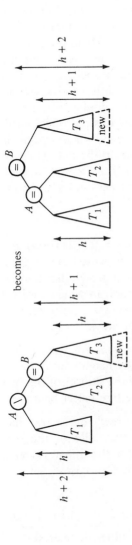

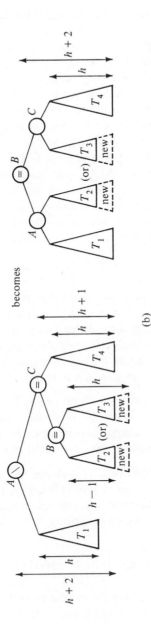

becomes

(a)

becomes

(b)

**Figure 7.15**

The transformations used to rebalance a height-balanced tree after the insertion of a new element: (a) rotation around $A$, (b) double rotation around $A$. The height condition codes in $A$ and $C$ in the right-hand drawing of (b) depend on whether the new element is at the bottom of $T_2$ or $T_3$. Both $T_2$ and $T_3$ are empty when $B$ is the new element (see also Exercise 9, page 324). Notice that in each transformation the inorder of the tree is unchanged and the height of the tree *after* the transformation is the same as the height of the tree *before* the insertion. In each case, there are corresponding mirror-image transformations.

the insertion the appropriate transformation needs to be applied *only* at the lowest unbalanced spot in the tree.

The insertion algorithm is thus as follows. Use Algorithm 7.7 to insert the new element into its proper place, setting its condition code to $=$ and storing the path followed down the tree, node by node, in a stack (the new element ends up as the top stack entry and the root is the bottom stack entry). Then, retrace that path backward, up the tree, popping nodes off the stack and correcting the height-condition codes until either the root is reached and its height-condition code corrected, or we reach a point when no more height-condition codes need to be corrected, or we reach a point at which a rotation or double rotation is necessary to rebalance the tree. More specifically, we follow this path backward, node by node, taking actions as defined by the following rules, where *current* is the current node on the path, *son* is the node before *current* on the path (that is, its son), and *grandson* is the node before *son* on the path (the *grandson* of *current*). Initially, *son* is the new element just inserted, *current* is its father, and *grandson* is **nil**:

1. If *current* has height condition $=$ , change it to $\searrow$ if *son* = RIGHT(*current*) and to $\diagup$ if *son* = LEFT(*current*). In this case the subtree rooted at *son* grew taller by one unit, causing the subtree rooted at *current* to grow taller by one unit, so we continue up the path, unless *current* is the root, in which case we are done. To continue up the path we set *grandson* ← *son*, *son* ← *current*, and *current* to the top stack entry, which is removed from the stack.

2. If *current* has height condition $\diagup$ and *son* = RIGHT(*current*) or *current* has height condition $\searrow$ and *son* = LEFT(*current*), change the height condition of *current* to $=$ , and the procedure terminates. In this case the shorter of the two subtrees of *current* has grown one unit taller, making the tree better balanced.

3. If *current* has height condition $\diagup$ and *son* = LEFT(*current*) or *current* has height condition $\searrow$ and *son* = RIGHT(*current*), then the taller of the two subtrees of *current* has become one unit taller, unbalancing the tree at *current*. A transformation is performed according to the following four cases:

|  | *grandson* = RIGHT(*son*) | *grandson* = LEFT(*son*) |
|---|---|---|
| *son* = RIGHT(*current*) | Rotate around *current* using Figure 7.15(a) | Double-rotate around *current* using Figure 7.15(b) |
| *son* = LEFT(*current*) | Double-rotate around *current* using the mirror image of Figure 7.15(b) | Rotate around *current* using the mirror image of Figure 7.15(a) |

In each case, the height conditions are set as shown in Figure 7.15. The procedure terminates, having rebalanced the tree at its lowest point of imbalance.

As an example of the insertion process, consider inserting the letter $T$ into the tree of Figure 7.16. Algorithm 7.7 inserts $T$ as the right son of $S$, and the path back up the tree is $T$, $S$, $R$, $P$, $K$, $U$, $H$. We thus begin the rebalancing with *current* $= S$, *son* $= T$, and *grandson* $=$ **nil**; the stack contents are then $R$, $P$, $K$, $U$, $H$. Case 1 applies to *current*, so we change the condition code of $S$ to $\setminus$ and set *grandson* $\leftarrow T$, *son* $\leftarrow S$, and *current* $\leftarrow R$ from the top of the stack. Again, case 1 applies to *current*, so we set the condition code of $R$ to $\setminus$ and *grandson* $\leftarrow S$, *son* $\leftarrow R$, and *current* $\leftarrow P$ from the top of the stack. Again, case 1 applies, so we set the condition code of $P$ to $\setminus$ and *grandson* $\leftarrow R$, *son* $\leftarrow P$, and *current* $\leftarrow K$ from the top of the stack. Now case 3 applies and, since *son* $=$ RIGHT(*current*) and *grandson* $=$ RIGHT(*son*), we apply the rotation of Figure 7.15(a) to the node *current* $= K$, setting the height-condition codes of $K$ and $P$ to $=$, as indicated in Figure 7.15(a). The procedure then terminates (ignoring the remainder of the stack contents). If we had been inserting $O$ instead, then upon reaching the node $K$ we would apply the double rotation of Figure 7.15(b).

The deletion process is more complex than insertion because it will not always be sufficient to apply a transformation only at the lowest point of imbalance; transformations may need to be applied at many levels between the site of the deletion and the root. To delete a node from a height-balanced tree, we proceed as for unconstrained trees: if the node is a leaf, just delete it; if it has one nonnil son, replace it with its son; if it has two nonnil sons, replace it by its inorder predecessor (successor), which will have a null right (left) son (see Figure 7.12). In the case of a height-balanced tree, however, if a node has only one

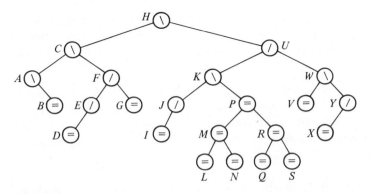

**Figure 7.16**
A height-balanced tree to illustrate the rebalancing algorithms after an insertion or a deletion

nonnil son, it must look like

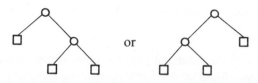

because of the height constraint. In either case, replacing the node by its only son has the identical effect on the heights as deleting the node with no sons. We thus need consider *only* the case of deleting a node with two null sons.

As in the insertion algorithm, we store on a stack the path followed down the tree to the site of the node to be deleted, then we retrace the path backward up the tree, popping nodes off the stack, correcting height-condition codes, and making transformations as needed. As we go back up the path, actions are taken as defined by the following rules, where (as before) *current* is the current node on the path and *son* is the node before *current*. Initially, *current* is the father of the node deleted, and *son* is the node deleted:

1. If *current* has height condition $=$ , then shortening either subtree does not affect the height of the tree rooted at *current*. The condition code of *current* is changed to $\setminus$ if *son* = LEFT(*current*) and to $/$ if *son* = RIGHT(*current*). The procedure then terminates.

2. If *current* has height condition $\setminus$ and *son* = RIGHT(*current*) or *current* has height condition $/$ and *son* = LEFT(*current*), the condition code of *current* is changed to $=$ . The subtree rooted at *current* has become shorter by one unit, so we continue up the path, unless *current* is the root, in which case we are done. To continue up the path we set *son* $\leftarrow$ *current* and *current* to the top stack entry, which is removed from the stack.

3. If *current* has height condition $\setminus$ and *son* = LEFT(*current*), then the height constraint is violated at *current*. There are three subcases, depending on the height-condition code at RIGHT(*current*), the brother of *son*. The subcases are as given in Figure 7.17.

4. If *current* has height condition $/$ and *son* = RIGHT(*current*), then the height constraint is violated at *current*. There are three subcases, depending on the height-condition code at LEFT(*current*), the brother of *son*. The subcases are the mirror images of those given in Figure 7.17, and we leave them as Exercise 12, page 325.

As an example of the deletion procedure, consider deleting the node $B$ from the tree of Figure 7.16. Case 2 applies to the node $A$ and then case 3(c) applies to the node $C$, so a double rotation is applied there. Then case 3(c) applies to the node $H$ and a rotation is applied there. The tree is then completely rebalanced. The reader can become familiar with the algorithm by going through the steps necessary to delete $D$ or $X$ from the tree of Figure 7.16.

What do height-balanced trees look like "on the average"—that is, when they are generated by a random sequence of insertions and deletions? There is no known mathematical analysis to answer this question, but empirical evidence strongly suggests that the average search time in such a tree is $\lg n + O(1)$ probes. This suggests that on the average height-balanced trees are almost as good as the completely balanced trees that correspond to binary search. Of course, unlike the case of binary search, height-balanced trees can be modified by insertions and deletions in logarithmic time.

The deletion algorithm will clearly require only time proportional to the height of the tree; thus deletion can be accomplished in $O(\log n)$ time, as can insertion. An insertion, however, will need at most one rotation/double rotation to rebalance the tree, while a deletion from a height-balanced tree of height $h$ can require as many as $\lfloor h/2 \rfloor$ rotations/double rotations, but no more (Exercise 14).

**Weight-Balanced Trees.** We now consider another class of balanced trees having the properties that the height of a tree is logarithmic in the number of nodes and that insertions and deletions can be accomplished in logarithmic time. The difference between the trees considered here and height-balanced trees is the constraint used to balance the trees. In addition, the algorithms for rebalancing the trees after an insertion or deletion are simpler, generalize to weighted trees, and, most important, allow an explicit trade-off to be made between search times and rebalancing times.

Let $T$ be an extended binary tree, so that $T$ either is a single external node $\square$ or consists of a root and left and right subtrees $T_l$ and $T_r$, respectively. Let $|T|$ be the number of external nodes in $T$. The *balance* of the root of a tree is defined as

$$\beta(T) = \begin{cases} \dfrac{1}{2}, & \text{if } T = \square, \\[2mm] \dfrac{|T_l|}{|T|}, & \text{otherwise.} \end{cases}$$

Notice that $\beta(T)$ roughly indicates the relative number of nodes (internal or external) in the left subtree of $T$. If the balance is $\frac{1}{2}$, then half the nodes are in each of the two subtrees, and this is the ideal balance; this explains our setting $\beta(\square) = \frac{1}{2}$, since the tree $\square$ is (trivially) well balanced. Obviously, $0 < \beta(T) < 1$ for any extended binary tree $T$.

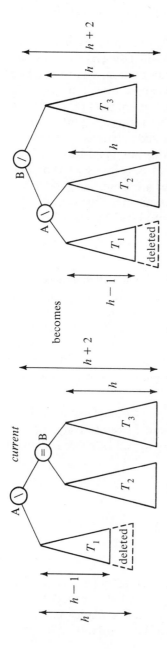

(a) Apply the rotation of Figure 7.15(a) to *current* and the procedure terminates, since the height-balance has been restored and the height of the tree after the transformation is the same as it was before the deletion. (See also Exercise 13.)

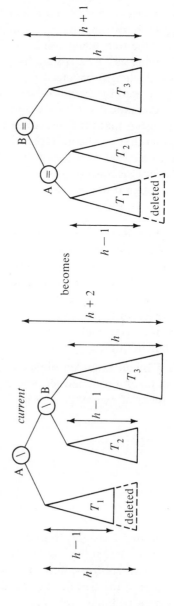

(b) Apply the rotation of Figure 7.15(a) to *current*. If *current* is the root, the procedure terminates; otherwise continue up the path toward the root because the height of the subtree is now one less than it was before the deletion.

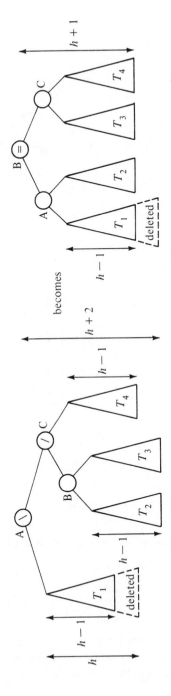

(c) Apply the double rotation of Figure 7.15(b) to *current*. If *current* is the root, the procedure terminates; otherwise continue up the path toward the root because the height of the subtree is now one less than it was before the deletion. The height-condition codes of A and C are both = if that of B was =. If B was \, then A is / and C is =. If B was /, then A is = and C is \.

**Figure 7.17**

The three subcases of case 3 in the algorithm for deletion from a height-balanced tree when *current* has height condition \ and *son* = LEFT(*current*). The subcases of Case 4 are obtained by taking the mirror images of those shown here (Exercise 12, page 325).

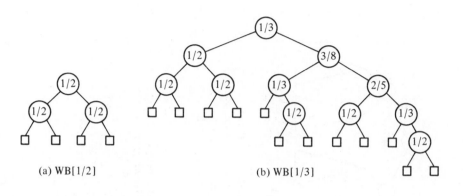

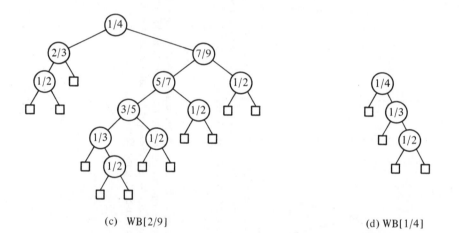

**Figure 7.18**
Examples of weight-balanced trees in WB[$\alpha$], for various values of $\alpha$

A tree $T$ is said to be of *weight-balance* $\alpha$, or in the set WB[$\alpha$], for $0 \leqslant \alpha \leqslant \frac{1}{2}$, if

1. $\alpha \leqslant \beta(T) \leqslant 1 - \alpha$.

2. If $T \neq \square$ then both $T_l$ and $T_r$ are in WB[$\alpha$].

In other words, at any node in a WB[$\alpha$] tree the balance factor lies between $\alpha$ and $1 - \alpha$. Figure 7.18 shows four examples of weight-balanced trees for various values of $\alpha$; in each case the balance of every subtree is given in the root of the subtree (except leaves).

Clearly $\text{WB}[\alpha_1] \subseteq \text{WB}[\alpha_2]$ for $\alpha_1 \geq \alpha_2$, hence the set $\text{WB}[\alpha]$ becomes more and more restricted as $\alpha$ goes from 0 to $\frac{1}{2}$. Since $\alpha = 0$ imposes no restriction at all on the trees, $\text{WB}[0]$ is the set of *all* extended binary trees. On the other hand, $\alpha = \frac{1}{2}$ forces the subtrees of every node to contain equal numbers of leaves, implying that the trees contain $2^h$ leaves and are completely balanced. For any $\alpha$, $\frac{1}{3} < \alpha < \frac{1}{2}$, $\text{WB}[\alpha] = \text{WB}[\frac{1}{2}]$. To prove this, suppose that a tree $T$ is not in $\text{WB}[\frac{1}{2}]$—that is, is not completely balanced. Since $T$ is not in $\text{WB}[\frac{1}{2}]$, it contains some subtree not in $\text{WB}[\frac{1}{2}]$; consider the smallest subtree $T'$ of $T$ not in $\text{WB}[\frac{1}{2}]$; both its left and right subtrees must be in $\text{WB}[\frac{1}{2}]$. Thus the number of leaves each contains must be a power of 2, say $2^l$ in the left subtree of $T'$ and $2^r$ in the right subtree of $T'$. Without loss of generality, suppose $l < r$. Then

$$\beta(T') = \frac{2^l}{2^l + 2^r} = \frac{1}{1 + 2^{r-l}} \leq \frac{1}{3},$$

and $T$ cannot be in $\text{WB}[\alpha]$ for $\alpha > \frac{1}{3}$.

Two observations are in order here, to demonstrate that weight-balanced trees bear no relation to height-balanced trees. First, consider the tree of Figure 7.18(b), which is in $\text{WB}[\frac{1}{3}]$. Since $\text{WB}[\alpha] = \text{WB}[\frac{1}{2}]$ for $\alpha > \frac{1}{3}$, the tree is as weight-balanced as it can be without being in $\text{WB}[\frac{1}{2}]$ and it is *not* height-balanced. Thus we cannot in general conclude anything about the height-balance of a weight-balanced tree. Second, there are height-balanced trees that are not weight-balanced, as can be seen by considering the tree with a completely balanced tree of $2^h$ leaves as its right subtree and the tree $T_h$ of (7.18) as its left subtree. Such a tree is height-balanced, but the weight-balance at the root goes to zero as $h \to \infty$ (why?). Thus we cannot conclude anything about the weight-balance of a height-balanced tree. The two concepts are therefore independent.

We need to prove that weight-balanced trees have only logarithmic height, and further that their weight-balanced state can be maintained efficiently under insertions and deletions. We will outline such a demonstration for $0 < \alpha < 1 - \sqrt{2}/2 \approx 0.29289$ with insertions but no deletions. For deletions we will require $\frac{2}{11} < \alpha < 1 - \sqrt{2}/2$, approximately $0.18182 < \alpha < 0.29289$.

What is the height of the tallest possible $\text{WB}[\alpha]$ tree with $n + 1$ external nodes (and hence $n$ internal nodes)? Let it be $h_\alpha(n + 1)$. By Exercise 20, $h_\alpha(n + 1) \geq h_\alpha(n)$, so the tallest $\text{WB}[\alpha]$ tree is the one formed by putting as many of the nodes as possible on one side, say the right side. Since the tree must be in $\text{WB}[\alpha]$, if it has $n + 1$ external nodes at most $\lfloor (1 - \alpha)(n + 1) \rfloor$ of them can be in either subtree and thus

$$h_\alpha(n + 1) = 1 + h_\alpha(\lfloor (1 - \alpha)(n + 1) \rfloor)$$

and trivially

$$h_\alpha(1) = 0.$$

The solution to this recurrence relation is

$$h_\alpha(n + 1) = \log_{1/(1-\alpha)}(n + 1) + O(1)$$

$$= \frac{1}{\lg \dfrac{1}{1 - \alpha}} \lg(n + 1) + O(1).$$

For $\alpha = \frac{1}{2}$, we have

$$h_{1/2}(n + 1) = \lg(n + 1) + O(1)$$

as we should, since WB[$\frac{1}{2}$] trees are the completely balanced trees in which $n + 1 = 2^t$, for some $t$. For $\alpha = 1 - \sqrt{2}/2$, the largest value for which the rebalancing algorithms will work, we have

$$h_{1-\sqrt{2}/2}(n + 1) = 2\lg(n + 1) + O(1),$$

so the worst-case number of probes is just twice as many as the minimum possible. For $\alpha = \frac{2}{11}$, the smallest value for which the rebalancing algorithms will work in the case of a deletion, we have

$$h_{2/11}(n + 1) \approx 3.45 \lg(n + 1) + O(1),$$

or about $3\frac{1}{2}$ times the optimal. In any case, the height of a WB[$\alpha$] tree is clearly logarithmic in its number of nodes, as desired.

In weight-balanced trees we can also say something about the average number of probes to search a WB[$\alpha$] tree, whereas in height-balanced trees no such analytical results are known. We leave it to Exercise 22, page 326, to verify that the internal path length of a tree in WB[$\alpha$] is at most

$$\frac{1}{H(\alpha)}(n + 1)\lg(n + 1) - 2n$$

where $H(\alpha) = -\alpha \lg \alpha - (1 - \alpha) \lg(1 - \alpha)$ is a simple instance of the entropy function introduced in Section 7.2.1. This bound tells us that the average number of probes for any WB[$\alpha$] will be at most

$$\frac{1}{H(\alpha)}\lg(n + 1) + O(1).$$

For $\alpha = 1 - \sqrt{2}/2$, $1/H(\alpha) \approx 1.15$ so that the average search time will be no more than 15 percent larger than the optimal. For $\alpha = \frac{2}{11}$, $1/H(\alpha) \approx 1.46$ so that the average search time will be no more than 46 percent larger than the optimal. These bounds hold for *any* WB[$\alpha$] tree, so the average search time in the *average* WB[$\alpha$] tree is surely much better, but there is no mathematical analysis known for this case.

When the insertion or deletion of an element in a WB[$\alpha$] tree causes the balances at various nodes to be outside the range [$\alpha, 1 - \alpha$], the tree can be rebalanced in logarithmic time using rotations and double rotations. Figure 7.19 shows the effects of these transformations on the balances of the subtrees involved. The formulas for the balances given in Figure 7.19(a) can be verified as follows. By definition,

$$\beta_A = \frac{|T_1|}{|T_1| + |T_2| + |T_3|} \quad \text{and} \quad \beta_B = \frac{|T_2|}{|T_2| + |T_3|}.$$

Since

$$|T_1| + |T_2| = (|T_1| + |T_2| + |T_3|) \times$$

$$\left[ \frac{|T_1|}{|T_1| + |T_2| + |T_3|} + \left(1 - \frac{|T_1|}{|T_1| + |T_2| + |T_3|}\right) \frac{|T_2|}{|T_2| + |T_3|} \right]$$

$$= (|T_1| + |T_2| + |T_3|)[\beta_A + (1 - \beta_A)\beta_B],$$

we have

$$\beta_A' = \frac{|T_1|}{|T_1| + |T_2|} = \frac{|T_1|}{|T_1| + |T_2| + |T_3|} \frac{1}{\beta_A + (1 - \beta_A)\beta_B}$$

$$= \frac{\beta_A}{\beta_A + (1 - \beta_A)\beta_B}$$

and

$$\beta_B' = \frac{|T_1| + |T_2|}{|T_1| + |T_2| + |T_3|} = \beta_A + (1 - \beta_A)\beta_B.$$

We leave the similar verification of the balances in Figure 7.19(b) as Exercise 23. An alternate derivation of these formulas is the object of Exercise 24.

We use rotations and double rotations as follows to rebalance a tree during an insertion or deletion. As in the case of height-balanced trees, we search the tree starting at the root for the site of the insertion or deletion. As each node on this downward path is visited, we can tell immediately whether the insertion or

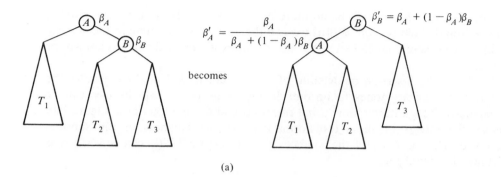

$$\beta'_A = \frac{\beta_A}{\beta_A + (1 - \beta_A)\beta_B}$$

becomes

$$\beta'_B = \beta_A + (1 - \beta_A)\beta_B$$

(a)

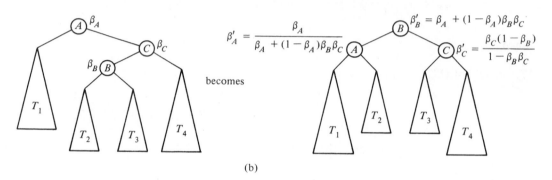

$$\beta'_A = \frac{\beta_A}{\beta_A + (1 - \beta_A)\beta_B\beta_C}$$

becomes

$$\beta'_B = \beta_A + (1 - \beta_A)\beta_B\beta_C$$

$$\beta'_C = \frac{\beta_C(1 - \beta_B)}{1 - \beta_B\beta_C}$$

(b)

**Figure 7.19**
The effects in weight-balanced trees of (a) a rotation and (b) a double rotation on the subtrees involved

deletion will disturb the balance at the node enough to throw the tree out of WB[$\alpha$]. We make this determination by a simple calculation on the SIZE fields that we maintain in every node: SIZE($p$) is one less than the number of leaves in the subtree rooted at $p$. (See also Exercise 27.) As each node is visited, its size information is updated, and if this results in an imbalance, a transformation is made to rebalance the tree according to the following rules. Let $T$ be a binary tree with left and right subtrees $T_l$ and $T_r$, respectively:

1. A rotation as shown in Figure 7.19(a) rebalances $T$ if $|T_l|/|T| < \alpha$, $\beta(T_r) \leq 1/(2 - \alpha)$, and either $|T_l|/(|T| - 1) \geq \alpha$ ($T$ was in WB[$\alpha$] and an insertion into $T_r$ threw it out of balance) or $(|T_l| + 1)/|T| + 1) \geq \alpha$ ($T$ was in WB[$\alpha$] and a deletion from $T_l$ threw it out of balance).

2. A double rotation as shown in Figure 7.19(b) rebalances $T$ if in case 1 we have instead that $\beta(T_r) > 1/(2 - \alpha)$.

3. The mirror image of the rotation in Figure 7.19(a) rebalances $T$ if $|T_l|/|T| > 1 - \alpha$, $\beta(T_l) \geqslant (1 - \alpha)/(2 - \alpha)$, and either $(|T_l| - 1)/(|T| - 1) \leqslant 1 - \alpha$ ($T$ was in WB[$\alpha$] and an insertion into $T_l$ threw it out of balance) or $|T_l|/(|T| + 1) \leqslant 1 - \alpha$ ($T$ was in WB[$\alpha$] and a deletion from $T$ threw it out of balance).

4. The mirror image of the double rotation in Figure 7.19(b) rebalances $T$ if in case 3 we have instead that $\beta(T_l) < (1 - \alpha)/(2 - \alpha)$.

The verification that these rules do indeed always rebalance the tree for $0 < \alpha \leqslant 1 - \sqrt{2}/2$ in the case of an insertion and $\frac{2}{11} < \alpha \leqslant 1 - \sqrt{2}/2$ in the case of a deletion is elementary but very tedious, and we leave it for Exercises 25 and 26, page 326. Notice the crucial fact that the SIZE fields are easy to maintain as the tree undergoes rotations and double rotations.

Of course, storing the SIZE fields will require much more than the two bits required for the height-condition codes in height-balanced trees. But, as we will see in the applications discussed below, that information is useful, and in certain cases we would have the SIZE fields (or something equivalent) even in height-balanced trees. The extra storage per node, therefore, need not be an impediment to using weight-balanced trees. However, the extra arithmetic required in the rebalancing operations may not be worth its expense. The primary advantage of weight-balanced trees is that the parameter $\alpha$ can be chosen to give some desired trade-off between search time and rebalancing effort.

There is one result known about weight-balanced trees for which no comparable result is known for height-balanced trees: given any sequence of $n$ insertions and deletions on an initially empty weight-balanced tree, the total number of rotations and double rotations required by the $n$ insertions and deletions is $O(n)$. In other words, on *any* single such sequence, the average number of transformations per insertion or deletion is constant. All that is known for height-balanced trees is that on the average over *many* different sequences of insertions and deletions, the number of transformations per insertion or deletion *seems* (on the basis of empirical evidence) to be constant. The statement made for weight-balanced trees is clearly much stronger than that made for height-balanced trees.

**Applications of Balanced Trees to Lists.** The size fields that we used for weight-balanced trees allow us to perform efficiently a search by *index position* of the elements, in addition to the search of Algorithm 7.5 which is by their alphabetic value. That is, we can find the $k$th element in inorder with the SIZE fields as follows. Compute the index position of the element at the root—it is $rank = 1 + \text{SIZE}(\text{left subtree})$. If this is equal to $k$, we are done. If it is greater than $k$, search the left subtree for the $k$th element; otherwise, search the right subtree for the $(k - rank)$th element. Algorithm 7.8 specifies this process more precisely.

$m \leftarrow k$    〚the index position for which we are searching〛
$p \leftarrow T$    〚the tree in which we are searching〛
*found* ← **false**
**while** $p \neq$ **nil and not** *found* **do**
          〚compute the relative index of the root $p$ in its subtree〛
          **if** LEFT($p$) = **nil then** *rank* ← 1
                                    **else** *rank* ← SIZE(LEFT($p$)) + 1
     **case**
          $m < rank$: $p \leftarrow$ LEFT($p$)
          $m = rank$: *found* ← **true**
          $m > rank$:
                         $m \leftarrow m - rank$
                         $p \leftarrow$ RIGHT($p$)
          **if** *found* **then**  〚found: $p$ points to the $k$th element〛
                    **else**  〚not found: $k < 1$ or $k >$ number of elements in
                              the tree〛

**Algorithm 7.8**
Tree search by inorder index position

Of course, we do not need weight-balanced trees to use Algorithm 7.8. It will work correctly in any binary tree that has correct SIZE fields. We have these fields already in weight-balanced trees because we need them to do the rebalancing after an insertion or deletion, but we could easily add them to nodes in height-balanced trees and maintain them just as in weight-balanced trees (in that case, however, it might be more natural to store the rank information directly, not the size—see Exercise 28, page 326). Algorithm 7.8 requires time proportional to the height of the tree, and for height- or weight-balanced trees this will be logarithmic in the number of elements in the tree. Furthermore, it is clear that Algorithm 7.8 can be combined with the insertion, deletion, and rebalancing algorithms, allowing us to make insertions and deletions by index position [for example, to insert a new element between the $k$th and $(k + 1)$st elements or to delete the $k$th element] in $O(\log n)$ time.

We can therefore use weight-balanced trees or height-balanced trees augmented with SIZE fields as a compromise between the linked and sequential storage structures for representing linear lists. The normal sequential storage structure (Section 4.1) could be easily searched by index position but was expensive to modify, while the normal linked structure (Section 4.2) was easy to modify once the location was known but expensive to search. Both searching by index position and insertions/deletions can be done with height- or weight-balanced trees in logarithmic time. This compromise is especially useful in implementing a priority queue (see the introduction to Section 4.3) which operates in a first-in, highest-priority-out-first order. As the elements arrive they are inserted into the tree according to their value by Algorithm 7.7 and the appropriate rebalancing scheme. The element deleted is always the one with the highest priority. In this case we do not even need the full power of Algorithm 7.8 to find the element to be deleted (why?).

Using balanced trees as a storage structure for linear lists suggests the need to be able to concatenate them together and to split them apart, just as we can do with linked lists. Can these operations also be done in logarithmic time? Yes, and in fact the concatenation algorithm is remarkably simple for height-balanced trees. It is similar but slightly more complicated for weight-balanced trees (see Exercise 31). Suppose we want to concatenate height-balanced tree $T_2$ to the right of height-balanced tree $T_1$ and have the result be a height-balanced tree. We proceed as follows. Compute the heights of $T_1$ and $T_2$ in logarithmic time (Exercise 17). Now, assume that height($T_1$) $\geqslant$ height($T_2$); the other case is essentially the mirror image. Delete the first element of $T_2$, call it $Q$, leaving $\hat{T}_2$. We will use $Q$ to paste $T_1$ and $\hat{T}_2$ together. Having kept track of the effect of the deletion of $Q$ from $T_2$, we know the height of $\hat{T}_2$. Starting at the root of $T_1$, we now follow zero or more RIGHT links, keeping track of the heights of the subtrees of the nodes encountered. We keep track of the heights by knowing the height of $T_1$ initially and subtracting either 1 or 2 depending on the height-condition code, 1 if the code was = or $\searrow$ and 2 if it was $\diagup$ (why?). We continue following RIGHT links until we find a node $P$ such that

$$0 \leqslant \text{height}(P) - \text{height}(\hat{T}_2) \leqslant 1$$

(see Exercise 32, page 326). The node

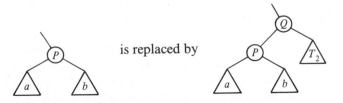

with the height-condition code of $Q$ being = or $\diagup$ depending on height($P$) − height($\hat{T}_2$). Having stored the nodes encountered along the right boundary of $T_1$ on a stack, we then go back up that boundary beginning at the original father of $P$, correcting height-condition codes and performing a rotation or double rotation as though we had done an insertion into the subtree rooted $Q$ and had thereby increased its height by one unit.

The problem of splitting a tree in two, corresponding to splitting a linked list into two pieces, is solved by an insertion followed with a sequence of concatenations. To understand the idea, consider the tree of Figure 7.20. The list represented is the inorder of the tree, $S_1 a S_2 b S_3 c S_4 d T_1 e T_2 f T_3$, and suppose this is to be split into two lists $S_1 a S_2 b S_3 c S_4 d$ and $T_1 e T_2 f T_3$; in other words, the list is to be split after node $d$, as shown in Figure 7.20 by the dashed line. Assume that in

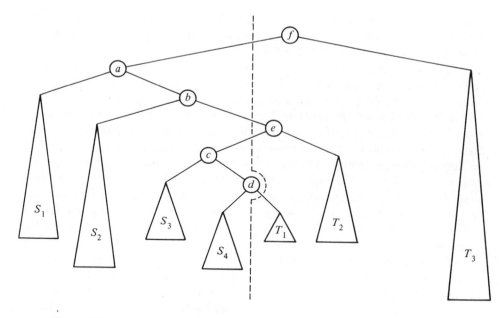

**Figure 7.20**
An example of splitting a binary tree into two pieces based on the inorder of the nodes

tracing the path from the root to $d$ the nodes have been stored on a stack, as in the other height-balanced tree algorithms. We now go back up that path toward the root, breaking the tree apart and concatenating the pieces together to form the desired lists. First $d$ is inserted at the extreme right of $S_4$, to give $S_4 d$. $S_3$ and $S_4 d$ are then concatenated to form $S_3 c S_4 d$ using $c$ as the paste node in the concatenation algorithm. The node $e$ is then used as the paste node in concatenating $T_1$ and $T_2$ to form $T_1 e T_2$. The node $b$ is then used as the paste node in concatenating $S_2$ to $S_3 c S_4 d$, giving $S_2 b S_3 c S_4 d$, which is in turn concatenated to $S_1$ using $a$ as the paste node, giving $S_1 a S_2 b S_3 c S_4 d$. On the last step, $f$ is used as the paste node in concatenating $T_3$ to $T_1 e T_2$, giving $T_1 e T_2 f T_3$.

Algorithm 7.9 gives an outline of the procedure in general. Notice that the algorithm as outlined will work to split any binary tree, not just a height-balanced tree. For example, using the insertion and concatenation procedures for weight-balanced trees, Algorithm 7.9 serves to split a weight-balanced tree, giving two weight-balanced trees as its result. The time required by Algorithm 7.9 will be proportional to the total required by the insertion and the sequence of concatenations. In height- or weight-balanced trees these are potentially $O(\log n)$, and the concatenation process requires $O(\log n)$ time, suggesting that the splitting algorithm might require time proportional to $(\log n)^2$ in the worst case for such balanced trees. Fortunately, however, the concatenation algorithm requires logarithmic time only to delete the node that will be used to paste the trees together. If given that node, as is the case in the concatenations done in the splitting process, the concatenation will require only time proportional to the difference in height of the two trees being concatenated. This leads to a logarithmic worst-case time for splitting balanced trees (Exercise 33).

$x \Leftarrow stack$ 〚stack contains the path from the node back up to the root; $x$ is the node on that path we are currently processing〛
$S \leftarrow$ tree formed by inserting $x$ at right of LEFT($x$)
$T \leftarrow$ RIGHT($x$)
**while** $stack \neq empty$ **do**
    $oldx \leftarrow x$
    $x \Leftarrow stack$
    **if** $oldx =$ RIGHT($x$) **then** $S \leftarrow$ tree formed by concatenating $S$ to the right of LEFT($x$) using $x$ as the paste node
                **else** $T \leftarrow$ tree formed by concatenating RIGHT($x$) to the right of $T$ using $x$ as the paste node
〚at this point $S$ is the left tree and $T$ is the right tree that results from the split〛

**Algorithm 7.9**
Splitting a binary tree in two pieces based on the inorder of the nodes

Two final remarks about representing lists by trees are in order. First, the algorithms described can be used *without* the rebalancing parts, essentially allowing the trees to grow randomly. If the insertions, deletions, concatenations, splittings, and searches were all random, the resulting trees would probably maintain logarithmic height on the average. *But* in most applications it is extremely unlikely that the sequences of operations would be truly random; rather, biases would occur that would cause the trees to deteriorate badly. Second, if possible when using binary trees to represent lists, FATHER pointers should be maintained in the nodes. This will greatly facilitate the algorithms that require retracing the path from a node to the root: insertion, deletion, concatenation, and splitting. Furthermore, it will allow the deletion of a node given only a pointer to the node; in this sense FATHER pointers give something of an analog to doubly linked lists.

**Exercises**

1. Use elementary integral calculus to show that $H_n = \sum_{i=1}^n 1/i \approx \ln n$. [*Hint:* Compare $H_n$ to $\int_1^n dx/x$ and use the rectangle rule.]

2. This exercise is an alternative way of deriving the result of Equation (7.16), which gives the external path length of a randomly constructed binary search tree of $n$ elements. Let $U_n$ be the average number of probes in an unsuccessful search in a given binary search tree $T$ of $n$ nodes. Let $U_{n+1}$ be the average number of probes in an unsuccessful search after a random insertion has been made in $T$.

(a) What is $U_1$?

(b) What is the relationship between $U_n$ and the $E_n$ of Equation (7.16)?

(c) Prove that $U_{n+1} = U_n + 2/(n + 2)$. [*Hint*: If the insertion is at level $i$, by how much does the external path length increase? What is the average level of a leaf in the tree?]

(d) Use the above results, along with Exercise 1, to obtain (7.16).

(e) Let $S_n$ be the average number of probes in a successful search of a randomly constructed binary search tree of $n$ elements. Prove that $S_n = 1 + (U_0 + U_1 + \cdots + U_{n-1})/n$.

3. Prove that if a node $X$ in a binary tree has a nonnil left (right) son, then its inorder predecessor (successor) has a null right (left) son.

4. Write out the details of the algorithm to delete an element from a binary search tree.

5. Derive a recurrence relation for $H_h$, the number of distinct height-balanced trees of height $h$. How fast does $H_h$ grow?

6. Compute the external path length of the tree $T_h$ as defined by (7.17) and (7.18).

7. Find height-balanced trees whose internal path length is close to $(1 - \sqrt{5}/20)/\lg[(1 + \sqrt{5})/2] \approx 1.44$ times that of the completely balanced tree of Section 5.3.

8. In a height-balanced tree of $n$ internal nodes, at least how many *must* have = as their height-condition code? (*Answer*: $[(3 - \sqrt{5})/2]n \approx 0.38n$.)

9. Can either or both of $T_1$ and $T_4$ in Figure 7.15(b) be empty?

10. Explain how the transformations of Figure 7.15 might be used to implement *self-organizing binary search trees*, in which the more an element is accessed the closer it moves to the root.

11. The backward scan up the tree from the site of the insertion in a height-balanced tree can be eliminated at the expense of retracing part of the path down the tree. This can be done roughly as follows. As we go down in the tree to the site of the insertion, keep track of the node $S$ that is the latest

one along the path to have height condition code $\searcrow$ or $\diagup$. When the insertion is made, each of the elements between $S$ and the newly inserted element has height-condition code $=$ and each must be changed to $\searrow$ or $\diagup$. It is at $S$ that a rotation or double rotation may be needed. Work out the details of this insertion algorithm.

12. Draw the trees for the subcases of case 4 in the algorithm for deletion from a height-balanced tree.

13. Under what conditions does a double rotation also rebalance the tree in case 3(a) in the algorithm for deletion from a height-balanced tree.

14. Prove that the deletion algorithm for height-balanced trees may require as many as $\lfloor h/2 \rfloor$ rotations and double rotations for a tree of height $h$, but no more.

15. Devise a technique for height-balanced trees that requires only one bit per element for the height-condition code. [*Hint*: If a node in the tree has two sons, then its two-bit height-condition code can be stored with one bit in each son. What if a node has zero or one son?]

16. As an alternative to the two-bit height-condition code or the complication of Exercise 15, we can insist that if one subtree is taller than its brother, it must be the right subtree. Thus a *one-sided, height-balanced tree* always has $0 \le h(T_l) - h(T_r) \le 1$. Obviously, a single bit suffices in this case for the height-condition code. Develop insertion and deletion algorithms for these trees.

17. Give an $O(\log n)$ algorithm to determine the height of a height-balanced tree.

18. Determine the exact shape of the height-balanced tree that results when $n$ elements are inserted in increasing order.

19. Develop, in parallel with the presentation given in the text for height-balanced trees, the notion of height-balanced trees in which the heights of the left and right subtrees can differ by at most 2.

20. Show that for all $\alpha$, $0 < \alpha \le \frac{1}{3}$, the tallest possible WB[$\alpha$] tree containing $n + 1$ internal nodes is at least as tall as the tallest possible WB[$\alpha$] tree containing $n$ internal nodes. (*Hint*: Show that, given a WB[$\alpha$] tree, a node can be added without throwing the tree out of WB[$\alpha$].)

21. What is the least number of nodes in a tree of height 10 in (a) WB[$\frac{1}{2}$], (b) WB[$\frac{1}{3}$], (c) WB[$1 - \sqrt{2}/2$], (d) WB[$\frac{1}{4}$], (e) WB[$\frac{2}{11}$]?

22. Use induction on the number of nodes to prove that the internal path length of a tree in WB[$\alpha$] is at most

$$\frac{1}{H(\alpha)}(n+1)\lg(n+1) - 2n,$$

where $H(\alpha) = -\alpha \lg \alpha - (1-\alpha)\lg(1-\alpha)$.

23. Verify the formulas for the balances $\beta'_A$, $\beta'_B$, and $\beta'_C$ in Figure 7.19(b).

★24. Interpreting $\beta_A$, $\beta_B$, and $\beta_C$ as probabilities, derive the formulas for $\beta'_A$, $\beta'_B$, and $\beta'_C$ in Figures 7.19 by using the law of *conditional probability*. [*Hint*: In Figure 7.19(a) $\beta_A$ is the probability that an external node is in $T_1$. What is the conditional probability that an external node is in $T_2$, given that it is in either $T_2$ or $T_3$?]

25. Does the rebalancing algorithm for weight-balanced trees work in the case of a deletion for $\alpha = \frac{2}{11}$?

26. Verify that the rebalancing algorithms for weight-balanced trees really do restore the balances properly for $0 < \alpha \leq 1 - \sqrt{2}/2$ in the case of an insertion and $\frac{2}{11} < \alpha \leq 1 - \sqrt{2}/2$ in the case of a deletion.

27. Suppose that, instead of using a field SIZE in a node giving the number of nodes in the subtree rooted at that node, we use a field RANK that gives the number of nodes in the left subtree of that node plus one. Suppose further that the tree is stored as the left son of the header node and that the header node contains a RANK field, the value of which would be, naturally, the number of nodes in the tree $T$ plus one. Show that these RANK fields are sufficient to maintain weight-balanced trees.

28. Show that both the SIZE and RANK fields can be properly maintained in logarithmic time in height- or weight-balanced trees under the various operations presented.

29. Give the details of the algorithms for insertion and deletion from a weight-balanced tree.

★30. Assuming that the balance of each node in a WB[$\alpha$] tree is uniformly distributed over the interval [$\alpha, 1 - \alpha$], independently of the balances of other nodes, show that the expected number of rotations and double rotations required to insert or delete a node is less than $2/(1 - 2\alpha)$.

31. Show that weight-balanced trees can be concatenated in logarithmic time.

32. In the algorithm to concatenate height-balanced trees we require that $0 \leq \text{height}(P) - \text{height}(\hat{T}_2) \leq 1$. Why don't we allow $\text{height}(P) - \text{height}(\hat{T}_2) = -1$?

33. Verify that Algorithm 7.9 for splitting a tree requires only $O(\log n)$ time for height-balanced trees. Is this also true for weight-balanced trees?

34. What changes would be necessary to Algorithm 7.9 if we wanted to split the tree *before* (rather than after) the splitting node?

35. In the application of representing a list as a binary tree, it might be more natural to keep a field INDEX rather than SIZE or RANK, where the INDEX field of the $i$th element in the list being represented is $i$. Comment on the advisability of this suggestion.

36. Consider the following memory management problem. Memory consists of an infinite number of locations that are to be allocated according to the first-fit rule: when $k$ locations are requested, they are to be allocated from the *leftmost* (lowest-address) block of locations currently available. Describe a data structure and algorithm for implementing reservations and releases so that each of these operations requires only time proportional to $\log n$, where $n$ is the number of blocks currently available.

37. Consider $n$ elements $x_1 < x_2 < \cdots < x_n$ organized into the following data structure. A subset of $l - 1$ elements $x_{i_1} < x_{i_2} < \cdots < x_{i_{l-1}}$ divides the rest

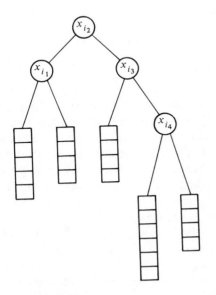

**Figure 7.21**
Exercise 37

of the keys into $l$ subfiles; the $j$th subfile consists of elements between $x_{i_{j-1}}$ and $x_{i_j}$. A *file tree* is the structure obtained linking these $l-1$ elements and $l$ subfiles together, where each subfile is organized as an ordered sequential list. The elements $x_{i_1}, x_{i_2}, \ldots, x_{i_{l-1}}$ are organized as a height-balanced binary tree. The resulting structure looks like the structure in Figure 7.21. There is an additional restriction that none of the subfiles can have more than $3m$ elements or less than $2m$ elements. (In the above example $m = 2$.) To search a *file tree* for $z$ we search as in binary search trees to find the proper subfile and then search that subfile sequentially.

(a) Describe a procedure to insert a new element into a file tree so that it remains a file tree.

(b) How many key comparisons are required in the worst case for your procedure?

38. Suppose we want to represent a priority queue that will never contain more than ten elements. Should we use the balanced-tree representation presented here?

39. Develop algorithms that keep trees balanced by bounding the amount by which the external path lengths of the two subtrees of a node can differ.

## 7.3  DIGITAL SEARCH TREES

We can use trees to organize tables based on the representation of the elements, rather than on the ordering of the elements as in the previous section. Actually, we have already seen a simple example of such an organization: Figure 5.6 shows how a binary tree could be used to store a table for decoding characters in Morse code. The principal idea there is to go down the tree starting at the root, going left or right based on whether the next bit is a dot or a dash, respectively. This idea is easily extended to larger alphabets, say decimal digits or alphabetic characters, in place of dots and dashes. If the alphabet contained $c$ characters, each node in the tree would be a $c$-way branch—one branch for each possible character. The structure thus obtained, called a *digital search tree* or *trie* (taken from the middle letters of the word "re*trie*val," but pronounced "try"), is illustrated in Figure 7.22 which shows 30 common English words.

To explain Figure 7.22 we will describe how to search it. Suppose we are given the word HAVE as the object of the search. We consider each of the letters H, A, V, E in turn, starting at the root of the tree and proceeding as follows. Follow the branch labeled H out of the root; at the next node follow the branch labeled A, then the branch labeled V, and finally the branch labeled E. At that point we are at the bottom of the tree and the letters of the search object are

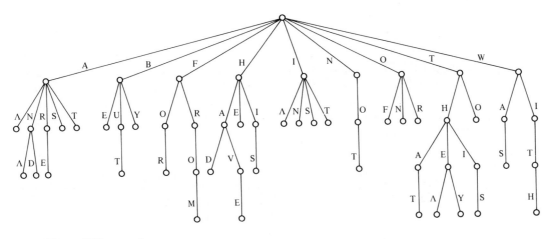

**Figure 7.22**
A digital search tree consisting of 30 common English words

exhausted, so we have successfully found HAVE in the tree. If the search object had been HAVING, we would have followed down branches in the tree corresponding to H, A, and V, but then there would be no branch labeled I, indicating that HAVING is not in the tree. If needed, we could store information about the word in a record pointed to by the last letter. A digital search tree is only a thumb-index arrangement in the extreme: at the top level there is a section for each initial character, within each initial letter there is a section for each possible second character, and so on.

Notice at the extreme lower left of Figure 7.22 there are branches of the tree labeled $\Lambda$. This special symbol is used to distinguish between words in the tree that are prefixes of one another. In Figure 7.22 for example, we have the words A, AN, and AND. The $\Lambda$ as a possible character after taking the branch labeled A from the root means that A itself is an element of the tree. We thus need to modify the search algorithm outlined above as follows. To search for an object $C_1 C_2 \cdots C_k$ in the tree, we follow the branch out of the root labeled $C_1$, then follow the branch labeled $C_2$, and so on. After following the branch for $C_k$, if the node reached has no outgoing branches or has a branch labeled $\Lambda$ then we have successfully found the object. Otherwise, the search fails. The search also fails if at any point along the way there was no branch labeled $C_i$ as required. When a search object is not in the tree, the search procedure finds the element in the tree that has the longest match of initial characters with the search object.

The advantage of a digital search tree is that in many circumstances the multiway branch required at every node of the tree will require little or no more time than a binary decision. In $c$-way branching, on the average only $\log_c n = \lg n / \lg c$ tests would be needed instead of, say, $\lg n$ for binary search. For example, in the case of alphabetic information $c = 27$ (why not 26?) and $1/(\lg 27) \approx 0.21$, indicating that only about a fifth as many tests would be made in a digital search tree based on letters as would be made by binary search. Of course, if the 27-way branch were five times as expensive as a two-way branch, there would be no savings. Typically, though, by using subscripts into an array it is possible to implement a $c$-way branch in only slightly more time than is required by a two-way branch. In the case of alphabetic characters each tree node could be of the form

Assuming the binary representations of the letters A, B, C, and so on are $00001, 00010, 00011, \ldots$, we can choose $00000$ as the binary representation for $\Lambda$ and use these "values" of the letters as indices into the particular record.

For the tree of Figure 7.22 we would need 27 such records, one for each node in the tree that has a descendant. Each record has room for 27 pointers, so the total storage requirements are about $27 \times 27 = 729$ pointers to represent a table of 30 words totaling only 79 characters. Unfortunately, this rather excessive storage requirement is typical of digital search trees! It arises from the fact that with records as shown above for tree nodes, there is an enormous amount of room for elements *not* in the table (why?). It is precisely this characteristic that allowed us the $c$-way branches that lead to $\log_c n$ average search times.

There is an obvious possibility for reducing the heavy storage requirements of digital search trees. We could use the natural correspondence (see Section 5.1.1) to represent the tree of Figure 7.22 as the binary tree shown in Figure 7.23. Searching the tree in this form means going to the RIGHT zero or more times until the desired character is found, going LEFT one step, and then repeating the process for the subsequent characters of the search object (Exercise 3). This process has replaced $c$-way branches by two-way branches, and we know from Section 5.3 that the average and worst-case number of such branches examined in searching such a tree cannot be less than $\lg n$, quite poor in comparison to $\log_c n$. This method for reducing the storage requirement of digital search trees is thus *not* to be recommended.

A second method of reducing storage requirements in digital search trees is to collapse sequences of nodes in which no branching occurs into a single node. Even in this small example the payoff for the additional complication is significant (we could make 14 such changes to the tree, saving 14 nodes, or more than half of the space); in many cases the savings would be even more dramatic (see

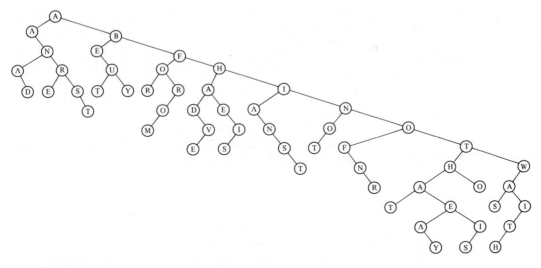

**Figure 7.23**
The binary tree resulting from the digital search of Figure 7.22 by the natural correspondence

Exercise 2). This collapsing need not be done just at the bottom of the tree. If we had a digital search tree of names of fish containing both "REDFIN" and "REDFISH", we could use

instead of

A third, even more useful strategy to reduce the storage requirements is to take a cue from the thumb-index analogy and use a digital search tree only for the few top levels, changing to some other technique when the number of elements is small enough.

Another weakness of digital search trees is that it would take a relatively long time to distinguish between almost identical words such as "IRRESPONSI-BLE" and "IRRESPONSIBILITY", while binary search would do so in only one comparison. In the nonrandom elements found in many practical applications such identical sequences of initial characters are common. Given a set of elements to be organized into a digital search tree, it may therefore be advantageous to base the tree on a different order of the characters in an element—for example, taking the characters of an element from right to left rather than left to right. In general the problem of determining the *optimal* order in which the characters should be considered (given some probability distribution for the search objects) is computationally intractible.

Finally, digital search trees can be used for dynamic tables, since insertions and deletions are not hard to accommodate. The time for such operations cannot be guaranteed to be logarithmic in the table size, however.

### Exercises

1. Give a digital search tree for the following approximations to well-known mathematical constants: $\sqrt{2} \approx 1.414$, $1/\ln 2 \approx 1.44$, $(1 + \sqrt{5})/2 \approx 1.6$, $\ln 10 \approx 2.3$, $e \approx 2.718$, $\pi \approx 3.14159$, $\gamma \approx 0.57722$, and $\cos 1 \approx 0.5403$. Do the same for these approximations in *octal*.

2. Give a digital search tree for the names of the books of the Old Testament of the Bible. Modify the tree by collapsing sequences of two or more nodes that have no branches and compute the amount of storage thus saved.

3. Give the complete algorithm to search a digital search tree that is represented by the binary tree arising from the natural correspondence.

4. Give the digital search tree for the same elements as in Figure 7.22, except with the letters of the elements in reverse order. Which of these two trees is more efficient? Consider space and average search time with some reasonable distribution of search objects.

★5. Prove or disprove: In an alphabet of $c$ characters the average length of $n$ distinct equally probable strings of characters cannot be much less than $\log_c n$ as $n$ gets large.

## 7.4  HASHING

We now examine a class of table organizations in which we attempt to store elements in locations that are easily computed from the value or representation of the elements. This contrasts markedly with the techniques presented in previous sections of this chapter; in those sections we based a search on comparisons (usually binary), and the location in which an element was stored depended on its position in an *ordered* arrangement of the elements. In this section we discuss techniques based on *directly* transforming the element into an address at which it will be stored.

A very simple version of this technique has already been presented in the discussion of multidimensional arrays in Section 3.1.1. There we considered the layout in memory of an array $A[l_1: u_1, l_2: u_2, \ldots, l_n: u_n]$ and gave a formula for the location of $A[i_1, i_2, \ldots, i_n]$ as a function of $i_1, i_2, \ldots,$ and $i_n$. Considering the $n$-tuple $(i_1, i_2, \ldots, i_n)$ as the key $z$ and the associated array value $A[i_1, i_2, \ldots, i_n]$ as some additional information to be carried in a field $A(z)$ in the corresponding record, the problem of finding $z$ reduces to the problem of computing the location from the $n$-tuple comprising $z$; that is where $A[z]$ will be found. Because there is a one-to-one correspondence between $n$-tuples $(i_1, i_2, \ldots, i_n)$ and locations, we do not need to store the $n$-tuple itself, only the array value. For the same reason, we need not worry about two different $n$-tuples yielding the same location.

More generally, we will suppose that we have an array of $m$ table locations $T[0], T[1], \ldots, T[m - 1]$, say, and given an element $z$ to be inserted we transform it to a location $h(z)$, $0 \leqslant h(z) < m$; $h$ is called a *hash function*. We then examine $T[h(z)]$ to see if it is empty. Most of the time it will be, so we set $T[h(z)] \leftarrow z$ and we are done. If $T[h(z)]$ is not empty, a *collision* has occurred and we must resolve it somehow. Taken together, the hash function and the collision resolution method are referred to as *hashing* or *scatter storage* schemes. In Section 7.4.1 we discuss various possible hash functions, and in Section 7.4.2 we discuss alternative strategies for resolving collisions.

For ease of presentation we will assume that the elements to be stored in the hash table are small enough to be stored in their entirety in the table locations. In practice, if the elements are large records, we would instead store only pointers to the actual records, supplied by **getcell**, in the table locations. This procedure slows down access into the table slightly because we must follow an extra pointer to reach the record contents. It also increases the storage overhead. It is advantageous, however, because it allows us to use a larger table size $m$ which in turn will lead to better performance, as we will see.

Under the proper conditions, hashing is unsurpassed in its efficiency as a table organization, since the average times for a search or an insertion are generally *constant, independent of the size of the table*. However, some important caveats are in order. First, hashing requires a strong belief in the law of averages, since in the worst case collisions occur every time and hashing degenerates into linear search! Second, while it is easy to make insertions into a hash table, the full size of the table must be specified *a priori*, because it is usually closely connected to the hash function used; this makes it extremely expensive to change dynamically: choose too small a size and the performance will be poor and the table may overflow, but choose too large a size and much memory will be wasted. Third, deletions from the table are not easily accommodated. Finally, the order of the elements in the table is unrelated to any natural order that may exist among the elements, and so an unsuccessful search results only in the knowledge that the element sought is not in the table, with no information about how it relates to the elements in the table.

### 7.4.1  Hash Functions

The hash function takes an element to be stored in a table and transforms it into a location in the table. If this transformation makes certain table locations more likely to occur than others, the chance of collision is increased and the efficiency of searches and insertions is decreased. The phenomenon of some table locations being more likely is called *primary clustering*. The ideal hash function spreads the elements uniformly throughout the table—that is, does not exhibit primary clustering. In fact, we would really like a hash function that, given any $z$, chooses a random location in the table in which to store $z$; this would minimize primary clustering. This is, of course, impossible, since the function $h$ cannot be probabilistic but must be deterministic, yielding the same location every time it is applied to the same element (otherwise, how would we ever find an element after it was inserted?!). The achievable ideal is to design hash functions that exhibit pseudorandom behavior—behavior that appears random but that is reproducible.

Unfortunately, there are no hard and fast rules for constructing hash functions. We will examine four basic techniques that can be used individually or in combination. The properties of any particular hash function are hard to determine because they depend so heavily on the set of elements that will be encountered in practice. Thus the construction of a good hash function from these basic techniques is more an art than anything amenable to analysis, but we will present general principles that usually prove successful, pointing out their pitfalls as well.

For convenience we will assume that an element to be hashed is encoded as a string of bits that can also be interpreted as an integer written in binary. The hash function has to take such a string of bits and produce an address from it. As a continuing example, we will use the 30 common English words in the trie of Figure 7.22. We will assume that each letter is encoded as a five-bit binary string, $A = 00001, B = 00010, C = 00011, \ldots, Z = 11010$. (This corresponds to the binary representation of integer($c$) $-$ integer("A") $+ 1$ if the characters are represented in USASCII—see Section 2.2.3.) Each word is thus a concatenation of such strings, so that, for example, THE would be represented as the fifteen-bit string

$$\underbrace{10100}_{T}\underbrace{0100000}_{H}\underbrace{0101}_{E} = 20{,}741_{10}$$

and OF would be represented as the ten-bit string

$$\underbrace{01111}_{O}\underbrace{00110}_{F} = 486_{10}.$$

**Extraction.** The simplest hash functions are those that merely extract a few scattered bits from an element, putting those bits together to form an address.

| Word | Binary Form | Third bit $\downarrow$ $\sqsubset$ Last two bits |
|------|-------------|--------------------------------|
| THE | 10100 01000 00101 | $(\overset{\frown}{1}01)_2 = 5$ |
| OF | 01111 00110 | $(110)_2 = 6$ |
| AND | 00001 01110 00100 | $(000)_2 = 0$ |
| TO | 10100 01111 | $(111)_2 = 7$ |
| A | 00001 | $(001)_2 = 1$ |
| IN | 01001 01110 | $(010)_2 = 2$ |
| THAT | 10100 01000 00001 10100 | $(100)_2 = 4$ |
| IS | 01001 10011 | $(011)_2 = 3$ |

**Table 7.1**
A simple hash function based on extraction.

Suppose we want to store the eight words THE, OF, AND, TO, A, IN, THAT, and IS in a table. We could take the third bit from the left and the last two bits on the right as in Table 7.1. This works out neatly and is a perfect way to fit the eight words into $T[0], T[1], \ldots, T[7]$. Of course, the same hash function would do rather poorly with the words THE, FROM, THEY, ONE, DRY, TEA, GLUM, and WE, each of which would give the result $(101)_2 = 5$.

Extraction is generally a poor way to do hashing except in ad hoc situations where the table contents are completely known in advance and the bits to be extracted can be carefully chosen to prevent primary clustering. The weakness of extraction as a technique for hashing is that the resulting location depends only on a small subset of the bits of an element. A first principle in the design of hash functions is thus that *the hash location should be a function of every bit of the element*.

**Compression.** A simple way to get a location from an element in such a way that every bit of the element participates is to compress the bits of the element into the number required for an address. We could, for example, break the bit string to be hashed into fixed-length segments and then add them up as binary numbers or take their exclusive-or. The function $h_1$ shown in Table 7.2 is just such a hash function, being formed by the exclusive-or of the bit strings of the letters in a word; thus $h_1(\text{THE}) = 10100$ **xor** $01000$ **xor** $00101 = 11001 = 25_{10}$. Exclusive-or may be somewhat preferable to addition because there is no need to worry about arithmetic overflow (also see Exercise 1).

One weakness of such a method of compression is that the operations of addition and exclusive-or are commutative, so that $a + b = b + a$ and $a$ **xor** $b = b$ **xor** $a$. This means that in our example different words formed from the same letters will hash to the same location: $h_1(\text{STEAL}) = h_1(\text{STALE}) = h_1(\text{TALES}) = h_1(\text{LEAST})$, and so on, suggesting another principle in the design of hash functions: *a hash function should break up naturally occurring clusters of*

| word | $h_1(word)$ | $h_2(word)$ | $h_3(word)$ |
|------|-------------|-------------|-------------|
| THE | 25 | 2 | 23 |
| OF | 9 | 21 | 21 |
| AND | 11 | 19 | 4 |
| TO | 27 | 4 | 7 |
| A | 1 | 1 | 19 |
| IN | 7 | 23 | 30 |
| THAT | 9 | 18 | 17 |
| IS | 26 | 28 | 2 |
| WAS | 5 | 12 | 26 |
| HE | 13 | 13 | 27 |
| FOR | 27 | 8 | 16 |
| IT | 29 | 29 | 20 |
| WITH | 2 | 29 | 7 |
| AS | 18 | 20 | 8 |
| HIS | 18 | 5 | 30 |
| ON | 1 | 29 | 18 |
| BE | 7 | 7 | 8 |
| AT | 21 | 21 | 26 |
| BY | 27 | 27 | 16 |
| I | 9 | 9 | 16 |
| THIS | 6 | 25 | 23 |
| HAD | 13 | 13 | 30 |
| NOT | 21 | 18 | 21 |
| ARE | 22 | 24 | 4 |
| BUT | 3 | 12 | 11 |
| FROM | 22 | 21 | 3 |
| OR | 29 | 2 | 2 |
| HAVE | 26 | 5 | 26 |
| AN | 15 | 15 | 6 |
| THEY | 0 | 27 | 1 |

**Table 7.2**

Examples of the three major types of hash functions for the set of words used in the trie of Figure 7.22. $h_1(word)$ = the **xor** of the characters; for example, $h_1(\text{THE}) = 10100 \textbf{ xor } 01000 \textbf{ xor } 00101 = 11001 = 25_{10}$. $h_2(word)$ = value of word mod 31; for example, $h_2(\text{THE}) = 101000100000101_2 \bmod 31 = 20741_{10} \bmod 31 = 2_{10}$. $h_3(word) = 1 + \lfloor 30(\text{fractional part of } 0.6125423371 \times \text{value of word})\rfloor$; for example, $h_3(\text{THE}) = 1 + \lfloor 30(\text{fractional part of } 0.6125423371 \times 20741)\rfloor = 1 + \lfloor 30 \times 0.74061 \rfloor = 1 + 22 = 23$.

*elements.* In the case of $h_1$, this can be done by shifting different segments circularly by different amounts. Instead of the function $h_1$ of Table 7.2 we might therefore use $\hat{h}_1$, in which the first segment is left alone, the second is circularly shifted one position, the third two positions, and so on (Exercise 2).

In general, compression techniques are most useful for converting multi-word elements into a single word, making it easy to apply either the division- or multiplication-based hash functions that we now describe. These two classes of hash functions have proven to be extremely reliable in practice.

**Division.** Given a table $T[0], T[1], \ldots, T[m-1]$, we can take

$$h(z) = z \bmod m,$$

that is, $h(z)$ is the remainder when $z$ is divided by $m$. The function $h_2$ in Table 7.2 is a hash function of this type with $m = 31$; therefore $h_2(\text{THE}) = 101000100000101_2 \bmod 31 = 20{,}741_{10} \bmod 31 = 2$. Such a hash function satisfies a third design principle: *a hash function should be very quick and easy to compute*. Of course, to use such a division hash function we must choose the value of $m$ carefully in order to satisfy our first two design principles, namely that the hash location depends on all the bits of the element and that naturally occurring clusters are broken up.

Suppose $m$ is even. Then even elements will hash to even locations and odd elements will hash to odd locations. More generally, if $d$ divides $m$, then $h(z) \equiv z \pmod{d}$, which would greatly increase collisions in the unfortunate event that the elements are not equally distributed across the equivalence classes modulo $d$. This strongly suggests that $m$ be prime, or at least have no "small" prime factors.

Let $r$ be the radix of the character set; in our example $r = 32$, because each letter is represented by a string of five bits, but $r$ would more usually be 256 (on a binary computer with eight-bit bytes) or 100 (on a decimal computer). If $r \equiv 1 \pmod{m}$, then $h(z)$ would degenerate into the sum of the individual characters modulo $m$ (why?), so that all permutations of given letters have the same hash location. Exactly this occurs in $h_2$ of Table 7.1, where $r = 32$ and $m = 31$ (Exercise 4). Similarly, if $r^k \equiv 1 \pmod{m}$ for some $k$, then $h(z)$ is a sum of blocks of $k$ characters modulo $m$ (again, why?). In general, $r^k \equiv \pm a \pmod{m}$ for "small" $a$ results in combinations of individual characters or groups of characters that can be problematic. This suggests that in addition to being a prime number, $m$ should *not* be of the form $r^k \pm a$ for small values of $a$. Getting back to our example of $r = 32$, we should choose as our value of $m$ some prime *not* in the vicinity of 32, $32^2 = 1024$, or $32^3 = 32{,}768$.

**Multiplication.** Given a real number $\theta, 0 < \theta < 1$, we can construct a hash function as follows. To find $h(z)$ compute $(z\theta \bmod 1)$—that is, the fractional part of $z\theta$. Multiplying this fractional part by $m$ and taking the floor gives a value

$$h(z) = \lfloor m(z\theta \bmod 1) \rfloor$$

satisfying $0 \leq h(z) < m$. The function $h_3$ in Table 7.2 is such a function with $\theta = 0.6125423371$ and $m = 30$; 1 has been added so that $h_3$ can be used for later examples that require a hash function with a range $1 \leq h(z) \leq m$.

Unlike the division hash function, we need not be concerned with the table size $m$, but we do need some guidelines in choosing $\theta$. We do not want $\theta$ to be too close to 0 or 1, since this would cause small elements to cluster at the ends of the table. For instance, if $\theta = 0.99999$, all one- and two-letter words would hash to the last 1 percent of the table. Similarly, we do not want $(r^k \theta \bmod 1)$ to be close to 0 or 1, where $r$ is the radix of the character set, for then elements of the form $ar^k + b$ will cluster for small values of $a$. For example, if $\theta = 0.3066407$, then all three-letter words ending in AT will have hash locations of the form $m[(32^2 a + 52)\theta \bmod 1]$ where $2 \leq a \leq 22$ (for example, BAT = 00010 00001 10100 = $32^2 \times 2 + 52$). But $(32^2 \theta \bmod 1) \approx 0.000077$, making the contribution of the initial letter of the word negligible, so that BAT, CAT, EAT, and so on will all hash to around $0.94m$.

Values approximately $i/(r - 1), 1 \leq i < r - 1$, are also problematic choices for $\theta$, since $z\theta \bmod 1$ will tend to cluster elements differing only by a permutation of characters. This occurs because

$$\frac{1}{r - 1} = \frac{1}{r} + \frac{1}{r^2} + \frac{1}{r^3} + \cdots,$$

written $(0.11\ldots)_r$, and the product of $z$ and such a $\theta$ will have base $r$ digits that are essentially sums of base $r$ digits of $z$—that is, sums of the individual characters of $z$. For example, consider $\theta = 6/31 \approx 0.193548387$ which, expressed as a decimal fraction, appears quite innocent. However, with our five-bit character encoding, EAT, TEA, ETA, and ATE all result in $z\theta \bmod 1$ being approximately 0.032258. Such values of $\theta$ are obviously to be avoided. In a similar fashion, a value of $\theta$ close to $i/(r^2 - 1), 1 \leq i < r^2 - 1$, will tend to cluster elements that differ only by independent permutations of the first, third, fifth,\ldots, positions and the second, fourth, sixth,\ldots positions (why?). For example, with $\theta = 479/1023 \approx 0.468230694$, MATE, META, and TAME give $z\theta \bmod 1$ as about 0.2610, while MEAT gives it as about 0.4731 and TEAM as about 0.0792. Generally, a value of $\theta$ close to $i/(r^k - 1), 1 \leq i < r^k - 1$, will exhibit a corresponding type of clustering, but in most cases we need only consider $k = 1$ or 2 [unless there is some special reason to suspect that the elements occurring in practice will differ from each other by independent permutations of the first, $(k + 1)$st, $(2k + 1)$st,\ldots positions, the second, $(k + 2)$nd, $(2k + 2)$nd,\ldots positions, the third, $(k + 3)$rd, $(2k + 3)$rd,\ldots positions, and so on, for some $k > 2$].

As a final comment on the choice of $\theta$, various theoretical results suggest that $\theta = (\sqrt{5} - 1)/2 \approx 0.6180339887$ or $\theta = 1 - (\sqrt{5} - 1)/2 \approx 0.3819660113$ tend to spread values out most uniformly of all choices for $\theta$. Care should be taken in using such a $\theta$ with $r = 100$, however, because the fractional part of $10^6 \times 0.3819660113$ is close to 0.

The computation of $h(z) = \lfloor m(z\theta \bmod 1) \rfloor$ can be simplified to a single-integer multiplication followed by the extraction of a block of bits from the product, if $m$ is properly chosen. For simplicity, assume that we are working on binary computer with $w$-bit words; the parallel case for a decimal computer is essentially identical. We write $\theta = q/2^w$ and choose a $w$-bit integer $q$ so that the resulting $\theta$ does not exhibit primary clustering as outlined in the preceding paragraphs. Choosing $m = 2^l$ causes $\lfloor m(z\theta \bmod 1) \rfloor$ to be a block of $l$ bits in the integer product $qz$:

$$z\theta = zq/2^w = (\longleftarrow t\text{ bits}\longrightarrow)(\longleftarrow w\text{ bits}\longrightarrow)/2^w$$

$$= (\longleftarrow\qquad t + w\text{ bits}\qquad\longrightarrow)/2^w$$

$$= \left(\underset{\substack{\text{radix}\\\text{point}}}{\longleftarrow t\text{ bits}\longrightarrow}\cdot\underset{z\theta\bmod 1}{\longleftarrow w\text{ bits}\longrightarrow}\right)$$

Clearly, $\lfloor 2^l(z\theta \bmod 1) \rfloor$ is just the leftmost $l$ bits of the segment labeled $z\theta \bmod 1$. Thus we have

$$zq = (\longleftarrow\qquad t + w\text{ bits}\qquad\longrightarrow).$$
$$\longleftarrow w\text{ bits}\longrightarrow$$
$$\underset{h(z)}{\longleftarrow l\text{ bits}\longrightarrow}$$

**Summary.** There is no such thing as a general-purpose hash function! A good hash function should be efficient to compute, requiring only a few machine instructions, and it should minimize primary clustering. Both of these criteria depend heavily on the particular computer and application, so it is best to think carefully about the nature of the elements being hashed, looking for any nonuniformities of the machine encoding of characters and any naturally occurring clusters of elements. We must bear in mind, however, that even an excellent hash function for a given application will occasionally be defeated in its attempt to minimize primary clustering.

## Exercises

1. Why should the operations of **and** and **or** not be considered for compression instead of **xor**?

2. Compute $\hat{h}_1$ for the words of Table 7.1 where $\hat{h}_1$ is like $h_1$ *except* that, as the **xor** of the characters is being accumulated, the accumulation is circularly right-shifted by one bit position before the next character is **xored** with it.

3. Under what conditions would compression hash functions be ill advised?

4. Verify that if two words differ from each other only by a permutation of letters, then $h_2$ of Table 7.1 maps them to the same location.

5. Under what conditions would a multiplicative hash function be ill advised?

6. In some applications elements tend to occur in arithmetic progressions (consider variable names *PARTA*, *PARTB*, *PARTC* or *T*1, *T*2, *T*3, for example). Which of the classes of hash functions described in the text avoid primary clustering in such a situation?

7. A hash function that has fallen into disrepute is the *middle-square* technique in which $h(z)$ is extracted from the middle bits (or digits) of $z^2$. This method is very similar to the multiplicative hash functions described, so we might think it acceptable or even desirable. Explain why it is definitely undesirable!

8. When might the hash function $h(z) = 0$ be useful?

### 7.4.2   Collision Resolution

Typically, the number of possible elements is so enormous compared to the relatively small number of table locations that no hash function, not even the most carefully designed, can prevent collisions from occurring in practice. A perfect illustration is the fact that the chances are better than fifty-fifty that among 23 people some two of them will have the same birthday. In terms of hashing, this means that if we have 365 table locations and only 23 elements to be stored in them, the chance of a collision is more than fifty-fifty. How much more likely, then, are collisions when we intend to fill the table to greater than $23/365 \approx 6.3$ percent of its capacity! In fact, the likelihood of collisions under even the most ideal circumstances (Exercise 1, page 361) suggests that the collision-resolution scheme is more critical to overall performance than the hash function, provided that at least minimum care is taken to avoid primary clustering.

When a collision occurs and the location $T[h(z)]$ is already filled at the time we try to insert $z$, we must have some method for specifying another location in the table where $z$ can be placed. A *collision-resolution scheme* is a method of specifying a list of table locations $\alpha_0 = h(z), \alpha_1, \alpha_2, \ldots, \alpha_{m-1}$ for an element $z$. To insert $z$, the locations are inspected in that order until an empty one is found. In parallel with the linear lists of Chapter 4, our two choices are to store pointers describing the sequence explicitly, or to specify the sequence implicitly by a fixed relationship between $z$, $\alpha_i$, and $i$. Techniques for collision resolution based on these two possibilities are explored in this section.

**Chaining.** In this scheme a sequence of pointers is built going from the hash location $h(z)$ to the location in which $z$ is ultimately stored. As above, we will assume that the table consists of $m$ locations $T[0], T[1], \ldots, T[m-1]$. In *separate chaining* each table location $T[i]$ is a list header, pointing to a linked list of those elements $z$ with $h(z) = i$. If the list is unordered, we insert $z$ just after the list header $T[h(z)]$, before the first element on that list. A search for $z$ in this case is done by applying Algorithm 7.1(b) to the list $T[h(z)]$. If the list is ordered, inserting $z$ is done by applying Algorithm 4.1 to the list $T[h(z)]$; a search for $z$ is done by applying Algorithm 7.2(b) to the list $T[h(z)]$. Figure 7.24, on page 342, shows the 31-position hash table built when the 30 words of Table 7.2, on page 336, are inserted by separate chaining with ordered lists using the hash function $h_2$.

Separate chaining requires the $m$ pointers of the table, in addition to the records stored in the lists. Frequently, however, the records are such that an extra link field can be packed into them without increasing the storage required per record. In this case, the overhead of the $m$ list headers needed by separate chaining can be eliminated by using coalesced chaining, in which each table location $T[i]$ is used to store a record, containing within it a field $LINK[i]$. When $T[h(z)]$ is found to contain another element on an attempted insertion of $z$, we follow the $LINK$ fields until we reach one that is null; then we take an empty table location $T[free]$, set that last null $LINK$ field to point to it, and store $z$ in $T[free]$. The search for empty locations originally begins at $T[m-1]$ and goes backward in the table toward $T[0]$. Each time an empty location is needed, we continue backward from where we stopped on the previous occasion; to stop this search, we introduce a dummy element $T[-1]$ that is always empty. The table will overflow when all locations are full. The details of such a "search and insert" process for coalesced chaining are given in Algorithm 7.10, which is on page 343; notice that the search can be easily separated from the insertion, if desired.

In separate chaining, elements with the same hash address stay in separate linked lists, hence the name. On the other hand, in coalesced chaining these lists can become intermingled, as an example will show.

Figure 7.25 shows how Algorithm 7.10 would behave when inserting the 30 words in the order shown in Table 7.2 into a 31-location table with $h_1$ as the hash function. $-1$ values for $LINK$'s have been omitted for clarity. The insertion of THE, OF, AND, TO, A, and IN results in no collisions, so in each case $z$ is simply stored in $T[h_1(z)]$. On the insertion of THAT, however, a collision occurs, because $h_1(\text{THAT}) = h_1(\text{OF}) = 9$. So, we begin scanning the table backward from $T[30]$, looking for an empty location. We find one at $T[30]$ and then set $LINK[9] \leftarrow 30$ and $T[30] \leftarrow \text{THAT}$, completing the insertion. Figure 7.25(a) shows the table at this point. Next, IS, WAS, and HE are inserted without

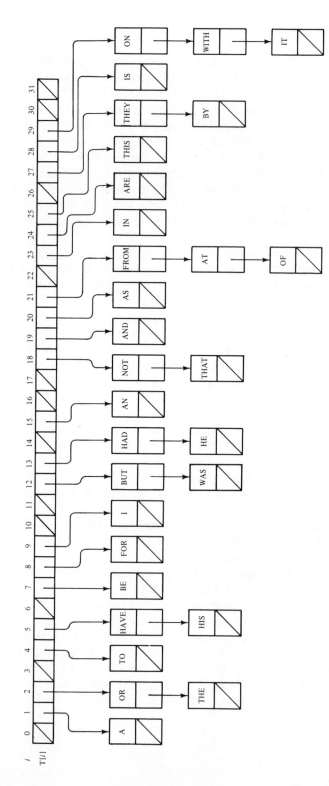

**Figure 7.24**
A hash table built by separate chaining with ordered lists and the hash function $h_2$ from Table 7.2

$loc \leftarrow h(z)$    $\llbracket 0 \leqslant h(z) < m \rrbracket$
*found* $\leftarrow$ **false**
**if** $T[loc]$ is not empty **then**
    $\llbracket$search linked list beginning at $T[loc]\rrbracket$
    $i \leftarrow loc$
    **repeat**
      **if** $T[i] = z$ **then** *found* $\leftarrow$ **true**
        **else**
          *previ* $\leftarrow i$    $\llbracket$trails behind $i$ for
              later insertion$\rrbracket$
         $i \leftarrow LINK[i]$
     **until** *found* **or** $i = -1$    $\llbracket -1$ corresponds to **nil**$\rrbracket$

$\llbracket$insert if necessary$\rrbracket$
**if not** *found*
  **then if** $T[loc]$ is empty **then** $T[loc] \leftarrow z$
  **else**
    $\llbracket$find empty spot$\rrbracket$
    **repeat** *free* $\leftarrow$ *free* $- 1$ **until** $T[free]$ is empty
    **if** *free* $= -1$ **then** $\llbracket$table overflow$\rrbracket$
        **else**
          $\llbracket$insert at *free*$\rrbracket$
          $LINK[previ] \leftarrow free$
          $T[free] \leftarrow z$
          $LINK[free] \leftarrow -1$

**Algorithm 7.10**
Search and insertion into a hash table with collisions resolved by coalesced chaining. The table consists of $T[-1], T[0], T[1], \ldots, T[m-1]$ where $T[-1]$ is a dummy location that is always empty. *free* is initialized to $m$ once, before the first insertion. Also, each $LINK[i]$ is initially set to $-1$ (in effect, the empty pointer).

collisions, but in the insertion of FOR we find $h_1(\text{FOR}) = 27$; $T[27]$ is filled already, so we scan backward looking for an empty location, this time from $T[28]$. We find $T[29]$ empty and store FOR in it, setting $LINK[27] \leftarrow 29$. The table is now as shown in Figure 7.25(b). Figures 7.25(c) and (d) show the table after more and more words have been inserted.

  How well does chaining do in practice? In other words, how time-consuming are the search and insertion processes on the average? The answer, of course, depends on how good the hash function is at avoiding primary clustering. As is customary, we therefore discuss the behavior of this (and other) collision-resolution schemes as though the hash function were truly random—that is, as though

**(a)** After the insertion of THE, OF, AND, TO, A, IN, and THAT. The table is $7/31 \approx 23$ percent full. A successful search requires an array of 1.14 probes and an unsuccessful search 1.03 probes.

| i: | 0 | 1 | 2 | 3 | 4 | 5 | 6 | 7 | 8 | 9 | 10 | 11 | 12 | 13 | 14 | 15 | 16 | 17 | 18 | 19 | 20 | 21 | 22 | 23 | 24 | 25 | 26 | 27 | 28 | 29 | 30 |
|---|---|---|---|---|---|---|---|---|---|---|---|---|---|---|---|---|---|---|---|---|---|---|---|---|---|---|---|---|---|---|---|
| T[i]: | | A | | | | | | IN | | OF | | AND | | | | | | | | | | | | | | THE | | TO | | | THAT |
| LINK[i]: | | | | | | | | | | 30 | | | | | | | | | | | | | | | | | | | | | |

**(b)** After the further insertion of IS, WAS, HE, and FOR. The table is $11/31 \approx 35$ percent full. A successful search requires an average of 1.18 probes and an unsuccessful search 1.06 probes.

| i: | 0 | 1 | 2 | 3 | 4 | 5 | 6 | 7 | 8 | 9 | 10 | 11 | 12 | 13 | 14 | 15 | 16 | 17 | 18 | 19 | 20 | 21 | 22 | 23 | 24 | 25 | 26 | 27 | 28 | 29 | 30 |
|---|---|---|---|---|---|---|---|---|---|---|---|---|---|---|---|---|---|---|---|---|---|---|---|---|---|---|---|---|---|---|---|
| T[i]: | | A | | | | WAS | | IN | | OF | | AND | | HE | | | | | | | | | | | | THE | | TO | | FOR | THAT |
| LINK[i]: | | | | | | | | | | 30 | | | | | | | | | | | | | | | | | | 29 | | | |

**(c)** After the further insertion of IT, WITH, AS, HIS, ON, BE, AT, BY, I, THIS, and HAD. The table is $22/31 \approx 71$ percent full. A successful search requires an average of 1.70 probes and an unsuccessful search 1.42 probes.

| i: | 0 | 1 | 2 | 3 | 4 | 5 | 6 | 7 | 8 | 9 | 10 | 11 | 12 | 13 | 14 | 15 | 16 | 17 | 18 | 19 | 20 | 21 | 22 | 23 | 24 | 25 | 26 | 27 | 28 | 29 | 30 |
|---|---|---|---|---|---|---|---|---|---|---|---|---|---|---|---|---|---|---|---|---|---|---|---|---|---|---|---|---|---|---|---|
| T[i]: | | A | WITH | | | WAS | THIS | IN | | OF | | AND | | HE | | | | HAD | AS | I | BY | AT | BE | ON | HIS | THE | IS | TO | IT | FOR | THAT |
| LINK[i]: | | 23 | | | | | | 22 | | 30 | | | | 17 | | | | | 24 | | | | | | | | | 29 | | 28 | 19 |

**(d)** After the further insertion of NOT, ARE, BUT, FROM, OR, HAVE, AN, and THEY. The table is $30/31 \approx 97$ percent full. A successful search requires an average of 1.73 probes and an unsuccessful search 1.74 probes.

| i: | 0 | 1 | 2 | 3 | 4 | 5 | 6 | 7 | 8 | 9 | 10 | 11 | 12 | 13 | 14 | 15 | 16 | 17 | 18 | 19 | 20 | 21 | 22 | 23 | 24 | 25 | 26 | 27 | 28 | 29 | 30 |
|---|---|---|---|---|---|---|---|---|---|---|---|---|---|---|---|---|---|---|---|---|---|---|---|---|---|---|---|---|---|---|---|
| T[i]: | THEY | A | WITH | BUT | | WAS | THIS | IN | AN | OF | HAVE | AND | OR | HE | FROM | ARE | NOT | HAD | AS | I | BY | AT | BE | ON | HIS | THE | IS | TO | IT | FOR | THAT |
| LINK[i]: | | 23 | | | | | | 22 | | 30 | | | | 17 | 8 | 14 | | | 24 | 12 | 12 | 16 | 15 | | | | 10 | 29 | 20 | 28 | 19 |

**Figure 7.25**
A hash table built by coalesced chaining (Algorithm 7.10) and the hash function $h_1$ from Table 7.2. *LINK*s with the value $-1$ are omitted for clarity

given any element $z$, $h(z)$ were chosen as a uniformly distributed random value $0 \leqslant h(z) < m$. If our hash function has been carefully designed, it will exhibit pseudorandom behavior and this assumption will not be far from the truth. This assumption amounts to our competely ignoring the question of primary clustering when discussing a collision-resolution scheme.

Examining Figure 7.25(a) we see that seven of the 31 table locations are filled, so the table is $7/31 \approx 23$ percent full. On a successful search in which each of the seven elements present is equally likely, the average number of locations probed in the **until** *found* **or** $i = -1$ loop of Algorithm 7.10 is 1.14, computed as follows: six of the seven elements are found on the first probe and one (THAT) is found on the second probe; the average number of probes is thus $8/7 \approx 1.14$. For an unsuccessful search we calculate the average number of probes by assuming that each of the 31 table locations is equally probable as the hash address. If that hash address is 0, 2 through 6, 8, 10, 12 through 24, 26, 28, or 29, then a single probe suffices, since those locations are empty. Similarly for locations 1, 7, 11, 25, 27, and 30—since the *LINK* fields are null, a single probe suffices. If the hash address is 9, however, two probes are required, first at $T[9]$ and then at $T[LINK[9]]$. The average number of probes in an unsuccessful search is thus $32/31 \approx 1.03$. The computations are similar for Figures 7.25(b), (c), and (d).

When collisions are resolved by separate chaining with unordered lists, the average number of probes in a successful search in a table of $m$ locations containing $n$ elements can be shown to be

$$S(\lambda) = 1 + \frac{n-1}{2m} \approx 1 + \frac{1}{2}\lambda$$

and in an unsuccessful search

$$U(\lambda) = \left(1 - \frac{1}{m}\right)^n + \frac{n}{m} \approx e^{-\lambda} + \lambda,$$

where $\lambda = n/m$ is called the *load factor* of the table. It is customary to express the behavior of the collision-resolution scheme in terms of $\lambda$ rather than $n$ and $m$ because the behavior of the algorithms is typically governed more by the fullness of the table in relative, rather than absolute terms. If the lists are kept in order, then the average number of probes for an unsuccessful search is decreased to

$$U(\lambda) = 1 + \frac{1}{2}\frac{n}{m} - \frac{m}{n+1}\left[1 - \left(1 - \frac{1}{m}\right)^{n+1}\right] + \left(1 - \frac{1}{m}\right)^n$$

$$\approx 1 + \frac{1}{2}\lambda - \frac{1}{\lambda}(1 - e^{-\lambda}) + e^{-\lambda}.$$

|  | | $\lambda = n/m$ | | | | | Full Table |
|---|---|---|---|---|---|---|---|
|  |  | 0.25 | 0.5 | 0.75 | 0.9 | 0.95 | ($\lambda \to 1$) |

**Separate Chaining**

$U(\lambda) = \left(1 - \dfrac{1}{m}\right)^n + \dfrac{n}{m}$ (unordered lists)
$\approx e^{-\lambda} + \lambda$

| 0.25 | 0.5 | 0.75 | 0.9 | 0.95 | Full |
|---|---|---|---|---|---|
| 1.03 | 1.11 | 1.22 | 1.31 | 1.34 | 1.37 |

$U(\lambda) = 1 + \dfrac{n}{2m} - \dfrac{m}{n+1}\left[1 - \left(1 - \dfrac{1}{m}\right)^{n+1}\right] + \left(1 - \dfrac{1}{m}\right)^n$ (ordered lists)
$\approx 1 + \dfrac{1}{2}\lambda - \dfrac{1}{\lambda}(1 - e^{-\lambda}) + e^{-\lambda}$

| 0.25 | 0.5 | 0.75 | 0.9 | 0.95 | Full |
|---|---|---|---|---|---|
| 1.02 | 1.07 | 1.14 | 1.20 | 1.22 | 1.24 |

$S(\lambda) = 1 + \dfrac{n-1}{2m} \approx 1 + \dfrac{1}{2}\lambda$

| 0.25 | 0.5 | 0.75 | 0.9 | 0.95 | Full |
|---|---|---|---|---|---|
| 1.12 | 1.25 | 1.38 | 1.45 | 1.48 | 1.50 |

**Coalesced Chaining**

$U(\lambda) = 1 + \dfrac{1}{4}\left[\left(1 + \dfrac{2}{m}\right)^n - 1 - \dfrac{2n}{m}\right]$
$\approx 1 + \dfrac{1}{4}(e^{2\lambda} - 1 - 2\lambda)$

| 0.25 | 0.5 | 0.75 | 0.9 | 0.95 | Full |
|---|---|---|---|---|---|
| 1.04 | 1.18 | 1.50 | 1.81 | 1.95 | 2.10 |

$S(\lambda) = 1 + \dfrac{1}{8}\dfrac{m}{n}\left[\left(1 + \dfrac{2}{m}\right)^n - 1 - \dfrac{2n}{m}\right] + \dfrac{1}{4}\dfrac{n-1}{m}$
$\approx 1 + \dfrac{1}{8\lambda}(e^{2\lambda} - 1 - 2\lambda) + \dfrac{1}{4}\lambda$

| 0.25 | 0.5 | 0.75 | 0.9 | 0.95 | Full |
|---|---|---|---|---|---|
| 1.14 | 1.30 | 1.52 | 1.68 | 1.74 | 1.80 |

**Open addressing with linear probing**

$U(\lambda) = \dfrac{1}{2}\left[1 + \displaystyle\sum_{k \geq 0} \dfrac{k+1}{m^k} \dfrac{n!}{(n-k)!}\right]$
$\approx \dfrac{1}{2}\left[1 + \dfrac{1}{(1-\lambda)^2}\right]$

| 0.25 | 0.5 | 0.75 | 0.9 | 0.95 | Full |
|---|---|---|---|---|---|
| 1.39 | 2.50 | 8.50 | 50.50 | 200.50 | $\dfrac{1}{2}m$ |

$S(\lambda) = \dfrac{1}{2}\left[1 + \displaystyle\sum_{k \geq 0} \dfrac{1}{m^k} \dfrac{n!}{(n-k)!}\right]$
$\approx \dfrac{1}{2}\left(1 + \dfrac{1}{1-\lambda}\right)$

| 0.25 | 0.5 | 0.75 | 0.9 | 0.95 | Full |
|---|---|---|---|---|---|
| 1.17 | 1.50 | 2.50 | 5.50 | 10.50 | $\sqrt{\dfrac{\pi}{8}\,m}$ |

| | | | formula | | 1.33 | 2.00 | 4.00 | 10.00 | 20.00 | $\frac{1}{2}m$ |
|---|---|---|---|---|---|---|---|---|---|---|

Open addressing with double hashing[a]

$$U(\lambda) = \frac{m+1}{m+1-n} \approx \frac{1}{1-\lambda}$$
  1.33  2.00  4.00  10.00  20.00  $\frac{1}{2}m$

$$S(\lambda) = \frac{m+1}{n} \sum_{k=m+2-n}^{m+1} \frac{1}{k} \approx \frac{1}{\lambda}\ln\frac{1}{1-\lambda}$$
  1.15  1.39  1.85  2.56  3.15  $\ln m$

Open addressing with double hashing in ordered hash tables[a]

$$U(\lambda) = \frac{m+1}{n+1} \sum_{k=m+2-n}^{m+1} \frac{1}{k} \approx \frac{1}{\lambda}\ln\frac{1}{1-\lambda}$$
  1.15  1.39  1.85  2.56  3.15  $\ln m$

$$S(\lambda) = \frac{m+1}{n} \sum_{k=m+2-n}^{m+1} \frac{1}{k} \approx \frac{1}{\lambda}\ln\frac{1}{1-\lambda}$$
  1.15  1.39  1.85  2.56  3.15  $\ln m$

[a] Based on the analysis of uniform hashing which, empirically, is how double hashing behaves.

**Table 7.3**

Expected numbers of probes for successful ($S$) and unsuccessful ($U$) searches in hash tables with various collision-resolution schemes and various load factors $\lambda = n/m$. Other than for chaining, both $S(\lambda)$ and $U(\lambda)$ grow unboundedly with $m$ as $\lambda \to 1$. For open addressing, the algorithms require $n \leq m - 1$, while coalesced chaining requires $n \leq m$. For separate chaining, the formulas are valid even for $n > m$ (that is, $\lambda > 1$).

When coalesced chaining is used, the corresponding functions are

$$U(\lambda) = 1 + \frac{1}{4}\left[\left(1 + \frac{2}{m}\right)^n - 1 - \frac{2n}{m}\right] \approx 1 + \frac{1}{4}(e^{2\lambda} - 1 - 2\lambda)$$

$$S(\lambda) = 1 + \frac{1}{8}\frac{m}{n}\left[\left(1 + \frac{2}{m}\right)^n - 1 - \frac{2n}{m}\right] + \frac{1}{4}\frac{n-1}{m}$$

$$\approx 1 + \frac{1}{8\lambda}(e^{2\lambda} - 1 - 2\lambda) + \frac{1}{4}\lambda.$$

The top lines of Table 7.3 give some numerical values for all of these functions $U(\lambda)$ and $S(\lambda)$ for various values of $\lambda$. The remaining lines of the table give corresponding values for other collision-resolution schemes.

   A brief examination of Table 7.3 indicates that separate chaining is superior to coalesced chaining which is, in turn, superior to other methods of collision resolution. In separate chaining we also have the advantage of ignoring the problem of the table overflowing its allocated storage since the usual **getcell** mechanism can be used to provide the actual storage for the table elements. This means that we can even have $n > m$, giving $\lambda > 1$; the formulas of Table 7.3 for separate chaining are also valid in this case. A further advantage of separate chaining is that it allows very easy deletion of elements, something difficult or impossible with other collision resolution schemes. The disadvantage of chaining compared to the other schemes represented in Table 7.3 is that it requires additional storage overhead for the link fields; this makes the other schemes more desirable in some circumstances.

   **Linear Probing.** The simplest alternative to chaining that does not require the storage of LINK fields is to resolve collisions by probing sequentially, one location at a time, starting from the hash address, until an empty location is found. This is called *open addressing with linear probing*, or simply *linear probing*. Algorithm 7.11 gives the details of such a procedure, and Figure 7.26 illustrates the use of this procedure in building a hash table with the 30 words and $h_1$ of Table 7.2 into a 31-location table; the words are added in the order given in Table 7.2.

   The insertion of THE, OF, AND, TO, A and IN results in no collisions, so in each case $z$ is stored in $T[h_1(z)]$. On the insertion of THAT a collision occurs, because $h_1(\text{THAT}) = h_1(\text{OF}) = 9$. So we begin scanning forward from $T[9]$, one location at a time, until we find an empty location. We find one immediately in $T[10]$, which is where THAT is stored, completing the insertion. The table now looks as shown in Figure 7.26(a). Figures 7.26(b), (c), and (d) show the table in various stages as the remaining words are inserted. Notice that in scanning forward to find an empty location we "wrap around" from $m - 1$ to 0; that is, the addition of one is computed modulo $m$, as shown in Algorithm 7.11.

$$i \leftarrow h(z) \quad \llbracket 0 \leqslant i < m \rrbracket$$

*found* ← **false**

**while** $T[i]$ is not empty **and not** *found* **do**

        **if** $T[i] = z$ **then** *found* ← **true**

            **else** $i \leftarrow (i + 1) \bmod m$

$\llbracket$insert if necessary$\rrbracket$

**if not** *found* **then**

        **if** $n = m - 1$ **then**  $\llbracket$table overflow$\rrbracket$

            **else**

                $n \leftarrow n + 1$

                $T[i] \leftarrow z$

**Algorithm 7.11**

Search and insertion into a hash table in which collisions are resolved by linear probing. The table consists of $T[0], T[1], \ldots, T[m-1]$. $n$ is the number of elements currently in the table, initialized once to 0 before the first insertion. The table overflows on trying to insert the $m$th element; this guarantees that one location is always empty to stop the **while** loop.

Examining Figure 7.26(a), we see that seven of the table locations are filled, so the table is $7/31 \approx 23$ percent full. On a successful search in which each of the seven elements present is equally likely, the average number of locations probed in the **while** loop of Algorithm 7.11 is 1.14, computed as follows: six of the seven elements are found on the first probe and one (THAT) is found on the second probe; the average number is thus $8/7 \approx 1.14$. For an unsuccessful search for $z$, assuming each of the 31 locations is equally likely as $h(z)$, the average number of probes is 1.32, computed as follows. For the 24 empty locations a single probe suffices. For locations 1, 7, 11, 25, and 27 two probes are needed, for 10 three probes, and for 9 four probes for an average of $(24 \times 1 + 5 \times 2 + 1 \times 3 + 1 \times 4)/31 \approx 1.32$. The computations for Figures 7.26(b), (c), and (d) are similar.

The number of probes expected on the average with linear probing when a table has load factor $\lambda = n/m$ is

$$S(\lambda) \approx \frac{1}{2}\left(1 + \frac{1}{1 - \lambda}\right)$$

for successful searches and

$$U(\lambda) \approx \frac{1}{2}\left(1 + \frac{1}{(1 - \lambda)^2}\right)$$

for unsuccessful searches. The exact formulas for $S(\lambda)$ and $U(\lambda)$ in terms of $n$ and $m$ are given in Table 7.3, along with a tabulation of their values for various $\lambda$.

(a) After the insertion of THE, OF, AND, TO, A, IN, and THAT. The table is $7/31 \approx 23$ percent full. A successful search requires an average of 1.14 probes and an unsuccessful search 1.32 probes.

| i: | 0 | 1 | 2 | 3 | 4 | 5 | 6 | 7 | 8 | 9 | 10 | 11 | 12 | 13 | 14 | 15 | 16 | 17 | 18 | 19 | 20 | 21 | 22 | 23 | 24 | 25 | 26 | 27 | 28 | 29 | 30 |
|---|---|---|---|---|---|---|---|---|---|---|---|---|---|---|---|---|---|---|---|---|---|---|---|---|---|---|---|---|---|---|---|
| T[i]: | | A | | | | | | IN | | OF | THAT | AND | | | | | | | | | | | | | | THE | | TO | | | |

(b) After the further insertion of IS, WAS, HE, FOR, IT, WITH, AS, HIS, and ON. The table if $16/31 \approx 52$ percent full. A successful search requires an average of 1.31 probes and an unsuccessful search 2.06 probes.

| i: | 0 | 1 | 2 | 3 | 4 | 5 | 6 | 7 | 8 | 9 | 10 | 11 | 12 | 13 | 14 | 15 | 16 | 17 | 18 | 19 | 20 | 21 | 22 | 23 | 24 | 25 | 26 | 27 | 28 | 29 | 30 |
|---|---|---|---|---|---|---|---|---|---|---|---|---|---|---|---|---|---|---|---|---|---|---|---|---|---|---|---|---|---|---|---|
| T[i]: | | A | WITH | ON | | WAS | | IN | | OF | THAT | AND | | HE | | | | | AS | HIS | | | | | | THE | IS | TO | FOR | IT | |

(c) After the further insertion of BE, AT, BY, I, THIS, and HAD. The table is $22/31 \approx 71$ percent full. A successful search requires an average of 1.59 probes and an unsuccessful search 3.45 probes.

| i: | 0 | 1 | 2 | 3 | 4 | 5 | 6 | 7 | 8 | 9 | 10 | 11 | 12 | 13 | 14 | 15 | 16 | 17 | 18 | 19 | 20 | 21 | 22 | 23 | 24 | 25 | 26 | 27 | 28 | 29 | 30 |
|---|---|---|---|---|---|---|---|---|---|---|---|---|---|---|---|---|---|---|---|---|---|---|---|---|---|---|---|---|---|---|---|
| T[i]: | | A | WITH | ON | | WAS | THIS | IN | BE | OF | THAT | AND | I | HE | HAD | | | | AS | HIS | | AT | | | | THE | IS | TO | FOR | IT | BY |

(d) After the further insertion of NOT, ARE, BUT, FROM, OR, HAVE, AN, and THEY. The table is $30/31 \approx 97$ percent full. A successful search requires an average of 2.93 probes and an unsuccessful search 16 probes.

| i: | 0 | 1 | 2 | 3 | 4 | 5 | 6 | 7 | 8 | 9 | 10 | 11 | 12 | 13 | 14 | 15 | 16 | 17 | 18 | 19 | 20 | 21 | 22 | 23 | 24 | 25 | 26 | 27 | 28 | 29 | 30 |
|---|---|---|---|---|---|---|---|---|---|---|---|---|---|---|---|---|---|---|---|---|---|---|---|---|---|---|---|---|---|---|---|
| T[i]: | OR | A | WITH | ON | BUT | WAS | THIS | IN | BE | OF | THAT | AND | I | HE | HAD | HAVE | AN | THEY | AS | HIS | | AT | NOT | ARE | FROM | THE | IS | TO | FOR | IT | BY |

**Figure 7.26**
A hash table built by linear probing (Algorithm 7.11) with $m = 31$ and the hash function $h_1$ from Table 7.2

We notice immediately that both $S(\lambda)$ and $U(\lambda)$ grow without bound as $\lambda \to 1$, that is, as the table occupancy increases and is filled to capacity. The values shown in Table 7.3 suggest that the behavior of linear probing is tolerable as long as the table is less than $3/4$ full, but beyond that it deteriorates rapidly. Experience confirms this observation.

It is instructive to compare coalesced chaining and linear probing to see why linear probing behaves so poorly in comparison. In both schemes collisions result in lists of items that have collided, but comparing Figure 7.25(c) with Figure 7.26(c) we see that in Figure 7.25(c) the 22 elements comprise thirteen lists with an average length of just $22/17 \approx 1.7$, while in Figure 7.26(c) there are five lists with an average length of $22/5 = 4.4$. The comparatively poor behavior of linear probing stems from a tendency to have fewer (and hence longer) lists while coalesced chaining results in more (and hence shorter) lists. Such a tendency would indeed produce poor behavior, because the average successful search will have to examine about half the elements on a list and the average unsuccessful search will have to examine all the elements on a list (unless, of course, the first probe is to an empty table location; as the table becomes fuller this will happen more infrequently, but if the table is less than 75 percent full it happens often enough to lower the average number of probes significantly). It is easy to see why linear probing results in fewer, longer lists: the longer a list is, the higher the chances that it will get longer! This is true in coalesced chaining also and is just a consequence of the fact that the more elements in a list, the more likely a collision is to occur in that list, thus increasing the length of the list by one. The key difference between coalesced chaining and linear probing is that in linear probing two existing lists can become concatenated: in Figure 7.26(c), for example, $T[4]$ will be filled with $z$ if $h(z)$ is any of the locations 1, 2, or 3. When that happens, the result is a single list of fourteen elements, $T[1], T[2], \ldots, T[14]$. The insertion of such a $z$ increases the average list length to $23/4 \approx 5.8$, quite a jump from 4.4.

**Double Hashing.** Part of the problem with linear probing is the phenomenon of *secondary clustering*: the tendency of two elements that have collided to follow the same sequence of locations in the resolution of the collision. Clearly, such a tendency will aggravate the unavoidable fact that long lists are more likely to grow than short lists. This suggests that the sequence of locations followed in resolving a collision of $z$ should be a function of the element $z$. This can be accomplished very easily by only a minor change to Algorithm 7.11: instead of incrementing $i$ by 1 in the **while** loop, we increment it by an amount $\Delta$, $1 \le \Delta < m$, where $\Delta$ is a function of $z$. In order to insure that every location in the table will be probed on a collision, we must have $\Delta$ and $m$ relatively prime (what happens if an integer $d > 1$ divides both $\Delta$ and $m$?). Since we want $\Delta$ to have pseudorandom behavior, we can use another hash function $\delta(z)$, $1 \le \delta(z) < m$, as our value for

$\Delta$. This means that we will now have to compute two hash functions instead of one, but the resulting improvement in behavior will more than compensate for the extra calculation. As a practical matter, it is easiest to guarantee that $\delta(z)$ and $m$ are relatively prime for all $z$ by insisting that $m$ be a prime number. Algorithm 7.12 gives this modification of linear probing, appropriately called *double hashing*. Figure 7.27 shows the application of this algorithm to our example using $h_1$ and $h_3$ from Table 7.2 as $h(z)$ and $\delta(z)$, respectively. Notice that in Figure 7.27(c) the average length of a list is $22/6 \approx 3.67$, compared with 4.4 in Figure 7.26(c).

$$i \leftarrow h(z) \qquad [\![0 \leqslant i < m]\!]$$
$$\Delta \leftarrow \delta(z) \qquad [\![1 \leqslant \Delta < m]\!]$$

*found* $\leftarrow$ **false**
**while** $T[i]$ is not empty **and not** *found* **do**
$\qquad$ **if** $T[i] = z$ **then** *found* $\leftarrow$ **true**
$\qquad\qquad$ **else** $i \leftarrow (i + \Delta) \bmod m$
$[\![$insert if necessary$]\!]$
**if not** *found* **then**
$\qquad$ **if** $n = m - 1$ **then** $\quad [\![$table overflow$]\!]$
$\qquad\qquad$ **else**
$\qquad\qquad\qquad$ $n \leftarrow n + 1$
$\qquad\qquad\qquad$ $T[i] \leftarrow z$

**Algorithm 7.12**
Search and insertion into a hash table in which collisions are resolved by double hashing. The algorithm is identical to Algorithm 7.11 in all respects *except* that $\Delta = \delta(z)$ is the increment in the **while** loop.

A complete analysis of the average behavior of double hashing has not yet been made, but both empirical results and some fragmentary theoretical results indicate that it behaves approximately like *uniform hashing*, an idealization of double hashing that we can analyze. We assume that the sequence of locations $\alpha_0 = h(z), \alpha_1, \alpha_2, \ldots$ used to insert an element $z$ into the table has the property that each location $\alpha_i$ is equally likely to be any of $0, 1, 2, \ldots, m - 1$ *independently* of the other $\alpha_i$'s. In other words, we assume that the probe sequence $(\alpha_0, \alpha_1, \ldots, \alpha_{m-1})$ is equally likely to be any of the $m!$ permutations of $(0, 1, \ldots, m - 1)$. This assumption implies that each of the $\binom{m}{n}$ possible configurations of empty and full locations is equally likely to occur (Exercise 8, page 362); it does not fully hold for double hashing, but it is close enough to the truth to give us a good approximation of the behavior of double hashing. In a table with load factor $\lambda$ the probability of at least $k$ probes being required in an unsuccessful search for $z$ is

$$\Pr(at\ least\ k\ probes) = \Pr(T[\alpha_0], T[\alpha_1], \ldots, T[\alpha_{k-2}]\ are\ full).$$

**(a)**

| i: | 0 | 1 | 2 | 3 | 4 | 5 | 6 | 7 | 8 | 9 | 10 | 11 | 12 | 13 | 14 | 15 | 16 | 17 | 18 | 19 | 20 | 21 | 22 | 23 | 24 | 25 | 26 | 27 | 28 | 29 | 30 |
|---|---|---|---|---|---|---|---|---|---|---|---|---|---|---|---|---|---|---|---|---|---|---|---|---|---|---|---|---|---|---|---|
| T[i]: |  | A |  |  |  |  |  | IN |  | OF |  | AND |  |  |  |  |  |  |  |  |  |  |  |  |  | THE | THAT | TO |  |  |  |

(a) After the insertion of THE, OF, AND, TO, A, IN, and THAT. The table is 7/31 ≈ 23 percent full. A successful search requires an average of 1.14 probes.

**(b)**

| i: | 0 | 1 | 2 | 3 | 4 | 5 | 6 | 7 | 8 | 9 | 10 | 11 | 12 | 13 | 14 | 15 | 16 | 17 | 18 | 19 | 20 | 21 | 22 | 23 | 24 | 25 | 26 | 27 | 28 | 29 | 30 |
|---|---|---|---|---|---|---|---|---|---|---|---|---|---|---|---|---|---|---|---|---|---|---|---|---|---|---|---|---|---|---|---|
| T[i]: |  | A | WITH |  |  | WAS |  | IN |  | OF |  | AND | FOR | HE |  |  |  | HIS | AS | ON |  |  |  |  |  | THE | THAT | TO | IS | IT |  |

(b) After the further insertion of IS, WAS, HE, FOR, IT, WITH, AS, HIS, and ON. The table is 16/31 ≈ 52 percent full. A successful search requires an average of 1.31 probes.

**(c)**

| i: | 0 | 1 | 2 | 3 | 4 | 5 | 6 | 7 | 8 | 9 | 10 | 11 | 12 | 13 | 14 | 15 | 16 | 17 | 18 | 19 | 20 | 21 | 22 | 23 | 24 | 25 | 26 | 27 | 28 | 29 | 30 |
|---|---|---|---|---|---|---|---|---|---|---|---|---|---|---|---|---|---|---|---|---|---|---|---|---|---|---|---|---|---|---|---|
| T[i]: |  | A | WITH | I |  | WAS | THIS | IN |  | OF | HAD | AND | FOR | HE | BY | BE |  | HIS | AS | ON |  | AT |  |  |  | THE | THAT | TO | IS | IT |  |

(c) After the further insertion of BE, AT, BY, I, THIS, and HAD. The table is 22/31 ≈ 71 percent full. A successful search requires an average of 1.68 probes.

**(d)**

| i: | 0 | 1 | 2 | 3 | 4 | 5 | 6 | 7 | 8 | 9 | 10 | 11 | 12 | 13 | 14 | 15 | 16 | 17 | 18 | 19 | 20 | 21 | 22 | 23 | 24 | 25 | 26 | 27 | 28 | 29 | 30 |
|---|---|---|---|---|---|---|---|---|---|---|---|---|---|---|---|---|---|---|---|---|---|---|---|---|---|---|---|---|---|---|---|
| T[i]: | OR | A | WITH | I | THEY | WAS | THIS | IN | ARE | OF | HAD | AND | FOR | HE | BY | BE | HAVE | HIS | AS | ON | AN | AT | BUT | NOT | FROM | THE | THAT | TO | IS | IT |  |

(d) After the further insertion of NOT, ARE, BUT, FROM, OR, HAVE, AN, and THEY. The table is 30/31 ≈ 97 percent full. A successful search requires an average of 3.23 probes.

**Figure 7.27**
A hash table built by double hashing (Algorithm 7.12) using $h_1$ and $h_3$ from Table 7.2 as $h(z)$ and $\delta(z)$, respectively. The computation of the average number of probes on an unsuccessful search is omitted because of the complexity of its computation (Exercise 6)

Since the probability of any given location being full is $\binom{m-1}{n-1} / \binom{m}{n} = n/m$ $= \lambda$ and of being empty is thus $1 - \lambda$, independently for all $m$ locations, we have

$$\Pr(\textit{at least } k \text{ probes}) = \lambda^{k-1}.$$

The expected number of probes for an unsuccessful search for $z$ in a table with load factor $\lambda$ is thus

$$
\begin{aligned}
U(\lambda) &= \sum_{k=1}^{\infty} k \Pr(\textit{exactly } k \text{ probes}) \\
&= \sum_{k=1}^{\infty} \left[ \sum_{i=k}^{\infty} \Pr(\textit{exactly } i \text{ probes}) \right] \\
&= \sum_{k=1}^{\infty} \Pr(\textit{at least } k \text{ probes}) \\
&= \sum_{k=1}^{\infty} \lambda^{k-1} \\
&= \frac{1}{1 - \lambda}.
\end{aligned}
$$

We can determine $S(\lambda)$, the expected number of probes for a successful search, by the following argument. In collision resolution by double hashing the number of probes needed to find $z$ after it is in the table is the same as the number of probes used on the unsuccessful search when it is inserted. This observation is also true for linear probing as well as various other table organizations we have considered [see Exercise 2(e) of Section 7.2.2]. If we consider the table as being constructed as elements are inserted one by one, the load factor increases in small discrete steps from 0 to its final value of $\lambda$. We approximate this situation by letting the load factor grow *continuously* from 0 to $\lambda$. $S(\lambda)$ is then the "average" value of $U(x)$ for $x$ in the range $0 \leqslant x \leqslant \lambda$:

$$
\begin{aligned}
S(\lambda) &= \frac{1}{\lambda} \int_0^{\lambda} U(x) \, dx \\
&= \frac{1}{\lambda} \int_0^{\lambda} \frac{dx}{1 - x} \\
&= \frac{1}{\lambda} \ln \frac{1}{1 - \lambda}.
\end{aligned}
\tag{7.20}
$$

Notice that, as expected, Equation (7.20) also holds for the formulas given for linear probing. The formulas derived above for $S(\lambda)$ and $U(\lambda)$ for uniform hashing are only approximate; the exact formulas are

$$U(\lambda) = \frac{m + 1}{m + 1 - n}$$

$$S(\lambda) = \frac{m + 1}{n} \sum_{k = m + 2 - n}^{m+1} \frac{1}{k},$$

whose derivation is left to Exercise 9. Examination of Table 7.3 indicates that double hashing (as approximated by uniform hashing) is much better than linear probing, performing reasonably well even for tables up to 90 percent full.

**Ordered Hash Tables.** In many, if not most, cases there is an ordering of the elements that may be useful in speeding up searches in hash tables just as it was for linear lists (Section 7.1). We will now investigate how such an ordering can be utilized in conjunction with a hashing scheme. The idea will be applicable to chaining, linear probing, or double hashing, but we will consider it only in the context of double hashing, because chaining is so efficient that it needs no improvement while linear probing is so inefficient we would probably never choose it over double hashing if economy were a factor.

If we had been extremely lucky in Algorithm 7.12 and the elements arrived in decreasing order to be inserted, then each of the lists built up through collisions would be in decreasing order by element. Assuming that an empty table location had a value less than that of any element in the table (for instance, in our continuing example we might choose the empty string, which is lexicographically less than any sequence of characters), we could do a search by Algorithm 7.13(a). This algorithm stops a search as soon as it reaches an element less than the search object $z$.

$$i \leftarrow h(z) \qquad [\![ 0 \leqslant i < m ]\!]$$
$$\Delta \leftarrow \delta(z) \qquad [\![ 1 \leqslant \Delta < m ]\!]$$
**while** $T[i] > z$ **do** $i \leftarrow (i + \Delta) \bmod m$
**if** $T[i] = z$ **then** $[\![$ found it $]\!]$
$\qquad\qquad\quad$ **else** $[\![$ not in the table $]\!]$

**Algorithm 7.13(a)**
Search of an ordered hash table

Of course, we cannot count on the elements being inserted into the table in decreasing order, making Algorithm 7.13(a) useless unless we can somehow keep the hash table "ordered" no matter in what order the elements are inserted. Consideration of the insertion part of Algorithm 7.12 leads us to an insertion algorithm that maintains an ordered hash table. When an insertion is made and there is no collision or when the element being inserted is less than the elements it collides with, then Algorithm 7.12 works fine and the hash table remains ordered. When an insertion leads to a collision with a smaller element, the algorithm must react as though the smaller element were not in the table! In such a collision, then, the idea is to have the (larger) element being inserted "bump" the (smaller) resident element it collided with temporarily out of the table; the larger element takes the location formerly occupied by the smaller. To reinsert the displaced element into the table, we simply apply the insertion algorithm to it; if that leads to a collision with a smaller element, that smaller element is bumped from its location and then reinserted. Each element thus bumped is smaller than the previous one, so this process must end. The result of such an insertion is that the table is in the same order that would have resulted from inserting the elements in decreasing order (Exercise 10.)

> **if** $n = m - 1$ **then**  [table overflow]
>            **else**
>                 $i \leftarrow h(z)$      $[0 \leqslant i < m]$
>                 $\Delta \leftarrow \delta(z)$      $[1 \leqslant \Delta < m]$
>                 **while** $T[i]$ is not empty **do**
>                     **if** $T[i] < z$ **then**
>                                 $T[i] \leftrightarrow z$
>                                 $\Delta \leftarrow \delta(z)$
>                     $i \leftarrow (i + \Delta) \bmod m$
>               $T[i] \leftarrow z$
>               $n \leftarrow n + 1$

**Algorithm 7.13(b)**
Insertion of $z$ into an ordered hash table

An example to illustrate the method will make it more understandable. Suppose the hash table looks as shown in Figure 7.28, $\delta(z) = h_3(z)$ from Table 7.2, and $h$ (HAVE) = 26. The insertion of HAVE causes the following sequence of events: $T[26]$ contains AS and AS is bumped. $\delta(\text{AS}) = 8$, so we try to insert AS into $26 + 8 = 34 \equiv 3$, but $T[3]$ contains FOR. Because AS is less than FOR, we continue, probing $3 + 8 = 11$. $T[11]$ contains AND, which is less than AS and therefore gets bumped. We now continue inserting AND; $\delta(\text{AND}) = 4$, so we try to insert it at $11 + 4 = 15$, but $T[15]$ contains I, which is bigger than AND. We next probe at $15 + 4 = 19$ and, finding $T[19]$ empty, insert AND there. Notice that every time an element is bumped the increment function $\delta$ must be recomputed. Algorithm 7.13(b) gives the details of the insertion of an element $z$ into an ordered hash table. Algorithms 7.13(a) and (b) can be com-

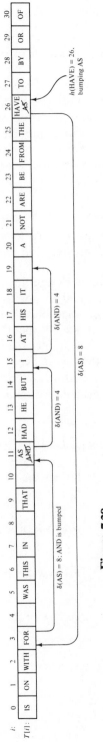

**Figure 7.28**
Insertion of HAVE into the ordered hash table shown, using Algorithm 7.13(b)

bined into a single search-and-insert algorithm in the style of Algorithms 7.10, 7.11, and 7.12.

We can give an approximate analysis of the number of probes needed on the average to search an ordered hash table; as in double hashing the analysis is based on the uniform hashing assumption: the sequence of locations $\alpha_0 = h(z), \alpha_1, \alpha_2, \ldots$ used to insert an element $z$ into the table has the property that each $\alpha_i$ is equally likely to be $0, 1, \ldots,$ or $m - 1$ independently of the other $\alpha_i$'s. In a table with load factor $\lambda$ the probability of at least $k$ probes in an unsuccessful search is $\lambda^{k-1}/k$, computed as follows:

$$\lambda^{k-1} = \text{probability that first } k - 1 \text{ locations probed will be full,}$$
$$1/k = \text{probability that of the } k - 1 \text{ elements thus probed the}$$
$$\text{search object will be smaller than all of them.}$$

The former probability is just as we saw in double hashing. The latter should be understood as stating that it is equally likely for the search object to be larger than all $k - 1$ elements probed, larger than only $k - 2$ of them,..., larger than only one of them, or smaller than all of them. Of these $k$ equally likely possibilities, only in the last case will more than $k - 1$ probes be needed. Assuming there is no significant correlation between $h(z)$ or $\delta(z)$ and the size of $z$, the relative sizes of the elements probed will be independent of the distribution of elements within the table. The probability of at least $k$ probes is thus the product $(1/k)\lambda^{k-1}$:

$$\Pr(\text{at least } k \text{ probes}) = \frac{\lambda^{k-1}}{k}.$$

The expected number of probes for an unsuccessful search in an ordered hash table can now be computed:

$$U(\lambda) = \sum_{k=1}^{\infty} k \Pr(\text{exactly } k \text{ probes})$$

$$= \sum_{k=1}^{\infty} \left( \sum_{i=k}^{\infty} \Pr(\text{exactly } i \text{ probes}) \right)$$

$$= \sum_{k=1}^{\infty} \Pr(\text{at least } k \text{ probes})$$

$$= \sum_{k=1}^{\infty} \frac{\lambda^{k-1}}{k}$$

$$= \frac{1}{\lambda} \sum_{k=1}^{\infty} \frac{\lambda^k}{k}.$$

This final summation is the Taylor series expansion for $\ln[1/(1 - \lambda)]$ so that

$$U(\lambda) = \frac{1}{\lambda} \ln \frac{1}{1 - \lambda}.$$

For $S(\lambda)$, we argue that it is exactly as for double hashing. Since, as Exercise 10 shows, the ultimate contents of the table are as if the elements had been inserted in decreasing order by double hashing, we may assume that they *were* so inserted. In this case, the expected number of probes for a successful search is $(1/\lambda) \ln[1/(1 - \lambda)]$ from our previous discussion (that discussion was independent of the order of insertion of the elements).

In fact, the exact values of $S(\lambda)$ and $U(\lambda)$ can be shown to be

$$S(\lambda) = \frac{m + 1}{n} \sum_{k=m+2-n}^{m+1} \frac{1}{k}$$

and

$$U(\lambda) = \frac{m + 1}{n + 1} \sum_{k=m+2-n}^{m+1} \frac{1}{k}$$

(see Exercise 11).

Comparing this to double hashing, we see that successful searches are no different, but unsuccessful searches require many fewer probes on the average as the table fills up (see Table 7.3). Is the savings worthwhile, considering that Algorithm 7.13(b) requires extra work per insertion compared to Algorithm 7.12? To answer, we need to analyze the extra work done to keep the table ordered. Two quantities are of interest—the average number of times the test "**if** $T[i] < z$" is made (the number of probes on an insertion) and the average number of times that test is true so that $\Delta$ must be recalculated. The number of probes on an insertion can be seen to be about $1/(1 - \lambda)$ using the argument of Exercise 12. This means that the average insertion into an ordered hash table is no more expensive than the average insertion into a conventional hash table. As to the number of times the function $\delta(z)$ must be reevaluated, its analysis is somewhat involved, but it can be shown to be about $(1/\lambda)\ln[1/(1 - \lambda)] - 1$ per insertion.

In summary, ordered hash tables are to be recommended over conventional hash tables when unsuccessful searches are common and $\delta(z)$ can be computed without much expense.

**Deletion and Rehashing.** Except in separate chaining deletion of an element from a hash table poses special problems. For example, consider Figure 7.26(c). The deletion of THAT by setting $T[10]$ to the value used for an empty table location makes it impossible to find I in the table: $h(I) = 9$, so we would probe $T[9]$, finding that it was not what we wanted, then probe $T[10]$, and finding it empty we would abandon the search. It is clear that we cannot simply remove an element from the table, because such a removal will disrupt the probe sequence for elements that collided with the one to be deleted. Our only choice is to mark the table location as containing an element that has been deleted. Such a location acts like an empty location with respect to insertions, but like a full location with respect to searches. This causes two distinct problems. First, most obvious, and most serious, is that search times will *not* change for the better after a deletion; for example, if we fill a table to 90 percent of its capacity and then delete half the elements, the table still behaves like a table 90 percent full as far as searching is concerned. The second and more subtle problem is how the value of $n$ should be treated in Algorithms 7.11, 7.12, and 7.13(b). Should $n$ be decremented following such a deletion? If it is decremented, then further insertions can fill *all* $m$ table locations, so that an unsuccessful search can go into an infinite loop. This is certainly not acceptable, but if $n$ is not decremented, the table may signal an overflow when there is plenty of room. This second problem can be solved by not

⟦mark every location as containing an element from the old table⟧
**for** $i = 0$ **to** $\max(m - 1, M - 1)$ **do** *new*$[i] \leftarrow$ **false**
*numelts* $\leftarrow 0$ ⟦number of elements relocated⟧
⟦relocate the elements⟧
**for** $i = 0$ **to** $m - 1$ **do**
        **while not** *new*$[i]$ **and** $T[i]$ is not empty **do**
                **if** $T[i]$ is deleted **then** $T[i] \leftarrow$ empty
                **else**
                        ⟦relocate $T[i]$ to a new spot⟧
                        *numelts* $\leftarrow$ *numelts* $+ 1$
                        **if** *numelts* $= M$ **then** table overflow
                        ⟦follow new probe sequence for $T[i]$⟧
                        $j \leftarrow h(T[i])$
                        $\Delta \leftarrow \delta(T[i])$
                        **while** *new*$[j]$ **do**
                              $j \leftarrow (j + \Delta) \bmod M$
                        ⟦bump current contents of $T[j]$⟧
                        $T[i] \leftrightarrow T[j]$
                        *new*$[j] \leftarrow$ **true**
                        ⟦$T[j]$ is now the relocated element⟧

**Algorithm 7.14**
Rehashing $T[0], T[1], \ldots, T[m - 1]$ to $T[0], T[1], \ldots, T[M - 1]$ and eliminating deleted elements

decrementing $n$ and incrementing $n$ only when inserting into a previously empty table location.

The only solution to the problem of degraded search times, however, is the ultimate total reconstruction of the table by a process called *rehashing*. In rehashing, each table location is scanned in turn and its contents, if any, are relocated as necessary. That is also precisely what is required to reduce or increase the storage allocated for a hash table, so we will present the rehashing algorithm in this slightly more general context. We give the algorithm only for conventional hash tables in which collisions are resolved by linear probing or by double hashing, leaving rehashing algorithms for coalesced chaining and ordered hash tables to the exercises.

Algorithm 7.14, the rehashing algorithm, rehashes the table from locations $T[0], T[1], \ldots, T[m-1]$ to locations $T[0], T[1], \ldots, T[M-1]$. We can have $M > m$ for expanding the hash table, $M < m$ for shrinking it, or $M = m$ for simply eliminating deleted elements. If $M > m$, the new locations $T[m]$, $T[m+1], \ldots, T[M-1]$ are assumed to be empty. The hash functions $h(z)$ and $\delta(z)$ used are for the new table size $M$, with $\delta(z) = 1$ for linear probing. The algorithm uses a bit table $new[0], new[1], \ldots, new[\max(m-1, M-1)]$ to record whether or not the corresponding table position contains an element of the new table.

The rehashing algorithm is clearly very expensive since it requires time proportional to $\max(m-1, M-1)$ in addition to the time required to insert all the elements into the new table. It should be used only when search times have deteriorated through many deletions, or when the size of the hash table must change.

## Exercises

1. In a hash table with load factor $\lambda$, show that out of $n$ elements inserted into the table, $\lambda(n-1)/2$ can be expected to have collided with a previously inserted element.

2. In some collision resolution schemes the number of probes used during an unsuccessful search followed by an insertion is the same number that will thereafter be required for a successful search for that element. Is that true of coalesced chaining? Is it true of ordered hash tables?

3. Suppose that in Algorithm 7.10 (coalesced chaining) we have *free* go forward from $i$, wrapping around from $m-1$ to 0 if needed, in searching for an empty table location. Does this improve or degrade the performance of the algorithm?

4. Explain why the poor performance of linear probing can be exacerbated by the use of division hashing if consecutive element values are likely to occur. Is this a problem with multiplicative hash functions?

5. What is wrong with having a 32-position table using double hashing with $h(z) = h_1$ and $\delta(z) = h_3$ of Table 7.2?

6. Explain how to compute the expected unsuccessful search time for the table configurations of Figure 7.27. Do the calculation for Figure 7.27(a).

7. Investigate the idea of using $\delta(z) = f(h(z))$ in double hashing.

★8. Show that the uniform hashing assumption that each of the $m!$ possible probe sequences is equally likely is stronger than the assumption that for all $n, 0 \leqslant n \leqslant m$, each of the $\binom{m}{n}$ possible configurations of $n$ full and $m - n$ empty locations is equally likely.

9. The object of this exercise is a more precise analysis of uniform hashing. As usual, let the table have $m$ locations and $n$ elements, assuming that the elements were inserted into random locations in the table so that each of the $\binom{m}{n}$ configurations of empty and nonempty locations is equally probable.
   (a) What is the probability that at least $k$ probes are needed to insert the $(n + 1)$st element?
   (b) Let $U_n$ be the average number of probes on an unsuccessful search in such a table, so that $U_n = \sum_{k=1}^{m} k \Pr(\textit{exactly } k \text{ probes are used})$. Compute $U_n$ using your result from (a). [*Hint*:

$$\sum_{k=1}^{m} \binom{m+1-k}{m-n} = \binom{m+1}{m-n+1}.]$$

   (c) Let $S_n$ be the average number of probes on a successful search in such a table. Prove that $S_n = (1/n)\sum_{k=0}^{n-1} U_k$ and use this to find $S_n$.

10. Prove that Algorithm 7.13(b), insertion into an ordered hash table, maintains the order of the elements in the table exactly as it would be if they were inserted in decreasing order. [*Hint*: Prove that the order of the elements in the table is unique by supposing that two different orderings were possible and arriving at a contradiction.]

11. Let $\hat{S}_n$ and $\hat{U}_n$ be, respectively, the analogs for ordered hash tables of $S_n$ and $U_n$ of Exercise 9.
    (a) Explain why $\hat{S}_n = (1/n)\sum_{k=0}^{n-1} U_k$. This means that $\hat{S}_n = S_n$.
    (b) Explain why $\hat{U}_n = \hat{S}_{n+1}$.

12. Let $C_n$ be the average number of probes made during an insertion with Algorithm 7.13(b)—that is, the average number of times that the test of the **while** loop is made. Show that $\sum_{k=1}^{n} C_k = n\hat{S}_n$. [*Hint*: Each time the loop is executed, the total number of probes to find one of the elements is

increased by one.] Combine this result with the previous exercise to show that $C_n = U_{n-1}$.

13. When linear probing is used to resolve collisions, true deletions can be accommodated, because all the elements that may have collided with the deleted element are localized. Design an algorithm for deletion in such a hash table.

14. As in the previous exercise, deletions with coalesced chaining are not too bad. Explain how to delete elements in such a hash table.

15. Give a rehashing algorithm analogous to Algorithm 7.14 for hashing with coalesced chaining.

16. Give a rehashing algorithm analogous to Algorithm 7.14 for ordered hash tables.

17. Comment on the appropriateness of using a hash-table scheme for storing sparse matrices.

18. Suppose we split the $n$ elements of a hash table into two tables, one of the $\lfloor \alpha n \rfloor$ high-frequency elements (to be searched first) that account for a fraction $\beta$ of all searches and the other of the remaining elements that account for the remaining $1 - \beta$ of all searches. For what values of $n$, $\alpha$, and $\beta$ is this worthwhile?

19. Discuss the merits of the following technique for making unsuccessful searches faster. Maintain a bit table $B[0]$, $B[1], \ldots, B[m-1]$, all initially **false**. $B[j]$ is set to **true** during an insertion in such a way that in general $B[j]$ is **true** if and only if some successful search passes through $T[j]$. Thus if a search ever reaches a $T[i]$ with $B[i] = $ **false**, the search must be unsuccessful.

20. Design a data structure based on hashing so that the following operations can be done efficiently on the average:

| | |
|---|---|
| $INSERT(a, b)$ | Adds the pair $(a, b)$ to the table |
| $FIND(a, b)$ | Finds the pair $(a, b)$ or indicates its absence |
| $FIND(a, *)$ | Returns a list of all pairs whose first element is $a$ |
| $FIND(*, b)$ | Returns a list of all pairs whose second element is $b$ |
| $DELETE(a, b)$ | Deletes the pair $(a, b)$ from the table |
| $DELETE(a, *)$ | Deletes all pairs whose first element is $a$ |
| $DELETE(*, b)$ | Deletes all pairs whose second element is $b$ |

21. Suppose we have a hash table in which collisions are resolved by coalesced chaining. Explain how the table locations can be shared by other linked

data structures *without* degrading the performance of the search and insert algorithms for the hash table (in other words, table locations that are being used by some other data structure must have the effect of empty locations in the search and insert algorithms). Assume that in addition to the fields for the key and a LINK for chaining, each table location has a two-bit TAG field. [*Hint*: Use the TAG field to distinguish four possible states of the table location—whether it is completely unused, part of another data structure, part of the hash table but not at the end of a chain, or part of the hash table and at the end of a chain. Assume that some minor changes will be needed in algorithms manipulating the *other* data structures sharing the space.]

22. We can improve the performance of Algorithm 7.10, collision resolution by coalesced chaining, by restricting the hash function so that $0 \leqslant h(z) < \hat{m} \leqslant m$. This means that elements *never* have home addresses in the last $m - \hat{m}$ table locations, called the *cellar*. Since Algorithm 7.10 is unchanged, these cellar locations are used *only* in the collision resolution phase. This technique delays, until collisions have filled the cellar, the tendency for long lists of collided elements to get longer. The factor $\hat{m}/m$ is called the *address factor*. By experiment or mathematical analysis, determine the optimal address factor as a function of the load factor. The value 0.86 has been proposed as a good compromise value for the address factor over a wide range of load factors; determine the range of load factors for which it is a good compromise.

## 7.5   TABLES IN EXTERNAL STORAGE

The methods of table organization that have been presented so far in this chapter are geared to internal memory—that is, memory that can be accessed randomly at speeds matching the speed of the computer itself. For small or medium-sized tables internal memory is fine, but for large tables, such as the telephone listings for a major city, social security records, and so on, we must rely on (relatively) slow external memory devices such as magnetic tapes, disks, or drums. This section briefly examines some of the issues involved in searching tables stored on such external memory devices.

**Magnetic Tapes.** Magnetic tape is a sequential storage medium. This means that to examine the $i$th record on the tape it is necessary to have examined or moved past the first $i - 1$ records. Essentially, then, organizing a table on magnetic tape is the same as using a linked list. We have already examined such table organizations earlier in this chapter, and nothing need be said here except a brief recapitulation. Such a table must be searched in a linear fashion, and the only possible refinement is to order the records on the tape so as to minimize search time. The two relevant orders are (1) the natural order of the records (alphabetical, numerical, etc.) or (2) so that $p_1/L_1 \geqslant p_2/L_2 \geqslant \cdots \geqslant p_n/L_n$, where $p_i$ is the probability that the $i$th record is the search object and $L_i$ is the length of the $i$th record—the cost of reading it into memory (see Exercise 4 of Section 7.1.1). In general, magnetic tapes are a poor choice for storing frequently accessed information unless that information will always be scanned in a linear fashion; the problem of sorting is an example of such an instance, and we will examine the use of magnetic tapes in sorting in Section 8.3.

**Disks and Drums.** Disks and drums do allow random access to all records stored, but the time required for such an access is very great compared to internal memory speeds. For purposes of this section we need not go into all the details of a disk drive or a drum. All we must know is that there is a high overhead in time to initiate an *access*—that is, a transfer of records from the disk or drum to internal memory. This overhead comes from the time required to position the reading mechanism at the proper location before reading the records into memory. Efficient organization of tables on disks and drums thus requires minimizing the number of times such an access is initiated and transferring large numbers of records on each access. We will now examine how this can be done within the framework of search trees and hash tables. For economy of style we will use the term "disk" exclusively, rather than "disk or drum."

### 7.5.1  Balanced Multiway Trees

The binary search tree techniques discussed in Section 7.2 could be used directly for organizing tables on disks; the LEFT and RIGHT pointers become addresses on the disk instead of addresses in internal memory. In this way the algorithms of Section 7.2 would require a disk access whenever a LEFT or RIGHT pointer was followed, essentially making one probe per disk access. Since disk accesses are costly compared to probes, it is preferable to make a number of probes for each disk access. We can do so if nodes in the tree contain $m$-way branches instead of two-way (binary) branches.

In analogy with binary trees, we define an *m-way tree*: such a tree $T$ either is empty or consists of a distinguished node called the root and $k$ subtrees $T_1, T_2, \ldots, T_k$, $2 \leqslant k \leqslant m$, each of which is an $m$-way tree. A node in an $m$-way tree specifies a $k$-way branch and thus contains $k - 1$ elements, for some $k$, $2 \leqslant k \leqslant m$, as shown in Figure 7.29. In an *m-way search tree* we require additionally that $x_1 < x_2 < \cdots < x_{k-1}$ and that elements in subtree $T_i$ be greater than $x_{i-1}$ and less than $x_i$ (with $x_0$ taken as $-\infty$ and $x_k$ taken as $\infty$). Figure 7.30 shows a five-way search tree. Notice that we do not insist that every node of an $m$-way tree contain $m - 1$ elements, for then the total number of elements stored

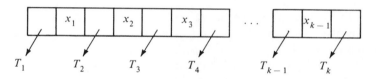

**Figure 7.29**
A degree-$k$ node in an $m$-way tree, $k \leqslant m$

in the tree would have to be a multiple of $m - 1$; this would be an unreasonable restriction, so we merely insist that each node contains *at most $m - 1$* elements.

To search an $m$-way search tree for $z$ we proceed similarly to the search of a binary search tree. Beginning at the root, we search for $z$ among the elements $x_1, x_2, \ldots, x_{k-1}$. If we find $z$, the search ends, if not, we continue the search in the subtree $T_i$ such that $x_{i-1} < z < x_i$. How should $m$ be chosen? If we have a *complete $m$-way tree* in which every node contains $m - 1$ elements and is an $m$-way branch, then searching a table of $n$ elements will require examining about $\log_m n$ tree nodes. The cost of examining a tree node is

$$I + Rm + S \lg m,$$

where

$\qquad I =$ time to initiate the disk access,
$\qquad Rm =$ time to read the $m$-way node from the disk,
$\qquad S \lg m =$ time to do a binary search on the elements of the $m$-way node.

The total time for a search is thus

$$(\log_m n)(I + Rm + S \lg m)$$

and, since $\log_m n = \lg n / \lg m$, this becomes

$$\left( \frac{I + Rm}{\lg m} + S \right) \lg n.$$

To minimize the search time we should choose $m$ so that $(I + Rm)/\lg m$ is minimized. Differentiating this with respect to $m$ and setting the derivative to zero, we find that the minimum occurs for $m$ such that

$$\lg m = \frac{I/R}{m} + 1.$$

The exact minimum depends on the value of $I/R$, but in most cases the cost of initiating a disk access is several thousand times the cost of reading a block of

**Figure 7.30**
A five-way search tree

words, so that $I/R$ will be something like 2000. The following table shows values of $m$ for various values of $I/R$ between 1000 and 3000.

| $I/R$ | 1000 | 1500 | 2000 | 2500 | 3000 |
|-------|------|------|------|------|------|
| $m$   | 159  | 221  | 280  | 338  | 394  |

Thus we expect to be dealing with multiway trees with a branching factor of a couple hundred or so. The best choice depends on the precise physical characteristics of the storage device (access and transfer times), the size of the elements in the table, and the amount of internal memory available to store elements.

Have we saved on disk accesses? Yes! If we choose $m = 256$, for example, we will be able to search a table of over 16 million elements with no more than two disk accesses, provided that the root node of the tree is always kept in internal memory. Had we used a binary tree, such a table would have required 24 accesses. Since the accesses dominate the cost of a search, a search in the 256-way tree will be twelve times faster than in the binary tree.

Our analysis above assumed that the $m$-way search tree was perfectly balanced with every node containing $m - 1$ elements. As we have seen with binary search trees, it is quite time-consuming to keep the tree perfectly balanced under insertions and deletions, so we need a compromise like that of height- or weight-balanced trees. A good compromise is the following: we insist that all paths from the root to an external node are of equal length and that each node except the root has at least $\lceil m/2 \rceil$ subtrees. (Exercise 9 considers the case in which each node has at least $\lceil (2m - 1)/3 \rceil$ subtrees.) Thus we define a *balanced multiway tree of order m* or *B-tree* as an $m$-way tree in which

1. All external nodes are at the same level.

2. The root has anywhere from two to $m$ subtrees.

3. Other internal nodes have anywhere from $\lceil m/2 \rceil$ to $m$ subtrees.

Returning to Figure 7.30, we see that the tree it depicts is a $B$-tree of order 5 because all the external nodes (not shown) are two levels down from the root and the number of subtrees of each internal node falls within the prescribed limits. The case of $m = 3$ is of interest as an alternative to height- or weight-balanced trees (Exercise 3), and that of $m = 2$ is related to height-balanced trees (Exercise 4). For purposes of the organization of tables on disks, we will be interested primarily in $m$ being around several hundred.

For $B$-trees to be useful, we need to verify that search times remain within bounds and that insertions and deletions can be accommodated. As to search times, suppose that there are $n$ elements and $n + 1$ external nodes. The root has at least two subtrees and the remaining nodes have at least $\lceil m/2 \rceil$ subtrees. After $l$ levels there are *at least* $2\lceil m/2 \rceil^{l-1}$ nodes. In a tree of height $h$ there are at least $2\lceil m/2 \rceil^{h-1}$ external nodes, so that

$$n + 1 \geqslant 2\lceil m/2 \rceil^{h-1}$$

or

$$h \leqslant 1 + \log_{\lceil m/2 \rceil}\left(\frac{n+1}{2}\right).$$

Thus in our example of a table with 16 million elements we need at most three disk accesses for a search if $m = 256$, so that such a search will be more than eight times faster than if we had used a height-balanced tree to organize the table (why?).

We now outline procedures to insert or delete elements in $B$-trees; the procedures require time comparable to that needed to search such a tree. To illustrate the process of insertion, consider inserting BY into the tree of Figure 7.30. We begin with an unsuccessful search for BY; as the search proceeds, a record is kept on a stack of the nodes visited—this lets us retrace the path up the tree. The search for BY fails at the bottom level of internal nodes in the tree. If the node at which it fails contains less than $m - 1$ elements, the new element is simply inserted into its proper place in that node (this would happen, for example, if we were inserting AN into the tree of Figure 7.30). BY cannot be so inserted because the node

$$\boxed{\text{AT} \mid \text{BE} \mid \text{BUT} \mid \text{FOR}}$$

already contains the maximum allowable number of elements. In such a case the $m$ elements, consisting of $m - 1$ in the node and the new element, are split into two nodes containing the smallest $\lceil m/2 \rceil - 1$ elements and the largest $\lfloor m/2 \rfloor$ elements; the median element is pushed up into the father node to be the separator element between the two halves. In our example, the split results in

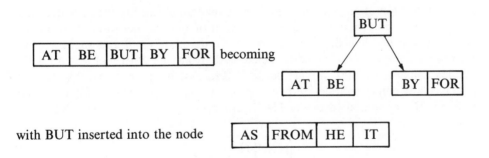

with BUT inserted into the node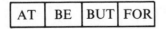

In general a split changes

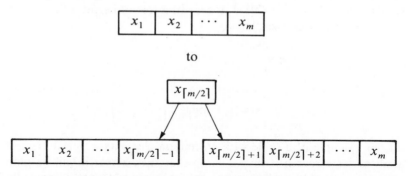

with $x_{\lceil m/2 \rceil}$ inserted into the father node. If the father node contains $m - 2$ or fewer elements, the addition of the new element causes no problem and the insertion ends. If the father node already contains $m - 1$ elements, as is the case of BUT inserted into

| AS | FROM | HE | IT |
|----|------|----|----|

,

then that node is split in turn; this process continues up the tree, as needed. When the root splits, a new node is created that becomes the root of the tree and the tree becomes one level taller. In our example,

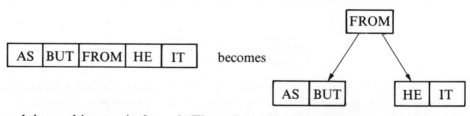

and the resulting tree is shown in Figure 7.31. The insertion process is remarkable in that trees grow taller by adding levels to the root, rather than to the leaves! This explains why we allow the root node a greater range of sizes than the other nodes in $B$-trees.

The deletion of an element is no harder than an insertion. We consider only the deletion of an element at the bottom level of the tree because, as in binary search trees, if an element is not at that level, its predecessor and successor are (Exercise 6). Suppose we want to delete HIS from the $B$-tree of Figure 7.31. Nothing could be simpler: we just delete HIS from its node in the tree; since that node has only empty subtrees and since it has enough elements, nothing else need be done. If we wanted to delete HAD, however, the deletion would leave the node

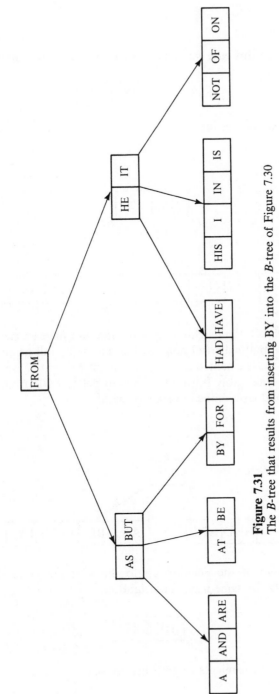

**Figure 7.31**
The *B*-tree that results from inserting BY into the *B*-tree of Figure 7.30

insufficiently full. In this case we could take an element from the neighboring

and use it to remake the tree into

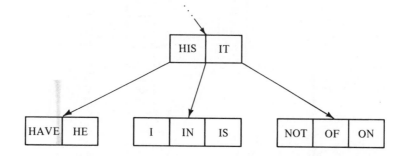

If we are deleting FOR, the neighboring node does not have an element to spare. In such a case of a minimally full node with a minimally full neighbor, the node, its neighbor, and the element that separates them in their father node can be combined into an acceptably full node. Thus to delete FOR from the *B*-tree of Figure 7.31 we would replace the left subtree with

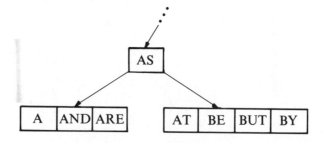

pushing the problem of the deletion up to the next higher level, where it is handled in precisely the same way: the neighbor

$$\boxed{\text{HE} \mid \text{IT}}$$

has insufficiently many elements to give one up, so

$$\boxed{\text{AS}} \quad \text{and} \quad \boxed{\text{HE} \mid \text{IT}}$$

and their separator are combined into a single node. The resulting *B*-tree is that shown in Figure 7.30, except BUT has been replaced with BY. Notice that as in an insertion, the height of the tree changes at the root, not at the leaves.

To close this section, a brief comment about storage utilization is in order. In the worst case, *B*-trees allow about 50 percent of their space to be wasted, since a minimally full node will have only $\lceil m/2 \rceil$ subtrees. On the average, however, the storage utilization can be shown to be about $\ln 2 \approx 69$ percent. The wasted space is the price we pay for fast search times involving few disk accesses, combined with the ability to make insertions and deletions.

## Exercises

1. Give a declaration for a node in a *B*-tree of order *m*.

2. We have seen that the best value for *m* is so that $\lg m = [(I/R)/m] + 1$. Verify that in general there is a fairly large *range* of values of *m* for which the value of $(I + Rm)/(\lg m)$ will be close to minimal.

3. A *3-2 tree* is a *B*-tree of order 3. Examine the possibility of using 3-2 trees as an alternative to height- or weight-balanced trees: What is the worst-case search time in a 3-2 tree of *n* elements? Can concatenation and splitting be done? Can 3-2 trees be maintained under insertions and deletions?

4. A *brother tree* is a *B*-tree of order 2 with the additional property that if *N* is a unary node of the tree, then *N* must have a binary brother. Prove that there is an exact correspondence between brother trees and height-balanced trees.

5. Show the tree that results from successively inserting THE, TO, THAT, WAS, WITH, THIS, OR, and THEY into the *B*-tree of order 5 shown in Figure 7.31.

6. In parallel with the discussion given in Section 7.2.2, prove that if an element is not at the lowest level of a *B*-tree, then its predecessor and successor are.

7. Show the tree that results from deleting BUT, HE, NOT, THE, WAS and OR from the tree that was the end result of Exercise 5.

8. Explain how the insertion and deletion processes (as outlined) would need to be modified if each level in the *B*-tree had its own limit on the number of elements in a node. What are the advantages and disadvantages of such a scheme?

9. Suppose we modify the definition of *B*-tree to (1) all leaves are at the same level, (2) the root has anywhere from 2 to $2\lfloor (2m - 2)/3 \rfloor + 1$ subtrees, (3) other internal nodes have anywhere from $\lfloor (2m - 1)/3 \rfloor$ to *m* subtrees.
   (a) What is the effect on the storage utilization?
   (b) Explain how insertions and deletions are done.

### 7.5.2  Bucket Hashing

We can adapt hashing schemes to disks by having the address computed by the hash function be a disk address. As in the previous section, we want to minimize the number of disk accesses needed on a search. There are two things we can do. First, we can enlarge the basic table component from a single element to a group of many elements. The elements would thus be grouped into *buckets* of $b$ elements each; a hash address would specify the disk address of a bucket and a disk access would retrieve all $b$ elements of the bucket. The elements in a bucket would then be searched one by one (the elements in a bucket would not ordinarily be kept in order). The second thing we can do is use a hash function that causes fewer collisions; even before, of course, we wanted a hash function with few collisions, but the cost of a collision is so enormous when disk accesses are at stake, that it becomes worthwhile to spend much more time in computing the hash function so as to minimize collisions.

Because we are hashing to buckets rather than individual table locations, we expect collisions to be no problem, so we can use a relatively simple-minded scheme to resolve them. Every bucket will have a pointer that will be used to point to the $(b + 1)$st element that hashed to that bucket. That element has a pointer to the $(b + 2)$nd element, and so on. Thus the first $b$ elements hashing to a given bucket are stored in that bucket; the remaining elements are formed into a linked list with the pointer to the start of the list kept in the bucket itself. All elements not in their proper bucket are kept in a special "overflow" area of the disk in which there are no buckets, only individual elements. If there are $n$ elements and $m$ buckets, we define the load factor as

$$\lambda = \frac{n}{mb}$$

(notice that this ignores the locations in the overflow area). The performance of such a hashing scheme is difficult to analyze, but it can be shown that for load factors $\lambda$ up to 90 percent and bucket sizes $b$ up to 50, an average of between one and two probes will be needed for either successful or unsuccessful searches. As $\lambda$ increases past 90 percent, the performance degrades somewhat, with the average successful search requiring about 1.5 probes and the average unsuccessful search requiring about $\sqrt{b/2\pi} + 1$ probes.

### Exercise

Explain why we might have a load factor $\lambda > 1$ in a hash table as described in this section.

## 7.6   REMARKS AND REFERENCES

In this chapter we have seen a diversity of table organizations, and we have only scratched the surface. Because of minor variations or combinations of table organizations, the possibilities are endless. The techniques of major importance have been presented, and many variations are covered in exercises. In choosing among the possible alternatives, some of the many things to ponder are

1. What are the elements? Are they simple keys, or complicated records of which the key is only a small part? Are the keys numerical? Alphabetical?

2. What kind of operations are necessary? Is the table static or dynamic? Are the elements going to be needed in any special order? Is the table size known a priori, or must the table be expandable? Will we have to merge tables together? Will we have to split one into pieces?

3. What is the nature of a search? Will we want a yes-or-no answer, the closest element, or a range of elements? Will searches be mainly successful or unsuccessful? Will an unsuccessful search always be followed by an insertion?

4. Are we concerned about average time or worst-case time?

5. What is the nature of the storage medium?

Further details and complete analyses for most of the table organizations discussed here can be found in Chapter 6 of

> Knuth, D. E., *The Art of Computer Programming*, Vol. 3, Sorting and Searching, Reading, Mass.: Addison-Wesley Publishing Co., 1972.

In two cases, more recent and/or detailed information can be found elsewhere. See

> Gonnet, G. H., L. D. Rogers, and J. A. George, "An Algorithmic and Complexity Analysis of Interpolation Search," *Acta Informatica* **13** (1980), 39–52.

for an extended discussion of interpolation search along with an extensive bibliography. For hashing techniques see

> Knott, G. D., "Hashing Functions," *Computer J.* **18** (1975), 265–278.
> Amble, O., and D. E. Knuth, "Ordered Hash Tables," *Computer J.* **17** (1974), 135–142.

Guibas, L. J., and E. Szemeredi, "The Analysis of Double Hashing," *J. Comput. Syst. Sci.*, **16** (1978), 226–274.

The first of these three papers gives a comprehensive discussion of hash functions; the second gives, in great detail, the analysis of ordered hash tables; the third gives the best analysis known of double hashing.

Finally, for the reader interested in pursuing the examples used in this chapter on letter and word distributions in the English language, we recommend

Kahn, D., *The Codebreakers*. New York: The Macmillan Company, 1969.

Kučera, H., and W. N. Francis, *Computational Analysis of Present-Day American English*. Providence, R.I.: Brown University Press, 1967.

# Chapter 8

# Sorting

The old order changeth, yielding place to new.

*The Passing of Arthur,*   Alfred Lord Tennyson

In this chapter we discuss methods for sorting a table of elements $x_1, x_2, \ldots, x_n$ into order, based on their arithmetic or lexicographic value. Our goal is to rearrange the elements so that $x_{\pi_1} \leqslant x_{\pi_2} \leqslant \cdots \leqslant x_{\pi_n}$, where $\Pi = (\pi_1, \pi_2, \ldots, \pi_n)$ is a permutation of the integers $1, 2, \ldots, n$. Normally we do not explicitly determine $\Pi$, but rather we rearrange the $x_i$ so that they are in order. Section 8.1 discusses various methods, their strengths and their weaknesses, for the case in which all the $x_i$ can fit into memory simultaneously for the sorting process. Section 8.2 discusses two problems closely related to sorting, that of selecting the $k$th largest among the unordered elements $x_1, x_2, \ldots, x_n$ and that of merging two ordered sequences of elements. Finally, in Section 8.3, we examine how sorting can be done when there are too many elements to be in memory together and it is possible only to examine the elements in piecemeal fashion, bringing them into memory in batches from an external storage device.

In the various algorithms presented in this chapter, the elements are stored in a sequence that we will denote by $x_1, x_2, \ldots, x_n$ *regardless of the data moves that occur*; thus the value of $x_i$ is whatever element is currently in the $i$th position of the sequence. Many of the algorithms are best performed on an array; in this case $x_i$ denotes the $i$th component of the array. Other algorithms are better suited to linked lists, in which case $x_i$ denotes the $i$th element of the list. In still other algorithms, $x_i$ will be the $i$th record in a file on tape or disk. Thus the $x_i$ are really variables in the programming sense so that

$x_i \leftrightarrow x_j$   means that elements $x_i$ and $x_j$ in the sequence are interchanged.

$x_i \leftarrow y$   means that the element being temporarily stored in $y$ becomes the element $x_i$.

$y \leftarrow x_i$   means that the element $x_i$ is stored as the value of $y$; this leaves $x_i$ temporarily undefined.

$x_i \leftarrow x_j$   means that the element $x_j$ replaces the element $x_i$, leaving $x_j$ temporarily undefined.

## 8.1   INTERNAL SORTS

In this section we will examine about a half-dozen different strategies for rearranging $x_1, x_2, \ldots, x_n$ into order. Each of the algorithms we consider requires, more or less, direct access to all of the elements being sorted: this makes them suitable only as internal sorting methods; even though they could be used directly with tape or disk storage, the result would be incredibly inefficient. Additionally, the algorithms considered here are all *in-place* sorting algorithms. In other words, the rearrangement process must occur entirely within the sequence $x_1, x_2, \ldots, x_n$ along with one or two additional locations in which to store elements temporarily. This in-place restriction is based on the assumption that the number of elements to be sorted is too large to permit copying them into a different storage area as they are sorted. If sufficient storage is available to allow such copying, then some of the algorithms presented here can be speeded up considerably.

Just as there are many internal table organizations for searching, so there are many internal sorting methods. And, as is the case for searching, there is no universal answer to the question, "What is the best method?" The answer varies with the particular circumstances—the type of elements being sorted, the amount of storage available, the relative emphasis on good average performance versus good worst-case performance, the relative speeds of various computer instructions, and so on. For this reason we will examine a variety of algorithms, each having its own assets and liabilities.

### 8.1.1   Insertion Sort

The following simple idea leads to a sorting algorithm that is useful for small numbers of elements: repeatedly insert another element into the sequence of already sorted elements. Thus at stage $i$ the $i$th element is inserted into its proper place among the previously sorted $i - 1$ elements. At stage 1, $x_1$ is inserted into place trivially, since there were no previously sorted elements; at stage 2, $x_2$ is inserted into place either before or after $x_1$; at stage 3, $x_3$ is inserted into place among the now sorted $x_1$ and $x_2$; and so on. Algorithm 8.1 describes this process precisely, with the $j$th stage inserting $x_j$ among $x_1, x_2, \ldots, x_{j-1}$. This is done repeatedly, for $j = 2, 3, \ldots, n$. In each case the insertion is done by storing $x_j$ temporarily in $t$ and scanning the elements $x_{j-1}, x_{j-2}, \ldots, x_1$, comparing each with $x_j$ and shifting it to the right if it is found to be greater than $t$. We thus have combined the finding of the spot where $x_j$ will be inserted together with the

| $j$ | $x_0$ | $x_1$ | $x_2$ | $x_3$ | $x_4$ | $x_5$ |
|---|---|---|---|---|---|---|
| 2 | $-\infty$ | 8 | 7 | 2 | 4 | 6 |
| 3 | $-\infty$ | 7 | 8 | 2 | 4 | 6 |
| 4 | $-\infty$ | 2 | 7 | 8 | 4 | 6 |
| 5 | $-\infty$ | 2 | 4 | 7 | 8 | 6 |
|   | $-\infty$ | 2 | 4 | 6 | 7 | 8 |

**Figure 8.1**
The insertion sort, Algorithm 8.1, applied to $n = 5$ elements. The dashed vertical lines separate the already sorted part of the table $x_1, x_2, \ldots, x_{j-1}$ from the unsorted part $x_j, x_{j+1}, \ldots, x_n$

process of making room at that spot for the insertion. For additional simplification, we have introduced a dummy element $x_0 = -\infty$ to stop the backward scan at the left end. Figure 8.1 illustrates this algorithm on five elements.

In this algorithm, as in most sorting algorithms, the performance depends on the number of element comparisons and the number of data moves made in the worst case, on the average (assuming that each of the $n!$ permutations of the elements is equally likely), and in the best case. The key to analyzing Algorithm 8.1 is the number of times the test "$t < x_i$" is made in the **while** loop, for once that number is known, the number of data moves "$x_{i+1} \leftarrow x_i$" can easily be deduced.

$$x_0 \leftarrow -\infty$$
**for** $j = 2$ **to** $n$ **do**
    ⟦insert $x_j$ into place among $x_1, x_2, \ldots, x_{j-1}$⟧
    $i \leftarrow j - 1$
    $t \leftarrow x_j$
    **while** $t < x_i$ **do**
        ⟦shift the "hole" one position left⟧
        $x_{i+1} \leftarrow x_i$
        $i \leftarrow i - 1$
    $x_{i+1} \leftarrow t$    ⟦drop it into the hole⟧

**Algorithm 8.1**
Insertion sort

The comparison "$t < x_i$" is made

$$\sum_{j=2}^{n} (1 + d_j) = n - 1 + \sum_{j=2}^{n} d_j$$

times, where $d_j$ is the number of elements larger than $x_j$ and to its left. We can understand this by observing that, for each $j$, the loop scanning backward from $j - 1$ will make the comparison "$t < x_i$" until it finds an element $t \geqslant x_i$; it thus makes the comparison (and executes the contents of the loop) once for each element to the left of $x_j$ that is larger than $x_j$ and once more when it fails, finding an element less than or equal to $x_j$. The number of data moves is thus $(n - 1) + (n - 1) + \sum_{j=2}^{n} d_j$ because "$t \leftarrow x_j$" and "$x_j \leftarrow t$" are done $n - 1$ times each, and "$x_{i+1} \leftarrow x_i$" is done $\sum_{j=2}^{n} d_j$ times.

It remains to analyze the quantity $\sum_{j=2}^{n} d_j$. In the best case, each $d_j = 0$; this occurs when the elements are already in sorted order. Clearly, the most number of elements to left of $x_j$ and larger than it is $j - 1$, so the worst case is $d_j = j - 1$; this occurs when the elements are in *reverse* order. The average value of $d_j$, assuming that each of the $n!$ permutations of the $x_i$ is equally likely, can be shown to be $(j - 1)/2$ (Exercise 1). The time required by Algorithm 8.1 is thus summarized by the following table:

| | Element Comparisons | Element Moves | Extra Storage |
|---|---|---|---|
| Best case | $n - 1$ | $2n - 2$ | |
| Average case | $\dfrac{(n-1)(n+4)}{4}$ | $\dfrac{(n-1)(n+8)}{4}$ | One extra element ($t$) and integer variables $i$ and $j$ in all cases |
| Worst case | $\dfrac{(n+1)(n-2)}{2}$ | $\dfrac{(n+1)(n-4)}{2}$ | |

Notice that we have not taken into account the overhead for the loop controls and the incrementing and decrementing of the counters. We have ignored these because we cannot determine their expense relative to the cost of manipulating the elements being sorted. If the $x_i$ are very long character strings, then what we have ignored is negligible. We have analyzed Algorithm 8.1 to the limit possible without knowledge of the elements being sorted or the machine instructions available.

Insertion sort has in its favor its elegant simplicity, which leads to very low overhead. Its disadvantage is that both on the average and in the worst case it requires time proportional to $n^2$, and we will find other sorting techniques that require only time proportional to $n \lg n$, a big improvement. Still, Algorithm 8.1 is the best choice for small numbers of elements (say, no more than fifteen to twenty) because of its low overhead, and it is the best choice when the elements are not in too much disarray, making $\sum_{j=2}^{n} d_j$ small.

Can the insertion sort be improved? We could do *binary insertion*, using binary search (Algorithm 7.3) and hence at most about $\lg j$ element comparisons to insert the $j$th element. A total of $\sum_{j=2}^{n} \lg j \approx n \lg n$ such comparisons would thus suffice. Unfortunately, after having found the spot in which to insert $x_j$, the in-place restriction causes the insertion itself to require moving $j/2$ names on the average, and the algorithm would still require proportional to $n^2$ operations. Alternatively, we could use a linked list for the sorted portion of the table, thereby making the insertion more efficient. Of course, in this case we could not use binary search to find the place at which $x_j$ should be inserted, and we would be stuck with a sequential search. The number of operations required by the resulting sorting algorithm would be proportional to $n^2$.

If we could combine the searching ease of a sequentially allocated list with the insertion ease of a linked list, we would be able to obtain an algorithm requiring $O(n \log n)$ time. This can be done by using the balanced tree schemes of Section 7.2.2, but the overhead involved makes such schemes prohibitive for sorting, since we will be able to achieve $O(n \log n)$ time sorting in simpler ways. It *is* possible to use the insertion idea to obtain an $O(n(\log n)^2)$ time sorting algorithm; see Exercises 3 and 4.

## Exercises

1. Given a permutation $\Pi = (\pi_1, \pi_2, \ldots, \pi_n)$ of $1, 2, \ldots, n$, let $d_i, 1 \leq i \leq n$, be the number of elements in $\Pi$ to the left of $\pi_i$ and greater than it.

   (a) What is the possible range of values of each $d_i$?

   (b) Show that the permutation can be reconstructed from the $d_i$'s, that is, that the $d_i$'s uniquely determine the permutation.

   ★(c) Show that if each of the $n!$ permutations is equally likely, then each of the possible values of each $d_i$ is equally likely and that the $d_i$'s are independent of each other.

2. (a) Modify Algorithm 8.1 by adding a test "$i > 0$" in the **while** loop so that the sentinel $x_0 = -\infty$ is not needed.

   ★(b) Analyze the modified algorithm. [*Hint:* The analysis is complicated by the fact that the number of times the comparison "$t < x_i$" is made is diminished by $l$, the number of times that $i = 0$, that is, the number of values $j, 1 < j \leq n$, such that $x_j < x_i$ for all $i, 1 \leq i < j$. Determine the minimum and maximum values of $l$. Show that the average value of $l$ is $H_n - 1$, where $H_n = \sum_{i=1}^{n} 1/i$.]

3. Rewrite Algorithm 8.1 so that in a table $x_1, x_2, \ldots, x_n$ it sorts only the subtables $x_k, x_{k+\delta}, x_{k+2\delta}, \ldots$, for $1 \leq k \leq \delta$. Call this $\delta$-*sorting* and show that, given any sequence of positive integers $\delta_t > \delta_{t-1} > \cdots > \delta_1 = 1$, successively $\delta_t$-sorting, $\delta_{t-1}$-sorting, $\ldots, \delta_1$-sorting leaves the table correctly

sorted. This sorting algorithm is known as the *diminishing-increment sort*. Why should we expect it to be better than Algorithm 8.1?

★4. (a) Show that, for any positive integers $\varepsilon$ and $\delta$, if a table is $\varepsilon$-sorted after being $\delta$-sorted it remains $\delta$-sorted.

(b) Show that if $l$ and $m$ are relatively prime positive integers, then the largest integer not representable in the form $ul + vm, u, v \geq 0$, is $(l - 1)(m - 1) - 1$.

(c) Let $t \geq s_2 > s_1 > s_0 \geq 1$. Use the results of parts (a) and (b) to show that if the values of $\delta_{s_2}$ and $\delta_{s_1}$ in Exercise 3 are relatively prime, then the $\delta_{s_0}$-sorting portion of the diminishing increment sort has running time $O(n \delta_{s_1} \delta_{s_2} / \delta_{s_0})$.

(d) Show that if the values of $\delta_t, \delta_{t-1}, \ldots$ in Exercise 3 are chosen as the set of all integers of the form $2^p 3^q < n$, then the diminishing-increment sort described there requires time $O(n(\log n)^2)$. [*Hint:* How many inversions are there in a table that has been both 2-sorted and 3-sorted? How many integral points $(p, q)$ are there in the triangular region $p \ln 2 + q \ln 3 < \ln n, p, q \geq 0$?]

5. When the elements being sorted consist of long records of which the key $x_i$ is only a small part, the data moves in Algorithm 8.1 may be too costly. Develop a version of Algorithm 8.1 that eliminates the data moves by linking the elements being sorted with an extra pointer field in each record. The end result of your algorithm should be the linked list of the elements in sorted order.

6. Does Algorithm 8.1 disturb the relative order of equal elements? A sorting algorithm that does *not* is called *stable*. Why might it be necessary to know whether or not a sorting algorithm is stable?

★7. Find the necessary loop-invariants to prove that Algorithm 8.1 terminates and upon termination leaves the elements properly sorted.

### 8.1.2  Transposition Sorts

Sorting methods based on transpositions systematically interchange pairs of elements that are out of order until no such pairs exist. In fact, Algorithm 8.1 can be considered a transposition sort in which the element $x_j$ is interchanged with its left-hand neighbor until it is in its correct place. In this section we discuss two transposition sorts: the well-known but inefficient bubble sort and quicksort, one of the best all-around internal sorting algorithms.

**Bubble Sort (Pilloried).** An obvious way to systematically interchange out-of-order pairs of elements is to scan adjacent pairs of elements from left to right repeatedly, interchanging those found to be out of order. This technique has come to be known as the *bubble sort*, since larger elements "bubble-up" to the top (that is, the right). Algorithm 8.2 shows how this simple idea is implemented with one slight improvement: there is no point in continuing the scan into the large names (at the right end) that are known to be in their final positions. Algorithm 8.2 uses the variable $k$, whose value at the beginning of the **while** loop is the largest index $t$ for which $x_t$ is *not* known to be in its final position. Figure 8.2 illustrates how the algorithm works on $n = 8$ elements.

$$k \leftarrow n$$
**while** $k \neq 0$ **do**
$$\quad t \leftarrow 0$$
$$\quad \textbf{for } j = 1 \textbf{ to } k - 1 \textbf{ do}$$
$$\quad\quad \textbf{if } x_j > x_{j+1} \textbf{ then}$$
$$\quad\quad\quad x_j \leftrightarrow x_{j+1}$$
$$\quad\quad\quad t \leftarrow j$$
$$\quad k \leftarrow t$$

**Algorithm 8.2**
Bubble sort

The analysis of the bubble sort depends on the number of passes (the number of times the body of the **while** loop is executed), the number of comparisons "$x_j > x_{j+1}$", and the number of interchanges "$x_j \leftrightarrow x_{j+1}$". The number of interchanges is $\sum_{j=2}^{n} d_j$ because each element must move past (be interchanged with) exactly those elements smaller and to the right. This is similar

| | $x_1$ $x_2$ $x_3$ $x_4$ $x_5$ $x_6$ $x_7$ $x_8$ | $d_1$ $d_2$ $d_3$ $d_4$ $d_5$ $d_6$ $d_7$ $d_8$ |
|---|---|---|
| | 4  7  3  1  5  8  2  6 | 0  0  2  3  1  0  5  2 |
| Pass 1 | | |
| | 4  3  1  5  7  2  6  8 | 0  1  2  0  0  4  1  0 |
| Pass 2 | | |
| | 3  1  4  5  2  6  7  8 | 0  1  0  0  3  0  0  0 |
| Pass 3 | | |
| | 1  3  4  2  5  6  7  8 | 0  0  0  2  0  0  0  0 |
| Pass 4 | | |
| | 1  3  2  4  5  6  7  8 | 0  0  1  0  0  0  0  0 |
| Pass 5 | | |
| | 1  2  3  4  5  6  7  8 | 0  0  0  0  0  0  0  0 |
| Pass 6 | | |
| | 1  2  3  4  5  6  7  8 | 0  0  0  0  0  0  0  0 |

**Figure 8.2**
The bubble sort (Algorithm 8.2) applied to $n = 8$ elements. Also shown are the $d_i$ values: $d_i$ is the number of elements larger than $x_i$ and to its left

to the number of data moves in Algorithm 8.1, *except* that in the bubble sort we are counting interchanges—each interchange $x_j \leftrightarrow x_{j+1}$ is shorthand for the three moves *temp* $\leftarrow x_j$, $x_j \leftarrow x_{j+1}$, and $x_{j+1} \leftarrow$ *temp*. The number of data moves in Algorithm 8.2 thus is $3\sum_{j=2}^{n} d_j$: 0 in the best case, $3n(n-1)/2$ in the worst case, and $3n(n-1)/4$ on the average.

Examination of the values of the $d_i$ in Figure 8.2 suggests that each pass of the bubble sort, except the last, decreases by one each nonzero $d_i$; this is indeed the case (Exercise 1). Thus the number of passes is one plus the largest $d_i$: 1 in the best case and $n$ in the worst case. The average number of passes can be shown to be $n - \sqrt{\pi n/2} + O(1)$. The number of comparisons "$x_j > x_{j+1}$" can be shown to be $n - 1$ at best, $n(n-1)/2$ at worst, and $[(n^2 - n\ln n)/2] + O(1)$ on the average.

The following table summarizes the behavior of the bubble sort:

| | Element Comparisons | Element Moves | Number of Passes | Extra Storage |
|---|---|---|---|---|
| *Best case* | $n - 1$ | $0$ | $1$ | |
| *Average case* | $\frac{1}{2}n(n - \ln n) + O(n)$ | $\frac{3}{4}n(n-1)$ | $n - \sqrt{\pi n/2} + O(1)$ | In all cases, one extra element for the interchanges and integer variables $k$ and $j$ |
| *Worst case* | $\frac{1}{2}n(n-1)$ | $\frac{3}{4}n(n-1)$ | $n$ | |

Comparison of this table with the corresponding table for Algorithm 8.1 makes it clear that *the bubble sort is inferior to the insertion sort in almost every regard*: the bubble sort requires many more element moves except under the luckiest circumstances when the elements are almost completely in order. It requires many more element comparisons on the average, the same number at best, and only slightly fewer at worst. Its extra storage requirements are the same. Furthermore, it is a more complex algorithm both in its technique and in its analysis. *Unless we are sorting elements that already are nearly in order, the bubble sort has nothing to recommend it as an internal sorting technique*! It is possible to improve the bubble sort slightly (Exercise 2), but not enough to make it competitive with other sorting algorithms.

In both the insertion sort and the bubble sort, a major source of inefficiency is the fact that the interchanges do very little work, since elements move only one position at a time. As Exercise 3 shows, such algorithms are doomed to require proportional to $n^2$ operations in both the average and the worst cases. Thus a promising improvement is to interchange elements that are far away from each other, which is why the diminishing-increment sort (Exercises 3 and 4 of Section 8.1.1) is asymptotically so efficient. Another way to make each interchange do more work is used in quicksort.

**Quicksort.** The idea in quicksort is to select one of the elements $x_1, x_2, \ldots, x_n$ and to use it to partition the remaining elements into two groups—those less than and those greater than the selected element—which are then sorted by applying quicksort recursively. The partitioning can be implemented by simultaneously scanning the elements from right to left and from left to right, interchanging elements in the wrong parts of the table. The element used to partition the table is then placed between the two subtables and the two subtables are sorted recursively.

Algorithm 8.3 gives the details of this method for sorting the table $(x_f, x_{f+1}, \ldots, x_l)$, using $x_f$ to partition the table into subtables. Figure 8.3 shows how Algorithm 8.3 uses the two pointers $i$ and $j$ to scan the table during partitioning. At the beginning of the loop "**while** $i < j$", $i$ and $j$ point, respectively, to the first and last elements known *not* to be in the correct parts of the file. When they cross, that is, when $i \geq j$, all elements are in the correct parts of the table, and $x_f$ is placed between the two parts by interchanging it with $x_j$. In order to simplify the central mechanism of the loop, the algorithm assumes that $x_{l+1}$ is defined and larger than or equal to $x_f, x_{f+1}, \ldots$, and $x_l$; as we will see, this can be easily arranged.

**procedure** $QUICKSORT(f, l)$
⟦sort $x_f, x_{f+1}, \ldots, x_l$ assuming that $x_{l+1}$ exists and is
greater than or equal to $x_f, x_{f+1}, \ldots, x_l$⟧
**if** $f < l$ **then**
      ⟦partition the elements⟧
      $i \leftarrow f + 1$
      **while** $x_i < x_f$ **do** $i \leftarrow i + 1$
      $j \leftarrow l$
      **while** $x_j > x_f$ **do** $j \leftarrow j - 1$
      **while** $i < j$ **do**
            ⟦at this point $i < j$, and $x_i \geq x_f \geq x_j$⟧
            $x_i \leftrightarrow x_j$
            **repeat** $i \leftarrow i + 1$ **until** $x_i \geq x_f$
            **repeat** $j \leftarrow j - 1$ **until** $x_j \leq x_f$
      $x_f \leftrightarrow x_j$    ⟦put $x_f$ into place between the subtables⟧
      ⟦sort the subtables recursively⟧
      $QUICKSORT(f, j - 1)$
      $QUICKSORT(j + 1, l)$

**Algorithm 8.3**
Recursive version of quicksort, using the first element to partition the remaining elements. The algorithm assumes that $x_{l+1}$ exists and is larger than or equal to $x_f, x_{f+1}, \ldots, x_l$.

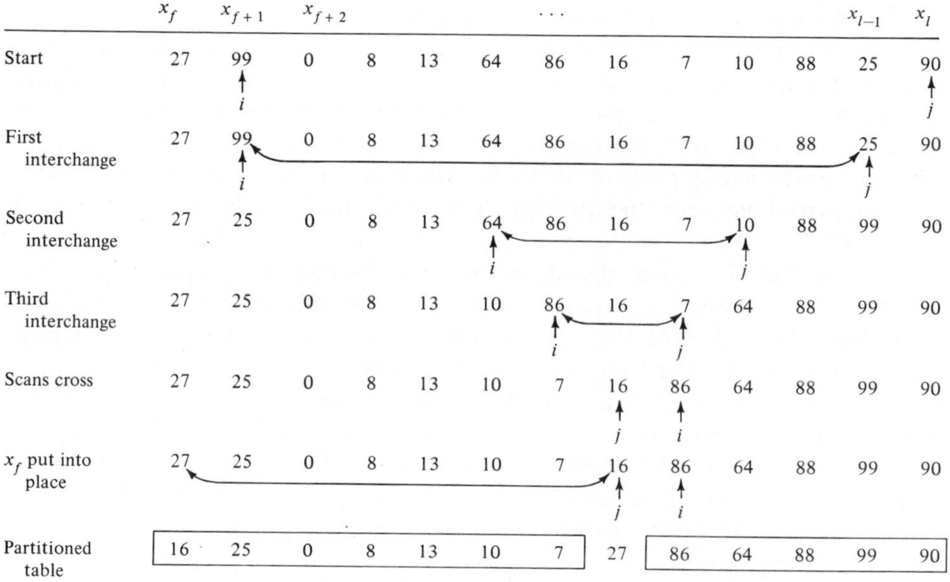

| | $x_f$ | $x_{f+1}$ | $x_{f+2}$ | | | ... | | | | | | $x_{l-1}$ | $x_l$ |
|---|---|---|---|---|---|---|---|---|---|---|---|---|---|
| Start | 27 | 99 | 0 | 8 | 13 | 64 | 86 | 16 | 7 | 10 | 88 | 25 | 90 |
| First interchange | 27 | 99 | 0 | 8 | 13 | 64 | 86 | 16 | 7 | 10 | 88 | 25 | 90 |
| Second interchange | 27 | 25 | 0 | 8 | 13 | 64 | 86 | 16 | 7 | 10 | 88 | 99 | 90 |
| Third interchange | 27 | 25 | 0 | 8 | 13 | 10 | 86 | 16 | 7 | 64 | 88 | 99 | 90 |
| Scans cross | 27 | 25 | 0 | 8 | 13 | 10 | 7 | 16 | 86 | 64 | 88 | 99 | 90 |
| $x_f$ put into place | 27 | 25 | 0 | 8 | 13 | 10 | 7 | 16 | 86 | 64 | 88 | 99 | 90 |
| Partitioned table | 16 | 25 | 0 | 8 | 13 | 10 | 7 | 27 | 86 | 64 | 88 | 99 | 90 |

**Figure 8.3**
The partitioning phase of quicksort using the first element to partition the table. The value of $x_{l+1}$, not shown, is assumed to be larger than or equal to the other values shown

To analyze the total number of element comparisons "$x_i < x_f$" and "$x_j > x_f$" in Algorithm 8.3, notice that at the end of the loop "**while** $i < j$" all the elements $x_{f+1}, \ldots, x_l$ have been compared once with $x_f$ except the elements $x_s$ and $x_{s+1}$ (where the scans cross), which have been compared twice with $x_f$. Let $\overline{C}_n$ be the average number of element comparisons to sort a table of $n$ distinct elements, assuming that each of the $n!$ permutations of the elements is equally likely. Obviously $\overline{C}_0 = \overline{C}_1 = 0$ and in general we have

$$\overline{C}_n = \sum_{s=1}^{n} p_s \left( n + 1 + \overline{C}_{s-1} + \overline{C}_{n-s} \right), \qquad n \geqslant 2, \tag{8.1}$$

where

$$p_s = \Pr\left( x_f \text{ is the } s\text{th smallest element} \right) = \frac{1}{n},$$

since the two subtables produced by the partitioning are random—that is, since each of the $(s-1)!$ permutations of the elements in the left subtable is equally likely and each of the $(n - s)!$ permutations of the elements in the right subtable is equally likely: the partitioning technique used in Algorithm 8.3 has been very carefully designed to insure the randomness of the two resulting subtables. Less subtle partitioning methods lead to nonrandom (that is, biased) subtables, causing a degradation of the performance of the algorithm (see Exercise 4).

The recurrence (8.1) simplifies to

$$\bar{C}_n = n + 1 + \frac{2}{n} \sum_{s=0}^{n-1} \bar{C}_s, \qquad n \geq 2.$$

This is an instance of the recurrence (7.9) of Section 7.2.2, the solution of which is given by Equation (7.14), page 300:

$$\bar{C}_n = 2(n + 1) \sum_{i=1}^{n} \frac{1}{i} - \frac{8}{3}n - \frac{2}{3}$$

$$\approx 1.386 n \lg n. \tag{8.2}$$

In other words, on the average, quicksort requires a total amount of work proportional to $n \lg n$, since the amount of work done *per comparison* is constant (why?). This makes quicksort *asymptotically* better on the average than the other sorting algorithms so far considered. It turns out that, with various refinements, the quicksort idea gives better performance on the average in practice than other sorting algorithms.

The worst-case performance of quicksort is another matter, however. Let $C_n$ be the number of element comparisons used by Algorithm 8.3 in the worst case on a table of $n$ elements. We claim that

$$C_n = \frac{1}{2}n^2 + \frac{3}{2}n + O(1), \tag{8.3}$$

making a quicksort's performance proportional to $n^2$ in the worst case—that is, no better than that of the insertion or bubble sorts. We verify (8.3) by showing an example on which quicksort requires $\frac{1}{2}n^2 + \frac{3}{2}n - 2$ comparisons and then showing that no matter what, $C_n \leq \frac{1}{2}n^2 + \frac{3}{2}n$. First, consider what happens when quicksort is applied to an *already sorted table*: it makes $(n + 1) + n + \cdots + 3$ element comparisons, a total of $\frac{1}{2}n^2 + \frac{3}{2}n - 2$. On the other hand, by inspection $C_0 = C_1 = 0$ and $C_2 = 3$. By induction, then, $C_n \leq \frac{1}{2}n^2 + \frac{3}{2}n$, for it holds when $n \leq 3$ and

$$C_n = n + 1 + \max_{1 \leq k \leq n} (C_{k-1} + C_{n-k})$$

$$\leq n + 1 + \max_{1 \leq k \leq n} \left[ \frac{1}{2}(k - 1)^2 + \frac{3}{2}(k - 1) + \frac{1}{2}(n - k)^2 + \frac{3}{2}(n - k) \right].$$

The maximum occurs at $k = 1$ or $k = n$ so that

$$C_n \le n + 1 + \tfrac{1}{2}(n-1)^2 + \tfrac{3}{2}(n-1) = \tfrac{1}{2}n^2 + \tfrac{3}{2}n,$$

as desired. A slightly more careful analysis shows that in fact for $n \ge 1$, $C_n = \tfrac{1}{2}n^2 + \tfrac{3}{2}n - 2$ (see Exercise 7).

Because $\bar{C}_n$ is $O(n \log n)$, the quadratic performance of the worst case would be extremely rare *on random tables*. In practice, however, the tables encountered are likely to be in rough order, or to have large segments that are in order; this makes the quadratic performance more probable, degrading the expected behavior. The difficulty is the choice of the element to partition the table; when the order of the elements is not random, the first element cannot be expected to split the table evenly, on the average. For this reason, we can improve quicksort for nonrandom tables by using a randomly chosen element to partition the table. This change is easily incorporated in Algorithm 8.3: add the statement "$x_f \leftrightarrow x_{\text{rand}(f, l)}$" just before the statement "$i \leftarrow f + 1$", where rand($f, l$) gives a random integer $r$, $f \le r \le l$. The use of a randomly chosen partitioning element guarantees that the average indicated by Equation (8.2) will be the average in practice *regardless of the expected distribution of the permutations of the $x_i$*.

There is a disadvantage to the use of a random partitioning element, however: the generation of the required pseudo-random numbers is time-consuming if the numbers generated are to seem random (see Exercise 10). Moreover, the effect we want (that of keeping the partitioning element near the middle) can be achieved by a different method that also improves the average behavior. We choose as the partitioning element the median of a small sample of elements. The median of three elements turns out to be an excellent choice, since there is a process of diminishing returns when larger samples are used. To eliminate the possibility of nearly ordered tables causing poor performance, we choose as our sample the first element, the middle element, and the last element. We can implement this modification to Algorithm 8.3 by inserting the following statements just before the statement "$i \leftarrow f + 1$":

$$x_{\lfloor (f+l)/2 \rfloor} \leftrightarrow x_{f+1}$$
$$\textbf{if } x_{f+1} > x_l \textbf{ then } x_{f+1} \leftrightarrow x_l$$
$$\textbf{if } x_f > x_l \textbf{ then } x_f \leftrightarrow x_l$$
$$\textbf{if } x_{f+1} > x_f \textbf{ then } x_{f+1} \leftrightarrow x_f$$

This makes $x_f$ the median of $x_f$, $x_{\lfloor (f+l)/2 \rfloor}$, and $x_l$ without introducing any bias in the expected distribution of the subtables. Furthermore, it leaves $x_{f+1} \le x_f \le x_l$ so that the initializations of $i$ and $j$ can be changed to "$i \leftarrow f + 2$" and "$j \leftarrow l - 1$", respectively. The complete partitioning part of the algorithm is thus as shown in Algorithm 8.4(a). With these changes, it can be shown that $\bar{C}_n \approx 1.188\, n \lg n$; of course, even in this case $C_n$ is proportional to $n^2$, but such behavior is *extremely* unlikely, even on tables occurring in practice.

〚find the median of the first, middle, and last elements〛
$x_{\lfloor (f+l)/2 \rfloor} \leftrightarrow x_{f+1}$
**if** $x_{f+1} > x_l$ **then** $x_{f+1} \leftrightarrow x_l$
**if** $x_f > x_l$ **then** $x_f \leftrightarrow x_l$
**if** $x_{f+1} > x_f$ **then** $x_{f+1} \leftrightarrow x_f$
〚$x_f$ is now the desired median and $x_{f+1} \leqslant x_f \leqslant x_l$〛
〚partition using $x_f$〛
$i \leftarrow f + 2$
**while** $x_i < x_f$ **do** $i \leftarrow i + 1$
$j \leftarrow l - 1$
**while** $x_j > x_f$ **do** $j \leftarrow j - 1$
**while** $i < j$ **do**
$\qquad$〚at this point $i < j$, and $x_i \geqslant x_f \geqslant x_j$〛
$\qquad x_i \leftrightarrow x_j$
$\qquad$**repeat** $i \leftarrow i + 1$ **until** $x_i \geqslant x_f$
$\qquad$**repeat** $j \leftarrow j - 1$ **until** $x_j \leqslant x_f$
$x_f \leftrightarrow x_j \qquad$〚put $x_f$ into place between the subtables〛

**Algorithm 8.4(a)**
Partitioning by the median-of-three method

Unfortunately, the changes outlined above for using the median-of-three partitioning mean that the algorithm as written will no longer work properly for subtables of two elements; the point at which the recursion "bottoms out" must thus be changed. This is a wise idea in any case, because the overhead in Algorithm 8.3 makes it less efficient than the insertion sort of Algorithm 8.1 for small numbers of elements. Depending on the particular instruction set of the computer, the optimum crossover point at which quicksort becomes worthwhile is between five and fifteen elements, with ten being typical. Thus we should select a crossover value $m$, $5 \leqslant m \leqslant 15$ (the exact optimum is *not* necessary), and change the beginning of Algorithm 8.3 to

$\qquad$**if** $l - f + 1 < m$ **then** apply insertion sort to $x_f, x_{f+1}, \ldots, x_l$
$\qquad\qquad$**else** apply quicksort to $x_f, x_{f+1}, \ldots, x_l$

In fact, there is a better way to accomplish the same thing. Ignore subtables of size less than $m$ during partitioning (that is, do not make a recursive call for such a subtable), leaving them unsorted; this is done in Algorithm 8.3 by simply changing the initial test from "**if** $f < l$" to "**if** $l - f + 1 \geqslant m$". Then, after all the partitioning has been finished, the table has all the elements used for partitioning in their final (sorted) locations with small groups of unsorted elements between them. A single application of Algorithm 8.1 completes the sorting of the table. The advantage of this approach to small subtables is seen as follows: it takes a single application of the insertion sort a bit longer to sort the entire table than it would have to apply it to each of the small subtables individually, but all the overhead of the multiple applications has been eliminated, more than making up the difference.

$m \leftarrow$ a value near the optimum crossover point from quicksort
     to insertion sort; probably around 10
$S \leftarrow$ empty stack
$S \Leftarrow (0,0)$    〖bottom stack entry, with values chosen to stop the loop〗
$f \leftarrow 1$
$l \leftarrow n$
$x_{n+1} \leftarrow \infty$    〖to stop the scan at the right end〗
$x_0 \leftarrow -\infty$    〖for the insertion sort applied below〗
**while** $f < l$ **do**
    partition as in Algorithm 8.4(a)
    **case**
        $j - f < m$ **and** $l - j < m$: $(f, l) \Leftarrow S$    〖ignore both
                                              small subtables〗
        $j - f < m$ **and** $l - j \geq m$: $f \leftarrow j + 1$    〖ignore small left
                                              subtable〗
        $j - f \geq m$ **and** $l - j < m$: $l \leftarrow j - 1$    〖ignore small
                                              right subtable〗
        $j - f \geq m$ **and** $l - j \geq m$:
                〖neither subtable small, put larger one on $S$〗
                **if** $j - f > l - j$ **then**
                        〖left subtable is larger〗
                        $S \Leftarrow (f, j - 1)$
                        $f \leftarrow j + 1$
                **else**
                        〖right subtable is larger〗
                        $S \Leftarrow (j + 1, l)$
                        $l \leftarrow j - 1$
apply the insertion sort of Algorithm 8.1 to the entire table
    **Algorithm 8.4(b)**
    Improved quicksort

Algorithm 8.4(b) presents a version of quicksort that contains all the improvements outlined above, plus one more. In the worst case, the stack that is implicit in the recursion of Algorithm 8.3 can reach depth $n$, consequently requiring extra storage proportional to the size of the subtable being sorted. For large $n$ this can be unacceptable, so we have written Algorithm 8.4(b) in an iterative manner, maintaining the stack explicitly (see Section 4.2). A stack entry is a pair $(f, l)$; when it is on the stack, it means that the subtable $x_f, \ldots, x_l$ is to be sorted. Algorithm 8.4(b) puts the larger of the two subtables on the stack and applies the algorithm immediately to the smaller subtable. This reduces the worst-case stack depth to about $\lg n$ (Exercise 12).

The behavior of Algorithm 8.4(b) is summarized by the following table:

| | Element Comparisons | Element Interchanges | Extra Storage |
|---|---|---|---|
| Best case | $n \lg n$ | | |
| Average case | $1.188\, n \lg n$ | Between $\frac{1}{2} n \lg n$ and $n \lg n$ | About $\lg n$ stack entries and several integer variables |
| Worst case | $\frac{1}{2} n^2$ | | |

## Exercises

1. Prove that each pass of the bubble sort except the last decreases by one each nonzero $d_i$, where $d_i$ is, as before, the number of elements larger than $x_i$ and to its left.

2. Implement the *cocktail shaker sort*: alternate passes in the bubble sort to go in opposite directions. Show that if $x_j$ and $x_{j+1}$ are not interchanged on two successive passes, then they are in their final positions.

★3. Let $\Pi = (\pi_1, \pi_2, \ldots, \pi_n)$ be a random permutation of $1, 2, \ldots, n$. What is the average value of

$$\frac{1}{n} \sum_{i=1}^{n} |\pi_i - i|,$$

the expected distance that an element will travel during sorting? What can be concluded about sorting algorithms that perform only adjacent interchanges?

4. Using mathematical analysis and/or extensive random number tables, compare the efficiency of the following partitioning methods for quicksort with that of Algorithm 8.3. For each of the methods below, first verify that

it correctly partitions the elements (especially in the presence of equal elements). Assume that $x_{f-1} = -\infty$ and that $x_{l+1} = \infty$ as needed.

(a)    $i \leftarrow f$
       $j \leftarrow l + 1$
       $v \leftarrow x_f$
       **while** $i < j$ **do**
                   **repeat** $j \leftarrow j - 1$ **until** $x_j \leqslant v$
                   **if** $i \geqslant j$ **then** $j \leftarrow i$
                       **else**
                               $x_i \leftarrow x_j$
                               **repeat** $i \leftarrow i + 1$ **until** $x_i \geqslant v$
                               **if** $i < j$ **then** $x_j \leftarrow x_i$
       $x_j \leftarrow v$


(b)    $i \leftarrow f - 1$
       $j \leftarrow l + 1$
       $p \leftarrow \left\lfloor \dfrac{f + l}{2} \right\rfloor$
       $v \leftarrow x_p$
       **while** $i < j$ **do**
                   **repeat** $i \leftarrow i + 1$ **until** $x_i > v$
                   **repeat** $j \leftarrow j - 1$ **until** $x_j < v$
                   **if** $i < j$ **then** $x_i \leftrightarrow x_j$
       **if** $i < p$ **then**
                       $x_i \leftrightarrow x_p$
                       $i \leftarrow i + 1$
       **if** $j > p$ **then**
                       $x_p \leftrightarrow x_j$
                       $j \leftarrow j - 1$


(c)    $i \leftarrow f - 1$
       $j \leftarrow l + 1$
       $v \leftarrow x_f$
       **repeat**
                   **repeat** $i \leftarrow i + 1$ **until** $x_i \geqslant v$
                   **repeat** $j \leftarrow j - 1$ **until** $x_j \leqslant v$
                   $x_i \leftrightarrow x_j$
       **until** $i \geqslant j$
       $x_j \leftrightarrow x_i$ [[undo extraneous exchange]]

★5. Prove by induction that $QUICKSORT$ $(1, n)$ using Algorithm 8.3 (and hence Algorithm 8.4) correctly sorts the table $x_1, x_2, \ldots, x_n$. [*Hint*: First prove by induction that the partitioning was done correctly.]

★6. Prove that the partitioning method of Algorithm 8.3 [and hence of Algorithm 8.4(a)] produces random subtables; that is, prove that if each of the $n!$ permutations of the table is equally likely, then each of the permutations of the left and right subtables is equally likely.

7. Do a more careful analysis of $C_n$, the number of element comparisons used by Algorithm 8.3 in the worst case, to show that

$$C_n = \begin{cases} 0, & n = 0, \\ \frac{1}{2}n^2 + \frac{3}{2}n - 2, & n \geq 1. \end{cases}$$

★8. Show that $\bar{I}_n$, the average number of interchanges used in Algorithm 8.3 called by $QUICKSORT$ $(1, n)$, satisfies

$$\bar{I}_n = \sum_{s=1}^{n} p_s \left( 1 + \sum_{t=0}^{s-1} q_{s,t} t + \bar{I}_{s-1} + \bar{I}_{n-s} \right), \qquad n \geq 2,$$

where $p_s$ is the probability that $x_1$ is the $s$th smallest element, $p_s = 1/n$, and $q_{s,t}$ is the probability that when $x_1$ is the $s$th smallest element, then among $x_2, \ldots, x_s$ there will be $t$ elements greater than $x_1$,

$$q_{s,t} = \frac{\binom{s-1}{t}\binom{n-s}{t}}{\binom{n-1}{s-1}}.$$

Establish the identity

$$\sum_{t=0}^{s-1} \binom{s-1}{t}\binom{n-s}{t} t = \binom{n-2}{s-2}(n-s)$$

and use it to show that for $n \geq 2$

$$\bar{I}_n = \frac{1}{6}n + \frac{2}{3} + \frac{2}{n}\sum_{i=0}^{n-1} \bar{I}_i$$

and thus that

$$\bar{I}_n = \frac{1}{3}(n+1)H_n - \frac{1}{9}n - \frac{5}{18},$$

where

$$H_n = \sum_{i=1}^{n} \frac{1}{i}.$$

9. Show that $I_n$, the worst-case number of interchanges used in Algorithm 8.3, is proportional to $n \lg n$.

10. Assuming that each of the $n!$ permutations of the elements in the table is equally likely, how many random numbers would be needed on the average to sort the elements with the random partitioning scheme described in the text?

11. Verify the claim in the text that, in the context of Algorithm 8.4, a single application of the insertion sort is, in total, more efficient than its application to the small subtables individually.

12. Analyze the worst-case stack depth required in Algorithm 8.4 as a function of $n$ and $m$.

13. Is Algorithm 8.3 stable (see Exercise 6 of Section 8.1.1 for the definition of stability)? Is Algorithm 8.4?

14. Explain the effect of changing the stack in Algorithm 8.4 to a queue.

15. Devise and implement a sorting algorithm analogous to quicksort in which the partitioning is done on the basis of the most significant bit of the elements; if that bit is 1, the element goes into the right-hand part of the table; if that bit is 0, the element goes into the left-hand part of the table. Subsequent partitions are based on the less significant bits of the elements. This method is known as *radix exchange* sorting. Analyze the best and worst cases for

    (a) The number of interchanges.
    (b) The number of bit inspections.
    (c) The maximum stack depth.

    Analyze the average cases for (a), (b), and (c), assuming that the elements to be sorted are $0, 1, 2, \ldots, 2^t - 1$ in random order.

### 8.1.3 Selection Sorts

In a selection sort the basic idea is to go through stages $i = 1, 2, \ldots, n$, selecting the $i$th largest (smallest) element and putting it in place on the $i$th stage.

The simplest form of selection sort is that of Algorithm 8.5: at the $i$th stage, the $i$th largest element is found in the obvious manner by scanning the remaining $n - i + 1$ elements. The number of element comparisons at the $i$th stage is $n - i$, leading to a total of $(n - 1) + (n - 2) + \cdots + 1 = \frac{1}{2}n(n - 1)$, regardless of the input. This is clearly not a very good way to sort. Its behavior is summarized by the following table:

| | Element Comparisons | Element Interchanges | Extra Storage |
|---|---|---|---|
| *In all cases* | $\frac{1}{2}n(n - 1)$ | $n - 1$ | Integer variables $i, j, k$, and $t$ |

Despite the inefficiency of Algorithm 8.5, the idea of selection can lead to an efficient sorting algorithm. The trick is to find a more efficient method of determining the $i$th largest element, which can be done by using the mechanism of a *knockout tournament*. Make comparisons $x_1 : x_2, x_3 : x_4, x_5 : x_6, \ldots, x_{n-1} : x_n$ and then compare the winners (the larger elements) of those comparisons in a like manner, and so on, as illustrated for $n = 16$ in Figure 8.4. Notice that this process requires $n - 1$ element comparisons to determine the largest element (see Exercise 3); but, having determined the largest element, we possess a great deal of information about the second largest element: it must be one of those that lost to the largest. Thus the second largest can now be determined by replacing the largest with $-\infty$ and remaking all the comparisions along the path from the largest element to the root. This is illustrated in Figure 8.5 for the tree in Figure 8.4.

> **for** $j = n$ **to** 2 **by** $-1$ **do**
>     ⟦stage $i = n - j + 1$: find the $i$th largest, that is, the $j$th smallest (see also Exercise 1)⟧
>     $t \leftarrow 1$
>     **for** $k = 2$ **to** $j$ **do if** $x_t < x_k$ **then** $t \leftarrow k$
>     ⟦put $i$th largest ($j$th smallest) into place⟧
>     $x_j \leftrightarrow x_t$

**Algorithm 8.5**
Simple selection sort

Since the tree has height $\lceil \lg n \rceil$ (why?), we can find the second largest element by remaking $\lceil \lg n \rceil - 1$ comparisons instead of the $n - 2$ used in the simple selection algorithm. This process can obviously be continued. On finding the second largest element, we replace it by $-\infty$ and remake another $\lceil \lg n \rceil - 1$ comparisons to find the third largest element, and so on. Thus the entire process

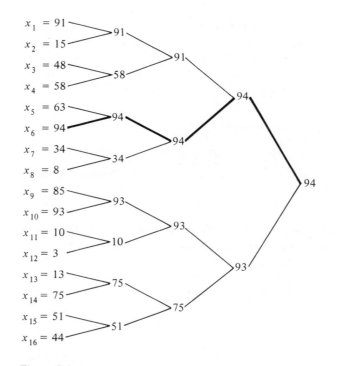

**Figure 8.4**
Using a knockout tournament to find the largest element. The path of the largest element is shown in boldface

uses

$$n - 1 + (n - 1)(\lceil \lg n \rceil - 1) \approx n \lceil \lg n \rceil$$

element comparisons. This *may* yield a reasonable sorting method, provided that the movement of the elements can be handled in an efficient manner (recall that binary insertion, mentioned at the end of Section 8.1.1, uses only $n \lg n$ element comparisons but proportional to $n^2$ interchanges).

A tournament process as described above is essentially a *priority queue*, mentioned in Sections 4.2 and 7.2.2, except that in this case all the elements arrive before the deletions begin. Thus, an efficient data structure for priority queues will lead to a tournament-like selection sort with the potential for $n \lg n$ worst-case behavior. In Section 7.2.2 we saw that balanced trees could be used to implement priority queues in which insertion, deletion, splitting and merging could all be done in logarithmic time. We hardly need such flexibility for a tournament sort, however, so we now introduce the *heap*, an efficient implementation for priority queues in which only insertions and deletions occur, not splittings or mergings.

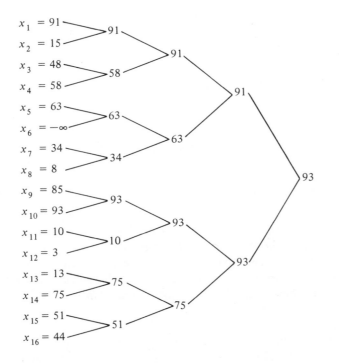

**Figure 8.5**
Finding the second largest element by replacing the largest with $-\infty$ and remaking the comparisons won by the largest element

A *heap* is a completely balanced binary tree of height $h$ in which all leaves are at distance $h$ or $h-1$ from the root (see Section 5.3) and all descendants of a node are smaller than it; furthermore all leaves at level $h$ are as far to the *left* as possible. Figure 8.6 shows a set of elements arranged into a heap. The advantage of a heap is that it can be stored compactly in an array, without the need for an explicit tree structure: the sons of the element in the $i$th position of the array are the elements in positions $2i$ and $2i+1$ (compare this with the ordering used in Figure 5.25). The heap of Figure 8.6 thus becomes

| $i$: | 1 | 2 | 3 | 4 | 5 | 6 | 7 | 8 | 9 | 10 | 11 | 12 |
|---|---|---|---|---|---|---|---|---|---|---|---|---|
| $x_i$: | 94 | 93 | 75 | 91 | 85 | 44 | 51 | 18 | 48 | 58 | 10 | 34 |

We will leave the description of general algorithms for insertion and deletion in heaps to Exercises 9 and 10; these algorithms are needed in order to use heaps as priority queues. Our attention here will be restricted to the manipulation of heaps as required by sorting.

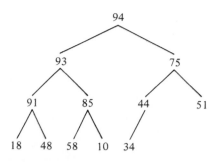

**Figure 8.6**
A heap containing twelve elements

Notice that in a heap the largest element must be at the root and thus always in the first position of the array representing the heap. Interchanging the first element with the $n$th element places the largest element into its correct position, but it destroys the heap property of the first $n - 1$ elements. If we can initially build a heap and then restore it efficiently, we are finished, for we can sort as follows:

$$\text{build a heap from } x_1, x_2, \ldots, x_n$$
$$\textbf{for } i = n \textbf{ to } 2 \textbf{ by } -1 \textbf{ do}$$
$$x_1 \leftrightarrow x_i$$
$$\text{restore the heap in } x_1, x_2, \ldots, x_{i-1}$$

This is an outline of *heapsort*.

Given a binary tree all of whose leaves are as far left as possible and both of whose subtrees are heaps, how do we transform the whole thing into a heap? We compare the root to the larger of the two sons. If the root is larger, the tree is already a heap; but if the root is smaller, we interchange it with the larger son and apply the restoration algorithm recursively to the subtree whose root has been interchanged. (See Figure 8.7.) The procedure to restore $x_j, x_{j+1}, \ldots, x_k$ to a heap, assuming that all the subtrees are heaps, is thus

$$\textbf{procedure } RESTORE(j, k)$$
$$\textbf{if } x_j \text{ is not a leaf } \textbf{then}$$
$$\text{let } x_m \text{ be the larger of the sons of } x_j$$
$$\textbf{if } x_m > x_j \textbf{ then}$$
$$x_m \leftrightarrow x_j$$
$$RESTORE(m, k)$$

Rewriting *RESTORE* in an iterative manner to eliminate the needless use of

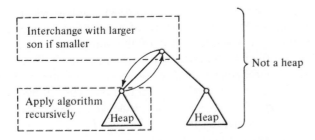

**Figure 8.7**
Recursive restoration of the heap

recursion and filling in the details, we obtain Algorithm 8.6. Notice that $x_j$ is a leaf if and only if $j > \lfloor l/2 \rfloor$ (Exercise 6).

**procedure** *RESTORE*($f, l$)
    $j \leftarrow f$
    **while** $j \leqslant \lfloor l/2 \rfloor$ **do**
          **if** $2j < l$ **and** $x_{2j} < x_{2j+1}$ **then** $m \leftarrow 2j + 1$
                        **else** $m \leftarrow 2j$
        $[\![x_m$ is the larger son of $x_j]\!]$
        **if** $x_m > x_j$ **then**
                $x_m \leftrightarrow x_j$
                $j \leftarrow m$
        **else** $j \leftarrow l$     $[\![$to terminate the loop$]\!]$

**Algorithm 8.6**
Restoration of the heap from a tree both of whose subtrees are heaps

    To build the heap initially, note that the heap property is already satisfied (vacuously) by each of the leaves ($x_i$, $\lfloor n/2 \rfloor < i \leqslant n$) and that calling *RESTORE*($i, n$) for $i = \lfloor n/2 \rfloor$, $\lfloor n/2 \rfloor - 1, \ldots, 1$ transforms the table into a heap at all higher levels. The heapsort procedure is thus as given in Algorithm 8.7: the first **for** loop builds the heap and is known as the *creation phase*; the second **for** loop is called the *sift-up phase*.

              **for** $i = \lfloor n/2 \rfloor$ **to** 1 **by** $-1$ **do** *RESTORE*($i, n$)
              **for** $i = n$ **to** 2 **by** $-1$ **do**
                    $x_1 \leftrightarrow x_i$
                    *RESTORE*($1, i - 1$)

**Algorithm 8.7**
Heapsort

The behavior of heapsort, on the average, is unknown, and so we consider only the worst case. We need to know the amount of work done during a call $RESTORE(f, l)$. In particular, if $h$ is the height of the subtree (a single node having height 0) rooted at $x_f$, then $RESTORE$ will perform at most $2h$ element comparisons and at most $h$ interchanges; thus we need to determine $h$. Notice that the left son of $x_f$ is $x_{2f}$, whose left son is $x_{4f}$, and so on. The subtree has a height of $h$, where $h$ is the largest integer such that the leftmost descendant $x_{2^h f}$ exists—that is, such that $2^h f \leq l$, which implies that $h = \lfloor \lg(l/f) \rfloor$.

The creation phase thus requires at most

$$\sum_{i=1}^{\lfloor n/2 \rfloor} \left\lfloor \lg \frac{n}{i} \right\rfloor \leq \sum_{i=1}^{\lfloor n/2 \rfloor} \lg \frac{n}{i} = \left\lfloor \frac{n}{2} \right\rfloor \lg n - \lg \left\lfloor \frac{n}{2} \right\rfloor ! = O(n)*$$

element interchanges and hence also $O(n)$ element comparisons. Similarly, the sift-up phase requires at most

$$(n - 1) + \sum_{i=2}^{n} \left\lfloor \lg \frac{i - 1}{1} \right\rfloor = n \lg n + O(n)*$$

element interchanges and hence $2n \lg n + O(n)$ element comparisons.

In summary we have

| | Element Comparisons | Element Interchanges | Extra Storage |
|---|---|---|---|
| Worst case | $2n \lg n + O(n)$ | $n \lg n$ | Integer variables $i, j, f, l,$ and $m$ |

Thus heapsort will require only proportional to $n \lg n$ operations for any table, in contrast to quicksort, which could require proportional to $n^2$ operations under extremely unlikely conditions. Despite this fact empirical evidence *strongly* suggests that quicksort is far superior to heapsort in practice (Exercise 8), unless worst-case performance is critical.

---

*This derivation requires the use of *Stirling's formula*, which states

$$n! = \sqrt{2\pi n} \left( \frac{n}{e} \right)^n \left( 1 + \frac{1}{12n} + \frac{1}{288n^2} - \frac{139}{51,840n^3} - \frac{571}{2,488,320n^4} + \cdots \right)$$

so that

$$\ln n! = \left( n + \frac{1}{2} \right) \ln n - n + \ln \sqrt{2\pi} + \frac{1}{12n} - \frac{1}{360n^3} + \cdots .$$

**Exercises**

1. Why is there no stage $i = n$ in Algorithm 8.5? In other words, why doesn't the loop go from $n$ **to** 1 **by** $-1$?

2. Is Algorithm 8.5 a stable sorting algorithm (see Exercise 6 of Section 8.1.1 for a definition of stability)?

3. Show that *any* method of determining the largest element in a table of $n$ elements requires $n - 1$ element comparisons. [*Hint:* At most how many of the elements can be eliminated from consideration by any such comparison?]

4. Prove a result similar to that in the previous exercise for determining the smallest element.

5. Give a method using only $\lceil 3n/2 \rceil - 2$ element comparisons to determine both the largest and smallest elements in a table of $n$ elements. Does this in any way contradict the results in the two previous exercises?

6. Show that a heap with $n$ elements has $\lceil n/2 \rceil$ leaves and that they are $x_{\lceil (n+1)/2 \rceil}, x_{\lceil (n+1)/2 \rceil + 1}, \ldots, x_n$; the last nonleaf is $x_{\lfloor n/2 \rfloor}$.

7. Is heapsort stable?

8. Perform experiments to support or refute the claim made that quicksort is superior to heapsort.

9. Design and analyze an algorithm for adding an arbitrary element to a heap of $n$ elements to form a heap of $n + 1$ elements.

10. Design and analyze an algorithm for deleting a specified element $x_i$ from a heap of $n$ elements, leaving a heap of $n - 1$ elements.

11. Explore the idea of a ternary heap and a sorting algorithm based on it.

12. A *biparental heap* is like a heap except that most of the elements have two parents, instead of just a father. Using an arrow to point from larger to smaller elements, the following is a biparental heap of ten elements:

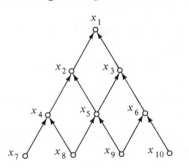

This can be neatly represented in a triangular array (see Exercise 4 of Section 3.3.1):

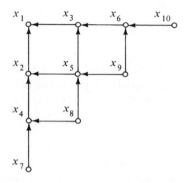

The elements of a biparental heap are thus regarded as being partitioned into *blocks*, the $i$th block consisting of the elements $x_{[i(i-1)/2]+1}$ through $x_{i(i+1)/2}$. The ordering imposed on the elements is that the $k$th element of the $p$th block is less than both the $k$th and $(k+1)$st elements of the $(p+1)$st block. In the triangular array representation shown above, a block is a positive slope diagonal.

★(a) Using the result of Exercise 4 of Section 3.3.1, we can encode the above triangular array in a linear array by the addressing function

$$\text{loc}(A[i,j]) = \frac{(i+j-2)^2 + i + 3j - 2}{2}.$$

It is possible with this encoding to move along rows, columns, and diagonals without explicitly calculating the addressing function. Assuming that a one-dimensional array $x[1], x[2],\ldots,x[m]$ is used to store a biparental heap of $n \leqslant m$ elements, find formulas (in terms of $k$) for the location in the array of the elements above, below, to the left, and to the right of $x[k]$.

(b) Design and analyze an algorithm to find the smallest element in a biparental heap.

(c) Design and analyze an algorithm to find the largest element in a biparental heap.

(d) Design and analyze an algorithm to insert a new element into a biparental heap.

(e) Design and analyze an algorithm to delete a specified element from a biparental heap.

### 8.1.4 Distributive Sorts

When faced with the real-life task of sorting a very large handful of index cards into alphabetical order, one might first attempt to make the task more manageable by dividing the cards into 26 piles—one of the a's, one of the b's, and so on, and then sorting each of the piles individually. This idea leads us to a class of sorting algorithms based on the distribution of the elements into "buckets."

In fact, quicksort can be viewed as a distributive sort in which there are two buckets, those elements below some chosen element and those above. Quicksort then recursively applies the same distribution idea to each of the two buckets. As we will see, the number of buckets and the way they are chosen is important to the efficiency of the resulting distribution sort, and in some cases we will be able to do much better than quicksort by choosing buckets based on the representation or the numerical distribution of the elements being sorted. This parallels the case in searching in which additional knowledge about the elements can allow us to use tries or interpolation search, methods not applicable to the general case. The benefit in the case of sorting is the possibility of sorting $n$ elements in time proportional to $n$.

**Radix Distribution.** The sorting algorithm we are about to discuss differs from those considered so far in that it is based *not* on comparisons between elements, but on the *representation* of the elements. We assume that the elements $x_1, x_2, \ldots, x_n$ each have the form

$$x_i = \left(x_{i,p}, x_{i,p-1}, \ldots, x_{i,1}\right)$$

and that they are to be sorted into increasing *lexicographic order*; that is,

$$x_i = \left(x_{i,p}, x_{i,p-1}, \ldots, x_{i,1}\right) < \left(x_{j,p}, x_{j,p-1}, \ldots, x_{j,1}\right) = x_j$$

if and only if, for some $t \leq p$, we have $x_{i,l} = x_{j,l}$ for $l > t$ and $x_{i,t} < x_{j,t}$. For simplicity, we will assume that $0 \leq x_{i,l} < r$, and so the elements can be viewed as integers represented in base $r$, each element having $p$ $r$-ary digits. If the elements are of different lengths, the short elements are padded with zeros to make the lengths uniform. In the common example of alphabetic elements, we would take $r = 27$ (26 letters and the blank character), padding the short elements on the right with blanks.

In the *radix distribution* sort the buckets correspond to the $r$ values of a base $r$ digit. The sort is based on the observation that if the elements have been sorted with respect to the $p - 1$ low-order positions, they can be completely sorted by sorting them according to the $p$th (highest) order position, being careful not to disturb the relative order of elements having equal values in their $p$th position. In effect, this means sorting the elements according to the lowest-order position, then according to the next-lowest-order position,..., and finally according to the highest-order position, never changing the relative order of equal values. This is the basis for old-fashioned mechanical card sorters: to sort cards on the field in, say, columns 76 through 80, the cards are sorted into increasing order on column 80, then on column 79, then on column 78, then on column 77, and finally on column 76. Each column sort is done by reading the column in each card and physically moving the card to the back of a pile (bucket) that corresponds to the digit punched in that column of the card. Once all the cards have been placed in the proper piles, the piles are stacked together (concatenated) in increasing order; the process is then repeated for the next column to the left. This procedure is illustrated in Figure 8.8 for three-digit decimal numbers.

Notice that both the buckets and the table itself are used in a first-in, first-out manner, and so it is best to represent them as queues. In particular, assume that a link field $LINK_i$ is associated with each element $x_i$; these link fields can then be used to hook all the elements in the table together in an input queue, $Q$. The link fields can also be used to hook the elements together into the queues used to represent the buckets, $Q_0, Q_1, \ldots, Q_{r-1}$. After the elements have been distributed into buckets, the queues representing those buckets are concatenated together to reform the table $Q$. The broad outline of this sorting algorithm is shown in Algorithm 8.8; the details of applying the techniques from Section 4.2.2 to build the queues are left as Exercise 1. The result of the algorithm is that the queue $Q$ will contain the elements in increasing order; that is, the elements will be linked in increasing order by the link fields, starting with the front of the queue $Q$.

> use the fields $LINK_1, LINK_2, \ldots, LINK_n$ to
> form $x_1, x_2, \ldots, x_n$ into an input queue, $Q$
> **for** $j = 1$ **to** $p$ **do**
> > initialize each of the queues $Q_0, \ldots, Q_{r-1}$ to be empty
> > **while** $Q$ **not** empty **do**
> > > $X \leftarrow Q$
> > > let $X = (x[p], x[p - 1], \ldots, x[1])$
> > > $Q_{x[j]} \leftarrow X$
> > Concatenate queues $Q_0, \ldots, Q_{r-1}$
> > together to form the new queue $Q$

**Algorithm 8.8**
Radix distribution sort

The analysis of Algorithm 8.8 must be different from those of the other sorting algorithms, since we cannot count the number of element comparisons and interchanges. Instead we will count the total number of queue operations.

207, 095, 646, 198, 809, 376, 917, 534, 310, 209, 181, 595, 799, 694, 334, 522, 139

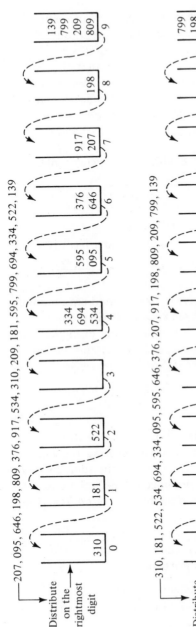

Distribute on the rightmost digit

310, 181, 522, 534, 694, 334, 095, 595, 646, 376, 207, 917, 198, 809, 209, 799, 139

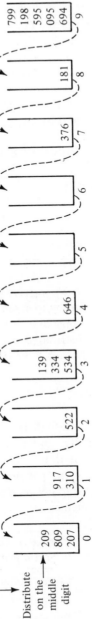

Distribute on the middle digit

207, 809, 209, 310, 917, 522, 534, 334, 139, 646, 376, 181, 694, 095, 595, 198, 799

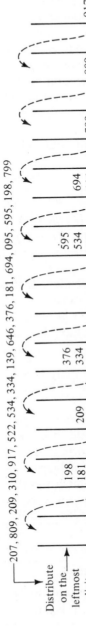

Distribute on the leftmost digit

Sorted table: 095, 139, 181, 198, 207, 209, 310, 334, 376, 522, 534, 595, 646, 694, 799, 809, 917

**Figure 8.8**
The radix distribution sort. The dashed arrows show how the buckets of elements are concatenated into a table

There are always $p$ passes over the table, and each pass requires removing each element from $Q$ and entering it on one of the $Q_i$. Hence there are a total of $2np$ insertion/deletion operations on the queues. Each pass also requires $r - 1$ concatenations to produce $Q$ from $Q_0, \ldots, Q_{r-1}$, and so there are a total of $(r - 1)p$ queue concatenation operations. It is clear from Section 4.2.2 that an insertion/deletion operation or a concatenation operation can be done in constant time. Therefore Algorithm 8.8 requires time proportional to $np + rp$ to sort elements $x_1, x_2, \ldots, x_n$, where $x_i$ has the form $(x_{i,p}, x_{i,p-1}, \ldots, x_{i,1})$ and $0 \leqslant x_{t,l} < r$.

We summarize the behavior of Algorithm 8.8 with the following table:

|  | Queue Operations | Extra Storage |
|---|---|---|
| In all cases | $np$ insertions<br>$np$ deletions<br>$(r - 1)p$ concatenations | $n$ link fields<br>$r$ queue header records |

**Value Distribution.** If the elements $x_1, x_2, \ldots, x_n$ to be sorted are randomly distributed in the range $(x_0, x_{n+1})$ according to some known distribution, then we can base our choice of buckets on this range and distribution. For example, if the elements $x_1, x_2, \ldots, x_n$ are uniformly distributed over $(x_0, x_{n+1})$ and we have $b$ buckets $B_1, B_2, \ldots, B_b$, we can let the bucket $B_j$ be defined by the range

$$x_0 + \frac{x_{n+1} - x_0}{b}(j - 1) < x_i \leqslant x_0 + \frac{x_{n+1} - x_0}{b}j.$$

We can sort by making a pass over the $x_i$, distributing them into their proper buckets, sorting the buckets individually, and then concatenating the buckets together. If the sorting of the buckets is done recursively by the same method, then the number of buckets determines the performance of the technique. The following table summarizes the relationship between the number of buckets $b$ and the average amount of work to sort $n$ elements:

| Number of Buckets b | Average Amount of Work to Sort n Elements |
|---|---|
| Constant | Proportional to $n \lg n$ |
| Proportional to $\sqrt{n}$ | Proportional to $n \lg \lg n$ |
| Proportional to $n$ | Proportional to $n$ |

The sorting technique outlined above is marred by the fact that at successive stages the numbers of elements in the buckets decrease to the point where the few elements in any bucket are better sorted by some more direct technique. Thus for the same reasons we made a hybrid of quicksort and the insertion sort, we should make a similar hybrid here: if there are $m$ or fewer elements in a bucket, ignore it, otherwise distribute the elements into suitably chosen buckets, process each bucket recursively, and finally concatenate the buckets together. After returning to the top level of recursion, apply the insertion sort once to the entire set of elements (see Exercise 11 of Section 8.1.2). We leave the details of this hybrid to Exercise 7. This hybrid will work well, outperforming even quicksort, for even moderate numbers of elements, *provided that those elements really satisfy the assumed distribution* over the range $(x_0, x_{n+1})$.

## Exercises

1. Using the techniques of Section 4.2.2 for implementing a queue with a linked list, fill in the details omitted from Algorithm 8.8 and implement it.

2. Use the bucket sort idea to sort a permutation of the elements $1, 2, 3, \ldots, n$ in time proportional to $n$.

3. Devise a bucket sort to sort $n$ elements of the form $(i, j)$, $1 \leqslant i \leqslant m$, $1 \leqslant j \leqslant m$ according to the value

$$\phi[(i, j)] = \begin{cases} 2i, & j \geqslant i, \\ 2j, & 2j > i > j, \\ 2j + 1, & i \geqslant 2j. \end{cases}$$

Your algorithm should require only time proportional to $n + m$.

4. Examine the suitability of the following hybrid sorting scheme: use a radix distribution sort on the first few high-order positions and then apply the insertion sort. The purpose of the radix distribution on the high-order positions is to reduce the amount of disorder among the elements so that the insertion sort becomes efficient.

5. Are the two distributive sorts described in the section stable? (See Exercise 6 of Section 8.1.1 for a definition of stability.)

★6. Given a matrix $A = (a_{ij})$, suppose that each row is sorted into increasing order and then each column is sorted into increasing order. Do the rows remain sorted into increasing order? Prove your answer.

7. (a) Give a detailed algorithm for the value-distribution/insertion-sort hybrid described in this section.

   (b) By analysis and/or extensive testing, determine the optimal crossover point $m$, below which it is more economical to sort by the insertion sort.

### 8.1.5 Lower Bounds

Now that we have examined various practical sorting algorithms; let us consider the problem of sorting from a theoretical point of view in order to get some idea of what kinds of improvements may (or may not) be possible. Except in Section 8.1.4 on distributive sorts, we measured the performance of the sorting algorithms by the number of element comparisons "$x_i : x_j$". The algorithms based on distribution do not make comparisons of this type, since they are based on the representation or numerical distribution of the $x_i$; the other algorithms are all based on an abstract ordering among the $x_i$, and the only way to get information about that ordering is to make comparisons "$x_i : x_j$". Of course, the number of comparisons is not the only determining factor in the performance of a sorting algorithm, but it usually does give a good indication of the amount of work being done. Consequently, the minimum possible number of element comparisons necessary to sort $n$ elements is of interest because it will provide us with a benchmark against which to compare the performance of many sorting algorithms. In this section we derive some lower bounds that are meaningful in that context.

In order to eliminate from this discussion sorting algorithms other than those based on comparisons of the elements, we will consider only the algorithms that are based on the abstract linear ordering of the elements: between every pair $x_i, x_j, i \neq j$, either $x_i < x_j$ or $x_i > x_j$. (It is not difficult to extend this discussion to the case in which equal elements are allowed; see Exercise 2.) Any such sorting algorithm can be represented by an extended binary tree (see Section 5.3) in which each internal node represents an element comparison and each leaf (external node) represents an outcome of the algorithm. This tree can be viewed as a flowchart of the sorting algorithm in which all loops have been "unwound" and only the element comparisons are shown. The two sons of a node thus represent the two possible outcomes of the comparison. For example, Figure 8.9 shows a binary tree for sorting three elements. In this section we will consider only sorting algorithms that can be written as such *decision trees*.

In any such decision tree each permutation defines a unique path from the root to a leaf. Since we are considering only algorithms that work correctly on all of the $n!$ permutations of the elements, the leaves corresponding to different permutations must be different. Clearly, then, there must be at least $n!$ leaves in a decision tree for sorting $n$ elements.

Notice that the height of the decision tree is the number of comparisons required by the algorithm for its worst-case input. Let $S(n)$ denote the minimum number of comparisons required by any sorting algorithm in the worst case; that is,

$$S(n) = \min_{\substack{\text{sorting} \\ \text{algorithms}}} \left[ \max_{\text{inputs}} (\text{number of comparisons}) \right].$$

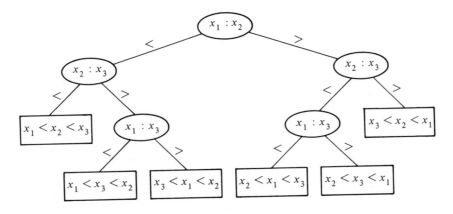

**Figure 8.9**
A binary decision tree for sorting three elements

Noticing that a binary tree of height $h$ can have at most $2^h$ leaves, we conclude that

$$2^{S(n)} \geqslant n!,$$

and so

$$S(n) \geqslant \lg n! \approx n \lg n.$$

(See the footnote on page 400.) Thus *any* sorting algorithm based solely on comparisons of the elements being sorted will require at least $n \lg n$ such comparisons for its worst-case input.

We can also derive a lower bound for $S(n)$, the minimum possible *average* number of element comparisons required by an algorithm that correctly sorts all the $n!$ permutations, assuming that each of these permutations is equally probable. That is,

$$\bar{S}(n) = \min_{\substack{\text{sorting} \\ \text{algorithms}}} \left[ \frac{1}{n!} \sum_{\text{permutations}} (\text{number of comparisons}) \right].$$

The external path length (see Section 5.3) of the decision tree is the sum of all distances from the root to the leaves; dividing it by $n!$ gives the average number of comparisons for the corresponding sorting algorithm. Equation (5.10) from the end of Section 5.3 states that in an extended binary tree with $N$ leaves the minimum possible external path length is $N \lg N + O(N)$; setting $N = n!$, we find that

$$\overline{S}(n) \geq \lg n! \approx n \lg n.$$

Therefore *any* sorting algorithm based on element comparisons will require an average of at least $n \lg n$ such comparisons.

Comparing these results with the sorting algorithms discussed in Sections 8.1.1 through 8.1.3, we find that heapsort, requiring about $2n \lg n$ comparisons, is within a small constant factor of being optimal in the worst case and hence also on the average. Similarly quicksort, requiring an average of $1.188n \lg n$ comparisons is within a very small constant factor of being optimal on the average, although it is not in the worst case. The remaining algorithms, being quadratic in performance, are far from optimal for large $n$.

**Exercises**

1. Show the decision trees that arise for sorting $x_1$, $x_2$, $x_3$, and $x_4$ by
   (a) The insertion sort.
   (b) The bubble sort.
   (c) Quicksort.
   (d) The simple selection sort.
   (e) Heapsort.

2. Extend the argument given in this section to show that even if equal elements are allowed, at least $\lg n!$ element comparisons are needed in the worst and average cases for sorting $n$ elements. (*Note:* Let $o_n$ be the number of outcomes when $n$ not necessarily distinct elements are sorted. The obvious extension would show only that $\log_3 o_n$ are needed, not $\lg n!$.)

3. (a) Suppose that all the elements are known to be zero or one. Prove that $n - 1$ element comparisons are necessary and sufficient in the worst case to sort $n$ elements.

   ★(b) Prove that the minimum average number of element comparisons for such a sorting algorithm is $2n/3 + O(1)$.

## 8.2  RELATED PROBLEMS

In the previous section we studied the problem of completely ordering a set of elements, given no a priori information about the abstract ordering of the elements. In this section we consider two special cases of this problem: instead of requiring determination of the complete ordering, the *selection problem* asks only for the $k$th largest element. The *merging problem* is to completely sort a set of elements, starting from two sorted subsets. For both selection and merging, we will measure efficiency entirely in terms of element comparisons. As we have seen in the previous section, this is not because other operations are negligible, but rather because the number of element comparisons usually determines, to within a small constant factor, the overall performance of the algorithm.

### 8.2.1  Selection

Given the (unordered) elements $x_1, x_2, \ldots, x_n$, how can we find the $k$th largest element? The problem is obviously symmetrical: finding the $(n - k + 1)$st largest (the $k$th smallest) can be done by using an algorithm for finding the $k$th largest but reversing the actions taken for the $<$ and $>$ results of element comparisons. Thus finding the largest element $(k = 1)$ is equivalent to finding the smallest element $(k = n)$; finding the second largest $(k = 2)$ is equivalent to finding the second smallest $(k = n - 1)$, and so on. Of special interest is the problem of finding quantile values $(k = \lceil \alpha n \rceil, 0 < \alpha < 1)$ and especially the median $(\alpha = \frac{1}{2})$.

Of course, all these cases of the selection problem can be solved by using any of the methods of Section 8.1 to sort the elements completely and then trivially accessing the $k$th largest. As we have seen, this will require proportional to $n \log n$ element comparisons *regardless of the value of* $k$. But we would be computing far more information than we need, and so there should be better ways. There are: first we examine the application of various sorting algorithms to the selection problem and then we describe an algorithm that requires only $O(n)$ comparisons regardless of the value of $k$.

In using a sorting algorithm for selection, the most apparent choice would be one of the algorithms based on selection, either the simple selection sort (Algorithm 8.5) or heapsort (Algorithm 8.7). In each case, we can stop after the first $k$ stages have been completed. For the simple selection sort, this means using

$$(n - 1) + (n - 2) + \cdots + (n - k) = kn - \frac{k(k + 1)}{2}$$

element comparisons, and for heapsort it means using proportional to $n + k \lg n$ element comparisons. In both cases, we are computing more information than we need because we are completely determining the order of the largest $k$ elements. This is not serious when $k$ is a constant, independent of $n$, since very little extra information is being computed; but when $k = \lceil \alpha n \rceil, 0 < \alpha \leq 1$, we are sorting a table of length $\alpha n$ and thus computing a great deal of unneeded information.

Although it does not seem so at first glance, the quicksort idea provides a reasonable method of selecting the $k$th largest element in $x_1, x_2, \ldots, x_n$. The table is partitioned into two subtables, those bigger than $x_1$, and those smaller than $x_1$, and then the appropriate subtable of the table is examined recursively. Assume that after partitioning, the original $x_1$ is in position $n - j + 1$ (that is, it is the $j$th largest element). If $k = j$, we are done; if $k < j$, we search for the $k$th largest element among $x_{n-j+2}, \ldots, x_n$; and if $k > j$, we search for the $(k - j)$th largest element among $x_1, \ldots, x_{n-j}$. (Compare this method with Algorithm 7.8) It is easy to modify Algorithms 8.3 and 8.4 (quicksort) so that they use this technique to find the $k$th largest element, and we leave that to the reader.

How efficient is this algorithm? If $\overline{C}_{n,k}$ is the number of element comparisons required on the average, we see that

$$\overline{C}_{n,k} = n + 1 + \frac{1}{n}\left( \sum_{j=1}^{k-1} \overline{C}_{n-j,\,k-j} + \sum_{j=k+1}^{n} \overline{C}_{j-1,\,k} \right)$$

and it is not difficult to show by induction that $\overline{C}_{n,k} = O(n)$ (see Exercise 1). Unfortunately, $C_{n,k}$, the number of element comparisons in the worst case, is proportional to $n^2$, and so, although good on the average, this algorithm can be *very* inefficient.

The inefficiency occurs because the element used to partition the table can be too close to either end of the table instead of close to the median, as we would like it (to split the table nearly in half). Thus the way to improve this algorithm is to discover a way of efficiently finding some element that is guaranteed to be near the median of the table and to use that element for partitioning the table into subtables. The following inductive technique shows how to do this. Assume that we have a selection algorithm that finds the $k$th largest of $n$ elements in $28n$ comparisons. This is certainly true for $n \leqslant 50$, since $28n \geqslant n(n - 1)/2$ element comparisons are sufficient to sort the elements completely, even using the bubble sort when $n \leqslant 50$. Suppose that $28t$ element comparisons are sufficient for $t < n$. First divide the table of $n$ elements into $\lceil n/7 \rceil \approx n/7$ subtables of seven elements each, adding some dummy $-\infty$ elements if needed to complete the last subtable (that is, when $n$ is not a multiple of 7). Then completely sort each of the $n/7$ subtables. Sorting a table with seven elements requires at most $7(7 - 1)/2 = 21$ element comparisons when using the bubble sort, and so sorting all these subtables requires at most $21(n/7) = 3n$ element comparisons in total. Next, apply the selection algorithm recursively to the $n/7$ *medians* of the $n/7$ sorted subtables to find the *median of the medians* in $28(n/7) = 4n$ element comparisons. At this point we have information about the elements as shown in Figure 8.10. (See Exercise 5.)

The elements in region $A$ are known to be less than the median of the medians, and the elements in region $B$ are thus known to be greater than the median of the medians; the remaining elements can be either less than or greater than the median of the medians. Clearly, there are about $2n/7$ elements in each of regions $A$ and $B$. Therefore the median of the medians is guaranteed to be

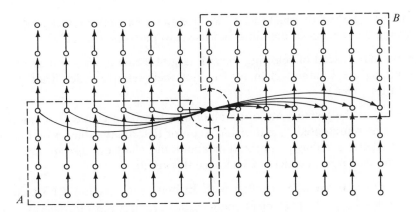

**Figure 8.10**
The information known about the elements after determining the median of the $n/7$ medians. An arrow o—→o means that the element at the arrow's head is *greater* than the element at the arrow's tail. All elements are shown as o except the median of the medians, which is shown as •.

"near" the middle of the table, and we can use it to partition the table when applying the quicksort idea to selection. After partitioning the table, at least the elements in one of $A$ or $B$ will not need to be examined further.

How many element comparisons are needed in total? We used $3n$ to sort the subtables, $4n$ to find the median of the medians, $n$ to partition the table, and at most $28(\frac{5}{7}n)$ to apply the selection algorithm recursively (since there can be as many as $5n/7$ elements in the subtable to which the selection algorithm is recursively applied). Thus at most $28n$ element comparisons are made; and since it is clear that the total amount of work is proportional to the number of element comparisons, we have an $O(n)$ selection algorithm. Of course, the algorithm is *not* as efficient as sorting by heapsort unless $28n < 2n\lg n$, that is, $n > 2^{14} = 16,384$. But we were quite sloppy; it is easy to bring the $28n$ down to $15n$ or less (Exercise 2), and, in fact, it can be reduced still further to $391n/72 \approx 5.431n$, so that the algorithm becomes reasonable for smaller values of $n$.

### Exercises

1. Use mathematical induction to verify that, as stated, $\overline{C}_{n,k}$ is proportional to $n$ for all $k$.

2. Show that the $k$th largest of $n$ elements can be determined in at most $27n/2$ element comparisons. How large must $n$ be for this to be more efficient than heapsort (in terms only of comparisons)? Try to lower the constant $27/2$ still further.

3. Is anything gained in quicksort if we use the $O(n)$ selection algorithm to determine the median of the table and use it to partition the table?

4. Suppose that instead of selecting the $k$th largest element, we are interested in determining the $k$ largest elements but not their relative order. Can this be done in time $O(n)$?

5. Prove that at the termination of an algorithm based on element comparisons (that is, one represented by a decision tree) for determining the second largest element, there is enough information about the elements *already known* to determine the largest element. Generalize this result.

6. Use the previous exercise, together with Exercise 3 of Section 8.1.3 to prove that $n - k + \left\lceil \lg \left( {n \atop k-1} \right) \right\rceil$ element comparisons are necessary to determine the $k$th largest of $n$ elements. [*Hint*: Use an argument similar to the one in Section 8.1.5 that shows that $S(n) \geq \lg n!$.]

7. Given sorted tables $x_1 \leq x_2 \leq \cdots \leq x_n$ and $y_1 \leq y_2 \leq \cdots \leq y_n$, how quickly can the $n$th smallest of the $2n$ elements in the combined sets be found? Does a lower bound (in the sense of Section 8.1.5) for this problem say anything about a lower bound for the problem of finding the median of a set of elements? Compare your result to that of the previous exercise.

## 8.2.2 Merging

The second sorting-related problem that we consider is that of merging two sorted tables $x_1 \leq x_2 \leq \cdots \leq x_n$ and $y_1 \leq y_2 \leq \cdots \leq y_m$ into a single sorted table $z_1 \leq z_2 \leq \cdots \leq z_{n+m}$. There is an obvious way to do this: scan the tables to be merged in parallel, at each stage selecting the smaller of the two elements and putting it into the output table. This process is simplified a little by adding sentinels $x_{n+1} = y_{m+1} = \infty$, as in Algorithm 8.9. In this algorithm $i$ and $j$ point, respectively, to the last elements in the two input tables that have not yet been put into the output table.

$$x_{n+1} \leftarrow y_{m+1} \leftarrow \infty$$
$$i \leftarrow j \leftarrow 1$$
$$\textbf{for } k = 1 \textbf{ to } n + m \textbf{ do}$$
$$\qquad \textbf{if } x_i < y_j \textbf{ then}$$
$$\qquad\qquad\qquad z_k \leftarrow x_i$$
$$\qquad\qquad\qquad i \leftarrow i + 1$$
$$\qquad \textbf{else}$$
$$\qquad\qquad\qquad z_k \leftarrow y_j$$
$$\qquad\qquad\qquad j \leftarrow j + 1$$

**Algorithm 8.9**
Straight merging

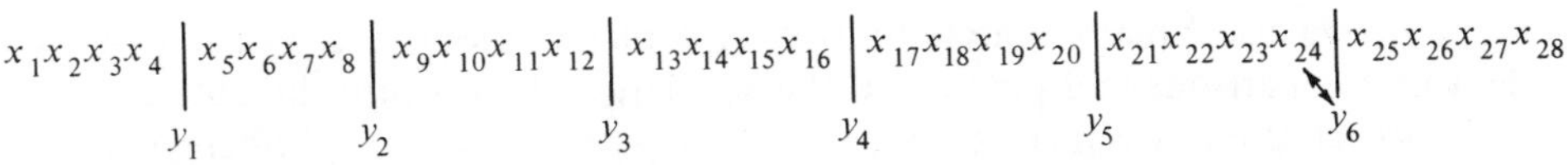

$$x_1 x_2 x_3 x_4 \;\Big|\; x_5 x_6 x_7 x_8 \;\Big|\; x_9 x_{10} x_{11} x_{12} \;\Big|\; x_{13} x_{14} x_{15} x_{16} \;\Big|\; x_{17} x_{18} x_{19} x_{20} \;\Big|\; x_{21} x_{22} x_{23} x_{24} \;\Big|\; x_{25} x_{26} x_{27} x_{28}$$

$$\underset{y_1}{\phantom{x}} \qquad \underset{y_2}{\phantom{x}} \qquad \underset{y_3}{\phantom{x}} \qquad \underset{y_4}{\phantom{x}} \qquad \underset{y_5}{\phantom{x}} \qquad \underset{y_6}{\phantom{x}}$$

**Figure 8.11**
The idea behind binary merging is this: assume that $n \geqslant m$ and divide the $x$'s into $m + 1$ subtables of $n/(m + 1)$ elements each. Then use binary search on the subtables. In this example, $n = 28$, $m = 6$, and $l = 24$.

The analysis of this algorithm is quite simple, for the comparison "$x_i < y_j$" is made exactly once for each element placed in the output table—that is, $n + m$ times. This can be reduced to $n + m - 1$ times by slightly complicating the algorithm (Exercise 1).

When $n \approx m$, this method of merging is quite good; in fact, when $n = m$, it is possible to show that in the worst case at least $n + m - 1 = 2n - 1$ element comparisons are always necessary to do the merging (Exercise 3). However, when $m = 1$, we can merge far more efficiently by using binary search to find the place in which $y_1$ should be inserted. We would like a method of merging that combines the best aspects of both binary search and Algorithm 8.9.

Figure 8.11 shows the central idea of binary merging, a scheme that behaves like binary search when $m = 1$ but like straight merging when $n \approx m$. It also provides a good compromise for other values of $m$. Assume that $n \geqslant m$; the idea is to divide the larger table into $m + 1$ subtables. We then compare the rightmost (largest) element of the smaller table, $y_m$, with the largest element of the next to the rightmost subtable of the larger table, say $x_l$ (see Figure 8.11). If $y_m < x_l$, then $x_l$ and all the rightmost subtable of the larger table can be put into the output table. If $y_m \geqslant x_l$, then $y_m$ is inserted into the rightmost subtable by using binary search; $y_m$ and the $x_i$'s in the subtable found to be greater than $y_m$ can now be put into the output table, and the algorithm continues recursively. Recall, however, that binary search works most efficiently for tables of size $2^k - 1$ (why?) and so, instead of having the last subtable of the larger table contain about $n/m$ elements, we do better if it has $2^{\lfloor \lg(n/m) \rfloor} - 1$ elements. Thus, for the case shown in Figure 8.11 we have $2^{\lfloor \lg(28/6) \rfloor} - 1 = 3$, and we would compare $y_6$ with $x_{25}$, as shown in Figure 8.12. If $x_{25} > y_6$, then $x_{25}, x_{26}, x_{27}$, and $x_{28}$ can be put into the output table, and we continue merging $x_1, x_2, \ldots, x_{24}$ with $y_1, y_2, \ldots, y_6$. If

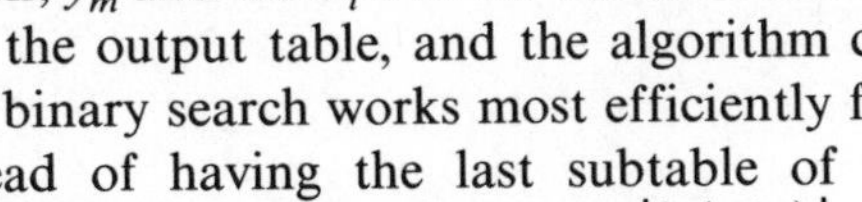

$$x_1 x_2 \cdots x_{21} x_{22} \;\Big|\; x_{23} x_{24} x_{25} \;\Big|\; x_{26} x_{27} x_{28}$$

$$\underset{y_1}{\phantom{x}} \;\; \cdots \qquad \underset{y_5}{\phantom{x}} \qquad \underset{y_6}{\phantom{x}}$$

**Figure 8.12**
The first comparison in binary merging

$x_{25} < y_6$, we use binary search to find $y_6$'s place among $x_{26}, x_{27}, x_{28}$ in two element comparisons and put $y_6$ and the $x_i$'s larger than it into the output table. Next, we continue by merging $x_1, x_2, \ldots, x_k$ with $y_1, y_2, \ldots, y_5$, where $k$ is the largest integer such that $y_6 > x_k$.

Assuming that $n \geqslant m$, Algorithm 8.10(a) gives an outline of the procedure that performs the first stage of the binary merge as described above. By interchanging the parameters, Algorithm 8.10(b) applies Algorithm 8.10(a) properly to the general case, using it iteratively until the tables are completely merged. We leave the omitted details of Algorithms 8.10(a) and (b) for the reader to fill in (Exercise 5).

How many element comparisons are made? Let $C_{m,n}$ be the number of element comparisons made in the worst case by Algorithms 8.10(a) and (b) when

**procedure** *MERGE*($x, n, y, m$)
⟦first stage of binary merging to merge $x_1, x_2, \ldots, x_n$
and $y_1, y_2, \ldots, y_m$ when $n \geqslant m$; the parameters $n$ and $m$ are
modified by the procedure and must be passed by reference⟧
$t \leftarrow \left\lfloor \lg \dfrac{n}{m} \right\rfloor$
**if** $y_m < x_{n+1-2^t}$ **then**
> Put $x_{n+1-2^t}, \ldots, x_n$ into the output table
> $n \leftarrow n - 2^t$

**else**
> Using $t$ element comparisons, insert $y_m$ into
> $x_{n+2-2^t}, \ldots, x_n$ by binary search
> Let $k$ be the largest integer such that $y_m > x_k$
> ($k$ is determined by the binary search)
> Put $y_m, x_{k+1}, x_{k+2}, \ldots, x_n$ into the output table
> $m \leftarrow m - 1$
> $n \leftarrow k$

**Algorithm 8.10(a)**
An outline of the first stage of binary merging

**while** $n \neq 0$ **and** $m \neq 0$ **do**
> **if** $m \leqslant n$ **then** *MERGE*($x, n, y, m$)
> > **else** *MERGE*($y, m, x, n$)

**if** $n = 0$ **then** put $y_1, \ldots, y_m$ into the output table
> **else** put $x_1, \ldots, x_n$ into the output table

**Algorithm 8.10(b)**
Binary merging, using the *MERGE* procedure of Algorithm 8.10(a) to do all the work

merging $x_1,\ldots,x_n$ and $y_1,\ldots,y_m$. It can be shown that

$$C_{m,n} = m + \left\lfloor \frac{n}{2^t} \right\rfloor - 1 + tm, \qquad \text{for } n \geqslant m, \ \ t = \left\lfloor \lg \frac{n}{m} \right\rfloor$$

(see Exercises 6, 7, and 8). When $m = n$, this gives

$$C_{n,n} = 2n - 1,$$

and when $m = 1$, it gives

$$C_{1,n} = 1 + \lfloor \lg n \rfloor = \lceil \lg(n + 1) \rceil.$$

These results mean that binary merging performs like straight merging at one extreme and like binary search at the other. Furthermore, it is also reasonably efficient for intermediate values of $m$ (Exercise 9).

**Exercises**

1. Modify Algorithm 8.9 (straight merging) so that the sentinel elements $x_{n+1} = y_{m+1} = \infty$ are not needed and thus give an algorithm that uses at most $n + m - 1$ element comparisons to merge $x_1 \leqslant x_2 \leqslant \cdots \leqslant x_n$ and $y_1 \leqslant y_2 \leqslant \cdots \leqslant y_m$.

2. (a) Give a version of your algorithm from the previous exercise for the case when the $x$'s and $y$'s are given as *linked lists*.

   (b) Use this version of the algorithm as the basis of a sorting algorithm in which the list to be sorted is split in half, each half is sorted recursively, and then the two are merged together. Analyze the worst-case behavior of this algorithm.

3. Show that $2n - 1$ element comparisons are necessary and sufficient to merge $x_1 \leqslant x_2 \leqslant \cdots \leqslant x_n$ and $y_1 \leqslant y_2 \leqslant \cdots \leqslant y_n$. [*Hint:* What if $x_i < y_j$ if $i < j$ and $x_i > y_j$ otherwise?]

4. Show the sequence of element comparisons made by binary merging [Algorithms 8.10(a) and (b)] when applied to merge the sorted tables

$$78 \leqslant 201 \leqslant 400 \leqslant 897$$

and

$$13 \leqslant 15 \leqslant 25 \leqslant 90 \leqslant 121 \leqslant 122 \leqslant 180 \leqslant 205 \leqslant 305 \leqslant 390 \leqslant 402.$$

5. Fill in all the missing details in Algorithms 8.10(a) and (b).

6. Prove that in the worst case binary merging uses $C_{n,m} = n + m - 1$ element comparisons when $m \leqslant n < 2m$.

★7. Prove that for binary merging $C_{m,n} = C_{m,\lfloor n/2 \rfloor}+m$ for $2m \leqslant n$. [*Hint:* Prove by induction on $m$ that $C_{m,n} \leqslant C_{m,n+1}$ for $n \geqslant m$ by observing that $C_{m,n} = \max(C_{m,n-2^t} + 1, C_{m-1,n} + t + 1)$, where $t = \lfloor \lg(n/m) \rfloor$.]

8. (a) By combining the results of the two previous exercises, show that $C_{m,n} = m + \lfloor n/2^t \rfloor - 1 + tm$ for $m \leqslant n$ where $t = \lfloor \lg(n/m) \rfloor$ as stated in the text.

   (b) Verify that this formula gives $C_{n,n} = 2n - 1$ and $C_{1,n} = 1 + \lfloor \lg n \rfloor$.

9. (a) In analogy with the discussion of $S(n)$ in Section 8.1.5, show that there must be at least $\lceil \lg \binom{m+n}{n} \rceil$ element comparisons used in the worst case for merging $x_1 \leqslant x_2 \leqslant \cdots \leqslant x_n$ and $y_1 \leqslant y_2 \leqslant \cdots \leqslant y_m$.

   (b) Use the formula for $C_{m,n}$ to show that $C_{m,n} < \lceil \lg \binom{m+n}{n} \rceil + m$, $n \geqslant m$. [*Hint:* First prove that $m! \leqslant m^m/2^{m-1}$.]

## 8.3   EXTERNAL SORTING

In the sorting methods discussed in Section 8.1 we assumed that the table fit in high-speed internal memory. This assumption is too strong for many real-life data processing problems. In this section we examine the sorting of very large tables that fit only on auxiliary storage devices such as magnetic tape, disks, or drums. In each case we will assume that we have a table of elements $x_1, x_2, x_3, \ldots, x_n$ stored externally and that the internal memory can hold only $m \ll n$ elements, along with other data, programs, and so forth.

The general strategy in external sorting is to use the internal memory to sort the elements in a piecemeal fashion so as to produce *initial runs* (also called *strings*) of elements in increasing order. As they are produced, these runs are stored externally. Later they are merged together, again in piecemeal fashion, until finally all the elements are in order. In our discussion we examine these two phases separately. We describe first how to generate the initial runs (Section 8.3.1), then how to do the merging both for tapes (Section 8.3.2) and for disks and drums (Section 8.3.3).

### 8.3.1   Initial Runs

Assuming that there is room in internal memory for $m \ll n$ elements, together with whatever buffers and other program variables are needed, the obvious method to generate initial runs is simply to read in $m$ elements, sort them in internal memory, and write them into external memory as a run, continuing in this fashion until all the *n* elements have been processed. The initial runs thus obtained all contain *m* elements (except perhaps for the last one). Since the number of initial runs ultimately determines the cost of the merging later, we would like to find some method of producing longer, and hence fewer, initial runs.

We can produce longer runs by organizing the *m* elements in internal memory into a priority queue that operates as follows. After initially being filled with the first *m* elements from the input file, the highest-priority element is deleted and put into the output file of the initial runs; the next element from the input file is then inserted in the priority queue. In this way the priority queue always contains *m* elements, except during its initialization and during the generation of the final run as it is being emptied. The priorities of the elements must be defined so that elements are deleted at the proper time and go into the proper run. Since each element selected from the priority queue to be added to the output is replaced by another element from the input, this method is called *replacement selection*.

To make certain that elements enter and leave the priority queue in the proper order, we will have the priority queue consist of ordered pairs $(r, x)$, where $r$ is the run number of the element $x$. A pair $(r, x)$ has higher priority than $(\hat{r}, \hat{x})$ if $(r, x) < (\hat{r}, \hat{x})$ lexicographically—that is, if $r < \hat{r}$ or if $r = \hat{r}$ and $x < \hat{x}$. After an element has been deleted from the priority queue and is being replaced by a new element $x$, we know the run number of $x$ by the following reasoning: if $x$ is smaller than the last element just added to the current run, then $x$ must be in the *next* run; otherwise, $x$ is the *current* run.

An outline of the above described replacement selection is given in Algorithm 8.11, and we leave it to Exercise 2 to fill in the missing details. The major unspecified aspect of Algorithm 8.11 is the implementation of the priority queue. The easiest way to implement it is with a heap as described in Section 8.1.3, except that the pair $(r, x)$ at the top of the heap is the *smallest* pair (lexicographically), not the largest, so the sense of the comparisons made in the *RESTORE* procedure (Algorithm 8.6) must be reversed. Furthermore, since the elements being sorted will in most cases be long records, there is no point in moving the actual elements around during the restoration of the heap—it is sufficient to maintain a heap of pairs $(r, p)$, where $r$ is the run number and $p$ is a *pointer* to the actual element.

How long are the runs produced by replacement selection? Clearly it does at least as well as the obvious method of reading in *m* elements, sorting them, and writing them—all runs generated by replacement selection, except possibly the last, contain at least *m* elements (why?). In fact it can be shown that on the average, assuming that the elements are in random order, the expected length of the runs is $2m$, double the length of the runs with the obvious method (see

$P \leftarrow$ empty priority queue with room for $m$ elements
⟦read in first $m$ elements and fill $P$ with them⟧
**for** $i = 1$ **to** $m$ **do**
      $x \leftarrow$ next element from the input file on external storage
      $P \Leftarrow (1, x)$
$R \leftarrow 0$     ⟦current run number⟧
**while** $P$ not empty **do**
      $(r, lastout) \Leftarrow P$
      **if** $r = R$ **then** add *lastout* to end of the current run
             **else**
                  ⟦begin a new run⟧
                  $R \leftarrow R + 1$
                  make *lastout* the first element in the new run
      ⟦replace the deleted element, if possible⟧
      **if** input file is not exhausted **then**
              $x \leftarrow$ next element from the input file
              **if** $x \geqslant lastout$ **then**
                        ⟦$x$ is part of the current run⟧
                        $P \Leftarrow (R, x)$
              **else**
                        ⟦$x$ is part of the next run⟧
                        $P \Leftarrow (R + 1, x)$

**Algorithm 8.11**
Replacement selection for generating initial runs

Exercise 8). Furthermore, in cases where the elements are not too out of order the length of the runs will be very long; in fact in the best case we would end up with only one initial run—that is, a sorted table!

We can improve on replacement selection with the following idea. The weakness of replacement selection is that sooner or later the $m$ locations in the priority queue get cluttered up with elements waiting to go into the next run. Suppose that instead of allowing this to happen we store such elements temporarily in external storage until, say, $M$ of them are accumulated. At that point, we dump the remaining elements in the priority queue into the current run and use the accumulated $M$ temporarily stored elements as the first part of the input for the next run. This leads to longer initial runs, but at the price of the extra input/output operations necessary to temporarily store the elements in external storage (see Exercise 9). The following table shows the expected initial run lengths with this technique; the size of $M$ is given in terms of $m$. There is no improvement, and hence no point in the method, for $M \leqslant 0.386m$.

| M | Expected Run Length |
|---|---|
| $0.386m$ | $2m$ |
| $m$ | $2.718m$ |
| $2m$ | $3.535m$ |
| $3m$ | $4.162m$ |
| $4m$ | $4.694m$ |
| $5m$ | $5.164m$ |

The fact that when $M = m$ the expected run length is $em$ ($e \approx 2.71828$, the base of the natural logarithms) has caused this technique to be called "natural selection."

**Exercises**

1. Elements from how many different runs can be in the priority queue at the same time in Algorithm 8.11?

2. Fill in the details of the priority-queue maintenance in Algorithm 8.11 using a heap of pointers as suggested in the text.

★3. Given an infinite "random" sequence of elements $x_1, x_2, \ldots$, where random means that each of the $n!$ possible relative orderings of the first $n$ elements is equally likely, compute the expected length of the first run in replacement selection.

4. Why would you expect the second run in replacement selection to be longer than the first?

5. Under what conditions will the last run in replacement selection be longer than $m$ elements?

6. Under exactly what conditions will replacement selection result in a single run? [*Hint*: Examine the analysis of Algorithm 8.1.]

7. What is the "worst-case" input file for replacement selection?

★8. We can describe the behavior of replacement selection with the following physical model. A snowplow is going round and round a circular road, plowing snow that is falling continuously and uniformly. Once a snowflake is plowed off the road, it is hauled away in a truck. The snowplow corresponds to the replacement-selection algorithm and the snowflakes correspond to the elements of the input file. Label the points on the road by real numbers $x$, $0 \leq x \leq 1$; a snowflake landing at $x$ represents an element $x$ from the input file. Assume the system is in a "steady state" in the sense that the speed of the snowplow is constant and the total amount of snow on the road is always $m$. Explain why the amount of snow removed by the plow on one revolution (the run length) is $2m$.

9. Under what conditions, if any, will natural selection put an element into the temporary external storage *twice*? Three times?

### 8.3.2   Tape Merge Patterns

After the initial runs have been generated, we have the problem of repeatedly merging them, piecemeal fashion, until we ultimately obtain the final, sorted table. The organization of this merging phase is heavily dependent on the type of external memory available. In this section we discuss merging the initial runs with tapes and in the next section we discuss the same problem with disks and drums. As pointed out at the beginning of Section 7.5, tapes are a completely sequential storage medium. Thus the entire problem of storing and merging the runs on tape is the organization of the tapes and the merges so that the runs are accessible on the tapes as they are needed.

We will assume that there are a total of $t + 1$ tapes; that is, initially we had $t$ scratch tapes and an input tape containing the table to be sorted. The initial runs have been generated from the table as described in the previous section, so that the situation now is that there are $r$ initial runs on one of the tapes and the remaining $t$ tapes are empty.

To begin, let us consider a simple-minded merging scheme. Distribute the runs as evenly as possible onto $t$ of the tapes and merge the runs together, one from each tape, to form longer runs on the $(t + 1)$st tape. The resulting (longer) runs are distributed onto the other $t$ tapes and merged together to form yet longer runs. This process continues until there is only one run, the sorted table. In comparing the behavior of various merging strategies the dominant factor is the number of times the external storage medium is accessed; we can thus get a good idea of the relative performance of merging strategies by comparing the number of times each element of the table is "examined"—that is, read into internal memory and then written out again. In the case of our simple-minded $t$-way merging, each merging decreases the number of runs by a factor of $1/t$ (why?). Consequently, since there are $r$ initial runs, $\lceil \log_t r \rceil$ mergings are required. Each of these mergings consists of first distributing the runs onto the $t$ tapes and then merging the runs together. In other words, each of the elements is examined twice, and hence in total each element of the table is examined $2\lceil \log_t r \rceil$ times: we describe this situation by saying that there are $2\lceil \log_t r \rceil \approx (2/\lg t) \lg r$ *passes* over the elements.

Thus for $t + 1$ tapes (that is, $t$-way merging), the number of passes over the elements in this simple-minded merging strategy is approximately

| $t + 1$ | *Number of Passes* |
|:---:|:---:|
| 3 | 2.000 lg $r$ |
| 4 | 1.262 lg $r$ |
| 5 | 1.000 lg $r$ |
| 6 | 0.861 lg $r$ |
| 7 | 0.774 lg $r$ |
| 8 | 0.712 lg $r$ |
| 9 | 0.667 lg $r$ |
| 10 | 0.631 lg $r$ |
| 20 | 0.471 lg $r$ |

Half the passes (the distribution passes) do nothing to reduce the number of runs; these passes are just copying the elements.

**Polyphase Merge.** The copy passes can be eliminated by being more clever. Suppose that 57 initial runs are distributed onto tapes 1, 2, 3, and 4 as shown in Figure 8.13. In this figure, $n \bullet i$ means $n$ runs of order $i$, an $i$th-order run being the result of merging $i$ initial runs together. The runs are merged as indicated by the arrows in the figure. The idea of this *polyphase merge* is to arrange the initial runs so that after each merge except the last there is exactly one empty tape—this tape will be the recipient on the next merge. Furthermore, we want the last merge to be merging only one run from each of $t$ nonempty tapes. A distribution of initial runs with such properties is called a *perfect* distribution.

Suppose that we had $t + 1$ tapes instead of four tapes. The polyphase merge idea generalizes easily, and we want to compute the perfect distributions for this case. We do this by working backward. Let $a_{k, j}$ be the number of runs on the $j$th tape when $k$ merge phases remain to be done. Moreover, we will assume that tape $t + 1$ is *always* the recipient of the merge and that tape 1 contains at least as many runs as tape 2, which contains at least as many runs as tape 3, and so on. This can be done by using "logical" tape numbers and switching them around appropriately. Thus we have the distribution

$$j: \quad 1 \quad 2 \quad 3 \quad \cdots \quad t-1 \quad t \quad t+1$$

Number of
runs on tape $j$: $\quad 1 \quad 0 \quad 0 \quad \cdots \quad 0 \quad 0 \quad 0$

when no more merge passes remain, that is, at the very end. If, when $k$ merge

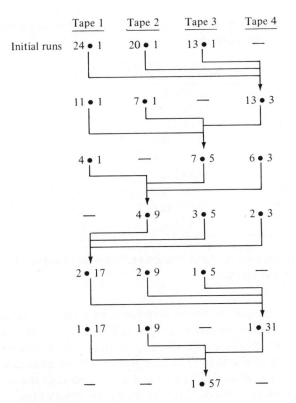

**Figure 8.13**
A merging pattern for 57 initial runs and four tapes. The notation $n \bullet i$ means $n$ $i$th-order runs.

passes remain, we have the distribution

$$j: \quad 1 \quad 2 \quad 3 \quad \cdots \quad t-1 \quad t \quad t+1$$

Number of
runs on tape $j$: $\quad a_{k,1} \quad a_{k,2} \quad a_{k,3} \quad \cdots \quad a_{k,t-1} \quad a_{k,t} \quad 0$

then the next merge leaves the distribution

$$j: \quad 1 \quad 2 \quad \cdots \quad t-1 \quad t \quad t+1$$

Number of
runs on tape $j$: $\quad a_{k,1} - a_{k,t} \quad a_{k,2} - a_{k,t} \quad \cdots \quad a_{k,t-1} - a_{k,t} \quad 0 \quad a_{k,t}$

It is easy to show that for $k \geq 1$, we have $2a_{k,t} \geq a_{k,1}$, and hence switching

logical tape numbers so that the runs are in decreasing order gives us

$$j: \quad 1 \qquad 2 \qquad 3 \qquad \cdots \qquad t \qquad t+1$$

Number of
runs on tape $j$: $\quad a_{k,t} \quad a_{k,1} - a_{k,t} \quad a_{k,2} - a_{k,t} \quad \cdots \quad a_{k,t-1} - a_{k,t} \quad 0$

Therefore

$$a_{k-1,1} = a_{k,t}$$
$$a_{k-1,j} = a_{k,j-1} - a_{k,t}, \qquad 2 \leqslant j \leqslant t,$$

or

$$a_{k+1,t} = a_{k,1}$$
$$a_{k+1,j} = a_{k,j+1} + a_{k,1}, \qquad 1 \leqslant j < t,$$

Consequently, the perfect distributions for $t + 1$ tapes are as shown in Table 8.1.

Of course, a perfect distribution for $t + 1$ tapes is possible only when $r$, the number of initial runs, is of a special form. For example, in the case of four tapes shown in Figure 8.13, $r$ must be one of 1, 3, 5, 9, 17, 31, 57, .... (What is the pattern here? See Exercise 2.) What do we do if we have, say, 40 runs, too few for the distribution (24, 20, 13) and too many for the distribution (13, 11, 7)? The easiest way to handle such cases is to add enough "virtual" or "dummy" runs to fill out the distribution; if we had four tapes and 40 initial runs, we would add seventeen dummy runs to allow us to have the (24, 20, 13) distribution of Figure 8.13. These dummy runs are only a bookkeeping device: we do not actually have such runs on tape, we just keep track of how many there are and where they are. This can be done by having counters $d_1, d_2, \ldots, d_{t+1}$, where $d_i$ is the number of dummy runs currently considered to be on tape $i$. In merging the runs together, it

| Pass | Tape 1 | Tape 2 | Tape 3 | | Tape $t-2$ | Tape $t-1$ | Tape $t$ | Tape $t+1$ |
|---|---|---|---|---|---|---|---|---|
| $k = 0$ | 1 | 0 | 0 | $\cdots$ | 0 | 0 | 0 | 0 |
| $k = 1$ | 1 | 1 | 1 | $\cdots$ | 1 | 1 | 1 | 0 |
| $k = 2$ | 2 | 2 | 2 | $\cdots$ | 2 | 2 | 1 | 0 |
| $k = 3$ | 4 | 4 | 4 | $\cdots$ | 4 | 3 | 2 | 0 |
| $\vdots$ | | | | | | | | |
| $k$ | $a_{k,1}$ | $a_{k,2}$ | $a_{k,3}$ | $\cdots$ | $a_{k,t-2}$ | $a_{k,t-1}$ | $a_{k,t}$ | 0 |
| $k+1$ | $a_{k,2} + a_{k,1}$ | $a_{k,3} + a_{k,1}$ | $a_{k,4} + a_{k,1}$ | $\cdots$ | $a_{k,t-1} + a_{k,1}$ | $a_{k,t} + a_{k,1}$ | $a_{k,1}$ | 0 |

**Table 8.1**
Perfect distributions for the polyphase merge

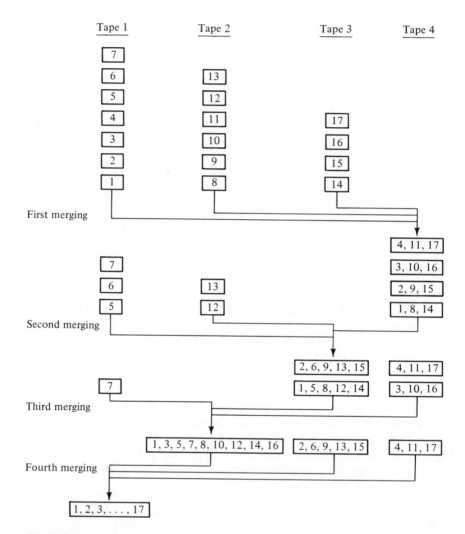

**Figure 8.14**
A four-tape polyphase merge of seventeen initial runs with the locations of
each initial run shown throughout

is easy to merge dummy runs! All we need do is change the values of the
appropriate $d_i$.

How should these dummy runs be distributed on the tapes? Some distributions will lead to more economical merging than others because some of the initial
runs will get handled more than others in the merging process, and we are better
off if those initial runs are dummy runs, not real runs. Suppose we have
seventeen initial runs and four tapes. The polyphase merge pattern is shown in

Figure 8.14, with the location of every initial run shown. Counting the number of times each initial run is handled, we find the following results:

Run number:                     1   2   3   4   5   6   7   8   9   10   11   12   13   14   15   16  17

Number of mergings
in which it participates:       4   3   3   2   3   2   2   4   3   3   2   3   2   4   3   3   2

Thus runs 1, 8, and 14 are handled more than the other runs, and so it would be wise to make them the dummy runs, if dummy runs are needed.

The optimal placement of the dummy runs is hard to determine, so instead we use the following heuristic of distributing the dummy runs as evenly as possible on the various tapes. This heuristic produces excellent results in general, and its simplicity makes it inexpensive to implement. Algorithm 8.12(a) describes an implementation in detail. It works by computing the next perfect distribution

⟦start at line $k = 0$ of Table 8.1 of perfect distributions⟧
$k \leftarrow j \leftarrow d_0 \leftarrow 0$
$a_1 \leftarrow d_1 \leftarrow tape_1 \leftarrow 1$
**for** $i = 2$ **to** $t + 1$ **do**
    $a_i \leftarrow 0$    ⟦actually $a_{k,i}$ of Table 8.1; the perfect
                distribution we are working toward⟧
    $d_i \leftarrow 0$    ⟦$d_i$ is the number of dummy runs on
                logical tape $i$⟧
    $tape_i \leftarrow i$    ⟦$tape_i$ is the number of the physical
                tape corresponding to logical tape $i$⟧
rewind all tapes
**while** initial runs not exhausted **do**
    ⟦at this point $d_1 = d_2 = \cdots = d_j, d_{j+1} \geqslant d_{j+2} \geqslant \cdots \geqslant d_{t+1}$ and
    $d_{j+1} = d_j$ or $d_j + 1$⟧
    **if** $d_j < d_{j+1}$ **then** $j \leftarrow j + 1$
             **else if** $d_j \neq 0$ **then** $j \leftarrow 1$
                  **else**
                        ⟦all $d_i$ are 0 so we have attained the distribution of
                        line $k$ in Table 8.1; we now go on to the next line⟧
                        $k \leftarrow k + 1$
                        $A \leftarrow a_1$
                        **for** $i = 1$ **to** $t$ **do**
                                $d_i \leftarrow A + a_{i+1} - a_i$    ⟦add dummy runs to
                                                        take up the slack
                                                        between the new dis-
                                                        tribution and the old⟧
                              $a_i \leftarrow A + a_{i+1}$    ⟦new distribution⟧
                      $j \leftarrow 1$
    read an initial run and write it on $tape_j$
    $d_j \leftarrow d_j - 1$    ⟦the real run replaces a dummy run⟧

**Algorithm 8.12(a)**
Even distribution of initial runs for the polyphase merge

(starting with the line $k = 0$ in Table 8.1), setting the $d_i$ so that the differences between the new distribution and the previous one just attained are filled by dummy initial runs. As real initial runs are processed, they replace the dummy initial runs until there are no more dummy runs; at that point the new distribution has been attained and we go on to the next.

Having distributed the initial runs, real and dummy, by Algorithm 8.12(a), we now face the problem of merging them according to the polyphase scheme as described above. The algorithm for doing this is given in Algorithm 8.12(b), which assumes that the values of variables left by Algorithm 8.12(a) are unchanged. The only subtle part of Algorithm 8.12(b) is that it proceeds by going up level by level in Table 8.1.

If there are $r$ initial runs and dummy runs are added as needed to form a perfect distribution, it can be shown that the number of passes over the elements

〚$k$, the $d_i$, and the *tape*$_i$ are assumed to have the values as left by Algorithm 8.12(a).〛
rewind all tapes
**while** $k \neq 0$ **do**

> 〚we now have the distribution shown in line $k$ of Table 8.1〛
> **while** *tape*$_t$ is not exhausted **or** $d_t > 0$ **do**
>> **if** $d_i > 0$ for all $i$, $1 \leq i \leq t$ **then**
>>> 〚merge $t$ dummy runs into a new dummy run on *tape*$_{t+1}$〛
>>> **for** $i = 1$ **to** $t$ **do** $d_i \leftarrow d_i - 1$
>>> $d_{t+1} \leftarrow d_{t+1} + 1$
>> **else**
>>> merge one run from each *tape*$_i$ with $d_i = 0$ onto *tape*$_{t+1}$
>>> **for** each $d_i > 0$ **do** $d_i \leftarrow d_i - 1$
>
> 〚go up a level in Table 8.1〛
> $k \leftarrow k - 1$
> rewind all tapes
> 〚renumber logical tape units〛
> $T \leftarrow tape_{t+1}$
> $D \leftarrow d_{t+1}$
> **for** $i = t + 1$ **to** $2$ **by** $-1$ **do**
>> $tape_i \leftarrow tape_{i-1}$
>> $d_i \leftarrow d_{i-1}$
>
> $tape_1 \leftarrow T$
> $d_1 \leftarrow D$

〚sorting is complete: *tape*$_1$ contains the sorted file〛

**Algorithm 8.12(b)**
Polyphase merge after the distribution of initial runs by Algorithm 8.12(a)

for $t + 1$ tapes is approximately

| $t + 1$ | Number of Passes |
|---|---|
| 3 | 1.042 lg $r$ |
| 4 | 0.704 lg $r$ |
| 5 | 0.598 lg $r$ |
| 6 | 0.551 lg $r$ |
| 7 | 0.528 lg $r$ |
| 8 | 0.516 lg $r$ |
| 9 | 0.509 lg $r$ |
| 10 | 0.505 lg $r$ |
| 20 | 0.500 lg $r$ |

Comparing this table with the similar one given for the simple-minded merging strategy, we see that the polyphase merge is far superior for ten or fewer tapes, but diminishing returns cause little to be gained by going beyond eight tapes.

**Cascade Merge.** For $t + 1 \geqslant 6$ there is an even better way to do the merging; it is most easily explained by an example. Suppose we have 190 initial runs distributed on six tapes as shown at the top of Figure 8.15. The *cascade merge* merges in the pattern shown in that figure. First, one initial run from each tape is merged, the resulting run being written on the empty tape. Second, runs from the four tapes still containing initial runs are merged onto the newly emptied tape. Third, runs from the three tapes still containing initial runs are merged onto the empty tape. Finally, runs from the two tapes still containing initial runs are merged onto the empty tape. This completes the first phase of the merge, the remaining phases being similar, as shown. The final phase merges one run from each tape onto the empty tape. It is surprising that this merge pattern is a reasonable choice, compared to the polyphase merge, because it does first a $t$-way merge, then a $(t - 1)$-way merge,..., and finally a two-way merge, while the polyphase merge always does $t$-way merges. Nevertheless, mathematical analysis shows that the number of passes over the elements is as follows:

| $t + 1$ | Number of Passes |
|---|---|
| 3 | 1.042 lg $r$ |
| 4 | 0.764 lg $r$ |
| 5 | 0.622 lg $r$ |
| 6 | 0.536 lg $r$ |
| 7 | 0.479 lg $r$ |
| 8 | 0.438 lg $r$ |
| 9 | 0.407 lg $r$ |
| 10 | 0.383 lg $r$ |
| 20 | 0.275 lg $r$ |

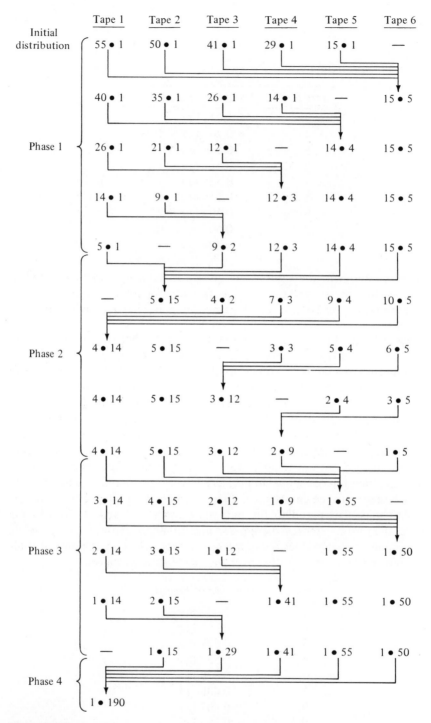

**Figure 8.15**
The cascade merge for 190 initial runs and six tapes

| Phase | Tape 1 | Tape 2 | Tape 3 | | Tape $t-2$ | Tape $t-1$ | Tape $t$ | Tape $t+1$ |
|---|---|---|---|---|---|---|---|---|
| $k=0$ | 1 | 0 | 0 | $\cdots$ | 0 | 0 | 0 | 0 |
| $k=1$ | 1 | 1 | 1 | $\cdots$ | 1 | 1 | 1 | 0 |
| $k=2$ | $t$ | $t-1$ | $t-2$ | $\cdots$ | 3 | 2 | 1 | 0 |
| $k=3$ | $\dfrac{t(t+1)}{2}$ | $\dfrac{t(t+1)}{2}-1$ | $\dfrac{t(t+1)}{2}-3$ | $\cdots$ | $3t-3$ | $2t-1$ | $t$ | 0 |
| $\cdots$ | | | | | | | | |
| $k$ | $a_{k,1}$ | $a_{k,2}$ | $a_{k,3}$ | $\cdots$ | $a_{k,t-2}$ | $a_{k,t-1}$ | $a_{k,t}$ | 0 |
| $k+1$ | $\displaystyle\sum_{i=1}^{t} a_{k,i}$ | $\displaystyle\sum_{i=1}^{t-1} a_{k,i}$ | $\displaystyle\sum_{i=1}^{t-2} a_{k,i}$ | $\cdots$ | $a_{k,1}+a_{k,2}+a_{k,3}$ | $a_{k,1}+a_{k,2}$ | $a_{k,1}$ | 0 |

**Table 8.2**
Perfect distributions for the cascade merge

This is identical to the polyphase merge for $t + 1 = 3$ (Exercise 3), somewhat inferior to the polyphase merge for $t + 1 = 4$ or 5, and superior to the polyphase merge for $t + 1 \geqslant 6$.

The perfect distributions for the cascade merge are given in Table 8.2, derived by an analysis similar to the one given for the polyphase merge. The distribution of initial runs and dummy runs for the cascade merge is similar to Algorithm 8.12(a) and the merging itself is similar to Algorithm 8.12(b); we leave both of these algorithms to Exercise 4.

**Buffering.** In examining the various ways of utilizing $t + 1$ tapes for sorting, we have ignored an important aspect of the problem: how information is actually transferred from tape to memory and back again. Since the nature of tape sorting is reading-processing-writing, we can save a good deal of time by fully overlapping these operations, after some initialization. In order to accomplish this feat, we need to be able to read from the input tapes *before* we have

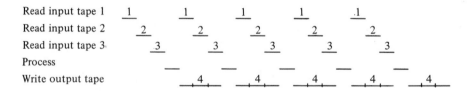

(a)  Single buffering: Buffers 1, 2, 3, 4 for the three input tapes and the output tape, respectively, with the possible overlap of reading, processing, and writing.

(b)  Double buffering: Buffers 1a, 1b, 2a, 2b, 3a, 3b, 4a, 4b for the three input tapes and the output tape, respectively, with full overlap of reading, processing, and writing.

**Figure 8.16**
Time charts for single and double buffering with four tapes showing the overlap possible in both cases. The relevant buffer number is given above each time-line segment in which it is read or written.

completely processed the elements read previously. Similarly, we need to write on the output tape before we have completely processed the elements. Although it sounds physically impossible to do, we can do it by *double buffering*, having two buffers for every tape. As we are processing elements from one of the input buffers, we fill the other; as we are putting elements into one of the output buffers, we write the other. A time-chart comparison of double buffering with single buffering is shown in Figure 8.16. Since the number of tapes is usually no more than ten, there is usually no problem in having enough internal memory to accommodate the extra buffers.

Figure 8.16 shows only one advantage of double buffering—the overlapping of not only read/write operations, but also the internal processing. Another significant improvement in Figure 8.16(b) compared to Figure 8.16(a) is that the writing operation goes on continuously. This is important, because one of the physical characteristics of tape is that once a read or write operation has finished and the tape is coasting to a stop, a good deal of time is saved if we can initiate the next read or write before the tape has slowed down too much. The time saved is that required by the tape to accelerate to its full reading or writing speed. Because double buffering allows the writing to go on continuously while single buffering does not, the savings in time can be much greater than Figure 8.16 suggests.

## Exercises

1. What is the relationship between the Fibonacci numbers $F_0 = 0$, $F_1 = 1$, $F_{i+2} = F_{i+1} + F_i$ and the polyphase merge?

2. Let $p_k = \sum_{i=1}^{t} a_{k,i}$, where the $a_{k,i}$ are as defined in Table 8.1, so that the $p_{k+t}$ are the numbers of runs in the perfect distribution on line $k$ of Table 8.1.

   (a) Prove that the $p_k$ satisfy the following recurrence relation:

$$p_1 = p_2 = \cdots = p_t = 1,$$
$$p_i = p_{i-1} + p_{i-2} + \cdots + p_{i-t}, \qquad i > t.$$

   ★(b) Determine the growth rate of $p_i$ as a function of $i$. [*Answer:* $p_i \approx C\alpha^i$ for some constant $C$, where $\alpha$ is the dominant root of the equation $x^t - x^{t-1} - x^{t-2} - \cdots - x - 1 = 0$, $\alpha \approx 2 - 1/(2^t - 1)$.]

3. Why do both the polyphase and cascade merges require the same number of passes over the elements for $t + 1 = 3$?

4. Design algorithms comparable to Algorithms 8.12(a) and (b) for the cascade merge.

5. Investigate the advantages of the following compromise between the poly-phase and cascade merges. Merge $t$ tapes until one is empty, next merge $t - 1$ tapes (excluding the one just created), next merge $t - 2$ tapes, then $t - 3$, and so on, down to a $(t - r + 1)$-way merge. At that point, the cycle of $t, t - 1, \ldots, (t - r + 1)$-way merges is repeated. When $r = 1$, this is the polyphase merge, and when $r = t - 1$, this is the cascade merge.

6. Examine the effect of rewind time on the merging strategies discussed in this section.

7. Suppose that the tapes can be read backward as well as forward. Show how this can be used to speed up the sorting process by doing mergings instead of rewinding the tapes. Be careful, because when merging runs while reading the tapes backward we get *descending* runs.

8. How should multireel tables of elements be sorted?

9. One way to describe a $t + 1$ tape merge pattern is by a $t$-way tree in which the leaves are initial runs and internal nodes represent merges. For example, the tree in Figure 8.17 represents the merge pattern corresponding to the polyphase merge of seventeen initial runs with $t + 1 = 4$ tapes shown in Figure 8.14. Each edge is labeled with the tape number on which the merged runs end up.

(a) What is the relationship between the external path length of such a tree and the merge?

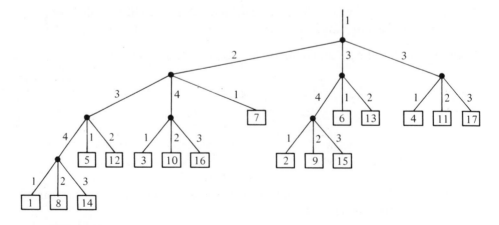

**Figure 8.17**
Exercise 9, the merge pattern tree corresponding to Figure 8.14.

    (b) Draw the tree corresponding to the cascade merge of fifteen initial runs and $t + 1 = 6$ tapes.

    (c) What trees correspond to $t + 1 = 3$ tapes for the polyphase and cascade merges?

    (d) Where should dummy initial runs be placed in the tree to minimize sorting time? Why?

10. To use either the polyphase or cascade merging schemes in tape sorting, we need at least three tapes. Suppose that only two tapes are available: find a method for tape sorting. [*Hint:* Combine the ideas of bubble sort and replacement selection.]

11. Investigate the advantages (if any) of double buffering during the generation of initial runs.

### 8.3.3  Disk Merge Techniques

As pointed out in Section 7.5, the key characteristic of disks (and drums) for external storage is direct access to anything on the disk, with a relatively high cost (in time) to initiate a transfer. When we do external sorting with tapes, after the generation of initial runs, all our energy goes into designing merge patterns so that runs are available as they are needed; this is necessitated by the sequential nature of access on tapes. The direct access to anything on a disk eliminates the need for such carefully arranged merge patterns—we can do any kind of merging we like! In particular, we can use the simple $t$-way merge because the copy phases, which are needed for tape sorting and which effectively double the input/output time, are not needed for disk merging because of the direct access.

With $r$ initial runs and $t$-way merging, the number of passes over the elements with repeated $t$-way merging is $\lceil \log_t r \rceil$, so it is to our advantage to minimize $r$ and to maximize $t$. In Section 8.3.1 we discussed the issue of minimizing $r$; the discussion there was independent of the external storage device, and it applies equally well to both tapes and disks. Maximizing $t$, however, is a new problem. For tape merging, $t$ was limited by the number of tape units available; but for disk merging the only constraint on $t$ is the amount of internal storage available for buffers. This is not typically a constraint for tape merging because, as pointed out in the previous section, the number of tapes is usually relatively small, with plenty of room even for the $2t + 2$ buffers needed for double buffering.

Suppose we have room for a total of $M$ elements in internal memory for all the buffers. The more buffers we have, the smaller they are and the more often they must be filled or emptied. Each filling of an input buffer or emptying of an output buffer requires the initiation of a transfer with the disk, so the more buffers we have, the more time we spend filling and emptying them. If there are $n$

elements and the buffers hold $b$ elements each, each pass of the merge will require

$$2n/b \text{ read/write initiations,}$$

$$2n \text{ read/write transfers of elements.}$$

If

$$I = \text{time to initiate an access,}$$
$$T = \text{time to transfer one element,}$$

then the total time per pass is

$$\frac{2n}{b}I + 2nT$$

if there is no overlap of reading and writing (say, if we are sorting with only one disk—most disks do not permit simultaneous reading and writing) and

$$\frac{n}{b}I + nT$$

if there is full overlap of reading and writing. In either case, the total time *per element* over the entire sort is

$$C\left(\frac{I}{b} + T\right)\lceil\log_t r\rceil \tag{8.4}$$

for some constant $C$ that also includes the internal processing time. The number of buffers would be $t + 1$ for single buffering and $2t + 2$ for double buffering, so that

$$b = \frac{M}{t+1} \qquad \text{for single buffering,}$$

$$b = \frac{1}{2}\frac{M}{t+1} \qquad \text{for double buffering.}$$

Formula (8.4) is thus of the form

$$C\left(\frac{I}{M}k(t+1) + T\right)\lceil\log_t r\rceil, \tag{8.5}$$

where $k = 1$ for single buffering and $k = 2$ for double buffering.

We would like to choose $t$ so as to minimize (8.5). This is difficult to do analytically, however, because of the ceiling function and the fact that $t$ can assume only integer values. An example will illustrate, however, how the optimum $t$ can be found by fairly simple calculations.

Suppose

$$M = 1000 \text{ elements,}$$
$$I = 75 \text{ ms/initiation,}$$
$$T = 0.5 \text{ ms/element,}$$
$$r = 100 \text{ initial runs,}$$

and we are doing single buffering. Without loss of generality, we can take $C = 1$ and compute the following table:

| $t$ | Time per Element | Number of Passes |
|----|-----|-----|
| 2 | 10.15 | 7 |
| 3 | 8.00 | 5 |
| 4 | 7.00 | 4 |
| 5 | 5.70 | 3 |
| 6 | 6.15 | 3 |
| 7 | 6.60 | 3 |
| 8 | 7.05 | 3 |
| 9 | 7.50 | 3 |
| 10 | 5.30 | 2 |
| 11 | 5.60 | 2 |
| 99 | 32.00 | 2 |
| 100 | 16.15 | 1 |

The minimum time occurs for $t = 10$-way merging. A similar table for $r = 60$ initial runs would have $t = 8$ as its optimum (Exercise 1), so we see that the number of initial runs has an effect on the optimum $t$.

Actually, we do not need to compute the entire table above to find the optimum $t$. Notice that the ceiling function in (8.4) causes five-way merging to be the most efficient way of finishing in three passes. There is no point in doing six-, seven-, eight-, or nine-way merging; they will be more costly than five-way merging, since the buffer size will be smaller, but the number of passes will be the same. Similarly, there is no point in considering eleven-, twelve-,..., or 99-way merging, since ten-way merging will require the same number of passes (only two), but with larger buffers. In general, for a fixed number of passes $p$, the most efficient $t$ is the smallest one that requires exactly that many passes; that value is

$t = \lceil r^{1/p} \rceil$ (why?). Of course, we will never require more passes than in two-way merging, $\lceil \lg r \rceil$. We can therefore find the optimum $t$ by the following simple program:

$$
\begin{aligned}
&min \leftarrow \infty \\
&\textbf{for } p = 1 \textbf{ to } \lceil \lg r \rceil \textbf{ do} \\
&\qquad t \leftarrow \lceil r^{1/p} \rceil \qquad [\![\text{smallest } t \text{ using } p \text{ passes}]\!] \\
&\qquad cost \leftarrow \text{from formula (8.4)} \\
&\qquad \textbf{if } cost < min \quad \textbf{then} \\
&\qquad\qquad\qquad min \leftarrow cost \\
&\qquad\qquad\qquad t_{\text{opt}} \leftarrow t
\end{aligned}
$$

Since $r$ will typically be at most a few thousand, the loop will be executed only about a dozen times, a trivial amount of time compared to what will be expended in the sorting itself.

**Exercises**

1. Compute the optimum $t$ for $M = 1000$, $r = 60$, $I = 75$ ms, and $T = 0.5$ ms with single buffers.

2. Compute the optimum $t$ for $M = 1000$, $r = 100$, $I = 75$ ms, and $T = 0.5$ ms with double buffering.

3. Suppose only a single disk is available for sorting. To prevent the irretrievable loss of data, it is wise to do a "read-back check" operation after writing. What is the effect of this on sorting time?

4. We have assumed that all buffers are the same size. Is there a way to reduce the number of disk initiations by using unequal buffer sizes?

## 8.4  REMARKS AND REFERENCES

Just about everything one might need to know about the sorting problems discussed in this chapter can be found in

> Knuth, D. E., *The Art of Computer Programming*, Vol. 3, *Sorting and Searching*. Reading, Mass.: Addison-Wesley Publishing Co., 1973.

This is an encyclopedic treatment of the entire subject. A lengthy bibliography of all the papers examined by Knuth in writing the section on sorting can be found in *Computing Reviews*, 13 (1972), 283–289. Section 5.5 of Knuth's book contains an excellent comparison of sorting methods and a short history of the subject of sorting.

A more complete treatment of quicksort and its variants can be found in

Sedgewick, R., "The Analysis of Quicksort Programs," *Acta Informatica* 7 (1977), 327–355.

Sedgewick, R., "Implementing Quicksort Programs," *Comm. ACM* 21 (1978), 847–857.

Sedgewick, R., "Quicksort with Equal Keys," *SIAM J. Comput.* 6 (1977) 240–267.

The linear time selection algorithm of Section 8.2.1 can be greatly improved. In

Schönhage, A., M. S. Paterson, and N. Pippenger, "Finding the Median," *J. Comput. Sys. Sci.* 13 (1976), 184–199.

it is shown that the median of $n$ elements can be found in $3n + O((n \log n)^{3/4})$ element comparisons; the $k$ th largest can be found in a similar number of element comparisons.

Ohe, iam satis est, ohe, libelle,
Iam pervenimus usque ad umbilicos.
Tu procedere adhuc et ire quaeris,
Nec summa potes in schida teneri,
Sic tamquam tibi res peracta non sit,
Quae prima quoque pagina peracta est.
Iam lector queriturque deficitque,
Iam librarius hoc et ipse dicit
"Ohe, iam satis est, ohe, libelle."

*Epigrams*, IV, 89,   Martial

# Index

**440**

"The review modules are so helpful because they give me bulleted highlights and concise, nursing information... To top it off, you also get critical thinking exercises! These books are fantastic! They have undoubtedly been the greatest review item I have found."

Kimberly Montgomery
Nursing student

**Terim Richards** *Nursing student*

"I immediately went to my nurse manager after I failed the NCLEX® and she referred me to ATI. I was able to discover the areas I was weak in, and focused on those areas in the review modules and online assessments. I was much more prepared the second time around!"

**Molly Obetz** *Nursing student*

"The ATI review books were very helpful in preparing me for the NCLEX®. I really utilized the review summaries and the critical thinking exercises at the end of each chapter. It was nice to review the key points in the areas I was weak in and not have to read the entire book."

**Lindsey Koeble** *Nursing student*

"I attribute my success totally to ATI. That is the one thing I used between my first and second attempt at the NCLEX®....with ATI I passed!"

**Danielle Platt** *Nurse Manager • Children's Mercy Hospital • Kansas City, MO*

"The year our hospital did not use the ATI program, we experienced a 15% decrease in the NCLEX® pass rates. We reinstated the ATI program the following year and had a 90% success rate."

"As a manager, I have witnessed graduate nurses fail the NCLEX® and the devastating effects it has on their morale. Once the nurses started using ATI, it was amazing to see the confidence they had in themselves and their ability to go forward and take the NCLEX® exam."

**Mary Moss** *Associate Dean of Nursing and Health Programs • Mid-State Technical College • Rapids, WI*

"I like that ATI lets students know what to expect from the NCLEX®, helps them plan their study time and tells them what to do in the days and weeks before the exam. It is different from most of the NCLEX® review books on the market."

## Contributors

**Jeanne Barrett, RN, BSN**

Nursing Education Author and Textbook Author
Jersey City, New Jersey

**Teri Scott, RN, MSN**

Associate Professor of Nursing
University of Missouri—Kansas City
Kansas City, Missouri

**Kim Tankle, RN, MSN**

Associate Professor of Nursing
University of Kansas
Kansas City, Missouri

**Barbara Kuhn Timby, RNC, MSN**

Professor Nursing
Glen Oaks Community College
Centerville, Michigan
Textbook Author

**Judith M. Wilkinson, RN, PhD, MSN, MS, ARNP**

Professor of Nursing
University of Missouri—Kansas City
Kansas City, Missouri

## Editor-in-Chief

**Leslie Schaaf Treas, RN, PhD(c), MSN, CNNP**

Director of Research and Development
Assessment Technologies Institute™, LLC
Overland Park, Kansas

## Editors

**Jim Hauschildt, RN, EdD, MA**

Director of Product Development
Assessment Technologies Institute™, LLC
Overland Park, Kansas

## Copyright Notice

## Important Notice to the Reader of this Publication

the content herein (with such content to include text and graphics), and the publishers, editors, advisors, and reviewers take no responsibility with respect to such content. The publishers, editors, advisors, and reviewers shall not be liable for any actual, incidental, special, consequential, punitive or exemplary damages (or any other type of damages) resulting, in whole or in part, from the reader's use of, or reliance upon, such content.

## Introduction to Assessment–Driven Review

To prepare candidates for the licensure exam, many different methods have been used. Assessment Technologies Institute™, LLC, (ATI) offers Assessment–Driven Review™ (ADR), a newer approach for customized board review based on candidate performance on a series of content-based assessments.

The ADR method is a four-part process that serves as a type of competency-assessment for preparation for the NCLEX®. The goal is to increase preparedness and subsequent pass rate on the licensure exam. Used as a comprehensive program, the ADR is designed to help learners focus their review and remediation efforts, thereby increasing their confidence and familiarity with the NCLEX® content. This type of program identifies learners at risk for failure in the early stages of nursing education and provides a path for prescriptive learning prior to the licensure examination.

The ADR approach may be preferable to a traditional "crash course" style of review for a variety of reasons. Time restriction is a fundamental barrier to comprehensive review. Because of the difficulty in keeping up with the expansiveness of information available today, a more efficient and directed approach is needed. Individualized review that starts with the areas of deficit helps the learner narrow the focus and begin customized remediation instead of a blanket A-to-Z approach. Additionally, review that occurs sequentially over time may be preferable to after-the-fact efforts after completion of a program when faculty are no longer available to assist with remediation.

Early identification of content weaknesses may prove advantageous to progressive program success. "Smaller bites" for content achievement and a shortened lapse of time between the introduction of course content and remediation efforts is likely to be more effective in catching the struggling learner before it is too late. Regular feedback keeps learners "on track" and reduce attrition rate by identifying the learner who is "slipping." This approach provides the opportunity to tutor or implement intensified instruction before the learner reaches a point of no return and drops out of the program.

### Step I: Proctored Assessment

The ADR program is a method using a prescriptive learning strategy that begins with a proctored, diagnostic assessment of the learner's mastery of nursing content. The topics covered within the ADR program are based on the current NCLEX® Test Plan. Proctored assessments are administered in paper-pencil and online formats. Scores are reported instantly with Internet testing or within 24 hours for paper-pencil testing. Individual performance profiles list areas of deficiencies and guide the learner's review and remediation of the missed topics. This road map serves as a starting point for self-directed study for NCLEX® success. Learners receive a cumulative Report Card showing scores from all assessments taken throughout the program—beginning to end. Like reading a transcript, the learner and educator can monitor the sequential progress, step-by-step, an assessment at a time.

## Step II: Modular Reviews

A good test is one that supports teaching and learning. The score report identifies areas of content mastery as well as a means for correction and improvement of weak content areas. Eight review modules contain concise summaries of topics with a clinical overview, therapeutic nursing management, and client teaching. Key concepts are provided to streamline the study process. The ATI modules are not intended to serve as a primary teaching source. Instead, they are designed to summarize the material relevant to the licensure exam and entry-level practice.

Learners are taught to integrate holistic care with a critical thinking approach into the review material to promote clinical application of course content. The learner constructs responses to open-ended questions to stimulate higher-order thinking. The learner may provide rationales for actions in various clinical scenarios and generate explanations of why the solution may be effective in similar clinical situations. These exercises serve as the venue to shift from traditional didactic memorization of facts toward the use of analytical and evaluative reason in a client-related situation. The clinical application scenarios involve the learner actively in the problem-solving process and stimulate an attitude of inquiry.

These exercises are designed to provoke creative problem solving for the individual learner as well as collaborative dialogue for groups of learners in the classroom. Through group discussion, learners discover the technique of elaboration. Learners use group dialogue to increase their understanding of nursing content. In study groups, they may pose questions to their peers or explain various topics in their own words, adding personal experiences with clients and examples from previously acquired knowledge of the topic. Together they learn to reframe problems and assemble evidence to support conclusions. Through the integration of multiple perspectives and the synergy involved in the exchange of ideas, this approach may also facilitate the development of effective working relationships and patterns for lifelong learning. Critical thinking exercises for each topic area situate instruction into a problem-solving environment that can capture learners' attention, increase motivation to learn, and frame the content into an application context. Additionally, the group involvement can model the process for effective team interaction.

## Step III: Non-Proctored Assessments

The third step is the use of online assessments that allow users to test from any site with an Internet connection. This online battery identifies specific areas of content weakness for further directed study. The interactive style provides the learner with immediate feedback on all response options. Rationales provide additional information about the correctness of an answer to supplement learners' understanding of the concept. Detailed explanations are provided for each incorrect response to clarify topics that learners often confuse, misunderstand, or fail to remember. Readiness to learn is often peaked when errors are uncovered; thus, immediate feedback is provided when learners are most motivated to find the answer. A Performance Profile summarizes learners' mastery of content. Question descriptors for each missed item are used to stimulate inquiry and further exploration of the topic. The online assessment is intended to extend the learners' preparation for NCLEX® in a way that is personally suited to their deficiencies.

### Step IV: ATI-PLAN™ DVD Series

This multi-disk set contains more than 28 hours of nursing review material. The DVD content is designed to complement ATI's Content Mastery Series™ review modules and online assessments. Using the ATI-PLAN™ navigational points, learners can easily find the content areas they want to review.

Recognizing that individuals process information in a variety of ways, ATI developed the ATI-PLAN™ DVD series to offer nursing review in a way that simulates the classroom. However, individuals viewing the ATI-PLAN™ DVDs can navigate through more than 28 hours of material to their topics of choice. Nursing review is available at the convenience of the learner and can be replayed as often as necessary to ensure mastery of content.

The regulation of personal learning goals and the ability to plan and pursue academic intentions are the keys to successful learning. The expert teacher is the one who can determine individual learning needs and appropriate strategies to master learning. The ADR program is an efficient method of helping students prepare for the nursing licensure exam using frequent and systematic content review directed by the identified areas of content weakness. The interactive approach for mastery of nursing content focused in the areas of greatest need is likely to increase student success on the licensure exam.

ATI's ADR method parallels the nursing process in concept and in design. Both provide a framework for solving actual and potential problems purposefully and methodically. Assessment ADR-style is accomplished with ATI's battery of proctored assessments. Diagnosis is facilitated by the individual and group score reports the proctored assessments generate. Planning for improving performance in identified areas of weakness incorporates ATI's modular review system. Implementation begins with modular review and culminates in use of ATI's online assessments to validate improvement. Evaluation is reflected in the score reports, and performance can then be strengthened or further improved with the ATI-PLAN™ DVD series. Just like the nursing process, ATI's ADR prescriptive learning method often leads to specific, measurable results and highly desirable outcomes.

# Table of Contents

# Health Care Delivery Systems

## Key Points

- Primary care: Promotion, prevention, early detection
- Secondary care: Diagnosis and treatment
- Tertiary care: Long-term, rehabilitation
- Health care is provided in a variety of settings, increasingly in ambulatory settings.
- Medicare provides insurance for adults over age 65 and disability income for disabled individuals.
- Medicaid provides health care for people with low incomes.
- Nursing care is organized in various ways, e.g., functional care, team nursing, and primary nursing.
- **Key Concepts/Terms**: Health care settings, health care providers, payment sources, nursing care delivery systems

## Overview

The health care system consists of all the agencies and professionals that are organized to provide health services. Its purposes are to provide (1) health promotion and illness prevention, (2) diagnosis and treatment of illness or injury, and (3) rehabilitation and health restoration. Health care services can be organized according to their complexity:

**Primary Care:** Health promotion and education, preventive care, early detection, and environmental protection

**Secondary Care:** Diagnosis and treatment, acute care, emergency care

**Tertiary Care:** Long-term care, rehabilitation, care of the terminally ill

## Health Care Settings

- **Public Health**: Government agencies—federal, state, or local—funded primarily by taxes and administered by elected or appointed officials. Local health departments develop and carry out programs to meet the health needs of groups within the community and the community as a whole. An example of a federal agency is the Centers for Disease Control and Prevention (CDC), which provides disease surveillance and epidemiological investigations.
- **Clinics**: May be inside or outside a hospital. May provide a variety of services or specialized services, such as infant immunizations. Some functions overlap with physicians' offices and ambulatory care centers.
- **Extended care facilities**: Includes skilled nursing (intermediate care) and extended-care (long-term care) facilities. May be inside or outside a hospital.

- **Ambulatory care centers:** (e.g., one-day surgery centers, diagnostic centers)
- **Primary care providers' offices**
- **Industrial clinics**
- **Hospitals**
- **Retirement and assisted-living centers**
- **Rehabilitation centers**
- **Home health agencies**
- **Hospice services**

## Health Care Providers

Also referred to as health professionals, health care givers, or the health care team, these are people from different disciplines who work together to promote client wellness and/or restore health.

- Nurses (RNs and LVNs/LPNs)
- Certified nursing assistants (CNAs)
- Unlicensed assistive personnel (UAPs) (e.g., nursing assistants)
- Physicians
- Physicians' assistants (PAs) diagnose and treat certain diseases and injuries under the direction of a primary care provider.
- Advanced Registered Nurse Practitioner (ARNP)
- Dentists
- Pharmacists
- Dietitians and nutritionists
- Physical therapists
- Respiratory therapists
- Occupational therapists
- Social workers
- Case managers
- Spiritual support persons
- Alternative care providers (e.g., chiropractors, acupuncturists)

## Factors Affecting Health Care Delivery

- Access to health care (e.g., eligibility for government programs/benefits/health insurance)
- Special populations (e.g., single-parent families, foreign born, homeless populations)
- Increasing number of older and very old adults
- Technologic advances (e.g., computers, organ transplants, minimally invasive surgical procedures)
- Cost of health care—Medical care costs have increased more than 400 percent since 1965.

- Uneven distribution of services (e.g., rural areas may not have enough health care professionals to provide needed services)
- Increased specialization—Contributes to fragmentation of care and increased cost of care. A client may receive care from up to 30 people during a hospital stay.

## Payment Sources

**Medicare**: A federal and state program to provide health insurance for adults over age 65. It includes hospital, extended care, and home health benefits. Medicare provides partial coverage of primary care provider services if clients choose to pay the premium for this coverage. Clients pay a deductible and co-payment for services. It does not cover dental care, eyeglasses, hearing aids, or most preventive care.

**Diagnosis-related groups (DRGs)**: In order to limit the amount paid to hospitals, Medicare has established a system that pays a predetermined amount for clients with a specific diagnosis (e.g., regardless of the length of stay or cost of services, a hospital would receive a set amount—say $1,500—for a client with asthma).

**Medicaid**: A federal public assistance program for people with low incomes. Coverage varies widely from state to state.

**Private (commercial) insurance**: (e.g., health maintenance organizations [HMO], preferred provider organizations [PPO]). In the U.S., about 80% of private health insurance is provided by employers.

## Nursing Care Delivery Systems

**Functional care**: Task oriented. Focuses on jobs to be done. Tasks are assigned by the nurse in charge (e.g., one nurse may administer medications for all clients on a unit; nursing assistant may take vital signs for all clients).

**Team nursing**: Led by a professional nurse, a team of 2 or 3 persons provides total care for a group of clients during an 8 or 12-hour shift.

**Primary nursing**: One nurse is responsible for total care of a group of clients. Associates provide care on some shifts; the primary nurse coordinates it.

# Critical Thinking Exercise: Health Care Delivery Systems

**Situation**: A hospital-based clinic serves a range of clients (e.g., children scheduled for routine immunizations, women seeking prenatal care, and individuals requiring treatment for a variety of chronic conditions). The following questions relate to health care delivered in this and other settings.

1. What factors distinguish primary care from secondary care? What levels of care would you expect to see provided in a clinic?

2. Define tertiary care. In what setting would you expect to see tertiary care provided? Give at least two examples.

3. Who is eligible for Medicare? Who is eligible for Medicaid? What are these programs intended to accomplish?

# Health Promotion and Disease Prevention

## Key Points

- **Primary care**: Promotion, prevention, early detection
- **Secondary care**: Diagnosis and treatment
- **Tertiary care**: Long-term, rehabilitation
- Health care is provided in a variety of settings, increasingly in ambulatory settings.
- Medicare provides insurance for adults over age 65 and income for disabled individuals.
- Medicaid provides health care for people with low incomes.
- Nursing care is organized in various ways (e.g., functional care, team nursing, and primary nursing)
- **Key Concepts/Terms**: *Healthy People 2010*, focus area, Pender's health promotion model, nursing interventions for wellness assessment, health promotion and disease prevention

## Overview

Health promotion and disease prevention are increasingly important aspects of nursing practice, as the public has become more interested in the relationship between lifestyle and illness. A healthy lifestyle involves adequate exercise, rest, good nutrition, and reducing the use of tobacco, alcohol, and other drugs. One way to talk about health promotion and disease prevention is by describing "levels of prevention":

- **Primary prevention** involves health promotion (e.g., exercise and nutrition) and protection against specific health problems (e.g., immunization against tetanus).
- **Secondary prevention** involves early identification of health problems (e.g., cancer screening) and prompt intervention to alleviate health problems (e.g., early identification of diabetes, with efforts to control blood sugar).
- **Tertiary prevention** involves restoration and rehabilitation (e.g., referring a grieving client to a support group; teaching foot care to a client with diabetes).

## Healthy People 2010

Although most of the U.S. health budget is spent on care and treatment of illnesses, government agencies are beginning to give more support to health-promotion initiatives. *Healthy People 2010* (2000) is a report by the U.S. Department of Health and Human Services that provides a framework for national health promotion, health protection, and preventive service strategies. National nursing organizations and individual nurses were involved in its development. This massive document outlines two broad goals for the next decade:

- Increase the quality and years of healthy life
- Eliminate health disparities

The *Healthy People 2010* established a list of leading health indicators to help communities and individuals target areas for improving the health of the individual or community. These indicators are behaviors, physical and social environmental factors, and health system issues that affect the health of communities and individuals. They are as follows:

- Access to quality health services
- Arthritis, osteoporosis, and chronic back conditions
- Cancer
- Chronic kidney disease
- Diabetes
- Disability and secondary conditions
- Educational and community-based programs
- Environmental health
- Family planning
- Food safety
- Health communication
- Heart disease and stroke
- Human immunodeficiency virus (HIV)
- Immunization and infectious diseases
- Injury and violence prevention
- Maternal, infant, and child health
- Medical product safety
- Mental health and mental disorders
- Nutrition and obesity
- Occupational safety and health
- Oral health
- Physical activity and fitness
- Public health infrastructure
- Respiratory disease
- Sexually transmitted disease
- Substance abuse
- Tobacco use
- Vision and hearing

*Healthy People 2010's* Focus Areas

## Pender's Health Promotion Model

Pender's Health Promotion Model is similar to the health beliefs model that is discussed in the section, "Concepts of Health and Illness." However, it explains the likelihood that a person will engage in **health promoting behaviors** rather than health protecting (illness) behaviors. Pender considers **cognitive-perceptual factors** to be the main motivators for engaging in healthy behaviors. These factors include (1) the importance the person places on health, (2) how much control a person feels he/she has over own health, (3) perceived self- efficacy (that is, whether the person believes that he/she can be successful in carrying out the behavior), (4) the person's definition of health, (5) the person's perception of his/her health status, (6) whether the person perceives that there will be benefits from the healthy behaviors, and (7) perceived barriers (e.g., how difficult the person thinks the behaviors/activities will be).

**Modifying factors** are those that affect the cognitive-perceptual factors. They include (1) **demographic factors**, such as age, sex, education, income; (2) **biologic characteristics**, such as body build; (3) **interpersonal characteristics**, such as expectations of significant others; (4) **situational factors**, such as ease or difficulty accessing healthy alternatives such as balanced meals; and (5) **behavioral factors**, such as previous experience, knowledge, and skill. **Cues to action** are those things that make the person aware of the need for action. They may be internal (e.g., personal awareness of the potential for growth) or external (e.g., conversations with others about their health behaviors).

### NANDA Wellness Diagnoses

- Effective Breastfeeding
- Anticipatory Grieving
- Effective Readiness for Enhanced Management of Therapeutic Regimen
- Family Coping
- Health Seeking Behavior (specify)
- Readiness for Enhanced Community Coping
- Readiness for Enhanced Organized Infant Behavior
- Readiness for Enhanced Spiritual Well-being.

(The phrase "Readiness for Enhanced" can be used to create wellness diagnoses from other NANDA labels.)

## Nursing Interventions for Wellness Assessment

### Health history and physical examination

**Physical fitness assessment**: This includes indicators of the body's physical functioning: muscle strength and endurance, flexibility, body composition (e.g., skin fold measurements), and cardiorespiratory endurance.

**Lifestyle assessment**: Lifestyle assessment is meant to help clients assess the impact of their personal lifestyle and habits on their health. Categories assessed include: physical activity, stress management, nutritional practices, smoking, alcohol consumption, and drug use.

**Health risk appraisal (HRA):** These assess a client's risk for disease or injury over the next 10 years by comparing the client's risk with the mortality risk of his/her age, sex, and racial group. A risk factor is a phenomenon (e.g., age or lifestyle behavior) that increases the chance that a person will acquire a specific disease such as cancer or heart disease.

**Health care beliefs:** This includes **locus of control** and other beliefs (see Pender's model).

**Life stress review:** Various instruments have been created to assess the amount of stress present in a person's life.

Most give numerical values to life events such as divorce, job loss, and pregnancy. High scores have been shown to be associated with increased possibility of illness in an individual.

## Nursing Interventions for Health Promotion and Disease Prevention

- Providing and facilitating support (e.g., individual and group counseling, telephone counseling, group support, facilitating social support from family and friends).
- Providing health education
- Enhancing behavior change
- Modeling healthy behaviors

# Critical Thinking Exercise: Health Promotion and Disease Prevention

**Situation**: A 45-year-old male with a family history of cardiovascular disease seeks advice about quitting smoking. The client, who smokes approximately one pack a day, admits that several previous attempts to stop smoking have been unsuccessful. Keep in mind the goals of *Healthy People 2010*, when answering the following questions.

1. In counseling this client, explain the risks associated with tobacco use. For a long-time smoker, what is the benefit of quitting?

2. What are the two-fold goals of Healthy People 2010? How might they apply to the client described?

3. Describe the three levels of prevention—primary, secondary, and tertiary. Provide an example of each.

# WORD SEARCH

Here are 12 terms associated with Healthy People 2010. Can you find them below?

| ARTHRITIS | FOOD SAFETY | NUTRITION |
|-----------|-------------|-----------|
| CANCER | HEARING | OSTEOPOROSIS |
| DIABETES | HIV | STROKE |
| DISABILITY | IMMUNIZATION | VISION |

| | | | | | | | | | | | |
|---|---|---|---|---|---|---|---|---|---|---|---|
| D | I | S | A | B | I | L | I | T | Y | L | R |
| I | M | L | R | P | M | L | H | I | V | N | N |
| A | C | T | T | F | M | P | S | I | H | U | Y |
| B | A | E | H | E | U | I | S | N | H | T | T |
| E | N | N | R | H | N | I | S | T | E | R | E |
| T | C | F | I | P | I | C | I | K | A | I | F |
| E | E | F | T | N | Z | V | O | I | R | T | A |
| S | R | J | I | T | A | R | I | L | I | I | S |
| C | T | O | S | R | T | I | O | O | N | O | D |
| L | E | R | U | S | I | P | U | R | G | N | O |
| S | I | S | O | R | O | P | O | E | T | S | O |
| V | I | S | I | O | N | S | T | L | M | P | F |

# Contemporary Nursing Practice

## Key Points

- Nursing is the diagnosis and treatment of human responses to actual or potential health problems and is both art and science. It involves interpersonal relationships.
- Nurses assume a variety of roles (e.g., teacher, leader, caregiver, advocate).
- The American Nurses Association (ANA) has developed standards of nursing practice. Official journal: *American Journal of Nursing (AJN)*
- The National League for Nursing (NLN) and Commission on Collegiate Nursing Education (CCNE) focus on nursing education and accrediting nursing schools. NLN official journal: *Nursing & Health Care Perspectives*. CCNE official journal: *Journal of Professional Nursing.*
- The National Student Nurses' Association (NSNA) is the official organization for nursing students. Official journal is the *Imprint.*
- Standards of practice provide a framework for evaluating nursing practice.
- There are two types of licensed nurses in the U.S.: the LPN (or LVN) and the RN.
- The 3 major educational routes for becoming an RN are the diploma, associate degree, and baccalaureate degree.
- Nurses participate in research of varying levels, but all nurses should use research findings in their practice.
- **Key Concepts/Terms:** Roles of the nurse, major nursing organizations, standards of practice, nursing education, nursing research

**Overview**

Many definitions of nursing exist. Over 100 years ago, Florence Nightingale said that nursing is "the act of utilizing the environment of the client to assist him in his recovery" (Nightingale, 1860). Virginia Henderson said, "The unique function of the nurse is to assist the individual, sick or well, in the performance of those activities contributing to health or its recovery (or to peaceful death) that he would perform unaided if he had the necessary strength, will, or knowledge, and to do this in such a way as to help him gain independence as rapidly as possible". Two more current definitions state that nursing is: (1) "the diagnosis and treatment of human responses to actual or potential health problems" and (2) "a unique blend of art and science applied within the context of interpersonal relationships for the purpose of promoting wellness, preventing illness, and restoring health in individuals, families, and communities". Theorists agree that nursing is different from medicine because it focuses on the person's physical, psychosocial, cultural, and spiritual responses to illness rather than just focusing on treating the disease.

## Roles of the Nurse

Nurses care for individuals, families, and communities. They are involved in promoting wellness, preventing illness, restoring health, and caring for the dying. In providing care, they assume a variety of roles, including the following:

- Caregiver
- Communicator
- Teacher
- Client advocate
- Counselor
- Change agent
- Leader
- Manager
- Case manager
- Research consumer

## Nursing Organizations

Professional nursing organizations provide leadership for many areas of nursing practice, education, and research. The primary organizations for U.S. nurses are the American Nurses Association, the National League for Nursing, the Commission on Collegiate Nursing Education, and the National Student Nurses Association. There are also many nursing specialty organizations, (e.g. for maternity, emergency department, operating room, pediatrics, and critical care nurses).

**American Nurses Association (ANA):** Founded in 1986, the ANA is the national professional organization for nursing in the United States, which develops standards of nursing practice and promotes the educational and professional advancement of nurses. Its official journal is the *American Journal of Nursing*.

**National League for Nursing (NLN):** Formed in 1952, its purpose is to foster the development and improvement of nursing education. Non-nurses (e.g., consumers) can become members of NLN. The NLN offers voluntary accreditation for educational programs in nursing, as well as preadmission and achievement testing for students. Its annual survey provides a primary source of data about nursing education in the U.S. Its official journal is *Nursing and Health Care Perspectives*.

**Commission on Collegiate Nursing Education (CCNE):** Established in 1996, CCNE was established by a majority vote of member schools of the American Association of Colleges of Nursing. CCNE is the only national accrediting body that focuses exclusively on baccalaureate and master's degree programs. Its official publication is the *Journal of Professional Nursing*.

**International Council of Nurses (ICN):** Established in 1899, the ICN is a federation of national nurses' associations (such as the ANA). Over 100 national nurses' associations are affiliated with ICN, representing almost 1.5 million nurses worldwide. It serves as a voice for nurses and nursing throughout the world. Its official journal is *International Nursing Review*.

**National Student Nurses Association (NSNA):** Formed in 1953 it is the official preprofessional organization for nursing students. It communicates with the ANA and the NLN, but is an autonomous body. Its official journal is *Imprint*.

**Sigma Theta Tau, International Honor Society for Nurses:** Founded in 1922, its purpose is professional rather than social. Students in baccalaureate, master's, doctoral, and postdoctoral programs are eligible to be selected for membership, based on their academic achievement. The official journal is Image: *Journal of Nursing Scholarship.*

## Standards of Practice

To protect the public, nursing practice is regulated legally by state nurse practice. However, standards of practice are set by professional nursing organizations to reflect the values and priorities of the nursing profession, provide direction for nursing practice, define nurses' accountability to the public, and provide a framework for evaluating nursing practice. The ANA has developed "generic" standards of practice that apply to all nurses regardless of area of specialization. Various specialty organizations have developed standards that apply specifically to nursing practice in their areas.

## Nursing Education

State laws in the U.S. recognize two types of nurses: the registered nurse (RN) and the licensed practical or vocational nurse (LPN or LVN). The three major educational routes for becoming an RN are: diploma (hospital-based, usually 3 years), associate degree (community college, 2 years), and baccalaureate programs (college or university, 4 or 5 years). Although nursing programs vary a great deal, all program graduates take the same licensing examination (NCLEX®).

The State Board of Nursing must approve all nursing programs. In addition, program accreditation by NLN or CCNE indicates excellence in nursing education.

## Nursing Research

All nurses should use research findings in practice; however, the level of participation in research depends on the nurse's educational level, position, experience, and employing agency. The Institute for Nursing Research (part of the National Institutes of Health) sets research priorities and recommends funding for nurses conducting research.

# Critical Thinking Exercise: Contemporary Nursing Practice

**Situation**: As a nursing student, you have been asked to address fellow students on the multifaceted roles related to contemporary nursing practice. In preparing your talk, consider the various ways nursing has been defined and the forces that have shaped the profession.

1. In what ways is nursing both an art and a science? Describe a situation from your own clinical experience to support your answer.

2. Nurses assume a variety of roles: from teacher, caregiver, communicator, counselor. Choose one of the many roles and describe how it might be carried out in contemporary practice?

# Theory-Based Practice

## Key Points

- **Nightingale**: Environmental theory, self-healing
- **Peplau**: Interpersonal relations; health as growth in human processes
- **Henderson**: Nursing separate from medicine; well and ill clients; "nursing needs"
- **Rogers**: Human as dynamic energy field; wholeness
- **Neuman**: Systems theory
- **Orem**: Self-care, self-care deficit; nursing systems are wholly or partly compensatory or supportive-educative
- **Roy**: Adaptation model
- **Watson**: Caring is the unifying focus for practice; interventions are "carative factors"
- **Key Concepts/Terms**: Nursing theories, conceptual framework, model

**Overview**

**Concepts** are abstract ideas or mental images of phenomena. They are words or other symbols (e.g., numbers) that evoke mental images and meanings of events, objects, or things. A **theory** is a set of concepts and ideas that explain or describe a given phenomenon (e.g., the theory of gravity). A theory expresses relationships between facts or concepts and may predict future events and relationships. **Nursing theories** describe the relationships among four major concepts: (1) person/client, (2) the environment, (3) health, and (4) nursing. They provide guidance for nursing practice, education and research; and they differentiate nursing from other professions. The terms **theory**, **model**, and **conceptual framework** are often used interchangeably.

## Purposes of Nursing Theories

- Help nurses to describe, explain, and predict everyday experiences
- Guide nursing assessments and interventions
- Help establish criteria to evaluate the quality of nursing care
- Build a common nursing language to use in communicating with other health professionals
- Promote nursing autonomy by defining independent nursing functions
- Guide curriculum decisions in schools of nursing
- Provide a framework for nursing research

### Environmental Theory (Florence Nightingale, 1860)

Nightingale is often referred to as the first nurse theorist. She thought that illness was caused by deficiencies in any of five environmental factors: (1) fresh, pure air, (2) pure water, (3) efficient drainage, (4) cleanliness, and (5) light (especially sunlight). She stressed keeping the client warm, maintaining a quiet environment, and monitoring the client's diet. Her concept of nursing was the provision of optimal environmental conditions to enhance the person's ability for self- healing.

### Interpersonal Relations Model (Hildegard Peplau, 1952, 1963, 1980)

Peplau, a psychiatric nurse, emphasized the use of a therapeutic relationship between the nurse and the client. She focused on the individual, rather than families or communities, and did not specifically talk about the environment. She described the client as a developing organism living in unstable equilibrium and striving to reduce anxiety. Health was more or less equated to growth in human processes.

For Peplau, nursing is one of the human processes needed to make health possible—a human relationship between a sick person with a felt need and a nurse who is educated to recognize and respond to the need.

### Definition of Nursing (Virginia Henderson, 1966)

Henderson's definition was a major influence in the emergence of nursing as a discipline separate from medicine. She described nursing in relation to the client and the environment; however, she saw the nurse as concerned with both well and ill persons (unlike Nightingale); and she mentioned the teaching and advocacy roles of the nurse. She conceptualized the nurse's role as assisting individuals to gain independence in meeting the following fundamental needs:

- Breathing
- Eating and drinking
- Eliminating body wastes
- Moving and maintaining a desirable position
- Selecting suitable clothing
- Maintaining body temperature within normal range
- Keeping the body clean and well- groomed to protect the skin
- Avoiding environmental dangers and not injuring others
- Communicating with others
- Worshiping according to one's faith
- Feeling a sense of accomplishment from one's work
- Playing or participating in recreation
- Learning, discovering, satisfying the curiosity, and using available health facilities
- Sleeping and resting

### Science of Unitary Human Beings (Martha Rogers, 1970)

This theory contains complex concepts from various disciplines (e.g., Von Bertalanffy's general systems theory, Einstein's theory of relativity). Rogers views the person (client) as a "whole" which is greater than the sum of all its parts. She sees humans as dynamic energy fields in continuous exchange with environmental energy fields. Nurses applying this theory would (1) focus on the person's wholeness, (2) promote/coordinate the human and environmental energy fields, and (3) direct and redirect patterns of interaction between the two energy fields to promote maximum health. Nurses' use of noncontact therapeutic touch is based on the idea of human energy fields.

### Neuman's Systems Model (Betty Neuman, 1974)

This model is based on the client's relationship to stress, the reaction to it, and the state of adaptation to stressors. Nursing interventions focus on retaining system stability and strengthening lines of defense and resistance.

### General Theory of Nursing (Dorothea Orem, 1971)

Orem's theory is based on three related concepts: self-care, self-care deficit, and nursing systems. **Self-care** refers to those activities a person performs independently to promote and maintain well-being. If the person cannot perform such activities, **a self-care agent** (e.g., a nurse or a parent) provides the care. **Self-care requisites** are actions taken to provide self-care (e.g., maintaining intake and elimination of air, water, and food; adjusting to loss of a spouse; seeking health care assistance when ill or injured). **Self-care deficit** occurs when the person is unable to meet self-care needs—this is when nursing is needed. Nurses help by: acting or doing for, guiding, teaching, supporting, and providing a supportive environment. **Wholly compensatory** systems are needed for clients who cannot control and monitor their environment or process information; **partly compensatory** systems are needed for clients who are able to perform some, but not all, self-care activities; **supportive-educative (developmental)** systems are meant for persons who need assistance in learning to perform self-care measures.

### Adaptation Model (Sister Callisat Roy, 1976)

Roy defines **adaptation** as a process and outcome in which a thinking/feeling person uses conscious awareness and choice to create human and environmental adaptation. The individual is a biopsychosocial adaptive system that uses a feedback cycle of input, throughput, and output. Both the person and the environment are sources of stimuli; both require modification to promote adaptation. Individuals respond to stimuli in one or more of four modes: (1) physiologic, (2) self-concept, (3) role function, and (4) interdependence. Nurses support the client's ability to enhance life processes through adaptation in the four adaptive modes.

### Human Caring Theory (Jean Watson, 1979)

Watson believes that caring is central to nursing and that it is the unifying focus for practice. Nursing interventions related to care are called **carative factors**. They are:

- Forming a humanistic-altruistic system of values
- Instilling faith and hope
- Cultivating sensitivity to one's self and to others
- Developing a helping-trust relationship
- Promoting and accepting the expression of positive and negative feelings
- Systematically using the scientific problem-solving method for decision making
- Promoting interpersonal teaching and learning
- Providing a supportive, protective, or corrective mental, physical, sociocultural, and spiritual environment

- Assisting with the gratification of human needs
- Allowing for existential-phenomenologic forces

# Critical Thinking Exercise: Theory-Based Practice

**Situation**: A classmate expresses difficulty in understanding the relationship between theoretical concepts and actual nursing practice. Before responding, consider the purpose of nursing theories.

1. Nursing theory is intended to guide practice. Choose one of the theories presented in this chapter and describe how it is reflected in your own practice or in a clinical setting you have observed.

2. Write a diary or journal entry based on a particularly memorable incident from your own clinical experience. Be sure to include elements that reflect how theory affects actual nursing practice.

# Values and Ethics

## Key Points

- **Moral principles**: Autonomy, nonmaleficence, beneficence, justice, fidelity, veracity
- **Consequence-based theories**: Action was right if it produced "good" results
- **Principles-based theories**: Action is moral if done according to an impartial principle, rights, duties, and obligations.
- **Relationships-based (caring) theories**: Actions are good that nurture and maintain relationships and the welfare of the group.
- *The ANA Code of Ethics for Nurses With Interpretive Statements* is a formal statement of values that are shared by nurses. A code of ethics provides guidance but does not guarantee that an ethical decision will be made.
- Nurses frequently encounter ethical problems in their practice (e.g., cost-containment issues that jeopardize client welfare, confidentiality, and end-of-life decisions).
- **Key Concepts/Terms**: Terminology, moral framework, moral principles, values, nursing code of ethics, ethics committee

**Overview**

In their day-to-day work, nurses often encounter ethical problems/situations in which the question is not, "What should be done?" but "Should I do this?" Moral problems are created as a result of technological advances, societal changes, and nurses' conflicting obligations to clients, families, employers, primary care providers, and other nurses. Sometimes an ethical problem exists because the nurse is not sure what is "right." In other instances, the nurse may feel sure of the right action, but cannot carry out that action without great personal risk (e.g., of losing her/his job).

When moral problems arise, the nurse's input is important. However, many people are usually involved in making an ethical decision. Therefore, collaboration, communication, and compromise are important skills. The goal of ethical reasoning in nursing is to reach a mutual, peaceful agreement that is in the best interests of the client. Nurses must maintain a nonjudgmental attitude, be honest, respect and support client decisions, and protect the client's privacy and confidentiality.

**Terminology**

**Morality**: Usually refers to private, personal standards of what is right and wrong. Moral issues involve important social values and norms, not trivial things. They can be recognized by an aroused conscience or feelings such as guilt, hope, or shame; or the tendency to respond to the situation with words such as ought, should, right, wrong, good, and bad. Morality and ethics are used interchangeably by lay people, although philosophers distinguish between them.

**Ethics:** Refers to: (1) a method of inquiry about the rightness/wrongness of human actions, (2) the practices or beliefs of a group (e.g., nursing ethics), and (3) the standards of moral behavior described in the group's formal code of ethics

**Bioethics:** Refers to ethics as applied to "life" (e.g., decisions about euthanasia, prolonging life, abortion)

**Nursing ethics:** Refers to ethical issues that occur in the practice of nursing—to questions that nurses must answer with regard to their own actions

**Values:** Freely chosen, long-lasting beliefs or attitudes about the worth of something (e.g., a person or an idea)

**Value system:** People organize a personal set of values (usually, unconsciously) on a continuum from most to least important. This forms their value system, which gives direction to their life and provides the basis for decisions and choices.

### Model Framework (Theories)

Nurses use moral theories to develop explanations for their ethical decisions and in discussing problem situations with others. Moral theories can be differentiated by their emphasis on either consequences of an action, principles and duties, or relationships.

**Consequence-based (teleological) theories:** Judge the "rightness" of an action based on the consequences it produces. Utilitarianism (one consequentialist theory) views a good act as one that brings the least harm and the most good to the most people.

**Principles-based (deontological) theories:** Emphasize individual rights, duties, and obligations. They judge an action independently of its results. An action is moral if it is done according to an impartial, objective principle.

**Relationships-based (caring) theories:** Stress courage, generosity, commitment, and the need to nurture and maintain relationships. Caring theories promote the common good, or the welfare of the group, rather than stressing individual rights.

A moral framework guides moral decisions, but does not determine the outcome of a decision. Two nurses using the same framework might make entirely different decisions about a situation.

## Moral Principles

Moral principles are statements about broad, general concepts such as autonomy and justice. Even though people disagree about the right action to take, they may be able to agree on the principles that apply—this can be the basis for discussion and an acceptable compromise. For example, most people would agree on the principle that nurses should tell the truth to clients (veracity), even if they disagree as to whether the nurse should deceive a particular client about his or her prognosis. Some basic principles are:

**Autonomy:** The right to make one's own decisions; the right to choose personal goals. This means that nurses respect a client's right to make decisions even when those choices seem not to be in the client's best interest. It also means treating clients with consideration and respect.

**Nonmaleficence:** The duty to do no harm. Harm includes "risk of harm," which is not always clear. For example, a nursing intervention that is meant to be helpful may have a harmful side effect.

**Beneficence**: "Doing well." Nurses are obligated to implement actions that benefit clients and families.

**Justice**: `ns that call for a sense of fairness. For example, they must weigh the facts carefully in order to divide their time fairly among the clients they care for during a workday.

**Fidelity**: Keeping promises and agreements

**Veracity**: Telling the truth

Although principles are important, they may conflict sometimes. For example telling the truth (veracity) to a client may cause the client to become anxious and hypertensive (violating the principle of nonmaleficence).

## Values

Values are important in nursing ethics because they influence both clients' and nurses' decisions and actions. Even though values may be unspoken, and even unconscious, they underlie all moral decisions. Values are learned through observation and experience of family, peer, cultural, ethnic, professional, and religious groups. Nurses acquire **professional values** from socialization in nursing school, from codes of ethics, and from nursing experiences.

According to the American Association of Colleges of Nursing, the essential values of a professional nurse are:

- Aesthetics (e.g., appreciation, creativity, sensitivity)
- Altruism (e.g., caring, commitment)
- Equality
- Freedom
- Human dignity
- Justice
- Truth

### Nursing Codes of Ethics and Standards of Professional Performance

A code of ethics is a formal statement of the ethical values and principles that are shared by members of a group. It provides a standard for professional actions. Nursing codes of ethics include the *International Council of Nurses' ICN Code of Ethics for Nurses* (2000) and the *ANA Code of Ethics for Nurses with Interpretative Statements* (2001). The present code reflects a commitment to the principles of respect for autonomy, beneficence, nonmaleficence, justice, veracity, fidelity, and confidentiality. Codes of ethics are not legally enforceable, and they are only as effective as are the nurses who use them. They are general, so even though they provide guidance for ethical decision-making, they do not guarantee the "right" answer in any given situation.

The *ANA Standards of Clinical Nursing Practice* now include standards of professional performance as well as standards of care. The fifth standard of professional performance states, "Ethics: the nurse's decisions and actions on behalf of clients are determined in an ethical manner."

## Ethics Committees

Many institutions have ethics committees that provide education, policy making, case review, and consultation. Usually they are multidisciplinary and provide a forum for widely differing views. Nurses' input into these committees is important, since they are often the ones with the most intimate knowledge of the client.

## Specific Bioethics Issues Involving Nurses

The ANA Center for Ethics and Human Rights has identified the following as some of the ethical problems that nurses encounter:

- Quality of life
- End-of-life care
- Human genetics
- Informed consent and advanced directives
- Privacy and confidentiality (e.g., computerized information management)
- Resource allocation and cost-containment issues that jeopardize client welfare and access to health care

# Critical Thinking Exercise: Values and Ethics

**Situation:** On an initial visit to a new client, a hospice nurse is greeted by family members with a barrage of questions ranging from advance directives to end-of-life care. The following questions relate to this and similar situations.

1. What is an advance directive? When is it implemented?

2. In many situations, a nurse cares for both client and family. How might it be possible to resolve conflicts that arise?

3. What ethical principles apply to nurses in caring for clients and their families? For example, what does the principle of veracity require of a nurse?

# Legal Responsibilities

## Key Points

- Nurses are legally responsible for nursing care.
- Satisfied clients usually do not sue.
- Knowledge of the law helps to maintain standards of practice and to protect the nurse from liability.
- Nursing practice is regulated primarily at the state level through licensing, nurse practice acts (legislation), and accreditation.
- Negligence is practice (1) below the expected standard, which (2) places the client at risk for harm.
- Malpractice is negligence by a person performing as a professional and must be "proven".
- Intentional torts include assault (threatening to touch), battery (touching), and false imprisonment.
- Informed consent is voluntary consent made by an informed, competent client of legal age.
- Areas of potential liability for the nurse include: breach of confidentiality, invasion of privacy, loss of client property, unprofessional conduct, primary care providers' orders, record keeping, incident reports, giving aid to someone in distress (e.g., at the scene of an accident).
- Good Samaritan laws are designed to protect health care workers who give emergency assistance. Providing aid is more of a moral than a legal duty.
- **Key Concepts/Terms:** Functions of law, sources of law, types of law, torts, potential liability, regulation of nursing practice

**Overview**

Nurses are legally responsible and accountable for assessing, diagnosing, planning, implementing, and evaluating nursing care. Knowledge of nursing law is important to ensure that the nurse's actions follow legal guidelines and to protect the nurse from liability. The best way to avoid legal conflicts is to practice within the legally defined scope of practice, identify and develop strategies to prevent potential liabilities in your practice, and develop good nurse-client relationships. Satisfied clients usually do not sue. Nurses may need to testify in a legal action for various reasons. The nurse may be the defendant in a lawsuit or may have provided care to a client who has filed a lawsuit. The nurse should seek the advice of an attorney before testifying.

## Functions of the Law in Nursing

- By making nurses accountable, it helps to maintain a standard of nursing practice.
- It differentiates nurses' responsibilities from those of other health care professionals.
- It specifies which nursing actions are legal in caring for clients.

## Sources of Law

**Constitution**: The Constitution is the supreme law of the United States. It creates legal rights and responsibilities and is the foundation for a system of justice.

**Legislation (statutes)**: These are laws enacted by federal, state, or other legislative bodies (e.g., the U. S. Senate). When federal and state laws conflict, federal law prevails. Regulation of nursing is provided by state law, that is, the Nurse Practice Acts.

**Common law**: These are laws that evolve from court decisions. Common law is often used to resolve disputes between two parties, as in a lawsuit.

## Types of Law

**Public law**: Is the body of law that deals with relationships between individuals and the government.

**Criminal law**: Is a type of public law, that deals with the safety and welfare of the public (e.g., homicide, rape, and theft).

**Private (civil law)**: The body of law that deals with relationships between private persons.

**Contract law**: Enforces agreements. (If a contract is written, it is considered expressed. An implied contract is a verbal or an understood responsibility in a specific relationship, involving services one can expect to receive from an individual in a job or role.)

**Tort law**: Defines duties and rights among individuals that are not based on contracts. Examples of tort law relevant to nurses are negligence, malpractice, invasion of privacy, and assault and battery.

## Unintentional Torts

**Negligence**: Practice that is below the standard expected of an ordinary, reasonable, and prudent practitioner, placing another person at risk for harm.

**Gross negligence**: Extreme lack of knowledge or skill that the practitioner clearly should have known would place another at risk.

**Malpractice**: Negligence that occurs while a person is performing as a professional. In order for malpractice to be proven, the following four elements must be present:

- The nurse has a relationship with the client that involves providing care (e.g., has been assigned to care for the client in the hospital).
- The nurse did not observe a standard of care that is expected in the specific situation.
- The client sustained harm, injury, or damage.
- The harm must have occurred as a direct result of the nurse's failure to observe the standard, and the nurse should have known that such failure could result in harm.

## Intentional Torts

**Assault:** An attempt or threat to touch another person without justification (e.g., a nurse who threatens, "If you don't eat, we'll have to force feed you.")

**Battery:** Intentional touching of a person or something the person is holding or wearing. To qualify as battery, the touching must be wrong in some way; for example, embarrassing, causing injury, or done without permission (e.g., continuing to bathe a client after the client has said, "Stop.").

**Informed consent:** This implies that (1) consent was voluntary (no coercion was involved), (2) the client was competent and of legal age, and (3) the client had enough information with which to make an informed choice.

**False imprisonment:** Restraining or detaining another person against his or her will. For example, a nurse cannot detain a client who insists upon leaving the hospital against medical advice. The client has a right to leave even though it may be harmful to him or her. This does not apply if the client is incompetent; instead of false imprisonment, then, it is a matter of protecting the client from injury.

## Other Areas of Potential Liability

- **Invasion of privacy:** Nurses may be liable if they breach confidentiality or intrude into a client's private domain (e.g., by taking photographs without permission). Clients have the right to privacy, which means that no client or his/her case should be discussed in public or with anyone other than those that "need to know" the information to care for the client.

- **Loss of client property**

- **Unprofessional conduct:** (e.g., falsification of client records, illegal use of controlled substances)

- **Good Samaritan:** Laws designed to protect health care workers who give assistance in an emergency. In the U.S., most state laws do not require a person to give aid to someone in distress—it is considered more of a moral than a legal duty. Most states do have legislation protecting a "Good Samaritan" from legal liability for injuries caused under these circumstances. Nurses who choose to render emergency care should (1) limit their actions to first aid, if possible, (2) not perform actions with which they are unfamiliar, (3) offer help, but do not insist, and (4) stay at the scene until the injured party leaves or until you are relieved by another qualified person.

- **Primary care providers' orders:** Nurses are expected to ask for clarification of ambiguous or seemingly erroneous orders. Otherwise, they are expected to carry out the order exactly as written. If the order reads "50 mg Demerol," for example, the nurse should not administer 25 or 75 mg. Nurses should question any order:

  - That the client questions (e.g., "I usually get only one pill").
  - If the client's condition has changed.
  - That is a verbal order. Read it back to the primary care provider and document that the primary care provider confirmed it as read.
  - That is illegible, unclear, or incomplete.

- **Record keeping:** The medical record is a legal document. Nurses must keep accurate and complete records of care provided and client responses.

Documentation should be objective, non-judgmental, include specific information (measurement in inches, cm, lbs or kg, time), timed, dated, and initialed by the person completing the charting. The client record (chart) is a legal document and should be treated as though it will be used as such. The nurse should never document any action until it has been completed. He/she should document according to the professional standards of nursing, following the institution's guidelines, and should limit redundant documentation (to limit mistakes). Documentation is legal protection for the nurse. Generally, if something has not been documented in the client's record it is considered to have not been done.

- **Incident reports**: This is an agency record of an unusual occurrence (e.g., a client fall, a medication error). They are used to make decisions about risk management and quality control. Incident reports should be completed by the person who identifies the incident as soon as possible, and filed according to agency policy.

- **Against Medical Advice (AMA)**: AMA is the term used when a client decides to leave the hospital without a primary care provider's order. The nurse should report a client's decision to leave the hospital to the physician and nursing supervisor promptly. The nurse should be prepared to offer education on the health, legal, and insurance risks to the client, family, and significant other(s) in the event the client leaves the hospital AMA.

## Regulation of Nursing Practice

Nursing practice is regulated at the state level primarily through licensing and nurse practice acts.

**Licenses**: These are legal permits that are granted by a government agency (the State Board of Nursing, in the case of nursing) to allow an individual to engage in the practice of a profession and use a particular title (e.g., registered nurse, licensed practical nurse, primary care provider). In the U.S., nurses must be licensed in order to practice (**mandatory licensure**). A nurse's license can be revoked for just cause (e.g., incompetent practice or conviction of a crime).

**Nurse practice acts** are specific to the state in which a nurse is licensed and practices. Nurse practice acts do several things and nurses in each state should be familiar with that state's provisions. Although they differ from state to state, each nurse practice act:

- Defines the scope of practice for the licensed nurse.
- Protects the practice of nursing exclusively to the licensed nurse.
- Protects the public by limiting the practice of nursing to the licensed nurse.
- Legally controls the practice of nursing through required licensing.

**Accreditation** is a process by which the state board of nursing or other private organization (e.g., the National League for Nursing or Commission on Collegiate Nursing Education) evaluates and approves educational programs or services that meet predetermined criteria. All schools of nursing in the U.S. must be accredited by the state board of nursing.

# Critical Thinking Exercise: Legal Responsibilities

**Situation:** A nurse who is present at the scene of an automobile accident provides assistance until emergency aid arrives. The following questions relate to nurses' responsibilities under the law.

1. What law governs a nurse's actions in a situation like this? With respect to the obligation to offer assistance, is it a legal one or a moral one? Explain by citing the law that applies to this situation.

2. What is the source of law that governs nursing practice?

3. Legal responsibility is intended to promote a high standard of nursing practice. Explain how it does this. Consider one area of focus, such as the right to privacy, professional conduct, or documentation, when giving your response.

# Concepts of Health and Illness

| Key Points |
| --- |

- Health is a "state of complete physical, mental, and social well-being, and not merely the absence of disease or infirmity" (World Health Organization, 2002).
- Wellness includes physical, emotional, social, intellectual, and spiritual well-being.
- The health belief model attempts to predict a person's **motivation** to engage in preventive behaviors, based on: perceptions (e.g., of the seriousness of the disease) and modifying factors (e.g., social pressure, availability of care).
- The health locus of control model refers to whether clients believe that their health is under their own control (internal LOC) or the control of others (external LOC).
- Disease is an alteration in body functions, causing reduced capacities or shortened life span.
- Illness is the person's perception of decreased functioning: it may or may not be related to disease.
- Acute illness is < 6 months; chronic illness > 6 months with remissions and exacerbations.
- Illness behavior is a coping mechanism. In general, sick people are excused from social roles and tasks, but are expected to seek help and get well as soon as possible.
- **Key Concepts/Terms**: Models of health, health belief models, illness terminology, illness behaviors.

**Overview**

According to the World Health Organization (WHO), **health** is "a state of complete physical, mental, and social well-being, and not merely the absence of disease or infirmity". Others believe that this is not realistic because "complete well- being" is not possible to achieve. Health is sometimes defined in terms of ability to perform role functions and as a creative process in which individuals are actively and continually adapting to their environments. The ANA defines health as "a dynamic state of being in which the developmental and behavioral potential of an individual is realized to the fullest extent possible". Most people describe health as being active and able to do what they want and need to do, being in good spirits, and being as free as possible from disease and pain.

**Wellness**: Maximizing personal potential and state of well-being. It involves attitudes and behaviors that enhance quality of life. **Well-being** is a subjective perception of vitality and feeling well. To realize optimal wellness, people must deal with the following dimensions:

**Physical**: The ability to carry out daily tasks, maintain adequate nutrition, achieve fitness, avoid drugs and alcohol, and practice a positive lifestyle

**Emotional**: Management of stress, expression of emotions and acceptance of personal limitations

**Social**: Interactions with people

**Intellectual**: Ability to learn and use information effectively; striving for continued growth; dealing with new challenges

**Spiritual**: The belief in a higher power that provides meaning and purpose to life

## Models of Health

Models of health are attempts to explain health, and in some instances, its relationship to illness and injury.

**Clinical Model**: In this narrow interpretation, health is the absence of signs and symptoms of disease or injury—the state of "not being sick."

**Role Performance Model**: Health is defined in terms of the person's ability to work and perform societal roles, even if the person appears clinically ill.

**Adaptive Model**: Health is defined as a creative process; disease is a failure to adapt. The aim of treatment is to help the person cope (adapt).

**Eudaemonistic Model**: Health is defined as a condition of actualization or realization of potential. Illness is a condition that prevents self-actualization.

## Health Belief Models

These are theories that are used to help determine how likely an individual is to participate in health-promotion and disease-prevention activities.

**Health Locus of Control (LOC) Model**: Locus of control refers to whether clients believe that their health status is under their own control or the control of others. Those with **internal LOC** believe that health is mostly self-determined, and are more likely to take the initiative for their health care and to be knowledgeable about it. People with **external LOC** believe that their health is controlled mostly by outside forces (e.g., chance or the primary care provider). They may need help to shift to an internal locus of control in order to achieve high-level wellness.

**Health Belief Models**: This model is based on motivational theory. It was intended to predict which individuals would or would not use preventive measures (e.g., screening for cancer). The following factors determine a person's motivation to use preventive measures:

- **Individual perceptions** include the person's **perception of susceptibility** (the likelihood that he will get the disease), **perceived seriousness** (whether the person thinks the disease will cause serious consequences or death), and **perceived threat** (perceived susceptibility and perceived seriousness combine to determine the total threat perceived by the individual).

- **Modifying factors** are factors that modify a person's perceptions about susceptibility and seriousness of disease. They include **demographic** variables (such as age, sex, and race), **sociopsychologic** variables (e.g., social pressure not to drive after drinking), **structural** variables (e.g., knowledge about the disease), and **cues to action** (e.g., feelings of fatigue, magazine articles, advice from others).

- **Likelihood of action** is the probability that a person will take preventive health action, and it depends on how he or she weighs the perceived benefits of the action against the perceived barriers to the action. **Perceived benefits** are such things as exercising and eating low-calorie foods because the person believes he/she will benefit (by losing weight). **Perceived barriers** are such things as cost, inconvenience, lifestyle changes, and unpleasantness (e.g., a colorectal examination).

## Illness Terminology

**Illness** is a state in which the person perceives physical, emotional, intellectual, social, developmental, or spiritual functioning to be decreased. It may or may not be related to disease. A person can have a disease (e.g., high blood pressure) and not feel ill. Or a person can feel discomfort and yet have no identifiable disease process. Illness is highly subjective.

**Disease** is an alteration in body functions which causes reduced capacities or a shortened life span.

**Acute illness** is characterized by severe symptoms and relatively short duration (e.g., colds, appendicitis, fractures).

**Chronic illness** lasts for an extended period (6 months or longer). It usually has a slow onset and periods of remission (symptoms disappear) and exacerbation (symptoms reappear). Examples are osteoarthritis and diabetes mellitus.

## Illness Behaviors

Illness behavior is a coping mechanism. It involves ways that individuals describe, monitor and interpret their symptoms, how they use the health care system, and what remedies they use. When people are sick they are:

- Not held responsible for their condition
- Excused from some social roles and tasks
- Obliged to try to get well as soon as possible
- Expected to seek competent help

## Five "Stages" of Illness

**Symptom experiences:** This is when the person begins to believe something is wrong (e.g., they feel unwell or get a rash). Usually they consult with support persons and they may try home remedies.

**Assumption of the sick role:** When self-management fails, the person seeks confirmation from family and friends, but may continue with self-treatment. He or she is often excused from normal role expectations.

**Medical care contact:** When symptoms persist, the person seeks the advice of a health professional, seeking validation that the illness is real, explanation of the symptoms, and prediction of the outcome.

**Dependent client role:** The client becomes dependent on the professional for help with the illness, even though maintaining varying control over his/her own life. People vary greatly in their ability to accept dependence and their need for independence.

**Recovery or rehabilitation:** During this stage the client is expected to gradually become independent and resume former roles and duties.

# Health Promotion and Disease Prevention Measures

**Healthy pregnancy**: Pregnant women can avoid infant disorders with early prenatal care, avoidance of tobacco and alcohol, and by taking dietary supplements of folic acid.

**Childhood injuries and illnesses**: Follow immunization programs and attend parenting classes.

**Dental health**: Regular preventive dental care is essential for all people.

**Vision care**: Preventive measures include regular check-ups for early detection of glaucoma and cataracts, and teaching proper eye hygiene.

**Heart health**: To prevent heart attack, a low-fat diet, get regular exercise, maintain ideal body weight, avoid tobacco, manage stress, blood pressure and cholesterol levels.

**Smoking**: Smoking cessation programs stress reasons to quit smoking and conditions aggravated or caused by tobacco use.

**Diabetes**: Early detection is the key to the prevention of related disorders and death.

**Hypertension**: Manage hypertension by increasing physical activity, reducing weight, limiting alcohol intake, and not adding extra salt to food.

**Cholesterol**: Manage cholesterol by eating low-fat foods and increasing activity. For every 10% reduction in cholesterol, heart disease risk is decreased by 20%.

**Cancer**: Follow a low-fat diet with adequate fiber. Adhere to screening recommendations for specific cancers.

**Alcohol and drugs**: Programs are available to detect and to treat abuse, and to educate children and adults.

**Mental health**: Manage stress. Recognize symptoms of depression, anxiety, panic disorder, and seasonal affective disorder.

**Regular physical examination and self-examination**: Adhere to physical guideline recommendations and maintain immunizations.

# Critical Thinking Exercise: Concepts of Health and Illness

**Situation**: As part of his cardiac rehabilitation, a 58-year-old male recovering from an acute myocardial infarction quits smoking and modifies his diet but is reluctant to adopt a program of moderate exercise. Overall, does this client exhibit an internal or an external locus of control?

1. Describe an individual with an internal locus of control regarding health beliefs. How would this individual behave differently from someone with an external locus of control? How might circumstances like a hospitalization affect one's locus of control?

2. Compare the definition of health offered by the World Health Organization with that of the American Nurses Association. Which best reflects your own definition of health? Draw on your own observations and experiences to support your response.

3. Distinguish disease from illness. In what ways might it be possible to have a disease without having an illness?

# Caring for Individuals, Families and Communities

| Key Points |
| --- |

- Homeostasis is maintaining equilibrium and balance during times of change.
- Family-centered nursing considers the health of the family.
- "Family" is the basic unit of society and is defined as, one or more individuals closely related by blood, marriage, or friendship.
- Family types include nuclear, extended, traditional, single-parent, blended, cohabiting.
- Needs theories include a focus on the individual (e.g., Maslow).
- Developmental stage theories focus on individuals and families.
- General systems theories focus on individuals, families, and communities.
- Structural-functional theories primarily focus on family structure and function.
- Community includes a group of people who share some attribute of their lives.
- Community health assessment includes physical environment, education, safety, transportation, politics, health services, communication, economics, and recreation.
- NANDA has nursing diagnoses appropriate for describing community health status.
- **Key Concepts/Terms:** Individual concepts, homeostasis, family concepts, theoretical framework, community health concepts

## Overview

Nurses care for individuals, families, and communities. Even when caring for an individual, it is important to remember that the person lives in a family and a community, which influence health and health behaviors. Therefore, nurses need to understand concepts related to all three types of clients.

## Concepts Related to Individuals

Each person is unique—with a different genetic makeup, life experiences, and environment. Nevertheless, all people have a common core of sameness, which allows nurses to make some general, tentative predictions about kinds of reactions to expect. For example, everyone feels pain and everyone has the same homeostatic responses, such as a pulse increase during anxiety. When caring for a client, the nurse should consider all the principles that apply to any client of that age and condition (e.g., all postoperative clients are at risk for infection); the nurse must also learn about the client as an individual and modify the general approach as needed (e.g., when hydrating a client, it is important to know what kinds of fluids he prefers). The following are important concepts in the care of individuals:

**Homeostasis** is the tendency of the body (person) to maintain a state of equilibrium (balance) while continually changing. This concept views the client as separate from the external environment and constantly striving to adapt to it.

**Physiologic homeostasis** refers to the stability of the internal environment of the body. Self- regulation is the automatic use of homeostatic mechanisms in a healthy person. When the person is ill, the mechanisms are compensatory. That is, they counteract abnormal conditions for the person. For example, when temperature drops, peripheral blood vessels constrict to divert blood internally.

**Psychologic homeostasis** refers to emotional balance or mental well-being. Needs for love, security, and self-esteem must be met in order to maintain psychologic homeostasis.

## Concepts Related to Families

**Family-centered nursing** considers the health of the family as a whole in addition to the health of individual family members. The **family**, a basic unit of society, is composed of one or more individuals closely related by blood, marriage, or friendship. A **nuclear family** consists of parents and their children. An **extended family** is composed of relatives of the nuclear family (e.g., grandparents, aunts, cousins).

## Additional Types of Families

**Nuclear**: a two generation unit consisting of husband, wife, and their immediate children (biologic, adopted, or both)

**Dual-career**: a nuclear family in which both parents work outside the home

**Nuclear dyad**: a married couple that lives together without children

**Extended family**: nuclear family and other related persons, such as grandparents, aunts, uncles and cousins

**Single-parent family**: an adult head of household with dependent children; the adult may be single as a result of separation, divorce, death, or never being married

**Binuclear family**: a separation or divorce of the adult person occurs, but both parents continue to assume a high level of childrearing responsibilities

**Reconstituted family**: formerly called a blended family, where existing family units join together to form a new family unit

**Alternative families**: include the following:

- **Cohabitation**: unmarried individuals in a committed partnership live together with or without children

- **Gay or lesbian family**: intimate partners of the same sex may live together, with or without children

- **Communal family**: several people living together who share financial resources, work, and child care

- **Foster family**: children live in a temporary arrangement with paid caregivers

## Assessing Family Health

The purpose of a family assessment is to provide an overview of the family process and help the nurse identify areas that need further assessment. It includes data about family structure, roles and functions, interaction/communication patterns, values, and coping resources, as well as about the physical health status of each family member.

### Family Functions

- Providing for physical health
- Providing for mental health
- Socializing members
- Reproducing
- Providing economic well-being

### Family Tasks

- Provision for physical needs: shelter, food, clothing, safety, and health care
- Allocation of resources: planning and use of family money, material goods, space and abilities
- Division of labor: assigning the workload
- Socialization: guiding towards acceptable standards of living
- Reproduction, recruitment, and release: bearing children, adding new members, and allowing members to leave
- Maintenance of order: providing interaction and communication opportunities
- Assistance with fitting into the larger society: community, school, and spiritual center
- Maintenance of motivation and morale: recognition, affection, encouragement, and family loyalty

## NANDA Nursing Diagnoses for Families

Interrupted Family Processes:

- Impaired Parenting
- Caregiver Role Strain
- Family Coping: Readiness for Enhanced
- Home Maintenance: Impaired
- Family Coping: Disabling
- Family Coping: Compromised

## Theoretical Frameworks for Individuals, Families, and Committees

**Needs theories** are useful for promoting the health of individuals. The best known is hierarchy of needs, which ranks human needs on five levels (in ascending order): physiological, safety and security, love and belonging, and self-actualization needs.

**Kalish** ranks needs on six levels, suggesting the category of "stimulation needs" to fall between the physiologic and safety/security needs. "Stimulation" includes sex, activity, novelty, and exploration.

**Needs theories** state that all people have the same basic needs, but that an individual's needs are shaped by that person's culture. For example, individual achievement may be important in one culture but not another.

**Developmental stage theories** are also useful in caring for individuals. They categorize a person's behaviors or tasks into age ranges or stages. A stage describes the characteristics associated with the majority of individuals at periods when developmental changes occur. Stage theories emphasize a predictable sequence of development that is orderly and continuous. Developmental theories have also been formulated for families.

**General systems theories** explain the breaking of whole entities into parts and describe the working together of those parts, explaining the relationship between wholes and parts (e.g., the relationship of cellular metabolism to the person's weight). Individuals, family units, and communities can be viewed as systems. Most are open systems. Some important systems terminology follows:

- A **system** is a set of interacting parts/components. It can be an individual, a family, or a community. Fundamental components of a system are matter, energy, and communication.

- The **boundary** of a system is a real or imaginary line that differentiates one system from another or from the environment (e.g., the skin separates a human being from the external environment).

- A **closed system** does not exchange energy, matter, or information with its environment (e.g., a chemical reaction in a test tube). It receives no input and gives no output.

- An **open system** moves energy, matter, and information in and out through the system boundary. All living systems (e.g., plants) are open systems and in a constant state of change.

**Structural-functional theories** focus on family structure and function. They describe the membership of the family and relationships among the members. They also examine the effects of relationships between various members on the functioning of the whole family. The main functions of a family are to develop a sense of family purpose and affiliation, to add and socialize new members, and to provide and distribute care and services to family members. A healthy family functions in harmony, working toward shared goals.

## Concepts Related to Community Health

A **community** is a collection of people who share some attribute of their lives. It can also be defined as a social system in which the members interact and form networks that benefit all within the community.

**Functions of a community** are to:

- Produce, distribute, and consume goods
- Socialize members
- Exert social control
- Promote social inter-participation
- Provide mutual support

**Community health nursing** can be defined as:

- Applying nursing and public health practice to the health of populations/groups
- Nursing in the community (e.g., home health care)

- **Community health assessment** includes information about the following:
  - The physical environment (e.g., population density, crime statistics)
  - Education (e.g., facilities, programs)
  - Safety and transportation (e.g., fire, police, air quality)
  - Politics and government (e.g., type of government, influential people)
  - Health and social services (e.g., existing health care facilities, immunization levels)
  - Communication (e.g., newspapers available)
  - Economics (e.g., industries and occupations, unemployment)
  - Recreation (e.g., theaters, playgrounds)

**Sources of data for community assessment** include: city maps, census data, employment statistics, health departments, city or regional health planning boards, telephone books, libraries, health facility administrators, recreational directors, police, teachers and school nurses, local newspapers, and on-line computer services.

## NANDA Nursing Diagnoses for Communities

- Ineffective Community Coping
- Ineffective Community Management of Therapeutic Regimen
- Potential for Enhanced Community Coping

# Critical Thinking Exercise: Caring for Individuals, Families and Communities

**Situation:** A community health nurse coordinates the care of a 75-year-old woman following hip replacement surgery. The client lives with her daughter and son-in-law, both of whom work full time and who admit to difficulty in coping with the care of an elderly parent. The following questions apply to the nursing care of individuals, families, and communities.

1. In caring for individuals, why is it important to include the family and the community in your assessment?

2. Select a nursing diagnosis that applies to the family described above. Based on this diagnosis, how might a nurse assist individuals and families in a similar situation?

3. Define homeostasis. Give an example of an individual's homeostatic response to the environment.

# Cultural Concepts

## Key Points

- The U.S. is a multicultural society.
- Transcultural care depends on nurses understanding their own cultural beliefs/ values.
- Culture is learned, transmitted, social and adaptive.
- Culture can be easily seen in art, clothes, and behaviors; values and beliefs are more difficult to identify.
- A cultural assessment should include information about the effect of the client's culture on his/her: health beliefs/practices, family patterns, verbal and nonverbal communication, space orientation, time orientation, nutritional pattern, pain responses, practices surrounding death.
- Nurses should accommodate clients' cultural beliefs as much as possible.
- Many clients use folk medicine and traditional healers in place of, or along with, Western medicine.
- **Key Concepts/Terms**: Terminology, characteristics of culture, cultural phenomena, health beliefs and practices, nursing interventions

## Overview

Our society is made up of widely diverse groups of people. By the year 2080, over half the population of the U.S. will consist of people who are from what have traditionally been considered minority groups. A transcultural care perspective is now considered essential for quality nursing care. A transcultural care perspective depends, in part, on nurses' understanding of their own cultural beliefs and biases. It is important to remember that individuals within a group vary greatly, and an individual will not share all the characteristics common to her/his cultural group.

## Terminology

**Culture** is, "the learned, shared, and transmitted values, beliefs, norms, and lifeway practices of a particular group that guide thinking, decisions, and actions in patterned ways".

**Subculture** is a group composed of people who have a distinct identity, but are also related to a larger cultural group (e.g., nurses, feminists).

**Ethnicity** is a sense of identification with a cultural group, largely based on the group's common heritage. One belongs to an ethnic group either through birth or through adopting the characteristics of that group. An ethnic group shares common language, migratory status, traditions, food preferences, religious practices, and so on.

**Race** is a way of categorizing people into subgroups according to physical characteristics, such as skin pigmentation and facial features. The three major classifications are Caucasian, Negroid, and Mongoloid. Clearly, people of the same race may belong to different subcultures and ethnic groups.

**Diversity** is "being different." Differences may be because of race, gender, sexual orientation, culture, ethnicity, social status, education, religion, and so on.

**Stereotyping** assumes that all members of a culture or ethnic group act alike. Not all stereotypes are negative.

**Ethnocentrism** is the view that the beliefs and values of one's own culture are superior to those of other cultures.

**Prejudice** is a strongly held opinion about some topic or group of people, usually stemming from a strong sense of ethnocentrism. Prejudice may result from ignorance, misinformation, past experience, or fear. Examples of negative prejudice are ageism, sexism, homophobia, and racism.

**Discrimination** occurs when treating individuals or groups differently based on categories such as race, gender, social class, or exceptionality.

**Racism** is a belief that race is the primary determinant of human traits and abilities, and in the inherent superiority of a particular race.

**Transcultural Nursing** is caring for clients while taking into consideration their religious and sociocultural backgrounds.

## Characteristics of Culture

- It is learned through life experiences from birth.
- It is transmitted from parents and peers over successive generations, by verbal and nonverbal communication patterns.
- It is social.
- It is adaptive—that is, it changes as people adapt to the social environment and their needs change.
- It is satisfying.
- It is difficult to articulate/explain, primarily because many of the values and behaviors are subconscious and habitual.
- It exists at many levels. For example, art, tools, and clothes of a culture are easy to identify. More abstract concepts, such as values and traditions, are more difficult to identify.

## Cultural Phenomena Significant to Nursing

**Health beliefs and practices**: The range of health and illness beliefs and practices is almost infinite. **Folk medicine** is the use of objects, substances, and religious practices to treat and prevent illness (e.g., the use of homeopathic remedies; using garlic to prevent the evil eye). Some people may wear a talisman or amulet to protect them. Foods may be used to prevent illness: in many cultures the body is kept in "harmony" by the types of food eaten, and many taboos exist. An example of a religious approach to preventing and treating illness is the use of prayer and burning of candles. In many cases, a client may consult a traditional healer instead of, or along with, a Western physician.

**Family patterns and gender roles**: The nurse must identify the person who has the authority to make decisions in the family. For example some families are matriarchal; that is, the oldest female is the leader and decision maker. In others, it is the man. Other family patterns revolve around the value placed on children, keeping personal information within the family, and to what extent the needs of the family are valued against the needs of individuals.

**Verbal communication**: Obviously, communication may be difficult in the U.S. if a client does not speak English. Other communication patterns involve the willingness to initiate conversation and the need to establish a social relationship before discussing business or personal matters.

**Nonverbal communication**: Always consider the possibility that nonverbal communication may have a different meaning for the client than it does for you. Some cultures view silence as a sign of respect; others may take it to indicate agreement. Touch, eye contact, facial expressions, and body postures all vary among cultures.

**Space orientation**: In Nomadic societies, space is merely occupied, not owned. In Western societies, people are more territorial. Special distances are defined as the intimate zone, the personal zone, and the public zone. The size of these zones varies among cultures. Some clients may withdraw or back away if the nurse is perceived as being too close physically.

**Time orientation**: This refers to a person's focus on past, present, or future. Although most cultures combine all three, usually one orientation dominates. In the U.S., orientation is directed to the future, emphasizing time and schedules. Past-oriented cultures view people and experiences in the past as influencing activities happening in the present. Present- oriented cultures focus on the here and now and may not understand the need for clocks or calendars; these cultures generally do not worry about consequences of actions that are not immediate. Not surprisingly, the culture of nursing and health care also values time and scheduling.

**Nutritional patterns**: This includes the way staple foods are prepared (e.g., the rice may be steamed instead of boiled), and the foods that are traditional for certain holidays. Food- related behaviors include whether to breast-feed or bottle-feed infants, and when to introduce solid foods. Some foods may be required or forbidden at certain times or circumstances (e.g., it may be forbidden to eat milk and meat at the same meal; or eating meat may be forbidden altogether).

**Pain responses**: Beliefs about and responses to pain vary among groups. For example, in some cultures pain may be viewed as punishment for sins. The individual is expected to tolerate pain without complaint in order to make amends. Some cultures teach that pain should be borne quietly; in others pain expressions elicit sympathy and attention.

**Death and dying**: Various cultural and religious practices help people cope with death and dying. Knowledge of a client's heritage helps nurses to provide individualized care even though they may not participate in the rituals. In most cultures, it is unacceptable to die alone. In some cultures, the family believes that a dying person's last days should be free of worry, so they prefer that the client not know the diagnosis. Beliefs about preparation of the body, autopsy, organ donation, cremation, and prolonging life are all culturally influenced.

## Therapeutic Nursing Interventions for Culture

- It is essential for nurses to recognize their own ethnicity and cultural values.
- Even when nurse and client share the same ethnic background, expect some differences in beliefs and values. Nurses need to understand that the meaning of behavior varies widely across cultures.
- Items under "Cultural Phenomena Significant to Nursing" should be used as the basis for assessing a client's culture.
- To obtain cultural assessment data, use broad statements and open-ended questions to encourage clients to express themselves completely.
- Before performing a cultural assessment, determine what language the client speaks and whether an interpreter will be needed.
- Accept and comply with the client's cultural beliefs insofar as possible (e.g., provide hot instead of cold water if the client wishes).
- Accommodate culturally specific food preferences, religious practices, child-care, and treatment practices.
- If the client's views/practices could lead to harm, attempt to shift the client's perspective to the scientific (traditional Western medical) view.

# Critical Thinking Exercise: Cultural Concepts

**Situation:** You have been assigned to care for a client whose language and culture are different from your own. Consider the principles that will guide a cultural assessment of your client.

1. What awareness must a nurse first develop in order to provide transcultural care?

2. Work together with a fellow student to refine your understanding of transcultural nursing. Below is a list of topics and questions to guide your discussion.

   • Choose a cultural concept, such as attitudes toward nonverbal communication, touch, or pain response. Reflect on variations among cultural groups. For example, how does your own culture view these concepts?

   • Do you believe it is possible for a nurse to remain culturally neutral? Why or why not?

   • Define diversity. In what ways might you expect to find evidence of diversity within a cultural group?

   • Although not all stereotyping is negative, what dangers are inherent in stereotyping individuals?

# Spiritual Concepts

- Spiritual beliefs and practices may provide strength and support during illness; or they may create anxiety when they conflict with a proposed medical therapy.
- Nurses should assess and respect clients' spiritual needs.
- Religious practices include observing holy days; sacred writings and symbols; prayer and meditation; and beliefs about diet and nutrition, dress, birth, and death.
- Spirituality is not the same thing as religion.
- Spirituality is characterized by: inner strength and self-reliance, a harmonious relationship with nature, positive relationships with others, and positive relationship with a religious or nonreligious deity (something "larger than self").
- Do not assume that a client follows all the practices of her/his stated religion.
- Pray with the client only if there is mutual agreement between you and the client about this.
- Support religious practices, and refer clients for spiritual counseling, if needed.
- **Key Concepts/Terms:** Religion, spirituality, faith, atheist, agnostic

## Overview

Spiritual beliefs and practices influence all aspects of a person's life, including health and illness (e.g., daily living habits, relationships with others, required and prohibited behaviors). Taking a holistic view, nurses should be sensitive to clients' spiritual, physical, and psychosocial needs. This includes being knowledgeable about and respectful of spiritual needs. A client's spiritual beliefs may provide strength and support during illness, but they may also create anxiety when they conflict with a proposed medical therapy.

**Religion** is an organized system of worship and offers:

- A sense of community bound by common beliefs
- The collective study of scripture
- The performance of ritual
- The use of disciplines and practices, commandments, and sacraments

**Agnostic** describes a person who doubts the existence of God or a supreme being, or believes the existence of God has not been proven.

**Atheist** denies the existence of God.

**Monotheism** is the belief in the existence of one God who created and rules the universe.

**Polytheism** is the belief in more than one god.

**Faith** is the complete and unquestioning acceptance of a belief that cannot be demonstrated or proven by the process of logical thought.

**Hope** is a multidimensional concept that includes perceiving realistic expectations and goals, having motivation to achieve goals, anticipating outcomes, establishing trust and interpersonal relationships, relying on internal and external resources, having determination to endure, and being oriented to the future.

## Westerhoff's Four Stages of Faith

| Stage | Age | Behavior |
|---|---|---|
| Experienced faith | Infancy and early adolescence | Experiences faith through interaction with others who are living a particular faith tradition |
| Affiliative faith | Late adolescence | Participates in activities that characterize a particular faith tradition; experiences awe and wonderment; feels a sense of belonging |
| Searching faith | Young adulthood | Through a process of questioning and doubting own faith, acquires a cognitive as well as an affective faith |
| Owned faith | Middle adulthood and old age | Puts faith into personal and social action and is willing to stand up for beliefs even against the nurturing community |

## Spiritual Distress

Spiritual distress is when a person questions or denies his/her belief or value system that generally provides meaning and answers to him/her. Spiritual distress may be initiated by physiologic problems, treatment-related concerns, and situational concerns. Characteristics of spiritual distress include:

- Inconsistency in personal belief system
- Questioning or denying spiritual beliefs
- Not wanting to live, or sensing hopelessness
- Asking for help with personal/spiritual beliefs

## Characteristics of Spirituality

**Inner strength and self-reliance** (includes self-knowledge, trust in self, trust in life and the future; and peace of mind)

Harmonious relationship with nature (e.g., knowing about plants and weather, gardening, preserving nature)

Positive relationships with others (e.g., sharing time, caring for children or the elderly, and reaffirming the living and dead)

Positive relationship with a religious or nonreligious deity (e.g., prayer, meditation, religious articles, church participation)

## Spiritual Well-Being

Spiritual well-being is a feeling of being alive, purposeful, and fulfilled, viewing and living life as pleasurable, rooted in spiritual values and/or religious beliefs. Indicators of spiritual well-being include:

- A sense of inner peace
- Compassion for others
- Gratitude
- Reverence for life
- Appreciation of both unity and diversity
- Humor
- Wisdom
- Generosity
- Transcending self
- Capacity for unconditional love

## Common Religious Practices

Holy days: Days set aside for special religious observance (e.g., observing Sunday as the day of worship). A client's religion may require fasting, extended prayer, or other rituals on holy days. Those who are seriously ill are usually exempt from such requirements.

Sacred writings: Each religion has its sacred writings, believed to be the word of the Supreme Being (e.g., for Jews it is the Torah; for Christians, the Bible).

Sacred symbols: Sacred symbols (e.g., religious medals, prayer beads) may be worn to demonstrate one's faith, as a source of comfort, or to provide spiritual protection.

Prayer and meditation: Prayer (communication to a Supreme Being) and/or meditation are a part of every religion. Meditation is the act of deeply focusing one's thoughts.

Beliefs about diet and nutrition: Many religions have rules about which foods and drinks are allowed and which are prohibited. For example, Mormons are forbidden to drink alcoholic beverages.

Beliefs about dress: As an example, Orthodox Jewish men wear a yarmulke because they believe it is important to have their head covered at all times. If the religion requires that the body be covered, hospital gowns may cause the client anxiety.

Beliefs about birth: Some religions have rituals that consecrate the newborn to God (e.g., baptism, circumcision, naming in the synagogue).

Beliefs about death: Many religious people believe that the person who dies transcends this life for a better place or state of existence. Many religions offer rituals to provide comfort to the dying person and family (e.g., the Sacrament of the Sick). There may be religious observances that must be followed after death, as well. For example, some religions specify that the body be touched only by members of that faith; others require a ritual bath.

## Related NANDA Nursing Diagnoses

- Readiness for Enhanced Spiritual Well-Being
- Risk for Spiritual Distress
- Spiritual Distress

## Therapeutic Nursing Interventions for Spirituality

- Include spiritual data as a part of the nursing history.
- Never assume that a client follows all the practices of her/his stated religion.
- Provide "presence." This means offering and sharing oneself—communicating a willingness to care, to listen, and to be available. It assumes compassion, kindness, honesty, love, gentleness, and patience.
- Support religious practices.
- Assist clients with prayer by providing a quiet environment and privacy.
- Prayers should be said with clients when there is mutual agreement between the clients and those praying with them.
- Tell the client you will say a private prayer for her/him if you are not comfortable praying aloud.
- Refer clients for spiritual counseling, if needed.
- During a terminal illness, ask the client and family if there are any special procedures to be followed upon death. Provide an environment conducive to performing the death rituals.

# Critical Thinking Exercise: Spiritual Concepts

**Situation**: During a home care visit, you are conducting an admission interview with an elderly client. In response to a question about spirituality, the client states, "Oh, I don't really get to church too often." How might you proceed in your assessment of this client?

1. What kinds of questions might you ask in obtaining a spiritual assessment?

2. As a nurse, what is your role in promoting a client's spiritual health? Once you have completed a spiritual assessment, how might this information direct your care?

3. What resources might you utilize in planning a client's spiritual care?

4. Explain the signs of spiritual distress. Under what circumstances might an individual experience spiritual distress? How might you respond?

# Holistic Healing (Alternative) Therapies

| Key Points |
| --- |

- Holistic nursing practice, (healing the whole person) incorporates mind-body and biobehavioral therapies to treat physiologic, psychologic, social, and spiritual responses to illness.
- Alternative medical therapies include acupuncture, chiropractic therapy, naturopathy, homeopathy, and herbal medicine.
- Touch therapies include therapeutic massage, acupressure, and foot reflexology.
- Mind-body therapies include biofeedback, progressive relaxation, imagery, yoga, meditation, music therapy, humor, and hypnosis.
- Pharmacologic and biologic therapies include aroma therapy.
- Transpersonal therapies include noncontact therapeutic touch, and intercessory prayer.
- Many clients today are using alternative methods of health care, and nurses are beginning to incorporate such healing techniques into their practice.
- Be aware that some alternative therapies (e.g., foot reflexology, therapeutic touch) require specialized training before they can be used by the health care professional
- **Key Concepts/Terms**: Holism, alternative therapy, complementary healing

**Overview**

Today, many people are pursuing alternative methods of health care, and nurses are beginning to incorporate alternative (complementary) healing techniques into their practice (e.g., therapeutic touch, massage, meditation, imagery, acupressure, art and music therapy, biofeedback, prayer, reflexology, and humor). **Holism** conceptualizes all living organisms, including human beings, as unified wholes that are more than the mere sum of their parts. Therefore, any disturbance in one part (e.g., the heart) affects the whole being. When assessing a client, the nurse must consider how the "disturbed" part relates to other parts and to the whole person. For example, what happens to the peripheral circulation when the heart is damaged? What is the person's emotional response?

**Holistic health** involves the total person, including her/his lifestyle. **Holistic health care** includes health promotion, health education, health maintenance, illness prevention, and restorative care. **Holistic nursing practice** has as its goal the healing of the whole person. It incorporates body-mind and biobehavioral therapies in all areas of nursing to treat physiologic, psychologic, social, and spiritual responses to illness.

## Some Beliefs Underlying Holistic Practice

- Mind, body, and spirit are interdependent and share one consciousness.
- The spirit is the core of the person.
- A person's attitudes and beliefs are major causal factors in health and illness.
- Human beings are energy fields, which can become unbalanced in response to stress in the body, mind, or spirit.
- Each individual is an open system with the environment.
- Changes in health occur through experiential learning that occurs as a result of living through an activity, situation, or event.
- The client-nurse relationship is a partnership, although the responsibilities of each partner differ.

## Common Alternative Therapies

Nurses do not create changes in others; it is the person receiving the treatment (whether medication or "alternative" therapy) who does the healing. It is important for nurses to care for themselves (e.g., learn to manage stress) in order to reveal "the healers within." Specialized education is needed to perform many of the following healing therapies:

**Acupuncture** is an ancient Chinese practice based on the principle that energy is channeled through the body along specific pathways. Fine-gauge needles are inserted into the body at specific points along these pathways (meridians). The internal organs are believed to be connected to those points; the acupuncture helps balance the energy flowing through the meridians.

**Chiropractic therapy** maintains that manipulation of the spinal column is effective in dealing with back pain and may have a positive effect on the nervous system and other body systems. Chiropractors emphasize allowing the body to heal itself.

**Naturopathy** makes use of nutritional counseling, acupuncture, homeopathy, and herbal medicine.

**Homeopathy** is the theory that the cure for the disease is found in the disease itself. Therefore, ill people are treated with very dilute substances that would produce the same symptoms of their disease if given in more concentrated doses. For example, minute quantities of an allergen are injected to treat an allergy (antigen therapy); belladonna is administered for a cold because it creates cold-like symptoms.

**Herbal medicine** has been used throughout history. Herbs are plants that are valued for their flavor, scent, or medicinal properties. Over 10,000 herbs have been identified as useful for medicinal purposes. The U. S. Food and Drug Administration considers herbs to be "foods" rather than drugs, so they are not regulated as long as the seller makes no claims that the product will cure a particular disease or condition. Nurses should be sure to ask about use of herbs when taking a medication history. Some popular herbal preparations are:

- Echinacea: Supports the immune system.
- St. John's wort: Treats depression.
- Milk thistle: An antioxidant; good for the liver.
- Ginseng: Reduces cholesterol; improves endurance.
- Ginkgo biloba: An antioxidant, thought to improve memory

## Touch Therapies

**Therapeutic massage** relaxes muscles, releases the buildup of lactic acid, and improves the flow of blood and lymph. It can relieve anxiety and provide a sense of relaxation.

**Acupressure** theory is similar to acupuncture. The therapist applies finger pressure to 657 designated points to restore balance in the flow of energy.

**Foot reflexology** (zone therapy) has its history in Egypt. The theory is that various "zones" of the body are connected to reflex areas on the hands and feet. Massaging a reflex area stimulates the corresponding organs in that zone. This should be used with caution when there are circulatory disorders in the extremities.

## Mind–Body Therapies

**Biofeedback** teaches clients to attain a state of relaxation and parasympathetic dominance. The client attempts to consciously control bodily processes such as muscle tension, heartbeat, blood flow, peristalsis, and skin temperature. Feedback is provided through temperature meters or electromyogram.

**Progressive relaxation** has been used successfully and widely to reduce stress and chronic pain. The client learns to control the body's responses to tension and anxiety. For relaxation, the client must have correct posture, a mind at rest, and a quiet environment.

The client is asked to tense and then relax successive muscle groups, concentrating on the difference between the feelings experienced when the muscle group is relaxed and when it is tense.

**Imagery** is the internal experience of memories, dreams, and fantasies that serve as a bridge between body, mind, and spirit. The idea is that people can create a mental picture of how they want things to be. The nurse might have the client visualize a tiger eating up his tumor cells, or have a client with a total knee replacement visualize playing tennis.

**Yoga** is the uniting of all the powers of the mind, body, and spirit. It is an approach to living a balanced life. Various schools of yoga use different techniques (e.g., breathing techniques, posturing, and so on), but all have mastery of the self as their goal.

**Meditation** is a combination of focused attention and relaxation that is used to quiet the mind, focus it in the present, and release fears and doubts concerning the past and the future. Techniques of breathing, progressive relaxation, and imagery are used to facilitate meditation. In the meditative state, a person is in a state of deep peace and rest, combined with mental alertness.

**Prayer** is similar to meditation but it is intended as a communication with a Higher Power.

**Music therapy** consists of listening, rhythm, body movement, and singing. The theory is that music that relates closely to the body's fundamental frequency pattern can help to restore regulatory functions that are altered during stress and illness.

**Humor** has physiologic benefits. Laughter stimulates respiratory and heart rates, muscle tension, and oxygen exchange. A state of relaxation follows, during which an opposite state exists. Humor stimulates the production of hormones and catecholamines, and the release of endorphins increases pain tolerance. Some health care agencies have "humor rooms."

**Hypnosis** is an altered state of concentration in which concentration is focused and distraction is minimized. It requires active client participation. Hypnosis is used to control pain (and other symptoms), alter body functions, and change habits.

## Pharmacologic and Biologic Therapies

**Aromatherapy** may help with some minor problems, but has not been shown to be effective in managing serious disease. Various aromas are thought to effect body functions, emotions, and memory. Essential oils (distilled from flowers, roots, bark, leaves, wood resins, and lemon and orange rinds) are inhaled, sprayed into the air, massaged on the body, applied as warm or cold compresses, or added to bath water.

## Transpersonal Therapies

**Noncontact therapeutic touch** uses the concepts that the human being is an energy field and that energy can be channeled from one person to another. Practitioners "feel" the energy field and then transmit energy to the ill or injured person to potentiate the healing process. The process consists of four steps:

- Centering (a meditative step)
- Assessing (scanning with the hands)
- Unruffling (making the client's energy field more receptive)
- Transferring energy through the nurse to the client

**Intercessory prayer** refers to prayer offered for another person. Research has shown that clients who were prayed for daily did better on average than those who were not.

# Critical Thinking Exercise: Holistic Healing (Alternative) Therapies

**Situation**: At a health fair, a middle-aged woman expresses interest in holistic healing. "But there are so many different therapies" she states. "I don't know where to start." How might you offer guidance?

1. What is the underlying principle or belief that governs all holistic practices? Explain this briefly in your own words.

2. Select a holistic healing therapy that interests you and design a brochure or poster to illustrate its benefits. In developing your advertisement, consider your target audience, describe the therapy, and outline its benefits.

# Nursing Process

## Key Points

- The nursing process is both a problem-solving process and a thinking/doing "approach" to client care.
- The nursing process provides a framework in which nurses can apply their knowledge and skills.
- The nursing process can be used to provide care for groups, families, and individuals of all ages and of any health status.
- The phases of the nursing process are sequential and overlapping—in actual practice, they do not always occur in neat, stepwise order.
- The nursing process phases (steps) are:
  - Assessment: Get the facts. Collect, organize, validate, and record client data.
  - Diagnosis: Identify the client's health status. Make the nursing diagnosis.
  - Planning (Outcomes): Identify the desired health status. Determine goals and outcomes.
  - Planning (Interventions): Identify actions to achieve the goals. Select nursing activities to promote wellness or prevent, correct, or relieve the problem(s).
  - Implementation: Doing, delegating, and documenting
  - Evaluation: Judging whether the plan worked.
- **Key Concepts/Terms**: Benefits of nursing process, phases of nursing process

## Overview

The nursing process is a thinking/doing approach that nurses use in their work. The following characteristics should be included in a definition of the nursing process.

**The nursing process is:**
- A special way of thinking and acting
- A systematic problem-solving approach; it is planned and outcome-directed
- Dynamic and cyclic
- Client centered
- Useful with any medical diagnosis, in any setting and any specialty area of practice
- A cognitive (thinking) process

**The nursing process:**
- Involves creativity and intuition
- Provides a framework in which nurses use their knowledge and skills to

express human caring
- Can be used to provide care for groups, families, and individuals of all ages, at any point on the wellness-illness continuum

## Benefits of the Nursing Process

The most important reasons for using the nursing process include benefits to clients, nurses and the nursing profession. The nursing process:

- Helps people understand what nurses do
- Helps nurses meet professional standards of practice
- Promotes client autonomy and increases client participation in care
- Promotes individualized care
- Promotes efficiency
- Promotes continuity and coordination of care
- Increases job satisfaction
- Promotes collaboration

## Critical Thinking

A complicated mix of inquiry, knowledge, intuition, logic, experience, and common sense is called critical thinking. This kind of thinking enables the nurse to grasp the meaning of multiple clues and to find quick answers when facing difficult problems.

**Characteristics of the critical thinker**

- Examines facts and compares these facts with known information
- Forms ideas or concepts that are mental pictures of reality
- Is reasonable and rationale and continuously searching to understand the entire situation
- Thinking may be random; however, conclusions are not reached quickly
- Forms own beliefs and ideas rather than automatically accepting the thoughts and ideas of someone else
- Is open-minded and flexible to alternatives

## Phases of the Nursing Process

The nursing process consists of five phases: assessment, diagnosis, planning, implementation, and evaluation. The planning phase consists of two distinctly different activities: goal-setting and planning nursing interventions. Those two activities are described separately here.

**Assessment. What is the relevant data?** In this phase the nurse collects, organizes, validates, and records data about the client's present health status. The nurse obtains data by examining clients, talking to them and their families, and reading charts and records.

- Subjective data is apparent only to the client, described or verified by the client. Any information supplied by the client directly is subjective. Any information supplied by anyone that is based on opinion rather than fact should be noted as subjective data.

- Objective data is that which is observable or obtained by testing against an accepted value or norm; it can be seen, felt, heard, or smelled through observation or physical assessment.

- Primary data is data provided by the client.

- Secondary data is any data provided by a source other than the client.

- Sources of data may include the client, family, significant other(s), caregivers, health care records, health care providers, and literature.

No conclusions are drawn about the data in this phase.

**Diagnosis. What is the client's present health status?** What is contributing to it? In this phase the nurse: (1) sorts, clusters, and analyzes client data in order to identify the client's present health status (actual and potential health problems and strengths), (2) writes a nursing diagnosis (a precise statement describing the client's present status and the factors contributing to it), (3) prioritizes the diagnoses, and (4) decides which diagnoses will respond to nursing care and which must be referred to another health care professional (e.g., a physician). The North American Nursing Diagnosis Association (NANDA) and the International Council of Nurses (ICN) have developed a set of standardized terms for nurses to use in writing nursing diagnoses.

**Planning outcomes. What is the desired health status of the client?** In this phase the nurse works with the client to choose the desired outcomes. The outcomes should be realistic and measurable goals specific to the client. That is, they should decide exactly how they want the client's health status to change, and within what period of time. The outcomes chosen in this phase are the criteria that the nurse will use in the evaluation phase. A group of nursing researchers has published a standardized vocabulary and classification for describing client outcomes thought to be sensitive to nursing interventions ("Nursing Outcomes Classification (NOC)". The International Council of Nurses is also developing a set of standardized terms for client outcomes. www.lcn.ch/lcnpupdate.htm

**Planning interventions. How can you help achieve the desired outcomes?** In this phase the nurse chooses interventions for promoting wellness or preventing, correcting, or relieving health problems. The interventions are meant to achieve the client outcomes associated with each nursing diagnosis. The interventions may be:

- Independent: Interventions that the nurse does independently

- Dependent: Interventions that the nurse depends upon the primary care provider's order or on an institutional protocol (a preprinted step-by-step list of actions to achieve a goal: for example admissions, dismissals, or blood administration)

- Collaborative: Interventions the nurse carries out in collaboration with other health care professionals

The end product of the planning phases is often a written care plan. However, in some cases, planning is simply the mental process of choosing what to do. **Nurses often act without a written plan, but never without a plan.** A group of nurse researchers has developed a standardized classification of nursing interventions, called the "Nursing Interventions Classification (NIC)". The International Council of Nurses (2002) also has a classification of terms to describe nursing interventions.

**Implementation. What actions are necessary to achieve the expected goals and outcomes of nursing care?** In this phase, the nurse communicates the plan of care to other members of the health care team and carries out the interventions indicated on the plan or delegates them to others. The final activity in this phase is to record the care that was given and the client's responses.

**Evaluation. Did it work?** In this phase, after implementing the plan, the nurse compares the client's health status with the desired outcomes identified in the Planning Outcomes phase. The nurse determines which interventions were helpful in achieving desired outcomes and revises the care plan as needed.

Because the nursing process is cyclic, all the phases (assessment, diagnosis, outcomes, interventions, and implementation) are continually reexamined in order to determine what is effective and what should be changed.

Nursing process phases are sequential. This means that each step depends on satisfactory completion of preceding steps. For example, the nurse must have accurate data (assessment) in order to make the correct diagnosis (diagnoses). However, the phases do not always occur in the assessment, diagnosis, planning, implementation and evaluation order, and nurses do not always perform one phase completely before proceeding to the next.

Although the phases of the nursing process are distinct and they are described separately in order to help you understand them, in practice they overlap a great deal. For instance, even though the first encounter with a client usually begins with some form of data collection, assessment is actually an ongoing process that occurs along with the other phases.

# Critical Thinking Exercise: Nursing Process

**Situation**: A classmate is about to begin a clinical assignment in a long-term care facility. Before reporting, she wonders whether the nursing process will apply in this facility as it did in an acute care setting. How might you respond?

1. Explain how the steps of the nursing process can be utilized in a variety of clinical situations and settings.

2. To reinforce your own appreciation for the nursing process, explain to a classmate how utilizing the process requires creativity and flexibility. Draw on your own experiences in formulating your response.

3. Identify two important benefits derived from using the nursing process?

# Developmental Theories

## Key Points

- Rates of growth and development are unique to each person; however, everyone follows the same pattern (sequence) of growth and development.
- Growth and development occur cephalocaudally, from proximal to distal, and from simple to complex.
- Psychosocial theories explain the development of the personality. Theorists include Freud and Erikson.
- Maslow's hierarchy of needs theory is commonly accepted and used in promoting health.
- Cognitive theories explain the development of the intellectual abilities (thinking, reasoning, and use of language). One theorist is Piaget.
- Moral development theories describe how the reasoning a person uses to decide what is right and wrong changes with age. Theorists include Kohlberg and Gilligan.
- Spiritual theories describe the development of faith. Theorists include Fowler and Westerhoff.
- **Key Concepts/Terms**: Growth, development, developmental task, personality

## Overview

Growth and development are interdependent and interrelated. **Growth** is physical change and increase in size. **Development**, the behavioral aspect of growth, is the progressive increase in complexity of skills and functions. The rates of growth and development are unique to each person; however, their sequence is predictable. Stages of growth usually parallel certain stages of development. Growth and development occur in five dimensions: physiological, psychosocial, cognitive, moral, and spiritual.

## Selected Principles of Growth and Development

- Everyone follows the same **pattern** of growth and development.
- Growth and development are continuous, orderly, sequential processes.
- Growth and development are influenced by maturation, environmental factors, and genetic makeup.
- Each developmental stage has its own set of characteristics (e.g., a child in Piaget's intuitive thought phase uses words to express thoughts.)
- Growth and development occur **cephalocaudally**. That is, they begin at the head and move toward the legs and feet (e.g., an infant lifts its head before lifting legs and feet).

- Growth and development occur from **proximal to distal**. That is, from the center of the body outward (e.g., an infant rolls over before it is able to grasp objects with its thumb and finger).
- Development proceeds from **simple to complex**.
- The pace of growth and development are uneven (e.g., growth is greater during infancy than in childhood).

## Developmental Task Theory (Havighurst, 1972)

A **developmental task** is one that arises at or about a certain period in a person's life which, if achieved, leads to happiness and success with later tasks. If not achieved, it leads to unhappiness, difficulty with later tasks, and social disapproval. People continue to learn throughout life. Growth and development occur in six stages, each associated with several tasks to be learned: (1) infancy and early childhood, (2) middle childhood, (3) adolescence, (4) early adulthood, (5) middle age, and (6) later maturity. Once a person masters a task, it is mastered for life.

## Psychosocial Theories

Psychosocial theories are meant to explain the development of the personality. **Personality** is the outward expression of the inner self. It includes feelings, temperament, character traits, self- esteem, behaviors, interactions with others, and ability to adapt to change. Many psychosocial theories exist.

### Freud (1923)

According to Freud's theory of psychosexual development, the personality develops in five stages from birth to adulthood, and a certain body area has special significance at each stage. The stages are: **oral** (birth-1 year), **anal** (2-3 years), **phallic** (4-5 years), **latency** (6-12 years), and **genital** (13 years and older). The **unconscious mind** is the part of the mind that the person is unaware of. **Defense mechanisms** result from conflicts from inner impulses and the anxiety they sometimes produce. The **id** is responsible for a person's instinctive and unconscious urges (which are sexual in nature). The **ego** serves to make effective contact with social and physical needs. The **superego** is the sense of right and wrong and serves to mediate the ego and id; it is the source of guilt, shame, and inhibition. The **libido** is an energy form or life instinct that underlies the motivation for human development.

### Erikson (1963)

Erikson, unlike Freud, believes that people continue to develop throughout life. He believes the ego to be the core of the personality. He describes eight stages of development. Each stage has a task that is achieved completely, partially, or unsuccessfully. The greater the task achievement, the healthier the personality. No stage can be "skipped," but a person can become "stuck" in one stage or regress to a previous stage. The stages and central tasks are:

| Developmental Stage | Approximate Age | Psychosocial |
| --- | --- | --- |
| Infancy | Birth-18 months | Trust vs. mistrust |
| Early childhood | 18 months-3 years | Autonomy vs. shame and doubt |
| Late childhood | 3-5 years | Initiative vs. guilt |

| Developmental Stage | Approximate Age | Psychosocial |
|---|---|---|
| School age | 6-12 years | Industry vs. inferiority |
| Adolescence | 12-20 years | Identity vs. role confusion |
| Young adulthood | 18-25 years | Intimacy vs. isolation |
| Adulthood | 25-65 years | Generativity vs. stagnation |
| Maturity | 65 years-death | Integrity vs. despair |

## Needs Theory

Maslow's Hierarchy of Needs is a commonly accepted needs theory used in promoting health. Maslow found that those who satisfy basic needs appropriately are healthier and happier than those individuals who do not.

### Maslow's Needs

- **Self-actualization**
- **Self-esteem**
- **Love and belonging**
- **Safety and security**
- **Physiologic**

Maslow's theory states that people will not progress successfully to the next level of needs without appropriately satisfying the lower level needs first.

**Physiologic** needs include air, food, water, shelter, rest, sleep, activity, and temperature. These are considered the lowest level needs which are crucial for survival.

**Safety and security** needs are both physical and psychological and are the second level of needs.

**Love and belonging** needs are the third level and include giving and receiving affection, and attaining and maintaining the feeling of belonging in a group.

**Self-esteem** needs include feelings of independence, competence, self-respect, and esteem which includes recognition, respect, and appreciation from others.

**Self-actualization** is the highest level in Maslow's hierarchy of needs. When the self-esteem need is satisfied one strives for self-actualization, the innate need to develop one's maximum potential and realize one's abilities and qualities is fulfilled.

## Cognitive Theory

### Piaget (1966)

Describes cognitive development as an orderly, sequential process in which a variety of new experiences must occur in order for intellectual abilities to develop. Cognitive development refers to thinking, reasoning, and use of language. Piaget's five stages, which are complete by age 15, are: sensorimotor stage (birth-2 years), preconceptual phase (2-4 years), intuitive thought phase (4-7 years), concrete operations phase (7-11 years), and formal operations phase (11-15 years). In the first stage, most actions are reflexes. By the end of the final stage, the person can use rational thinking and use deductive, futuristic thinking.

## Moral Development Theories

Moral developmental theories are based on the work of Piaget. They describe the reasoning a person uses to decide what is right and wrong. **Morality** refers to the necessary requirements for people to live together in a society. **Moral behavior** is the way people interpret and respond to moral requirements. **Moral development** is the pattern of change in moral behavior that occurs with age.

### Kohlberg (1981)

Describes moral development in children and adults. Although moral development tends to parallel cognitive development because it includes reasoning, they are not always linked because some people progress to a higher level of moral development than others do. At the first **premoral/preconventional level**, actions are taken to satisfy one's own needs and the action is judged to be wrong if it is punished and right if not punished. At the **conventional level**, actions are taken to please others and gain approval; an action is right if it follows the rules (obeys the law). At the **postconventional level**, actions are based on obeying laws/rules that protect the welfare of others. The person recognizes personal values and opinions, avoids violating the rights of others, and believes that relationships are based on trust.

### Gilligan (1982)

Gilligan found that Kohlberg's stages did not fit well with women because it was limited to reasoning and justice, and did not include the concepts of caring and responsibility. Gilligan's three stages are:

- Caring for oneself
- Caring for others
- Caring for self and others

Gilligan believes that because women see morality in caring relational integrity, their moral problems are different from those of men. Men tend to consider that what is just (fair) is right; women see "right" as choosing to take responsibility for others.

## Spiritual Theories

Spiritual theories describe the development of faith.

### James Fowler (1981)

Fowler describes the development of faith as a process of interaction between the person and environment. **Faith** is a relational phenomenon, a way of being in relation with "another or others in which we invest commitment, belief, love, risk and hope." Fowler describes six stages that progress from an infant (who is unable to formulate concepts about self or environment) through the **paradoxical-consolidative** stage at about 30 years of age (when the person comes to an understanding of truth from a variety of perspectives), and finally to the universalizing stage (in which the person adapts an incarnation of the principles of love and justice). Fowler says that the final stage is not always achieved. Fowler's theory is based on the work of Kohlberg.

### Westerhoff (1976)

Westerhoff describes faith as a way of being and behaving. It evolves through experience and guidance by others during infancy and childhood, and progresses to an internalized, owned faith that serves as a directive for personal action in adulthood.

Westerhoff stages are:
- Experience faith
- Affinitive faith
- Searching faith
- Owned faith

# Critical Thinking Exercise: Developmental Theories

**Situation**: A fellow student enlists your help in understanding the concepts of growth and development. Use the questions below to guide your discussion.

1. How is the process of growth different from that of development? Respond by defining and providing an example of each term.

2. Identify the five dimensions of growth and development. What factors influence growth and development?

3. Select one of the various psychosocial, cognitive, spiritual, and moral development theories. What does this theory seek to explain? In what ways might this theory either complement or conflict with other developmental theories?

4. Why is it important for a nurse to understand the theoretical basis for growth and development?

# Infant Through Adolescence

## Key Points

**Conception and Prenatal Development**

- Embryonic phase: First 8 weeks of pregnancy
- Fetal phase: Remaining 32 weeks of pregnancy
- Health promotion: Essential for prenatal development

**Neonates and Infants (birth-1 year)**

- Physical: Birth weight triples by one year; begin to walk at about one year, teeth begin to erupt at 6-7 months with the first teeth to erupt generally the two lower central incisors
- Psychosocial: Erikson stage of trust vs. mistrust; Freud, oral stage
- Cognitive: Piaget's sensorimotor phase
- Health problems: Colic, failure-to-thrive, child abuse, sudden infant death syndrome
- Nursing interventions: Teach parents about developmental norms, safety measures, sensory stimulation, and the need for immunizations.

**Toddlers (1-3 years)**

- Physical: Lose baby look; height growth slows after age 2; tactile sense important; uses spoon at 2 years; toilet trained at 3 years
- Psychosocial: Erikson, autonomy vs. shame and doubt: Freud, anal stage; separation anxiety
- Cognitive: Piaget, preconceptual phase begins at 2 years; egocentric
- Moral: Responds to punishment and reward
- Health problems: Accidents, respiratory and ear infections, dental caries
- Nursing interventions: Teach parents to avoid saying "No" too much; how to set limits; to offer safe choices; about accident prevention and suitable toys. Recommend dental visits by age 3.

**Preschoolers (4-5 years)**

- Physical: Growth slows; taller and thinner; extremities grow more quickly than body
- Psychosocial: Erikson, initiative vs. guilt; Freud, phallic stage
- Cognitive: Piaget, phase of intuitive thought; begins to develop reading skills
- Moral: Model parental behaviors; seeks approval
- Health problems: Communicable diseases, accidents, dental caries
- Nursing interventions: Teach parents that this is a time of testing independence.

Health promotion includes accident prevention, dental health, nutrition, cognitive stimulation, and adequate sleep.

### School-age (6-12 years)

- Physical: This period begins when the deciduous teeth are shed, includes prepuberty, and ends with onset of puberty. Weight gain is rapid. Growth spurt occurs just before puberty.

- Psychosocial: Erikson, industry vs. inferiority; Freud, latent stage

- Moral: Preconventional (act to avoid punishment), instrumental-relativist (do things to benefit self), or conventional (law-and-order)

- Health Problems: Same as preschoolers, plus violence, scabies, impetigo and lice

- Nursing Interventions: Promote exercise and physical fitness; teach hygiene to prevent infections.

### Adolescence (12-18 years)

- Matures physically and psychologically; acquires a personal identity

- Physical: Rapid growth; dramatic and sudden physical changes; sexual characteristics develop

- Psychosocial: Erikson, identity vs. role confusion

- Cognitive: Piaget, formal operations stage

- Moral: Kohlberg's conventional level (Golden Rule)

- Developmental Tasks: Havighurst, achieving independence from parental domination and accepting individual responsibility for oneself

- Health Problems: Substance abuse, sexually transmitted diseases, motor vehicle accidents, athletic injuries, eating disorders

- **Key Concepts/Terms:** Physical development, psychosocial development, cognitive development, moral development, health problems, therapeutic nursing interventions

## Conception and Prenatal Development

**Embryonic phase** is the first 8 weeks of pregnancy. The fertilized ovum is implanted, placental function starts, and fetal membranes differentiate. Most organs are formed.

**Fetal phase** is the remaining 32 weeks of pregnancy. Fetus grows rapidly in size, and at the end of the 6th lunar month looks like a small baby. At about 5 months, the mother can feel movement, and the fetal heartbeat may be auscultated. Few developmental changes occur in the final 2 months of pregnancy, as the fetus primarily accumulates weight.

**Health promotion** is essential for prenatal development because the embryo or fetus relies on the maternal blood flow through the placenta to meet its needs for nutrition, elimination, and oxygenation. The nurse should promote a well-balanced diet with additional calories. Pregnant women can provide a safe environment for their fetus by avoiding alcohol, street drugs, or other medications not ordered by a health professional.

## Neonates and Infants (Birth–1 year)

### Physical Development

**Basic task** is survival—breathing, sleeping, sucking, eating, swallowing, digesting, and eliminating.

**Weight at birth (most babies):** 6.0 to 9 lb. Doubles birth weight by 5 months; triples birth weight by age 12 months.

**Length at birth (average):** European American newborns—20 in. African American infants tend to be shorter; females are typically smaller than males. Infants add about 8.5 in. of height in the first year of life.

**Vision:** Can follow large moving objects at birth, and blinks in response to light and sound. By 6 months, can perceive colors. After 9 months, will smile at a familiar face. Depth perception develops by 12 months.

**Hearing:** Newborns will react to a loud noise with a startle (Moro reflex). They can recognize their mother's voice, versus that of another woman, within just a few days. By 1 year, listens to sounds, distinguishes some words, and responds to simple commands.

**Motor development:**

- 1 month: When prone, can turn head and lift head momentarily
- 6 months: Sits without support
- 9 months: Reaches, grasps rattle and transfers from hand to hand
- 12 months: Turns book pages, puts objects in a container, walks with help

### Psychosocial Development

**Erikson stage:** Trust vs. mistrust. Trust is facilitated by having needs fulfilled consistently during the first year of life. Newborns react socially by paying attention to the caregiver's voice and face and by cuddling. By age 8 months, most infants show displeasure when left with strangers. Crying is the infant's way of communicating stress.

**Freud stage:** Oral stage. Infants reduce tension by sucking and mouthing objects.

### Cognitive Development

**Piaget:** Sensorimotor phase

- 4-8 months: Perceptual recognition begins
- 6 months: Responds to new stimuli, remembers some objects and will look for them for a short time
- 12 months: Has concepts of space and time; experiments to reach a goal (e.g., trying to reach a toy on a chair)

### Moral Development

Right and wrong is associated with pleasure and pain. From the parents' smiles and vocal tones of approval, children learn that certain behaviors are good; they learn that pain is the result of other ("bad") behaviors.

### Health Problems

- Failure-to-thrive

- Colic
- Child abuse
- Sudden infant death syndrome

### Therapeutic Nursing Interventions for Neonates and Infants

- Apgar Scoring—At birth, provides a numerical indicator of the baby's ability to adapt to extrauterine life. Maximum score is 10; infants with very low scores require resuscitation.
- Denver Developmental Screening Test (DDST)—Used to assess the infant's development. Used for screening children from birth to 6 years of age. Screens personal- social, fine motor adaptive, language, and gross motor development.
- Teach caregiver normal behaviors, responses, and activities of the infant.
- Listen for caregiver cues to problem areas.
- Observe parent-infant interactions and assess attachment.
- Remind caregiver to schedule health examinations at 2 weeks and at 2, 4, 6, and 12 months.
- Encourage immunizations, fluoride supplements, screening for tuberculosis and phenylketonuria.
- Stress the importance of supervision.
- Teach caregiver to avoid propping bottle, time to introduce solid foods and indication for iron supplements at 4-6 months of age.
- Teach about car seat, crib, playpen, bath, and home safety measures.
- Toys should not have small parts or sharp edges.
- Teach the importance of sensory stimulation (touch, vision, hearing, play).

## Toddlers (1–3 years)

### Physical Development

**Loses the baby look**: Is chubby, with short legs and large head. Has lumbar lordosis and protruding abdomen.

**Weight**: At two years should weigh about 4 times its birth weight. A 3-year-old weighs about 30 lb.

**Height**: Between ages 1 and 2 years, the average toddler gains 4-5 inches in height. Between 2 and 3 years, growth in height slows to 2-1/2 to 3-1/2 inches.

**Head circumference**: By 2 years, the head is 80% of adult size and the brain is 70% of its adult size.

**Sensory abilities**: Visual acuity and accommodation to near and far objects are fairly well established by 18 months. The other senses become increasingly developed. Touch is important and toddlers can often be soothed by tactile sensations (e.g., a soft toy).

**Motor abilities**: Both fine muscle coordination and gross motor skills improve. At 18 months, can hold a spoon and cup and walk up stairs with help. At 2 years, can put a spoon into the mouth correctly; can run with steady gait and balance on one foot. By age 3, most are toilet trained. By age 3, can copy a circle, jump, and walk on toes.

### Psychosocial Development

**Freud**: Anal phase

**Erikson**: Autonomy vs. shame and doubt. Toddlers begin to develop their sense of autonomy by using the word "no" frequently or by dawdling. An attempt to restrain their behavior frustrates them and they may have tantrums. Gradually, with caregiver guidance, they gain control over their emotions. Separation anxiety (fear of abandonment when the parent leaves them) is common. Experience with separation helps the child cope with parental absences.

### Cognitive Development (Piaget)

- Completes sensorimotor phase and starts preconceptual phase at about 2 years of age
- Solves problems by trial-and-error method, and then begins to do it mentally
- Has some symbolic thought (e.g., a blanket may represent comfort)
- Learns about the sequence of time
- Begins to form concepts at about age 3 (e.g., recognizes "chair" even though there are several kinds of chairs in the house)

### Moral Development (Kohlberg)

In the pre-conventional level, the child responds to punishment and reward. Learns what attitudes parents have about moral matters.

### Spiritual Development (Fowler)

Spiritual development is undifferentiated. The child is primarily involved in learning knowledge and emotional reactions; repeats bedtime prayers, for example, because the parents give praise or affection.

### Health Problems

- Accidents
- Visual problems (amblyopia, strabismus)
- Dental caries (as a result of excessive sweets or use of bottle during naps and bedtime)
- Respiratory tract and ear infections
- Sleep disturbances

### Therapeutic Nursing Interventions for Toddlers

- Teach parents about accident prevention.
- Provide suitable toys: challenging, but not so difficult that the toddler will fail.
- Teach parents to make positive suggestions rather than give commands. Avoid saying "no" too much.
- Give the toddler safe choices (limit to 2 or 3).
- Set and enforce reasonable, consistent limits.
- Praise toddlers' accomplishments.
- Encourage dental visits at age 3.
- Schedule hearing tests by 18 months or sooner.

- Screen for tuberculosis and lead poisoning.
- Stress home safety (e.g., locked medicine cabinet, gun locks).
- Stress outdoor safety (e.g., close supervision near water, car seats).

# Preschoolers (4-5 years)

## Physical Development

- Physical growth slows; body control and coordination increase greatly.
- Preschoolers appear taller and thinner than toddlers do.
- Extremities grow more quickly than body.
- By 5 years, the brain reaches adult size.
- **Weight:** Age 5, weight is 40-45 lb.
- **Height:** Grows about 2.0-2.5 inches per year; doubles birth length at age 5 (about 40 inches tall).
- **Vision:** Usually farsighted. Normal vision is about 20/30.
- **Motor Abilities:** By age 5, can wash hands and face and brush own teeth; runs skillfully and can jump three steps; dresses without assistance.

## Psychosocial Development

- **Freud:** In this phallic stage, oedipus or Electra complex occurs. That is, the child focuses feelings of love chiefly on the opposite sex parent, and may be hostile to the same-sex parent.
- **Erikson:** Initiative vs. guilt
- Imitates behavior
- Imagination and creativity are active
- Self-concept based, in part, on gender identification; aware of two sexes and identify with the correct one
- Becomes interested in clothes and hairstyles
- Learns adaptive mechanisms: identification, introjection, imagination, and repression
- Becomes increasingly aware of himself or herself
- Can draw a person with all the features, by age 5
- Has words for feelings ("sad"); begins to control feelings
- Tests limits by refusing to cooperate and ignoring parental requests

## Cognitive Development (Piaget)

- This is the phase of intuitive thought. Egocentrism begins to subside as they encounter different experiences.
- Learns through trial and error
- Forms concepts
- Becomes concerned about death, which he or she associates with others rather than himself or herself.
- Counts pennies

- Begins to develop reading skills
- Likes books about animals, children, and fairy tales

### Moral Development

- Capable of prosocial behavior (e.g., kindness, sharing, helping, protecting, befriending, and giving affection and encouragement)
- Conscience not fully formed, but does have some internal controls
- Preschoolers control their behavior because they seek approval from their parents.
- Moral behavior is learned by modeling parents.
- Usually well-behaved in social settings

### Spiritual Development

- Usually enjoys Sunday school classes because of the social interaction
- Fowler: In his intuitive-projective stage, faith is a result of the teaching of significant others. Children imitate religious behavior. They require simple explanations of spiritual matters.

### Health Problems

Similar to toddlers: Respiratory tract problems and communicable diseases; accidents; dental caries. Congenital anomalies (e.g., hernias) are often corrected at this age.

### Therapeutic Nursing Interventions for Preschoolers

- The child can answer some of the questions during an interview (e.g., What do you usually have for lunch? What do you like to do for fun?).
- Health promotion includes accident prevention, dental health, nutrition, cognitive stimulation, and adequate sleep.
- Teach parents that preschoolers need to feel they are loved and are an important part of the family; they need time to adjust to a new baby.
- Teach parents to give extra attention to help the preschooler adjust to a new baby.
- Parents should provide guidance and discipline during this time of testing independence.
- Encourage parents to give recognition for actions such as sharing.
- Encourage parents to answer "why" questions and discuss values with the child.

## School-Age Children (6–12 years)

This period begins at about the time the deciduous teeth are shed. It includes the prepubescent period and ends with the onset of puberty (at about age 12). Weight gain is rapid and the child appears less chubby than preschoolers.

### Physical Development

**Weight**: Average gain of 7 lb./year. Major weight gain occurs from age 10-12 for boys and 9- 12 for girls. By age 12, the average weight for boys is 88-95 lb.; girls are heavier.

**Height**: At 6 years, both boys and girls are about 46 in. tall. By 12 years, they are about 60 inches tall. A growth spurt occurs just before puberty (girls at age 10-12, boys at

12-14 years). The extremities grow more quickly than the trunk, thus the "gangly" appearance.

**Vision**: Depth and distance perception are accurate; full binocular vision is present. Vision should be 20/20.

**Hearing and Touch**: Hearing is fully developed. School-age children can identify fine differences in sound and pitch. Sense of touch is well developed. Stereognosis is present (they can identify an unseen object simply by touch).

**Prepubertal Changes**: At ages 9 to 13, endocrine functions gradually increase, resulting in increased perspiration and sebaceous gland activity.

**Motor Abilities**: Muscle skills and coordination are perfected. By age nine, most have fine motor control for activities such as working puzzles or sewing.

### Psychosocial Development

**Erikson**: This is the stage of industry vs. inferiority. Children begin to develop a sense of competence and perseverance. If they successfully mastered tasks of previous stages, they are motivated to be industrious and work with others to achieve a goal.

**Freud**: In the latent stage, the focus is on physical and intellectual activities; sexual tendencies seem to be repressed. The focus of interest moves to school, peers, and other activities; however, the home is the most important factor for developing high self-esteem.

### Cognitive Development

**Piaget**: This is the phase of concrete operations. Children progress from egocentric to cooperative interactions. They progress from intuitive to logical reasoning (e.g., they can add and subtract). They also learn about cause-and-effect relationships. Most know the value of coins by age 7 or 8. They learn time periods and begin to understand about "the past." Reading skills are well developed in later childhood. By age 9, most children are self- motivated; they compete with themselves, to talk, discuss different subjects, and debate.

### Moral Development

**Kohlberg**: Some are at the pre-conventional level (e.g., they act to avoid punishment). Some are still at the instrumental-relativist level (e.g., they do things to benefit themselves). Fairness becomes an important concept. At about ages 10-13 most progress to the conventional, law-and-order level (they want to be thought of as a "good boy" or "nice girl"). The motivation for moral action is to live up to what significant others think of them.

### Spiritual Development

**Fowler**: This is his mythic-literal stage. Children learn to distinguish fantasy from fact. Spiritual "facts" are those beliefs that are accepted by their religious group. When they do not understand events (e.g., creation), they use fantasy to explain them. Concepts such as prayer must be presented in concrete terms.

### Health Problems

- Communicable diseases, dental caries, and accident rates are the same as preschoolers
- Homicide and violence

- Scabies, impetigo, and lice
- Alcohol and drug abuse (not as common as in adolescence)

### Therapeutic Nursing Interventions for School-Age Children

- Demonstrate interest and enthusiasm for the child's strengths and achievements.
- Encourage and teach dental hygiene and regular dental examinations.
- Teach safety measures for accident prevention.
- Promote exercise and physical fitness.
- Support autonomy and self-esteem.
- Teach hygiene measures to prevent infections and infestations.

## Adolescence (Age 12–18 years)

**Adolescence** is the period in which the person matures physically and psychologically and acquires a personal identity. At the end of adolescence (which may be later than age 18), the person is ready to assume adult responsibilities. **Puberty** is the first stage of adolescence (sexual organs begin to grow and mature. **Menarche** (onset of menstruation) occurs in girls. **Ejaculation** (expulsion of semen) occurs in boys. Onset of puberty is age 10-14 for girls and 12-16 for boys.

### Physical Development

- Growth is rapid; physical changes are dramatic and sudden. This period is called the "adolescent growth spurt." The period of most rapid growth occurs at about age 12 for girls, and age 14 for boys.
- Boys reach their maximum height at 18-19 years. During adolescence, boys gain about 72 lb. and grow about 16 inches in height.
- Girls reach their maximum height at about 15-16 years. During adolescence, they gain about 55 lb. and grow about 9 inches.
- Growth is first noted in the musculoskeletal system. Growth to adult status occurs in this sequence: (1) head, hands and feet, (2) extremities, (3) trunk, (4) shoulders, chest, and hips.
- Skull and facial bones change: the forehead becomes more prominent and the jawbones develop.

### Glandular Changes

- The **eccrine** and **apocrine** glands become fully functional. The eccrine glands produce sweat over most of the body; the apocrine glands produce sweat (in response to emotional stimuli) in the axillae, anal and genital areas, external auditory canals, around the umbilicus, and the areolae of the breasts.
- **Sebaceous glands** become active on the face, neck, shoulders, back, chest, and genitals.
- **Sexual characteristics:** Primary sexual characteristics relate to reproductive organs (e.g., the vagina, ovaries, uterus, testes, and penis). Secondary sexual characteristics differentiate the male from the female but do not relate directly to reproduction (e.g., growth of pubic hair, voice changes, breast development). In males, the first sign of puberty is the appearance of pubic hair. In females,

the first sign is either hair along the labia or the appearance of a "breast bud." Menarche occurs about 2 years after appearance of the breast bud.

## Psychosocial Development

- **Erikson**: In this stage of identity vs. role confusion the adolescent is asking, "Who am I?" and "What will I be/do?" They help each other through this crisis by forming cliques, excluding those who are "different," and a separate youth culture. Their intolerance is a temporary defense against identity confusion.

- Adolescents are concerned about their body and appearance.

- Adolescents form their self-concept largely from the way in which others perceive them and relate to them, especially their peers.

- Sexual identification is a significant part of adolescence.

- Adolescents are sexually active and may engage in masturbation, heterosexual, and homosexual activity.

- Homosexual activity in adolescence is not necessarily an indicator of permanent sexual preference. Gay and non-gay adolescents may experiment sexually with persons of the same and opposite sex.

- The need for independence causes the adolescent to pull away from the family; the continued need for family support may create conflict.

- Peer groups are very important. They provide a sense of belonging, social learning, and sexual roles.

## Cognitive Development

**Piaget**: This is his formal operations stage. Adolescents can think beyond the present and the world of reality. Their thinking includes logic, organization, and consistency. Adolescents have a great capacity to absorb and use information.

## Moral Development

**Kohlberg**: The young adolescent is at the conventional level; they still want to abide by the Golden Rule ("Do unto others...") and the social order and existing laws. They examine their values and morals and may discard parental values for ones they find more suitable. As they move into the post-conventional (principled) level, they begin to question rules and laws and consider changing the laws. What is right/wrong becomes a matter of personal values and opinions. Not everyone proceeds to this level.

## Spiritual Development

**Fowler**: At this synthetic-conventional stage, adolescents are exposed to a wide variety of religious beliefs. They may reconcile their differences by (a) deciding that the "different" belief is wrong, (b) compartmentalizing the differences, or (c) asking for advice from a parent or minister.

## Health Problems

- Alcohol, tobacco, and drug use/abuse
- Sexually transmitted diseases
- Motor vehicle accidents
- Athletic injuries
- Eating disorders: obesity, anorexia nervosa, bulimia

### Therapeutic Nursing Interventions for Adolescence

- Encourage parents to provide love and support and help the adolescent feel valued by the family.
- Provide (and teach parents) consistent guidance, but give adolescents as much independence as they can handle. They need to know that parents are there to help if they need help.
- Promote peer understanding and acceptance of teenagers with handicaps (who are particularly vulnerable to rejection); discuss the specific problems with the peer group.
- Be open and accepting of adolescents' questions and statements.
- In spiritual matters, encourage adolescents to talk with members of their church or a peer group for support.
- Provide an environment in which adolescents can practice the rituals of their faith.
- Encourage adolescents to establish relationships in which they can discuss feelings, concerns, and fears.
- Teach parents to act as role models for social interactions.
- Encourage parents to make the home a comfortable place for adolescent peer group activities.
- Teach about periodic dental check-ups and vision and hearing screenings.
- Provide information about sexual issues.
- Be alert for signs of substance abuse and/or emotional disturbances.
- Teach and encourage healthy eating and regular exercise.

# Critical Thinking Exercise: Infant Through Adolescence

1. In working with parents to promote the health of children at all stages of development, nurses address a variety of issues and concerns. Develop a chart highlighting the major teaching concerns unique to each stage of childhood development.

   - Neonates and infants

   - Toddlers

   - Preschoolers

   - School-age children

2. Select one of the theories described in this chapter, and briefly outline the developmental tasks accomplished at each stage from birth to adolescence.

# Young Adult (20-40 years)

## Key Points

- Criteria for young adulthood includes: maturity, financial independence, "legal" age, living outside parents' home.
- Physically, young adults are in their prime.
- Psychosocial: Erikson, intimacy vs. isolation; Freud, genital stage
- Cognitive: Piaget, formal operations
- Moral: Kohlberg, postconventional; Gilligan, ethic of care, obligation
- Developmental tasks: Havighurst, relationships, career choices, and family establishment
- Spiritual: Rational examination of past religious teachings
- Health problems: Motor vehicle accidents are the leading cause of mortality; suicide, hypertension, substance abuse, STDs, spousal abuse, testicular cancer, cervical cancer
- Nursing interventions: Recommend regular physical examinations, immunizations, screening for major health risks (e.g., Pap smears), adequate iron in diet, sun protection measures, and automobile safety
- **Key Concepts/Terms**: Adulthood, maturity, physical development, cognitive development, moral development, spiritual development, health problems, therapeutic nursing interventions

**Overview**

For purposes of this review, young adults are people 20-40 years old. The developmental tasks of young adults differ from those of middle-aged and older adults. Young adults are quite busy. They must assume new roles at work and at home and develop related interests and values.

**Adulthood** is described in various ways. Some commonly used criteria of adulthood include:

- Financial independence
- "Legal" age (e.g., for consuming alcohol, voting, driving a car)
- Establishing a home separate from one's parents

**Maturity** means being fully developed. Characteristics representative of maturity include:

- Having a philosophy of life
- Being tolerant of the views of others
- Making sense out of life
- Maintaining a sense of hope and purpose in the face of misfortunes

- Continued growth and openness to new experiences
- Ability to tolerate ambiguity
- Flexibility, able to adapt to change
- Self-acceptance
- Making decisions and accepting responsibility for them

## Physical Development

Young adults are in their physical prime. All body systems are functioning at peak efficiency.

## Psychosocial Development

**Freud**: Genital stage (energy is directed toward attaining a mature sexual relationship)

**Erikson**: During intimacy vs. isolation, choices are made about education, employment, choosing a mate, having and rearing children, and assuming community activities.

## Cognitive Development (Piaget)

Piaget believes that cognitive development is complete at about age 15 years, and that adult thinking is characterized by "formal operations" (e.g., solving abstract problems, generating hypotheses).

## Moral Development

**Kohlberg**: This is the post-conventional level. The person can separate self from the rules, laws, and expectations of others, and defines morality in terms of personal principles. If the person is in conflict with a law, he/she acts according to own principles.

**Gilligan**: He calls this the ethic of care. Women often define moral problems in terms of obligation to care and avoid hurting others.

## Developmental Tasks

Havighurst, early adulthood major tasks include selecting a mate, starting a family, raising children, managing a home, beginning an occupation, becoming involved in civic and religious group activities, and forming social groups.

## Spiritual Development (Fowler)

Young adults enter the individuating-reflective period sometime after 18 years of age. They begin to focus on reality and ask philosophic questions, and may be self-conscious about spiritual matters. Past religious teaching is accepted, not accepted, or redefined after rational thought.

## Health Problems

This is usually a healthy time of life. Some problems of this age group include:

- Accidents: Motor vehicle accidents are the leading cause of mortality.
- Suicide
- Hypertension, especially for African-American men
- Substance abuse
- Sexually transmitted disease
- Abuse of women

- Malignancies, especially testicular cancer for men and breast and cervical cancer for women

## Therapeutic Nursing Interventions

### Young Adults primarily teaching and guidance regarding:

- Routine physical examinations every 1-3 years for women (5 years for males)
- Recommended immunizations
- Annual dental assessments
- Periodic vision and hearing screening
- Monthly breast self-examination (professional exam every 1-3 years)
- Annual Pap smear
- Monthly testicular self-examination
- Screening for cardiovascular disease every 5 years
- Tuberculosis skin test every 2 years
- Workplace safety
- Sun protection measures
- Motor vehicle safety
- Importance of adequate iron intake in diet
- Balanced nutrition and exercise

# Critical Thinking Exercise: Young Adult (20–40 years)

**Situation:** As part of a routine pre-employment screening, a nurse interviews a healthy 25-year-old male. How might the interview be tailored to the specific health concerns of this age group?

1. What are the major health problems of young adults? What nursing interventions are appropriate to address in these areas? With these concerns and interventions in mind, design a one-page handout to promote healthful behaviors among young adults.

2. One of the defining characteristics of young adulthood is maturity. What guidelines signal that an individual has reached maturity?

3. Identify and explain one or two of the major developmental tasks of young adulthood.

# Middle Adult (40-65 years)

## Key Points

- A period of stability and consolidation
- Physical: Important terms include menopause and climacteric.
- Psychosocial: Erikson, generativity vs. stagnation; important term is midlife crisis.
- Cognitive: Intellectual abilities change little; however, reaction time may be slower in late middle age.
- Moral: Kohlberg, postconventional level
- Spiritual: Many view "truth" from a variety of viewpoints (paradoxical-consolidative stage); less dogmatic about religious beliefs
- Developmental tasks: Havighurst, self-awareness and personal fulfillment
- Health problems more likely to occur than in young adulthood. They include chronic diseases, cancer, coronary heart disease, obesity, alcoholism, and mental health problems.
- Nursing interventions: Stress routine physical exams; tetanus booster every 10 years; screening for cancer, glaucoma, and cardiovascular disease; home safety; exercise and diet to control obesity; and retirement planning
- **Key Concepts/Terms**: Physical changes, psychosocial changes, cognitive development, moral development, spiritual development, health problems, therapeutic nursing interventions

**Overview**

This is the period of stability and consolidation. Children have grown and moved away from home, and partners have time for other interests and for each other.

## Physical Changes

- Hair begins to thin; gray appears
- Skin turgor and moisture decrease and wrinkling occurs
- Fat deposits in the abdominal area
- Skeletal muscle bulk decreases at about age 60
- Thinning of the intervertebral discs causes a 1-inch loss in height
- Visual acuity declines, especially for near vision
- Auditory acuity for high-frequency sounds decreases
- Metabolism slows, resulting in weight gain
- Tone decreases in large intestine, predisposing to constipation
- **Menopause** is the cessation of menstruation; it occurs sometime between the

ages of 40 and 55 (average is about 47 years). Ovarian activity declines and ovulation stops. Common symptoms are hot flashes, "shrinking" of the breasts, weight gain, insomnia, and headaches. These changes may produce anxiety in some women.

- The **climacteric** refers to the "change of life" in men, when sexual activity decreases as a result of diminishing androgen levels. However, men can still father children, even late in life.

## Psychosocial Development

**Erikson**: Generativity vs. stagnation: Generativity is the concern for establishing and guiding the next generation. Young adults tend to be self and family-centered; middle-aged persons seem more altruistic and interested in service to others and in love and compassion. People who are unable to expand their interests at this time suffer boredom and have difficulty accepting their aging bodies. They become preoccupied with self and unable to give to others. In these years of middle adulthood, developing a meaningful and lasting relationship (this can be a new relationship or a mending/refinement of a current relationship) can mean that the individual has moved from stagnation to generativity.

### Midlife Crisis

This crisis occurs when individuals recognize that they have reached the halfway point in their life. They begin to realize that they can no longer take youthfulness and physical strength for granted and that "time is running out."

### Cognitive Development (Piaget)

This is the phase of formal and/or post-formal operations. The cognitive and intellectual abilities change little in middle age; however, reaction time may be slightly slower during late middle age.

### Moral Development (Kohlberg)

The adult can move beyond the conventional level to the post-conventional level, after benefiting from extensive experience of personal moral choice and responsibility. Moving into the post-conventional level requires that the person believes in the importance of the rights of others and takes steps to support another's rights. Not everyone reaches this level.

### Spiritual Development (Fowler)

Some, but not all, adults progress to the paradoxical-consolidative stage, in which they view "truth" from a variety of viewpoints. This stage corresponds to Kohlberg's post-conventional level. Middle-aged people tend to be less dogmatic about religious beliefs; they often rely on spiritual beliefs to help them deal with illness, death, and other misfortunes.

### Developmental Tasks (Havighurst)

The middle adult should focus on the following developmental tasks:

- Assist children to become responsible adults
- Achieve social and civic responsibility
- Attain satisfying career

- Develop leisure activities and hobbies appropriate to age
- Strengthen relationship with partner
- Accept and adjust to physical status and deal with aging parents

## Health Problems

The risk of developing a health problem is greater than in young adulthood. Health problems of the middle adult include:

- Motor vehicle and occupational accidents
- Chronic disease (e.g., cancer, cardiovascular diseases)
- Health problems often caused by lifestyle patterns (e.g., smoking and chronic respiratory disease; overeating and obesity, diabetes, and coronary artery disease)
- Cancer is second leading cause of death between the ages of 25 and 65
- Coronary heart disease is the leading cause of death in the U.S. Physical inactivity is the greatest risk factor for CHD.
- Obesity related to decreased metabolic activity and decreased physical activity, without a decrease in calorie intake
- Alcoholism
- Mental health problems

## Therapeutic Nursing Interventions for Middle Adults

Health promotion topics include:

- Routine physical examinations
- Tetanus booster every 10 years; annual influenza and pneumococcal vaccinations
- Yearly dental assessments
- Screening for glaucoma
- Monthly breast self-examination for women
- Monthly testicular self-examination for men
- Teach postmenopausal women to report any vaginal bleeding
- Screenings for cardiovascular disease (e.g., cholesterol)
- Screenings for colorectal, breast, cervical, uterine, and prostate cancer
- Screenings for tuberculosis every 2 years
- Workplace and home safety measures
- Importance of adequate protein, calcium, and vitamin D in diet
- Importance of monitoring fat intake for cardiovascular fitness
- Need for exercise and diet to control obesity
- Retirement planning
- Encourage discussion of feelings, concerns, and fears regarding life stage

# Critical Thinking Exercise: Middle Adult (40–65 years)

**Situation**: A 48-year-old male inquires about recommended health screening for his age group. Base your response on your knowledge of the specific health concerns of middle adulthood.

1. As individuals move into middle adulthood, what health problems are they likely to experience? Identify the most appropriate nursing interventions to address these concerns. Use this information to develop a chart of recommended health screenings of middle adults.

2. What is meant by a "midlife crisis"? What do individuals begin to realize at this stage of life?

# Older Adult (over 65 years)

## Key Points

- 15% of those in the U.S. are 65 years of age or older.
- Those over 85 are rapidly increasing in number.
- Important terms: frail elderly, osteoporosis, presbyopia, cataracts
- Physical changes are many and involve all body systems.
- Psychosocial: Erikson, ego integrity vs. despair. Psychosocial changes include retirement, economic changes, and relocation of family home, inability to maintain independence, and facing death and grieving.
- Cognitive: Senses become less acute; more a loss of speed rather than ability; intellectual loss reflects disease process, not normal aging; and short-term memory declines.
- Moral and spiritual: Similar to middle adulthood
- Developmental tasks: Havighurst, necessary adjustments that arise from inevitable physical and social changes associated with aging
- Health problems: Leading causes of death are heart disease, stroke, pneumonia/influenza, obstructive lung disease, and cancer. Other problems include: accidents, chronic diseases (e.g., arthritis), drug misuse, alcoholism, dementia, and elder abuse.
- Nursing interventions: Be alert for depression, changes in cognitive function, and signs of abuse or neglect; be aware of the costs of health care.
- **Key Concepts/Terms**: Frail elderly, physical changes, psychosocial development, cognitive development, moral development, spiritual development, health promotion, therapeutic nursing interventions

## Overview

The number of Americans age 65 and older is about 15% of the total population. The group of people over 85 years old is increasing rapidly. The following are two systems for categorizing the older adult population.

- 55-64 years: the older population
- 65-74 years: the elderly
- 75-84 years: the aged

- 60-74 years: the young-old
- 75-84 years: the middle-old
- 85 and older: the old-old

The term **frail elderly** is used to describe the old-old, the extreme aged, and elderly individuals with significant physical and functional impairment, whatever their age.

## Physical Changes

**Integument:** Skin is dry, pale, and fragile, with wrinkling and sagging (loss of subcutaneous fat). Brown "age" spots appear on face, hands, and arms. Hair thins and grays. Nails thicken and grow more slowly.

**Neuromuscular:** Slowed reaction times, impaired balance, joint stiffness (degenerative joint changes), loss of height (atrophy of intervertebral discs). **Osteoporosis** is a decrease in bone density, occurring more often in people with insufficient calcium intake, postmenopausal women, and physically inactive individuals. There is a steady decrease in muscle fibers after age 50; therefore, elderly adults often complain about lack of strength and tiring quickly.

**Sensory/Perceptual:** All five of the senses become less efficient. Visually, there is increased sensitivity to glare and decreased ability to adjust to darkness. The eye lens thickens, becomes opaque (**cataract**), and inelastic (**presbyopia**). May become less sensitive to pain, touch, and temperature.

**Pulmonary:** Respiratory efficiency is reduced. There is decreased lung expansion, less effective exhalation, and increased residual volume (because the rib cage becomes more rigid, thoracic muscles weaken, and alveoli are less elastic).

**Cardiovascular:** Reduced cardiac output and stroke volume; may result in shortness of breath on exertion and pooling of blood in the extremities. Arteries lose elasticity; blood pressure increases. Orthostatic hypotension is common and is characterized by faintness due to a sudden drop in blood pressure when a person rises from a supine position to a standing or sitting position.

**Gastrointestinal:**

- Because of decreased enzyme production, loss of muscle tone in the intestines, and loss of absorbing cells in the intestinal tract, there is increased tendency for indigestion and constipation.

- Due to an increase in adipose tissue along with a decrease in muscle tone, the older client's abdomen may be more rounded. This is considered normal.

**Urinary:** Reduced filtering ability of the kidneys, waste products filtered and excreted more slowly. Tendency for nocturnal frequency and retention of residual urine. Other complaints include urinary urgency and frequency related to weakened supporting muscles or sphincter.

**Genitals:** Mild prostate enlargement in men. In women, shrinkage and atrophy of the vulva, cervix and other reproductive organs, reduced secretions, and changes in vaginal flora. Reduced vaginal secretions may cause painful intercourse, necessitating the use of lubricating jellies.

## Psychosocial Development

**Erikson:** This is the stage of ego integrity vs. despair. Ideally, ego integrity brings serenity and wisdom. Those who are successful with this task derive satisfaction from past accomplishments and view death as an acceptable completion of life. Those who do not achieve ego integrity wish they could live life over, they feel frustrated and discouraged, and that life has not been worthwhile.

**Retirement:** Most people over 65 are unemployed. However, many do continue to work, either for self-satisfaction or for economic reasons. The lifestyle after retirement is formulated in youth. Those who have always lived well-balanced, fulfilling lives are more likely to continue to do so in retirement.

**Economic change**: Costs continue to rise, along with inflation, making it difficult for some to manage on a fixed income. Food and medicine are major expenses for many. Adequate financial resources make it possible to remain independent. Elderly women of all ages have lower incomes than do men.

**Relocation**: The family home may be too large or too expensive to keep, or it may present difficulties for those with mobility problems. Some people must relocate to nursing homes when they can no longer care for themselves (e.g., because of memory impairment). For most, relocation is difficult and stressful. For some it is voluntary (e.g., the person may be seeking a more moderate climate).

**Maintaining independence and self-esteem**: Aging people thrive on independence. They should be allowed to do as much as possible for themselves, even though they may be slower and less meticulous than in the past.

**Facing death and grieving**: When a partner dies, the remaining partner experiences great loss and loneliness. More women than men face this loss because women usually live longer. Those with meaningful friendships, good relationships with children and grandchildren, and a peaceful philosophy of life cope better with bereavement.

## Cognitive Development

**Perception** (the ability to interpret the environment) changes as the senses become less acute. Changes in the nervous system also affect perception.

**Blood flow to the brain decreases**, brain metabolism slows, and there is progressive loss of neurons.

**Cognitive changes** are more often a difference in speed rather than in ability. Intellectual loss reflects a disease process (e.g., atherosclerosis, which decreases nutrients to the brain). Most older adults do not experience cognitive impairment.

**Memory**: Older adults tend to forget the recent past (short-term memory).

**Information retrieval** is slower, so older people need additional time for learning.

### Moral Development (Kohlberg)

Most old people remain at the conventional level and some are at the preconventional level. At the conventional level, they follow society's rules and the expectations of others. Because they were formed at a different time, the values of older adults may have little significance to younger people.

### Spiritual Development

Many older people have strong religious convictions. Religious involvement often helps the person resolve issues about the meaning of life and adversity. The older person derives a sense of worth by sharing spiritual views and experiences. One who has not matured spiritually may feel despair as economic and professional successes diminish.

### Developmental Tasks (Havighurst)

- Adjust to decreased physical strength and declining health
- Adjust to retirement and fixed income
- Adjust to death of spouse or companion

- Establish appropriate living arrangements if needed
- Accept one's own mortality
- Accept oneself as an aging person

## Health Problems

**Leading causes of death** in people over age 65 are heart disease, stroke, pneumonia/influenza, obstructive lung disease, and cancer.

**Accidents:** Vision is impaired, reflexes are slowed, and bones are brittle. Night driving requires caution because of impaired accommodation to light and diminished peripheral vision. A lowered metabolism and loss of subcutaneous tissue increases the risk for hypothermia.

**Chronic disabling illness:** Examples are: arthritis, osteoporosis, heart disease, lung disease, and cognitive dysfunctions. Chronic illness creates many difficulties: the person may need help with activities of daily living, health care expenses increase, family roles must be altered, and so on.

**Drug use and misuse:** Older adults often take many prescribed (and over-the-counter) medications. They may take combinations that interact, take too much or too little, take someone else's medication, and so on. Also, there are changes in the ability to absorb, metabolize, and excrete drugs.

**Alcoholism:** Those who begin drinking later in life do so to help them cope with the challenges of aging. Many late-onset alcoholics are widowers.

**Dementia:** This is a general term for a permanent or progressive organic mental disorder that is characterized by personality changes, confusion, disorientation, and confusion. Control of memory, judgment, and impulses is impaired. The most common cause of dementia is Alzheimer's disease, which affects more than 3 million people in the United States.

**Elder abuse:** Victims are most often women over the age of 75. It may occur in homes or health care facilities. Older adults may fail to report abuse because they fear being sent to a nursing home, because they fear retaliation, or because they are ashamed to admit that a family member has abused them.

## Therapeutic Nursing Interventions for Older Adults

- Physical assessment should include measuring weight, height, and vital signs, examination of the skin for hydration and lesions, examination of visual and hearing acuity.

- Assessment interviews should include questions about usual dietary pattern, bowel or urinary elimination problems, activity and sleep patterns, family and social activities, adjustment to loss of a partner, and any problems with reading, writing, or problem solving.

- Be aware of the costs of health care (e.g., consider which foods a client can afford to buy; ask the primary care provider to prescribe generic medications; use supplies carefully).

- Respect the values of the older person (e.g., heat soup on the stove instead of the microwave if the client prefers).

- Be alert for symptoms of depression; assess suicide risk.

- Be alert for changes in cognitive function.
- Assess for signs of physical abuse or neglect.
- Assess for tooth decay, loose teeth, or gingivitis.

# Critical Thinking Exercise: Older Adult (over 65 years)

**Situation**: In planning the discharge of an 82-year-old client from an acute care setting to home, a nurse takes into account the special needs of older adults.

1. Based on your knowledge of older adults, how might you adapt your discharge instructions to their specific needs?

2. Devise a safety plan for older adults living at home. Take into consideration limitations they are likely to experience and specific safety challenges encountered in the home.

# Therapeutic Relationships

## Key Points

- A helping relationship focuses on the client's needs (although the nurse may benefit).
- Helping relationships depend on the nurse's ability to communicate and provide comfort.
- Trust is essential to the helping relationship.
- Helping relationships are goal-directed (not just social chatter).
- Phases of the relationship: preinteraction, orientation, working, terminal
- Therapeutic nursing interventions for the helping relationship need to be varied, and include active listening.
- **Key Concepts/Terms**: Helping relationships, preinteraction phase, orientation phase, working phase, termination phase

**Overview**

The nurse-client relationship involves a collaborative effort of the nurse and the client to resolve a problem and/or promote the client's health and adaptation. A therapeutic relationship promotes a psychological climate that supports positive change and growth. Although the nurse may benefit from the relationship, the relationship focuses on the client's needs. The relationship depends on the nurse's ability to communicate and to provide physical and psychosocial comfort to the client. Only then will the client trust and be willing to share concerns with the nurse. The nurse is expected to identify the boundaries of the therapeutic relationship with each client. This means being able to accept the client's trust while maintaining a professional standard and limiting the relationship to focus on the well-being of the client.

## Characteristics of a Helping Relationship

A helping relationship:

- Involves caring, empathy, and sympathy on the part of the nurse
- Focuses on client needs
- Is goal-directed (e.g., has a purpose, is not just social chatter)
- Requires trust on the part of the client. This means that the nurse must be reliable, consistent, and competent.
- Respects client confidentiality
- Respects client individuality
- Considers family and cultural values

- Is based on mutual trust, respect, and acceptance
- Maximizes the client's abilities to participate in decision-making

## Swanson's Theory of Caring

Swanson defines caring as a nurturing way of relating to a valued other toward whom one feels a personable sense of commitment and responsibility. Caring consists of five processes:

- Knowing: Striving to understand an event as it has meaning to the life of others
- Being with: Being emotionally present to the other
- Doing for: Performing tasks/activities or providing care in a manner the person would most prefer
- Enabling: Facilitating the other's passage through life transitions
- Maintaining belief: Sustaining faith in the other's capacity to get through an event and face a future with meaning.

## Phases of a Helping Relationship

**Preinteraction phase**: Planning for the interaction (e.g., choosing a time and place, reviewing client data)

**Orientation phase**:

- Opening the relationship
- Clarifying the problem
- Structuring and formulating the contract (obligations to be met by both nurse and client)

Orientation begins when nurse and client first meet. It sets the tone for the rest of the relationship. After introductions, the nurse may engage in social "chit-chat" to put the client at ease. Clients may "test" the nurse in this phase, because trust is not yet developed and because they are reluctant to admit that they need help. By the end of this phase clients should feel comfortable talking with the nurse about sensitive issues, develop trust in the nurse, and view the nurse as able to help.

The primary task of the nurse in this phase is to build trust and to set goals with the client for the rest of the relationship.

### Working phase

- Exploring and understanding thoughts and feelings
- Facilitating and taking action

During this phase, nurse and client begin to view each other as unique individuals. This is when caring and empathy develop. The nurse provides support, but the client is responsible for decisions and actions.

### Termination phase

- Nurse and client accept feelings of loss
- Relationship ends

During this phase the nurse prepares the client for the end of the relationship. Together they evaluate goal achievement. The nurse may summarize or review the process; both nurse and client may share feelings about termination. Termination discussions should begin before the final interaction.

## Therapeutic Nursing Interventions for the Helping Relationship

You can facilitate development of a helping relationship by:

- Listening actively
- Helping the person to identify what he/she is feeling (e.g., "You seem angry....")
- Putting yourself in the client's shoes (e.g., empathy)
- Being honest
- Being genuine
- Being creative
- Being aware and respectful of cultural differences
- Maintaining client confidentiality
- Knowing your role and your strengths/limitations

# Critical Thinking Exercise: Therapeutic Relationships

**Situation**: In an effort to assist an anxious client, a nurse utilizes the principles of therapeutic communication. Keep these principles in mind as you respond to the following exercises.

1. What is the purpose of the therapeutic relationship?

2. What distinguishes therapeutic communication from other forms of communication?

3. Describe the phases of the therapeutic relationship. Do you believe it is possible to skip any of the phases without compromising the relationship?

# Basic Communication

## Key Points

- Communication is a complex, ongoing, and dynamic process in which meaning is generated and transmitted.
- Factors influencing communication: Developmental stage, perceptions, values, emotions, sociocultural background, gender, knowledge, roles/relationships, environment, space and territoriality
- Communication techniques: Attentive listening, physical attending, silence, providing general leads, being specific and tentative, open-ended questions, touch, restating/ paraphrasing, clarifying, offering self, giving information, acknowledging, clarifying sequence, presenting reality, focusing, reflecting, summarizing
- Communication barriers: Stereotyping, agreeing or disagreeing
- **Key Concepts/Terms**: Communication, nonverbal communication, verbal communication, therapeutic communication

**Overview**

Communication skills are important for assessing and meeting client needs, as well as for building relationships with other professionals. Effective communication is essential to helping relationships. **Communication** is the process of sharing information or of generating and transmitting means. **Verbal communication** is the use of both the spoken and written word. **Nonverbal communication** (body language) is the exchange of information without using words (e.g., touch, facial expression, posture, gait, gestures, sounds, voice tone, silence, and general physical appearance). **Therapeutic communication** is goal-directed and client-centered. The nurse needs to respond not only to the content of the client's communication, but also to the underlying feelings.

## Components of Communication

- **Sender**: the originator or source of the idea
- **Message**: the idea
- **Medium or channel**: a means of transmitting the idea, which can be verbal or nonverbal
- **Receiver**: the person who receives and interprets the message
- **Interaction**: the receiver's response to the message through feedback

## Factors Influencing Communication

- **Perceptions (the personal view of events)**: The client's perception of the nurse's intent affects her/his willingness to talk. Differences in perceptions can be a barrier to communication.

- **Values**: The nurse must be aware and respectful of client values. Nurses must not let their own personal values interfere with client relationships; they should not be judgmental.

- **Emotions**: are subjective feelings about events. A client who is angry, for example, will respond differently to the nurse's comments than one who is frightened. Nurses should not take it personally when clients ventilate their emotions on them. Nurses must also be aware of their own emotions.

- **Sociocultural background**: Culture is reflected in language, gestures, attitudes, and personal space. It sets limits for the way people act and communicate. Language differences can be a barrier to communication, and an interpreter may be necessary. A hospital interpreter is usually preferable to a family member.

- **Gender**: Men and women have different communication styles. Active listening and seeking clarification should help to prevent misperceptions.

- **Knowledge**: Nurses and clients usually have different levels of knowledge. Nurses need to assess the client's knowledge and use terms and phrases that they understand, without "talking down" to them.

- **Roles and relationships**: (e.g., students talk differently with friends than they do with instructors)

- **Developmental stage**: Period whereby an individual experiences physical, social, cognitive, moral or spiritual maturation. Fundamental tasks are achieved.

- **Environment**: (e.g., temperature, privacy)

- **Space and territoriality**: (e.g., if the nurse sits too close, the client may feel intimidated).

## Communication Techniques

**Attentive listening** means listening actively, using all the senses—paying attention to both the verbal and nonverbal message and noting whether the two are congruent. It means being sensitive to the feeling the person is conveying and not selectively listening only to what the nurse **wants** to hear.

**Physical attending** is a way of "being present" while listening. Five actions which indicate "involvement" are:

- Facing the other person squarely
- Adopting an open posture (do not cross arms or legs)
- Leaning toward the person
- Maintaining eye contact
- Trying to be relaxed (e.g., allow pauses, use natural gestures)

**Using silence** (e.g., accepting silences of several minutes without making any verbal response)

**Providing general leads** (e.g., "Perhaps you would like to talk about..." "And then what?") These are statements that encourage the client to verbalize and choose the topic of conversation.

**Being specific and tentative**: Example of specific statement: "You spilled the water." (not "You are clumsy.") Example of tentative statement: "You seem upset about..." (not "You're upset today!")

**Using open-ended questions** (e.g., "Tell me about..." "How are you feeling?") Such questions specify the topic to be discussed and lead the client to elaborate, clarify, and describe thoughts and feelings.

**Using touch** (e.g., placing a hand over the client's hand) Be sensitive to differences in clients' need for, and acceptance, of touch.

**Restating or paraphrasing** (e.g., The client says, "I didn't sleep a wink all night; just tossed and turned." Nurse replies, "You had trouble sleeping.") Repeating the client's statement in similar words conveys that the nurse has listened and understood the message.

**Seeking clarification** (e.g., "I'm not sure I understand," or "Would you say that again?") This is a way to make the client's broad overall meaning more understandable.

**Offering self** (e.g., "I'll stay with you until you feel better.") This conveys caring.

**Giving information** (e.g., "Your surgery is scheduled for 9 a.m.") Providing specific, factual information that the client may or may not have requested.

**Acknowledging** (e.g., "You walked twice as far today.") Giving recognition, non-judgmentally, of an effort or a change in behavior.

**Clarifying time sequence** (e.g., client says, "I took my pills already." Nurse replies, "Was that before or after lunch?")

**Presenting reality** (e.g., "That is not an elephant; it is your television.") Some clients need help differentiating the real from the unreal.

**Focusing** (e.g., helping the client expand on a topic of importance) The focus may be an idea or a feeling; however, the nurse often emphasizes a feeling to help the client recognize an emotion behind the words. For example, the client says, "My husband says he loves me, but I look so ugly now, I don't think he will even want to touch me, much less stay with me." The nurse replies, "You're worried about how your husband will react to your scars?"

**Reflecting**: This technique directs ideas, feelings, or questions back to clients to enable them to explore their own ideas or feelings. For example, the client says, "What should I do?" and the nurse replies, "What do you think might help?"

**Summarizing** (e.g., "We've talked about...") Stating the main points of the conversation helps to clarify any necessary points and introduce future planning.

## Communication Barriers

**Stereotyping** (e.g., "Men don't cry," "Women are whiners.")

**Agreeing or disagreeing** (e.g., client says, "That night nurse doesn't know what she's doing." Nurse replies, "Oh, she's pretty knowledgeable, actually.") Agreeing and disagreeing imply that the nurse is in a position to judge whether the client is right or wrong. It may cause the client to become defensive.

**Being defensive** (e.g., client says, "The food here is terrible," and the nurse responds, "They do the best they can; we don't have a huge budget to work with.") Such responses are attempts to protect professionals or the agency from negative comments. They imply that the client has no right to complain and keep the client from expressing true concerns.

**Challenging** (e.g., "It can't hurt that bad!") Such responses suggest that the client needs to prove his/her point of view. They do not consider the client's feelings.

**Asking "Why"** (e.g., "Why didn't you go to the doctor before this got out of control?") "Why" questions may be considered prying, may violate privacy, and can cause clients to be defensive. Nurses often ask them more out of curiosity than from intent to help the client.

**Testing** (e.g., "You're throwing around some big words; what is your medical background?") Such questions meet the nurse's needs, not the client's. They are a way of forcing the client to admit to something, such as the fact that they are "only" a client.

**Rejecting** (e.g., "I can't talk now; I'm too busy" or "Let's not talk about that.") This may make clients feel that the nurse is rejecting them, as well as their communication. By changing the subject and not following the client's leads, the nurse would be exhibiting rejecting behavior.

**Unwarranted reassurance** (e.g., "You'll feel better soon." or "Don't worry.") Using such clichés blocks the client's further expression of fears, feelings, and other thoughts.

**Passing judgment** (e.g., "You did the right thing," or at a birth, saying, "What a beautiful baby you have.") When nurses give opinions, approve/disapprove, moralize, or state their own values, it implies that the client must think as the nurse does. This fosters client dependence.

**Giving advice** (e.g., "You should have a regular bedtime for your children.") Such comments deny the client's right to be an equal partner. There are times, however, when expert advice is needed. But giving advice does not promote communication.

# Critical Thinking Exercise: Basic Communication

**Situation**: An elderly woman experiencing shortness of breath is accompanied by her adult children to an emergency room. The client, who speaks limited English, relies on her children as interpreters. In order to meet the needs of the client and her family, a nurse utilizes techniques of effective communication.

1. When a client speaks limited English, what strategies might a nurse employ to facilitate communication?

2. Working with a fellow student, identify and discuss the major factors influencing communication. Use the following questions to guide your discussion.

   • In addition to the spoken or written word, by what means does communication occur?

   • Identify attitudes or actions that serve as barriers to effective communication.

   • Describe techniques and nursing interventions that can be used to foster effective communication.

# Teaching and Learning

## Key Points

- **Learning** is a change in human disposition or behavior that occurs as a consequence of experience.
- Learning theories attempt to explain how and why learning occurs. They include: **behaviorism**, **cognitive theory**, and **humanism**.
- Bloom's "domains" of learning: cognitive, affective, and psychomotor.
- Factors that facilitate learning include: learner readiness, active involvement, and motivation; sufficient client energy, feedback, repetition, and timing.
- Factors that inhibit learning include: emotions (e.g., fear), pain, fatigue, cultural barriers, and ability.
- Before teaching, assess the learner's knowledge level and factors that might affect learning.
- Use repetition and reinforcement; summarize periodically.
- Use techniques to increase motivation (e.g., relate the content to something the client values).
- Sequence content thoughtfully (e.g., review what the learner knows and then proceed to new material).
- A teaching plan is devised and documented based on the clients learning needs; learning needs are prioritized.
- Collaboration with other health care professionals involved in the client's care is essential to the uniformity and the continuity of teaching.
- **Key Concepts/Terms**: Learning theories, factors that facilitate learning, factors that inhibit learning, guidelines for teaching, increasing motivation, teaching plans, content sequencing

**Overview**

Client education is a major part of nursing practice, a right of all clients, and a legal and professional responsibility of all nurses. **Learning** is a change in human disposition or behavior that is a consequence of experience; it is an intellectual, emotional, internal, collaborative, and cooperative process.

## Learning Theories

**Behaviorism** believes that the environment is the essential factor in determining human action. Where there is an environmental **stimulus**, it causes a **response** (behavior). Learning is a matter of **conditioning** and providing stimuli that cause the desired response/behavior. **Positive reinforcement** fosters repetition of the desired action. **Behaviorism** includes the concepts of imitation and modeling. **Imitation** is the process by which individuals copy what they observe; **modeling** is the process by which a person learns by observing the behavior of others.

**Cognitive theory** depicts learning as largely a mental or thinking process in which the learner structures and processes information. Developmental readiness and individual readiness are thought to be key factors in learning. Bloom identified three areas ("domains") of learning:

- **Cognitive** includes intellectual skills such as knowing, comprehending, applying, and analyzing
- **Affective** includes changes in feelings, emotions, interests, attitudes, and appreciation
- **Psychomotor** includes motor skills such as learning to walk with crutches

**Humanism** focuses on both the cognitive and affective qualities of the learner. Learning focuses on self-development and achieving full potential. Autonomy and self-determination are important: the learner identifies the learning needs and takes responsibility for meeting them. Learning occurs best when it is relevant to the learner.

## Factors that Facilitate Learning

- Motivation of the learner
- Readiness of the learner
- Energy level of client
- Active involvement by the learner (e.g., listening to a lecture is passive; discussion is active)
- Mild degree of anxiety that increases motivation to learn
- Relevance of the content to the learner
- Feedback that is meaningful to the learner
- Nonjudgmental support
- Simple to complex: that is, logical organization of the material that proceeds from simple to complex
- Repetition
- Timing (e.g., information is retained best when it can be used soon after it is presented)
- Environment (e.g., minimal distraction, physical comfort)

## Factors that Inhibit Learning

- Emotions such as fear, anger, or depression
- Physiologic events such as illness, pain or fatigue
- Cultural barriers
- Psychomotor ability (e.g., a client needs enough strength to learn to use a walker)

## Therapeutic Nursing Interventions: Guidelines for Teaching

- Use teaching activities that help the learner meet self-chosen objectives.
- Rapport between teacher and learner is essential.
- Make use of the client's previous learning in the present situation.
- Communicate clearly and concisely.
- Before writing a teaching plan or doing client education, assess the client's knowledge level and knowledge deficit as well as other factors that may affect the ability and incentive for the client to learn. Target the client's identified strengths. Utilize a method of teaching that meets the client's knowledge level and addresses the specific knowledge deficit in a way that is most meaningful to the client (use examples that the client can relate to, use facilities and references that are available and meaningful to the client).
- Involve the learner in planning the goals and content.
- Use strategies that involve a variety of the senses (e.g., use discussion and demonstration).
- Gear the pace of the teaching session to the client's needs. If the client loses interest, the pace may be too slow, or the client may be tired.
- Create a facilitative environment—free from noise or interruptions.
- Repetition reinforces learning. Summarize content periodically.
- Use vocabulary that the client understands. Words such as "OR" (operating room) and "vital signs" may be misunderstood by clients.

## Therapeutic Nursing Interventions: Increasing Motivation

- Relate learning to something the client values.
- Help the client make the learning situation pleasant and non-threatening.
- Encourage self-direction and independence.
- Show a positive attitude about the client's ability to learn.
- Offer support and encouragement (positive reinforcement).
- Create a learning situation in which the client is likely to succeed. Break the learning up into small tasks or steps.
- Use reading material that is at the client's reading level.

## Teaching Plans

A teaching plan should include cognitive, affective, and behavioral objectives (statements of what the client is expected to be able to know, feel, and do), an outline of the content to be learned, and a list of the teaching methods to be used. The following is an example of a portion of a teaching plan:

| Behavioral Objectives | Content Outline | Teaching Methods |
| --- | --- | --- |
| 1. Describe normal wound healing | I. Normal wound healing | • Use audiovisuals; describe wound healing |

| Behavioral Objectives | Content Outline | Teaching Methods |
|---|---|---|
| 2. Describe signs and symptoms of infection | II. Infection<br>• Signs and symptoms | • Give handout to show infected wound appearance<br>• Discuss mechanism of wound infection |

## Sequencing the Teaching Content

- Start with something the client is concerned about (e.g., how the disease will affect his lifestyle, not the pathophysiology of the disease).
- Review what the learner knows, then proceed to new material.
- Address early any area that is creating client anxiety. Too much anxiety can impair concentration (although slight anxiety can be a motivating factor).
- Teach the basics (simple) first, before teaching variations or adjustments (complex). For example, teach how to insert a urinary catheter before teaching how to deal with problems such as leaking or blockage.
- Schedule time for review and question.

## Evaluating the Effectiveness of Teaching

After the education/teaching program, the nurse should reevaluate the client's knowledge level and answer these questions:

- Have the objectives of the program been met by the client?
- Can the client express the information verbally or through demonstration?
- Can the client explain the information to another person?

# Critical Thinking Exercise: Teaching and Learning

**Situation:** You are responsible for teaching a 60-year-old client about a new medication. Before formulating your teaching plan, consider the various factors that influence learning.

1. Before you begin client teaching, what must you first do?

2. What strategies promote effective learning? How might you overcome obstacles to learning?

3. Describe the role of the nurse in client education.

# Leadership and Management

## Key Points

- Leading and managing are not the same thing; although leaders often manage, and managers must have leadership abilities.
- Effective leaders must have vision, influence, and power.
- Leadership styles: autocratic, democratic, "laissez-faire", bureaucratic, charismatic, transactional, transformational, connective, and shared
- Management functions: planning, organizing, directing, and controlling the work of others
- Leadership is the art of getting others to want to do something you are convinced should be done.
- Leaders are the ones who show the way and have a grasp of the big picture.
- **Key Concepts/Terms**: leader, manager, leadership style, power, reward power, coercive power, legitimate power, expert power, referent power, vision, influence, first- level manager, middle-level manager, upper-level manager

## Overview

Professional nurses frequently function as leaders and managers. **Leaders** influence others to work together to accomplish goals. They have vision and the ability to motivate and inspire confidence; they inspire change, facilitate behavior, and mentor others. They may or may not have an official appointment, but people listen to them and follow their ideas. **Managers** are employees to whom the organization has given authority, power, and responsibility for directing the work of others, establishing standards, and evaluating that work. They implement organizational policies, control resources, set goals, make decisions, and solve problems. They initiate and implement change. Good managers must have leadership abilities, and leaders often manage.

## Leadership Styles

**Autocratic (authoritarian, directive):** Leader makes decisions for the group. Leader assumes people are externally motivated and incapable of independent decision-making. Most effective in emergencies (e.g., a fire on the unit).

**Democratic (participative, consultative):** Leader encourages group discussion and decision- making. Leader assumes individuals are internally motivated and capable of making decisions.

**"Laissez-faire" (nondirective, permissive, ultraliberal):** Leader assumes the group is internally motivated and needs autonomy and self-regulation. The leader assumes a "hands-off" approach.

**Bureaucratic**: Leader presumes the group is externally motivated, but does not trust self or others to make decisions. Instead, the leader relies on organizational rules and policies, taking an inflexible approach.

**Charismatic**: Characterized by an emotional relationship between the leader and group members. The leader's personality evokes strong feelings of commitment.

**Transactional**: Represents the traditional manager focused on the day-to-day tasks of achieving organizational goals. The leader provides incentives to promote loyalty and performance (e.g., gives a nurse a weekend off to get her to work a night shift).

**Transformational**: Leader fosters creativity, risk-taking, commitment, and collaboration by empowering the group to share in the organization's vision. The leader facilitates individual independence, growth, and change, and converts followers into leaders through shared values, honesty, trust, and continued learning.

**Connective**: Leader promotes collaboration and teamwork within the organization and among other organizations in the community.

**Shared**: Leader assumes that a professional work force contains many leaders. No one person is considered to have more knowledge or ability than any others do. In effect, all are leaders.

## Effective Leadership

**Leaders** must have vision, influence, and power.

**Vision** is the mental image of a desirable and possible future state. It is the basic ingredient of leadership.

**Influence** is an informal strategy used to gain cooperation without exercising formal authority. It is based on a trusting relationship and often involves persuasion and good communication skills.

**Power** is the ability to influence—that is, to exert actions that result in changes in attitudes or behaviors of followers.

- **Reward power** is based on incentives the leader can offer.
- **Coercive power** is based on fear of retribution or withholding of rewards.
- **Legitimate power** is granted by a specific position or role (e.g., a manager).
- **Expert power** comes from the respect others have for the leader's abilities, knowledge, or skills.
- **Referent power** refers to the admiration and respect for the leader's charisma or success.

## Management

The manager's job is to accomplish the work of the organization. Roles and functions vary with the type of organization and the level of management.

## Levels of Management

**First-level managers** manage the work of non-managerial staff and the day-to-day activities of a specific work group.

**Middle-level managers** supervise several first-level managers and serve as liaisons between them and upper-level managers.

Upper-level managers are organizational executives who are responsible for establishing goals and strategic plans for the organization.

## Management Functions

- Planning (assessing a situation, establishing goals, developing a plan of action, writing policies and procedures to address situations in a consistent and acceptable manner)
- Organizing (the work of others)
- Directing (the work of others)
- Controlling (includes evaluating and rewarding the work of others)

## Skills and Competencies Needed by Nurse Managers

- Critical thinking
- Communication skills
- Networking
- Managing resources (e.g., budgeting)
- Enhancing employee performance (e.g., mentoring)
- Team building
- Evaluating effectiveness and efficiency
- Delegating

## The Five Rights of Delegation

- Right task: the task is delegable for a specific client, such as tasks that are repetitive, require little supervision, and are relatively non-invasive
- Right circumstances: appropriate client setting, and available resources
- Right person: the right person is delegating the right task to the right person to be performed on the right person
- Right direction/communication: a clear, concise description of the task including its objective limits, and expectations is given
- Right supervision: appropriate monitoring, evaluation, intervention are needed and feedback is provided

# Critical Thinking Exercise: Leadership and Management

**Situation**: Imagine that you are about to interview for your first position as a staff nurse. In preparing for this meeting, consider the qualities you will look for in a prospective manager. Enlist the help of a classmate in formulating questions that will guide you in assessing the leadership potential of this individual.

1. With your classmate assuming the role of the manager and you that of the prospective employee, role play the interview. Focus on questions that will help you discern the manager's leadership style and its effect on the workplace environment.

2. What distinguishes a leader from a manager? What qualities must both possess? What is unique to each?

3. Consider the advantages and the disadvantages of various styles of leadership. When, for example, might an autocratic style be necessary and appropriate? Under what circumstances might it be detrimental to the good of individuals or of the organization? Do you consider any one style of management superior to another? Be prepared to support your answer.

4. You have been hired by an advertising agency to write a 25-to-50 word job description for a nurse manager on a medical-surgical unit. Describe the ideal candidate.

# Change Process

| **Key Points** |
| :--- |

- Change is the process of making something different.
- Change can be unplanned, planned, covert, overt, or developmental.
- Characteristics of effective change agents include: communication and interpersonal skills, project expertise, knowledge of available resources, problem-solving skills, self-confidence, decisiveness, trust in self and others, and respect of others.
- Change occurs gradually and in stages.
- The change agent must identify the forces that motivate change and the forces that inhibit change (restraining forces).
- Techniques for dealing with resistance include:
  - Asking resisters about the reasons for their opposition
  - Clarifying and correcting inaccurate information
  - Emphasizing the positive results of the change
  - Being open to revisions, but clear about what must be maintained
- **Key Concepts/Terms:** Change agent

**Overview**

Change is the process of making something different. Change is a constant in nursing, and nurses are often **change agents** (individuals who initiate, motivate, and implement change). Accepting change takes time, especially when the change conflicts with a person's values and attitudes.

## Types of Change

**Unplanned change** occurs without any control or effort by a person or a group (e.g., a natural disaster).

**Planned change** is intended, a purposeful attempt by individual or groups to influence its own status quo or that of another system or person.

**Covert change** occurs without the individual's awareness (e.g., developing hypertension).

**Overt change** occurs with the person's awareness (e.g., the development of abdominal pain).

**Developmental change** occurs as a result of the physical and psychosocial changes that occur during the life cycle.

## Characteristics of Effective Change Agents

- Excellent communication and interpersonal skills

- Project expertise
- Knowledge of available resources
- Skilled in problem-solving
- Skilled in teaching
- Respect from those involved
- Self-confidence, risk-taking
- Inspires trust in themselves and others
- Ability to make decisions
- Broad knowledge base
- Good sense of timing

## Stages in Accepting Change

The individual:

- Becomes aware of the new idea or practice
- Seeks more information about the change
- Evaluates the information in relation to the present situation
- Mentally tries out the proposed change
- Actually tries out the change (e.g., a pilot study)
- Adopts and integrates the change into the present system

## Forces that Motivate Change

- Belief that it will bring economic gain
- Belief that the change will improve the situation
- Perception of the change as a challenge
- Visualization of the future impact the change will have
- Potential for growth, recognition, achievement, and/or improved relationships

## Forces that Inhibit Change ("Restraining forces")

- Intellectual and emotional security (low tolerance for change)
- Lack of time or energy
- Fear that something of value will be lost (e.g., threat to job security)
- Misunderstanding of the meaning and implications of the change
- Failure to see the "big picture"
- Perception that the change will not have any benefits ("won't work")
- Perceived loss of freedom to engage in desired behaviors/activities

## Therapeutic Nursing Interventions for Implementing Change

- Communicate with resisters to find out their reasons for opposing the change.
- Clarify information; correct inaccurate information.
- Present the negative consequences of resistance (e.g., poor client care).

- Emphasize the positive consequences of the change—how the person or group will benefit.
- Be open to suggestions/revisions, but be clear about what must be maintained.
- Maintain a climate of trust and support.
- Encourage interaction between supporters and resisters.
- Encourage supporters to empathize with resisters and relieve unnecessary fears.

# Critical Thinking Exercise: Change Process

**Situation:** In an effort to improve staffing, a community hospital has hired several new nurses. How might current staff and management work together with new nurses to promote effective change?

1. Imagine that, as a new member of a nursing staff, you encounter a lack of support and recognition from senior staff. What steps might you take to overcome resistance and implement positive change?

2. In your own experience, have you ever worked to promote a needed change in your family, school, or community? If so, what qualities enabled you to act as an agent of change? What resistance did you encounter? How were you able to respond to or perhaps even overcome this resistance?

3. Interview an experienced nurse on the topic of change. In preparing for your interview, you may either adapt the questions below or develop your own.

   - How long have you been a nurse, and what patterns of change have you seen during your career?

   - Have most of the changes you've witnessed been more positive than negative? Explain why you think so.

   - As a nurse, how have you participated in the process of change? What advice would you give a new nurse about implementing positive change?

# General Health Assessment

## Key Points

- Techniques for a general health assessment include assessment of general appearance and behavior and measurement of vital signs, height and weight.
- Techniques of physical assessment include inspection, palpation, percussion, and auscultation.
- Avoid chilling the client; expose only the areas of the body you are examining.
- Provide privacy.
- Explain each step of the examination.
- Warn the client before you touch her/him.
- Observe for expressions of discomfort.
- Measure height and weight without shoes.
- A complete health assessment is the process of making an evaluation of the client's condition.
- The review of systems is a systematic method for collecting data on all body systems.
- **Key Concepts/Terms**: Health assessment, inspection, palpation, percussion, auscultation, general survey, appearance, behavior, nursing interventions

**Overview**

A complete health assessment includes a nursing health history and a physical assessment. Physical assessment begins with a general assessment of appearance and behavior and measurement of vital signs, height, and weight. Assessment is conducted in a deliberate, purposeful, and systematic manner (e.g., head-to-toe). Physical assessment data is collected through techniques of inspection, palpation, percussion, and auscultation.

## Techniques of Physical Assessment

**Inspection**: Visual examination, either with the naked eye (e.g., assessing skin color) or with a lighted instrument (e.g., otoscope).

**Palpation**: Using the sense of touch to examine the body (e.g., palpating for distended bladder or pain on pressure). Use finger pads, as a rule. Use back of hand and fingers to test skin temperature. Palpate areas of tenderness last.

**Percussion**: Striking the body surface to elicit sounds or vibrations. In indirect percussion, the middle finger of the nondominant hand (called the "pleximeter") is placed firmly on the client's skin. The tip of the flexed middle finger of the other hand strikes the pleximeter. To obtain a clear sound, blows must be firm, rapid, and short.

**Auscultation**: Listening to sounds produced within the body. Direct auscultation uses the unaided ear (e.g., hearing the grating of a moving joint); indirect auscultation uses a stethoscope to amplify the sounds.

## Assessment of General Appearance

- Sex and race
- Vital signs
  - Temperature (normal 96.6°–99.3° F or 35.9–37.4° C)
  - Pulse (normal 60–80 bpm for adults; 120–160 bpm for infants)
  - Respirations (normal 12-18 breaths per minute for adults; 40-60 breaths per minute for infants)
  - Blood pressure (normal 120/80 for young and middle-aged adults; 140/80 for older adults)
- Body build, height and weight
- Odors (body, breath)
- Hygiene, grooming
- Signs of distress (e.g., bending over in pain)
- Facial expression (e.g., wincing in pain)
- Obvious signs of illness (e.g., pallor, labored breathing)

## Assessment of Behavior Includes:

- Overall attitude (e.g., cooperative, negative, withdrawn)
- State of arousal
- Affect/mood, appropriateness of responses, level of orientation
- Quantity, quality, and organization of speech
- Relevance and organization of thoughts
- Activity level (e.g., restlessness, pacing, wringing hands)
- Gait

## Therapeutic Nursing Interventions for General Health Assessment

- For frail or elderly clients, perform several short assessments in order not to tire them.
- Expose only areas to be examined, to avoid chilling (especially infants and elderly clients).
- Provide privacy (gowning, draping).
- Position client comfortably.
- Explain each step of the examination; warn the client, for example, "I'm going to touch your back now."
- Be sensitive to client's verbal and nonverbal expressions of discomfort.
- Ask clients about their height and weight before measuring them in order to get an idea of the person's self-image.

- Measure height and weight without shoes (always set the heavy weight first on a balance scale, then move the smaller weight until the scale balances at the client's weight).
- Wash hands before beginning.

# Critical Thinking Exercise: General Health Assessment

**Situation**: In caring for clients across the life span, a family nurse practitioner in a small clinic routinely performs health assessments. Use the following exercises to guide you in understanding the purpose and components of a general health assessment.

1. Enlist the help of a family member or close friend in refining your health assessment skills. With this individual's permission, obtain a comprehensive nursing health history. Structure your interview to include all essential elements related to your subject's health status.

2. What kind of information are you attempting to ascertain in a general health assessment? How will you integrate this information with the findings of a physical assessment?

# Vital Signs: Temperature

---

### Key Points

- Body temperature is regulated by the hypothalamus; the body produces heat by metabolism.
- Temperature reflects the balance between heat produced and lost from the body, measured in degrees.
- Normal temperature range: 96.6°-99.3° F (35.9°-37.4° C)
- Factors affecting temperature: age, circadian rhythms, environmental temperature, exercise, illness, injury, hormone changes, dehydration
- Oral temperature: Do not use for clients who might bite the thermometer (e.g., confused). Be sure client has not ingested hot or cold liquids in the 20 minutes preceding.
- Rectal temperature: Stool in rectum can cause inaccurate reading. Do not use for newborns or clients who have a rectal disorder.
- Axillary temperature: Be sure axillary area is dry.
- Tympanic temperature: Place snugly into client's outer ear canal.
- Temperature variations: Rectal temperatures are a few tenths of a degree higher, and axillary temperatures are a few tenths of a degree lower than oral temperatures. Tympanic temperatures are usually charted without conversion, although it is possible to convert to oral, rectal, or core equivalents.
- **Key Concepts/Terms**: Temperature regulation, radiation, conduction, convection, evaporation, fever

## Overview

**Vital signs** are an indication of body function necessary to sustain life, and are used to measure a person's health status. Temperature, pulse, respiration, and blood pressure make up a "set" of vital signs. **Temperature** reflects the balance between heat produced and heat lost from the body, measured in degrees.

**Temperature regulation** and heat production is affected by the basal metabolic rate, muscle activity, thyroxine output, sympathetic stimulation, and fever. When the skin becomes chilled: (1) shivering increases heat production, while (2) vasoconstriction and decreased sweating occur to decrease heat loss. Heat is lost from the body through:

- **Radiation**: Transfer of heat from one object to another object without contact between them (e.g., heat lost from the body to a cold room)
- **Conduction**: Transfer of heat from the body to another surface (e.g., when the body is immersed in cold water)
- **Convection**: Dispersion of heat by air currents

- **Evaporation:** When water vaporizes heat, energy is used to change it from a liquid to a gas (e.g., sweating cools the body).

**Normal temperature range** is 96.6°–99.3° Fahrenheit, or 35.9°–-37.4° Celsius. Rectal temperatures are a few tenths of a degree higher and axillary temperatures are a few tenths of a degree lower than oral temperatures. The tympanic membrane is also used.

**Fever** is an elevation in temperature that is caused by disease. Fever is usually not harmful unless it goes above 39° C (102° F).

### Factors Affecting Body Temperature

- **Age:** Older adults have a loss of subcutaneous fat that results in lower body temperatures and feeling cold. Because of changes in temperature regulation, it takes longer for body temperature to register on the thermometer. Newborns have a large surface-to-mass ratio and therefore lose heat rapidly to the environment.

- **Circadian (diurnal) rhythms** have to do with the 24-hour, day-night cycle. Temperature is usually lowest between 1 and 4 a.m. and peaks between 4 and 7 p.m.

- **Environmental temperature** (e.g., heat stroke, hypothermia)

- **Exercise and activity:** Increase metabolism and produce heat

- **Illness or injury:** Pyrogens alter the hypothalamic set point.

- **Hormone changes** (e.g., during ovulation)

- **Dehydration**

## Nursing Interventions for Taking Temperature

### Sites

- **Oral:** Place under tongue in posterior sublingual pocket lateral to center of lower jaw. Leave in place 2 min. or according to agency policy, using mercury thermometer. Do not use for clients who might bite thermometer (e.g., small children, confused clients), mouth breathers, clients with trauma to the face or mouth.

- **Rectal:** Provide privacy. Use Sims' position with upper leg flexed. If using glass thermometer, hold in place for 2 min. or according to agency policy. Stool in rectum can cause inaccurate reading. Do not use for newborns or clients who have a rectal disorder.

- **Axillary:** Be sure axilla is dry. This is the site that is routinely used for newborns. If using glass thermometer, hold in place for 5-10 min. or according to agency policy.

- **Tympanic:** The tympanic probe should surround the skin of the outer ear, rather than the mucous membrane to minimize the risk of spreading infection. Temperature will be recorded in 1 to 2 seconds.

**Thermometers** can be glass, disposable/single-use, electronic. For glass, shake down to 35.5° C (96° F) before using.

**To Convert Celsius (centigrade) to Fahrenheit**: Multiply by 9/5 and add 32°.

**To Convert Fahrenheit to Celsius (centigrade)**: Subtract 32° and multiply by 5/9.

- When taking oral temperature, be sure client has not smoked or ingested hot or cold liquids in the past 20 minutes. Use of oxygen can also alter readings.

- For fever, provide fluids, rest, and antipyretics (e.g., aspirin, acetaminophen). A cooling blanket or tepid bath may also be used. Offer blankets during chills; remove when client feels warm.

- Provide oral hygiene and dry clothing and linens.

- If hypothermic, provide a warm environmental temperature, warming blanket, friction to extremities, or warmed oral or IV fluids.

# Critical Thinking Exercise: Vital Signs: Temperature

**Situation**: An 82-year-old male arrives at an emergency room with a body temperature of 101° F in combination with an elevated heart and respiratory rate, restlessness, and skin that is warm to touch. Describe specific nursing interventions you will use in caring for this and other febrile clients.

1. Discuss with a classmate how an understanding of aging and its effects on body temperature can guide your assessments. For example, how does the baseline body temperature of an elderly individual compare with that of a young or middle-aged adult?

2. The selection of a site for assessing body temperature may be guided by agency policy or left to the discretion of individual nurses. In the absence of a specific policy, identify the site you would select to assess temperature in these individuals and describe the rationale for your choice.

   • A newborn

   • A confused 82-year-old client

   • A healthy 60-year-old female

# Vital Signs: Pulse

| Key Points |
| --- |

- The normal pulse reflects an equal amount of time between beats.
- The volume of the pulse refers to the amount of blood pushing against the artery wall with each beat.
- Pulse is the wave-like sensation felt in a peripheral artery; it is a measure of how often the heart beats.
- Normal range: 60-100 bpm (adults); 120-160 bpm (infants)
- Factors affecting pulse: age, gender, exercise, fever, hemorrhage, stress, position changes
- Sites: temporal, carotid, apical, brachial, radial, femoral, popliteal, posterior tibial, pedal
- Assess rate, rhythm, volume, and arterial wall elasticity.
- Palpate using middle three fingertips; excess pressure can obliterate the pulse.
- Compare pulses bilaterally
- **Key Concepts/Terms**: Tachycardia, bradycardia, palpation, auscultation

**Overview**

**Vital signs** are an indication of body function necessary to sustain life, and are used to measure a person's health status. Temperature, pulse, respiration, and blood pressure make up a "set" of vital signs. **Pulse** is the wave-like sensation that can be felt in a peripheral artery. Pulses can be palpated in several areas of the body. The pulse rate is a measure of how often the heart beats.

> **Tachycardia**: Abnormally fast pulse
> **Bradycardia**: Abnormally slow pulse
> **Normal Range**: 60-100 beats per minute (at rest) for adults; 120-160 beats per minute for infants

## Factors Affecting Pulse

**Age**: Rate gradually decreases with age. In addition, when a person's arteries lose their ability to contract and expand (as in old age), greater pressure is required to pump the blood into the arteries. Therefore, the pulse may feel weak when palpated.

**Sex**: Males have a slightly lower rate than females.

**Exercise**: Rate increases with activity.

**Fever**: Rate increases because of increased metabolic rate and peripheral vasodilation.

### Medications

**Hemorrhage:** A small blood loss (e.g., 500 mL) results in compensatory increase in heart rate.

**Stress (e.g., in fear and anxiety):** Sympathetic nervous stimulation increases rate and force of heartbeat.

**Position changes:** Rate is faster when sitting and standing than when lying down.

## Nursing Interventions for Taking Pulse

- Sites
    - **Temporal:** Use when radial not accessible.
    - **Carotid:** Use in infants, cardiac arrest, and to determine circulation to brain.
    - **Apical:** 3 in. left of sternum at 4th, 5th, or 6th intercostal space; use for cardiac disease, irregular pulse.
    - **Brachial:** Use to measure blood pressure, for cardiac arrest in infants.
    - **Radial:** Thumb side of inner aspect of wrist is the most commonly used site. Not used for children up to 2-3 years, in very obese or elderly clients (unless using Doppler equipment), or persons with heart disease.
    - **Femoral:** Use in cardiac arrest, for infants/children, and to determine circulation to legs.
    - **Popliteal:** Use to determine circulation to lower leg.
    - **Posterior tibial:** Use to determine circulation to foot.
    - **Pedal (dorsalis pedis):** Use to determine circulation to foot.
    - **Apical-radial pulse:** A-R measurement is ordered when it is suspected that the client's heart is not effectively pumping blood. If the apical and radial measurements are not the same, a pulse deficit exists.
- **Palpation:** Use middle three fingertips to palpate all pulse sites except apical. Excessive pressure can obliterate pulses.
- **Auscultation:** Use stethoscope for apical pulses and fetal heart tones. Use Doppler ultrasound stethoscope for pulses that are difficult to hear.
- **Collect the following data:** Rate, rhythm (regular, irregular), volume (full, weak, thready, bounding, or use a scale of 0 to 3), and arterial wall elasticity
- Compare pulses bilaterally.
- If the pulse is regular, count for 30 seconds and multiply by 2. If irregular, count for one minute.
- For apical pulse, count for one minute.
- Assess for clinical signs of cardiovascular alterations (e.g., dyspnea, fatigue, pallor, and cyanosis).
- Assess for factors that may alter pulse rate (e.g., emotional status, medications such as digoxin).
- **Pharmacology:** Do not administer cardiac glycosides if apical pulse is below 60 bpm.

# Critical Thinking Exercise: Vital Signs: Pulse

**Situation:** Before administering digoxin, a nurse assesses a client's heart rate and rhythm. "Why are you doing that?" the client asks. How would you explain the purpose of your assessment? Under what circumstances is it appropriate to withhold digoxin? Describe your teaching plan for a client who will continue taking digoxin after discharge from an acute care setting.

1. Using a simple drawing or diagram of the adult human body, indicate the location of each of the nine sites where pulse is commonly assessed. Using a friend or classmate as a model, demonstrate how you would locate and assess each of these pulse sites.

2. In addition to pulse rate, rhythm, and volume, what other factors are most useful in evaluating a client's cardiovascular status?

3. In an acute care setting, as you auscultate a client's apical pulse while simultaneously palpating the radial pulse, you discover a disparity between these two rates. What are some possible causes of such a disparity, or pulse deficit?

# Vital Signs: Respirations

| Key Points |
| --- |

- Inspiration is inhaling air with oxygen into the lungs.
- Expiration is exhaling air with carbon dioxide.
- Respiration is the act of breathing; it includes the exchange of oxygen and carbon dioxide between the lungs and the pulmonary blood.
- Normal range: 12-20 breaths/minute (adult); 40-60 breaths/minute (infant)
- Procedure: Count for 30 seconds if regular; 60 seconds if irregular
- Observe for depth, rhythm, and character of respirations.
- Be aware that respirations, unlike other vital signs, can be controlled voluntarily (to some degree) by the client.
- **Key Concepts/Terms**: External respiration, internal respiration, inspiration, expiration, ventilation

## Overview

**Vital signs** are an indication of body function necessary to sustain life, and are used to measure a person's health status. Temperature, pulse, respiration, and blood pressure make up a "set" of vital signs. **Respiration** is the act of breathing. It includes the exchange of oxygen and carbon dioxide between the lungs and the pulmonary blood. The respiratory rate is the number of breaths in one minute.

**Normal Range**: 12-20 breaths per minute (at rest) for adults; 40-60 breaths per minute for infants

## Terminology

**External respiration**: The interchange of oxygen and carbon dioxide between the alveoli of the lungs and the pulmonary blood.

**Internal respiration**: The interchange of oxygen and carbon dioxide between the circulating blood and the cells of the body tissue.

**Inhalation (inspiration)**: Intake of air into the lungs.

**Exhalation (expiration)**: Breathing air out of the lungs.

**Ventilation**: Movement of air in and out of the lungs.

**Hyperventilation**: Very deep, rapid respirations.

**Hypoventilation**: Very shallow respirations.

**Eupnea**: Breathing that is normal in rate and depth.

**Bradypnea**: Abnormally slow respirations below 12 breaths a minute at rest.

**Tachypnea**: Abnormally fast respirations above 18 breaths per minute at rest.

**Apnea:** Absence of breathing.

**Dyspnea:** Painful or difficult breathing.

**Cheyne-Stokes respirations:** Slow and shallow at first, gradually grow faster and deeper, then taper off until they stop entirely.

## Factors Affecting Respirations

**Anxiety, Stress:** Increases rate and depth

**Activity:** Increases rate and depth

**Increased environmental temperature:** Increases rate

**Decreased environmental temperature:** Decreases rate

**Increased altitude:** Increases rate

**Certain medications (e.g., narcotic analgesics):** Decrease rate

**Increased intracranial pressure:** Decreases rate

**Age:** Lung capacity and gas exchange decrease with age, resulting in more rapid and shallow respirations.

## Nursing Interventions for Taking Respirations

- Choose a time when client is relaxed.
- Grasp the client's wrist and place the client's arm across the chest; observe the chest movements while appearing to take the radial pulse.
- Administer medications for respiratory conditions (e.g., bronchodilators, antibiotics).
- **Inspect** chest movements (e.g., intercostal and substernal retraction).
- **Palpate** and count the respiratory rate for 30 seconds if rate is regular; count for 60 seconds if irregular. An inhalation and an exhalation count as one respiration.
- **Assess:**
  - Depth, rhythm, and character of the respirations
  - Skin and mucous membrane color
  - Position assumed for breathing
  - Signs of cerebral anoxia (e.g., restlessness, drowsiness)
  - Activity tolerance
  - Associated symptoms such as chest pain and dyspnea
  - Medications affecting respiratory rate
  - Cough (e.g., productive, nonproductive)
  - Respirations following administration of respiratory depressants such as morphine

# Critical Thinking Exercise: Vital Signs: Respirations

**Situation:** You are caring for an elderly postoperative client who is receiving IV Demerol for pain management by means of patient-controlled analgesia. Within an hour of starting the PCA, the client's respiratory rate declines from 14 breaths/minute to 8 breaths/minute. Consider the action you will take.

1. Identify the parameters of normal respiration. When assessing respiration, what special considerations apply to clients receiving narcotic analgesia? Specifically, what adverse effects might these medications have on a client's respiration? What medications can be used to reverse postoperative opioid effects?

2. Imagine you are caring for a client with chronic obstructive pulmonary disease (COPD). What physical manifestations of COPD might you observe? When assessing respiration, what deviations from normal would you expect to see? Describe the way in which individuals with COPD typically breathe. How might you promote a more efficient pattern of breathing?

3. One of the best ways to learn a skill is by teaching it to someone else. To refine your own assessment skills, teach a classmate how to perform a respiratory assessment. Before you begin, consider the components of the assessment. What techniques will you use? How can you distinguish normal from abnormal patterns of respiration? Encourage your classmate to ask any questions that come to mind. When you have finished, reverse roles and allow your classmate to teach you how to assess respiration.

# Vital Signs: Blood Pressure

## Key Points

- Blood pressure (BP) is the force exerted by the blood in the arteries during heart contraction and relaxation.
- The systolic pressure is the higher number and represents the ventricles contracting.
- The diastolic pressure is the lower number and represents the pressure within the artery between beats.
- The difference between the two readings is called the pulse pressure.
- Arterial blood pressure is a result of the following factors: cardiac output, peripheral vascular resistance, blood volume, and blood viscosity.
- Normal range: 120/80 for young and middle-aged adults; 140/80 for older adults
- Factors affecting BP: age, exercise, stress, race, obesity, gender, medications, diurnal variations, dehydration, disease processes (e.g., kidney disease, heart failure)
- Usual site: brachial artery
- Usual position: Sitting, legs not crossed, arms not above heart level
- **Key Concepts/Terms**: Systolic pressure, diastolic pressure, hypertension, hypotension, orthostatic hypotension.

**Overview**

**Vital signs** are an indication of body function necessary to sustain life, and are used to measure a person's health status. Temperature, pulse, respiration, and blood pressure make up a "set" of vital signs. **Blood pressure** is the force exerted by the blood in the arteries during heart contraction (systole) and relaxation (diastole). **Systolic pressure** is the pressure exerted by the blood during systolethe higher pressure. **Diastolic pressure** is the pressure exerted by the blood during diastolethe lower pressure. Blood pressure varies considerably among individuals, and it also varies in a given individual, depending on a number of factors:

**Hypertension** is a blood pressure that is persistently above normal. Hypertension is usually asymptomatic.

**Hypotension** is a blood pressure that is below normal (e.g., 85-100 mm Hg systolic).

**Orthostatic hypotension** is a blood pressure that falls when a client stands or sits, usually as a result of peripheral vasodilation.

**Arterial blood pressure** is a result of several factors:

- **Cardiac output (the volume of blood pumped into the arteries by the heart)**: Strong heart action increases blood pressure.

- **Peripheral vascular resistance (e.g., the size and elasticity of the arterioles and capillaries):** Increased resistance causes increased blood pressure.
- **Blood volume (the total amount of blood circulating through the vascular system:** Blood pressure increases when volume is greater.
- **Blood viscosity ("thickness" of the blood):** Increased viscosity raises blood pressure. When the hematocrit is more than 60-65%, blood pressure increases markedly.
- **Normal Range:** Normal measurement is 120/80 for young and middle-aged adults. For adults over age 60, 140/80 is considered normal.

## Factors Affecting Blood Pressure

**Age:** Pressure rises with age because the elasticity of arterial walls decreases. Older adults also have higher incidence of chronic cardiovascular disorders, which affect pulse and blood pressure.

**Exercise** increases blood pressure by increasing cardiac output.

**Stress:** Due to stimulation of the sympathetic nervous system, BP increases.

**Race:** For example, African-American males over age 35 have higher blood pressures than European American males of that age.

**Obesity** raises blood pressure, in general.

**Sex:** Adult women usually have lower blood pressures than adult males.

**Medications** (e.g., morphine)

**Diurnal variations:** Pressure is usually lowest early in the morning.

**Dehydration** decreases blood volume.

**Disease processes:** Any condition affecting cardiac output or blood volume affects the blood pressure. Some examples are: stroke, seizures, myocardial infarction, kidney disease, and heart failure.

## Therapeutic Nursing Interventions

### Taking Blood Pressure

- Blood pressure is measured in millimeters of mercury (mm Hg). The systolic pressure is written over the diastolic pressure (e.g., 120/80).
- **Equipment** (a) blood pressure cuff, sphygmomanometer, and stethoscope, (b) Doppler ultrasound stethoscope. Cuff width should be 40 percent of the circumference of the arm. If cuff is too narrow, blood pressure will be erroneously high; if too wide, erroneously low.
- **Sites:** Usually over brachial artery in client's arm.
- **Position:** Usually sitting, elbow slightly flexed, palm of hand facing up; support client's forearm at heart level. Specify if reading is done in any other position.
- On initial BP, perform a preliminary palpatory determination of systolic pressure before auscultating.
- Use bell-shaped stethoscope diaphragm.

- Pump up to about 30 mm Hg above the point where the brachial pulse disappeared on palpation. To get an accurate reading, release valve on cuff so that pressure decreases at 2-3 mm Hg per second.

- If necessary to repeat measurement, wait 2 minutes before pumping up cuff again.

- Make sure that client has not smoked or ingested caffeine in the 30 minutes before taking the BP.

- Advise hypertensive clients to limit salt, exercise, limit stress, lose weight, and have regular blood pressure checks.

- For hypotensive clients, advise to rise from chair or bed slowly, promote safety if dizziness is present.

- **Electronic blood pressure apparatus**: Apply and manipulate the cuff the same as an anaeroid manometer: blood pressure and pulse will print out on the screen within a few seconds. The stethoscope may be used to double-check if needed.

- **Medications**: antihypertensives (e.g., clonidine, propranolol, hydralazine, nifedipine, captopril), vasopressors (e.g., norepinephrine, phenylephrine)

# Critical Thinking Exercise: Vital Signs: Blood Pressure

**Situation**: You are an adult nurse practitioner working in a collaborative practice with a physician. A healthy 40-year-old female arrives for a yearly physical examination. Although all other findings are normal, the client's blood pressure is measured as 145/90. How do you interpret this measurement?

1. Describe several factors affecting blood pressure. Is a single elevated blood pressure measurement sufficient to support a diagnosis of hypertension? Why or why not?

2. Design an information packet for clients called "Tips for Maintaining Normal Blood Pressure." In your packet, include the following information:

   • Why maintaining normal blood pressure is so important

   • Foods to choose and foods to avoid

   • Adopting an exercise program that works

   • Proven methods for reducing stress

   • When and why medical management might be needed

   Can you think of any other information to include in your packet? Consult the American Heart Association for additional ideas.

# Physical Assessment of Body Systems

## Key Points

- A complete physical assessment includes assessment of the integument, head, neck, upper and lower extremities, chest and back, abdomen, genitals, anus, and rectum.

- A health assessment is performed in addition to vital signs and a general assessment in order to obtain baseline data and make clinical judgments about the client's health status.

- Physical assessment techniques include inspection, auscultation, percussion, and palpation.

- **Neurologic**: Includes level of consciousness, orientation, ability to follow commands, behavior and appearance, language, memory, knowledge, abstract thinking, cranial nerve function, and sensory function

- **Skin**: Assess color, moisture, temperature, texture, turgor, edema, and lesions.

- **Hair**: Assess for amount, alopecia, distribution, texture, oiliness, scalp lesions, infections, or infestations.

- **Nails**: Assess shape, texture, color of nail bed, tissues surrounding nails, and capillary refill.

- **Skull and Face**: Inspect for size, shape, and symmetry; palpate for nodules, masses, and depressions.

- **Eyes**: Assess visual acuity, visual fields; pupils for reaction and accommodation.

- **Ears**: Inspect and palpate external ear; use otoscope to examine tympanic membrane; test auditory acuity.

- **Nose and Sinuses**: Assess shape, color, flaring, discharge, tenderness, masses, symmetry, and patency.

- **Mouth and Oropharynx**:
  - Equipment needed: Disposable gloves, tongue blade, gauze squares, penlight
  - Inspect for symmetry, color, hydration, caries, and lesions.

- **Neck**: Inspect for range of motion, stiffness, and discomfort. Palpate trachea, thyroid gland, and lymph nodes.

- **Thorax and Lungs**:
  - Auscultate both the anterior and posterior thorax.
  - There are landmarks for the anterior, posterior, and lateral chest, which help identify the position of the lobes of the lungs.
  - Use the flat-disc diaphragm of the stethoscope.

- **Heart**: Auscultate for $S_1$, $S_2$, systole, and diastole; palpate carotid pulse; inspect jugular veins.

- **Peripheral Circulation:**
  - Refer to "Vital Signs—Pulse."
  - Inspect skin color and temperature, varicosities, edema, and phlebitis; check Homan's sign.
- **Abdomen:**
  - Begin with inspection, then follow with auscultation, palpation, and percussion.
  - Auscultate in all four quadrants; listen for 3-5 minutes before concluding no bowel sounds.
- **Musculoskeletal:** Inspect for range of motion, inflammation, swelling, deformity, tenderness, warmth, stiffness, joint instability, and skin changes.
- **Key Concepts/Terms:** Functions of skin, appearance of skin, hair, nails, skull and face, eyes, ears, nose and sinuses, mouth and oropharynx, neck, thorax and lungs, heart, peripheral circulation, abdomen, musculoskeletal, neurologic

## Overview

A complete physical assessment includes assessment of the integument (skin, hair, and nails), head (skull and face, eyes, ears, nose and sinuses, mouth and oropharynx), neck, upper and lower extremities, chest and back, abdomen, genitals, anus and rectum.

Systems assessment is performed in order to obtain baseline data, confirm, or refute data obtained in the nursing history, and to make clinical judgments about the client's health status.

## The Skin

**Function:** Skin (integument) protects the body, regulates temperature, and acts as a sensory organ. Skin condition reflects the client's general state of health.

**In older adults,** skin loses its elasticity. It appears thin and translucent because of loss of dermis and subcutaneous fat. It is dry and flaky because sebaceous and sweat glands are less active.

**Procedure and technique** includes **inspection, palpation,** and olfaction to detect unusual odors. The skin is usually assessed gradually as other body systems are being examined.

**Color** varies among individuals, but is usually uniform over the body.

- **Pallor:** Indicates reduced tissue circulation or oxygenation
- **Cyanosis:** A bluish tinge, most easily seen in the nail beds, lips, and buccal mucosa (in dark-skinned clients the best site to assess for cyanosis is the nail beds)
- **Jaundice:** A yellowish tinge, first seen in the sclera of the eyes and then in mucous membranes and skin
- **Erythema:** Redness (associated with rashes)
- **Vitiligo:** Patches of hypopigmented skin
- **Moisture:** Palpate for wetness and oiliness (wear disposable gloves).
- **Temperature:** Palpate with back of the hand; reflects circulation adequacy.

- **Texture**: Palpate with fingertips, for smoothness or roughness.
- **Turgor**: Gently pinch skin on back of the hand and release. The skin stays pinched when turgor is poor. Skin is less elastic with age or in dehydration.
- **Edema**: Buildup of fluid in the tissues. Edematous skin is swollen, stretched, and shiny. Palpate to determine consistency and tenderness. When the finger leaves an indentation, it is called pitting edema. (1+ = barely detectable; 2+ = indentation of <5 mm; 3+ = indentation of <5-10 cm; 4+ indentation of >10 cm)

**Vascularity**: Look for localized pressure spots, superficial blood vessels, and petechiae (tiny, pinpoint-sized, red spots on the skin that may indicate blood-clotting disorders).

**Lesions**: Inspect, palpate and describe (size, shape and texture, color; distribution, and configuration):

- **Vesicle, bulla**: Circumscribed, round or oval, lesion filled with serous fluid or blood. Vesicles are smaller than 0.5 cm; bullae are larger than 0.5 cm. (e.g., burn blister)
- **Pustule**: Vesicle or bulla filled with pus (e.g., acne)
- **Macule**: Flat, unelevated change in color (e.g., freckle)
- **Papule, plaque**: Circumscribed, solid elevation of skin (e.g., warts). Papules are <1cm; plaques are >1cm
- **Nodule**: Elevated, solid, hard mass that extends deeper into the dermis than a papule
- **Tumors**: Larger than 2 cm and may have an irregular border (e.g., hemangioma)
- **Cyst**: A 1-cm or larger, elevated, encapsulated, fluid-filled or semisolid mass (e.g., chalazion of the eyelid)
- **Wheal**: A reddened, localized collection of edema fluid, irregular in shape (e.g., hives, mosquito bites)
- **Furuncle**: Deep inflamed pustular area usually surrounding a hair follicle

# Hair

### General Information

Normal hair is evenly distributed. Some therapies (e.g., chemotherapy) cause **alopecia** (hair loss). Hypothyroidism can cause thin, brittle hair. In severe protein deficiency, the hair appears reddish or bleached, and it is coarse and dry.

In **older adults** there is loss of scalp, pubic, and axillary hair. Hair of the eyebrows, ears, and nostrils becomes coarse; some women develop coarse facial hair.

### Procedure and Technique for Inspection:

- Even growth over the scalp, patches of hair loss
- Thickness or thinness
- Texture and oiliness
- Amount of body hair
- Infections or infestations (e.g., of lice or ringworm)
- Scalp lesions

## Nails

### Normal Findings

The nail plate should be colorless, smooth, and convexly curved. The angle between the nail and nail bed should be 160 degrees. If the angle is >180 degrees, the condition is called **clubbing**. The nail bed should be pink in color.

**Paronychia** is an inflammation of the tissues surrounding the nail.

**In older adults**, nails thicken and grow more slowly. Longitudinal bands commonly develop; nails often split. Bands across the nails may indicate nutritional deficiencies.

### Procedure and Technique for Inspection:

- Nail plate shape (curvature, angle)
- Texture (should be smooth, not grooved or furrowed)
- Nail bed color
- Tissues surrounding nails

**Perform** blanch test for capillary refill.

## Skull and Face

**Normocephalic** describes normal head size.

### Procedure and Technique for Inspection and Palpation:

Inspect

- **Skull** for size, shape, and symmetry
- **Facial features** (palpebral fissures, nasolabial folds) for symmetry
- **Eyes** for edema or hollowness
- Note **symmetry** of facial movements
- **Palpate** for nodules, masses, and depressions

## Eyes

### General Information

People under age 40 should have their eyes tested every 3 to 5 years. After age 40, examination is recommended every 2 years to rule out glaucoma.

**In older adults**, visual acuity decreases; the lens becomes more opaque and loses elasticity. Peripheral vision diminishes; night vision decreases. Color vision declines; accommodation to near objects decreases. Pupil reaction is less brisk, but should be symmetrical. The eyeballs may seem sunken because of decreased orbital fat.

### Terminology

**Myopia**: Nearsightedness (cannot see distant objects well)

**Hyperopia**: Farsightedness (cannot see near objects well, as in reading)

**Presbyopia**: Loss of elasticity of the lens

**Astigmatism**: Uneven curvature of the cornea that causes distorted images

**Cataract**: Opacity of the lens or its capsule, which blocks light rays

**Glaucoma**: Disturbance in the circulation of aqueous fluid, which causes increased intraocular pressure

### Procedure and Technique for Inspection and Assessment

#### Inspect

- **Eyebrows**: Hair distribution, alignment, and movement
- **Eyelashes**: Even distribution and direction of curl
- **Eyelids**: Skin quality and texture, symmetrical closing, blinking
- **Bulbar conjunctiva**: Color, texture, and lesions (sclera should be white)
- **Palpebral conjunctiva**: Evert the lids to inspect. Note color, texture, and lesions.
- **Lacrimal gland**: Inspect and palpate.
- **Lacrimal sac**: Palpate and note edema or tearing.
- **Nasolacrimal duct**: Palpate and note edema or tearing.
- **Cornea**: Use penlight at oblique angle; inspect for clarity and texture (details of iris should be visible).
- **Anterior chamber**: Transparency and depth
- **Pupils**: Color, shape, and symmetry of size
  - Direct and consensual reaction to light
  - Reaction to accommodation

#### Assess

- Peripheral vision
- Visual acuity with Snellen chart
- Ability to see light, note hand movements, and count fingers
- Six ocular movements to determine eye alignment and coordination (extraocular muscle tests). **Nystagmus** is rapid involuntary movement of the eyeball on the extreme lateral gaze. **Strabismus** is crossed eyes or a deviation of the eye that the patient cannot overcome.

## Ears

### Procedure and Technique

- Inspection and palpation of the external ear
  - Inspect auricles for symmetry, color, and position.
  - Palpate auricles for elasticity and tenderness.
- Otoscopic examination (In some settings, the nurse does not perform otoscopic examinations.)
  - Using otoscope, inspect external canal for cerumen, lesions, and drainage.
  - Using otoscope, inspect tympanic membrane (should be pearly gray and shiny).
- Testing auditory acuity
  - Check ability to hear normal voice tones and a watch ticking.
  - Use Weber's test to assess bone conduction.
  - Use Rinne test to compare air to bone conduction.

## Nose and Sinuses

### Procedure and Technique

- Inspect external nose for shape, color, and flaring or discharge from the nares.
- Palpate external nose for tenderness, masses, and symmetry.
- Assess patency of nares (occlude one naris and ask client to breathe through the other).
- Inspect the nasal cavities using a flashlight/penlight.
- Palpate maxillary and frontal sinuses for tenderness.

## Mouth and Oropharynx

**Plaque** is an invisible film on the enamel of the teeth. When it is allowed to remain, tartar forms.

**Tartar** is a visible, hard deposit that forms at the gum line. **Tartar buildup** can cause periodontal disease, which is characterized by **gingivitis** (red, swollen gums), bleeding, and receding gum lines. In time, the teeth become loose and the gums infected.

### Procedure and Technique

- The nurse should wear disposable gloves for physical examination of the mouth, which includes both inspection and palpation.
- Other equipment needed includes: tongue blade, gauze squares, and penlight or flashlight.

### Inspect

- Outer lips for symmetry, color, hydration (dry, cracked)
- Inner lips and buccal mucosa for color, moisture, textures, and lesions; also palpate
- Inspect dentition for missing teeth, caries, discoloration
- Gums for color, bleeding, tenderness, moisture, retraction/atrophy, swelling
- Dentures—ask client to remove them—check condition. (e.g. broken, worn areas)
- Tongue position, color, texture of surface, and free movement
- Base of tongue, mouth floor, and frenulum (also palpate tongue and floor of the mouth for nodules, lumps, or lesions.)
- Salivary ducts for swelling or redness
- Hard and soft palate (color, shape, texture, lesions)
- Uvula (position in midline); ask client to say "ah" to check for mobility.
- Oropharynx color and texture; use tongue blade and penlight.
- Tonsils for size, color, and discharge

## Neck

### Procedure and Technique

- Ask client to move the head through full range of motion; check coordination, stiffness, and discomfort.
- Palpate neck for enlarged lymph nodes (face client and have client bend the head forward and toward the side being examined).
- Palpate trachea (should be midline).
- Inspect and palpate thyroid gland (can be done from anterior or posterior approach); if enlarged, auscultate for a bruit.

## Thorax and Lungs

### General Information

**The right lung has three lobes.** Anteriorly the nurse can auscultate the upper, middle, and a small part of the lower lobe. Posteriorly, the middle lobe is not heard.

**The left lung has two lobes.** The upper and a small portion of the lower lobe can be auscultated anteriorly; both the upper and lower can be heard posteriorly.

**Landmarks** are a series of imaginary lines to help the nurse identify the position of the lobes of the lung. There are landmarks for the anterior, posterior, and lateral chest.

- **Lateral chest landmarks**
  - Anterior axillary line is a vertical line through the anterior axillary folds.
  - Midaxillary line is a vertical line through the apex of the axilla.
  - Posterior axillary line is a vertical line through the posterior axillary fold.
- **Anterior chest landmarks**
  - Anterior axillary line is a vertical line through the anterior axillary folds.
  - Midclavicular line is a vertical line through the midpoint of the clavicle.
  - Midsternal line is a vertical line through the center of the sternum.
- **Posterior chest landmarks**
  - Vertebral line is a vertical line along the center of the spine.
  - Scapular line is a vertical line from the inferior angle of the scapula.
- **Intercostal spaces** also function as landmarks; the intercostal space is numbered according to the number of the rib above the space.

### Procedure and Technique for Assessing Posterior Thorax

**Inspect** for shape, symmetry, spinal alignment and deformities.

**Palpate** for bulges, tenderness or abnormal movement, respiratory excursion and vocal (tactile) fremitus (vibration felt through the chest wall when client speaks).

**Percuss** for resonance and diaphragmatic excursion. Percussion determines whether underlying lung tissue is filled with air, liquid, or solid material.

**Auscultate**, using flat-disc diaphragm of the stethoscope; ask client to take slow, deep breaths through the mouth. Identify the following breath sounds:

- Vesicular: soft, low-pitched, normal sounds of air moving through small airways
- Bronchovesicular: moderate intensity "blowing" sounds of air moving through larger airways

- Adventitious: abnormal sounds
  - Crackles/rales: a fine to coarse popping as air passes through fluid in the small airways
  - Bronchi: low-pitched gurgling as the large airways fill with fluid
  - Wheezes: high-pitched whistling as air passes through narrowed airways
  - Friction rubs: grating or scratching as the inflamed pleura rubs

### Procedure and Technique for Assessing Anterior Thorax

- Assess breathing pattern (rate, rhythm).
- Inspect the costal angle (should be less than 90 degrees).
- Palpate, percuss, and auscultate anterior chest (see posterior chest).

## Heart

### Procedure and Technique

- The heart is usually examined with the client in a semi-reclining position.
- Inspection, palpation, and auscultation should be performed in that sequence.
- Perform assessment in a quiet room.
- Use both the flat-disc and the bell-shaped diaphragm to listen to all areas.
- Auscultate with client supine and head elevated to about 45 degrees. When finished, listen again with the client sitting upright.
- Inspect and palpate the chest simultaneously for abnormal pulsations or heaves from forceful movement.
- **Auscultate over the following areas, each of which is associated with the closure of a heart valve:**
  - **Aortic:** Right of sternum at 2$^{nd}$ intercostal space
  - **Pulmonic:** Left of sternum at 2$^{nd}$ intercostal space
  - **Tricuspid:** Left of sternum at 5$^{th}$ intercostal space
  - **Apical:** Medial to midclavicular line at 5$^{th}$ left intercostal space
- **Auscultate in each of the areas for the following sounds:**
  - $S_1$ is the first heart sound ("lub"). It is dull and low-pitched. It occurs when the atrioventricular valves close.
  - $S_2$, the second heart sound ("dub"), is higher pitched and shorter. It occurs when the semilunar valves close.
  - **Systole** is the period in which the ventricles contract. It begins with $S_1$ and ends when $S_2$ begins. **Diastole** is the period in which the ventricles relax. It starts with $S_2$ and ends at the following $S_1$.

**Carotid pulses:** (1) palpate one side at a time; do not massage; should be easily palpated; (2) auscultate for a buzzing or humming sound that would indicate a bruit.

**Jugular veins:** Inspect, in semi-Fowler's position, for pulsations and distention (which should not be present). The jugular veins should be distended and visible when the client is lying down; they should be flat and less visible when the client is standing.

# Peripheral Circulation

## Procedure and Technique for Assessing Peripheral Arteries

- Palpate pulses for adequacy of blood flow: rate, rhythm, and strength.
- Palpate for elasticity of vessel wall.
- Palpate for equality and symmetry of bilateral pulses (use fingertips).
  - **Radial pulse:** On radial side of forearm at the wrist
  - **Ulnar pulse:** On opposite side of wrist, not as strong, usually
  - **Brachial pulse:** Between triceps and biceps muscle, in the antecubital fossa above the elbow
  - **Femoral pulse:** Found best with client lying down, below the inguinal ligament, midway between the symphysis pubis and anterosuperior iliac spine. May require deep palpation.
  - **Popliteal pulse:** Palpate behind the knee. Have client flex the knee slightly with foot resting on bed; or have client assume prone position with knee slightly flexed. Requires deep palpation in the popliteal fossa, just lateral to the midline.
  - **Dorsalis pedis pulse:** Palpate along the top of the foot in a line with the groove between the first toe and the extensor tendons of the great toe.
  - **Posterior tibial pulse:** Locate on the inner side of each ankle; have client extend foot slightly; place fingers behind and below the medial malleolus (anklebone).

## Procedure and Technique for Assessing Peripheral Veins

- Inspect skin color and temperature.
- Inspect and palpate extremities for varicosities, peripheral edema, and phlebitis (inflammation of a vein wall). Look for localized redness, tenderness, and swelling.
- Check Homan's sign. (support leg and flex the foot upward; calf pain may indicate phlebitis or DVT.)

# Abdomen

**Landmarks:** The abdomen is divided into four imaginary quadrants. The tip of the sternum marks the upper boundary; the symphysis pubis marks the lower boundary; and a vertical line through the umbilicus divides the abdomen left and right.

**Procedure and technique:** The order of abdominal assessment is important. Begin with inspection, and then follow with auscultation, palpation, and percussion. The client must relax the abdominal muscles. Have the client void and then position in supine position with arms at the sides.

**Inspect:**

- Scars, venous patterns, lesions and striae (stretch marks)
- Shape and symmetry of abdomen; masses
- Distention (abdomen protrudes, skin may be taut); if present, measure with tape measure
- **"Shifting:"** Have client roll onto one side; if abdomen is distended with fluid, it will shift to the dependent side.

- Abdominal movement on breathing
- Peristaltic movement
- Aortic pulsation or bounding (abnormal)
- Umbilicus, for position, shape, color, inflammation, discharge, or masses

- **Auscultate for bowel sounds** (intestinal motility). Listen in all four quadrants; it may take up to a minute to hear a bowel sound. Describe sounds as normal or audible, absent, hypoactive, or hyperactive. Listen for 3 to 5 minutes before concluding that bowel sounds are absent.

- **Auscultate for vascular sounds.** Use bell of stethoscope to detect **bruits** (blowing sounds) over abdominal aorta (anteriorly) and renal arteries (listen over the costovertebral angle posteriorly). If bruits are present, **do not palpate**. Consult a primary care provider.

- **Palpate** to detect areas of tenderness and to assess any masses seen on inspection.

- **Percuss** to determine whether the liver is enlarged and to determine presence of air or fluid in stomach and intestines. **Tympany** (a hollow, drum-like sound) indicates air; dull sounds indicate fluid or solid masses. Percuss kidneys (posteriorly) for tenderness.

## Musculoskeletal

### Procedure and Technique

- The examination uses **inspection** *and* **palpation**.
- Client must assume a variety of positions: sitting, standing, supine, and prone.
- Gait and posture are checked in the general assessment (e.g., elderly clients walk with smaller steps and a wider base of support).
- Inspect for **lordosis** ("swayback" or increased lumbar curvature); **kyphosis** ("hunchback," exaggerated posterior curvature of the thoracic spine); and **scoliosis** (lateral spinal curvature).
- Ask the client to put each joint through its full **range of motion**; assess passively if client is too weak to perform; never force a joint into a painful position.
- During range of motion:
  - **Inspect** for swelling, deformity, tenderness, stiffness, or instability of the joint.
  - **Palpate** for unusual joint movement during range of motion.
  - **Palpate** for warmth in tissue surrounding joints.
  - **Assess** muscle tone and strength.
- **Inspect** joints for inflammation, muscle atrophy, and skin changes. Normal joints are non- tender, freely moveable, and without swelling or redness.

## Neurologic

Neurological assessment includes data about cranial nerve function, reflexes, and mental/emotional status. Data about mental capacities can be gathered by interacting with the client, although some specific questions may be needed. Take the client's educational and cultural background into consideration when assessing mental/emotional status.

## Procedure and Technique

- **Mental and emotional status assessment** includes:
  - Level of consciousness (fully awake, alert, cooperative, responsive to stimuli). The Glasgow coma scale is commonly used in medical or critical care areas.
  - Orientation to time, place, and person
  - Ability to follow simple commands (e.g., "Squeeze my finger.")
  - Behavior and appearance (e.g., grooming, speech)
  - Language (as an indicator of cerebral function). In receptive aphasia, the person cannot understand written or verbal speech; in expressive aphasia, the person understands but cannot write or speak appropriately.
  - Memory (e.g., have the client repeat a series of 5 or 6 numbers)
  - Knowledge (e.g., "Why did you come to the hospital?")
  - Abstract thinking (e.g., have client interpret clichés such as, "A rolling stone gathers no moss.")
- **Cranial nerve function** of the 12 cranial nerves is assessed in order of their number.
- **Sensory function** is usually just a quick screening. Have the client close his/her eyes and apply light touch to random locations.
- **Motor function** includes coordination, balance, and reflexes. Reflexes should be compared on both sides of the body for symmetry. They are graded as follows: 0=no response; 1+=low normal or diminished response; 2+=normal; 3+=more brisk than normal; 4+=very brisk and hyperactive.

# Critical Thinking Exercise: Physical Assessment of Body Systems

**Situation**: A nurse working in a hospital-based clinic is about to perform a physical assessment on an 80-year-old female. System by system, what changes might the nurse expect to find in this client compared to a younger client?

1. Describe a logical sequence for performing a head-to-toe assessment. How might this sequence be adapted to a client's needs? What measures might you take to assure a client's comfort during the assessment?

2. Identify the four basic techniques used in physical assessment. When you are assessing a client's cardiac status, for instance, what is the usual order in which these techniques are performed? Can you think of a specific situation in which it would be necessary to vary this order?

# General Information about Infection

## Key Points

- The Centers for Disease Control and Prevention (CDC) is the national public health agency in the U.S. that is concerned with prevention and control of disease.
- There are six links in the chain of infection: etiologic agent, reservoir, portal of exit, method of transmission, portal of entry to a susceptible host, and susceptible host.
- Nonspecific defenses include the inflammatory response and anatomic and physiologic barriers such as intact skin, stomach acidity and nasal cilia.
- Specific defenses are immune responses directed against a specific pathogen.
- Nosocomial infections are hospital or health care-acquired infections. This means that they are not present prior to the client receiving health care or being hospitalized.
- Hand washing and good medical/surgical aseptic techniques are essential to preventing nosocomial infections.
- Signs and symptoms of generalized infection: fever, malaise, leukocyte count >11,000/cu mm, enlarged lymph nodes, anorexia, nausea/vomiting
- Signs and symptoms of localized infection: redness, edema, pain, heat, loss of function
- Infection occurs when pathogenic microorganisms invade the client and multiply.
- **Key Concepts/Terms**: Resident flora, subclinical/asymptomatic infection, disease, pathogen, communicable disease, inflammatory response, immune response, active immunity, passive immunity, nosocomial infections, endogenous infection, iatrogenic infection, vehicle, fomite vector

**Overview**

Nurses are responsible for the biological safety of their clients' environment. Microorganisms are everywhere (e.g., on the skin, in the air and water), but most are harmless. **Resident flora** are the microorganisms normally present in a part/area of the body, but produce infection in another part (e.g., E. coli, a resident flora of the intestine is a common cause of urinary tract infections). An **infection** is the invasion and proliferation of microorganisms in body tissue. A person can be infected without having any symptoms or alterations in tissue function (**subclinical** or **asymptomatic infection**). **Disease** occurs when there are symptoms or alterations in tissue function. **Pathogens** are microorganisms that cause disease.

**Communicable diseases** are caused by microorganisms that can be spread by transmission to an individual by direct or indirect contact, through a vector or vehicle, or through the air. Infectious diseases are the major cause of death worldwide. The Centers for Disease Control and Prevention (CDC) is the principal national public health agency in the United States that is concerned with disease prevention and control.

## Types of Infection-Producing Organisms

- Bacteria (e.g., *Staphylococcus aureus, Escherichia coli*)
- Viruses (e.g., rhinovirus, influenza, hepatitis, herpes, human immunodeficiency virus [HIV])
- Fungi (e.g., *Candida albicans*)
- Parasites (e.g., protozoa, worms, mites, fleas, ticks)

## Chain of Infection

### Six "links" in the chain of infection

**Etiologic agent**: The microorganism causing the disease. The likelihood of a microorganism causing an infection depends on the number of organisms present, the virulence and potency of the organisms, their ability to enter the body, the susceptibility of the host, and the ability of the microorganism to live in the body. A **carrier** is a person or animal who harbors an infectious organism and can infect others, but does not have any symptoms or signs of disease.

**Reservoir**: Sources of microorganisms (e.g., humans, the client's own body, plants, animals, the environment, food, water, milk). For example, influenza is usually spread among humans; the Anopheles mosquito carries the malaria parasite.

**Portal of exit from reservoir**: The site from which the organism leaves the reservoir. In humans, this is commonly the: respiratory tract, gastrointestinal tract, urinary tract, reproductive tract, blood, and tissue.

**Method of transmission**: After leaving the reservoir (source), the organism requires a means of transmission to reach another host. There are four methods of transmission:

- **Direct transmission**: Is transfer from person to person through touching, biting, kissing, or sexual intercourse. Droplet transmission can occur if the source and the host are within 3 feet of each other (e.g., through coughing, talking, sneezing, and singing).
- **Vehicle-borne transmission**: A vehicle is any substance that provides a means of transport and introduces a pathogen into a host. Fomites are inanimate objects (e.g., handkerchiefs, toys, soiled clothes, eating utensils, surgical instruments, water, food, milk, and blood) that can act as vehicles.
- **Vector-borne transmission**: A vector is an animal or a flying or crawling insect that serves as a means of transport. Transmission may occur by biting or by depositing feces on the skin.
- **Airborne transmission**: May involve droplets or dust. Air currents transport the material to a suitable portal of entry, usually the respiratory tract. Tuberculosis and *Clostridium difficile* are airborne.

**Portal of entry to a susceptible host**: The microorganisms must enter the body before a person can become infected. Any break in the skin can serve as a portal of entry. Microorganisms can also enter the body through any of the routes they use to leave the body.

Susceptible host: A susceptible host is anyone who is at risk for infection. A **compromised host** is a person who is at increased risk; that is, more likely than others to acquire an infection (e.g., someone with a break in the skin or someone who is immunosuppressed).

## Body Defenses Against Infection

**Nonspecific defenses** protect against all microorganisms, regardless of prior exposure. **Specific (immune) defenses** are directed against a specific bacterium, virus, fungus, or other infectious agent.

**Anatomic and physiologic barriers** are nonspecific defenses.

They include:

- Intact skin and mucous membranes
- Cilia in the nasal passages
- Macrophages and phagocytes (cells that ingest microorganisms)
- Saliva in the oral cavity
- Tears in the eye
- Acidity in the stomach; resident flora of the large intestine
- Low pH of the vagina

**Inflammatory response**: This is a nonspecific adaptive mechanism that destroys or dilutes the infectious agent, prevents further spread, and promotes tissue repair. It is characterized by pain, swelling, redness, heat, and impaired function of the part. The suffix "–itis" means inflammation. Inflammation can be caused by physical agents (e.g., excessive heat), chemical agents (e.g., poisons), and microorganisms.

The three stages of the inflammatory response are:

- **Vascular and cellular responses** (e.g., increase in blood supply to the area, leukocytosis)
- **Exudate production**: Exudate consists of fluid that "leaked" from the blood vessels, dead phagocytes, and dead tissue cells. Exudate can be serous, purulent, or sanguineous. Fibrinogen, thromboplastin, and platelets make a barrier to wall off the area.
- **Reparative phase**: In this stage, injured tissues are repaired by regeneration or by replacement with scar tissue.

**Immune responses**: Immunity is the resistance of the body to a specific infection (pathogens or their toxins). In **active immunity**, the host produces its own antibodies upon exposure to antigens (e.g., an infection or a vaccine). In **passive immunity**, the host receives natural or artificial antibodies produced by another source (e.g., milk from a nursing mother or an injection of immune serum).

## Nosocomial Infections

Nosocomial infection occurs while the client is receiving health care. Nosocomial infections can either be **exogenous** (infection acquired from the hospital environment or other people), **endogenous** (infection acquired by microorganisms that the client himself harbors), or **iatrogenic** (infection received as the direct result of a treatment or diagnostic procedure, for example bacteremia in an intravenous line). The urinary tract is the most common site for nosocomial infections. The most common infections are caused by E. coli and *Staphylococcus aureus*. Hand washing is one of the most effective measures for preventing nosocomial infections.

### Risk factors increasing the susceptibility to nosocomial infections

- Inadequate hand washing (caregivers)
- Compromised hosts—that is, clients whose normal defenses are low because of disease, poor nutrition, immunosuppression, or surgery
- Poor medical and/or surgical aseptic techniques of caregivers (e.g., for catheterization, dressing changes, parenteral medications, suctioning, closed drainage systems, surgery)
- Presence of invasive devices
- Poor personal hygiene (client)
- Skin breakdown
- Decreased oxygenation
- Impaired circulation
- Stress
- Lifestyle/health habits
- Aging (older clients are more susceptible)

### Organisms most commonly causing nosocomial infections

- **Urinary tract**: *E. coli, Enterococcus, Pseudomonas aeruginosa*
- **Surgical sites**: *Staphylococcus aureus, Enterococcus, Pseudomonas aeruginosa*
- **Bloodstream**: *Coagulase-negative staphylococcus, Staphylococcus aureus, Enterococcus*
- **Pneumonia**: *Staphylococcus aureus, Pseudomonas aeruginosa, Enterobacter*

## Normal Course of Infection

### Infection usually follows a progressive course

- **Incubation period**: First stage of infection is the time from when the pathogen enters the body to the appearance of the first symptoms of ulcers.
- **Prodromal stage**: The period from the onset of initial symptoms to more severe symptoms; many illnesses are the most contagious during the prodromal stag.
- **Full stage of illness**: The third phase is when the symptoms are acute and specific to the type of infection.
- **Convalescence stage**: Final stage during which the acute symptoms of the infection subside and the person recovers.

## Signs and Symptoms of Infection

### Generalized/systemic infection

- Fever
- Increased pulse and respiratory rate (if fever is high)
- Loss of energy, malaise

- Anorexia or nausea and vomiting
- Enlarged lymph nodes

### Localized infection of the skin and mucous membranes

- Redness (from dilation of arterioles which deliver blood to the area)
- Edema/swelling
- Pain or tenderness with movement or palpation
- Palpable heat at the area
- Possible loss of function of the affected part

### Laboratory data indicating presence of an infection

- Leukocyte (white blood cell) count greater than 11,000/cu mm (leukocytosis)
- Increases or decreases in specific types of leukocytes (as revealed in the "differential" count); for example, the normal types and percentages of leukocytes are neutrophils 40% to 60%, eosinophils 1-3%, basophils .05% to 1%, lymphocytes 20% to 40%, and monocytes 4% to 8%. A client with leukocytosis with shift to the left is experiencing an increase in neutrophils.
- Elevated erythrocyte sedimentation rate (ESR)
- Growth of microorganisms in urine, blood, sputum, or other drainage cultures

### Related NANDA Nursing Diagnoses

- Impaired Tissue Integrity
- Risk for Infection
- Risk for Impaired Skin Integrity

### Pharmacology

- Antipyretics
- Anti-infectives—antibiotics, antifungal, antiviral
- Analgesics
- Corticosteroids

# Critical Thinking Exercise: General Information about Infection

**Situation**: You have been invited by the infection control specialist of an acute care facility to participate in developing an in-service for new employees on the prevention of nosocomial infections. What essential information do you want to convey about this important topic?

1. As a nurse, what specific measures can you take to prevent the spread of infection? Why are these measures so important?

2. With respect to a specific pathogen, such as Mycobacterium tuberculosis, identify the six links in the chain of infection. What interventions might be effective in breaking the chain of infection?

3. What are the goals of the Centers for Disease Control and Prevention (CDC) and the Occupational Safety and Health Administration (OSHA) regarding protecting the public from the spread of infectious disease?

# Medical Asepsis

## Key Points

- Asepsis is freedom from disease-producing microorganisms.
- Hand hygiene is the most effective way to prevent the spread of microorganisms.
- The CDC supports two methods of hand hygiene: In addition to traditional hand washing with soap and water, the CDC recommends using alcohol-based hand rubs for patient care. The product is applied to the palm of one hand and the hands rubbed together to cover all hand and finger surfaces until the hands are dry. When using the traditional method, the CDC recommends vigorous hand washing under a stream of water for at least 10 seconds, using soap (not bar soap) or antimicrobial soap.
- When washing hands, do not touch faucet handles with clean hands.
- The amount of time spent handwashing should be adjusted to the amount of likely contamination of the hands.
- Basic practices of medical asepsis include:
  - Hand washing and using gloves to prevent contact with blood or body fluids
- **Key Concepts/Terms**: Medical asepsis, clean, dirty, sepsis, antimicrobial agent, disinfectant, bactericidal, bacteriostatic, immunocompromised

**Overview**

**Asepsis** is freedom from disease-producing microorganisms. **Medical asepsis** refers to the protective practices intended to limit the number, growth, and transmission of microorganisms. Clean implies the absence of nearly all microorganisms. Dirty (soiled, contaminated) indicates the likely presence of microorganisms. **Sepsis** is the opposite of asepsis. It is the state of being infected.

## Basic Practices of Medical Asepsis

- Wash hands often, especially before eating or handling foods or after personal hygiene.
- Carry soiled linens and equipment so that they do not touch the clothing.
- Do not place soiled linens or other items on the floor, which is considered to be "grossly" contaminated.
- Teach clients to cover their mouth and nose when they cough or sneeze; provide them with disposable tissues.
- Avoid raising dust (e.g., do not shake linens).
- Clean areas from "clean to dirty"—that is, clean the least soiled areas first.
- Dispose of soiled items directly into appropriate containers. Wrap moist, soiled items in plastic bags before discarding.

- Pour liquids (e.g., mouth rinse) directly into the drain; avoid spattering.
- Follow agency policies for isolation and barrier techniques.
- Shampoo your hair regularly; keep it short.
- Keep fingernails short; use hand lotion.
- Do not wear rings with grooves and stones that may harbor microorganisms.
- Wear latex gloves for "dirty" procedures.

## Procedure and Technique for Handwashing

### Handwashing is the best way to prevent the spread of microorganisms.

- The Centers for Disease Control (CDC) recommends rubbing on fast-drying alcohol gels and solutions to kill antimicrobial agents.
- Remove all jewelry.
- Check hands for breaks in the skin (use lotion to prevent cracked, dry skin).
- Squirt a small amount of alcohol gel in palm of one hand, or alcohol solution may be available.
- Rub hands together, covering all surfaces until the hands are dry.
- Hand cleansing agents:
  - The CDC recommends fast-drying alcohol gels and solutions that contain antimicrobial agents:
    - In areas where there are known multiple resistant bacteria
    - Before invasive procedures
    - In special care units such as nurseries and ICUs
    - Before caring for clients who are immunocompromised
    - In cases of gastrointestinal, respiratory, skin, or wound infections
    - In cases of colonization with multidrug-resistant bacteria judged to be of special clinical and epidemiologic significance
    - In cases of enteric infections with a low infectious dose or prolonged environmental survival (e.g., *Clostridium difficile*, *E. coli*, *Shigella*, hepatitis A, or rotavirus in diapered or incontinent clients)

## Procedure and Technique

### Cleaning

Wear gloves when cleaning visibly soiled objects. Most objects (e.g., forceps, linens) can be cleaned using the following steps:

- Rinse article with cold water. (Hot water coagulates proteins of organic materials such as blood and makes it adhere.)
- Wash the article in hot water and soap.
- Use an abrasive (e.g., a stiff brush) to clean grooves and corners.
- Rinse article well with warm or hot water.
- Dry the article, which is considered clean at this point.
- Clean the brush and sink (they are still considered soiled until they are cleaned), preferably with a disinfectant.

### Disinfecting

A **disinfectant** is a chemical preparation (e.g., iodine, chlorine, phenol, alcohol, and chlorhexidine gluconate) used in inanimate objects. Disinfectants are often caustic to skin. A **bactericidal** preparation destroys bacteria. A **bacteriostatic** preparation prevents the growth and reproduction of some bacteria. Whether or not bacteria are destroyed depends on the type and number of infectious organisms on the object, the concentration of the disinfectant, and the duration of the contact. Most disinfectants are meant to be used at room temperature; some are not effective in the presence of soap or detergent. Organic materials (e.g., pus, blood, excretions) can inactivate many disinfectants, so the object should be cleaned before it is disinfected. The disinfectant must come into contact with all surfaces and areas of the object.

# Critical Thinking Exercise: Medical Asepsis

**Situation:** As a nurse in a skilled nursing facility, you are responsible for supervising patient care assistants. In response to a recent increase in the incidence of infection among clients, you have designed a 20-minute presentation on the importance of medical asepsis. Consider the key points you wish to communicate and the visual aids that will enhance your talk. The following are a few topics to consider including in your presentation:

- The best way to prevent the spread of infection.
- The importance of personal hygiene.
- The handling of bed linen and other equipment.
- Cleaning soiled objects.

Once you have developed a presentation that is clear and comprehensive, practice delivering it to a friend, classmate, or family member.

# Surgical Asepsis

## Key Points

- Surgical asepsis (sterile technique) is used to keep an area/object free of all microorganisms.
- Surgical asepsis is used in areas other than surgery; for example, in inserting an indwelling urinary catheter, inserting an intravenous catheter, and drawing up medications, and working with clients whose skin integrity is interrupted.
- All objects used in a sterile field must be sterile.
- Thorough handwashing must always be performed before and after contact with a client, and before and after contact with sterile equipment.
- Objects are considered unsterile:
    - After prolonged exposure to air
    - When touched by unsterile objects
    - When out of vision or below the waist level
- A wet sterile field becomes unsterile.
- The edges of a sterile field are considered unsterile (1 inch around the perimeter).
- The skin is unsterile; it cannot be sterilized.
- Pathogens can be killed or inactivated by disinfection, by sterilization, or by the use of anti-infective drugs.
- Alertness, conscientiousness, and honesty are essential to maintaining surgical asepsis.
- **Key Concepts/Terms**: Handwashing, sterile field, contaminated, dominant hand, sterile antiseptic solution, sterile gloves, closed gloving technique

## Overview

**Surgical asepsis** (also called sterile technique) includes methods and practices used to keep an area or object free of **all** microorganisms. It destroys microorganisms and spores before they can enter the body. Surgical asepsis is accomplished by first sterilizing articles and then preventing their contact with any unsterile articles. Sterile technique is practiced in surgery and for many procedures in general care areas (e.g., urinary catheterizations, injections).

A **sterile field** is an area that is free of microorganisms. Often this is the inner side of a sterile wrapper, or perhaps a sterile drape. Sterile forceps or sterile gloves are used to transfer sterile objects from their wrappings to the sterile field. Thorough handwashing must be performed before donning sterile gloves and after discarding the gloves.

## Principles of Surgical Asepsis

- Sterile objects become unsterile when touched by unsterile objects.
- Sterile items are considered unsterile when they are out of vision or below the waist level of the nurse.
- All objects used in a sterile field must be sterile.
- A sterile field is considered contaminated if anyone reaches across it who is considered contaminated (is not wearing a sterile gown and sterile gloves). Therefore, do not reach across the sterile field unless you are wearing a sterile gown and sterile gloves.
- Sterile objects are considered unsterile after prolonged exposure to the air.
- Fluids flow in the direction of gravity (e.g., "downhill").
- A wet sterile field becomes unsterile unless it has a waterproof barrier underneath.
- The edges of a sterile field are considered unsterile.
- The skin is unsterile and cannot be sterilized, and thus it is essential to wear sterile gloves when performing a sterile procedure.
- Thorough handwashing must be performed before donning sterile gloves.
- Alertness, conscientiousness, and honesty are essential to maintaining sterile asepsis.
- When a sterile object becomes unsterile, it does not necessarily change appearance.

## Procedure and Technique for Setting Up a Sterile Field

- Examine the sterile packages for signs of contamination.
- Review the institution's protocols for the procedure.
- Thoroughly wash your hands.
- Drape the sterile field by opening the package and letting the edges of the wrapping drop down and away. Touch only the outside surface of the drape.
- If you add additional items to the field, do NOT reach across the field.
- Open the additional items and carefully drop them onto the field.
- Do NOT allow the packaging to touch the sterile field.
- Open packages away from the field to avoid contamination.

## Procedure and Technique for Pouring Sterile Solution to a Sterile Bowl

- Thoroughly wash your hands.
- Obtain the exact amount of sterile solution, if possible.
- Read the label on the bottle three times to make certain it is the correct solution and strength, and the product has not expired.
- Remove the lid/cap from the bottle; invert the lid if placing it on an unsterile surface.
- Hold the bottle at a slight angle so that the label is on the upper side.
- Hold the bottle about 4 to 6 inches above the bowl and to the side of the sterile field—so that as little of the bottle as possible is over the field.

- Pour the fluid gently to avoid splashing.
- Replace the lid securely if you plan to use the solution again.
- Label the bottle with the date and time of use and any other information required by the institution.

## Procedure and Technique for Donning Sterile Gloves (Open Method)

- Thoroughly wash your hands.
- Open the sterile package; place on a clean, dry surface.
- Open the outer package.
- Remove the inner package from the outer package.
- Open the inner package, using procedure for opening sterile packages. Do not touch the inner surface of the package that is next to the sterile gloves. It must remain sterile.
- Put the first glove on the dominant hand. To do so, grasp the glove by the cuff with the non-dominant hand, touching only the inside of the glove. Insert the dominant hand into the glove, keeping the thumb against the palm while inserting.
- Put the second glove on the nondominant hand. To do so, use the gloved hand and insert the gloved fingers under the cuff of the second glove, holding the thumb close to the palm. Pull on the second glove, holding the thumb of the dominant hand as far as possible from the palm.
- When removing the gloves, if they are soiled with secretions, remove them by turning them inside out. Otherwise there is no special procedure for removal.

Procedure and Technique for Donning a Sterile Gown

- Open the package of sterile gloves.
- Unwrap the sterile gown pack.
- Wash and dry hands carefully again.
- Put on the sterile gown.
    - Grasp gown at the crease near the neck; unfold without permitting it to touch anything.
    - Put hands inside the shoulders of the gown and work arms partway into the sleeves without touching the outside of the gown.
    - If donning gloves by closed method, work hands down the sleeves just to the proximal edge of the cuffs; if donning gloves by open method, work the hands down and through the cuffs.
- Have a masked and gloved coworker:
    - Grasp the neckties without touching the outside of the gown and pull the gown up to cover the neck of your uniform in front and back.
    - Tie the neckties.
    - Hold the waist tie, using sterile gloves or forceps; make a three-quarter turn, then take the tie and secure it in front of the gown.
- A sterile gown is considered sterile in front from the waist to the shoulder. The sleeves are sterile from 2 inches above the elbow to the cuff.

# Critical Thinking Exercise: Surgical Asepsis

**Situation**: A nursing student accompanies a home care nurse on a series of visits. The student is surprised to learn how often procedures requiring sterile technique are performed in the home. Outside the acute care setting, in what situations might it be necessary to maintain surgical asepsis?

1. With respect to the importance of establishing and maintaining a sterile field, what specific challenges might nurses encounter in serving clients in the community? Explain resourceful strategies a nurse working outside acute care might use to meet these challenges.

2. Imagine that, as a nurse educator, you are preparing an inservice for nurses on the principles of surgical asepsis. Ask a group of your classmates to assume the role of participants and plan an interactive session. If possible, obtain permission from your school to obtain various supplies that may be needed, such as packages of sterile gloves, a sterile field, and perhaps a urinary catheter insertion kit. Once you have described the rationale for using sterile technique, practice the skills involved: open a sterile package, don sterile gloves, and set up a sterile field. Critique each other's technique throughout the practice session until all participants are confident of their skills.

# Isolation Precautions

## Key Points

- Isolation describes measures taken to prevent the spread of infectious microorganisms to clients, visitors, and health care workers.

- Standard (Tier One): CDC precautions assume that everyone has the potential for infection transmission; therefore, they protect against exposure to all body fluids, blood, secretions, and excretions.

- Transmission-based (Tier Two): CDC precautions are used in addition to standard precautions for clients with known or suspected infections. They differ depending upon whether the infection is spread by airborne transmission, droplet transmission, or direct contact.

- Gloves are worn to protect:
  - The nurse
  - The client from the nurse
  - Other clients

- Wash hands each time gloves are removed; gloves may have imperfections.

- Always change gloves between client contacts.

- Place contaminated articles in a sturdy bag that is impervious to microorganisms; or double-bag them.

- Complete the following steps if a worker is exposed to blood-borne pathogens:
  - Report incident immediately.
  - Complete injury/incident report.
  - Get appropriate evaluation and follow-up (e.g., medical consultation, blood testing).
  - For a puncture or laceration: encourage bleeding, clean with soap and water, initiate first aid or seek treatment.

- **Key Concepts/Terms**: Isolation, standard precautions, transmission based precautions, gloves, gown, face mask, removing soiled clothing, disposal of soiled supplies, federal protective regulations

**Overview**

Isolation describes measures taken to prevent the spread of infectious microorganisms to clients, visitors, and health care workers. The Centers for Disease Control and Prevention (CDC) guidelines for hospitals specify two types of precautions: (1) **standard precautions** (used in the care of all hospitalized clients), and (2) **transmission-based precautions** (used for clients with known or suspected infections.

## Standard ("Tier One") Precautions

These precautions are used in the care of all hospitalized clients regardless of their actual or potential infection status. They assume that everyone has the potential for infection transmission. Standard precautions (previously called universal precautions and body substance isolation) apply to all body fluids, blood, nonintact skin, mucous membranes, and all body secretions and excretions except sweat. They are:

- Wash hands after contact with body fluids, blood, secretions, excretions, and contaminated objects, even if gloves were worn.

- Wear clean (not sterile) gloves when touching body fluids, blood, secretions, excretions, and contaminated items (e.g., bedpans). Remove gloves before touching non- contaminated surfaces. Wash hands immediately after removing gloves.

- Wear a mask or other eye/face protection if splashes or sprays of body fluids, blood, secretions, or excretions are expected.

- Wear a clean, unsterile gown to protect the clothing if client care is likely to result in splashes of body fluids, blood, secretions, or excretions.

- Clean and disinfect/sterilize equipment that is soiled with body fluids, blood, secretions, or excretions; dispose of single-use equipment and supplies correctly.

- Handle, transport, and process linen that is soiled with body fluids, blood, secretions, or excretions in a manner to prevent contamination of clothing and transfer of microorganisms to others and the environment.

- Prevent injuries from used scalpels, needles, or other equipment; place in puncture-resistant containers.

## Transmission-Based ("Tier Two") Precautions

These precautions are used (alone or in combination with each other) in addition to Standard Precautions for clients with known or suspected infections. They are used for infections that are spread by airborne or droplet transmission or by direct contact.

**Airborne precautions:** For clients known or suspected of having serious illnesses transmitted by airborne droplet nuclei smaller than 5 microns (e.g., measles, varicella, and tuberculosis):

- Place client in a private room with negative air pressure, 6-12 air changes per hour, and either a filtration system or discharge of air to the outside.

- If no private room is available, place client with another client who is infected with the same microorganism.

- Wear a respiratory device (N95 respirator) when entering the room of clients known or suspected of having primary tuberculosis.

- Susceptible people should not enter the room of a client who has rubella (measles) or varicella (chicken pox). If they must enter, they should wear a respiratory device.

- Move client outside the room only for essential purposes. Place a surgical mask on the client during transport.

**Droplet precautions:** Used for clients known or suspected to have serious illnesses transmitted by particle droplets larger than 5 microns (e.g., diphtheria, mycoplasma pneumonia, pertussis, rubella, mumps, streptococcal pharyngitis, scarlet fever in infants and young children, and pneumonic plague) are as follows:

- Place client in a private room.
- If no private room is available, place client with another client who is infected with the same microorganism.
- Wear a mask when working within 3 feet of the client.
- Move client outside the room only for essential purposes. Place a surgical mask on the client during transport.

**Contact precautions:** Used for clients known or suspected to have serious illnesses transmitted by direct client contact or by contact with items in the client's environment (e.g., gastrointestinal, respiratory, skin, or wound infections; colonization with multidrug-resistant bacteria; enteric Clostridium difficile; enterohemorrhagic Escherichia coli ; Shigella; hepatitis A for diapered and incontinent clients; respiratory syncytial virus; parainfluenza virus; enteroviral infections in infants and young children; herpes simplex virus; impetigo; pediculosis; and scabies) are as follows:

- Place client in private room.
- If no private room is available, place client with another client who is infected with the same microorganism.
- Wear gloves as described in Standard Precautions. Change gloves after contact with infectious material; remove gloves before leaving client's room; wash hands with antimicrobial agent immediately after removing gloves.

**Vancomycin-resistant enterococci (VRE):** In addition to Standard and Contact precautions, the CDC recommends hand washing with antimicrobial soap for clients with VRE. They should not share equipment with clients who do not have VRE (e.g., blood pressure cuffs, stethoscopes). Place the client in a private room or in a room with other clients who have VRE. These precautions should be used until at least three negative cultures, taken one week apart, are obtained.

## Gloves

### Reasons to Use Gloves:

- **To protect the nurse** when the nurse is likely to come into contact with any body substances
- **To protect the client from the nurse**—that is, to prevent transmission of the nurse's own microorganisms to the client
- **To protect other clients** from transmission of microorganisms from the client or fomites
  - For most activities clean (not sterile) disposable gloves are used.
  - Gloves are always changed between client contacts.
  - Nurses with open sores or cuts on the hands must wear gloves.
  - Hands are washed each time gloves are removed because the gloves may have imperfections or the hands may become contaminated during glove removal.
  - Most gloves are made of latex. Long-term exposure to latex predisposes nurses to latex allergies.

## Gowns

Most agencies use single-use gown technique. After use, discard the gown (if it is disposable) or place it in a linen hamper. When removing a grossly soiled gown, avoid touching the soiled parts, if possible. Roll the gown up with the soiled part inside and discard it in the appropriate container. Wash hands before leaving the client's room.

## Face Masks

**For large-particle aerosol transmission (droplets)** (e.g., measles, mumps, acute respiratory diseases in children) face masks must be worn only by those close to the client. Large- particle droplets travel only short distances (about 3 feet) and must be transmitted by close contact.

**For small-particle aerosol transmission (droplet nuclei)** (e.g., tuberculosis) face masks must be worn by all persons entering the room. It is best to use special masks with a tighter face seal and filtration for these infections. Small-particle aerosols remain suspended in air and can travel great distances.

**Single-use disposable surgical masks are effective for most clients**: They should be changed if they become wet or soiled.

**Disposable particulate masks** may be effective for droplet transmission, splatters, and airborne transmission. The National Institute for Occupational Safety and Health (NIOSH) certifies N95 respirators to be effective in preventing inhalation of tuberculosis organisms.

**Caregivers who are especially susceptible** to a specific airborne disease should not be assigned to clients with that disease.

### When using disposable masks:

- Be sure the mask covers both the mouth and nose.
- Adjust the metal strip (if there is one) firmly over the bridge of the nose.
- If wearing glasses, place the upper edge of the mask under the glasses.
- Wear the mask only once, and only for the time recommended by the manufacturer.
- Change the mask if it becomes wet.
- Avoid unnecessary talking, sneezing, and coughing when caring for an at-risk client.
- When removing a mask, untie the lower strings first.
- Discard disposable masks in the waste container.
- Wash hands if they have touched the soiled part of the mask.

## Removing Soiled Personal Protective Equipment/Clothing

- **Remove gloves first**, except if wearing a gown tied at the waist in front, undo the ties before removing the gloves.
- **Remove the mask**, holding it by the strings.
- **Remove the gown.**
- **Remove the eye wear.**

## Disposal of Soiled Supplies and Equipment

Proper disposal and handling are important to (a) prevent health care workers from being exposed to contaminated articles, and (b) prevent contamination of the environment.

**Bagging**: Bag articles contaminated with infective material (e.g., pus, blood, body fluids, feces, or respiratory secretions). CDC guidelines recommend: A single bag, if (a) it is sturdy and impervious to microorganisms, and (b) the contaminated articles can be placed in the bag without contaminating its outside. Double-bag if the conditions in (a) and (b) are not met.

**Disposable equipment (including dressings and tissues)**: Dispose in the plastic bag that lines the wastebasket. If it is not contaminated, no special precautions are required.

**Non-disposable or reusable equipment that is visibly soiled**: Place in a labeled bag before removing from client's room; send to a central processing area for decontamination.

**Special procedure trays (e.g., epidural tray)**: Disassemble them into component parts; some parts are disposable and others need to be sent to a central processing area for cleaning and decontaminating.

**Soiled client personal clothing**: Bag before sending home or to the laundry.

**Linens**: Handle linens as little as possible and do not shake them. Place in hamper and close bag before sending to laundry.

**Laboratory specimens**: Need no special precautions if placed in a leakproof container and if the outside of the container was not contaminated during specimen collection. If visibly contaminated, place the container in a sealable plastic bag before sending to the laboratory.

**Dishes**: No special precautions required. Paper dishes may be placed in the wastebasket.

**Blood pressure equipment**: No special precautions unless contaminated with infective material. In that case, follow agency procedures.

**Thermometers**: Usually disposable or have disposable covers. Otherwise, follow agency guidelines.

**Disposable needles, syringes, and other sharps**: Place "sharps" (e.g., scalpels, broken glass, needles) in special, puncture-resistant containers. Do not detach needles from the syringe or recap the needle before disposal.

**Toys**: If visibly contaminated, bag and send them home. If they are agency toys, follow agency practice for cleaning. Special precautions may be required in some situations.

## Federal Regulations to Protect Health Care Workers

The Occupational Safety and Health Administration (OSHA) provides regulations to protect health care workers from occupational exposure to pathogens in the workplace. **Occupational exposure** is reasonably anticipated skin, eye, mucous membrane, or parenteral contact with blood or other potentially infectious materials that may result from the performance of an employee's duties. Primary modes of transmission of infectious fluids in the health care setting are: puncture wounds, skin contact, and mucous membrane contact (e.g., splashes in the eyes). OSHA requires health care employers to make hepatitis B vaccine available to all employees.

Steps that must be taken if a worker is exposed to blood-borne pathogens:

- Report the incident immediately.
- Complete an injury (incident) report.
- Get appropriate evaluation and follow-up, including: identification of the source, blood testing for source (if consent is obtained) and worker, medical counseling, post-exposure prophylaxis.
- For a puncture or laceration: encourage bleeding, clean with soap and water, initiate first aid or seek treatment.

# Critical Thinking Exercise: Isolation Precautions

**Situation:** You are caring for a 40-year-old male diagnosed with pulmonary tuberculosis. From your interactions with the client, you assess that both he and his family have limited knowledge about how TB is spread. Consider how you might best respond to questions about the reason for isolation in the hospital and the kind of follow-up that will be needed when he returns home. Develop a teaching plan for the client and his family with the goal of preventing the transmission of the TB microorganism.

1. Soon after a client with vancomycin-resistant enterococcus (VRE) is admitted to a medical unit, a patient care assistant asks about specific measures to reduce risk of transmission. Explain what VRE is and how it is transmitted, and describe the special precautions that apply in this situation.

2. Investigate the policies and procedures regarding isolation precautions in one of the clinical settings from your own experience. Who developed these policies, and where can they be found? How are nurses and other facility personnel made aware of these policies? In what specific situations have you observed nurses applying these principles?

# Client Safety

## Key Points

- Factors affecting client safety include: age/development, lifestyle, mobility, health status, sensory-perceptual alterations, cognitive awareness, emotional state, safety awareness/knowledge, and environmental hazards.
- Common hazards in the home include: storage of medications and toxic substances, slippery floors, throw rugs, slippery tubs/showers, frayed electrical cords, lack of smoke detectors.
- When a fire occurs in the health care setting, follow this sequence:
  - Protect and evacuate clients in immediate danger.
  - Report the fire.
  - Contain the fire.
  - Extinguish the fire.
- Infants and older adults are at special risk for falls; use a risk assessment tool; keep bed in low position; provide nonskid footwear; use safety sensors as needed to monitor patient activity.
- The main causes of poisoning in children are improper storage of toxic substances and inadequate supervision.
- The universal sign of choking is for the victim to grasp and point to the neck and throat without speaking. Abdominal thrusts should be used to dislodge the foreign object.
- Firearms should be in locked cabinets, stored separately from the bullets.
- Federal regulations require that: restraints be applied only as a last resort and with a physician's written order; the order must specify reason for and anticipated duration of restraint; the client must agree to be restrained; the client must not have any physical restraints except those required to treat her/his medical symptoms.
- If restraints are applied, document the situation thoroughly.
- **Key Concepts/Terms:** Clinical overview, assessing safety needs, scalds, burns, fires, falls, poisoning, suffocation, noise, electrical hazards, firearm safety, restraints

## Overview

Nurses are involved in accident prevention regardless of whether care is provided in the home, the community, or the hospital. The following factors affect clients' ability to protect themselves from injury:

**Age and development:** Children lack the knowledge and experience to maintain their safety. Older adults may have sensory and mobility deficits that increase their risk for injury.

**Lifestyle:** Factors that increase risk are, for example, living in a high-crime neighborhood; access to guns; lack of income to make repairs; illicit drug use; and risk-taking behavior (e.g., driving while intoxicated, not wearing seat belt).

**Mobility and health status:** People with muscle weakness and poor balance or coordination are at higher risk for injury, as are people with paralysis.

**Sensory-perceptual alterations:** For example, impaired touch perception, impaired vision, or impaired hearing.

**Cognitive awareness:** Impaired awareness may be caused by lack of sleep, unconscious states, disorientation, confusion, or altered judgment. These may be caused by disease processes or medications such as narcotics.

**Emotional state:** Any extreme emotion can decrease the ability to perceive environmental hazards. Stress reduces concentration and awareness of external stimuli; depression causes slower thinking and reaction to environmental stimuli.

**Safety awareness:** Clients in new environments may need specific safety information. For example, they may need to be told not to disconnect the intravenous tubing when changing clothes.

**Environmental factors:** Safety hazards can be found in the home (e.g., slippery bathtubs), the workplace (e.g., machinery, chemical hazards), and in the community (e.g., unsafe water, poor street lighting).

## Assessing Safety Needs in the Home

The preceding safety factors should be assessed routinely as a part of the history and physical examination. Special risk assessment tools are used to identify clients at risk for specific injuries, such as falls and hazards in the home. Common hazards in the home include:

- Medications (storage and disposal)
- Toxic substances (improperly stored or not labeled)
- Uneven, broken walkways; broken or loose stairs
- Slippery floors, unanchored rugs
- Furniture with sharp corners
- Bathrooms with slippery tubs and shower stalls; absence of grab bars; water temperature too hot
- Pilot lights in need of repair (kitchen and furnace)
- Inadequate lighting, especially night lights
- Unanchored or frayed electrical cords
- Absence of smoke detectors, fire extinguishers; improper storage of gasoline or corrosives
- No method to call for help, such as a telephone or intercom

## Related NANDA Nursing Diagnoses

- Knowledge Deficit (accident prevention)
- Risk for: Injury, poisoning, suffocation, trauma, and aspiration

## Scalds and Burns

Clients at special risk are those whose skin has decreased ability to sense temperature. Nurses must be cautious about temperature of bath water and about therapeutic applications of heat. In the home, scalds are caused by pot handles protruding over the edge of the stove, electric appliances used to heat liquids (dangling cords cause spills), and excessively high temperature in the water heater.

## Fires

**The correct fire extinguisher** must be used to extinguish a fire. Fires are classified according to the material that is burning:

- **Class A:** Paper, wood, upholstery, rags, ordinary rubbish
- **Class B:** Flammable liquids and gases
- **Class C:** Electrical fires

**In the health care setting** when a fire occurs, follow this sequence of action:

- Protect and evacuate clients who are in immediate danger.
- Report the fire.
- Contain the fire.
- Extinguish the fire.

**In the home, preventive measures include:**

- Keeping emergency numbers near the telephone (e.g., 9-1-1)
- Having a family exit plan; holding fire drills
- Maintaining operable smoke alarms
- Keeping fire extinguishers, be sure they are in working order
- If a fire occurs, close windows and doors, if possible. Cover mouth and nose with damp cloth when leaving through a smoky area. Bend over and keep head as close to the floor as possible to avoid the heaviest smoke.

## Falls

**Incidence:** Infants and older adults are at special risk for falls. Falls are the leading cause of accidents among older adults, and a frequent cause of hospital and nursing home admissions. Most falls occur in the home.

### Risk Factors for Falls

- Poor vision
- Cognitive dysfunction (e.g., confusion, disorientation, impaired memory, impaired judgment)
- Gait, balance, or walking problems (e.g., arthritis)
- Impaired transfer mobility (e.g., getting in and out of chair or bed)
- Orthostatic hypotension
- Urinary frequency
- Weakness (e.g., from disease process or therapies)
- Medications such as sedatives, narcotic analgesics, and hypnotics

### Therapeutic Nursing Actions to Prevent Falls

- On admission, orient clients to the surroundings (including use of grab bars, especially in the bathroom).
- Use a risk assessment tool to assess potential for falling; communicate to all staff.
- Keep call bell within reach; remind client and family to use it. Answer it promptly.
- If client is at risk for falls, assign to a room near the nursing station.
- Keep bed in low position when not providing care.
- Keep side rails up and bed in low position for sedated and unconscious clients when unattended.
- Lock wheels on wheelchairs, stretchers, and beds.
- Provide client with nonskid footwear.
- Use bed or chair safety sensors or leg bands as needed to monitor client activity.
- Check batteries of sensing devices regularly.
- Keep the floor free of clutter, with a clear path to the bathroom.
- Provide adequate lighting (leave a light on at night).

## Poisoning

The main causes of poisoning in children are improper storage of toxic household substances and inadequate supervision. Adolescents and adults are usually poisoned by recreational drugs, suicide attempts, insect bites, or snakebites. Older adults are usually poisoned by accidental ingestion of a toxic substance or an overdose of a prescribed medication (e.g., because of poor memory).

### Therapeutic Nursing Interventions for Poisoning

- Teach parents to "childproof" the home (e.g., dispose of medicines by flushing down a drain).
- Teach parents to place poison warning stickers on containers of bleach, lye, and other toxic substances.
- Teach parents to keep syrup of ipecac on hand at all times. It is a nonprescription emetic. Use it only on advice from the poison control center or a primary care provider.
- Provide information and counseling to adolescents and young adults about risky behaviors.
- For older adults, safeguard (or teaching to safeguard) the environment and monitoring underlying problems (such as memory loss).
- If poisoning occurs: (1) Identify the poison by searching for the container, (2) Contact the poison control center, (3) Keep the person as quiet as possible and lying on the side to prevent aspiration of vomitus, (4) Follow instructions from the poison control center.

## Suffocation or Choking

- A common cause of suffocation is food or a foreign object lodged in the throat.

The universal sign of choking is for the victim to grasp and point to his/her neck and throat without speaking. For this emergency, the Heimlich maneuver (or abdominal thrust) should be used to dislodge the foreign object.

- Other causes of suffocation are drowning, becoming entangled in a piece of plastic on the face (children, usually), and being trapped in a confined space (e.g., a discarded refrigerator).

## Excessive Noise

- Sound levels above 120 decibels are painful and may cause hearing damage even with only brief exposure. Levels of 85-95 decibels for several hours a day can lead to progressive or permanent hearing loss.
- When ill or injured, people are sometimes especially disturbed by noise (e.g., loud talking at the nurses' station).
- Physiologic effects of noise include (a) increased heart and respiratory rates, (b) nausea, and (c) increased muscle tension and activity, and d) hearing impairment.

## Electrical Hazards

- Electrical equipment must be grounded (a plug with three prongs).
- Faulty equipment (e.g., with a frayed cord) may cause electric shock or start a fire.
- **Electric shock** occurs when a current travels through the body to the ground. It can cause burns, muscle contractions and cardiac and respiratory arrest.
- When someone has been shocked, the rescuer should not touch the victim until he/she is removed from contact with the current or the electricity is shut off.

### Therapeutic Nursing Interventions for Electrical Safety

- Check for frayed cords.
- Do not overload outlets and fuse boxes with too many appliances.
- Use only grounded outlets and plugs.
- Do not use electric appliances near sinks or other wet areas.
- Use protective covers over wall outlets.
- Pull plugs from a wall outlet by grasping the plug, not the cord.
- Disconnect appliances before cleaning or repairing them.

## Firearms Safety

- Store firearms in locked cabinets, with keys inaccessible to children.
- Store guns separately from bullets.
- Teach children never to touch a gun.
- Teach children to leave if they discover that there is a gun at a friend's house.
- When handing a gun to someone else, first be sure that it is unloaded and the action is open.
- Don't handle guns when using alcohol or drugs of any kind (including prescribed drugs).
- Have firearms inspected by a gunsmith every two years.

# Restraints

**Restraints** are devices which limit a client's activity and are primarily used to prevent injury to the client, staff, another person, or property. Intended as protective devices, they can also create a safety hazard. **Physical restraints** are any physical, manual, or mechanical method or device attached to the body to restrict movement. Chemical restraints are medications (e.g., neuroleptics, sedatives, anxiolytics) used to control disruptive behavior. **Chemical restraints** should only be used as a last resort when caring for disruptive clients who are elderly. Chemical restraints can cause elderly clients to become confused and disoriented, possibly resulting in falls and injuries. The U. S. government has regulated the use of mechanical restraints in long- term facilities.

Regulations require that:

- Restraints be applied only as a last resort
- Restraints be applied only under a primary care provider's written order. Use of restraints without an order is considered false imprisonment and is illegal.
- The order must specify why the restraint is used and for how long it will be used.
- The client must agree to be restrained.
- The client must not have any physical restraints except those required to treat her/his medical symptoms.

## Safety Measures When Using Restraints

- Remove or replace the restraint frequently.
- Pad bony prominences under or near the restraint.
- Perform neurological and circulatory evaluations of color, sensation, temperature, mobility, and capillary refill of the area distal to the restraint at least every two hours.
- Allow enough slack on the straps of the restraint to enable the client to move both arms and both legs to do range-of-motion exercises.
- Use knots that are easily released and do not tighten when force is applied, such as a clove hitch or a half-bow knot.
- Keep the restraint loose enough to fit two fingers between the device and the client.
- Never tie straps to the bed's side rails. An upper body restraint must be tied to a portion of the bed that moves with the client when the bed position is altered.
- Explain the need for restraints to the client and family.
- Evaluate the continued need for the restraints on a regular, ongoing basis according to the institution's policy.
- Never leave the client alone when the restraint is temporarily removed for evaluation.
- Assess a client in restraints at least every 30 minutes and remove the restraints for a complete evaluation and range-of-motion exercises every two hours.

## Criteria for Selecting Restraints

- Restricts movement as little as possible
- Does not interfere with the client's treatment

- Is easily changed (e.g., when soiled)
- Is safe for the client
- Restrains the client only to the extent necessary
- Fits properly

### Documentation

- The behavior that made the restraint necessary
- The type of restraint used
- Explanations given to client and family
- The client's consent
- Exact times of application and removal
- Client behavior while restrained
- Type and frequency of care given while restrained (e.g., circulation assessment, range-of-motion exercises)

### Gerontological Considerations

- Older adults are at greater risk for falls and injuries as a result of diminished hearing, sight, reflexes, bone density and flexibility.
- Using restraints with older clients to prevent injury is not without risk. Skin that is fragile due to loss of adipose tissue and poor circulation can be damaged by the friction and pressure of restraints.

### Types of Physical Restraints

**Vest restraints**: For confused or sedated clients in beds and wheelchairs. Be sure not to put the restraint on backward. Vest restraints must be sized to fit the client. Tie with a half-bow knot to bed frame or chair leg. Do not tie to the head of the bed. Do not tie so tightly as to restrict breathing.

**Belt or safety strap body restraints**: For all clients being moved by stretcher or wheelchair. Apply even when side rails are up.

**Mitt or hand restraint**: To prevent confused clients and children from scratching and injuring themselves (e.g., by pulling out tubing). Remove mitts every two hours to wash and exercise the hands and check circulation.

- **Limb restraints**: Cloth restraints primarily used to maintain an intravenous infusion. Pad bony prominence (e.g., wrist or ankle); tie with half-bow knot, secure other end to the moveable bed frame.
- **Elbow restraints**: Rigid restraint used to keep infants and children from flexing their elbows to touch or scratch the face or head. Be sure none of the tongue depressors in the restraint are broken; be sure the ends are padded. Do not restrict circulation.
- **Mummy restraint**: Special folding of a blanket or sheet around a child's body to prevent movement during procedures such as eye irrigation.

# Critical Thinking Exercise: Client Safety

**Situation:** A nurse in a rehabilitation facility is planning the discharge of a 75-year-old client who had a left total hip replacement several weeks earlier. The client, who lives with his son and daughter-in-law, can walk approximately 50 feet with the use of a walker and will continue to receive physical therapy in the home. On the day of discharge, the client's son asks about specific measures to promote a safe environment in the home. Before you respond, consider the specific needs and abilities of this client.

1. Working with a classmate, prepare a basic checklist to be used by home care or community health nurses to promote safety in the home. Or, if possible, prepare an audio or video presentation called "Hidden Dangers in the Home." Develop a script for your presentation. You might even want to ask your classmate to videotape you walking through the home, pointing out common dangers and suggesting ways to promote safety.

2. In both acute and subacute settings, restraints may at certain times be used. Describe the basic rationale for applying restraints and discuss the following topics related to restraint use:

   • Alternatives to restraints

   • When restraints may be used

   • Guidelines governing their use

   • Basic measures that must be taken to ensure the safety and well-being of clients in restraints

# Body Mechanics

| Key Points |
| --- |

- A wide base of support and a low center of gravity increase stability and balance.
- Uncoordinated movements can sprain the ligaments, strain the muscles, rupture the disks, and irritate the joints.
- Widen the base of support by spreading the feet farther apart.
- Use the legs, abdomen, and arms—not the back—to lift.
- Pull objects rather than pushing them, when possible.
- Sliding, rolling, or pushing requires less energy than lifting.
- Make use of body weight by rocking forward or backward to move an object.
- When moving an object, avoid twisting the spine or bending over at the waist.
- Exercise regularly to strengthen the pelvic, abdominal, and lumbar muscles.
- Get help whenever possible to lift a heavy load rather than trying to lift a client or object alone.
- **Key Concepts/Terms**: Center of gravity, line of gravity, base of support, assistive devices, back strain or injuries

**Overview**

Body mechanics are the efficient, coordinated, and safe use of the body as a machine and as a means of locomotion. Good body mechanics is important to both clients and nurses in order to decrease the risk of injury and reduce fatigue. Important concepts are: center of gravity, line of gravity, and base of support.

The **center of gravity** of an object is the center of its mass. In humans, it is at the center of the pelvis about midway between the umbilicus and the symphysis pubis.

The **line of gravity** is the vertical line passing through the center of gravity.

The **base of support** is the foundation that provides the object/person's stability.

## Some Principles of Body Mechanics

- The wider the base of support and the lower the center of gravity, the greater the stability and balance of the object or person.
- Spreading the feet farther apart can widen the base of support.
- When a person moves, the center of gravity shifts in the direction of the movement. The closer the line of gravity is to the center of the base of support, the greater the stability. If the line of gravity falls outside the base of support, the person falls.

- The center of gravity can be lowered by flexing the hips and knees, assuming a squatting position.
  - Movements to avoid:
    - Twisting the thoracolumbar spine
    - Acute flexion of the back with the hips and knees straight (bending over at waist)

## Lifting

- Use major muscle groups of the thighs, knees, upper and lower arms, abdomen, and pelvis to prevent back strain.
- Tighten abdominal muscles when assisting with lifting clients to increase back support.
- When lifting, distribute the weight between the large muscles of the legs and arms.
- When lifting an object from the floor to waist level, flex the back and knees until the load is at thigh level, then keep the knees bent and straighten the back.
- Hold the lifted object as close as possible to the center of gravity.
- Keep the feet at least 12 inches apart and keep the load close to the body, especially when it is at waist level.
- Request assistance if the client is over 85 pounds.
- Use assistive devices (e.g., mechanical lifts) as needed.

### Pushing and Pulling

- When pushing an object, enlarge the base of support by moving the front foot forward.
- When pulling an object, move the rear leg back if facing the object, or move the front foot forward if facing away from the object (when moving a client, always face the direction you are moving them).
- It is easier and safer to pull an object toward one's center of gravity than to push it away.
- Use your body weight to push an object by rocking forward, and to pull an object by rocking backward (e.g., when sliding a client up in bed).
- Sliding, rolling, or pushing requires less energy than lifting.

### Other Strategies to Prevent Back Injuries

- Plan ahead for activities that might result in muscle or joint injury.
- Stop and rest as needed when performing high-risk movements.
- Maintain good posture.
- Exercise regularly to strengthen your back, arm, leg and abdominal muscles.
- Get help and offer help to others to lift or move a heavy client or object.
- Wear low-heeled shoes that provide good foot support.
- When standing for long periods, occasionally flex one hip and knee and rest your foot on an object if possible.

- Sit with knees slightly higher than hips.
- Exercise regularly, including exercises to strengthen the pelvic, abdominal, and lumbar muscles.
- Sleep on a firm mattress.
- Use smooth, even movements when lifting, never sudden or jerky movements.

## Complications Related to Poor Body Mechanics

- Muscle fatigue
- Joint strain
- Lower back injuries are the most common injury among nurses.
- Repetitive motion injuries

# Critical Thinking Exercise: Body Mechanics

**Situation**: A nursing colleague confides that she can no longer tolerate "all this heavy lifting and pulling". In observing this nurse with clients, what principles do you keep in mind as you assess the proper use of body mechanics? Consider how you might use this information to assist your colleague.

1. You are participating in a health fair. As a result of a number of questions about back pain, you decide to create a handout called "The Key to Preventing Back Injuries". What essential information will you include in this teaching tool? Here are a few topics to get you started.

   • Maintaining good posture

   • The importance of exercise

   • Choosing clothing and shoes

   • The right mattress

   • How to move heavy objects

2. Each day your best friend carries a schoolbag filled with heavy books over her left shoulder. You notice that this practice is beginning to affect her posture. What advice might you offer your friend?

# Hygiene Care: Bathing, Oral Hygiene, and Linen Change

| Key Points |
| --- |

- Bathing is done to cleanse and condition the skin, relax the client, promote circulation, stimulate respirations, relax the muscles, provide sensory input, and provide opportunity to foster the nurse-client relationship.
- Provide the amount of assistance needed.
- Work efficiently (e.g., organize materials, save steps).
- Use good body mechanics (e.g., elevate bed for bed bath).
- Ensure privacy.
- Ensure safety (e.g., lower bed after bath).
- Keep patient warm.
- Work from "clean to dirty, to prevent the spread of microorganisms.
- Promote circulation (e.g., wash and dry from distal to proximal on limbs).
- Oral hygiene should be provided with attention to:
  - Organization
  - Efficiency
  - Patient safety (e.g., position to prevent aspiration; check dentures for damage and fit)
  - Nurse safety (e.g., protect from contact with body fluids)
  - Privacy
  - Oral hygiene includes daily brushing and flossing of the teeth.
  - Remember that dentures are easily broken.
  - Purpose: clean the teeth; stimulate the gums
- The elderly client may not need a full bath every day due to decreased sebaceous and sweat glands.
- **Key Concepts/Terms**: Clinical overview, shower bath, tub bath, bed bath, towel bath, bag bath, morning care, HS (at the hour of sleep) care, gingiva, oral mucosa, sulcular technique, plaque, draw sheet, occupied bed, body mechanics, unlicensed assistive personnel, fanfold

**Overview**

When clients are unable to perform personal hygiene, the nurse assists or delegates hygiene care to other personnel. In hospitals, certain routines for hygienic care are usually followed.

**Early morning care** is given just after the client awakens. The nurse helps the client to the toilet or provides a bedpan or urinal, washes the face and hands, and provides mouth care.

**Morning care (a.m. care).** After breakfast, the nurse again assists as needed with toileting and assists with oral care, bathing, back massage, hair care, dressing, positioning for comfort, and possibly a linen change. Morning care is usually referred to as "self-care," "partial care," or "complete care."

**Afternoon care** consists of toileting, washing the hands and face, and oral care.

**Hour of sleep (HS) care** is provided before clients retire. It involves providing for elimination needs, washing face and hands, oral care, and a back massage.

**As-needed (PRN) care** is provided to meet special needs. For example, a diaphoretic client may need more frequent bathing and linen changes.

**Bathing** is done to cleanse and condition the skin, relax the client, promote circulation, stimulate the rate and depth of respirations, promote comfort through muscle relaxation and skin stimulation, provide sensory input, improve the client's self-image, and provide an opportunity to strengthen the nurse-client relationship.

## Bathing

### Shower and Tub Bath

- Preferred method for ambulatory clients
- Be sure the bathroom is available, clean, and safe (e.g., shower mats).
- See that the necessary articles are available (e.g., soap, washcloth, towel, gown).
- Provide a shower stool or chair for weak or physically disabled clients.
- Assist into and out of tub/shower as needed for safety.
- Ensure privacy (e.g., place a sign on the door).
- If client bathes independently, see that a call device is handy.
- Keep the bathroom door unlocked.
- Wash and dry areas (e.g., the back) that the client cannot reach.

### Bed Bath

- Water should be comfortable to the client (usually 110-115° F). It should be changed at least once during the bath.
- Some clients require a complete bed bath by the nurse.
  - Prepare the client, the environment, supplies, and equipment (e.g., offer bedpan, provide privacy, close windows to prevent drafts).
  - Place bed in high position, remove top sheet and replace with bath blanket; remove client's gown.
  - Make a bath mitt with the washcloth.
  - To keep client warm, uncover only parts being bathed; dry each area before washing the next area.
  - Wash the face.
  - Wash the arms and hands.

- Wash the chest and abdomen.
- Wash the legs and feet; soak feet in a basin, if possible. Use long strokes on the legs, washing from ankle to knee to thigh.
- Wash the back and then the perineum. For females wash "front-to-back" and use a separate area of the washcloth for each stroke.
- Assist the client with deodorant, lotion, and so on.
- Document pertinent data.

- Some clients who must remain in bed can, nevertheless, bathe themselves. The nurse should still help by:
  - Placing articles for bathing conveniently on the over-the-bed table
  - Providing privacy
  - Removing the top linens on the bed and replacing them with a bath blanket
  - Placing cosmetics within reach; providing a mirror and light
  - Providing hot water and a razor for clients who wish to shave
  - Bathing any parts of the body the client cannot reach

## Towel Bath

This is an in-bed bath using a quick-drying solution containing a cleaning agent, a disinfectant, and a softening agent mixed with water at a temperature of 110-120° F. It takes only about 10 minutes and is readily accepted by most clients. It minimizes client fatigue, and dry, itchy skin.

- Place large towel, about 3 x 7 feet, in a plastic bag and saturate it with the warmed cleaning solution.
- Wring out towel, then, unroll it over the client while simultaneously removing the top sheet off the client.
- Fold part of the towel under the client's chin for later use.
- Begin with the client's feet and work up the body, using a massaging motion to clean the body.
- Fold the towel upward as the bath proceeds, at the same time unfolding a clean sheet over the client.
- Use the folded-under part of the towel to wash the face, neck, and ears.
- Remove the towel, roll the client to one side, and apply the clean side of the towel to the back of the neck, back, and buttocks.
- Remove the towel.
- Place clean linen on the bed, dress the client, and position appropriately.
- No towel drying is needed because the cleaning solution dries immediately.

## Bag Bath

This is a variation of the towel bath. It is standard bathing practice in some facilities. Ten or twelve washcloths are placed in a plastic bag with a mixture of water and a non-rinsable cleaner. The solution and cloths are warmed in a microwave for about one minute. Each area of the body is cleaned with a different cloth and then air-dried. This permits the emollient ingredient to remain on the skin.

# Oral Hygiene

## Overview

Oral hygiene includes daily brushing and flossing of the teeth. The nurse's role includes teaching clients about oral hygiene and identifying problems that require referral to a dentist. Brushing removes food particles that can harbor bacteria and stimulates circulation in the gums. Like natural teeth, dentures need to be cleaned at least once a day to remove food and microorganisms. A client may have an upper plate, (a complete set of upper and lower teeth), or just a lower plate. Dentures are fitted to the individual and usually will not fit another person. Some clients have only a few artificial teeth called a bridge. These may be fixed or removable.

## Oral Hygiene for Dependent Clients

- Gather equipment: disposable gloves, emesis basin, towel, toothbrush, cup of water, toothpaste, dental floss, and mouthwash.
- Assist client to sitting position. If client cannot sit up, position on his or her side, laying with head on a pillow, so the client can spit out the water.
- Place towel under client's chin.
- Don gloves.
- Moisten brush and apply toothpaste.
- Hand toothbrush to the client or brush the client's teeth. Brush all surfaces, using a vibrating or jiggling motion.
- In the sulcular technique, the brush bristles are placed at a 45-degree angle against the teeth, with the tips of the outer bristles resting against and penetrating under the gingival sulcus.
- This technique removes plaque and cleans under the gingival margins.
- Floss the teeth (or provide floss to the client).
- Assist client to rinse the mouth with water and/or mouthwash.
- Assist client to wipe the mouth.
- Document assessment of the mouth.
- Foam swabs may be used to remove debris from the teeth and mouth, but they are not effective in removing plaque from the teeth and should not be used exclusively.

## Cleaning Dentures

Remove dentures, scrub with a toothbrush, rinse, and reinsert. The client will probably want privacy when taking out dentures; many people are self-conscious about being seen without their teeth. Gather equipment: disposable gloves, tissue or piece of gauze, washcloth, toothbrush, toothpaste or denture cleaner, water, emesis basin, towel, mouthwash, container for dentures.

- Assist client to sitting or side-lying position.
- Don gloves.

- Remove (or have client remove) dentures:
  - **Remove upper plate** by grasping with tissue or gauze and moving the denture up and down slightly to break suction, place in denture container.
  - **Remove lower plate** by turning it so that one side is slightly lower than the other and lifting it up and out of the mouth, place in denture container.
- Clean the dentures:
  - Place washcloth in bowl or sink to prevent damage if the dentures are dropped. Scrub with toothbrush.
  - Do not use hot water (it may change the shape of some dentures). Rinse. Alternately, soak dentures in a commercial cleaner.
- Inspect the dentures for rough or sharp areas.
- Inspect the mouth for redness and irritated areas.
- Return dentures to the mouth.
- Assess the fit of the dentures.

## Special Mouth Care for Debilitated or Unconscious Clients

- Unconscious clients or those with mouth problems need frequent oral hygiene (perhaps every two hours) to combat mouth dryness and prevent infections.
- Clients at risk for altered/dry oral mucous membranes include those who are unconscious, those receiving oxygen, those who cannot take fluids by mouth, and those who must breathe through the mouth.
- Special applicators of lemon juice and oil can be used; however, long-term use can lead to further drying of the mucosa and be harmful to tooth enamel.
- Do not use mineral oil (aspiration can cause lipid pneumonia).
- Do not use hydrogen peroxide (it irritates healthy mucosa and may alter the normal flora of the mouth).
- Normal saline solution is the recommended solution.

## Procedure for Unconscious Clients

- Prepare equipment: towel, emesis basin, disposable gloves, bite-block, toothbrush, cup of water, toothpaste, mouthwash, rubber-tipped bulb syringe, applicators and cleaning solution for mucous membranes, petroleum jelly.
- Position unconscious client in side-lying position; lower head of bed (to allow saliva and water to run out of the mouth by gravity).
- Place towel under client's chin.
- Place emesis basin against chin and lower cheek to catch fluid.
- Clean teeth or dentures.
- Fill syringes with about 10 mL of water or mouthwash and rinse client's mouth.
- Be sure all the rinsing solution runs out of the mouth into the basin.
- Inspect and clean oral tissues with applicators or gauze. Use a fresh applicator for each area of the mouth.

# Linen Change

## Overview

Although bed making is often delegated to unlicensed assistive personnel, the nurse may need to evaluate and teach the procedure. Traditional practice is to change bed linens after the bath, but some agencies change linens only when soiled. Most agencies use a fitted bottom sheet and a flat top sheet. A **draw sheet** may be used across the center of the bed, under the client's midsection, either to protect the bottom sheet or to use in turning and repositioning the client in bed.

## Principles

**Use good body mechanics** (e.g., adjust bed to high position and drop side rails).

**Prevent the spread of microorganisms:**

- Wash your hands before and after procedure.
- Roll all soiled linen inside of the bottom sheet and place directly in hamper; do not place them on the floor or furniture.
- Do not hold soiled linens against your uniform.
- Do not shake linens.

**Work efficiently:**

- Place all bottom linens (mattress pad, sheet, draw sheet, protective pad) before securing under mattress, then tuck all at once.
- Follow the same procedure with top linens (sheet, blanket, spread).
- Move around the bed from head to foot to head on far side; do not reach across bed.

**Consider client safety** (e.g., when finished, leave bed in low position and put call light within reach of client).

**When making an occupied bed:**

- Turn client toward far side of the bed (away from you).
- Loosen and roll soiled linen, tucking under or as close to client as possible.
- Open and fanfold clean linen as close to client as possible, and slightly under the soiled linen.
- Tuck clean bottom sheet and draw sheet on your side of the bed.
- Raise side rail and help client roll over the folded linen toward you; reposition client on that side.
- Move to the other side of bed.
- Loosen and remove all bottom linen.
- Ease the clean linen from under the client and finish tucking the bottom linen.
- Reposition client in center of bed and apply top linen.
- Be sure to finish with side rails up and bed in low position.

# Critical Thinking Exercise: Hygiene Care: Bathing, Oral Hygiene, and Linen Change

1. You are caring for an 82-year-old postoperative client who is now able to ambulate with the assistance of a nurse. At home, the client is independent with all activities of daily living. The client is reluctant to accept your help and wants to take a shower "on my own". What would you do?

2. Nurses in all practice settings encounter a variety of factors, ranging from culture to religious practice to personal preference, influencing clients' hygienic practices. For example. how might an individual's culture affect his or her attitude to hygiene? Begin by considering your own culture. What is the standard of hygiene accepted in your culture? In what ways might this differ from the standard of other cultures? How might you adapt this knowledge to promote the goal of optimal hygiene?

3. Consider how a nurse might integrate ongoing assessment into the process of bathing a client. What kinds of information can a nurse gather about a client's health status during this process?

# Application of Heat and Cold

| Key Points |
| --- |

- The nurse assesses the client's physical condition for signs of potential intolerance to heat and cold prior to their application.
- Heat and cold can have both local and systemic effects.
- When applying heat and cold, apply these principles:
  - Protect from localized injury (e.g., burns from a heating pad).
  - Protect from extreme systemic effects (e.g., overheating, over cooling, fainting).
  - Prevent chilling (by lowering body temperature gradually).
  - Prevent rebound phenomenon (by limiting the length of the application to 20 to 30 minutes).
- Methods of application include: hot water bag/bottle, aquathermia pad, commercial hot/ cold packs, compresses, ice bag/glove/collar, soaks, electric heating pads, sitz bath, cooling sponge bath, hypothermia and hyperthermia blankets.
- Use heat and cold cautiously with clients with neurosensory, mental, or circulatory impairment; clients with open wounds; and clients in the immediate post-op period.
- Never place an electric pad under a client.
- **Key Concepts/Terms**: Heat, cold, rebound phenomena, methods of application, therapeutic procedures, precautions, nursing interventions

**Overview**

Heat and cold may be applied for either local or systemic effects. Heat is most often used for clients with joint stiffness (e.g., arthritis, contractures) and low back pain. Cold is most often used for musculoskeletal injuries (e.g., sprains, fractures) to limit post- injury swelling and bleeding.

**Heat**

**Local effects**: Vasodilation and increased blood flow to the area; promotion of tissue healing (vasodilation increases blood delivery to the area which contains oxygen and nutrients for healing); increased suppuration (through increased vascular permeability allowing for absorption of the debris of infection/inflammation); muscle relaxation; and a reduction in joint stiffness.

**Systemic effects**: Applied over a large area, may cause excessive peripheral vasodilation and a drop in blood pressure. The client may faint.

# Cold

**Local effects:** Vasoconstriction and reduced blood flow to the area; produces skin pallor and coolness. Prolonged exposure can damage tissues from lack of oxygen and nourishment.

**Systemic effects:** With extensive applications, excessive vasoconstriction causes blood to shunt from peripheral circulation to internal blood vessels; this causes the blood pressure to increase. Shivering is a normal response to prolonged cold.

## Rebound Phenomenon

After the maximum therapeutic effect of the hot or cold application is achieved, the opposite effect will begin.

**Heat** produces maximum vasodilation in 20 to 30 minutes. If the application continues beyond 30 to 45 minutes, the blood vessels **constrict**. The constricted vessels cannot dissipate the heat adequately, and the client is at risk for burns.

**Cold** produces maximum vasoconstriction when the skin reaches a temperature of 60° F. Below that, vasodilation occurs as a protective mechanism to help prevent freezing of tissues.

## Methods of Application

Either dry or moist forms of heat and cold can be used.

**Hot water bag or bottle:** Provides dry heat; frequently used in the home. Burns may occur with improper use. Wrap bag in a towel or cover to apply. Recommended temperatures:

- Normal adults and children older than 2 years, 125° F
- Debilitated or unconscious adults and children under 2 years, 105-115° F

**Aquathermia pad (K-pad):** Disposable pad made with tubes through which water circulates. It is attached by tubing to an electric control unit that heats and circulates the water through the pad. The reservoir of the unit must be filled with distilled water, and it provides dry heat.

**Commercially prepared hot and cold packs:** These are activated by striking, squeezing, or kneading the pack. They provide dry heat/cold for a limited period of time.

**Compresses:** A compress is a moist dressing that can be either warm or cold. The temperature of a hot compress is indicated by medical order or agency protocol (about 105° F). For open wounds, sterile technique is used to apply the compress; and the compress and wetting solution are also sterile.

**Ice bags, gloves, and collars:** These provide dry cold to a localized area. They are filled with ice chips or an alcohol-based solution, wrapped in a towel or cover, and applied to the body.

**Soaks:** A body part (e.g., hand, leg) is immersed in a solution or wrapped in gauze dressings and then saturated with a solution. For open wounds (e.g., burns), sterile technique is used. Agency protocol determines the temperature of the solution.

**Electric heating pads** supply dry heat. Some have waterproof covers for use when pad is placed over a moist dressing. Use pads with a preset heating switch so the client cannot increase the heat. There is a risk of burns with this method.

Sitz bath: The client sits in a special tub, chair, or pan to soak the perineal or pelvic area. Because the legs are not immersed (as they would be in a regular tub), circulation to the perineum is increased. Water temperature should be 105-110° F. The duration of the sitz bath is 15 to 20 minutes.

Cooling sponge bath: Used to reduce a fever by promoting heat loss through conduction and evaporation. A fan may be used to increase air movement around the client. The temperature of a **tepid** bath is 90° F throughout the bath. A cool bath begins with a 90° water temperature and is gradually lowered to 65° F by adding ice chips during the bath. A **cool** bath must be used cautiously.

## Procedure and Technique for Cooling Bath

- Sponge face, arms, legs, back, and buttocks (usually not the chest and abdomen). Leave each area wet and cover with a damp towel.
- Place ice bags or a cool cloth on the forehead, in each axilla, and at the groin, where large superficial blood vessels are located.
- The bath should take about 30 minutes; if given too rapidly, the client will shiver and produce even more heat.
- If the client shivers, becomes pale or cyanotic, or if the pulse becomes rapid or irregular, stop the bath.
- When finished, pat the client dry. Assess the vital signs 15 minutes into the bath and after completing it.

Hypothermia and hyperthermia blankets: These have a control panel for setting the desired temperature and registering the client's core temperature. They should be used according to agency protocols and the instruction manuals accompanying them.

## Precautions

- Use heat and cold with caution in the following situations:
  - Clients with neurosensory impairment: They cannot perceive discomfort from heat or cold and are at risk for tissue injury.
  - Clients with impaired mental status (e.g., those who are confused or who have an altered level of consciousness) must be monitored closely during applications.
  - Clients with impaired circulation (e.g., peripheral vascular disease, diabetes, congestive heart failure) lack the ability to dissipate heat well, increasing their risk for tissue damage.
  - Immediately after surgery or injury, heat increases bleeding and swelling.
  - Open wounds: Cold decreases blood flow and inhibits healing.

## Therapeutic Nursing Interventions for Application of Heat or Cold

- Assess client's ability to tolerate the therapy (e.g., identify contraindications such as circulatory impairment).
- Inform the client to report any discomfort.
- Use heat applications cautiously for clients who have heart disease, pulmonary

disease, or circulatory disturbances such as arteriosclerosis (see "Systemic Effects of Heat").

- When using cold applications, observe for bluish purple, mottled skin, numbness, pain, and blisters.

- Explain to clients that heat and cold receptors adapt to temperature changes. After the heat (or cold) has been applied for several minutes, the strong stimulation disappears and it will feel less hot (or cold) to the client. The temperature of the application should not be changed as the feeling decreases.

- Apply heat (or cold) intermittently to prevent rebound phenomenon. Return to assess 15 minutes after starting the application.

- Do not insert sharp objects (e.g., safety pin) into electric heating pads.

- Do not place an electric pad under the client; heat will not dissipate and the client may be burned.

- Observe client in a sitz bath for faintness, dizziness, and pallor.

# Critical Thinking Exercise: Application of Heat and Cold

1. You are caring for a febrile 75-year-old client with peripheral neuropathy and a healing Stage II pressure ulcer on her sacrum. A cooling bath has been ordered. What special precautions apply in this situation?

2. A 16-year-old soccer player sprains his ankle as a result of a twisting movement on a fall. With respect to the application of heat or cold, what is the best way to manage the swelling of the ankle and foot? Explain the rationale for your action.

3. Conduct a literature search on the therapeutic uses of heat and cold applications. Focusing on the findings of one study, share and discuss its implications with a classmate.

4. Develop a one-page "Safety Sheet" to guide you in the appropriate application of heat and cold. Share these guidelines with classmates.

# General Medication Information

| Key Points |
| --- |

- A formula for computing drug dosage:
    - Dose ordered
    - Dose on hand x Amount on hand = Amount to administer
- The primary care provider prescribes, the pharmacist prepares, and the nurse administers to the client.
- The nurse should question any order suspected of being in error.
- Effects of a medication are influenced by weight, sex, biological factors, psychological factors, pathology, environment, and timing of administration.
- The "Five Rights" are the correct: Medication, Client, Dosage, Route, and Time.
- To identify a client, check the ID bracelet and ask the client to state his/her name.
- Do not administer medications prepared by another nurse.
- Return the medication to the pharmacy if the label is missing or illegible
- The nurse is ethically and legally responsible for ensuring that the client receives the correct medication ordered by the physician.
- **Key Concepts/Terms:** Drug/medication, generic name, trade name, brand name, absorption, distribution, metabolism/biotransformation, excretion, therapeutic effect, adverse effect, allergy, anaphylactic reaction, drug tolerance, cumulative effect, idiosyncratic effect, drug interaction, agnostic effect, antagonistic effect, placebo effect

**Overview**

A **drug** or **medication** is a substance that modifies body functions. The primary care provider is legally responsible for prescribing medications; the pharmacist prepares them; the nurse administers them to clients. The **generic name** (e.g., acetaminophen) of a drug is the name assigned by the manufacturer that first develops the drug; the **trade name** (e.g., Tylenol, Tempra, Liquiprin) is the **brand name** chosen by the company that sells the drug.

**Absorption** is the process by which a drug gets from its site of entry into the bloodstream.

**Distribution** is the movement of the drug, via the bloodstream, to the specific tissues for its action, where it accumulates.

**Metabolism** (biotransformation) is the breakdown of the drug into an inactive form. Usually this occurs in the liver.

**Excretion** occurs after the drug is metabolized. Most drugs leave the body via the kidneys, but excretion also occurs through the sweat, salivary, and mammary glands, the lungs, and the intestines.

## Factors Influencing the Effects of a Medication

- Weight
- Sex (because of the distribution of body fat and fluids)
- Genetic and biological factors
- Psychological factors (e.g., placebo effect)
- Pathology (e.g., liver disease slows drug metabolism)
- Environment (e.g., quiet environment enhances effect of pain medication)
- Timing of administration (e.g., some drugs are best absorbed from an empty stomach)

## Adverse Effects

The **therapeutic effect** is the desired effect of the drug (e.g., the therapeutic effect of morphine is pain relief). **Adverse effects** are those that are neither intended nor desired.

**Drug allergy** occurs if the person has previously had the drug and has developed antibodies. Allergic reactions range from minor to serious (e.g., rash, hives, diarrhea).

**Anaphylactic reaction** is a life-threatening immediate allergic reaction that causes respiratory distress, severe bronchospasm, and cardiovascular collapse. It is treated with epinephrine, bronchodilators, and antihistamines.

**Drug tolerance** occurs when the body becomes accustomed to a drug over a period of time, so that larger and larger doses must be given to the client to produce the same effects.

A **cumulative effect** occurs when the body cannot metabolize one dose of the drug before another is administered; so each new dose adds to the total quantity in the body.

An **idiosyncratic effect** is an abnormal or peculiar response. It may be an over-response, an under-response, or simply an unexpected response. They may be the result of genetic enzyme deficiencies. Older clients, for example, often have erratic responses to medications.

A **drug interaction** occurs when the combined effect of two or more drugs produces a different effect than that of each drug alone. An agonistic effect is greater than that of each drug alone (e.g., alcohol and barbiturates are synergistic). An antagonistic effect is less than that of each drug alone.

## Medication Orders

The nurse is legally responsible for drugs administered; any order suspected of being in error should be questioned. A medication order consists of seven parts:

- Client's name
- Date and time order is written
- Name of drug to be given
- Dosage of the drug
- Route of administration
- Frequency of administration
- Signature of the person writing the order

## Safety Measures for Preparing and Administering Drugs

- Check the label on the medication container three times: (1) when reaching for the package/container, (2) immediately before pouring or opening the package, and (3) when replacing the container or giving the unit dose to the client.
- Return medications to the pharmacy if the label is missing or illegible.
- Follow agency procedures for accounting for controlled substances (e.g., narcotics).
- Notify the nurse manager if there appears to be tampering with any medication.
- Never administer medications prepared by another nurse.
- Observe the Five Rights:
  - Right medication
  - Right client
  - Right dosage
  - Right route
  - Right time
- To identify a client, check his/her identification bracelet and ask the client to state his/her name. It is not safe to just call the client by name.
- Remain at the bedside until the medication is taken.
- Administer scheduled medications within 30 minutes of scheduled time.
- Recheck medications if the client questions their appearance or time of administration.
- Monitor the effects of the medications.

## Age-Related Changes—Gerontological Considerations

- Older adults have a decline in liver function and enzyme production needed for drug metabolism.
- Decreased kidney function results in diminished filtration and excretion.
- The older adult has altered peripheral venous tone—important if client is taking antihypertensive medications.
- Older adults often take multiple drugs that interact with each other.
- Because of sensory deficits such as hearing and sight, the older adult may not comprehend medication instructions or be able to see and read labels.
- Older adults often are dehydrated, which affects distribution of medication.
- Dosages should be re-evaluated frequently because of changes in weight.

# Critical Thinking Exercise: General Medication Information

1. As you are preparing to administer an oral medication to a 78-year-old nursing home resident who has dementia, the client states "I don't remember taking that before." What action is appropriate in this situation.

2. Discuss with a classmate the nature and extent of a nurse's responsibility with regard to medication administration. What part do the Five Rights play in this responsibility? For example, does the responsibility begin and end with the actual administration of the medication? Explain.

3. You are reviewing the medications of a home care client, who is currently taking an oral hypoglycemic, two cardiac medications and a nonsteroidal anti-inflammatory drug for arthritic pain. The client points to several bottles of herbal preparations and states "My wife has been taking these for months, so I decided to give them a try." How might you best respond?

# Oral Medication Preparations

## Key Points

- The oral (PO) route is the most common method of drug administration.
- Oral medications may be pills, capsules, tablets, or liquid preparations.
- Usually the safest route
- Some clients have difficulty swallowing oral medications.
- Some clients are at risk for choking and aspiration.
- Ensure that the client has adequate swallowing and gag reflexes.
- Pour liquid medications into a medicine cup at eye level to measure.
- Remember that older adults have special needs related to medication administration and metabolism.
- **Key Concepts/Terms**: Sublingual, buccal, crushed tablets, enteric-coated tablets, extended release capsules, elixir, spirit, suspension, syrup

## Overview

The **oral** (PO) route (through the mouth) is the most common because it is usually the most convenient, comfortable, and safe. Oral medications are intended for absorption in the gastrointestinal tract. Some are administered via gastric or intestinal tube. Compared to other routes, medications have a slower onset and a more prolonged but less potent effect.

**Sublingual** drugs (e.g., nitroglycerin) are placed under the tongue to dissolve and are readily absorbed. Drugs given by the **buccal** route are placed against the mucous membranes of the cheek until the drug dissolves. They act locally on the mucosa or systemically when swallowed in the saliva. The desired effect is not achieved if these drugs are swallowed or if the client drinks liquids before the drug is dissolved.

## Types of Oral Medications

- Oral medications may be solid or liquid.
  - Solid preparations are: tablets, capsules, and pills. **Enteric-coated** tablets have a coating that delays absorption until after the tablet leaves the stomach. Because the active ingredient irritates stomach mucosa, enteric-coated tablets should not be crushed or chewed. **Extended release** capsules also should not be crushed or chewed.
  - Liquid preparations include **elixirs, spirits, suspensions, and syrups.** They may be water or alcohol-based. They are administered in a specially calibrated, disposable cup.

## Safety Factors for Oral Drug Administration

- Oral medications can be difficult, and even dangerous, to give in some situations. The following are examples:
  - Children under age 5 years find it difficult to swallow tablets and capsules.
- Clients who are "NPO" cannot take oral medications.
- Clients with dysphagia or altered level of consciousness are at risk for choking and aspiration.
- Clients who have experienced oral trauma or surgery usually cannot take PO medications.

## Therapeutic Nursing Interventions for Oral Preparations

- Assess to be sure that the client has adequate swallowing and gag reflexes.
- Consult with the primary care provider about substituting liquid drugs for solid medications that must be instilled via gastric or intestinal tube.
- Pour liquid medication into medicine cup at eye level.
- Clients that may have trouble swallowing whole capsules or tablets may need to have the medication crushed and mixed with food or put down the feeding tube (e.g., nasogastric or NG tube). Not all tablets and capsules can be crushed. For safe administration, always obtain an order from the primary care provider before crushing a tablet. Medications that are slow release, long acting, or sustained release are coated to control the release of the drug. This coating must remain intact to prevent uncontrolled release or "dumping" of the dose. Tablets should not be crushed if they have been coated; capsules generally can be opened, but the beads cannot be crushed because the beads are generally the part that have been coated to control release of the medication. The nurse should follow the manufacturers' guidelines with regard to crushing and administering the medications with food or via the tube. Note: Before administering medications via a nasogastric tube, confirm proper positioning of the tube.
- For medications with objectionable taste, offer oral hygiene immediately after administering.
- Never put any of your clean items which are used in administering medications directly on the surface of the counter or medication cart/tray. Always provide a clean barrier to prevent contamination of clean items from unknown material on a surface.

## Age-Related Changes—Gerontological Considerations

- Older adults have decreased gastric emptying time, decreased gastric motility, and an increase in stomach pH.
- Older adults, because of arthritis, may have difficulty removing childproof caps.
- Aging causes a decrease in saliva production, which makes swallowing medication more difficult (increased risk of choking).
- Older adults need extra time to take medications related to slower reflexes and a decreased understanding of treatment.

Also review *"Medication Administration: General Information."*

# Critical Thinking Exercise: Oral Medication Preparations

**Situation**: A 55-year-old client refuses to take his antihypertensive medications. "I feel just fine," the client states. "Anyway, my blood pressure has been okay for at least a month. I just don't want to take any medication I don't need." Consider your response to this client.

1. What do you tell your client about hypertension to address his statement "I feel just fine."?

2. An elderly client is being discharged from your facility with several new medications. The client's family is concerned that the client, who lives alone, may have difficulty remembering to take so many medications. How might you assist this client and family?

# Topical and Inhalation Medications

| Key Points |
| --- |

- Topical medications are given for their local effects; inhalation medications are given for both local and systemic effects.
- Wear gloves when applying topical medications.
- Methods of inhalation: nasal inhalation, oral inhalation, endotracheal or tracheal administration (requires special training)
- **Key Concepts/Terms:** Transdermal disc/patch, instillation, irrigation, vasoconstriction, nasal sprays, nose drops, inhaler, metered dose inhaler

**Overview**

**Topical medications** are given for their local effects. **Inhalation** medications are given for both local and systemic effects.

## Topical Administration

- The medication is applied to the skin by painting or spreading it over an area, applying moist dressings, soaking in a solution, or by medicated bath.
- Systemic effects can occur if drug concentration is high or contact is prolonged.
- A **transdermal disc/patch** may be used to secure the medicated ointment to the skin (e.g., nitroglycerin, estrogen). This ensures a continuous delivery of the medication into the blood. Application sites should be rotated each time the patch is changed to avoid problems with absorption or skin breakdown.
- Topical drugs can be applied to mucous membranes by:
  - Direct application (e.g., having the client gargle)
  - Inserting the drug into a body cavity (e.g., inserting a suppository into the rectum)
  - Instillation into a body cavity (e.g., nose drops, eardrops)
  - Irrigation (washing out) of a body cavity (e.g., flushing the eye, vaginal douche)
  - Spraying (e.g., nose spray, throat spray)
- Nasal instillation of medications is a route used most often for rapid absorption of certain medications.
- Because these medications can cause systemic and local effects, the nurse should wear gloves or use an applicator to apply them.

# Inhalation Medications

## Nasal Drug Administration

- Medications given by the nasal route include:
    - Decongestants to reduce nasal and sinus congestion, such as phenylephrine (Neo-Synephrine) which causes local vasoconstriction in the nasal passages
    - Saline preparations to loosen secretions and help clear nasal passages
    - Antiseptics to treat local conditions
    - Corticosteroids to reduce inflammation from nasal polyps or allergic conditions
    - Anesthetic agents to ease the discomfort of transnasal examinations and treatments
    - A mast cell stabilizer to treat asthma
    - Nasal mist influenza vaccine
- When you are preparing to administer nasal drops to a client, follow the same initial steps as you would for any medication administration:
    - Screen the client for allergies to the medication.
    - Explain the procedure, along with what to expect during administration, the desired effects, and the possible adverse affects of the drug.
    - Check that you are administering the prescribed amount and strength of the drug to the right client by the right route at the right time.
    - Cross-check the medication container with the medication order and check the drug's expiration date.
    - Use medical asepsis, or clean technique since the nasal cavity is not considered a sterile area.
    - Assess the client before administering nasal medications, document your findings, and contact provider if necessary. Note:
        - Are the client's nares irritated?
        - Are the nasal passages congested?
        - Does the client have a nasal discharge, pain over the eyes, fever, anorexia or cough?
    - Have an assistant or family member on hand to assist you in administering nose medications for a young child or a confused client.
    - Assemble equipment for the procedure:
        - The medication container
        - Gloves
        - Facial tissues
        - A pillow, if needed, for positioning
    - Instruct the client to clear or blow nose gently (unless contraindicated) to remove any mucus or secretions that could interfere with the distribution and absorption of the medication.
    - Wash your hands and put on clean gloves.
    - Position the client, usually in the supine position.

- Position the client's head so that the individual receives maximum benefit from the medication.
- For most nasal instillations, tilt the head backward and aim the dropper toward the client's eye which will direct the solution toward the posterior pharynx.
- To treat the ethmoid and sphenoid sinuses, hyperextend the client's head by having him tilt his head over the edge of the bed, while you support his head with one hand to prevent neck strain.
- To treat the maxillary and frontal sinuses, have the client hyperextend his neck and turn his head slightly toward the side to be treated. Support the client's head with your hand. If the client cannot tolerate this position, place a pillow under his shoulder to tilt the head backward.
- Instruct the client to breathe through his mouth.
- With your dominant hand, hold the dropper one half inch (that's a little over one centimeter) above the appropriate nostril and instill the prescribed number of drops.
- Prevent contamination of the medication container and equipment during the procedure.
- Avoid touching the nasal mucosa with the tip of the medication dropper or bottle.
- If the dropper becomes contaminated, do NOT place it back in the container because the dropper will contaminate the remaining solution.
- Instruct the client to remain supine for at least five minutes following the procedure.
- Offer client a facial tissue to blot any seepage.
- Remind client to avoid blowing his nose for at least several minutes.
- Discard your gloves, wash your hands, and document the procedure.
- Evaluate and record how the client tolerated the procedure as well as the client's response to the medication, including therapeutic as well as adverse effects.
- When teaching clients to administer nasal drops to themselves or others, instruct them to:
  - Wash their hands thoroughly before and after the procedure.
  - Assume a supine or sitting position with the head tilted back.
  - Insert the tip of the dropper just inside the nares without touching the nasal mucosa.
  - Squeeze the dropper or bottle to release the appropriate number of drops.
  - Keep the head tilted back for at least one minute.
  - Avoid blowing the nose for several minutes after instilling the drops.
  - Check the nares regularly for irritation or inflammation.
  - Report any problems to the provider.

### Nasal Inhalation

- Nasal sprays are often self-administered, as it is easier for the client to control the timing of inhalation.

- The client should inhale as the spray enters the nasal passages.
- Assess the client's nasal passages for signs of irritation, and evaluate the client's ability to administer the nasal spray effectively.
- Nasal inhalation is the most common method for oxygen administration. The deep passages of the respiratory tract provide a large surface area for drug absorption.
- Nasal sprays should be administered into the nose during inspiration.
- Examples are phenylephrine (Neo-Synephrine), which causes local vasoconstriction in the nasal passages.
- Nasal inhalation is the common site for oxygen administration.
- The deep passages of the respiratory tract provide a large surface area for drug absorption.

## Oral Inhalation

- Hand-held inhalers that dispense an aerosol spray, mist, or powder that penetrates the lung airways
- Advantages/rationale
  - Protection from "first-pass" metabolism by the liver
  - Very rapid absorption for fast action (e.g., bronchodilators)
  - Alternate route for the NPO or unconscious client
- Metered-dose inhalers (MDI) deliver a pre-measured dose of medication in a mist and then air, oxygen, and/or vapor carry the drug into the lungs. The MDI could be thought of as a hand-held nebulizer. The nurse must know and instruct the client in the proper use of the MDI. The procedure for use is as follows:
  - After confirming the order, insert the canister of ordered medication into the dispenser.
  - Hold the canister upright and shake the medication (before each puff).
  - Remove the cap on the dispenser.
  - While sitting or standing, hold the dispenser 2 inches from the mouth.
  - Exhale, slowly and fully, through pursed lips.
  - Activate the dispenser while inhaling deeply through the mouth over 2-5 seconds.
  - Hold the breath for 5-10 seconds, then exhale slowly, again through pursed lips.
  - Wait 1-3 minutes between puffs, if more than one puff is ordered.
  - Rinse the mouth with water and blow the nose to reduce irritation from residual medication.
  - Remove the canister from the dispenser and clean the dispenser with soap and water, air dry.
  - Follow the primary care provider's orders regarding frequency of dosing and number of puffs of medication.
  - Instruct client to report any adverse effects to the primary care provider.
  - If the medication is a dry powder, hyperextend the neck during administration and breathe in quickly and deeply.

The nurse should assess the client's ability to use the MDI properly prior to administration or home use. The client should be observed using the MDI after instruction to assess the effectiveness of the teaching. The client should also be assessed for effectiveness and adverse effects after administration of the medication (the nurse must educate the client and family/ significant other about the potential side effects as well as the desired effects of the medication).

Newer MDIs are breath-activated; the pre-measured dose of medication is released when the client inhales. Spacers (extenders) are also available to ease the use of the MDI. The spacer is a holding area for the medication until the client inhales, especially for children, making it unnecessary to activate the dispenser and inhale at the same time. The spacer also decreases the loss of medication by exhalation and through large droplet absorption via the oral mucosa.

**Nebulizers**: Generally nebulizers are used for very ill clients (e.g., confused, very dyspneic, weak) or those who cannot use an MDI. The machine is not as portable; it is more expensive and larger than the MDI, and the dose delivered is not as accurate. This procedure also releases both medication and any airborne pathogens from the client into the environment, exposing others to both (one-way valves or filters can reduce this risk). The procedure is as follows:

- After confirming the order, add the medication to the nebulizer chamber as ordered by the health care provider. If necessary, dilute the medication with sterile 0.9% sodium chloride solution to a total volume of 3-5 mL.
- Connect the chamber to the aerosol mask or t-piece (keep the chamber upright and vertical). If using a mask, fit the mask well to prevent leakage around the mask.
- Attach the tubing to one end of the t-piece and the mouthpiece to the other.
- Attach air flow (compressed air or oxygen [8 L/min.]) to the nebulizer.
- Turn on air flow and wait for mist.
- Have the client breathe normally through the mouthpiece or mask.
- Do not stop until the mist is gone. The duration of treatment is usually about 5 to 15 minutes.
- Turn off air flow and/or reset oxygen to the rate of delivery ordered by the health care provider.
- Clean and replace equipment according to institutional policy (mouthpiece cleaned after use).
- Monitor the client for adverse effects and desired effects of the medication after administration.
- The nurse must monitor for correct use of the technique as these devices have several complex steps. The MDI, for example, requires that the client have adequate hand strength.

**Endotracheal or Tracheal Administration**: Used in an emergency, when a client does not have an intravenous line; requires special training to deliver by this route.

# Critical Thinking Exercise: Topical and Inhalation Medications

**Situation**: You are caring for a hospice client who is experiencing severe chronic pain associated with metastatic cancer. A transdermal fentalyl (Duragesic) patch has been ordered for pain management. On your visit to this client, you apply the patch and instruct the caregiver and other family members on the use of this medication. Consider how you might organize your teaching in order to anticipate any questions the family may have. For example, how would you respond to a caregiver who poses the following questions?

1. How does the patch "work"?

2. What is the advantage of using a patch? How is it different from just taking oral medication?

3. What do I need to know to apply the patch correctly?

4. Will the patch work immediately, or will other pain medication be needed until the effects of the patch are felt?

5. How long should the patch be left in place?

6. What are the best sites for applying the patch?

7. When I am ready to change the patch, can I apply it to the same place?

8. What adverse effects are associated with this medication?

9. What safety considerations should I be aware of?

10. How is the medication discarded?

# Parenteral Medications

## Key Points

- Parenteral medications are given by injection into body tissues.
- Drug absorption is fastest from the IV route; absorption is slowest from the SQ route.
- All injections require sterile technique when drawing up and administering the medication.
- The smaller the gauge, the larger the needle diameter.
- Long bevel needles are sharper and minimize discomfort.
- Subcutaneous injections: As a rule, dose should not exceed 1 mL; use 5/8", 25-27 gauge needle at a 45° angle.
- Intradermal injections: As a rule, use tuberculin syringe, 3/8," 25-27 gauge needle, and 5-15° angle.
- IM injections: As a rule, dose should not exceed 3 mL; use 1-1/2" needle and 90° angle; aspirate before injecting.
- Take precautions to avoid needlestick injuries (e.g., never recap a used needle).
- Avoid injections in limbs that are paralyzed or inactive.
- **Key Concepts/Terms**: Subcutaneous (SQ), intradermal (ID), intramuscular (IM), intravenous (IV), syringe, Luer-lok, insulin syringe, tuberculin syringe, needle length, needle gauge, needle bevel, vastus lateralis site, ventrogluteal site, dorsogluteal site, deltoid site, z-track method, lipotrophy, needlestick injuries

**Overview**

**Parenteral medications** are given by injection into body tissues. Depending on the rate of drug absorption and the route used, effects of a parenterally administered drug can occur almost instantaneously with some intravenous medications. Nurses are usually responsible for administering the following types of injections:

**Subcutaneous (SQ):** injected just below the dermis of the skin

**Intradermal (ID):** injected into the dermis, just below the epidermis

**Intramuscular (IM):** injected into a muscle

**Intravenous (IV):** injected into a vein

Depending on institutional policy and state Nurse Practice Acts, nurses may or may not be responsible for administration of medications through other parenteral routes. Even so, the nurse often is responsible for monitoring the delivering system and evaluating client responses to medications delivered by the following advanced techniques: epidural, intrathecal, intraosseous, intraperitoneal, intra-arterial.

## Preventing Infection

- All injections are invasive procedures and must be performed using sterile technique when drawing up and injecting the medication.
  - If the client's skin is noticeably soiled, wash it first.
  - Wash your hands thoroughly before preparing the injection and after administering the injection.
  - Wear gloves, because contact with blood is a possibility whenever you administer an injection.
    - Draw medication from ampules immediately; do not allow ampules to stand open.
    - Do not let needle touch unsterile surfaces (e.g., outer edges of ampule, outer surface of needle cap, hands, table top).
    - Do not touch the long part of the plunger or inner part of the barrel; keep syringe tip covered with cap or needle.
    - Prepare client's skin by washing (if noticeably soiled with drainage, etc.); then use an antiseptic swab and use friction while cleaning in a circular motion. Swab from center of site and move outward.
    - Promptly discard any items that become contaminated and begin again with sterile equipment.

## Equipment

- **Syringes** consist of a barrel, a tip on which to place a hypodermic needle, and a close-fitting plunger. One type of syringe contains a **Luer-Lok** connection, which is the adaptor designed to maintain a secure attachment between the syringe and the needle, stopcock, or tubing. The other type of syringe does not contain a Lure-Lok connection, whereby the syringe is directly attached to the needle, stopcock, or tubing. They come in a variety of sizes, holding from 0.5 to 60 mL of fluid. **Insulin syringes** are low-dose syringes, holding 0.5 to 1.0 mL. In the United States, they are marked in "units" and designed for use with U-100 strength insulin; therefore, each mL contains 100 units of insulin. A **tuberculin syringe** is also a low-dose syringe. It has a capacity of 1 mL and is calibrated in sixteenths of a minim and hundredths of a mL. It is useful for preparing small amounts of potent drugs and for small, precise doses for infants and children.
- **Needles** are packaged individually. A needle has three parts: the hub (which fits onto the syringe tip), the shaft (the long part of the needle), and the bevel (the slanted tip). Needle shafts vary in **length** from 1/4 to 5 inches. They vary in diameter as well, usually from 27- to 16-gauge. The smaller the **gauge**, the larger the needle diameter. Long **bevels** are sharper, minimizing the discomfort of SQ and IM injections.
- **Disposable prefilled injection units**: Many medications are available in disposable, single-dose, prefilled syringes. The Tubex and Carpuject systems include reusable plastic holders. The nurse secures the medication syringe in the plastic holder and uses it in the same manner as other syringes.

## Subcutaneous Route

- Drug absorption is slower than from IM and IV routes.
    - This is an appropriate route for slow, systemic absorption of small amounts of non- irritating aqueous solutions, such as insulin and heparin.
- **Syringe**: 1-2 mL
- **Needle**: 1 inch, 25-27 gauge (needle length depends on the client's amount of SQ tissue; needle length should be half the width of the skin fold of pinched skin). A tuberculin syringe is acceptable.
- Only small **amounts** (0.5 to 1 mL) of water-soluble medication should be given SQ, as the tissues are sensitive to irritating solutions and large volumes of medication.
- **Angle of insertion**: Inject at 45-degree angle (depending on client's weight and amount of SQ tissue, a 90-degree angle may be used).

### Sites

- The best sites are vascular areas around the outer aspect of the upper arms, the abdomen, and the anterior aspect of the thighs.
- Other safe sites are the scapular areas and the upper ventrodorsal gluteal areas.
- For obese clients, pinch the tissue and use a needle long enough to insert through the fatty tissue.
- For very thin clients, the upper abdomen is the best site for injection.
- Insulin and heparin injection sites should be rotated and recorded on an "injection diagram" to be sure that sites are rotated. When clients self-inject, they should check off the specific site used for the injection on the diagram and note the date and time. Orderly rotation of injection sites is beneficial because it:
    - Minimizes tissue damage.
    - Keeps the medication absorption rate consistent.
    - Reduces discomfort.
    - Prevents lipodystrophy, which is the wasting of subcutaneous fat.

## Intradermal Route

- Intradermal methods are typically used for tuberculin and allergy tests.
- **Syringe**: Use tuberculin or small hypodermic syringe.
- **Needle**: Fine, short (e.g., 3/8 in., 25-27 gauge)
- **Angle of insertion**: 5 to 15 degrees: a small bleb resembling a mosquito bite will appear on the skin if the medication goes into the dermis. If the site bleeds, the medication probably entered the SQ tissues and the test will not be valid.

## Intramuscular Route

- Injecting medication by the intramuscular route deposits the drug deep into muscle tissue beneath the dermal and subcutaneous layers. It allows rapid absorption of relatively large volumes of medication.
- Irritating and viscous drugs are better tolerated when inserted into muscle because muscle tissue has fewer sensory nerves than other tissues.

- IM medications are absorbed faster than SQ because of the greater vascularity in muscles.
- Muscle is less sensitive than SQ tissue to irritating and viscous drugs.
- It is safer to inject into deep muscle because there is less risk of injecting directly into a blood vessel.

- Weight influences needle size, angle of injection, and amount of fluid to inject in one site. Choose a two-inch needle for an obese client. Choose a one-inch needle for an emaciated or thin client, or for an older client who has less subcutaneous fat and more muscle wasting.
- As a general rule, do not give IM injections into the upper arm or upper thigh if the client's limbs have been inactive or if they are paralyzed.
- **Syringe:** The syringe should be selected to closely accommodate the amount of medication to be injected. Generally, a 3 cc syringe is used for an adult, and a smaller syringe in a child, small or underdeveloped adult (to prevent injection of too large an amount of medication).
- **Needle:** The needle should be selected dependent upon the size of the client, the development and size of the injection site, and the amount and type of medication to be injected. Generally, a 1-1/2 inch, 21, or 22 gauge needle is used for adults. A 1-inch needle is suitable for elderly clients. A smaller, 5/8-inch, 23-25 gauge needle should be used for a child or small or underdeveloped or emaciated adult. If the client is very obese a longer needle may be required. If the medication is more viscous a larger gauge needle may be required.
- **Angle of insertion:** 90°

## Sites

- Choose a site that is located away from large blood vessels, bones, and nerves.
- Make certain that the site is free of inflammation, edema, irritation, moles, birthmarks, scar tissue, abscesses, and lesions of any kind.
- The most common sites for IM injections are the ventrogluteal site, the vastus lateralis site, the dorsogluteal site, and the deltoid site.
  - **Ventrogluteal site (Hochsteter's site)** is the gluteus medius muscle (over the gluteus minimus). This is the preferred site due to the lack of large nerves, blood vessels, and fat (decreasing the probability of injecting subcutaneously). This site is not recommended for infants under 7 months of age. The client should be prone or side lying. The nurse should place a palm (right palm on the client's left hip or left palm on the client's right hip) on the greater trochanter, fingers towards the client's head, index finger pointing toward the anterior superior iliac spine, and the middle finger pressing below the crest of the ilium. A triangle is formed by the index finger, middle finger, and the crest of the (client's) ilium. The injection should be administered within this triangle. In a well-developed adult up to 4 mL of medication can be injected into the ventrogluteal site. For children or small or underdeveloped adults, 1-2 mL is the maximum amount recommended to be injected at this site.
  - **Vastus lateralis site:** Generally well developed in both adults and children. This is the recommended site for infants under 7 months of age. Again, this site has a relative lack of large blood vessels and nerves. The client

should be supine or sitting. The site is the anterior lateral area of the thigh, one hand width below the greater trochanter at the top and one hand width above the knee at the bottom. The area begins at mid- thigh vertically and goes to the mid-lateral thigh.

- **Dorsogluteal site**: This site is difficult to locate and the least preferred intramuscular site. The sciatic nerve is in this area, as are large vessels. This area, in most people, contains a relatively thick fat layer which increases the probability of injecting the medication subcutaneously. Generally, this site should not be used for children under 3 years of age (unless they have walked for one year to develop the muscle in this area). The client should be prone, toes pointed inward (if they are able to do so) to relax the muscles. The nurse must identify and locate the greater trochanter and the posterior-superior iliac spine, imagine a line going from one to the other (which would parallel and lateral to the sciatic nerve). The injection site is lateral and superior to this site. The nurse MUST palpate the ilium and greater trochanter to prevent injection in an incorrect area which could result in injury to underlying structures. Another technique for locating the dorsogluteal site is to divide the client's buttock into quadrants. The upper outer quadrant, just about two to three inches (that's about five to seven and a half centimeters) below the iliac crest, is the appropriate injection site.

- **Deltoid site** is a small muscle on the upper arm. This site is not generally recommended because of the small size of the muscle and close proximity to the radial nerve and radial artery. However, some nurses prefer the deltoid site for larger adults due to the relatively rapid absorption and convenience of administration. The client can be standing, sitting, or supine for this injection. If the deltoid muscle is used, the medication injected should be non-irritating. Small volume (1 mL) should be administered per dose. The nurse locates the lowest edge of the acromion process and places two or three fingers (about 2 inches) just below it. The thumb should be at a line opposite the axilla. The area between the fingers and the thumb identifies the injection site (at least 2 inches below the acromion).

- The client should be instructed to relax the area to be used for an intramuscular injection, and the nurse should assess for relaxation. A relaxed muscle decreases the pain of the intramuscular injection.

- Some experts recommend using a 0.2 mL "air-lock" to prevent medication from leaking back up into SQ tissues.

- Rotate injection sites if the client requires multiple or frequent IM injections. Mark the date, time and site of the injection on the medication administration record or on an injection rotation chart. An orderly rotation of injections speeds medication absorption, lessens discomfort, and helps prevent tissue damage.

- Access each site for inflammation, swelling, erythema, or tissue damage from previous injections.

## Z-Track Method

- Used for irritating preparations (e.g., iron, Vistaril)
- Minimizes tissue irritation by sealing the drug within the muscle tissues

- Apply new needle after drawing up the medication
- Draw up 0.2 mL of air to create an air lock
- Refer to a textbook for this technique, which involves displacing the skin and SQ tissues laterally before inserting the needle and holding them until after injecting the medication.

### Intravenous Route

- See *"Medication Administration: Intravenous Medications."*

## Preventing Needlestick Injuries

- The two most problematic pathogens involved in needlestick injuries are hepatitis B virus and the human immunodeficiency virus.
- Never recap a used needle. If it must be done, use a one-handed technique (refer to a textbook for procedure).
- Dispose of needles and other "sharps" into clearly marked, puncture- and leak-proof containers (often they are red plastic).
- Never force a needle into a full receptacle.
- Never place a used needle in a wastebasket, in your pocket, or at a client's bedside.
- Many institutions now have "needleless" systems or "safety syringes."

## Potential Complications of Parenteral Injections

- Nerve damage/paralysis of the extremity
- **Lipotrophy**: Subcutaneous (SQ) fat breaks down at the site of repeated injections.
- Infection/abscess formation at site
- Damage to the blood vessels
- Behavioral changes due to drug toxicity

## Procedure and Technique for Administration of Parenteral Medications

- Before administering an intramuscular injection, ask the client about any drug allergies.
- Review information about the medication you are going to administer, including the action of the drug, and any side effects, contraindications, or potential adverse reactions, and share the appropriate information with the client.
- Evaluate the client's age, weight, degree of muscle development (if an IM injection), and the condition of the tissue you plan to inject.
- Evaluate the client's ability to cooperate during the injection and make sure others are available to help as needed.
- Select the appropriate needle size and syringe, and check the medication record for the appropriate drug, dosage, route and time.
- Examine the medication bottle or vial to check the expiration date and to make

certain that contents are not discolored or thickened and that no sediment has formed.

- Assemble equipment:
    - The medication bottle, ambule or vial
    - The needle and syringe
    - Gloves
    - An antiseptic pad
- If the client's skin is noticeably soiled, wash the skin first.
- Wash your hands before medication preparation, and don gloves.
- Use a sharp, beveled needle in the smallest suitable length and gauge.
- When withdrawing a drug from an ampule, a filter needle may be required.
- When withdrawing a drug from a vial, inject air to permit easy withdrawal.
- Check for compatibility when mixing medications.
- When mixing medications from two vials: (a) do not contaminate one medication with the other (e.g., if using a multidose vial, draw from it first, or change needle before drawing): (b) eject all air before drawing up second drug, to ensure that dose is accurate.
- When mixing drugs from one vial and one ampule, draw from the vial first.
- When determining site for injection, consider:
    - Size and development of muscle tissue
    - Volume of injected medication
    - Depth to which the medication is administered
    - Integrity of the skin and tissue
    - Drug characteristics (e.g., irritating to tissues)
    - Sites should be free of infection, inflammation, swelling, erythema, or tissue damage from previous injections, bruises, skin lesions, scars, bony prominences, and large underlying muscles or nerves.
- Apply ice to numb the site before needle insertion.
- Cleanse the selected site with an antiseptic pad according to agency policy. Start at the center of the site, rotating in a widening circle to about two inches, or five centimeters.
- Allow the site to air dry, and position the swab on the client's skin above the intended site.
- Remove the needle cover and insert needle quickly to minimize discomfort. Withdraw the needle quickly, apply gentle pressure at the site with the antiseptic pad, then remove it and observe the site for bleeding.
- Exert pressure over the injection site to control bleeding; massage if not contraindicated. Have a colleague witness the wasting of narcotic medications.
- Wash your hands and discard gloves.
- Assess client carefully for any adverse effects. Medications given IM are absorbed rapidly and can produce such effects quickly.
- Observe for behavioral changes that could be a result of drug toxicity.

- Report any problems to the provider.
- Report if the medication is having the desired effect, such as pain control.
- Watch for potential complications in clients who receive repeated injections. These include nerve damage, infection or abscess formation at the site, and damage to blood vessels.
- Insert needle quickly to minimize discomfort.
- Hold the syringe steady while aspirating and injecting.
- Exert pressure over the injection site to control bleeding; massage if not contraindicated.
- Have a colleague witness the wasting of narcotic medications.

## Age-Related Changes—Gerontological Considerations

- Older adults have less subcutaneous fat, which affects needle length, and amount of medication injected.
- Because of sensory deficits, older adults may have difficulty self-administering their own insulin.
- Injections should be avoided in limbs that are paralyzed or inactive.
- Arthritis affects injection techniques if the client is self-administering medications.
- Behavior changes that are a result of drug toxicity or adverse effects may be misdiagnosed as senile dementia (Also review *Medication Administration: General Information.*").

# Critical Thinking Exercise: Parenteral Medications

**Situation:** An oncology nurse is preparing a client for discharge from an inpatient unit. At home, the client will self-administer Epoetin alfa (Epogen) subcutaneously in an initial dose of 150 U/kg three times a week to treat chemotherapy-induced anemia. The client, a 56-year-old woman, is motivated to participate in her own care and has several questions related to self-administration of this drug. Work with a classmate to develop a teaching plan that responds to a variety of potential concerns about self-administration of subcutaneous medications and about this medication in particular. You might even want to role play the scenario with your classmate assuming the role of the client and you the role of the nurse. Here are a sample of client questions to guide your teaching.

1. Why am I taking Epogen?

2. What does the medication do?

3. Why can't I take this medication orally?

4. How do I draw up and inject this drug?

5. Should I use the same site each time?

6. What should I know about handling the vials?

7. How do I store the drug?

8. What about the used syringes?  Can I throw them out with the trash, or is there a particular way to dispose of them?

9. What kind of monitoring will I need while I'm taking the medication?

# Intravenous Medications

## Key Points

- The IV route is used for medication administration and fluid replacement.
- Intravenous route may be used to administer blood and blood products, to administer cancer chemotherapy, and to keep a vein open for quick access.
- Medications may be given through an IV infusion line or an IV access device (e.g., saline lock).
- Before administering an IV bolus or piggyback, check for blood return; be sure the site is not infiltrated or inflamed.
- Label the IV container immediately after adding a medication.
- Potential complications include: fluid overload, sepsis, and phlebitis.
- All IV procedures require meticulous sterile technique.
- **Key Concepts/Terms:** Venipuncture, large-volume infusion, IV bolus (IV push), volume-controlled infusion, piggyback administration, intermittent venous access, infiltration

## Overview

The **intravenous (IV) route** is used for administering medications as well as for fluid replacement therapy. Medications may be given through an IV infusion line or an IV access device such as a heparin or saline lock. IV drugs are administered directly into the bloodstream. They begin acting immediately and there is no way to stop the action unless the drug has an antidote. Therefore, the nurse must observe clients closely for adverse reactions. The IV route is preferred for emergency medications, when constant blood levels must be maintained, and for highly alkaline drugs, which are irritating to muscle and subcutaneous tissues. (**Venipuncture** is discussed in the section on "Intravenous Fluid Therapy.") Many IV medications are mixed in the pharmacy; however, nurses also may add medication to existing IV fluids.

## Large-Volume Infusion

This is the safest and easiest method for IV administration. The drug is diluted in 500 to 1000 mL of compatible IV fluids (e.g., normal saline). Usually the medication is added to the primary fluid bag in the pharmacy. Because the drug is in a dilute form, the risk of serious reactions is decreased. The most serious complication is circulatory fluid overload if the IV fluid is infused too rapidly. This complication is less likely if the fluid is administered by pump.

## IV Bolus ("IV Push")

An IV bolus is used when rapid and predictable responses are necessary. Because only a small amount of fluid is used, it is useful for clients who are on fluid restrictions. Disadvantages are:

- It is dangerous because there is no time to correct errors.
- The concentrated drug may irritate the vein.

The concentrated dose is drawn up in a syringe and then "pushed" directly into a vein, into an existing IV fluids line, through an injection port, or through a heparin lock. The standard rate is to administer 1 mL/min; however, this differs widely. The rate may or may not be prescribed; if not, look it up.

## Volume-Controlled Infusions (e.g., Volutrol, Pediatrol)

In this method, the drug is mixed with a small amount of fluid (30-100 mL) in a secondary fluid container that is separate from, but connected directly to, the primary IV line. These small containers attach just below the primary infusion bag/bottle. This method allows for a more dilute dose than IV bolus and for infusion over a longer time (e.g., 30-60 minutes). It also allows for limiting IV fluid intake.

## Piggyback IV Administration

In this method, the drug is mixed in a small (50-100 mL) bag or bottle of fluid. It is connected to the upper Y-port of the primary infusion line or to a port for intermittent venous access (e.g., heparin lock). Most agencies use safety connectors to prevent needle sticks. Advantages and disadvantages are similar to those for volume-controlled infusions.

## Intermittent Venous Access

When intravenous fluids are not needed, an intermittent IV access device (also referred to as **sterile injection cap**, **heparin lock**, **saline lock**, or **med-lock**) is used. This is a special connector covered by a rubber diaphragm that is inserted directly into the venous catheter and left in place. After administering a bolus or piggyback medication, the nurse must flush the lock to keep it free of clots. Some agencies use saline flushes; others use heparin. Advantages of the use of intermittent venous access include increased mobility, increased safety, and increased comfort for the client.

## Procedure & Technique for Administration of Intravenous Medications

- Before administering an IV bolus or piggyback, check for blood return.
- Never give an IV bolus or piggyback if the insertion site appears puffy, or if the IV fluid cannot flow at the proper rate.
- Be sure to check medication and fluid compatibility when adding medications to IV fluids.
- Always label the IV container immediately after adding a medication. Include: name and dose of medication, date, time, and your initials.
- Always clean IV tubing ports and saline/heparin lock ports with antiseptic swab before inserting needle and after administering the medication.
- Insert needle through the center of the port diaphragm.

## Age-Related Changes—Gerontological Considerations

The veins of older adults are fragile, and the risk of infiltration is higher. Older adults are more prone to fluid volume overload related to their cardiovascular status and decreased cardiac output.

# Critical Thinking Exercise: Intravenous Medications

1. You are caring for a 62-year-old client who is receiving intravenous antibiotics following orthopedic surgery. As you assess the client's left forearm, you note that the area is cool to touch, there is swelling around the insertion site, and the flow rate has decreased significantly. Based on these findings, what do you think the problem might be? What action is appropriate?

2. A nurse working for an infusion company is instructing the parents of a 12-year-old who will be receiving short-term intravenous therapy at home. The parents are concerned about their ability to manage the therapy successfully. Develop a teaching plan to support the caregivers and address any potential concerns they might have.

3. A client complains of tenderness around his IV insertion site. You assess the site and detect redness, swelling and warmth in the area. What does your assessment suggest? What will you do next?

# Eye Medications

## Key Points

- Eye medications include ointments, liquid drops and medication disks
- The cornea is very sensitive; do not instill medication directly onto the cornea
- Wear disposable gloves to administer eye medications
- Infection can easily be transmitted from one eye to the other; do not touch the eyelid or eye with the eye dropper or ointment tube
- If the drug can cause systemic effects, apply gentle pressure to the nasolacrimal duct for 30-60 seconds after administration
- **Key Concepts/Terms:** OS, OD, OU

## Overview

Eye medications include ointments, liquid drops and medication disks that are inserted much like a contact lense. Common eye medications include:

- Antibiotics to treat infections
- Glucocorticoids to reduce inflammation
- Lubricants to reduce irritation from dryness
- Beta blockers and prostaglandin analogs to treat glaucoma
- A wide variety of drugs to treat allergic conjunctivitis including ocular decongestants, antihistamines, mast-cell stabilizers and NSAIDs
- Anesthetics
- Anticholinergics to dilate the pupil and paralyze the ciliary muscle
- Dyes such as florescein to facilitate examination.

Many clients use over-the-counter preparations such as artificial tears, lubricants, and vasoconstrictors (e.g., Visine). Also, contact lens wearers self-administer various preparations to enhance comfort while wearing their lenses. Ophthalmic drugs are prescribed for conditions such as glaucoma and cataract surgery. Many of the clients who receive ophthalmic medications are older adults, who may have difficulty grasping the containers and self-administering the medications. The cornea is very sensitive, and infection can easily be transmitted from one eye to the other.

## Procedure and Technique for Administration of Eye Medications

- Accurately identify the correct eye for instillation (OS is left eye; OD is right eye, and OU is both eyes).
- Carefully observe the Five Rights of medication administration prior to administration.

- Screen the client for allergies to the medication.
- Explain the procedure to the client, and tell the client what to expect during administration, the desired effects and any possible adverse reactions to the medication.
- Assess the client's eyes and surrounding structures prior to medication administration for:
  - Lesions
  - Exudate
  - Swelling
  - Inflammation, discoloration, or discharge from the lacrimal gland
- Ask the client if he or she is experiencing any itching, burning, blurred vision, pain in or around the eyes, or sensitivity to light.
- Note if the client is squinting, blinking excessively, frowning or rubbing the eyes.
- Withhold the medication and immediately notify the primary health care provider if the client is experiencing any serious signs or symptoms.
- Ensure that the client will be able to cooperate during the procedure. A confused client or a small child could suddenly move and cause injury or contamination. Obtain adequate help to assist with the procedure if needed.
- Obtain equipment for the procedure which includes:
  - The medication container
  - Clean gloves
  - Sterile absorbent sponges or cotton balls soaked in sterile saline irrigation solution
  - A dry, sterile absorbent sponge
- Thoroughly wash your hands and put on clean disposable gloves. If you are putting eye medication into both eyes, wash your hands before and after you medicate each eye.
- Help the client into a comfortable sitting or lying position
- Do not instill any medication directly onto the cornea.
- Do not touch the eyelids or other parts of the eye with the eyedropper or ointment tube.
- Never allow one person to use another's eye medications.
- Have the client tilt the head back comfortably and look upward.
- Uncap the medication and place the lid on its side on a clean barrier (not directly on the surface of the counter or medication cart/tray).
- Stabilize the hand with the dropper (dominant hand) on the client's forehead and with the other hand press on the cheek under the eye and gently pull downward to expose the lower conjunctival sac.
- Approaching from the side and with the dropper 1/2-3/4 inch above the cornea, instill the correct number of drops onto the outer third of the lower conjunctival sac.
- Ask the client to close the eye.

- With liquid eye medications (drops) press firmly on the lacrimal duct for about 30 seconds or according to institutional policy.
- Clean the eyelid as needed; use an eyepad if ordered or if necessary.
- Wash your hands thoroughly following the procedure and discard the gloves.
- Monitor the client for any adverse effects and/or desired effects of the medication.
- To administer eye ointment, squeeze 2 cm of ointment into the lower conjunctival sac from the inner canthus to the outer canthus.
- When administering drugs that cause systemic effects, apply gentle pressure to client's nasolacrimal duct for 30 to 60 seconds.
- Teach clients and families the proper techniques for administering eye drops and/or ointments. Instruct them to:
    - Wash their hands thoroughly before and after the procedure.
    - Use a separate towel or tissue for each eye if only one eye is infected.
    - Avoid sharing face towels or tissues with other household members.
    - Store eye medications away from direct heat and freezing temperatures.
    - Cap medications tightly.
    - Discard old, cloudy, or discolored ophthalmic solutions.
    - Discard unused eye medications once the treatment regimen is completed.
    - Never use another person's eye medications.
    - Report any problems to the primary care provider.

# Critical Thinking Exercise: Eye Medications

**Situation**: You are preparing a 70-year-old client for cataract surgery. The client asks, "Will I be able to drive home after the surgery, or should I arrange for a ride?" Based on his current knowledge, formulate an appropriate teaching plan for your client following surgery. Be sure to include the following topics in your discussion. You might even choose to develop a one-page handout of self- care activities following cataract surgery.

- The need for eye medication

- Instilling eye drops

- Activities to avoid

- Signs and symptoms to report

- Scheduling follow-up visits

After reviewing these topics, the client's wife remarks "I have eye drops left over from the cataract surgery I had last year. Maybe we can use these." What additional teaching might be needed?

# Otic Medications

## Key Points

- Otic medications are medications that are instilled in the ear(s).

- Otic medications are used primarily to treat ear infections, to relieve pain, to soften ear wax prior to removal, and to remove insects that are trapped in the ears.

- Sterile technique is recommended if the tympanic membrane is damaged in order to avoid the introduction of pathogens into the tissues and structures of the middle and inner ear; otherwise aseptic technique is used when instilling otic medications.

- Abbreviations relating to the ear(s): **AD**: right ear, **AS**: left ear, **AU**: each ear or both ear(s), TM: tympanic membrane.

- For infants: To ensure movement of the medication as far down the external auditory canal as possible, gently pull the pinna down and back.

- For adults and children over 3 years of age: To ensure movement of the medication as far down the external auditory canal as possible gently pull the pinna up and back.

- Always warm medication prior to instilling into the auditory canal. Otic instillation of fluids at extreme temperatures can cause vertigo, dizziness, nausea, and pain.

- **Key Concepts/Terms**: Auricle, cerumen, external auditory canal, tympanic membrane

## Overview

The instillation of otic medications, as with the administration of any medication, first requires a review of the medication order in the client's chart. Prior to administration of the medication the nurse should: 1) assess the client for any allergy to the medication, and 2) assess the ear(s) for signs of irritation or abrasion, any otic discharge (amount and type), or any discomfort relative to the ear or the procedure. Medication to be instilled in the ear should be warm (at or just above room temperature for the comfort of the client). Otic medications may be used to: 1) soften cerumen for removal, 2) treat inflammation and/or infection of the external canal, or 3) relieve discomfort of the ear. Because this is a topical administration, the medication should have maximal contact with the tissue of the external auditory canal. Thus, to administer otic medications properly, straighten and drop the medication down the side of the canal.

## Terms

**Auricle**, also known as the pinna, is the external part of the ear. The pinna catches the sound waves and transmits them into the external auditory canal.

**Cerumen**, also known as earwax, is a sticky yellowish-brown, thick protective drainage secreted from the ceruminous glands (that line the external canal). Its purpose is to catch small foreign bodies (bugs, dust, or bacteria) and keep them from entering the middle ear.

**External auditory canal** is the canal that transmits sound waves from the pinna to the middle ear (the tympanic membrane). In the adult the canal is about 2.5 cm or 1 inch long and is S- shaped; in the infant (child under 3) the canal is directed upward. The canal should be pinkish in color when viewed through the otoscope.

**Tympanic membrane (TM)**, also known as the eardrum, is the point between the outer and middle ear. The TM transmits sound waves to the small bones of the middle ear (auditory ossicles) by vibrating. The TM should be pearly gray in color, semitransparent, and shiny with a cone of light at the posterior inferior aspect when viewed through the otoscope.

## Therapeutic Nursing Interventions

- Review the medication order and check the medication to be sure it is an otic preparation.
- Take care to determine that the medication is to be instilled into one or both ears, and that it is the prescribed amount and strength.
- Screen the client for allergies to the medication.
- As with all medications, ensure the Five Rights of medication administration prior to instillation.
- Explain the procedure to the client along with what to expect during administration, the desired effects, and the possible adverse affects of the drug.
- Assess the client's ears prior to administration of the medication.
- Check the pinna for signs of inflammation or injury, and note any discharge or drainage.
- Observe any drainage for amount, color, odor, pus, or blood.
- Use a light source to examine the condition of the external canal and the tympanic membrane.
- Ask if the client is experiencing any discomfort, ear pain, or loss of hearing.
- Check the client for elevated temperature, nausea, vomiting, headache, and malaise if the client has an ear infection.
- Document your findings, compare your assessment with the client's baseline, and report any new deviations to the provider.
- Ensure that the client will be able to cooperate during the procedure. A confused client or a small child could suddenly move and cause injury or contamination. Obtain adequate help to assist with the procedure if needed.
- Obtain the following equipment for the procedure:
  - The medication container
  - Clean gloves

- • Moistened cotton-tipped applicators
- • Cotton fluff
- Assist the client to a side-lying position with the ear to be treated facing upward.
- Wash your hands and put on clean gloves.
- Clean the pinna and meatus of the external canal with warm water and a cotton-tipped applicator to remove any drainage or debris (this prevents flushing of any material into the canal with the medication).
- Warm the medication to, or just above, room temperature (run it under warm water for a short period or warm it in your hand).
- Fill the dropper partially with the medication unless the otic medication is available in an administration bottle with a dropper on the bottle itself.
- Straighten the canal as instructed above, according to the age of the client.
- Rest the hand holding the dropper gently against the client's head. This stabilizes your hand, and ensures that your hand will move with the client if the client moves suddenly. This precaution reduces the risk of striking the client's ear, or inserting the dropper too far into the canal and causing injury.
- Take care that the tip of the dropper does not touch the structures of the ear.
- Instill the ordered number of drops of medication down the side of the canal.
- To assist the movement of the medication down the canal, press gently on the tragus of the ear two or three times.
- Ask the client to remain on her/ his side for about five minutes to give the medication time to move down the canal and make maximal contact with the tissue of the external canal. Rising immediately after instillation of the drops will allow the medication to flow out of the canal into the pinna.
- Gently insert a piece of loose cotton into the meatus of the external canal and instruct the client to leave it in place 15 to 20 minutes.
- Discard gloves following the procedure and wash your hands.
- Assess the client after administration of the medication for:
  - • Amount and description of discharge
  - • Relief of symptoms or discomfort
  - • Change in the tissue of the ear canal as expected through administration of the medication
  - • Any adverse effects to the medication (including indications of allergic reaction)
  - • Any change in the texture and color of cerumen if instillation is to soften cerumen for removal
- Document the assessment done prior to instillation of the drops, administration of the medication, and the assessment done after instillation of the medication.
- Report any adverse effects or untoward responses to the ordering health care provider promptly.
- Provide home instruction to the client or home caregiver prior to discharge. If possible, have the client or home caregiver demonstrate administration of the otic medication to ensure adequate understanding of the procedure.

# Critical Thinking Exercise: Otic Medications

**Situation:** As you inspect the ear of a 4-year-old child, you note the absence of cerumen. The child's parent mentions that she uses a cotton swab to clean deep within the ear. How would you respond to this information?

1. You are instructing a parent on how to instill ear drops. Describe how the technique is adapted to a child under 3 years of age and an older child. What other information would you convey to the parent about the correct way to instill ear drops?

2. Unscramble the following words related to the ear.

- TOCI _ _ _ _
- INNPA _ _ _ _ _
- PATYNICM _ _ _ _ _ _ _ _
- TOPCOSOE _ _ _ _ _ _ _ _
- RUCEMEN _ _ _ _ _ _ _

# Insulin

## Key Points

- Insulin is a pancreatic hormone used to treat diabetes.
- Insulin is only given IV or SQ; it is a protein and would be destroyed in the GI tract.
- Some types of insulin cannot be mixed with others; however, regular insulin can be mixed with any other type of insulin.
- Do not shake insulin vials; rotate for 1 minute between both hands before mixing them.
- Administer insulin within 5 minutes after mixing them.
- Insulin can be stored at room temperature for about a month; longer, if refrigerated.
- Rotate injection sites; record location of injection.
- Have another professional nurse double-check the insulin vial and dosage you have prepared.
- **Key Concepts/Terms**: Sliding scale, rapid-acting insulin, intermediate-acting insulin, long-acting insulin, regular insulin, zinc suspensions, Lente insulin, insulin glargine (Lantus)

## Overview

Insulin is used to treat diabetes. It is a hormone produced by the pancreas that enables the body cells to use carbohydrates. Insulin can only be given intravenously (IV) or subcutaneously (SQ). Because it is a protein, it would be digested and destroyed in the gastrointestinal tract. Most clients learn to self-administer their insulin injections. Insulin is prescribed by specific dose at select times, or by "sliding scale." With a "sliding scale" order, the primary care provider orders different doses based on the client's blood glucose reading throughout the day. In the U.S. and Canada, insulin comes in a solution of 100 units (U) per mL. A small (usually 1 mL) syringe and a 3/8-5/8 inch needle are used to administer it (see *"Parenteral Medications, Subcutaneous Route"*).

## Types of Insulin

Insulins are classified according to onset and duration of action. Insulin may be **rapid-acting** (onset of 1-2 hours), **intermediate-acting** (onset of 1-2.5 hours), and **long-acting** (onset of 6-8 hours). Each type has a different peak and duration of action.

- Regular (unmodified), rapid-acting insulin is a clear solution; it can be given IV or SQ.
- All other types of insulin are cloudy solutions with a protein added; they can only be given SQ.

## Mixing Insulins

- Regular insulin can be mixed with any other type of insulin.
- Insulin zinc suspensions (Lente insulin) can be mixed with each other and with regular insulin. Do not mix with any other type.
- Insulin glargine (Lantus) must not be mixed with any other insulin.
- Before mixing different types of insulin, rotate each vial for at least 1 minute between both hands (to suspend modified preparations); **do not** shake the vials.
- Always draw up the unmodified (regular) insulin first (the clear vial). If two modified forms are mixed, it does not matter which is drawn up first.
- Administer insulins within five minutes after mixing them.
- Mixing reduces the action of the regular insulin.

## Insulin Storage

- Insulin can be stored safely at room temperature for about a month; it requires refrigeration for longer periods.
- Do not administer cold insulin; allow to warm to room temperature first.

## Procedure and Technique for Administration of Insulin

- Monitor (or teach client to monitor) blood glucose. Refer to medical-surgical text for in- depth information about diabetes management.
- Monitor for signs of hypoglycemia and hyperglycemia.
- Rotate injection sites and record location of injection.
- Have another licensed nurse double-check the insulin vial and the dosage in the syringe before administering.

# Critical Thinking Exercise: Insulin

**Situation:** A diabetes educator is working with a 12-year-old boy recently diagnosed with Type 1 diabetes mellitus and his family to promote adherence to an insulin regimen and maintain normal blood glucose levels. Before meeting with his clients, the educator will gather written and visual materials to enhance learning. With this in mind, design a brochure introducing basic concepts related to diabetes management. Here are topics to include in your brochure:

- The goals of insulin therapy

- Glucose monitoring with ease

- Changing insulin requirements

- Injecting insulin

- Rotating sites for optimal absorption (develop a simple chart for illustration)

Edit your brochure so that it is clear and concise, focusing on essential concepts of diabetes management.

# Heparin

## Key Points

- Heparin is an anticoagulant. It is metabolized by the liver and excreted by the kidneys.
- The antidote is protamine sulfate.
- Administration is by IV or SQ route; an infusion pump should be used for IV administration because dosage is critical.
- Heparin is available in different strengths; carefully check vial against order.
- Do not mix heparin with any other drug.
- Apply pressure for one minute after SQ injection; do not aspirate or rub the site.
- Potential complications include hemorrhage, hypovolemic shock, and allergic reactions.
- Associated diagnostic tests are partial thromboplastin time (PTT) and platelet count.
- **Key Concepts/Terms**: Low-dose prophylactic anticoagulation, full-dose anticoagulation

## Overview

Heparin is an anticoagulant. It affects the clotting factors, specifically thromboplastin, necessary for clotting. Heparin is used (1) to treat pulmonary emboli and deep vein thrombosis, (2) to reduce thrombus formation during cardiovascular surgery and after heart attacks, and (3) as an IV flush to maintain patency of indwelling IV catheters. Heparin is metabolized by the liver and excreted by the kidneys.

### Antidote

Protamine sulfate

## Administration

- **Routes**: Intravenous or subcutaneous (SQ). Heparin is not well absorbed from the gastrointestinal tract, so it is not given orally. It should not be given intramuscularly because it can cause bleeding, irritation, and pain at the injection site, and it is absorbed erratically from muscle tissue.
- For SQ injections use a 1 inch, 25-27 gauge needle and inject at a 45° angle. A tuberculin syringe is acceptable. (These are general guidelines; always check the primary care provider's orders and the manufacturer's recommendations.)
- IV heparin should be administered by infusion pump to carefully regulate dosage.

## Dosages

- **Low-dose, prophylactic anticoagulation (e.g., preoperatively)**: Usually involves giving 5,000 Units SQ just before surgery and repeating every eight to 12 hours postoperatively for 24 hours to a week.

- **Full-dose anticoagulation (e.g., treatment of myocardial infarction)**: May be given as slow, intermittent IV injections of 5,000-10,000 Units. May also be given by slow infusion: the client gets an initial IV injection of 5,000 U, followed by a continuous infusion rate of about 1,000 IU per hour.

## Risk Factors

- Heparin should be used with caution with clients who have other risks for hemorrhage such as severe hypertension, aneurysms, hemophilia, ulcers, ulcerative colitis, threatened abortion, and shock.

- Heparin may be contraindicated for clients having renal or liver disease.

- Certain drugs such as the antihistamines and the salicylates (aspirin) interfere with other steps in the clotting mechanism. When given with heparin, the risk of bleeding is increased.

## Potential Complications

- Bleeding, hemorrhage
- Hypovolemic shock
- Allergic reactions
- Tissue damage related to heparin administration

## Therapeutic Nursing Interventions for Administration of Heparin

- Obtain baseline measurements of complete blood count and coagulation studies (e.g., partial thromboplastin time [PTT] and platelet count).

- Check PTT each day before administering the drug (the therapeutic level for PTT is 1.5 to 2.5 times the "normal" (control) value.

- Observe for signs of bleeding (bleeding gums; bruises on arms or legs; petechiae; nosebleeds; black, tarry stools; hematuria; bloody sputum; bloody emesis; and prolonged menstruation in females).

- Double-check heparin concentration and dosage before administering (it is available in different strengths).

- Do not mix heparin with any other drug.

- Apply gentle pressure for one minute after administering SQ.

- Do NOT aspirate or rub the injection site.

- The abdomen is the preferred SQ site.

- Rotate sites and record location of injection.

- Avoid giving other injections during heparin therapy.

# Critical Thinking Exercise: Heparin

1. After checking medication orders, you are preparing to administer subcutaneous heparin to a 68-year-old postoperative client. The client asks, "I used to take warfarin (Coumadin). Isn't this the same thing?" How do you respond?

2. As part of a preadministration assessment, what baseline data should you obtain for a client on heparin therapy?

3. With a classmate, identify those clients who are at greatest risk for deep venous thrombosis (DVT). Describe your assessment of these clients. What measures can be taken to prevent thrombus formation?

# Wounds, Wound Healing and Wound Dressings

## Key Points

- A wound is a disruption of the normal anatomical structure and function of the skin.
- Inflammation is a localized protective response brought on by injury or destruction of tissues.
- Wounds heal in stages, by the process of primary, secondary, or tertiary intention.
- Factors affecting wound healing: age, circulation, oxygenation, depth and extent of the wound, client's overall state of wellness, malnutrition, dehydration, obesity, infection, chronic illness (such as diabetes and alcoholism)
- Complications of wounds are: infection, hemorrhage, delayed closure, dehiscence, evisceration, and fistula.
- **"RYB" Color Code:**
  - **Red:** protect (cover)
  - **Yellow:** cleanse
  - **Black:** debride
- Normal saline is the "rule" as a cleansing agent.
- Assess wound for appearance, character of drainage, wound closures, and pain.
- Wound healing is slower in older adults, as a rule.
- A dressing should:
  - Absorb drainage
  - Protect from contamination
  - Keep the wound slightly moist
- Types of dressings include: woven gauze sponges, wet-to-dry dressings, self adhesive transparent film (e.g., Op-Site), hydrocolloid dressings (e.g., DuoDERM), hydrogel (e.g., Aquasorb).
- Transparent wound barriers act as a temporary second skin, and are suitable for superficial wounds as well as partial-thickness ulcers and burns.
- Controversy exists over whether clean or sterile technique should be followed for dressing changes.
- The nurse should wear disposable gloves to avoid exposure to body fluids from the wound.
- Do not touch a fresh or open wound without wearing sterile gloves.
- Administer pain medication so that peak effect occurs during the dressing change.
- Be alert for tape allergy; non-allergenic tapes are available.

- Do not apply tape over irritated or broken skin.
- When removing tape, pull parallel with skin surface, toward the wound, not across the wound.
- **Key Concepts/Terms**: Primary, secondary, and tertiary healing; stages of wound healing, debridement, woven gauze sponges, wet-to-dry dressings, self-adhesive, transparent film, hydrocolloid (HCD) dressings, hydrogel dressings

**Overview**

A **wound** is a disruption of the normal anatomical structure and function of skin. Wounds heal by one of three processes: primary, secondary, or tertiary intention.

**Primary Healing**: Wound with little tissue loss, edges approximated (close together), heals rapidly with minimal scarring, low risk of infection (e.g., surgical incisions). Healing occurs in three stages:

- **Inflammatory Phase (Reaction)**: Begins within minutes and lasts about three days. Reparative processes control bleeding, deliver blood and cells (e.g., leukocytes) to the area, and form epithelial cells at the site of the wound.
- **Proliferative Phase (Regeneration)**: Begins with the appearance of new blood vessels; lasts from 3 to 24 days. The wound fills in with connective or granulation tissue and the top is closed by epithelization. Fibroblasts synthesize collagen, which closes the wound defect.
- **Maturation (Remodeling)**: The collagen scar continues to gain strength. The collagen fibers undergo remodeling (reorganization) before assuming their normal appearance. This process may take more than a year, depending on the extent of the wound.

**Secondary Healing**: Wounds involving loss of tissue; wound edges widely separated; wound appears pink to dark red, healing occurs by granulation, resulting in a large scar; increased likelihood of infection; healing time is longer (e.g., burns, pressure ulcers).

**Tertiary Healing**: Occurs when a widely separated wound is later brought together with some type of closure material. This type of wound usually is fairly deep and likely to contain extensive drainage and tissue debris. These wounds have a high risk of infection (e.g., wound dehiscence).

## Stages of Wound Healing

**Stage 1**: Inflammatory process occurs; blood vessels constrict, providing a clot; vasodilation brings more nutrients and WBCs to wound; blood flow is re-established after epithelial cells begin to grow. A slight fever is normal.

**Stage 2**: Collagen and granulation tissue forms in the wound.

**Stage 3**: Collagen fibers strengthen the wound; scar is pink and raised.

**Stage 4**: Scar becomes smaller, flatter and white.

## Factors Affecting Wound Healing

- Developmental considerations (age)
- Circulation and Oxygenation

- Wound condition (e.g., depth and extent)
- Overall client wellness
- Malnutrition/dehydration
- Obesity
- Infection
- Chronic Illness (diabetes, alcoholism)

## Potential Complications

- Infection
- Hemorrhage
- Delayed wound closure (third-intention wound healing)
- **Dehiscence** (separation of a wound's edges, revealing underlying tissues)
- **Evisceration** (protrusion of visceral organs through a dehisced surgical wound)
- **Fistula** (abnormal passage/opening between two organs or between an organ and the outside of the body)
- Psychological effects (pain, anxiety, fear, impaired self-image)

## The "RYB" Color Code to Guide Wound Care

Developed by Marion Laboratories, this approach is based on the color of an open wound (red, yellow, or black), rather than the depth, size, or stage of the wound.

**Red protect (cover):** Red wounds are usually in the late regeneration phase of tissue repair. Protect them by cleansing gently, avoiding dry gauze or wet-to-dry dressings, applying topical antimicrobial agent, using a transparent film or hydrocolloid dressing, and changing the dressing as infrequently as possible.

**Yellow cleanse:** Yellow wounds usually contain liquid or semiliquid "slough" and purulent drainage. Wet-to-damp dressings, wound irrigation, absorbent dressings, and hydrogel or other exudate absorbers are used to remove dead tissue. Antimicrobial agents may be used to impede bacterial growth.

**Black debride:** Black wounds are covered with **eschar** (thick, necrotic material). They must be debrided in order for healing to occur.

- **Sharp debridement** uses a scalpel or scissors; special training is needed.
- **Mechanical debridement** uses scrubbing or wet-to-damp dressings.
- **Chemical debridement** uses collagenase enzymes to remove dead tissue.
- **Autolytic debridement** utilizes dressings that contain wound moisture which makes use of the body's own enzymes to break down necrotic tissue.

## Therapeutic Nursing Interventions for Wounds/Wound Care

- Assessment includes inspection and palpation of wound for: appearance, character of drainage, wound closures (e.g., sutures), and pain.
- Assess for signs of complications: redness, warmth, edema, fever, chills, and purulent, odorous discharge.
- Prevent infection by hand washing and sterile or clean procedures.

- Normal saline is the preferred cleansing agent.
- Keep wound clean and change dressings according to medical orders or agency procedures.

## Age-Related Changes—Gerontological Considerations

- In general, healing time is slower because:
    - The skin loses turgor and is more fragile (loose epidermal attachment).
    - Peripheral circulation and oxygenation are decreased.
    - There is decreased absorption of nutrients.
    - Epidermal cell renewal is about 1/3 slower.
    - There is a decreased amount of collagen, so the healing wound is more likely to shear or tear.
    - Aging causes impaired immune function of cells in the skin and, thus, increased susceptibility to infection.

# Wound Dressings

**Overview**

The choice of dressing and the method of dressing influence wound healing. A dressing should not allow the wound to become too dry; ideally, it should leave the wound slightly moist to promote epithelial cell migration. The dressing should also absorb drainage to prevent maceration of the surrounding skin.

## Functions/Purposes of Dressings

- Protect the wound from bacterial contamination and trauma.
- Aid hemostasis (pressure dressing).
- Absorb drainage.
- Debride the wound (wet-to-dry dressing).
- Support/splint the wound site.
- Keep the wound moist.
- Prevent the client from seeing the wound (in some situations).
- Provide thermal insulation of the wound. Promote healing
- Improve the adherence of skin in a skin graft.

## Types of Dressings

**Woven gauze sponges**: These are absorbent and wick away the wound exudate. They can be used to pack wounds, as well. They come in various textures and sizes.

**Non-adherent dressings (e.g., Telfa)**: These are used over clean wounds. They have a shiny surface that does not stick to wounds, but does allow drainage to pass through to the gauze in the upper layer.

**Wet-to-dry dressings (e.g., saline-soaked gauze)**: These are used for mechanical debridement of necrotic tissue. Gauze moistened with the prescribed solution is placed over or into the wound and allowed to dry. When the dry dressing is pulled off, the debris adhering to it is also removed.

**Self-adhesive, transparent film (e.g., Acu-Derm, Op-Site, Tegaderm)**: These act as a temporary second skin and are ideal for small, superficial wounds, as well as partial-thickness ulcers and burns. They adhere well, serve as a barrier to bacteria, allow the wound to "breathe," keep the wound moist, seal in wound fluid which allows the enzymes in the fluid to digest the necrotic tissue, permit viewing the wound, do not require a secondary dressing, and can be removed without damaging underlying tissues and risking bacterial contamination. Transparent film dressings are also used to cover intravenous catheter insertion sites because they allow viewing of the site without removing the dressing and risking bacterial contamination. No secondary dressing is required and they eliminate the need for dressing changes, as long as the dressing remains intact. Because they are waterproof, the client can bathe or shower without affecting the wound or the dressing.

**Hydrocolloid (HCD) dressings (e.g., DuoDERM, Dermiflex, Biofilm)**: These are occlusive dressings. The wound-side of these dressings swells in the presence of exudate. They can be used to heal granulating wounds and to debride necrotic wounds autolytically. HCD dressings absorb drainage, maintain wound humidity, liquefy necrotic debris, cushion the wound, provide a barrier to contaminants, are self-adhesive, and may be left in place for three to five days.

**Hydrogel dressings (e.g., Aquasorb, IntraSite Gel, Transorb):** These are water or glycerin- based amorphous-gel-impregnated gauze or sheet dressings. They are used on partial to full thickness wounds, deep wounds with exudate, necrotic wounds, and burns. They have a high water content, are soothing to the wound, can debride the wound, are easily removed without sticking, and can be used in infected wounds.

**Other types include** foam dressings, alginate dressings, and exudate absorbers. After surgery, primary care providers frequently order "reinforce dressing PRN"—that is, add dressings without removing the original one. This is done to prevent accidental disruption of the suture line. The primary care provider's order for a dressing change should include the dressing type, frequency of changing, and any solutions/ointments to be applied. Controversy remains over whether clean or sterile technique should be followed for dressing changes. Observe primary care provider orders or agency procedures. After the original dressing is removed, surgical incisions are frequently left open to air.

## Procedure and Techniques for Dressing Changes

- Administer pain medication so that peak effect occurs during the dressing change.
- Wash hands thoroughly before and after wound care.
- Remove tape and dressing materials carefully.
- Do not touch a fresh or open wound without wearing sterile gloves.
- Dressings may be changed without gloves, if the wound is "sealed."
- Dressings over closed wounds should be removed or changed when they become wet.

## Procedure and Techniques for Application of a Sterile Dressing to a Sutured Wound

- A sutured wound is a wound that has been closed surgically with stitches, wire, or staples.
- Maintain surgical asepsis as you cleanse a surgical wound and change the dressing. Infection control is a high priority with a surgical wound.
- Check the provider's orders.
- Screen the client for allergies.
- Explain the procedure to the client.
- Assemble the equipment:
  - A towel or waterproof pad
  - A bath blanket
  - Clean gloves
  - A moisture-proof biohazard bag
  - Adhesive remover
  - A sterile dressing kit or the appropriate dressing materials
  - Sterile drapes
  - The prescribed cleansing solution

- Sterile gloves
- Other materials as ordered such as ointment
- Provide privacy and assist the client into a comfortable position that allows good access to the surgical wound.
- Protect the bed linens with a towel or pad.
- Cover the client with the bath blanket except for the area where the dressing is.
- Set up the disposal bag for easy access.
- Wash your hands and put on clean gloves.
- Gently remove the existing dressing and place it in the disposal bag without touching the outside of the bag.
- Note the characteristics of any drainage on the dressing.
- Remove the clean gloves and wash your hands.
- Assess the wound so that findings can be documented later.
- Open all sterile supplies and pour the cleansing solution into a sterile basin.
- Put on the sterile gloves and carefully cleanse the wound, following the provider's orders or the institution's policy. Cleanse the wound from the area of least contamination to the area of most contamination, using one sterile swab, gauze, or cotton ball for each stroke.
- Dry the area completely with gauze pads.
- Apply medication if prescribed.
- If the wound has a drain, cleanse the drain and the surrounding skin last. Apply pre- cut gauze around the drain, extra gauze below it, and add dressing material over it to collect drainage.
- Apply the sterile dressing over the wound.
- Place an abdominal pad over larger wounds as a final covering to absorb drainage, support the wound and protect it from trauma.
- Secure the dressing with tape or straps. If the client has fragile or sensitive skin, use paper tape.
- Remove the gloves and dispose of them in the proper receptacle.
- Make sure the client is comfortable and then wash your hands.
- Document the dressing change in the client's record. Note:
  - Assessment findings concerning the wound
  - Assessment of drainage or fluids on the old dressing including color and odor
  - Wound care provided
  - How the client tolerated the procedure

## Procedure and Techniques for Packing Wounds

- Use appropriate material (e.g., woven gauze).
- Use sterile technique.
- Fluff gauze before packing.
- Moisten gauze with normal saline only.

- Don't let sterile packing material drag across surrounding skin and wound.
- Pack the wound loosely and do not overfill the space. Wound packing should not extend higher than the wound surface or overlap onto the wound edges.

## Procedure and Techniques for Securing Dressings

- Dressings may be secured with tape, ties, bandages, secondary dressings, or cloth binders, depending on the wound size, location, amount of drainage, frequency of dressing changes, and client's level of activity.
- Tape is used most often. Non-allergenic paper and plastic tapes are available. It is available in various widths.
- Never apply tape over irritated or broken skin.
- When removing tape, loosen the ends and gently pull the end parallel with the skin surface, toward the wound (not across the wound). Apply light traction to the skin away from the wound as the tape is removed. Pull in the direction of hair growth if possible.
- Assess for inflammation and excoriation indicating skin sensitivity to tape.
- If dressing changes are frequent or skin is sensitive to tape, dressings can be secured with Montgomery ties. These are a pair of cloth strips: half of each strip has an adhesive backing while the other half folds back and contains a cloth tie to fasten across the dressing which can then be untied at dressing changes.

## Procedure and Techniques for Applying a Transparent Wound Barrier

- Check the primary health provider's orders.
- Explain the procedure to the client
- Review the manufacturer's instructions for applying the dressing.
- Assemble the equipment for the procedure:
  - Clean or sterile gloves
  - Alcohol or acetone
  - Wound cleaning supplies as needed or prescribed
  - The wound barrier dressing
  - Paper tape
- Assist the client into a comfortable position that allows access to the wound.
- Assess the wound, any drainage, and the stage of healing.
- Wash hands and put on clean gloves.
- Clip or shave any hair around the wound that might keep the dressing from adhering.
- Clean the wound and surrounding area as prescribed, and dry the surrounding skin with sterile gauze. Alcohol or acetone removes oils from the surrounding skin, which helps the dressing to adhere.
- Remove part of the paper backing on the dressing, and apply the dressing to at least one inch (that's 2 $^1/_2$) of surrounding skin at one edge of the wound.
- Gently lay the rest of the dressing over the wound.
- Frame the dressing with paper tape to keep the edges from curling up or

becoming loose.

- After applying the dressing, remove gloves and dispose of them in the proper receptacle, make sure the client is comfortable and wash your hands.
- Document the dressing change, your assessment of the wound, the wound care provided, how the client tolerated the procedure, and any other concerns or observations, in the client's record.

# Critical Thinking Exercise: Wounds/Wound Healing

1. While changing a surgical dressing, you use the opportunity to assess your client's readiness to manage dressing changes and incision care after discharge. What signs might indicate a client's readiness to manage his or her own wound care?

2. As a teaching tool to provide postoperative clients on discharge, develop a set of instructions on wound care. Keep your instructions concise but be sure to cover areas of potential concern to the client, such as:

   • Signs of infection

   • Managing soreness or discomfort

   • Reducing swelling

   • Signs to report to health care provider

3. On a visit to a home care client, a caregiver expresses concern about a sacral pressure ulcer that is slow to heal. The client, an 86-year-old female diagnosed with Alzheimer's disease, ambulates with assistance but spends the majority of the day in a wheelchair. The client is continent of urine and feces. The client is repositioned frequently and pressure-relieving devices are used regularly. From what the caregiver tells you, you suspect that the delay in healing may be related to the client's nutritional status. How might you educate the caregiver on the importance of adequate nutrition for wound healing?

# Incision and Drain Care

## Key Points

- Drains are tubes that provide a means for removing blood and drainage from a wound.
- Cleanse from least to most contaminated area.
- Never use the same piece of gauze to cleanse across a wound twice.
- As a rule, the incision site is considered less contaminated than the surrounding skin.
- A drain site is considered highly contaminated (more so than the surrounding skin).
- Observe for skin irritation and breakdown around the drain insertion site.
- Use sterile technique to irrigate open wounds.
- Use dressings to absorb drainage; change when damp.
- **Key Concepts/Terms:** Open drains, closed drains (drainage evacuators), irrigation, noncytotoxic solutions, antiseptics

**Overview**

A wound or drain site may need to be cleansed if the dressing does not absorb the drainage well or if an open drain deposits drainage onto the skin. Surgical or traumatic wounds are cleansed by applying noncytotoxic solutions (usually sterile saline) with sterile gauze or by irrigation. As a rule, the wound is considered to be less contaminated than the surrounding skin.

## Types of Drains

**Drains** are tubes that provide a means for removing blood and drainage from a wound.

**Open drains** are flat, flexible tubes that provide a pathway for drainage from the wound bed toward the dressing. Sometimes a safety pin or long clip is attached to the drain as it extends from the wound to prevent the drain from slipping within the tissue.

**Closed drains** (also called **drainage evacuators**) are tubes that lie in the wound bed and terminate in a receptacle (e.g., Jackson-Pratt, Hemovac). These receptacles pull fluid from the wound by creating a vacuum or negative pressure.

## Therapeutic Nursing Interventions

### Cleansing Wounds (Key Procedural Points)

- Cleanse in a direction from the least contaminated area toward the most contaminated area (e.g., begin at the incision and cleanse toward the surrounding skin; or begin at the site of an isolated drain and cleanse toward the surrounding skin).

- Use gentle friction when applying solutions to the skin.
- Never use the same piece of gauze to cleanse across a wound twice.
- A drain site is considered highly contaminated because the moist drainage harbors microorganisms. Use friction when applying antiseptics locally to the skin around a drain site.
- If there is a dry incision with a nearby moist drain site, clean from the incision toward the drain. However, if the drain is isolated away from the incision (or if there is no incision), begin at the drain and swab around it, moving in circular motions outward from the drain.

### Wound Irrigations (Key Points)

- In addition to removing exudate and debris, irrigation may be used to apply heat or locally acting medications to the area.
- Use sterile technique to irrigate open wounds; use a 35-cc irrigating syringe with a 19-gauge needle; hold tip about 1 inch above upper end of the wound and flush with slow, continuous pressure.
- Hold syringe tip over, but not in, the drainage site. Fluid should flow directly into the wound.
- Allow the solution to flow from the least to the most contaminated area.

### Therapeutic Nursing Interventions (Other)

- Change dressing and cleanse incision per primary care provider's orders.
- When emptying a closed drain (drainage evacuator), measure the output and reset the evacuator to apply suction.
- After cleansing around a drain, place a precut drain gauze around the base of the drain. An open drain may require additional layers of gauze material to absorb the drainage.
- Observe for skin irritation and breakdown around the drain insertion site.

# Critical Thinking Exercise: Incision and Drain Care

**Situation:** You are caring for a 50-year-old woman who has just undergone a mastectomy. Immediately after surgery, and as you prepare your client for discharge, what special considerations will apply with respect to incision and drain care?

1. Suppose the client described in this scenario is discharged with drains still in place. How will you assure continuity of care between the hospital and home?

2. In your clinical setting, familiarize yourself with the drainage systems in use, such as the Hemovac or Jackson-Pratt. Ask your clinical instructor if you might be assigned to a client requiring incision and drain care.

# Pressure Ulcers

## Key Points

- Excellent nursing care is the main factor in the prevention of pressure ulcers.
- A pressure ulcer is a specific tissue injury caused by unrelieved pressure that results in ischemia in and damage to the underlying tissue.
- Pressure ulcers occur most commonly over bony prominences.
- Risk factors include immobility, malnutrition, incontinence, and compromised peripheral circulation.
- The elderly are especially at risk because of a loss of lean body mass and changes in body tissues and peripheral circulation.
- Symptoms: Stage 1: redness of intact skin; Stage 2: abrasion, crater, or blister; Stage 3: deep crater, possibly with drainage; Stage 4: small or large surface wound, but with extensive tunneling, and foul smelling discharge
- The general focus of prevention and treatment are to relieve the pressure (e.g., by frequent turning of immobile patients) and provide for good nutrition and hydration.
- Complications include localized and systemic infection.
- Restore circulation to a deprived area by rubbing around a reddened area. Do not massage reddened skin as it has already suffered temporary damage.
- **Key Concepts/Terms**: Ischemia, friction, shearing force, maceration, dermis, epidermis, subcutaneous tissue, erythema, Braden Scale, Norton's Pressure Area Risk

## Overview

A pressure ulcer is a specific tissue injury caused by unrelieved pressure that results in damage to underlying tissue. When tissue is caught between two hard surfaces (e.g., the mattress and a bony prominence), localized ischemia (a deficiency in blood supply to the tissues) occurs. Oxygen and nutrients cannot reach the cells, metabolic waste products accumulate, and the tissue dies. Friction (e.g., sheets rubbing against the skin) and shearing force (a combination of friction and pressure, for example, when the body slides downward toward the foot of the bed) also contribute to pressure sores.

## Risk Factors

- Immobility (e.g., paralysis, muscle weakness)
- Malnutrition (especially protein deficiency)
- Fecal and urinary incontinence (moisture promotes skin **maceration** softening of the tissues)

- Impaired mental status (e.g., sedation)
- Diminished sensation (e.g., paralysis, neurological disease)
- Elevated body temperature
- Peripheral vascular disease
- Localized edema (decreases skin elasticity)
- Chronic illness

## Signs and Symptoms

**Stage I**: Nonblanchable erythema (redness) of intact, lightly-pigmented skin. With darker skin tones, the ulcer may have red, blue, or purple tones. Only the epidermis is involved. Reversible if pressure is relieved.

**Stage II**: Partial-thickness skin loss involving epidermis and/or dermis. The lesion is superficial and presents as an abrasion, shallow crater, or blister. It may be swollen and painful. Takes several weeks to heal when pressure is relieved.

**Stage III**: Full-thickness skin loss, including subcutaneous tissues. May extend down to the underlying fascia. Presents as a deep crater with or without undermining of adjacent tissue. May have a foul-smelling drainage if infected. May require months to heal after pressure is relieved.

**Stage IV**: Extensive damage to underlying structures including tendons, muscles, and bones. Wound can appear small on surface but have extensive tunneling (sinus tracts) underneath. Usually includes foul smelling discharge, and local infection can easily spread, causing sepsis. May take months or years to heal.

Wounds cannot be staged if they are covered with eschar (a dark, leathery scab or crust made of dead tissue) because the wound base would not be visible.

## Stage–Related Treatments

**Stage I**: Relieve pressure by frequent turning, using pressure-relieving devices (e.g., air-fluidized bed, low-air-loss bed), using pressure-reduction surfaces (air mattress, 4-inch foam overlay), and repositioning often. Keep client clean, dry and well nourished.

**Stage II**: Maintain a moist healing environment: use saline or occlusive dressing that promotes natural healing but prevents formation of a scar.

**Stage III**: Debride by using wet-to-dry dressings, surgical intervention, or proteolytic enzymes.

**Stage IV**: The wound should be covered with non-adherent dressings which are changed every eight-12 hours; may require skin grafts.

### Diagnostic Tests to Determine Secondary Infection

- Wound culture
- Complete blood count (CBC)

### Complications

- Localized infection
- Systemic infection (sepsis)
- Psychological effects (pain, anxiety, fear, impaired self-image)

## Therapeutic Nursing Interventions

### Assessment

- Use good lighting for assessing, preferably natural or fluorescent light.
- Consider environmental temperature: heat can cause the skin to flush; cold can cause the skin to blanch.
- Palpate surface temperature of skin over pressure areas.
- Palpate over bony prominences for edema.
- When a client first enters the agency, use a specially developed tool (e.g., the Braden Scale for Predicting Pressure Sore Risk or Norton's Pressure Area Risk Assessment Form) for predicting the risk for developing pressure ulcers.
- Note: Size and location of the lesion, stage of the ulcer, color of the wound bed, location of necrosis or eschar, condition of wound margins, integrity of surrounding skin, and signs of infection.

### Prevention

- Use proper lifting techniques to prevent friction and shearing.
- Rotate injection sites.
- Avoid placing clients on hard support surfaces.
- Be sure to apply pressure-relieving devices (e.g., bed cradles) according to instructions.
- Keep clean and dry.
- Avoid hot water, soap, or alcohol on the skin; use mild cleansing agents.
- Avoid massage over bony prominences.
- Provide smooth, firm, wrinkle-free bed and chair surfaces.
- Reposition at least every two hours; teach client to shift weight every 15 to 30 minutes.

### Support of Wound Healing

- Provide adequate fluids–at least 2,500 mL/day.
- Provide adequate nutrition– especially protein, vitamins A, B1, B5, and zinc.
- Prevent infection (e.g., use standard precautions and provide wound care according to agency policy).
- Position to keep pressure off the ulcer area.

## Age-Related Changes—Gerontological Considerations

The following changes in the skin and supporting structures increase the older adult's risk for pressure ulcers:

- Lean body mass is lost.
- The epidermis thins.
- Changes in collagen fibers cause the skin to lose strength and elasticity.
- Decreased oil production by the sebaceous glands creates skin dryness and scaliness.

- Pain perception is decreased.
- Peripheral circulation and oxygenation may be compromised (e.g., as in peripheral arterial disease and varicose veins).
- Cells regenerate more slowly, causing delayed healing.

# Critical Thinking Exercise: Pressure Ulcers

**Situation:** In accompanying a homecare nurse on a series of visits, you encounter a high percentage of clients with one or more risk factors for pressure ulcers, including immobility or limited mobility, incontinence, vascular disease, diabetes mellitus, and advanced age. Providing education on appropriate preventative measures is a major aspect of nursing responsibility. What kind of instructions will you provide to assist these clients and their caregivers?

1. Develop a brochure to help clients and caregivers cope with the challenge of preventing pressure ulcers. Include written and visual materials as part of your instructions. In concise, direct language, address the following questions:

   - What causes pressure ulcers?

   - Who is at risk?

   - How to recognize early signs of pressure formation

   - Where to assess the skin

   - What can be done to prevent pressure ulcers?

   - What measures can be taken to promote healing once a pressure ulcer has formed?

   Think of this brochure as a teaching tool to reinforce your instructions on pressure ulcer prevention and care.

2. Consider the clients you have cared for during your most recent clinical experiences. Choose one particular client who either had a pressure ulcer or was at greatest risk for developing pressure ulcers. What risk factors were present in this individual? What interventions were used either to prevent pressure ulcer formation or to promote healing? Were the interventions successful? What nursing diagnosis applied to this client?

# Wound Infection

## Key Points

- Wound infection is an invasion of the wound by pathogenic microorganisms. The most common is staphylococcus aureus.
- Drainage (exudate) results from the inflammatory process.
- Drainage is described as serous, sanguineous, and/or purulent.
- Symptoms occur within two to seven days of the injury or surgery.
- Risk factors include: extremes of age, decreased tissue oxygenation or circulation, chronic illness, malnutrition, and immunosuppression.
- The primary complication is sepsis.
- Good hand-washing and standard (tier one) precautions are necessary.
- The best way to prevent wound infection is to maintain strict asepsis when performing wound care.
- **Key Concepts/Terms**: Wound drainage descriptions, inflammatory process signs and symptoms, nursing interventions.

## Overview

A wound infection is an invasion of the wound by pathogenic microorganisms. It can occur at the time of trauma, during surgery, or at any time after the initial wound. Wound drainage (exudate) results from the inflammatory process that occurs in the first two stages of wound healing.

**Serous drainage** is clear or slightly yellow and thin. It consists mostly of the clear, serous portion of the blood.

**Sanguineous drainage** is thick and reddish. It consists of serum and red blood cells, and is the most common type of drainage from an uncomplicated surgical incision.

**Serosanguineous drainage** is blood-streaked or blood-tinged serous drainage (contains both serum and blood) and is generally watery.

**Purulent drainage** is the result of infection. It is composed of white blood cells, tissue debris, and bacteria. It is thick and the color is specific to the type of infectious organism present. It may have a foul odor.

## Risk Factors

- Extremes of age
- Decreased circulation and oxygenation of tissues
- Chronic illnesses (e.g., diabetes)
- Malnutrition

- Suppressed immune system
- Breaches in sterile technique
- Extent and nature of the wound

## Signs and Symptoms

- Symptoms usually become apparent within two to seven days after the injury or surgery.
- Purulent drainage. Drainage may be a mixture of three types: serous, sanguineous, and purulent.
- Pain
- Redness and edema in and around the wound
- Fever, chills
- Increased pulse and respiratory rates
- Increased white blood cell count

## Diagnostic Tests

- Complete blood count (CBC)
- Wound culture
  - Review the primary care provider's order for the proper site and test.
  - Wearing gloves, remove the old dressing, document drainage character and amount (after procedure is complete). Dispose of the old dressing material and gloves per institutional policy.
  - With new gloves, irrigate the wound with sterile normal saline to remove exudates. Place sterile gauze over the wound to "dry." Discard dirty gloves and gauze per institutional policy.
  - Remove the swab from the tube. Do not touch the swab or the inside of the cap to any surface. Swab the edges of the wound, avoid pus or exudates and any intact skin. Roll the swab as it is moved across the wound.
  - Put the swab into the tube without touching the top or sides of the tube, crush the ampule at the bottom of the tube. Label the culture and store according to instructions for the specific test or take the specimen to the lab promptly.
  - Redress the wound as ordered.
  - As with any procedure, wash your hands before and after the procedure, or after any steps that go from dirty to clean or from one area to another. Use waterless, bactericidal hand cleanser after washing hands. Change gloves when moving from a dirty to clean procedure.

## Complications

- Systemic infection (sepsis)

## Therapeutic Nursing Interventions

- Maintain medical asepsis (e.g., good hand-washing technique, standard precautions).

- Observe sterile technique during dressing changes and when handling tubes and drains.
- Maintain the client's hydration and nutritional status.
- Culture the wound before beginning antibiotic therapy.
- Administer antipyretics and antibiotics per order.
- Dressing changes as ordered.
- Cleanse and irrigate wound as prescribed.
- When obtaining wound culture, keep inside of culturette and lid sterile.

## Age-Related Changes—Gerontological Considerations

- Decreased immunological functioning, increasing susceptibility to infection.
- Reduced inflammatory response.

# Critical Thinking Exercise: Wound Infection

1. During your clinical experience on a medical-surgical unit, a classmate who is caring for a client with a fresh surgical incision asks you the most important measures for preventing wound infection. How do you respond?

2. A number of hospital-acquired infections are related to surgical wounds. What is the function of the inflammatory response that occurs with wound infection? As you assess postoperative clients, what signs of infection would typically accompany the inflammatory response? How long after surgery would you expect to see these signs?

3. Not all surgical clients experience the same risk of wound infection. What particular factors increase a client's risk of wound infection? From your own clinical experience, describe a client who is at increased risk of developing a wound infection.

# Wound Dehiscence/Eviscceration

## Key Points

- Dehiscence or evisceration are suspected when there is an increased flow of serosanguineous fluid between postoperative days five and 12.
- Immediate actions:
  - Cover wound immediately with sterile towels or dressings soaked in saline solution.
  - Place client in bed with knees bent.
  - Notify physician
- **Key Concepts/Terms**: Complications of wound healing include dehiscence and evisceration.

**Overview**

**Dehiscence** is a partial or total rupture of a sutured wound. Usually the layers below the skin also separate. **Evisceration** is a protrusion of visceral organs through the surgical incision. Both require immediate intervention.

## Risk Factors

- Chronic diseases (e.g., diabetes)
- Being elderly
- Morbidly obese individuals
- Persons with invasive abdominal cancer
- Vomiting
- Dehydration
- Malnutrition
- Failure of the sutures

## Signs and Symptoms

- Appreciable increase in the flow of serosanguineous fluid into the wound dressings between the 5th and 12th postoperative days.
- Preceded by a sudden straining, such as coughing, sneezing, or vomiting.
- Client may comment that "something has suddenly given way."

## Therapeutic Nursing Interventions

- Cover the wound area immediately with sterile towels or dressings soaked in saline solution.

- Place client in bed with knees bent.
- Notify the primary care provider immediately.
- Maintain a calm environment.
- Surgical repair may be necessary, although minor dehiscence may be allowed to heal by secondary intention.

# Critical Thinking Exercise: Wound Dehiscence and Evisceration

**Situation**: You are caring for a 38-year-old male client who underwent an emergency appendectomy five days earlier. As you enter the client's room, he holds his hands across the incision site and tells you "I think something has given way here." What has most likely occurred?

1. What warning signs suggest impending dehiscence? What immediate action is necessary if dehiscence or evisceration occurs?

2. What factors predispose a surgical client to dehiscence or evisceration? When are these events most likely to occur?

# Sensory Functioning and Sensoriperceptual Alterations

| Key Points |
| --- |

- The four components of the sensory experience are: stimulus, receptor, impulse conduction, and perception (awareness).
- Factors affecting sensory function are: illness, age, stress, culture, pain, medications, excess or insufficient environmental stimuli, and smoking.
- Sensoriperceptual function often decreases in older adults
- Awareness (consciousness) exists on a continuum from fully alert to comatose.
- Consiousness can be altered by physiological (e.g., medications) and psychological factors (e.g., grieving).
- Nursing goals for clients with altered consciousness are to provide for safety and to stimulate and reorient the client.
- Assess for sensory deprivation and overload; manipulate the environment to provide an appropriate amount of stimuli (e.g., provide a clock, allow pets to visit, minimize unnecessary light and noise).
- The nurse can help prevent sensory deficits, for example, by recommending eye examinations every three to five years after age 50; recommending immunizations against rubella, mumps, and measles; and teaching eye and ear protective measures.
- When communicating with visually impaired clients, identify yourself, explain what you are doing before touching the client, speak in a normal tone of voice.
- When communicating with hearing impaired clients, stand so the client can see your face, and speak in a normal tone of voice.
- **Key Concepts/Terms**: Sensory reception, sensory perception, presbyopia, tinnitu, awareness, level of consciousness, orientation, confusion, disorientation, solmnolence, comatose, sensory deprivation, sensory overload, sensory deficit

**Overview**

The nervous system receives massive amounts of information from sensory organs (e.g., ears, eyes). If the nervous system is intact, the sensory stimuli travel to the appropriate brain centers where the individual perceives, interprets, and reacts to it. Sensory reception begins with the stimulation of a nerve cell. The nerve impulse travels to the spinal cord or to the brain where it is interpreted as a sensation. Sensory perception is awareness of the sensation. It involves the organization and translation of stimuli into meaningful information. There are four components of the sensory experience:

**Stimulus**: Any agent act or influence that produces a reaction within a nerve receptor (e.g., sound waves, light).

**Receptor:** A specialized nerve cell that converts the stimulus to a nerve impulse. Receptors are specialized for only one type of stimulus. Some are grouped together in specialized organs such as the retina of the eye.

**Impulse conduction:** After the nerve impulse is created it travels to the spinal cord or directly to the brain.

**Perception (awareness):** Specialized brain cells in the cerebral cortex interpret the nature and quality of the sensory stimuli. When the person becomes conscious of the stimuli, perception takes place. Perception is affected by a person's level of consciousness that is, the brain must be alert in order for the person to receive and interpret stimuli.

## Factors Affecting Sensory Function

**Disease/Illness:** For example, peripheral vascular disease can cause reduced sensation in the extremities; diabetes can cause blindness and peripheral neuropathy.

**Age:** Infants have immature nerve pathways; older adults have changes in sensory organs.

**Stress:** Contributes to sensory overload.

**Culture:** Determines the amount of stimulation that a person considers "normal" (e.g., in some cultures touch is comforting; in others it may be offensive).

**Pain:** Affects the way a person perceives and reacts to stimuli.

**Medications:** Narcotic analgesics, antidepressants, and sedatives can change the level of consciousness and alter perception in other ways as well. Some antibiotics (e.g., streptomycin, gentamicin) can permanently damage the auditory nerve.

**Environment:** Excessive environmental stimuli (e.g., staff conversation) or inadequate stimuli (e.g., poor lighting, isolation) affect reception and perception.

**Smoking** can cause the taste buds to atrophy.

**Constant exposure to high noise levels** can cause hearing loss.

## Therapeutic Nursing Interventions

- Determine whether the client's sensory aids (e.g., eyeglasses, hearing aids) are functioning properly.
- Assess the environment for amount and type of stimuli (e.g., bright lights, noise, frequency of procedures).
- Assess level of consciousness and orientation.

## Age–Related Changes—Gerontological Considerations

- Middle-aged adults may develop presbyopia (inability to focus on near objects).
- Hearing changes include decreased acuity and tinnitus ("ringing" in the ears), and difficulty hearing conversation over background noise.
- Vision changes include sensitivity to glare, impaired night vision, reduced peripheral vision, reduced accommodation and depth perception, and reduced color discrimination.
- There is a delayed reception and reaction to speech.
- There is a decrease in the number of taste buds.

- Decreased sensitivity to odors and tastes are common.
- Tactile changes include decreased sensitivity to pain, pressure, and temperature.
- After age 60, proprioceptive changes create difficulty with balance and coordination.

# Sensoriperceptual Alterations

**Overview**      A variety of factors (e.g., disease, aging) can interfere with the reception and perception of stimuli. Sensoriperceptual alterations can be categorized as confusion/ disorientation, decreased level of consciousness, sensory deprivation, sensory overload, and sensory deficits.

## Related NANDA Nursing Diagnoses

- Acute Confusion
- Chronic Confusion
- Impaired Memory
- Risk for Disuse Syndrome
- Sensory/Perceptual Alteration (auditory, gustatory, kinesthetic, olfactory, tactile, visual)
- Unilateral Neglect

Sensoriperceptual problems can also be the etiology (cause) of other problems/diagnoses, for example:

- Impaired home maintenance management (related to declining visual abilities)
- Risk for impaired skin integrity (related to reduced tactile sensation)
- Risk for injury (related to sudden onset of blindness)
- Social isolation related to impaired hearing and decreased awareness

## Changes in State of Awareness

- **States of awareness exist on a continuum from fully conscious to comatose:**
    - **Fully conscious:** Alert; oriented to time, place, and person; understands written and verbal communication
    - **Disoriented:** Not oriented to time, place, or person (one or all three)
    - **Confused:** Decreased awareness; easily bewildered; memory deficits; misinterprets stimuli; judgment impaired
    - **Somnolent:** Extremely drowsy, but responds to stimuli
    - **Semicomatose:** Can be aroused by extreme (e.g., painful) or repeated stimuli
    - **Comatose:** Will not respond to stimuli
- **Awareness can be altered by physiological or situational factors. The most common are:**
    - **Effects of drugs and medications** (e.g., drug abuse, potentiating effects of multiple prescription drugs)
    - **Multiple losses** in a short time span
    - **Psychological trauma** (e.g., the abrupt loss of a significant person)
    - **Physiological disturbances** (e.g., hypoxia, metabolic imbalances, infectious processes, fluid

- Imbalance, neurologic disorders (e.g., Alzheimer's disease, other forms of dementia):
  - **Confusion** is a mental state in which stimuli are not interpreted accurately. The person becomes bewildered, misunderstands speech and behaviors, and may make inappropriate or nonsensical statements. Disorientation is a state of confusion in which the person does not accurately perceive time, place and person. The nurse assesses by asking questions such as, "Where are you? What day is this?"
  - **Decreased consciousness** is a mental state in which the person is less aware, or unaware, of self, others, and the environment. The person may not respond to any sensory stimuli, such as light, sound, or touch. Frequently, though, an unconscious person is able to hear speech and other sounds, even though unable to respond to them. Assessment includes observing the client's response to stimuli such as verbal commands, bright light, light touch, and painful touch.

## Therapeutic Nursing Actions for Changes in States of Awareness

- Maintain a consistent daily schedule. Routine gives the client a sense of security.
- Orient the client to time, place and person as often as needed.
- Touch and stroke the unconscious client.
- Wear a name tag with large print.
- Address the client by name and introduce yourself frequently.
- Place a calendar and clock in the room; refer to it often.
- Speak clearly and calmly to the client; allow time for the client to process and respond to your words.
- Interpret unfamiliar sights, sounds, and smells (e.g., buzzers, telephone ringing); correct any misinterpretations of situations.
- Assign the same caregivers, if possible.
- Place familiar items in the room (e.g., photographs).
- Keep the area uncluttered.

## Sensory Deprivation

**Sensory deprivation**: A decrease in or lack of meaningful stimuli. Because of the reduced stimulation, the person becomes more acutely aware of the few stimuli that are received and often perceives them in a distorted manner.

**Risk factors**: Isolation, a non-stimulating, monotonous environment (such as being confined to a room for long periods), inability to process environmental stimuli (e.g., a client who has brain changes), and inability to receive environmental stimuli (e.g., clients with impaired vision or hearing)

**Clinical signs**: Excessive yawning or drowsiness; reduced attention span; impaired memory; periodic disorientation and confusion (e.g., at night); hallucinations; apathy; emotional lability; depression

## Therapeutic Nursing Actions to Prevent Sensory Deprivation

- Be sure glasses are clean and hearing aids work properly.
- Communicate frequently with client.
- Use touch, as appropriate.
- Provide telephone, radio, TV, clock, and calendar.
- Have family bring flowers, plants, wall hangings, etc., from home.
- If possible, allow pets to visit, or consider having a pet in the agency.
- Provide tactile stimulation through physical care (e.g., back massage).
- Encourage changes in environment (e.g., take the client off the unit for a walk).

## Sensory Overload

**Sensory overload** occurs when a person receives multiple sensory stimuli and cannot selectively ignore some of the stimuli.

**Risk factors:** Increase in quantity or intensity of internal stimuli (e.g., pain, dyspnea, anxiety); increase in external stimuli (e.g., the client in a critical care unit is subject to constant, around-the-clock activity, the lights are always on, and there are sounds from monitors, alarms, and staff conversations); inability to disregard stimuli (e.g., as a result of medications or nervous system disorders)

**Clinical signs:** Fatigue, sleeplessness, irritability, anxiety, restlessness, disorientation, decreased ability to perform tasks or solve problems, muscle tension, difficulty concentrating

## Therapeutic Nursing Actions to Prevent Sensory Overload

- Minimize unnecessary light and noise. Provide earplugs, if necessary.
- Provide adequate pain control.
- Provide a private room; limit visitors.
- Plan care to allow for uninterrupted periods of rest and sleep.
- Speak in a low tone of voice and in an unhurried manner.
- Remove sources of odors (e.g., empty bedpan immediately after use, provide ventilation, use a room deodorizer, keep client clean).
- Give new information gradually so that the client can process the meaning.

## Sensory Deficits

A **sensory deficit** is the impaired reception, perception, or both, of one or more of the senses (e.g., blindness, deafness, changes in tactile perceptions). Sensory deficits create a risk for both sensory deprivation and sensory overload.

## Therapeutic Nursing Actions for Preventing Sensory Deficits

Teach clients to:

- Have regular eye examinations (every three to five years after age 50).
- Seek early medical attention if there is (1) reduced eye contact from infant; (2) persistent earache; (3) persistent eye redness or discharge; (4) increased tearing

of the eye; (5) pupil asymmetry; or (6) pain in the eyes.

- Have children immunized against rubella, mumps, and measles.
- Keep sharp instruments (e.g., scissors, pencils) away from infants and toddlers.
- Be sure that school-age children and adolescents use sports equipment properly.
- Wear sunglasses; never look directly at the sun.
- Wear protective goggles when working with power tools, spraying chemicals, or riding motorcycles.
- Wear ear protectors in environments with high noise levels (e.g., rock concerts).

## Therapeutic Nursing Actions for Managing Sensory Deficits

- Encourage client to use sensory aids, such as eyeglasses and hearing aids; assure that they function properly.
- When one sense is lost, assist the client to supplement with other senses (e.g., for a visually impaired client, provide audio tapes of music and books, chiming clocks, scented candles, etc.).
- When communicating with visually impaired clients: (1) identify yourself when entering the room; (2) speak in your normal tone of voice; (3) explain what you are about to do before touching the client; and (4) indicate when you are leaving the room.
- When communicating with hearing impaired clients: (1) move to a position where you can be seen or touch the person before beginning a conversation; (2) decrease background noises (e.g., TV); (3) talk in a normal tone of voice; (4) be sure the person can see your face well; (5) do not talk with gum in your mouth; (6) do not over-articulate or mouth words; and (7) pantomime or write ideas.

# Critical Thinking Exercise: Sensory Functioning and Sensoriperceptual Alterations

1. You are caring for a 65-year-old client with peripheral neuropathy secondary to diabetes mellitus. As you assess the client, what changes in the sensory functioning would you expect to find? What particular risks does the client face related to the disease process? How might you focus your assessment and teaching to promote the client's safety and well being?

2. In a skilled nursing facility, you are assigned to care for a 78-year-old woman with a significant hearing impairment. Describe the specific measure you will take in your interaction with this client to promote optimal communication.

3. A neighbor who is caring for an elderly parent tells you that "Dad just doesn't seem as sharp these days as he used to be." She wonders whether the changes she has observed are a normal part of aging or indicate an actual disease process. How do you respond to her concern?

**Situation**: You are caring for a 71-year-old client in an ICU who demonstrates an increasing level of anxiety and disorientation to time and place. Based on a nursing diagnosis of risk for sensoriperceptual alterations related to sensory overload, you develop a plan of care.

4. What nursing interventions might be most effective to reduce the potential for sensory overload in this client?

5. You accompany a home care nurse on a visit to a 78-year-old woman who has recently undergone orthopedic surgery. The client lives alone following the death of her husband four months ago. Although the client is progressing well with physical therapy and is now able to walk short distances without an assistive device, she expresses a lack of interest in activities she once found pleasurable and is becoming increasingly isolated. Based on these findings, formulate a plan to reduce the client's risk of social isolation.

6. A colleague working in a nursing home facility expresses concern for clients who become increasingly agitated in the evening. Based on her description, you recognize the signs of sundowner's syndrome. What are the possible causes of sundowning? What interventions might be useful to assist individuals who may wander or become disoriented around sunset?

7. A neighbor tells you that his father has recently been diagnosed with Alzheimer's disease. Although the family has met with their father's physician, your neighbor asks you to clarify some questions about the disease process. What causes Alzheimer's? Why does his father have difficulty concentrating? How might family members respond when their father seems confused? What can he and other family members do to assist their father and promote a safe environment?

# Self-Concept & Sexuality

## Key Points

- Through the process of growth and development, the individual accumulates and processes information that helps form a basic perception of who he is.
- Self-concept is a person's knowledge about and subjective image of the self—a mixture of conscious and unconscious feelings, attitudes, and perceptions.
- Components of self-concept are: body image, identity, role performance, self-esteem.
- Self-concept can be affected by such factors as: loss of a body part or function, inability to achieve goals, unrealistic expectations of self, illness, loss, and lack of positive feedback from significant others.
- Signs of low self-esteem include: avoiding eye contact, frequent apologizing, stooped posture, poor grooming, hesitant speech, and expressions of powerlessness/ helplessness.
- Help clients identify strengths; avoid criticism; teach positive self-talk; encourage to participate in decisions about their care; call the client by name.
- Sexuality is a part of a person's identity.
- Important dimensions of sexual health are: self-concept, body image, and sexual identity.
- Sexuality is influenced by health status, culture, religion, and personal ethics.
- Illness, medications, and hospitalization affect a person's sexuality (e.g., by producing physical, psychological, and social changes).
- Important facts:
  - Alcohol is not a sexual stimulant.
  - Older adults do have sexual intercourse.
  - Masturbation does not cause mental illness or physical harm.
- Sexual health teaching should include sex education, responsible sexual behavior, and monthly breast and testicular self-examination. Clients are usually reluctant to initiate a discussion about sex; the nurse must take the lead.
- It is essential for the nurse to be nonjudgmental, knowledgeable, and comfortable with her/his own sexuality.
- The nurse should learn methods for dealing with clients who act out sexually.
- **Key Concepts/Terms:** Self-knowledge (subjective self), self-expectation (ideal self), social evaluation, sexual identity, biologic sex, gender identity, gender-role behavior, sexual orientation, heterosexual, homosexual, erectile dysfunction, impotence, premature ejaculation, retarded ejaculation, orgasmic dysfunction, vaginismus, dyspareunia, PLISSIT model for sexual counseling

**Overview**

Self-concept is the person's knowledge about and subjective image of the self. It is a mixture of conscious and unconscious feelings, attitudes, and perceptions. Self-concept and perception of health are closely related, and a positive self-concept is necessary for personal growth and self-care behaviors. Four dimensions of self-concept formation are:

**Self-knowledge (subjective self):** "Who am I?" The knowledge that one has about oneself, including basic facts (e.g., age, race, sex), insight as to good and bad qualities, abilities and limitations, and characteristics (e.g., generous, intelligent).

**Self-expectation (ideal self):** "Who do I want to be?" What one expects of oneself or how one would prefer to be; may be a realistic or unrealistic expectation.

**Social self:** "What do others think of me?" How one believes one is perceived by others and by society.

**Social evaluation:** "How satisfied am I with myself?" How one judges oneself in relationship to others, events, or situations (e.g., "I am not as smart as most people.")

# Self-Concept

## Components of Self-Concept

Self-concept exists on a continuum from positive to negative, depending on the strengths of each of the four components: body image, identity, self-esteem, and role performance.

**Body image** is the person's perception of his/her body, both positive and negative. It is affected by cognitive and physical development, as well as by cultural attitudes and values. Body image depends only partly on the reality of the body. For example, those who have recently lost a great deal of weight may still perceive themselves as obese.

**Identity** is the internal sense of individuality. It implies being a whole and unique self, distinct and separate from others. Sexuality is a part of one's identity—that is, the image of self as a man or a woman.

**Role performance** relates to how the behaviors expected of a particular role compare to what the person actually does in that role.

**Self-esteem** is the judgment of one's own worth. If the person's self-concept does not match with the "ideal self," then low self-esteem results. Self-esteem is derived both from self and from others. Its foundation is laid in childhood and family experiences; however as an adult, self-esteem changes with the person's ongoing evaluation of self in relation to others. Severe stress (e.g., a long illness) can lower a person's self-esteem.

## Factors Affecting Self-Concept

### Body Image Factors

- Loss of a body part (e.g., mastectomy)
- Disfigurement (e.g., burns, colostomy, pregnancy)
- Loss of body function (e.g., spinal cord injury, arthritis, hearing loss)

### Identity Factors

- Achievement of goals
- Membership in a minority group
- Decreasing physical or mental abilities
- Concerns about sexuality
- Concerns about relationships
- Unrealistic expectations of self

### Role Performance Factors

- Loss of or change in job
- Retirement
- Divorce
- Illness, hospitalization
- Loss of parent, child, spouse, or close friend

### Self-Esteem Factors

- Repeated failures (e.g., career, parenting)
- Unrealistic expectations (self and others)
- Ineffective coping mechanisms
- Abusive relationships
- Lack of financial security
- Lack of positive feedback from significant others (e.g., parents to child)

## Related NANDA Nursing Diagnoses

- Altered role performance
- Body image disturbance
- Chronic low self esteem
- Self-esteem disturbance
- Situational low self esteem
- Personal identity disturbance

## Therapeutic Nursing Interventions for Self-Concept

- **Recognize signs of low self-esteem:**
  - Avoids eye contact
  - Overly critical of self
  - Overly critical of others
  - Frequently apologizes
  - Stooped posture
  - Poor grooming
  - Verbalizes feelings of helplessness and powerlessness ("Whatever happens, happens," or "I don't care; let's just get it over with.")

- Help clients to identify personal strengths (e.g., provide honest, positive feedback; acknowledge goals that have been attained).
- Help clients maintain a sense of self (e.g., respect privacy, provide explanations before starting a procedure, address the client by name, listen attentively).
- Enhance self-esteem by:
  - Encouraging clients to participate in decisions about their care
  - Encouraging clients to take part in activities in which they can be successful
  - Communicating that the client is valued: use client's name; ask for advice
- Provide accurate information.
- Avoid criticism.
- Encourage clients to express their feelings and ask questions.
- Teach clients to use positive self-talk ("I can tie my shoes.") instead of negative self-talk ("I can't dress myself any more.").

# Sexuality

## Overview

Sexuality is a part of a person's identity. In holistic nursing practice, the need to acknowledge and deal with issues of sexuality cannot be ignored. The World Health Organization has defined **sexual health** as "the integration of the somatic, emotional, intellectual, and social aspects of sexual being, in ways that are positively enriching and that enhance personality, communication, and love". Three important dimensions of sexual health are sexual self-concept, body image, and sexual identity.

**Body image** is how one perceives one's body, both negatively and positively. People who feel good about their bodies are more likely to be comfortable with sexual activity. Media focus on attractiveness and large breasts, for example, creates body image problems for some women. Likewise, men may worry about penis size, having bought into the myth that "bigger is better."

**Sexual self-concept** is how one perceives and values oneself as a sexual being. A healthy sexual self-concept is essential for forming intimate relationships.

**Sexual identity** consists of biologic sex (gender), gender identity, gender-role behavior/orientation, and sexual orientation/preference. **Biologic sex** refers to chromosomal makeup, external and internal genitalia, hormonal states, and secondary sex characteristics (e.g., facial hair). **Gender identity** is one's belief or awareness of being male or female. **Gender-role behavior** is the behavioral expression of gender identity as well as the expression of what is considered acceptable behavior. Each culture establishes expectations for "male" and "female" behavior. **Sexual orientation** is the preference of a person for one sex or the other as a sexual partner (e.g., a **heterosexual** person is sexually attracted to persons of the opposite gender; a **homosexual** person is sexually attracted to persons of the same gender). There is controversy over the determiners of sexual orientation.

## Factors Influencing Sexuality

**Health status**: For example, after a myocardial infarction, clients may be anxious about sexual activity and the risk of another heart attack. Other conditions affecting sexuality include diabetes mellitus, spinal cord injury, joint disease (e.g., arthritis), chronic pain, chronic lung disease, sexually transmitted diseases, mental disorders, and various medical and surgical procedures.

**Culture** influences expectations about sex role behaviors and specific sex practices (e.g., premarital coitus, polygamy, and homosexuality may or may not be acceptable or tolerated).

**Religion**: Religious values provide guidelines for determining what sexual behavior is acceptable, and under what circumstances (e.g., only in marriage).

**Personal ethics**: Many people have codes of conduct that they have developed outside of religion (e.g., views about masturbation).

## Effects of Illness on Sexuality

**Physiological changes** that may interfere with sexual functioning:

- Interruption of neural pathways and/or blood supply to the genitals and other soft body tissues that respond to sexual arousal
- Hormone changes
- Changes in joints and muscles that must bend and stretch during sexual activity
- Anything (e.g., chronic illness or providing home care to an elderly parent) that leaves the person with little energy for sexual feelings or activity

**Psychological changes** include problems with gender identity and low self-esteem.

**Medications** that can interfere with sexual desire and sexual response include: alcohol, alpha- blockers, antianxiety agents, anticonvulsants, antidepressants, antihistamines, antihypertensives, antipsychotics, barbiturates, beta-blockers, cardiotonics, cocaine, diuretics, and narcotics.

**Hospitalization:** The client lacks privacy and control over the environment. Some clients act out sexually (e.g., by using obscene language) as an attempt to validate their identity as a sexual being. The nurse should reinforce desirable behavior.

## Sexual Myths (Misconceptions)

- It is immoral to use positions for coitus other than the face-to-face, male superior position (Fact: Partners should use the position that provides the most pleasure and is acceptable to both.)
- Alcohol is a sexual stimulant. (Fact: It is associated with impotence.)
- Older adults do not have sexual intercourse. (Fact: Sexual desire and ability decrease very little after middle age.)
- Masturbation causes mental illness. (Fact: It is harmless.)
- A man with a large penis provides greater sexual satisfaction to a woman than one with a small penis. (Fact: There is no evidence that this is so.)
- Women who experience orgasm are more likely to become pregnant. (Fact: This is not true; conception is unrelated to orgasm.)

## Altered Sexual Function

Sexual dysfunctions may have physical or psychological origins.

### Male dysfunction includes:

- Erectile dysfunction/impotence (inability to achieve or maintain an erection)
- Premature ejaculation (inability to delay ejaculation long enough to satisfy the partner)
- Retarded ejaculation (inability to ejaculate into the vagina, or delayed ejaculation)

### Female dysfunction includes:

- Orgasmic dysfunction (inability to achieve orgasm)
- Vaginismus (irregular, involuntary contractions of vaginal muscles before penile penetration can be achieved)
- Dyspareunia (pain during intercourse)

NANDA nursing diagnoses include:

- Altered Sexuality Patterns
- Sexual Dysfunction
- (Sexual problems can be the source of other nursing diagnoses, such as Pain, Anxiety, Body Image Disturbance, and Low Self-Esteem)

## Therapeutic Nursing Interventions Client Education for Sexual Health

Many sexual problems occur as a result of ignorance. Therefore, sexual health teaching is important. Three areas to include are:

**Sex education**: e.g., Kegel exercises, sexual anatomy and physiology, effects of developmental stages (such as puberty, menopause), effects of illness and medications, and ways that parents can answer their children's questions

**Responsible sexual behavior**: e.g., the prevention of sexually transmitted diseases (STDs) and unwanted pregnancies

Teaching monthly breast self-examination for women and testicular self-examination for men

## Therapeutic Nursing Interventions for Sexual Counseling

Nurses vary in their knowledge and ability to help clients with sexual problems. The PLISSIT model describes four progressive levels of knowledge and skill.

Nurses should be able to function at the first three levels in most situations. For clients needing fourth level skills, the nurse should have special training or should refer the client to a counselor.

P: Permission Giving (e.g., "Many people worry about how this surgery will affect their sex life. Have you thought about it?")

LI: Limited Information (e.g., "After a heart attack, most people can resume intercourse in four to six weeks; but check with your primary care provider first.")

SS: Specific Suggestions (This level requires specialized knowledge about how sexual functioning may be affected by a disease or therapy and how the client can adapt sexual activity—for example, ways to handle a cast during sexual activity, or safe and unsafe practices following a heart attack.)

IT: Intensive Therapy (Should be provided by a clinical nurse specialist or sex therapist. Intensive therapy is used when the first three levels are ineffective. It may deal with issues such as self-concept and marriage.)

## Therapeutic Nursing Interventions Other

- Generally speaking, clients are reluctant to initiate a discussion about sexual concerns; the nurse will usually need to take the lead.
- Make assessments with a non-judgmental attitude.
- Encourage clients to discuss their concerns.
- Effective helping requires nurses to be knowledgeable and comfortable with their own sexuality.
- Develop and convey a nonjudgmental attitude when caring for clients with a sexual orientation differing from your own.

## Dealing with Inappropriate Behavior

Clients may act out sexually for a variety of reasons (e.g., fear of loss of sexual functioning, unmet intimacy needs, need for reassurance that they are still sexually attractive, misinterpretation of the nurse's behavior as sexually provocative). Examples of behaviors include, exposing themselves, touching or grabbing the nurse, asking the nurse to bathe the genital area when they are able to do so themselves, and making sexual comments to or about the nurse.

The following strategies are suggested:

- Directly state that the behavior is not acceptable ("I do not like the things you are saying," or "You do not have permission to touch me.").

- State your feelings ("When you do that, I am embarrassed. It makes me not want to come into your room.").

- State the behavior you expect ("Please keep yourself covered when I am in the room.").

- Set limits (use direct eye contact and say, "Don't do that.").

- Refocus the client from the inappropriate behavior to the underlying concerns. Offer to discuss concerns about sexuality and sexual functioning ("Sometimes people do that when they wonder how their illness will affect them sexually. Are there some things you would like to talk about?").

- Report the incident to the charge nurse or supervisor.

# Critical Thinking Exercise: Self-Concept & Sexuality

**Situation**: A school nurse shares her concerns with the parents of a 13-year-old boy who is experiencing extreme anxiety over school performance. The boy is an honor roll student who works hard at his studies but lacks close friendships or outside interests. In an interview with the nurse, the boy states "I have no time for friends or sports because I have to get straight A's so that I can get into a good school and make my Mom and Dad happy.".

1. Based on the child's presentation, what nursing diagnosis might apply? How might the nurse assist this family to help the child develop a more positive self-concept?

2. Body image is an important component in an individual's self-concept. Think of a client you have cared for who has experienced an altered self-concept related to a change in body image or function. How were you able to assist this individual maintain or recover a more positive self-image?

3. Discuss with a classmate how your own self-concept as a nurse affects your relationship with clients. As a caregiver, what measures can you take to promote a positive personal self image?

4. You are caring for a 54-year-old client who has undergone a modified radical mastectomy. Your client is reluctant to look at the incision and expresses concern about how her husband will view her altered appearance and the effect it will have on their relationship. How might you respond to her concerns about her health status and any factors that may result in altered sexual functioning?

5. A nurse working in a clinic performs a routine physical examination on an 18-year-old male. In response to the nurse's question about testicular self-exam, the client asks: What has that got to do with me? I'm too young to have cancer." How might a nurse educate this client about the importance of testicular self-examination?

6. Obtaining information about a client's sexual health history is an important component of a nursing assessment. How might a nurse use this information to identify factors that can place an individual at risk for sexual dysfunction?

# Stress, Adaptation, Grief and Loss

## Key Points

- Regardless of the cause of stress, the same chain of nonspecific physiologic events occurs—the release of epinephrine and cortisone, with subsequent changes in organ systems.

- Physiological signs/symptoms are mostly a result of sympathetic nervous system and adrenal activity (e.g., dilated pupils, diaphoresis, increased heart rate, decreased peristalsis, increased muscle tension).

- Psychological signs/symptoms include anxiety, anger, depression, and unconscious ego defense mechanisms.

- Cognitive signs/symptoms include: problem solving, structuring, self-control, suppression, fantasy, and daydreaming.

- A person can cope by (1) altering the stressor, (2) adapting to the stressor, or (3) avoiding the stressor; coping can be adaptive or maladaptive.

- The nurse should teach anxiety reduction measures, encourage health-promotion (e.g., exercise), and provide information about tests, treatments, etc.

- Relaxation techniques (e.g., visualization) can be used to counteract the "fight-or-flight" response of the General Adaptation (stress) System (GAS).

- Kubler-Ross's "stages of dying": denial and isolation, anger, bargaining, depression, acceptance.

- Physical manifestations of grief and stress are varied; e.g., fatigue, headache, altered libido, loss of appetite, difficulty sleeping, chest pain, palpitations, and menstrual disturbances.

- In dysfunctional grieving, the symptoms are more severe and the grieving is longer. The person may deny the loss and/or have difficulty expressing grief.

- Factors influencing the grief process: culture, spiritual beliefs, age, age of the deceased, available resources, significance/importance of the loss, expected versus unexpected loss, socioeconomic factors, cause of loss/death (e.g., preventable, socially acceptable).

- Older adults experience multiple losses, thus taxing their ability to cope.

- Determine the client's stage of grieving/dying: this affects how he/she perceives messages.

- During early stages, provide opportunity to talk about the loss.

- Do not offer advice or give unwarranted reassurance (e.g., do not say, "It's darkest just before dawn," or "He's in a better place now.").

- **Key Concepts/Terms**: Stress, stressor, internal stressor, external stressor, general adaptation syndrome (GAS), local adaptation syndrome (LAS), alarm reaction, fight or flight, stage of resistance, stage of exhaustion, coping, ego defense mechanisms, loss, grief, bereavement, mourning, anticipatory grieving, dysfunctional grieving

# Stress and Adaptation

**Overview**

Everyone experiences stress. **Stress** can result from both positive and negative experiences. Stress is any situation in which a nonspecific demand creates the need for a person to respond; to take action. Stress involves both physiological and psychological responses. A **stressor** is any stimulus (e.g., an event) that causes a person to experience stress. A stressor represents an unmet need. **Internal stressors** originate from within (e.g., a fever, pregnancy); **external stressors** originate outside a person (e.g., peer pressure, a death in the family). Stress affects the whole person: physically, emotionally, intellectually, socially, and spiritually.

## Selye's (1956, 1976) Model

Selye defined stress as a response: "the nonspecific response of the body to any kind of demand made upon it".

A **stressor** is any stimulus that produces stress and disturbs the body's equilibrium.

The **general adaptation (stress) syndrome (GAS)** is the body's nonspecific response to stress. Regardless of the cause of the stress, the same chain of physiologic events occurs (i.e., the GAS). The GAS is a result of the release of adrenal hormones, with subsequent changes in organ systems. The GAS occurs in three stages: (1) **alarm reaction** (when epinephrine and cortisone are released to prepare for "fight or flight"), (2) **stage of resistance** (when adaptation takes place), and (3) **stage of exhaustion** (when the adaptation made in stage 2 cannot be maintained).

The **local adaptation syndrome (LAS)** is the reaction of one organ or one part of the body. One example of the LAS is inflammation.

## Assessing for Indicators of Stress

Assessment should include physiological, psychological and cognitive signs and symptoms.

**Physiological signs/symptoms** occur mostly as a result of sympathetic nervous system and adrenal activity:

- Dilated pupils
- Diaphoresis (to control increased body heat due to increased metabolism)
- Increased heart rate
- Increased sodium and water retention with resulting increase in blood volume and increased cardiac output
- Increased rate and depth of respirations (hyperventilation)
- Skin pallor (because of peripheral vasoconstriction in response to norepinephrine)
- Decreased urinary output
- Dry mouth
- Decreased peristalsis, resulting in flatus and constipation

- Improved mental alertness (in serious threats)
- Increased muscle tension
- Increased blood sugar (in response to release of glucocorticoids and gluconeogenesis)

**Psychological signs and symptoms**

- Anxiety (ranging from mild anxiety to panic)
- Fear
- Anger
- Depression
- Unconscious ego defense mechanisms
  - Denial—ignoring the truth
  - Displacement—taking out one's anger or emotion on an object or subject that is not the real cause
  - Projection—placing blame on other persons or things when one does not get what she/he wanted or something happens that she/he did not want to happen
  - Rationalization—justifying actions in an illogical manner
  - Regression—moving to a previous state that is more comfortable, less stressful
  - Sublimation—adapting socially unacceptable behaviors/desires into a more socially acceptable action

These are psychological adaptations that serve to relieve inner conflicts and tensions. They are the unconscious mind's efforts to defend against anxiety. Defense mechanisms may be adaptive or maladaptive.

**Cognitive Signs and Symptoms**

- **Problem solving**: Thinking through the situation, using specific steps to find a solution.
- **Structuring**: Manipulating the situation so that threatening events do not occur (e.g., structuring an interview by asking only direct, closed questions in order to avoid threats to the nurse's knowledge or values).
- **Self-control (discipline)**: Assuming facial expression and calm behavior to convey a sense of being in control. This may help to prevent panic. Carried to an extreme, however, self-control can keep the person from receiving support from others, who may think the person is handling the situation well.
- **Suppression**: Consciously putting a thought or feeling out of the mind
- **Fantasy and daydreaming**: For example, fantasizing the positive outcome of chemotherapy. If used to excess, fantasy can keep the person from dealing with reality.

## Coping with Stress

**Coping** consists of thought processes and behaviors used to successfully manage demands that tax or exceed the person's resources (dealing with stress). There are three approaches to coping with stress:

- Alter/change the stressor

- Adapt to the stressor
- Avoid the stressor

**Coping can be adaptive or maladaptive. The effectiveness of coping depends on the:**

- Number, duration, and intensity of the stressors
- Person's past experiences
- Availability of support systems
- Personal qualities of the individual

## Related NANDA Diagnoses

- Anxiety
- Caregiver Role Strain
- Defensive Coping
- Fear
- Impaired Adjustment
- Ineffective Denial
- Ineffective Family Coping (Compromised)
- Ineffective Family Coping (Disabling)
- Ineffective Individual Coping
- Post-Trauma Response
- Relocation Stress Syndrome

## Therapeutic Nursing Interventions for Stress and Adaptation

- Encourage health-promotion strategies such as regular exercise, good nutrition, adequate rest and sleep, and effective time management.
- Teach anxiety reduction measures, for example, taking deep breaths during a painful procedure.
- Provide factual information about tests, treatments, and so on.
- Be prepared to deal with clients' anger (e.g., refusing treatment, being verbally abusive). Respond in a way that reduces the client's stress rather than your own.
- Use relaxation techniques (e.g., visualization, progressive muscle relaxation) to counteract the "fight-or-flight" responses of the GAS.
- Manage your own stress; for example:
  - Have a plan for daily relaxation (e.g., meditation, massage).
  - Exercise regularly to direct energy outward.
  - Develop assertiveness to overcome feelings of powerlessness.
  - Learn to say "No."
  - Accept what cannot be changed.
  - Develop peer support groups to deal with stressors in the work setting.

# Grief and Loss

## Overview

Loss is a situation in which something of value is no longer available (e.g., loss of a job, a loved one, or a body part). Loss can be actual, perceived or anticipatory (experienced before the loss actually occurs). Death is a fundamental loss, both to the dying person and for the survivors.

Grief is the emotional response to loss. It is associated with overwhelming sorrow or distress that is manifested in thoughts, feelings, and behaviors. Bereavement is a type of grief specifically experienced by survivors after the death of a loved one. Mourning is the process through which grief is resolved. Those who lose a loved one experience grief and mourning. This process occurs with those who suffer other kinds of losses. Grieving is a normal process that is essential for good health. Grieving enables the individual to cope with the loss and to accept it as reality.

## Stages of Grieving and Dying

Many authors have described stages of grieving. However, there is no single correct way for a person to progress through the grief process, nor a correct length of time in which to do it. Kubler-Ross (1969) has described the stages of dying, which are similar to the stages of grief:

Denial and isolation: Client refuses to believe that he or she will die, isolates self from reality, and represses what is discussed. The person may think, "They must have mixed my records up with someone else's," and may be artificially cheerful in order to prolong the denial.

Anger: Client or family express rage and hostility, sometimes at nurse and staff, about things that normally would not upset them. The person may think, "Why me? I didn't do anything to deserve this."

Bargaining: Client seeks to bargain for more time: "If I can just live until my daughter's wedding, I'll be satisfied." Client may begin putting affairs in order, for example, making wills and giving away personal items. May express guilt for real or imagined past sins. Bargaining helps the client to move into the final stages of dying.

Depression: Client accepts that death is a reality and goes through a period of grief. The client may cry and talk freely about the loss, or may withdraw: "I wanted so badly to see my grandchild, but I won't even be here when he is born."

Acceptance: Client comes to terms with dying. He or she has accepted death and is prepared to die. Client may have decreased interest in surroundings and significant others: "Everything is taken care of and I'm tired of fighting it. I'm ready to go."

## Signs and Symptoms of Grief

Physiologically, the body responds to loss with a stress reaction. Physical manifestations of grieving include: fatigue, headache, dizziness, fainting, blurred vision, menstrual disturbances, chest pain, palpitations, alterations in libido, difficulty sleeping, and loss of appetite. Unresolved, dysfunctional grieving is longer and the symptoms are more severe than in normal grieving. The person may deny the loss, have difficulty expressing grief, or grieve beyond the expected time.

### Normal Manifestations of Grief

- Verbalization of the loss
  - Crying
  - Sleep disturbances
  - Loss of appetite
  - Difficulty concentrating
  - Fatigue

### Signs of Dysfunctional Grieving

- Depression
  - Severe physiological symptoms
  - Suicidal thoughts
  - Extended period of grieving
  - Prolonged time in stage of denial
  - Prolonged or severe social isolation
  - Persistent guilt
  - Low self-esteem
  - Drug abuse

## Factors Influencing the Grief Process

**Culture** defines perception and expression of loss.

**Spiritual beliefs** influence both the perception of the loss and subsequent behaviors (e.g., funeral rituals).

**Age**: As they age, people experience losses and gain better understanding and acceptance of life, loss, and death.

**Age of the deceased (when loss is due to death)**: Because it seems "unnatural," it is usually more difficult to cope with the death of a young person than that of an older adult.

**Available resources** (e.g., support persons, physical and mental health)

**Multiple losses (e.g., death of family member, loss of employment, financial change or moving residence)**

**Significance/importance of the loss** (e.g., can the lost object be replaced?)

**Expected or unexpected loss**, that is, was there time to prepare for the loss?

**Socioeconomic factors** (e.g., an insurance plan can support a person with a disability while he or she pursues vocational reeducation)

**Cause of death or loss** (e.g., a disease such as a heart attack is considered "more acceptable" than AIDS; a death that is beyond the person's control is more acceptable than one, such as drunk driving, that is preventable)

## Related NANDA Nursing Diagnoses

- Anticipatory Grieving
- Anxiety
- Dysfunctional Grieving

- Grieving may be the etiology (cause) of other nursing diagnoses:
  - Altered Family Processes
  - Altered Nutrition
  - Impaired Adjustment
  - Risk for Loneliness

## Therapeutic Nursing Interventions for Grief and Loss

- Assure the terminal client and his/her family (or significant others) that effective pain management will be provided to the client throughout the illness.
- Deal with the person's perception of the loss.
- Determine the client's stage of grieving/dying. Whether the client is angry, depressed, or withdrawn affects how he or she perceives messages and how the nurse interprets the client's statements.
- Use communication skills such as active listening, silence, clarifying, reflecting feelings and summarizing. Do not offer advice, evaluate, pass judgment, or give unwarranted reassurance, such as "It's darkest just before dawn. You'll feel better soon."
- Help the client to identify the loss; it may be necessary to verbalize it for him/her.
- During the early stages, provide opportunities for the client or family to talk about the lost person/object/ability.
- Provide some way of filling the "gap" left by the loss—for example, occupational therapy to help a client learn to dress after losing the use of an arm.
- Offer choices that promote client autonomy and promote a sense of control over her/his life at a time when not much control is possible.
- Meet physiological needs of the dying client: hygiene, nutrition/fluids, elimination, movement, respiratory care, positioning for comfort and breathing, pain control.
- Meet psychological needs of the dying client: fear, loss of control, separation from family, and overwhelming pain.
- Meet spiritual needs of the dying client: obtain clergy, be present and available to listen to spiritual concerns.
- Obtain consent for autopsy and organ donation if appropriate.
- Encourage family to participate in care of dying client.
- Understand that a client's spiritual beliefs, customs and rituals may interfere with nursing routines.
- Make referrals to agencies such as the Grief Recovery Institute, Choice in Dying, and the American Association of Retired Persons (AARP).

## Age-Related Considerations—Gerontological Considerations

Changes in physical and mental capabilities during old age are generally perceived as losses. Older adults often experience other losses, such as loss of employment, independence, health, and significant others, thus taxing their coping capacity.

# Critical Thinking Exercise: Stress, Adaptation, Grief and Loss

**Situation:** During a routine physical examination, a 50-year-old client reports chronic stress related to the many demands of his business. He also reports skipping breakfast on a regular basis, a lack of exercise, an increasing irritability with his family, and an inability to relax at the end of the work day. Physical examination reveals a blood pressure of 144/90 and a 10-pound weight gain over the past year. The client acknowledges his unhealthful lifestyle, but does not know how to implement change. What might you suggest to help this individual cope with stress and promote good health?

1. Stress is an unavoidable part of everyday living. It cannot be entirely eliminated. However, there are many ways to minimize the effects of stress, such as adopting a program of regular exercise and a healthful diet. Develop a pamphlet that can be used as a simple teaching tool for clients who wish to incorporate stress-reducing activities into their daily lives.

2. A client who has been experiencing a number of stressors asks you to explain the mechanism that mediates the stress response. In simple terms, identify the adaptive hormones that are involved in the stress response and the changes they produce in the body. What measures can the individual take to cope with unavoidable stressors?

**Situation:** A hospice home care nurse is making an initial visit to a client diagnosed with lung cancer. At the door, the client's family greets the nurse and states "Dad doesn't know his diagnosis. For his own good, we don't want him to know." The nurse finds the client alert, oriented, and experiencing good pain control. The client asks no questions and doesn't mention his prognosis in the presence of the family.

3. What is the nurse's responsibility in this case? How might the nurse best respond to this potential conflict?

4. Interview a staff member of a hospice, whether a chaplain or a nurse or a volunteer. Ask this person to explain the benefits derived from participating in hospice care. Your questions might include:

- What is hospice care?

- What distinguishes hospice from other forms of care offered at the end-of life?

- How is the role of "team" emphasized in the delivery of hospice care?

5. Although death represents the ultimate loss, each of us experiences a variety of losses daily, both large and small. Think of a loss you have experienced—perhaps the loss of a pet, or a friend who moved away, or a failure to meet a goal. Consider your own grieving process as you came to terms with your loss. In a brief diary entry, identify the loss and explore its impact on your life.

6. A 65-year-old hospice client starts taking inventory of a number of keepsakes collected over the years and begins giving them to friends and family. How might a nurse encourage life review in such a process? At the same time, the client expresses a desire to buy a new winter coat. Based on this information, what can a nurse infer about the client's grieving process?

7. Read "Tuesdays with Morrie," Mitch Albom's best-selling chronicle of a young man's weekly visits to his former college professor and mentor who has been diagnosed with amyotrophic lateral sclerosis and is coming to terms with impending death. If possible, form a discussion group with classmates to share your reactions to this book. In particular, what lessons might nurses and future nurses learn from this book?

# Death and Post Mortem Care

## Key Points

- World Medical Assembly guidelines for defining death:
  - Coldness of the skin and extremities
  - Total lack of response to external stimuli
  - No muscular movement; no breathing
  - No reflexes
  - Flat encephalogram
- Clinical signs of impending death: muscle weakness (e.g., difficulty speaking, loss of sphincter control), respiration change, sensory changes (e.g., blurred vision), circulatory changes (e.g., mottling and cyanosis of extremities; slow, weak pulse; decreased blood pressure; loss of sensation)
- Post mortem care varies among institutions and is guided by institutional policies and procedures. The nurse's primary responsibility is care of the body and emotional support of the family/significant others. This generally includes cleaning the area for viewing of the body, closing the eyes and mouth (before rigor mortis), cleaning and caring for the body so it is as "normal" as possible (combing the hair, removing any medical equipment that is not invasive, adjusting the lighting in the room, covering the client as though she/he were sleeping). DO NOT REMOVE invasive tubing until it is determined that a post-mortem examination will not be done or per the institution's policy
- **Key Concepts/Terms:** Clinical signs of death, cerebral (higher brain) death, rigor mortis, algor mortis, livor mortix, shroud, DNR order, advance directive, living will, durable power of attorney, health care proxy, death certificate, autopsy, inquest, Cheynne Stokes breathing, death rattle

## Overview

Because we have the technology to maintain respirations and blood circulation, it is difficult to determine when death has actually occurred. Various criteria are used. The traditional **clinical signs of death** are cessation of the apical pulse, blood pressure, and respirations. **The World Medical Assembly guidelines are**:

- Total lack of response to external stimuli
- No muscular movement; no breathing
- No reflexes
- Flat encephalogram

**When the client is on artificial life support** electric currents from the brain (as shown by an electroencephalogram) must be absent for at least 24 hours before a primary care provider can pronounce death. Using the cerebral death (higher brain death) definition, death occurs when the cerebral cortex is irreversibly destroyed. The person may still be breathing, but is irreversibly unconscious.

**The nurse's role** in caring for dying clients includes some activities prescribed by law (e.g., advance directives), providing hospice and home care, meeting the physiological needs, providing spiritual support, supporting the family, and providing post mortem care. Preparation of the body after death involves cleaning and preparing the body, ensuring proper identification and releasing the body to the morgue.

## Terminology

**Rigor mortis:** Approximately two to four hours after death, the body stiffens when it stops synthesizing adenosine triphosphate (ATP), which is necessary for muscle fiber relaxation. Rigor mortis occurs first in the involuntary muscles (e.g., the heart), and then progresses to the head, neck, trunk, and extremities, in that order. Rigor mortis usually leaves the body about 96 hours after death.

**Algor mortis:** The decrease in body temperature that occurs after death. When blood circulation and hypothalamic function cease, body temperature drops about 1-2° F per hour until it reaches room temperature.

**Livor mortis:** The bluish-purple discoloration that appears in the dependent areas of the body (e.g., on the posterior surface when the body is in supine position). The red blood cells break down, releasing hemoglobin, which discolors the tissues.

**Shroud:** A large piece of plastic or cotton material used to enclose a body after death

## Legal Issues

Some nursing activities are prescribed by local and state laws as well as institutional policies. For example, some agencies do not permit do-not-resuscitate (DNR) orders; others require a primary care provider's signature for DNR, even though there may be an advance directive stating the client's wishes for DNR status.

**Advance directives:** All health care facilities receiving Medicare and Medicaid reimbursement must recognize advance directives. Ask clients whether they have advance directives, and provide educational materials regarding clients' rights to be involved in treatment decisions, including the right to refuse treatment. A living will (durable power of attorney for health care) provides specific instructions about what the client chooses to permit or refuse (e.g., mechanical ventilation) in the event the client later becomes unable to make those decisions. A health care proxy is a notarized statement appointing another person to manage health care decisions when the client is unable to do so.

**Death certificate:** Formal determination of death (pronouncement) must be made by a primary care provider or a coroner (and in some instances a nurse). By law, a death certificate (usually signed by the attending primary care provider) must be filed with a local health or other government office; the family usually receives a copy (e.g., to use for insurance claims).

**Autopsy (postmortem examination):** An examination of the organs and tissues of the body after death, to determine the exact cause of death, learn more about a disease, and accumulate statistical data. The law determines whether an autopsy must be performed (e.g., when a death is sudden or unexpected). Consent must be obtained from the decedent (before death) or from family members. Laws may differ as to who may give consent.

**Inquest**: A legal inquiry into the cause of death. For example, when the death appears to have been the result of a crime or an accident. Agency policy specifies who is responsible for reporting deaths to the coroner or medical examiner.

**Do-Not-Resuscitate (DNR, "no code") orders**: DNR orders are written by the primary care provider for clients who are in the terminal stage of irreversible illness and who have expressed the wish to not be resuscitated in the event of a respiratory or cardiac arrest. State laws and institutional policies vary with regard to DNR orders.

## Clinical Signs of Impending Death

**Muscle weakness** is manifested by difficulty speaking and swallowing, minimal body movement, incontinence due to loss of sphincter control, abdominal distention, and sagging of the jaw.

**Respiration change** is manifested by mouth breathing and/or rapid, shallow, irregular, or very slow respirations. **Cheyne-Stokes breathing** is a rhythmic increase and decrease in respirations from very deep to very shallow and temporary apnea. **Death rattle** is very noisy breathing caused by mucus collecting in the throat.

**Sensory changes**: Impaired taste and smell, blurred vision

**Circulatory changes**, including:

- Coldness of the skin of the extremities (first the feet, later the hands) and the nose. Subjectively, however, the client may feel warm because of elevated body temperature.
- Mottling and cyanosis of the extremities
- Slow, weak pulse
- Decreased blood pressure
- Loss of sensation

## Therapeutic Nursing Interventions for Post Mortem Care

- Determine that the person is dead (obtain a formal determination of death).
- Place body supine with arms folded over abdomen or extended at the side.
- Remove all medical equipment except drains, central lines or chest tubes (follow agency policy regarding removal of drains, central lines, and chest tubes; some agencies require that they be left in; others allow for removal).
- Make the area as clean and pleasant as possible (e.g., remove soiled linen, wipe up spills) before family and friends view the body.
- Place dentures in the mouth and close the eyes and mouth before rigor mortis sets in.
- Determine whether there are religious or cultural requirements for care of the body.
- Wash the body, placing disposable pads under the buttocks and between the legs.
- Allow the family private time to view the body and ask questions.

- Identify body by placing a tag on the ankle and on the shroud. In the hospital, the deceased's wrist identification tag is left on, as well.
- Wrap body in a shroud.
- Transport the body to the morgue or leave the body in the room until the mortuary personnel come for the body.

# Critical Thinking Exercise: Death and Post Mortem Care

1. A woman who is caring for her terminally ill mother at home asks what to expect as death approaches. In your attempt to prepare the family for an impending death, explain some of the common signs and symptoms that often accompany the dying process.

2. You are at the bedside of an 85-year-old client whose death appears imminent. Although the client is comatose, his son is present and asks if his father might still be able to hear. You encourage the son to speak to his father. Explain the appropriateness of this action.

3. The daughter of a critically-ill client has earlier met with her father's physician to discuss implementing a DNR. Later in the day, the daughter approaches you and expresses confusion about how having a DNR will affect the nursing care her father receives. How would you respond to her concern?

# Mobility - Exercise, Range of Motion and Positioning

### Key Points

- Exercise is physical activity for improving health, maintaining fitness, and conditioning the body.

- As a rule, people should engage in moderate intensity exercise for 30 minutes every day (e.g., walking, biking, and swimming).

- Immobility can create problems in all body systems; for example, contractures, thrombus formation, diminished cardiac reserve, pooling of respiratory secretions, negative nitrogen balance, urinary stasis, renal calculi, constipation, skin breakdown, exaggerated emotional reactions, impaired problem-solving.

- Factors affecting activity level and tolerance: growth and development, physical health, mental health, nutritional status, personal values/attitudes, environmental factors (e.g., a high-humidity climate), prescribed limitations (e.g., bed rest).

- When assisting with exercise, stop activity immediately if client experiences symptoms such as sudden facial pallor, dizziness, dyspnea, or chest pain.

- The range of motion (ROM) is the maximum movement that is possible for a given joint.

- Body movement involves four basic elements: posture (body alignment), joint mobility, balance (stability), and coordination.

- The flexor muscles are stronger than the extensor muscles.

- Active and/or passive range of motion exercises for immobile patients can prevent permanent shortening of the muscles (contractures).

- While exercising a limb, support it above and below the joint.

- Do not force a joint beyond its comfortable ROM; ROM should not be painful.

- If muscle spasms occur, stop the exercise temporarily, but continue applying gentle pressure on the limb until the muscle relaxes.

- **Key Concepts/Terms**: Isotonic (dynamic) exercises, isometric (static or sitting) exercises, isokinetic (resistive) exercises, aerobic exercise, anaerobic exercise, activity tolerance, osteoporosis, atrophy, contractures, joint ankylosis, Valsalva maneuver, thrombophlebitis, atelectasis, active ROM, passive ROM, active-assistive ROM

**Overview**

**Physical activity** is bodily movement produced by the skeletal muscles. It requires energy expenditure and improves health and fitness.

**Exercise** is physical activity for improving health, maintaining fitness, and conditioning the body.

Activity tolerance is the kind and amount of work or exercise a person is able to perform. The nurse assesses activity tolerance, for instance, when planning client activities such as walking, exercises, and activities of daily living.

# Exercise

## Benefits of Exercise

### Cardiovascular System

- Strengthens cardiac muscle
- Increases cardiac output
- Decreases resting heart rate
- Improves venous return

### Musculoskeletal System

- Maintains muscle tone and strength (mild exercise)
- Improves muscle tone and strength (strenuous exercise)
- Increases joint flexibility and range of motion
- Maintains bone density (through weight bearing)

### Respiratory System

- Improves alveolar ventilation
- Decreases breathing effort
- Improves diaphragmatic excursion
- Oxygen intake increases during strenuous exercise

### Gastrointestinal (GI) System

- Improves appetite
- Increases gastrointestinal tract tone, improving digestion and elimination

### Metabolic System

- Elevates basal metabolic rate
- Reduces serum triglycerides and cholesterol levels (by increasing the use of triglycerides and fatty acids)
- Increases use of glucose
- Increases production of body heat
- Burns excess calories

### Urinary System

- Promotes effective excretion of wastes
- Helps prevent urine stasis in the bladder

### Activity Tolerance

- Reduces fatigue
- Improves activity tolerance

### Psychosocial Factors

- Improves stress tolerance
- Produces a sense of well-being
- Reduces depression
- Improves body image
- Enhances quality of sleep
- Increases energy level

## Types of Exercise

**Isotonic (dynamic) exercises** are those in which the muscle shortens to produce muscle contraction and active movement (e.g., running, walking, activities of daily living, or using a trapeze to lift the body off the bed).

**Isometric (static or sitting) exercises** are those in which there is no joint or muscle movement; there is a change in muscle tension, but not in length. Isometric exercises are useful for strengthening abdominal, gluteal and quadricep muscles for use in ambulation.

**Isokinetic (resistive) exercises** are those in which the muscle is tensed or contracted against resistance (e.g., weight-lifting). They can be either isotonic or isometric.

**Aerobic exercise** is activity in which the body takes in oxygen equal to or more than the amount the body requires (e.g., walking, jogging, running, dancing, swimming). Aerobic exercises improve cardiovascular functioning and physical fitness.

**Anaerobic exercise** is activity in which the muscles cannot draw enough oxygen from the bloodstream; used for endurance training for athletes.

## Assistive Devices

Prior to selecting the device, the client must be assessed to identify if an assistive device is needed and which one may be indicated for that specific client.

- **Canes**—may be used to assist the client in ambulation. The client should be assessed for the type of cane she/he may need (straight-legged or standard, tripod or quad). The quad cane gives more stable support. All canes should have rubber tips or caps on the bottom(s) to prevent slipping of the cane. Canes may be the best choice for a client with problems affecting only one side of his/her body (but client must have the stability, balance, and strength to support him/herself safely).

- **Walkers**—used with clients who may need more support than a cane or canes may provide. A walker offers a wider base of support and may have wheels so the client with decreased upper body strength may push the device rather than have to lift and move it.

- **Crutches**—generally used by those with a temporary need or injury. Clients must be instructed in crutch walking to ensure their safety while using the devices. Clients should be instructed to support their weight with their arms, not their axillae, as this can cause radial nerve injury.

## Factors Affecting Activity Level and Tolerance

**Growth and development**: Newborn movements are reflexive and random. Gross motor development occurs in head-to-toe fashion. Between the ages of 6 and 12 years, motor skills are refined and exercise patterns for later life are generally determined. With advanced age, muscle tone and bone density decrease and reaction time slows.

**Physical health**: Any disorder that impairs the nervous system, musculoskeletal system and inner ear (vestibular apparatus) affects mobility and activity tolerance (e.g., muscular dystrophy, brain tumors, strokes, meningitis, head injury, inner ear infection, joint replacements). Illnesses that limit the supply of oxygen and nutrients to the muscles can decrease activity tolerance (e.g., congestive heart failure, angina, anemia, chronic obstructive pulmonary disease).

**Mental health**: Mental disorders such as depression or chronic stress may reduce the person's desire for activity. Chronic stress depletes energy reserves.

**Nutrition**: Poor nutrition can cause muscle weakness and fatigue. An obese person must expend extra energy to move, thus decreasing activity tolerance.

**Personal values/attitudes**: When a family incorporates regular exercise into their daily routine, children come to value physical activity. If the family takes part only in spectator sports or other sedentary activities, this lifestyle is usually transmitted to the children.

**External factors**: Such factors as a high-temperature, high-humidity climate and an unsafe neighborhood discourage outdoor activity. The availability of recreational facilities is also a factor, as is the financial ability to take part in certain programs.

**Prescribed limitations**: Some medical therapies limit the ability to move (e.g., casts, braces, splints, traction, and bed rest). Bed rest may vary in meaning. It may mean strict confinement to bed or it may mean that the client is allowed to use a bedside commode.

## Potential Effects of Immobility

Whether immobility causes problems for the client depends on the client's health status, duration of the inactivity, and the client's sensory awareness.

**Musculoskeletal system**: Disuse osteoporosis (bone demineralization with calcium loss), disuse atrophy (decrease in muscle size), contractures (permanent shortening of the muscle) and eventual joint deformity, stiffness and joint pain (from calcium deposits), and joint ankylosis (permanently "frozen" joint) can result from immobility.

**Cardiovascular system**

- Diminished cardiac reserve (resting heart rate increases approximately 1/2 beat per minute for each day of immobilization): rapid heart rate, reduced diastolic pressure, and reduced coronary blood flow may produce tachycardia and angina even with minimal exertion.

- Increased use of the Valsalva maneuver (holding the breath and straining against a closed glottis while moving, as when attempting to pull one's self up in bed). This increases intrathoracic pressure and interferes with blood return to the heart and can result in arrhythmias in clients with cardiac disease.

- Orthostatic (postural) hypotension

- Peripheral venous vasodilation and stasis

- Dependent edema, resulting from peripheral venous stasis
- Thrombus formation: Thrombophlebitis is a clot that is loosely attached to an inflamed vein wall. Predisposing factors are: impaired venous return to the heart, hypercoagulability of the blood, and injury to a vessel wall.

**Respiratory system**: Decreased respiratory movement (an immobile, paralyzed client can lose up to 50% of normal vital capacity), pooling of respiratory secretions (which interferes with normal diffusion of oxygen and carbon dioxide in the alveoli), inability to cough up secretions (due to loss of respiratory muscle tone and dehydration), atelectasis, and hypostatic pneumonia

**Metabolic system**: Decreased metabolic rate, negative nitrogen balance, anorexia, and negative calcium balance (loss from bones). Immobility creates an increase in catabolism (protein breakdown) over anabolism (protein synthesis).

**Urinary system**

- Urinary stasis (in a horizontal position, gravity impedes the emptying of urine from the kidneys and the urinary bladder)
- Urinary retention and occasionally urinary incontinence
- Urinary infection
- Renal calculi (immobility results in excessive amounts of calcium and phosphate in the urine; the urine is more alkaline and the calcium salts precipitate out as crystals to form calculi)

**Gastrointestinal system**: Constipation due to decreased peristalsis and weakness of abdominal and perineal muscles.

**Integumentary system**: Reduced skin turgor (elasticity) and skin breakdown (see "*Pressure Ulcers*").

**Neuropsychological system**: Low self-esteem, frustration, exaggerated emotional reactions (e.g., apathy, withdrawal, regression, anger, aggression); difficulty perceiving time intervals; impaired ability to make decisions and solve problems; anxiety; impaired social and motor development of young children

## Related NANDA Nursing Diagnoses

- Activity Intolerance
- Risk for Activity Intolerance
- Impaired Physical Mobility
- Impaired Bed Mobility
- Impaired Walking
- Impaired Wheelchair Mobility
- Impaired Transfer Mobility
- Risk for Disuse Syndrome

**NOTE: Impaired Physical Mobility is often the etiology of other problems, such as:**

- Fear of Falling
- Risk for Injury (falls)
- Self-Care Deficit
- Impaired Home Maintenance Management

## Therapeutic Nursing Interventions for Mobility and Exercise

### Assessments

- Assess body alignment, posture, gait, and joint movement.
- Identify factors that may hinder mobility (e.g., mental alertness, balance and coordination, pain, encumbrances such as intravenous lines or a cast, effect of the client's illness, vision, and medications such as narcotics).
- Obtain data regarding heart rate, strength, and rhythm; respiratory rate, depth, and rhythm; and blood pressure at the following times: before the activity, during the activity, immediately after the activity, and three minutes after the activity has stopped.
- Stop activity immediately if the client experiences: (1) sudden facial pallor, (2) dizziness or weakness, (3) heart rate or respiratory rate that is significantly higher than baseline, (4) change in heart or respiratory rhythm from regular to irregular, (5) dyspnea or shortness of breath, (6) chest pain, (7) weak, thready pulse, or (8) decrease of 10 mm/Hg in diastolic blood pressure.

### Client Teaching

- Teach clients guidelines for physical activity:
  - **Frequency**: Preferably every day
  - **Duration**: 30 minutes per day (can be divided throughout the day)
  - **Intensity**: "Moderate" intensity (e.g., the person should be able to carry on a conversation, but with some labored breathing)

**Types of exercise**: Walking, biking, and swimming (recommended for beginners and older adults). More strenuous exercise includes jogging, running, and jumping rope.

- Teach safe transfer and ambulation techniques.
- Teach safety measures to avoid falls (e.g., locking wheelchairs, using a raised toilet seat, installing grab bars, using rubber crutch tips, wearing appropriate footwear).
- Discuss ways to minimize fatigue (e.g., performing activities more slowly, resting more often).
- Teach ways to increase energy (e.g., intake of high-energy foods, adequate rest and sleep, pain control).
- Encourage client to use a progress graph or chart to promote adherence with the exercise regimen or therapy.

### Other

- Intersperse rest periods with activity periods to increase tolerance.
- Discuss pain control measures required before exercise.

# Range of Motion

## Overview

Mobility is the ability to move freely and purposefully. Mobility is essential in order for people to meet their basic needs, maintain independence, and have a good self-concept. Normal movement and stability require intact musculoskeletal and nervous systems and intact inner ear structures that control equilibrium. Body movement involves four basic elements: (1) posture (body alignment), (2) joint mobility, (3) balance (stability), and (4) coordination.

The flexor muscles are stronger than the extensor muscles. Therefore, when a client is inactive, the flexors pull the joints into a bent position. Active and/or passive range of motion exercises must be used to counteract this tendency and prevent permanent shortening of the muscles, which would "freeze" the joint into a flexed position. Joint flexibility can be maintained by activities of daily living (e.g., walking, bathing, dressing) and by active and passive range of motion exercises.

## Joint Movements (Range of Motion)

The **range of motion (ROM)** is the maximum movement that is possible for a given joint. The usual ranges of joint movement are as follows:

**Neck** (pivot joint):

- Flexion = 45° from midline
- Extension = 45° from midline
- Lateral flexion = 40° from midline
- Rotation = 70° from midline

**Shoulder** (ball and socket joint):

- Flexion = 180° from the side
- Extension = 180° from vertical position beside the head
- Hyperextension = 50° from side position
- Abduction = 180° from a neutral position
- Adduction (anterior) = 230°
- Circumduction = 360°
- External rotation = 90°
- Internal rotation = 90°

**Elbow** (hinge joint):

- Flexion = 150°
- Extension = 150°
- Rotation for supination = 70-90°
- Rotation for pronation = 70-90°

**Wrist** (condyloid joint):

- Flexion = 80-90°
- Extension = 80-90°

- Hyperextension = 70-90°
- Radial flexion (abduction) = 0-20°
- Ulnar flexion (adduction) = 30-50°

**Hand and Fingers** (metacarpophalangeal joints are condyloid; interphalangeal joints are hinge):

- Flexion = 90°
- Extension = 90°
- Hyperextension = 30°
- Abduction = 20°
- Adduction = 20°

**Thumb** (saddle joint):

- Flexion = 90°
- Extension = 90°
- Abduction = 30°
- Adduction = 30°
- Opposition (touch thumb to top of each finger of the same hand)

**Hip** (ball and socket joint):

- Flexion with knee extended = 90°
- Flexion with knee flexed =120°
- Extension = 90-120°
- Hyperextension = 30-50°
- Abduction = 45-50°
- Adduction = 20-30° beyond other leg
- Circumduction = 360°
- Internal rotation = 90°
- External rotation = 90°

**Knee** (hinge joint):

- Flexion = 135-145°
- Extension = 0° (Return leg to floor)

**Ankle** (hinge joint):

- Extension (plantar flexion) = 45-50°
- Flexion (dorsiflexion) = 20°

**Foot** (gliding):

- Eversion = 5°
- Inversion = 5°

**Toes** (interphalangeal and metatarsophalangeal joints are hinge; intertarsal joints are gliding):

- Flexion = 35-60°
- Extension = 35-60°

- Abduction = 0-15°
- Adduction = 0-15°

**Trunk** (gliding):

- Flexion = 70-90°
- Extension = 70-90°
- Hyperextension = 20-30°
- Lateral flexion = 35° on each side
- Rotation = 35-40°

## Range of Motion Exercises

**Active ROM exercises** are isotonic (muscle tension is constant and the muscle shortens to produce muscle contraction and active movement). The client moves each joint through its complete range of motion. In addition to preventing loss of joint movement, active ROM exercises increase muscle strength and endurance. The nurse instructs the client to perform each exercise three times and to do the entire series twice a day. Joints should be moved to the point of slight resistance, but not to the point of discomfort.

**Passive ROM exercises** are those in which the nurse or therapist moves each of the client's joints through its complete range of motion. Passive ROM will maintain joint flexibility but is of no value in maintaining muscle strength. Therefore, it should be used only for movements that the client cannot achieve independently. Passive ROM exercises prevent shortening of muscles, ligaments, tendons, and joint capsules, thus preventing joint deformities or contractures. These exercises also help prevent hypostatic pneumonia, thrombophlebitis, footdrop, circulatory problems, skin breakdown, and fecal impaction. They stimulate circulation and sensory nerve endings, increase endurance, and promote a sense of well being. Possible contraindications for passive ROM for some joints are acute arthritis, septic joints, a bone fracture, a head injury, a torn ligament, joint dislocation or subluxation, acute cardiovascular pathology, and acute thrombophlebitis.

**Active-assistive ROM exercises** are those in which the client uses a stronger, opposite arm or leg to move of the joints of an immobile limb. The client moves the joint as much as he or she is able and the nurse then continues the movement passively to its maximal degree.

## Procedure and Technique for Passive ROM Exercises

### General Guidelines

- Use proper body mechanics to avoid injury to self and client.
- Work systematically from the client's head to toes.
- While exercising a limb, support it above and below the joint to prevent muscle strain.
- Move the body parts slowly and smoothly. Jerky movements can cause discomfort and injury. Fast movements may cause muscle spasms.
- Do not force a joint beyond its comfortable range of motion—ROM should not be painful.

- Complete each exercise maneuver 3 times, completing each set of repetitions before proceeding to the next joint.
- If muscle spasms occur, stop the exercise temporarily, but continue to apply gentle pressure on the part until the muscle relaxes.
- If the client becomes tired, provide rest periods between exercises.

## Procedure

- Wash your hands
- Explain to the clients the reasons for the ROM exercises.
- Dress the client in a loose gown and cover with a bath blanket.
- Use proper body mechanics to avoid injury to self and client.
- Position the bed at a comfortable height.
- Expose only the limb being exercised.
- Support the client's head and flex his neck so that his chin moves toward his chest, then tilt his head backward, and then toward each shoulder.
- Exercise the shoulders. Support the client's elbow with one hand and hold his wrist with the other. Then raise the client's arm above his head, move his arm across his chest, and externally rotate his shoulder by moving his arm away from him.
- Exercise the client's elbows by placing the client's arm at his side with his palm upward, then flexing and extending his forearm.
- Exercise the forearm by stabilizing the elbow, then turning the palm downward and then upward.
- Exercise the wrist. Stabilizing his forearm, flex and extend the client's wrist, move it from side to side, then rotate it.
- Exercise the fingers, extending and flexing them, abducting and adducting them, and rotating the thumb.
- Exercise the client's hips and legs. Flex and extend the hip while suppporting the leg and bringing the knee toward the chest, allowing full flexion of the knee.
- Adduct and abduct the client's hip.
- Exercise the client's leg by turning the leg inward and then outward.
- Exercise the ankles by bending the client's foot upward for dorsiflexion and downward for plantar flexion. Rotate the ankle in a circular motion.
- Flex and extend each toe, then abduct and adduct each one.
- Move to the other side of the client's body and starting at the shoulder, repeat each exercise that has been performed for each joint.
- After completing passive ROM exercises, ensure that the client is comfortable, check vital signs, lower the bed, and place the client's signaling device within reach.
- Wash your hands.
- Document your assessment of the client's range of motion and response to the exercises.

- Note whether or not the client experienced joint stiffness, joint pain, spasms, or extreme fatigue.
- Record client's vital signs.
- Compare your assessment with the client's baseline data and earlier assessments.
- Arrange for home health care or physical therapy following discharge if ordered.
- Teach the client's family to perform passive ROM exercises safely.

# Positioning

**Overview**

Normally, people automatically reposition themselves for comfort. However, people who are paralyzed, unconscious, weak, injured, or in pain require assistance with turning, repositioning, and maintaining proper body alignment. Position changes and proper body alignment are essential in order to maintain muscle tone and stimulate postural reflexes, as well as to prevent muscle discomfort, injuries to the skin, and contractures. For clients who cannot move independently, it is best to have two or more people move and turn the client in order to reduce the risk of muscle strain and injury to both client and nurse.

**Proper body alignment:** When a client is lying down, the vertebrae should be in a straight line with no noticeable curves; the extremities should be aligned and not crossed over one another; the head and neck should be aligned without excessive flexion or extension.

## Common Alignment Problems

These problems can be prevented or corrected with proper positioning and use of support devices:

- Flexion of the neck
- Internal rotation of the shoulder
- Adduction of the shoulder
- Flexion of the wrist
- Anterior convexity of the lumbar spine
- External rotation of the hips
- Hyperextension of the knees
- Plantar flexion of the ankles

### Associated NANDA Nursing Diagnoses

- Potential for Injury (e.g., muscle strain, contractures)
- Risk for Impaired Skin Integrity
- Impaired Physical Mobility

## Positioning Devices

**Pillows:** Pillow must be the proper size: a thick pillow under the head may cause too much cervical flexion; a thin pillow under bony prominences may not adequately protect against tissue damage.

**Footboard:** Prevents foot drop by keeping foot in dorsiflexion; be sure it is placed so the client's feet reach and touch the board. A foot boot may be used instead.

**Trochanter roll:** Prevents external rotation of the hips when the client is in supine position. Fold a bath blanket lengthwise to extend from the waist to the lower border of the knee. Place under the client and roll it counterclockwise until the thigh is in neutral position or in inward rotation.

**Sandbags:** Sand-filled plastic tubes that can be used in place of a trochanter roll or to immobilize an extremity.

**Hand rolls:** Maintains the thumb in slight adduction and in opposition to the fingers (in "functional" position). Make from a washcloth or a roll of gauze.

**Hand-wrist splints:** Maintain the thumb and wrist in proper alignment (slight adduction and slight dorsiflexion, respectively). They are molded specifically for each client and cannot be used on other clients.

**Trapeze bar:** Allows the client to use upper extremities to raise the trunk off the bed or perform upper arm-strengthening exercises.

**Bed boards:** Plywood boards placed under the entire mattress to provide back support and alignment.

## Positioning

**Fowler's Position:** Head of bed elevated 45-90° (semi-Fowler's or low Fowler's is about a 30° elevation of the head of the bed); knees slightly elevated. This position allows for better chest expansion, lung ventilation, and (at 60-90° of elevation) provides better dependent drainage postop abdominal surgeries.

| Common Problems | Prevention |
|---|---|
| • Flexion of lumbar spine<br>• Increased cervical flexion<br>• Hyperextension of the knees<br>• External rotation of the hips<br>• Arms hanging unsupported at sides causing shoulder strain and edema of hands and arms<br>• Foot drop (plantar flexion)<br>• Pressure points at sacrum and heels | • Pillow or pad under lumbar region<br>• Be sure pillow under head is not too thick.<br>• Small pillow or pad under thighs<br>• Trochanter roll; use sandbags if necessary<br>• Place arm and hand on over bed table in front of client or place arm away from client's side and place pillow under elbow (**orthopneic** position).<br>• Footboard at bottom of feet<br>• Small pillow or roll under ankles |

**Supine Position (Dorsal recumbent):** Client lies on back with head and shoulders slightly elevated on a small pillow.

| Common Problems | Prevention |
|---|---|
| • Cervical flexion<br>• Hyperextension of neck<br>• Shoulders internally rotated<br>• Elbows extended<br>• Thumb not in functional position<br>• Hyperextension of knees<br>• External rotation of the hips<br>• Foot drop<br>• Unsupported lumbar area<br>• Unprotected pressure points | • Be sure pillow is not too thick.<br>• Pillow under upper shoulders, neck, and head<br>• Pillows under pronated forearms, keeping upper arms parallel to client's body<br>• Use hand rolls.<br>• Small pillow under thighs<br>• Use trochanter rolls or sandbags at thighs.<br>• Use pillows or footboard against feet.<br>• Small rolled towel under lumbar area<br>• Place small pillow or roll under ankle to elevate heels. |

**Prone Position:** Client lies on abdomen with head turned to one side. Hips are not flexed. Allows for full extension of the hip and knee; promotes drainage from the mouth (e.g., for clients recovering from throat surgery). Chest expansion is inhibited in this position; it is not practical for clients with cardiac or respiratory problems or for those with spinal abnormalities

| Common Problems | Prevention |
|---|---|
| • Cervical flexion/hyperextension <br> • Pressure on breasts (females) <br> • Plantar flexion of feet and pressure <br> • Hyperextension of lumbar vertebrae | • Small pillow under head unless there is a need to promote drainage from the mouth <br> • Small pillow under abdomen below diaphragm <br> • Allow toes to hang over end of mattress or place pillow under lower legs so toes do not touch the bed <br> • Small pillow under abdomen below diaphragm |

**Lateral (Side-Lying) Position:** Bed is flat; client rests on side with most of body weight on the dependent hip and shoulder; usually the top hip and knee are flexed and placed in front of the body. Both arms are flexed in front of the body. A pillow may need to be tucked behind the back to prevent client from rolling onto his or her back. A good sleeping position.

| Common Problems | Prevention |
|---|---|
| • Lordosis of lumbar spine <br> • Internal rotation and adduction <br> • Strain on sternocleidomastoid muscle <br> • Internal rotation and adduction <br> • Foot drop | • Flex top hip and knee and place in front of body. <br> • Pillow under upper arm to keep it level with shoulder <br> • Place pillow under top leg from groin to foot so femur, hips and shoulders are aligned <br> • Sandbag parallel to plantar surface of dependent foot |

**Sims (Semi-Prone) Position:** Client is on side, halfway between the lateral and prone positions; weight is on anterior ileum, humerus, and clavicle. The lower arm is positioned behind the client and the upper arm is flexed in front. The upper leg is flexed. This position facilitates drainage from the mouth and reduces pressure over the sacrum and greater trochanter. A comfortable sleeping position for many people.

| Common Problems | Prevention |
|---|---|
| • Lateral neck flexion <br> • Internal rotation of shoulder <br> • Internal rotation of hip and adduction of leg <br> • Foot drop | • Small pillow under head <br> • Pillow under upper flexed leg, keeping it level with hip <br> • Sandbag parallel to plantar surface of lower foot |

# Critical Thinking Exercise: Mobility - Exercise, Range of Motion and Positioning

1. You are preparing a client with a total hip replacement for discharge from an acute care facility. With respect to mobility, what home care considerations apply to this client?

2. A healthy 45-year-old woman asks your advice about starting an exercise program. Because of the demands of caring for family and work, she admits to being "a real couch potato" and wonders if a membership at a local gym might help her conquer her sedentary habits. How might you respond?

3. A client whose physician has prescribed antilipidemics has also been advised on the importance of regular exercise. The client asks you: "I really don't have time to exercise. Isn't it enough that I take the medication?" What might you say to this client about the importance of exercise?

# Sleep and Rest

## Key Points

- Factors affecting sleep: developmental stage (amount of sleep needed decreases gradually with age), emotional stress, sleep patterns (e.g., shift work), illness, environmental factors (e.g., noise), drugs, and other substances.
- L-tryptophan in milk, poultry, and meat promotes sleep.
- Hypnotics provide only temporary increase in sleep and have a "hangover" effect.
- Nursing interventions include:
  - Preparing a restful environment
  - Offering non-caffeinated bedtime snacks (e.g., milk)
  - Continuing client's usual bedtime rituals
  - Advising restful, quiet activities (e.g., a warm bath, reading) prior to bedtime
  - Older adults sleep about six hours per night; they have decreased REM and Stage 4 sleep
- Adequate rest and sleep are important factors in general health and recovery from illness.
- A person who is rested shows both increased pain tolerance and a greater response to analgesia.
- Normal sleep requirements vary from individual to individual.
- **Key Concepts/Terms**: Rest, sleep, circadian (diurnal) rhythm, biological clock, NREM sleep, REM sleep, sleep disorder, sleep deprivation, insomnia, sleep apnea (obstructive and central), narcolepsy, soomnambulism, night terrors, nocturnal enuresis, polysomnograph

## Overview

Rest is a state of decreased physical and mental activity that leaves a person feeling refreshed. **Sleep** is a sustained altered state of consciousness that involves a rhythmic pattern of repeated cycles. During sleep, the person's energy and well-being are restored. Sleep can be disturbed by illness and by the environment and disruptions in the health care institution. There is no specific number of hours of sleep needed to assure adequate rest, it varies with the individual. Sleep disorders can cause daytime sleepiness, fatigue, depression, and anxiety.

**Circadian (diurnal) rhythm** is the 24-hour, day-night cycle that influences major biological and behavioral functions (e.g., fluctuations of temperature, heart rate, blood pressure, hormone secretion, sensory acuity, and mood). Each individual has a **biological clock** that synchronizes the sleep cycle, the time of day they fall asleep and the time of day they function best.

## Sleep Cycle

Sleep is characterized by two phases: **NREM-nonrapid eye movement** and **REM-rapid eye movement**. A person cycles several times during these phases while sleeping:

**NREM sleep**, the lightest sleep, consists of four stages:

- **Stage 1**: Light sleep; easily aroused; lasts a few minutes
- **Stage 2**: Deeper relaxation; lasts 10-20 minutes
- **Stage 3**: Early phase of deep sleep; difficult to arouse; little or no movement; lasts 5-30 minutes
- **Stage 4**: Deep sleep; shortens toward morning; lasts 15-30 minutes

**REM sleep**, the period of sound sleep, cycles every 50-90 minutes after sleep occurs; averages 20 minutes, but duration increases with each cycle. Sleeper is hard to arouse; has darting eye movements, changes in vital signs, loss of skeletal muscle tone, and pauses in breathing for 15-20 seconds. This is the stage of vivid, full-color dreaming.

## Factors Affecting Sleep

**Developmental considerations**: The amount of total sleep a person needs decreases gradually from infancy to old age.

**Emotional stress/anxiety**, including loss and grieving, can cause loss of sleep or oversleep.

**Sleep patterns**: Examples are changes in circadian rhythm, and sleep deprivation from social activities or performing shift work.

**Illness**: Examples are nocturia (urination at night), leg cramps, pain, anxiety, depression, thyroid disease, and chronic lung disease.

**Environmental factors**: For example, noise, light, ventilation, room temperature, uncomfortable bed, presence of a bed partner.

**Drugs and other substances**: L-tryptophan (in milk products and meat) promotes sleep. Hypnotics provide only temporary increase in sleep and have a "hangover" effect (excess drowsiness, confusion, decreased energy); they may worsen sleep apnea. Diuretics cause nocturia. Antidepressants and stimulants suppress REM sleep. Alcohol speeds onset of sleep, but disrupts REM sleep. Caffeine prevents the person from falling asleep and may cause awakening during the night. Digoxin causes nightmares. Beta-blockers cause nightmares, insomnia, and make it difficult to stay asleep.

## Sleep Disorders

**Sleep disorder**: A condition that causes continued disruption in the pattern of sleep

**Sleep deprivation**: Involves decreases in the amount, quality, and consistency of sleep. Common in hospitalized clients, it is not a true sleep disorder. The two types of sleep deprivation are REM deprivation and NREM deprivation.

- **REM deprivation** is caused by alcohol, barbiturates, morphine, Demerol, jet lag, shift work, and extended time as a client in an intensive care unit. It is characterized by excitability, irritability and emotional lability.
- **NREM deprivation** can be caused by all of the preceding, as well as diazepam (Valium), flurazepam hydrochloride (Dalmane), hypothyroidism, depression, respiratory disorders, sleep apnea, and advanced age. It is characterized by apathy, hyporesponsiveness, excessive sleepiness, and speech deterioration.

**Insomnia**: Chronic difficulty falling asleep, remaining asleep, or inability to go back to sleep after awakening

**Sleep apnea** is cessation of breathing for short periods of time during sleep, causing the blood oxygen level to fall and creating a risk for cardiac dysrhythmias, hypertension, and right heart failure. **Obstructive sleep apnea** consists of upper airway blockage (lasting as long as a minute), but with continued chest and abdominal breathing movements; it is usually accompanied by snoring. This type of apnea can be treated by weight reduction, surgical removal of the tonsils and uvula, and use of a continuous positive airway pressure mask (CPAP). **Central sleep apnea** is caused by a defect in the brain's respiratory center, causing failure of the impulse to breathe; this results in cessation of both nasal airflow and chest movements.

**Narcolepsy** is sudden onset of sleep, during the day, which cannot be controlled. The person may feel an overwhelming wave of sleepiness. Sleep attacks can be mistaken for laziness or drunkenness. Brief daytime naps, abstinence from alcohol and avoiding exhausting activities may help to reduce the attacks. Medical treatment includes stimulants and medications that suppress REM sleep.

**Somnambulism**: Sleepwalking occurs during stages 3 and/or 4 of NREM sleep. Sleepwalkers should be gently awakened and led back to bed.

**Night terrors**: During stage 3 sleep, a child sits up, shakes, screams, seems very frightened, and is difficult to calm. (Rare in children over the age of six years.)

**Nocturnal enuresis**: Involuntary urination during the night. It occurs before REM sleep.

## Diagnostic Test for Sleep Disturbance Disorders

Polysomnograph evaluations uses electrodes to measure a sleeping client's brain waves, extraocular eye movements, and chin and facial muscle movements.

## Assessments

It may be helpful to have the client keep a sleep log that documents bed times and activities before bedtime, etc. A sleep history should include:

- A description of the sleeping problem (e.g., difficulty falling asleep at night)
- The nature of the sleep disturbance (e.g., the source of the problem, for example worrying about money)
- Severity of the problem
- Daytime symptoms
- The client's usual sleep pattern
- Current life events and stressors
- Medical history
- Emotional and mental status
- Bedtime rituals (e.g., does the client take anything to help him sleep?)
- Bedtime environment (e.g., is the bedroom dark and quiet?)
- Sleeping partner
- Symptoms of sleep deprivation (e.g., irritability, slurred speech, disorientation, delusions, paranoia); chronic sleep problems usually cause milder symptoms.

## Associated NANDA Nursing Diagnoses

- Activity Intolerance
- Altered Thought Processes (related to sleep deprivation)
- Fatigue
- Risk for Injury (related to sleepwalking or narcolepsy)
- Sleep Deprivation
- Sleep Pattern Disturbance

## Therapeutic Nursing Interventions for Sleep and Rest

- Prepare a restful environment (e.g., keep room temperature at 65-69.8° F, control noise and light).
- Promote bedtime rituals (e.g., physical exercise should be performed at least two hours before bedtime; relaxation may be promoted by a bedtime snack).
- Offer non-caffeinated bedtime snacks. Milk products have a sedative effect.
- Promote relaxation and comfort (e.g., give the client a massage; give analgesics for pain; at home take a warm bath, read, or do a quiet activity until becoming drowsy).
- Schedule nursing care to avoid unnecessary disturbances.
- Eliminate caffeine, smoking, alcohol and exercise before bedtime.
- Observe client's usual bedtime hours.
- Avoid long naps during the day.
- Give sleep medications only as a last resort because many alter REM or NREM sleep and cause morning hangover and daytime drowsiness. Tolerance can develop if these medications are used for several weeks; can cause rebound insomnia.
- If barbiturate sedative-hypnotics are used to promote sleep, they must be withdrawn gradually.
- Abrupt withdrawal can create severe withdrawal symptoms (including seizures) and even death.

## Age-Related Changes—Gerontological Considerations

- Older adults sleep about six hours a night.
- REM sleep decreases slightly and the first REM period is longer.
- Older adults have more nightmares related to a decrease in REM sleep.
- Stage 4 sleep is markedly decreased and in some cases absent. This means that older adults have less restorative sleep.
- Older adults do not sleep as deeply and are more likely to be disturbed by noise.
- Older adults have greater difficulty falling asleep.
- Older adults have more side effects from sedatives and hypnotics related to decreased kidney and liver function. For example, they are more prone to drowsiness and sedation the morning after receiving a hypnotic.

# Critical Thinking Exercise: Sleep and Rest

**Situation:** Following the recent death of her spouse, a healthy 70-year-old woman expresses concern about her inability to sleep restfully at night. After several hours of lying awake, she eventually falls asleep. By the next morning, she is so exhausted that she remains in bed until approximately 10 or 11 a.m. What advice might you offer?

1. What nursing diagnosis applies in this situation?

2. What interventions might be most helpful with this woman?

# Pain Management

## Key Points

- Pain is an unpleasant sensation caused by noxious stimulation of sensory nerve endings, usually as a result of disease or injury.
- Pain signals that something is wrong.
- Pain is subjective; it is whatever the experiencing person says it is.
- The pain experience consists of pain perception, pain reception and modulation, and pain reaction/responses.
- Moderate, superficial pain produces a sympathetic response.
- Severe, deep pain produces a parasympathetic response.
- Infants do experience pain; children are less able than adults to articulate their pain and are often undertreated.
- Assess location, intensity, quality, pattern, precipitating factors, alleviating factors, associated symptoms, effects on activities of daily living, coping resources and strategies, effective responses, behavior, and physiological responses.
- Provide pain relief measures before pain becomes severe.
- Regular administration of narcotic analgesics for pain relief does not lead to drug dependence.
- Choose analgesic on the basis of severity of pain—administer the mildest analgesic and the lowest dose that is effective.
- **Key Concepts/Terms**: Acute pain, chronic pain, the body's analgesia system, gate control theory, pain threshold, pain tolerance, pain reaction, pain response/behavior, analgesic, NSAIDs, opioid (narcotic) analgesics, AHCPR guidelines for acute pain management

**Overview**

**Pain** is an unpleasant sensation caused by noxious stimulation of sensory nerve endings, usually as a result of disease or injury. Pain is adaptive in that it signals that something is wrong. Pain is subjective and highly individualized: "Pain is whatever the experiencing person says it is, existing whenever he says it is and does". Pain is a leading cause of disability and is often the reason for seeking health care.

**Acute pain** (e.g., that caused by a sore throat or surgical incision) is rapid in onset, varies in intensity from mild to severe, and lasts for up to six months. Once the underlying cause is resolved, acute pain disappears.

**Chronic pain** (e.g., the pain of cancer or arthritis) lasts for six months or longer and interferes with normal functioning. It may be intermittent or persistent. Unlike acute pain, it seems to have no protective purpose and may lead to depression, anger, withdrawal and dependency.

## The Pain Experience

The pain experience includes reception and transmission of the pain stimulus, perception/ modulation of the pain, and the person's responses to the pain.

### Reception

In the presence of noxious stimuli, the injured tissue releases **bradykinin**, **prostaglandins**, and **substance P**, which sensitize the pain receptors and increase the inflammatory response. Pain impulses travel to the dorsal horn of the spinal cord by two types of fibers (the large, myelinated **A-delta fibers** relay localized impulses rapidly; they are associated with sharp, pricking pain; the smaller, unmyelinated **C fibers** transmit impulses more slowly and create long-lasting, burning or aching pain).

### Perception

Pain impulses travel up the spinal cord to the thalamus, and then on to the cortex, where pain is perceived. Perception occurs when the person becomes aware of the pain.

- **Modulation, "The Body's Analgesia System"**: When regions of the midbrain are stimulated, endogenous opioids (enkephalins, dynorphins, and beta endorphins) are released. Morphine-like in their actions, these neuromodulators bind with opiate receptor sites in the nervous system, decreasing or blocking pain impulses and pain perception. In chronic pain, endorphins often do not function.

- **Modulation, "The Gate Control Theory"**: This theory suggests that pain impulses can be blocked by a "gate" between transmission cells in the dorsal horn of the spinal cord. Sensory impulses traveling by large A and small C fibers pass through this gate. Only a certain amount of information can be processed by the central nervous system at a given moment; therefore, when too much information is sent through, the gate "closes." High activity in the large A fibers, which are stimulated by heat, cold, and touch, closes the gate to the small C fibers, which are stimulated by pain. Gating mechanisms may also be activated by thoughts, feelings, and distraction (e.g., by auditory or visual stimuli).

### Reaction/Responses

The **pain threshold** is the amount of stimulation the person needs in order to feel pain; the pain threshold is similar in most people. The degree to which pain is perceived depends on the interaction between the nervous system's transmission of stimuli and the body's analgesic system. **Pain tolerance** is the maximum amount and duration of pain that a person is willing to endure. Pain tolerance varies greatly among people. **Pain reaction** includes the physiological, behavioral, and affective responses to pain. The severity and duration of pain affect responses to pain (e.g., mild, brief pain may produce little or no behavioral responses; intense pain usually causes reflex action to escape the cause).

### Physiological responses

**When pain is moderate and superficial, a sympathetic response occurs**:

- Increased respiratory rate and dilation of bronchial tubes
- Increased heart rate
- Pallor and increased blood pressure (due to peripheral vasoconstriction)
- Increased blood glucose
- Increased output of adrenaline

- Diaphoresis
- Muscle tension
- Dilated pupils
- Decreased gastric motility

**Physiological Responses**

**When pain is severe and deep, a parasympathetic response occurs:**

- Pallor
- Rapid, irregular breathing
- Nausea and vomiting
- Weakness, exhaustion, fatigue
- Fainting, unconsciousness
- Prostration
- Decreased heart rate, decreased blood pressure

**Behavioral Responses**

- Moving away from the painful stimuli
- Clenching the teeth
- Holding the painful part
- Grimacing
- Bending over
- Tensing the abdominal muscles
- Crying, moaning
- Refusing to move
- Restlessness
- Making frequent requests of the nurse

**Affective Responses**

- Withdrawal
- Anxiety, fear
- Depression
- Anger
- Anorexia
- Hopelessness
- Powerlessness
- Stoicism

## Factors Affecting Pain Response and Pain Tolerance

**Culture** influences the meaning of the pain to the individual, as well as appropriate behaviors. For example, in some cultures, tolerance of pain signifies strength and endurance; in other cultures, family and friends are sympathetic to those who are vocal about their pain.

**Developmental stage** (see *"Age-Related Changes"*)

- Pain transmission is present in newborns.
- Children are less able than adults to articulate their pain, and are often undertreated.

**Meaning of pain**: If the client associates pain with a positive outcome (e.g., the pain of childbirth), it is viewed as a temporary inconvenience rather than a catastrophe, so the suffering is less. Unrelenting, chronic pain may cause greater suffering because the person cannot attach a positive purpose to it. Religious beliefs influence the meaning the individual attaches to pain (e.g., in some religions suffering is a means of purification or of atoning for one's sins). Faith can help the client cope with pain; on the other hand, pain can shake the client's faith.

**Environment**: Perception of pain can be increased by the strangeness, noise, constant activity, and sense of powerlessness that accompany admission to a hospital.

**Support people**: A lonely person may perceive more pain than one with supportive people around (e.g., a toddler will tolerate pain better when a parent is nearby). Expectations of significant others can also affect perceptions of and responses to pain (e.g., girls may be permitted to express pain more openly than boys).

**Anxiety, fatigue, and other stressors**: Pain almost always causes anxiety, and anxiety tends to increase the perceived intensity of pain. Pain is also increased when fatigue and muscle tension are present. A rested, relaxed person can cope better with pain.

**Past pain experiences:**

- Someone who has experienced pain may be more anxious about anticipated pain than someone with no pain experience, who may not realize how intense it can be.
- Past effectiveness or lack of effectiveness of pain relief measures influence the client's attitude and expectations of pain relief.
- In general, those who have experienced an unusual amount of pain during their lifetimes tend to have increased sensitivity to pain.

**Psychological variables**: Feelings of helplessness or lack of control tend to increase anxiety, thereby increasing pain perception.

## Assessments

- Location
- Intensity: The client's report of pain is the single most important indicator of intensity. Health care professionals are notoriously inaccurate in rating clients' pain. Pain intensity scales are easy and reliable. Most scales use a 0-5 or 0-10 range with 0 indicating "no pain" and the highest number indicating "the worst pain possible."
- Quality: For example, sharp, dull, piercing like a knife. Use client's exact words.
- Pattern: Includes time of onset, duration, and intervals without pain.
- Precipitating factors
- Alleviating factors
- Associated symptoms (e.g., nausea)
- Effects on activities of daily living

- Coping resources and strategies (e.g., support from family, prayer)
- Affective responses (e.g., anxiety, depression)
- Behavior and physiological responses (e.g., vital signs, grimacing, moaning)

### Associated NANDA Nursing Diagnoses

- Chronic Pain
- Pain
- (Note that pain can be the etiology of many other nursing diagnoses, such as Impaired Physical Mobility, Powerlessness, Self-Care Deficit, and Altered Sexuality Patterns).

## Pain Management—Pharmacologic

An **analgesic** is a medication that relieves pain. The three classes of analgesics are NSAIDs, opiods (narcotics), and adjuvant drugs (e.g., corticosteroids, anticonvulsants, antidepressants, and antihistamines) that enhance the effect of opioids, reduce side effects, or lessen anxiety.

**Nonsteroidal Anti-inflammatory Drugs (NSAIDs)** (e.g., aspirin, acetaminophen, ibuprofen)

- **Action:** Peripheral, outside the nervous system; inhibit biosynthesis of prostaglandins, analgesic and anti-inflammatory (with the exception of acetaminophen, which has minimal anti-inflammatory effect).
- **Indication:** Drug of choice for mild-to-moderate pain; combined with opioids, provide more analgesia than either drug taken alone.
- **Route:** Oral and injectable; many are available without prescription.
- **Side Effects:** Some have gastric side effects, but these may be alleviated if the drug is taken with food or antiacids. Contraindicated in those with bleeding disorders (NSAIDs interfere with platelet function) and those with probable infections (the anti-inflammatory effects can mask the signs of infection).

**Opioid (Narcotic) Analgesics** (e.g., morphine, codeine, meperidine [Demerol], propoxyphene [Darvon], hydromorphine [Dilaudid], butorphanol [Stadol], and fentanyl citrate [Sublimaze])

- **Action:** Bind to opiate receptors and activate endogenous pain suppression in the central nervous system
- **Indication:** Drug of choice for moderate-to-severe pain
- **Route:** Subcutaneous, intramuscular, intravenous, nasal, transdermal, epidural and oral. Narcotics are commonly administered by **client-controlled analgesia (PCA)**, in which an infusion pump is set to deliver the medication and the client pushes a button to initiate the dose. A lockout interval (usually 10-15 minutes) is set so that the client cannot give an additional dose too soon. Many pumps can deliver a low-dose continuous infusion to provide sustained analgesia while the client is sleeping. Teach the client to report signs of infiltration to the nurse. Monitoring the site for IV-related complications is often neglected because the client regulates drug administration.
- **Side Effects:** Drowsiness/sedation, respiratory depression, constipation, nausea and vomiting, pruritus

## Therapeutic Nursing Interventions for Pain

- Become familiar with the guidelines for acute pain management and management of cancer pain developed by the Agency for Health Care Policy and Research, which are the standards for effective pain management by nurses.
- AHCPR guidelines specify that the least invasive pain management modalities and the simplest dosage schedules should be first.
- Pain cannot be objectively measured; the only way to assess it is to rely on the client's words and behaviors.
- Do not stereotype clients as complainers or "difficult" just because an obvious organic cause of pain cannot be found.
- Do not assume that drug abusers and alcoholics overreact to discomforts.
- Understand that regular administration of narcotic analgesics does not necessarily lead to client drug dependence.
- Do not assume that the amount of tissue damage accurately indicates pain intensity.
- Do not assume that a minor illness causes less pain than severe illness/alteration.
- Individualize pain therapy: try different types of pain relief measures.
- Provide pain relief measures before pain becomes severe.
- Use measures that the client believes to be effective.
- Choose the analgesic on the basis of the severity of pain—never administer a potent analgesic for mild pain.
- Obtain baseline level of alertness and respiratory rate before administering narcotics.
- Assess for and prevent constipation (narcotic analgesics): increase intake of fluids and fiber, increase exercise, administer stool softeners and laxatives if necessary.
- For pruritus (narcotic analgesics), apply cool packs and lotion; use diversional activity; administer an antihistamine if necessary.
- Caution about use of alcohol with analgesics.
- Caution about driving when taking analgesics.

## Age-Related Changes—Gerontological Considerations

- Incidence of pain is higher among older adults because of the prevalence of acute and chronic diseases.
- Pain threshold does not change.
- Lower doses of analgesics, especially opioids, may be required due to changes in the body's metabolism of drugs. Monitor carefully for oversedation, confusion, respiratory depression, and toxicity.
- Older adults may under-report pain because they expect it as a result of aging and believe nothing can be done about it. Emphasize the importance of reporting pain to health care professionals.

# Critical Thinking Exercise: Pain Management

1. You are caring for a 75-year-old male who is experiencing severe pain related to a cancer diagnosis. Along with other medications, a narcotic analgesic has been prescribed to treat the pain. Your client expresses reluctance to take the medication and states "I don't want to become addicted." How might you respond to your client's concern?

2. You are preparing to administer a scheduled dose of narcotic analgesic to a client on an inpatient hospice unit. As you enter the room, you find the client sleeping. His daughter suggests that you skip the scheduled medication by explaining "My father can't be in any pain; he's sleeping." How do you respond?

3. McCaffery advanced a definition of pain that emphasized its subjective nature. "Pain" she told us, "is whatever the patient says it is." What implications does this definition have for nursing practice? Discuss this issue with classmates.

# Metabolic Function and Basic Nutrition: Glucose Monitoring

## Key Points

- The glucose value measures the effectiveness of the treatment of the client with diabetes.
- Capillary blood glucose is monitored by obtaining a small drop of blood from the finger and using a commercial glucose meter, such as an Accu-Chek or Glucometer.
- Normal blood glucose is 60-120 mg/dL.
- Measurement is done about 30 minutes before meals.
- The nurse should wear disposable gloves to prevent contact with bodily fluids or blood.
- Protect test strips from exposure to light.
- The fingertips are the most common site for skin puncture for adults and older children. Use the lateral aspect of the fingertip. Punctures in the pads of the fingertips can damage nerve endings and are also painful.
- The earlobes are another site for skin puncture if the fingertips are calloused, or if the client's hands are edematous, or the client is in shock.
- The glycosylated hemoglobin (HbA1c) level reports the client's average blood glucose level over the previous six to 10 weeks. The HbA1c indicates the amount of glucose that is attached to a portion of hemoglobin in the client's erythrocytes; HbA1c levels should be within a range of 5% to 8%.
- Measurement is done about 30 minutes before meals.
- The nurse should wear disposable gloves.
- Massage and/or hold finger in dependent position to encourage blood flow.
- Energy requirements vary and are influenced by age, body size and temperature, activity level, environmental temperature, gender, growth, emotional state, and food intake.
- Calorie content of nutrients:
  - 1 gram carbohydrate = 4 calories
  - 1 gram protein = 4 calories
  - 1 gram fat = 9 calories
- The six nutrients are: water, carbohydrates, proteins, lipids, vitamins, and minerals. The nurse should know the required daily allowance for each.
- For good nutrition:
  - Eat a variety of foods.
  - Maintain a desirable weight.
  - Limit fats; avoid saturated fats and cholesterol.
  - Eat foods with adequate starch and fiber.

- - Limit intake of sugar and sodium.
  - Drink alcohol only in moderation.
- Stimulate client's appetite by relieving illness symptoms and providing a clean, odor- free environment.
- Provide oral hygiene before mealtime.
- Community nutritional services may be available for those who are impoverished and/ or cannot prepare meals (e.g., Meals-on-Wheels, food stamps).
- Older adults need fewer calories because of a slower metabolic rate and decrease in physical activity.
- **Key Concepts/Terms:** Commercial blood glucose meter, signs and symptoms of hypoglycemia and hyperglycemia, metabolism, anabolism, catabolism, basal metabolic rate (BMR), caloric value, nutrients, complete proteins, incomplete proteins, complementary proteins, saturated fatty acids, unsaturated fatty acids, polyunsaturated fatty acids, linoleic acid, fat-soluble vitamins, water-soluble vitamins, macrominerals, microminerals, anthropometric measurements, body mass index (BMI)

# Basic Nutrition

## Overview

Food is ingested, digested, and absorbed to provide the energy needed for organ function, body movement, and work; and for enzyme function, growth, repair, and replacement of cells. **Metabolism** consists of the physical and chemical processes by which energy is made available for use by the body. **Anabolic** reactions build substances and body tissue; **catabolic** reactions break down substances. The **basal metabolic rate (BMR)** is the energy requirement of a person who is awake, but at rest the energy needed at the lowest level of cellular function. Energy requirements vary and are influenced by age, body size and temperature, activity level, environmental temperature, gender, growth, emotional states, and food intake. The amount of energy that nutrients or foods supply is their **caloric value**. There are 4 calories in a gram of carbohydrate or protein, and 9 calories in a gram of fat.

## Nutrients

**Nutrients** are foods that contain the elements necessary for body function. The six nutrients are water, carbohydrates, proteins, lipids, vitamins, and minerals.

- **Water:** The function of cells depends on a fluid environment. Fluids are found in liquids and in fruits and vegetables. Illness can create an increased need for fluids (e.g., as with a fever or hypermetabolic state). Conversely, illness can create a need to ingest less fluids (e.g., as with cardiac or renal disease). Thirst is a protective mechanism that alerts the person to the need for fluids.
- **Carbohydrates:** Complex carbohydrates are usually starches and fibers; simple carbohydrates are sugars, and are composed of carbon, hydrogen, and oxygen. They are the body's preferred energy source. They are needed to metabolize fats and to "spare" protein—that is, to keep the body from using protein for energy. At least 50-100 grams of carbohydrate are needed daily. Carbohydrates are found mainly in plant foods; the only important animal source is the lactose in milk. Monosaccharides are simple sugars (e.g., glucose, dextrose, fructose).

Disaccharides consist of two sugar units (e.g., sucrose, lactose, maltose); during digestion they are split into simple sugars. Polysaccharides are composed of many sugar units and are insoluble in water (e.g., glycogen, stored in the liver). Some polysaccharides, such as fiber, cannot be digested.

- **Proteins** are made of hydrogen, oxygen, carbon, and nitrogen. Protein is the body's only source of nitrogen. The amino acids in proteins are essential for synthesis of body tissue in growth, maintenance, and repair. If carbohydrate intake is inadequate, proteins can also be used for energy.

  - **Complete proteins (high-biological value proteins)** contain all essential amino acids in sufficient quantity to support growth and maintain nitrogen balance. They are found in meat, fish, poultry, milk, cheese, and eggs.
  - **Incomplete proteins** do not contain all the essential amino acids, or do not have them in sufficient quantities. They are found, for example, in cereals, legumes, and vegetables.
  - **Complementary proteins** are incomplete proteins that, when eaten together, combine to act as complete proteins (e.g., grains and legumes).

The required daily allowance of protein is about 45-60 grams for adults—an additional 30 grams is needed for pregnant women; lactating women need an additional 20 more than the required daily amount. Complete proteins contain significant amounts of saturated fatty acids and cholesterol. Most Americans ingest more protein than required.

- **Lipids** are composed of carbon, hydrogen, and oxygen, but in different proportions than carbohydrates. They are insoluble in water, but soluble in alcohol, ethanol and acetone. Fat is the body's form of stored energy, and it is needed for the absorption of fat-soluble vitamins. Metabolism of 1 gram of lipid yields 9 kcal—more than twice the energy provided by carbohydrates or proteins. Approximately 98% of the lipids in foods are in the form of triglycerides, which have been linked to arteriosclerosis. Most animal fats have a high proportion of **saturated fatty acids**, which tend to increase blood cholesterol levels. Most vegetable fats have high proportions of **unsaturated and polyunsaturated fatty acids** (e.g., safflower, corn, soybean, and peanut oils). **Linoleic acid** is the only essential fatty acid (it cannot be synthesized by the body, so it depends on adequate dietary intake). In general, American diets are too high in fat. Nutritional guidelines recommend that fats be limited to 30% of the total caloric intake.

**Vitamins**: These are organic substances essential to metabolism, that are found in small amounts in foods. The body cannot synthesize vitamins adequately and depends on dietary intake. Although contained in many foods, vitamins are destroyed by processing, storage, and preparation. Vitamin content is highest in fresh foods.

- **Fat-soluble vitamins (A, D, E, and K)** can be stored in the body; so daily intake may not be necessary. Large (megadoses) can be toxic. Many foods (e.g., milk, bread) are fortified with vitamins A and D.

  - Vitamin A sources include: liver, carrots, egg yolk, and fortified milk.
  - Vitamin D sources include: sunlight, fish liver oils, and fortified milk.
  - Vitamin E sources include: vegetable oils, wheat germ, and whole grains.
  - Vitamin K sources include: dark, green leafy vegetables, and synthesis from bacteria in the intestines.

- **Water-soluble vitamins (C and B complex)** cannot be stored in the body and therefore must be ingested daily. The **B complex** includes thiamine, riboflavin, niacin, pyridoxine, folic acid, cobalamin, pantothenic acid, and biotin.
    - Vitamin C sources include: citrus fruits, broccoli, green pepper, greens, and strawberries.
    - Vitamin B complex sources include: organ meats, pork, seafood, egg yolk, yeast, green leafy vegetables, grains, nuts, banana, and cantaloupe.
- **Minerals**: These are inorganic elements found in all body fluids and tissues as organic compounds, inorganic compounds, and free ions. Some function to provide structure in the body; others help to regulate body processes. Calcium and phosphorous make up 80% of all the mineral elements in the body. **Macrominerals** are those needed in amounts of 100 mg/day (calcium, phosphorus, sodium, potassium, magnesium, chloride, and sulfur). **Microminerals** are those needed in amounts of less than 100 mg/day (iron, zinc, manganese, iodine, fluoride, copper, cobalt, chromium, and selenium). Common problems associated with mineral deficiencies are iron deficiency anemia and osteoporosis, which results from loss of bone calcium.

## Food Guide Pyramid

The 2005 Food Guide Pyramid, developed by the U. S. Department of Agriculture, was designed to represent the foods needed daily in a healthy diet, which is based on a foundation of daily exercise and weight control. It is composed of six groups, represented as color stripes within the pyramid. The USDA chose to avoid offering serving recommendations because they differ depending on weight, gender, activity level, and physical condition of the individual.

Orange: Whole grains

Green: Vegetables

Red: FruitsYellow: Oils

Blue: Dairy

Purple: Meat and beans

## Risk Factors for Nutritional Problems

- Conditions that interfere with the ability to ingest, digest, or absorb nutrients
- Congenital anomalies
- Surgical revisions of the gastrointestinal tract
- Clients fed only by intravenous infusion for more than a week
- Pregnant clients
- Those with poor dietary habits
- Obesity
- Anorexia nervosa or bulimia
- Cancer and radiotherapy
- Immobilization
- Living alone
- Alcohol or substance abuse

## Nutritional Assessment

**Nursing History**–To identify clients who are at risk for malnutrition, assess:

- Age and sex
- Activity level
- Difficulty eating (e.g., missing teeth, impaired swallowing)
- Weight changes
- Changes in appetite
- Physical/functional changes that can affect purchasing and preparing food
- Economic status (e.g., for purchasing food)
- Cultural and religious beliefs that influence food choices
- Living arrangements (e.g., living alone)
- Usual and current pattern of dietary intake
- Primary medical diagnosis and any metabolic demands created by the disease

**Physical Examination**–Signs and symptoms of nutritional deficiencies:

- Loss of subcutaneous fat
- Muscle wasting
- Edema and ascites
- Dry, flaky, pale skin; bruising
- Brittle, ridged, or spoon-shaped nails
- Dry, dull, sparse hair
- Pale or red conjunctiva
- Swollen, red cracks at side of mouth (angular stomatitis), vertical fissures (cheilosis)
- Swollen, beefy red, or smooth tongue
- Swollen, inflamed, or bleeding gums
- Enlarged liver
- Sensory loss, burning and tingling of hands and feet (paresthesias)

**Anthropometric measurements** are used to determine body dimensions. The most common anthropometric measurements are height and weight. Others are **triceps skin fold measurements**, which measure subcutaneous fat stores; **midarm circumference**, which measures skeletal muscle mass. The **body mass index (BMI)** indicates whether a person's weight is appropriate for height. It is calculated as follows:

$$BMI = \frac{\text{weight in kilograms}}{(\text{height in meters})^2}$$

| BMI | Weight Status |
|---|---|
| Below 18.5 | Underweight |
| 18.5-24.9 | Normal |
| 25.0-29.9 | Overweight |
| 30.0 and Above | Obese |

Laboratory Data: Most commonly used tests are serum proteins, hemoglobin, urinary urea nitrogen and creatinine, and total lymphocyte count.

## Related NANDA Nursing Diagnoses

- Altered Nutrition: Less than Body Requirements
- Altered Nutrition: More than Body Requirements
- Risk for Altered Nutrition: More than Body Requirements
- Feeding Self-Care Deficit
- Impaired Swallowing
- Fluid volume excess, fluid volume deficit, and risk for fluid volume deficit (note that nutritional problems may be the etiology of other nursing diagnoses, such as activity intolerance, constipation, and risk for infection.)

## Guidelines for Avoiding Nutrition-Related Problems

- Eat a variety of foods.
- Maintain a desirable weight.
- Avoid fat, saturated fats, and cholesterol.
- Eat foods with adequate starch and fiber.
- Limit intake of sugar.
- Limit intake of sodium.
- Drink alcohol in moderation.

## Therapeutic Nursing Interventions

- Stimulate the appetite by relieving illness symptoms (e.g., give an analgesic to relieve pain before mealtime).
- Provide a clean, odor-free environment (e.g., remove used bedpans, soiled dressings, used dishes; pull the curtain between beds if necessary).
- Provide oral hygiene before mealtime.
- Provide food the client likes, encourage family and friends to bring food that is within the parameters of special diet requirements.
- When feeding a client, ask in which order the person would like to eat the food.
- If the client cannot see, tell her/him which food is being offered.
- When feeding a client, if the client cannot communicate, offer fluids after every three or four bites of food.
- Special community services may be available to help meet the nutritional needs of people who cannot prepare meals or leave their homes (e.g., Meals-on-Wheels).
- For the impoverished, the U. S. Department of Agriculture has a food stamp program, which helps low-income people to purchase food.

## Age-Related Changes—Gerontological Considerations

- Fewer calories are needed because of slower metabolic rate and decrease in physical activity. Approximately 1,200 kcal per day are needed, on average.

- Because of age-related changes, skin fold measurements should be taken from several body sites.
- Serum albumin level may be lower (<3.5 mg/dL) as a result of the aging process.
- Older adults may have lower than normal hemoglobin levels.
- Some may need more carbohydrates for fiber and bulk (fruits, vegetables, grains).
- Physical changes (e.g., tooth loss and impaired senses of taste and smell) may affect eating habits.
- People who live alone may not like cooking for themselves or eating alone.
- Anxiety, depression, loss of a spouse, dependence on others, and low income may all affect eating habits.
- Thirst is not a reliable guide to fluid needs for older adults and confused clients because they may not be able to communicate that they are thirsty.

# Glucose Monitoring

**Overview**

Capillary blood glucose is measured by obtaining a small drop of blood from the finger and using a glucose meter (e.g., Accu-Chek, Glucometer) for determination. Normal blood glucose levels are between 60 to120 mg/dL. Blood glucose is usually measured about a half-hour before meals for diabetic clients. The glucose value measures the effectiveness of the treatment of the client with diabetes.

All food that is ingested must be converted to glucose in order to be used by the cells; however, carbohydrates are more easily and quickly digested than protein and fat. When the supply of glucose exceeds what is needed for energy, it is stored as glycogen in the muscles and liver, or as fat in adipose tissue. Clients with diabetes cannot properly utilize glucose, so it accumulates in the blood, raising the blood glucose levels.

## Signs and Symptoms of Abnormal Blood Glucose Levels

**Hypoglycemia: Blood glucose <60 mg/dL**

- Hunger
- Shakiness
- Irritability
- Loss of concentration
- Rapid pulse
- Hypotension
- Pale, cool, clammy skin
- Nausea
- Seizures
- Coma

**Hyperglycemia: Blood glucose >120 mg/dL**

- Weakness
- Thirst
- Flushed cheeks
- Dry skin and mouth
- Nausea and vomiting
- Glucosuria
- Fruity odor to breath (late sign)
- Coma

## Procedure and Techniques for Glucose Monitoring

**Equipment:** Blood glucose meter, testing strips, sterile lancet, cotton balls, alcohol swab, disposable gloves

**Procedure:**

- Check the primary health care provider's orders, verify the client's identity and explain the procedure to the client.
- Review the institution's protocols for this procedure, as well as the manufacturer's instructions for using the blood glucose meter.
- Wash your hands and put on disposable gloves to prevent the spread of microorganisms to the client and to protect yourself from exposure to the client's blood and bodily fluids.
- Protect test strips from exposure to light.
- Place the selected finger in a dependent position, and massage the finger gently toward the site to increase blood flow to the area.
- Clean the site according to the institution's protocols and allow the site to dry completely.
- While using your nondominant hand to apply gentle pressure proximal to the site, use your dominant hand to puncture the client's skin at a 90-degree angle with the lancet injector.
- Wipe away the first drop of blood with a cotton ball because the first drop usually contains more serous fluid, which can alter the result.
- Gently squeeze the proximal area to produce a large drop of blood.
- Apply the blood to the reagent strip by touching the skin just below the drop until blood covers the indicator area.
- Instruct the client to keep pressure on the puncture site with a cotton ball until it no longer bleeds.
- Follow the manufacturer's directions for how long to expose the strip to the blood, whether or not to wipe the strip, and then how to insert the strip and obtain the reading.
- Remove and dispose of your gloves after obtaining the reading, and then wash your hands.
- Document the reading on the client's record. Record:
  - How the client tolerated the procedure
  - Any concerns or observations
  - If insulin was administered based on the result
- Compare the current result with previous glucose testing results, taking into consideration any treatments and medications that might influence glucose levels.
- Prepare the client who is newly diagnosed with diabetes for discharge. Teach the client and family members how to perform blood glucose testing at home. Ask the client to perform a return demonstration.
- Advise the client and family members to immediately notify the primary health care provider if the client experiences abnormal blood glucose levels.

## Age-Related Changes—Gerontological Considerations

- Blood glucose levels increase with age, and the use of some medications.
- Because dehydration reduces plasma volume, the older adult's blood glucose levels may be elevated even more.

# Critical Thinking Exercise: Metabolic Function and Basic Nutrition: Glucose Monitoring

Situation: A homecare nurse is instructing a client in the use of a glucometer. The client, a 59-year-old woman recently diagnosed with Type II diabetes mellitus, poses this question: "Isn't it enough that I follow up with my doctor? Do I really need to do this every day?" Based on this question, how might you convince your client of the importance of regular glucose monitoring? Keep these factors in mind when formulating your response:

- Self-management strategies in the treatment of diabetes

- Regular glucose monitoring

- Preventing diabetes-related complications

1. During a routine physical assessment, a healthy 28-year-old female tells you that she has recently become a "total vegetarian." Based on your knowledge of basic nutrition, what potential deficits might be associated with this diet?

2. Take stock of your own nutritional status. Using the food pyramid and your understanding of basic nutritional needs, make a careful assessment of your own diet. What are your nutritional strengths? What deficits might you work to overcome?

3. After you've completed your nutritional self-assessment, create a weekly sample menu based on foods from the food pyramid. Make sure your menu represents foods that are both balanced and nutritious.

4. A homecare nurse visits an 85-year-old client who is independent in all activities of daily living (ADLs) and until recently, loved to cook for herself. Her children, however, have asked her to stop cooking because she occasionally forgets to turn off the oven or burners. Your client is now concerned that her nutritional status will decline. What suggestions might you offer?

# Nasogastric Tubes & Gastroenteral Feedings

## Key Points

- A nasogastric (NG) tube is a flexible tube with a hollow lumen that is inserted through one nostril, down the nasopharynx and esophagus and into the stomach.
- One of the greatest challenges is maintaining client comfort.
- The NG tube may or may not be attached to suction– depending on its purpose.
- Procedure highlights:
  - Place client in high-Fowler's position.
  - To determine how far to insert the tube, measure from tip of nose to earlobe, and down to the xiphoid process.
  - Lubricate the tip with water-soluble lubricant.
  - Encourage client to tilt head forward and drink or swallow as tube advances.
  - If tube meets resistance, withdraw it, relubricate, and insert in other nostril.
  - Discontinue and remove if client cannot talk, is coughing, or becomes cyanotic.
  - Check for placement, to be sure tube is in stomach and not in airway.
  - Secure the tube to the bridge of the nose; secure the tube to the client's gown.
- Provide daily tube and nose care.
- Enteral nutrition involves administration of formula through a tube placed into the gastrointestinal tract, either by the nasogastric route or surgically through the abdominal wall.
- **Potential Complications**: Aspiration pneumonia, obstructed tubing, diarrhea, nausea, vomiting, abdominal distention, stoma infection, skin breakdown, hyperglycemia, dumping syndrome.
- After initial placement of an NG tube is confirmed by radiography, mark it with indelible ink or tape at its exit point from the nose.
- Monitor bowel sounds and aspirate residual before each feeding.
- Check tube placement prior to feeding (e.g., by injecting air).
- Elevate the head before, during, and 30 minutes after each feeding.
- Replace gastrostomy tube immediatly if it becomes dislodged.
- Administer formula at room temperature unless order specifies otherwise.
- **Key Concepts/Terms**: Assessment of client before and after nasogastric tube insertion, procedural techniques for nasogastric tube insertion, intermittent feedings, bolus intermittent feedings, continuous feedings, cyclic feedings, gastrostomy, jejunostomy, nasogastric

# Nasogastric Tubes

**Overview**

Nasogastric tubes have several uses including nourishment, administration of oral medications that cannot be swallowed, obtaining gastric aspirate for diagnostic testing, removing toxic or poisonous materials that have been ingested, decompressing the stomach, or controlling gastric bleeding. A nasogastric tube is placed through the nose and advanced into the stomach.

Prior to inserting a nasogastric tube the nurse should review the primary care provider's order to determine the type of tube to use, the suction settings, and any irrigation instructions. Intermittent suction is used with single lumen NG tubes to provide protection against damage of the mucous membranes at the tip of the suction and low suction is 80 to 100 mm of suction (these numbers are on most suction regulators; it may be indicated merely as "high" or "low"). Irrigation of the NG tube may be required if the lumen(s) become clogged. NG tubes are generally irrigated before and after tube feedings, when giving medications via the tube, or if they become clogged. Unless the provider orders a specific irrigation fluid, sterile normal saline should be used.

## Assessment

**Before insertion**, assess for factors that may make insertion difficult:

- Altered level of consciousness
- Presence of nasal polyps, deviated septum, or nasal trauma
- Inability to swallow, cough, and gag
- Impaired integrity of nasal and oral mucosa
- Presence of nausea and vomiting
- History of nasal surgery
- Inability to cooperate with NG tube insertion

**After insertion**, assess for complications:

- Improper placement into the lungs, esophagus or small intestine
- Dehydration
- Electrolyte imbalance
- Enteritis
- Aspiration
- Nasal erosion
- Gastric distention

## Procedure and Techniques for Nasogastric Tubes

- Before inserting, **determine the size of tube** to be inserted and whether or not it is to be attached to a suction device.
- Assemble equipment
  - Clean gloves
  - A cup of ice or water and a straw
  - A towel

- Facial tissues
- Flashlight or penlight
- The NG tube
- A basin of warm water to soften a plastic tube or a basin of ice if a rubber tube is used
- Water-soluble lubricant
- Tongue blade
- 50-milliliter syringe with a tip that fits the tube
- Stethoscope
- Pad or towel to protect clothing and bedding
- Clamp or plug
- pH strip
- Paper tape
- Rubber band
- Safety pin
- Wash your hands.
- Place client in **high-Fowler's position** with towel across chest.
- Explain the procedure: it is not painful, but is unpleasant because the gag reflex will be activated.
- Using a flashlight, **inspect the nares for patency.**
- **Place rubber tube in ice** for 5 to10 minutes to stiffen it; for a small-bore tube, insert a stylet or guidewire into the tube and secure it in position.
- **Determine how far to insert the tube**: Measure the tube from the tip of the nose, to the earlobe, and down to the xiphoid process.
- Put on clean gloves.
- **Lubricate the tip** with a water-soluble lubricant.
- Insert the tube through the one nostril.
- If the client is conscious, ask client to tilt the head forward and encourage to drink and swallow.
- **Advance tube** to the mark that was previously measured.
- **If tube meets resistance, withdraw it**, relubricate, and insert in other nostril.
- Discontinue and remove if the client cannot talk, is coughing, or becomes cyanotic.
- **Check for placement** after the tube has been advanced by (a) auscultating the stomach while injecting a small amount of air and (b) checking the pH of the aspirated fluid (gastric aspirate pH is usually 1-4, but may be as high as 6 in the presence of some medications; there is a chance that respiratory placement has occurred when the pH reading is 6; it is almost certain if the pH is 7 or above). Tube placement will be absolutely confirmed later by x-ray.
- **Secure the tube** to bridge of the nose with tape.
- Secure the tube to the client's gown with an elastic band and safety pin to prevent dislodging or movement of the tube which could cause discomfort and irritation/ulceration of the nare.

- Remove your gloves and dispose of them and other materials in the proper receptacle, and make certain the client is comfortable.
- Wash your hands
- Document the insertion and placement, check how the client tolerated the procedure, and any other concerns or observations in the client's record.
- Assess and provide care to the nare each shift (clean nostril and apply water-soluble lubricant), change the tape as needed, and give frequent mouth care.
- Xylocaine gel in nostril or spray in throat may be used to decrease discomfort.
- Before removing an NG tube, check the provider's order, assess the client for bowel sounds, and check for the presence of nausea or vomiting when the tube is clamped.
- Explain the procedure to the client.
- Assemble equipment
  - Pad or towel
  - Facial tissues
  - Clean gloves
  - Large syringe with a tip that fits the tube
  - A plastic bag for disposal
- Provide privacy and place a pad or a towel across the client's chest.
- Provide client with tissues to clean his/her mouth and nose after the tube is removed.
- Wash hands and then detach the tube from any suction or feeding apparatus, and clamps or plugs at the distal end of the NG tube.
- Unpin the tube from the client's gown and remove the tape securing the tube to the client's nose. Put on the gloves, unplug the end of the tube and instill 50 milliliters of air into the tube to clear it of any contents.
- Ask the client to take a deep breath and hold it to keep the glottis closed, thus preventing aspiration of gastric contents.
- Pinch the tube, and with a quick, smooth motion, withdraw the tube and place it in the disposal bag.
- Remove your gloves and dispose of them and all materials in the proper receptacle.
- Help the client clean his nose and mouth if needed, provide oral hygiene, and make sure client is comfortable.
- Wash your hands.
- Document the tube removal, how the client tolerated the procedure, and any other concerns or observations in the client's record.

## Age-Related Changes—Gerontological Considerations

- Gag reflex may be diminished in older clients.

# Gastroenteral Feedings

## Overview

Enteral nutrition involves placing a tube into the gastrointestinal tract to administer formula. Formula may be given by bolus, intermittent, continuous, or cyclic method. For short-term use (6 weeks or less) a nasogastric tube is placed through the nose into the stomach. For long-term nutritional support, a gastrostomy or jejunostomy tube is placed surgically through the abdominal wall into the stomach or jejunum. The surgical opening is sutured around the tube to prevent leaking. Before feedings begin, tube placement is confirmed by radiography. While the opening heals, tube care requires surgical asepsis. After the opening is healed, the tube can be removed and reinserted for each feeding.

**Intermittent feedings**: 300-500 mL of enteral formula is administered several times a day as ordered by the physician, dietician, or nurse practitioner.

**Bolus intermittent feedings**: A bag hanging by gravity or a syringe is used to deliver the formula into the stomach. Feeding is delivered rapidly by this method, so it may not be well tolerated. The client must be monitored closely for aspiration and distention.

**Continuous feedings**: An infusion pump administers feedings at a constant flow rate over a 24-hour period. Continuous feedings are essential when the tube is placed in the small bowel, when small-bore tubes are in place, or when gravity flow is not successful in instilling the feedings.

**Cyclic feedings**: These are continuous feedings that are administered in less than 24 hours, often at night. The client may eat regular meals throughout the day.

## Potential Complications of Tube Feedings

- Aspiration pneumonia
- Obstructed tubing
- Nasal erosion if being fed by NG
- Diarrhea
- Nausea, vomiting and distention
- Tubing inadvertently pulled out
- Stoma infection and skin breakdown
- Hyperglycemia
- **Dumping syndrome** may occur with bolus feedings in jejunostomy clients. Symptoms are nausea, vomiting, diarrhea, cramps, pallor, sweating, heart palpitations, increased pulse rate, and fainting after a feeding. These symptoms may occur when hypertonic foods and liquids suddenly distend the jejunum; body fluids then shift rapidly from the client's vascular system in an attempt to make the intestinal fluids isotonic.

## Assessments

- Before each feeding, aspirate to assess residual feeding contents and evaluate absorption of the last feeding to prevent overfilling of the stomach. If more than 100 mL (or more than half the last feeding) is withdrawn, the feeding may be withheld or the amount may be changed.

- Monitor bowel sounds prior to each feeding; every 4-8 hours for continuous feedings.
- Monitor daily for abdominal distention; measure abdominal girth at the umbilicus.
- Observe for regurgitation and feelings of fullness after feedings.
- Monitor for signs of dumping syndrome.
- Assess for diarrhea, constipation, or flatulence.
- Monitor blood glucose levels for the first 24-48 hours when clients are beginning enteral feeds; after that, monitor urine for sugar and acetone.
- Check hematocrit and urine specific gravity (signs of dehydration).
- Monitor serum BUN and sodium levels (feeding formula may have high protein content; kidneys may not be able to excrete nitrogen adequately).

## Therapeutic Nursing Interventions for Gastroenteral Feedings

- After initial placement of a nasogastric tube is confirmed by radiography, mark the tube with indelible ink or tape at its exit point from the nose.
- Check tube placement prior to feeding by (a) injecting a small amount of air through the tube and listening to the stomach, (b) aspirating 20-30 mL of gastric secretions (gastric secretions tend to be green, off-white, or tan), or (c) measuring the pH of aspirated fluid.
- Elevate the head before, during and 30 minutes after each feeding.
- Hang no more than four hours of formula at a time, and replace formula container every 24 hours.
- Rinse feeding tubes just before all of the formula runs through the tubing by instilling 50-100 mL of water through the tube. Clamp the tube before all the water is instilled to prevent leakage and air from entering the tube.
- Alternate nasal placement or use a small diameter flexible tube for nasal insertion.
- Replace gastrostomy tube immediately if it becomes dislodged.
- Always check for food allergies and check the expiration date of the tube feeding formula before administering.
- Tube feedings are administered at room temperature unless the order specifies otherwise.
- Gastronomy or jejunostomy feedings procedure:
  - Determine the type and amount of the tube feeding and the frequency of feedings by checking the nutrition order in the client's chart.
  - Determine from the chart (or from report) any pertinent information regarding the client's previous feedings.
  - Assess and prepare the client: auscultate for bowel sounds, inspect for abdominal distention or GI distress, and elevate the head of the bed 30° or more.

- Check the migration of the tube.
- Aspirate for residual feeding in the stomach. Generally, if it is greater than the volume of the previous feeding, the next feeding should be held. Check the primary care provider's order for any specific interventions related to residual feedings.
- Return the aspirated amount to the stomach to prevent electrolyte imbalance.
- Administer the feeding as it was ordered; if not otherwise ordered, the feeding should be administered slowly, by gravity (over about 15 minutes).
- Ensure the client's comfort and safety.
  - Keep the head of the bed elevated at least 30° for 2 hours after the feeding.
  - Assess the stoma and peristomal area at each feeding.
  - Clean and dress the peristomal area as ordered, or with mild soap and water at least once per day.
  - Observe for complications from the feeding or at the peristomal site.
  - Provide education and supplies for oral care at least once per shift.

## Age-Related Changes—Gerontological Considerations

- Gag reflex may become diminished.
- Hyperglycemia develops quicker in clients who receive tube feedings.
- Older adults tolerate smaller, continuous feedings better.

# Critical Thinking Exercise: Nasogastric Tubes & Gastroenteral Feedings

**Situation:** You are caring for a postoperative client who has a nasogastric tube in place. Describe your ongoing assessment of this client along with appropriate interventions to manage such potential complications as:

- Fluid volume deficit

- Pulmonary complications

- Irritations related to the presence of the tube

1. As you remove the tube, what measures are appropriate to assure the client's comfort?

# Bowel Elimination - Ostomies and Enemas

## Key Points

- Normal feces are about 75% water and 25% solid materials; they are soft and formed; they are normally brown; normal amount is 100 to 400 grams/day (varies with diet).

- Frequency of bowel movements varies with individuals; it is not necessary for everyone to have a daily bowel movement (although some clients have this mistaken idea).

- Common elimination problems include: constipation, fecal impaction, diarrhea, bowel incontinence, and flatulence.

- Diagnostic tests: anoscopy, proctoscopy, proctosigmoidoscopy, colonoscopy, x-rays (e.g., barium enema, upper GI, stool specimens [e.g., for ova and parasites], occult blood [Guaiac test, Hemoccult test])

- Certain foods (e.g., red meat, raw fruits and vegetables) and medications (e.g., aspirin, vitamin C) should be avoided for up to three days before testing for occult blood, in order to avoid obtaining a false positive or false negative result. In addition, specimens should not be collected during a woman's menstrual period or while there are bleeding hemorrhoids.

- Promote regular defecation by:
  - Providing privacy for defecation
  - Encouraging client to defecate when the urge is recognized; establish a time and routine
  - Assuring adequate intake of fluids and fiber
  - Providing as normal a position as possible for clients who must use a bedpan

- When placing a bedpan, wear disposable gloves.

- Document color, odor, amount, and consistency of feces, and condition of the perianal area.

- Habitual use of laxatives inhibits natural defecation reflexes and causes even more constipation.

- An ostomy is a surgical opening made on the abdominal wall for the purpose of providing an alternate feeding route or for the elimination of feces or urine.

- Potential complications: Skin breakdown at stoma site, infection, constipation, fluid and electrolyte imbalances, stomal prolapse/retraction

- Assess stoma and peristomal skin each time the appliance is changed.

- Report pale or dark-colored stoma and stomal bleeding.

- Empty ostomy bag frequently to keep odor-free.

- Apply a barrier (e.g., karaya gum) over the skin around the stoma to prevent contact with any excretions.

- Keep stoma site clean and dry; be sure there is no leakage and that bag fits properly. Teach clients self-care. Control odors by eating dark green vegetables, using bismuth subgallots, placing a deodorizer in the pouch, using an appliance with a charcoal filter disc.

- Initially, avoid high-fiber and gas-producing foods.

- Avoid heavy lifting and contact sports.

- Laxatives and enemas may cause severe fluid and electrolyte imbalance.

- Enemas act by distending the intestine, and sometimes by irritating the intestinal mucosa; this increases peristalsis and the expulsion of feces and flatus.

- When administering an enema, protect against contact with body fluids.

- Minimize trauma to the sphincters (e.g., lubricate insertion tube).

- Place client in left lateral position.

- Administer fluid slowly to minimize discomfort.

- Raise container to increase force of flow; lower container to reduce flow; usually, begin with solution about 12 inches above the rectum.

- Some clients cannot retain even small volumes of enema solutions. In this case, place client on a bedpan for the procedure and press client's buttocks together to help retain the solution.

- **Key Concepts/Terms:** Colon, rectum, anal canal, internal sphincter, external sphincter, peristalsis, mass peristalsis, defecation, laxative, stool softner, paralytic ileus, anatomy of bowel, physiology of defecation, common elimination problems, diagnostic tests, nursing diagnoses, and nursing interventions, gastrostomy, jejunostomy, ileostomy, colostomy, stoma, enema, cleansing enema (high, low large-volume, small-volume), carminative enema, oil retention enema, return-flow enema, antibiotic enema, anthelmintic enema, nutritive enema, hypertone solution, hypotonic solution, isotonic solution

# Bowel Elimination

## Overview

Elimination problems can be embarrassing to clients and can cause considerable discomfort. Nurses are frequently involved in promoting normal bowel elimination. Waste products of digestion, referred to as **feces** or **stool**, are eliminated through the bowel. Normal feces are about 75% water and 25% solid materials; they are soft and formed. If fluid intake is insufficient, feces may be hard and difficult to expel. Feces are normally brown, due to the presence of byproducts of bilirubin. The normal amount is about 100-400 g per day, but that varies with the diet. The action of microorganisms on the food is responsible for the odor of feces.

## Physiology of Defecation

The colon (**or large intestine**) is about 125-150 cm (50-60 in.) long. It is a muscular tube lined with mucous membrane. Its main functions are absorption of water and nutrients, secretion of mucous to protect the intestinal wall, and fecal elimination. Waste products leave the stomach; pass through the small intestine, and then through the ileocecal valve into the colon. **Peristalsis** (a wavelike movement produced by the muscle fibers of the intestinal walls) moves the intestinal contents (flatus and feces) forward. Colon peristalsis is very sluggish and moves the chyme only a little. **Mass peristalsis** is a wave of powerful muscle contraction that moves over large areas of the colon. It occurs after eating, when it is stimulated by the presence of food in the stomach and small intestine.

The **rectum** is about 4-6 in. (10-15 cm) long. The most distal 1-2 in. (2.5-5 cm) is the anal canal. The **anal canal** has an **internal sphincter** (which is under involuntary control) and an **external sphincter** (which is under voluntary control).

**Defecation**, also called a "bowel movement," is the expulsion of feces from the anus and rectum. When peristaltic waves move the feces into the sigmoid colon and the rectum, the sensory nerves are stimulated and the person becomes aware of the need to defecate. When the internal sphincter relaxes, feces moves into the anal canal. When the person voluntarily relaxes the external anal sphincter and contracts the abdominal muscles and the diaphragm, abdominal pressure is increased. Together with the contraction of the levator and muscles of the pelvic floor, this moves the feces through the anal canal. Repeated inhibition of the urge to defecate can result in distention of the rectum and loss of sensitivity to the need to defecate, resulting ultimately in constipation.

## Factors Affecting Defecation

**Age/development**: Infants do not have control of defecation. Some control begins at around two years of age. School-age children have bowel habits similar to those of adults.

**Diet**: Sufficient fiber/bulk in the diet is needed to provide fecal volume. Bland diets lack bulk and move more slowly through the intestinal tract. Increasing fluid intake increases the rate of movement. Spicy foods can cause diarrhea and flatus; excessive sugar can also cause diarrhea. Foods such as cabbage, onions, cauliflower, bananas, and apples are gas producing. Foods such as bran, prunes, figs, chocolate, and alcohol act as laxatives. Foods such as cheese, pasta, eggs, and lean meat are constipating.

**Activity**: Stimulates peristalsis and helps keep abdominal muscles strong.

**Psychologic factors**: People who are depressed may have decreased intestinal motility and constipation.

**Diagnostic procedures**: Before certain diagnostic procedures (such as a sigmoidoscopy), the client is restricted from eating food and fluids. In addition, a cleansing enema is given prior to the procedure. The client will usually not defecate normally until eating is resumed.

**Bowel habits**: Many people defecate after breakfast, when the gastrocolic reflex causes mass peristalsis. When normal defecation reflexes are continually ignored (e.g., because the client is "too busy"), the urge to defecate is lost. Hospitalized clients may suppress the urge to defecate because they are embarrassed about the lack of privacy or about using a bedpan.

**Medications**: Some medications have gastrointestinal side effects. Some (e.g., certain antibiotics) cause diarrhea; others (e.g., morphine, codeine, and iron tablets) cause constipation. Laxatives are medications that are given for the purpose of stimulating bowel activity and assisting fecal elimination. Stool softeners facilitate defecation by increasing the ability of the fecal material to absorb water. Some medications, given to treat diarrhea, are given because of their ability to suppress peristalsis.

**Anesthesia and surgery**: General anesthetics block parasympathetic stimulation to the muscles of the colon, thereby slowing or stopping normal colonic movements. Surgery that involves handling of the intestines can cause paralytic ileus, a cessation of intestinal movement that lasts 24 to 48 hours.

**Pathologic conditions** (e.g., spinal cord and head injuries, poorly functioning anal sphincters).

**Pain**: Clients who have pain when defecating (e.g., a person with hemorrhoids) may suppress the urge to defecate, resulting in constipation.

## Common Elimination Problems

**Constipation**: Fewer than three bowel movements per week. Symptoms are small, dry, hard stool; or no passage of stool. Causes include: insufficient intake of fiber and fluids, insufficient activity, irregular defecation habits, changes in daily routine, lack of privacy, chronic laxative or enema use, medications, and depression.

**Fecal impaction**: A collection of hardened, puttylike feces in the rectum. It occurs as a result of prolonged retention of fecal material. In severe cases, the feces accumulate far up into the sigmoid colon. The hardened mass can often be palpated by digital examination of the rectum. Other symptoms include: the passage of liquid fecal seepage, but no passage of normal stool; a frequent desire, but inability to defecate; and rectal pain. The client may experience a generalized feeling of illness, anorexia, abdominal distention, nausea, and vomiting. The cause of impaction is usually poor defecation habits. Fecal impaction is treated with an oil retention enema followed in two to four hours by a cleansing enema. This is followed with daily cleansing enemas, suppositories, or stool softeners. If these measures are ineffective, manual removal of the impacted feces may be necessary.

**Diarrhea**: Passage of liquid feces and increased frequency of defecation. Rapid movement of fecal contents through the large intestine reduces the time available for the large intestine to reabsorb water and electrolytes. Diarrhea is often accompanied by spasmodic abdominal cramps, increased bowel sounds, and vomiting. The feces may contain blood and excessive mucus. Prolonged and persistent diarrhea can result in irritation of the tissue in the anal region,perineum and buttocks. Fatigue, weakness and emaciation also commonly occur with severe diarrhea.

**Bowel incontinence**: Loss of the ability to control discharges of feces and flatus through the anal sphincter. It is usually associated with impaired functioning of the anal sphincter (as in spinal cord trauma or tumors of the external sphincter muscle). Because feces contain digestive enzymes, the area around the anal region may become excoriated.

**Flatulence**: The presence of excessive flatus in the intestines, which leads to stretching and inflation of the intestines. The three primary causes of flatus are: (a) action of bacteria on the chyme in the large intestine, (b) swallowed air, and (c) gas that diffuses from the bloodstream into the intestine. The gases that form in the large intestine are mostly absorbed through the intestinal capillaries into the circulation. However, flatulence can occur from a variety of causes, such as foods (e.g., cabbage), abdominal surgery, or narcotics. When gas is propelled through the colon faster than it can be absorbed, it may be expelled through the anus.

## Diagnostic Tests

**Direct visualization:** Anoscopy, proctoscopy, proctosigmoidoscopy, and colonoscopy allow the viewing of the bowel with a flexible instrument.

**Indirect visualization** is accomplished with x-rays, enhanced by introduction of a radiopaque substance such as barium. For an upper gastrointestinal or small bowel examination, the client drinks barium sulfate. For a lower gastrointestinal examination, the client is given a barium enema. Cleansing enemas must be given after the procedure to prevent impaction.

**Stool specimen collection:** Tell the client to defecate in a clean bedpan or commode and to void before defecating. Tell the client to not place toilet tissue in the bedpan and to notify the nurse immediately after defecation. When handling the bedpan, transferring the stool sample to a specimen container, and when disposing of the bedpan contents, wear disposable gloves and take care not to contaminate the outside of the specimen container. Use tongue blades to transfer the specimen to the container, then wrap them in a paper towel and place them in a waste container. The amount of stool needed depends on the purpose of the test—usually about 1 inch of formed stool or 15-30 mL of liquid stool is enough. Specimens should be sent to the laboratory immediately.

**Testing feces for occult blood (Guaiac Test):** A commonly used test is the Hemoccult test, which detects the presence of the enzyme peroxidase in the hemoglobin molecule. False positive results can occur if the client has recently ingested red meat, raw vegetables or fruits (especially radishes, turnips, horseradish, and melons); medications, such as aspirin, that irritate the gastric mucosa, and anticoagulants. False negative results can occur if the client has taken more than 250 mg/day of vitamin C per day up to three days before the test, even if bleeding is present.

## Related NANDA Nursing Diagnoses

- Bowel Incontinence
- Constipation
- Diarrhea
- Perceived Constipation
- Risk for Constipation
- Risk for Impaired Skin Integrity (related to diarrhea or incontinence)

## Therapeutic Nursing Interventions for Bowel Elimination

- Promote regular defecation by providing as much privacy as possible during defecation.
- Encourage the client to defecate when the urge is recognized, and to establish a time and routine for defecation.
- Assure an adequate intake of fluids and fiber (e.g., prunes, raw fruit, bran and whole-grain cereals and bread).
- For constipation, increase fluid intake and include hot liquids and fruit juices, especially prune juice.
- For diarrhea, encourage intake of fluids and small amounts of bland foods. Assess for potassium losses. Avoid hot or cold fluids because they stimulate

peristalsis; avoid spicy foods and fatty foods.

- For flatulence, limit carbonated beverages and chewing gum. Avoid use of drinking straws. Avoid gas-forming foods.
- Digital examination for fecal impaction should be done carefully to avoid stimulating the vagus nerve and slowing the client's heart.
- For clients with diarrhea and fecal incontinence, keep the anal region clean and dry and protect it with zinc oxide or other ointment.
- If excessive gas cannot be expelled, it may be necessary to insert a rectal tube to remove it.
- For fecal tests for occult blood, teach women to not collect specimens during their menstrual period or for three days afterward.
- For fecal tests for occult blood, teach clients to not collect specimens while they have bleeding hemorrhoids.
- Elevated toilet seats can be purchased for clients who have difficulty moving themselves to and from the toilet.
- When placing a client on a bedpan, wear disposable gloves; warm metal pans by rinsing with warm water. After placing the pan under the client, provide a more normal position by elevating the bed to semi-Fowler's position. Cover the client to maintain comfort and privacy.
- Document color, odor, amount, and consistency of feces, and the condition of the perineal area.

## Age-Related Changes—Gerontological Considerations

- Constipation is a common problem.
- Many older people believe that "being regular," means they should have a bowel movement every day, so they may use over-the-counter preparations. Consistent use of laxatives inhibits natural defecation reflexes and causes even more constipation.
- Older adults may be incontinent because of the inability to recognize the urge to defecate.
- Older adults with impaired mobility may be incontinent because they are unable to get to the toilet soon enough.

# Intestinal Ostomies

**Overview**

An ostomy is a surgical opening made on the abdominal wall for the purpose of providing an alternate feeding route or for the elimination of feces or urine. There are many types of ostomies. The location of the ostomy within the intestine determines the consistency of the excreted stool and the management of the discharge. When an ostomy is placed lower in the intestine, the stool will be more formed. Additionally, stomal discharge is easier to control and this site can be managed with less difficulty.

**Gastrostomy:** An opening through the abdominal wall into the stomach (for feeding)

**Jejunostomy:** An opening through the abdominal wall into the jejunum (for feeding)

**Ileostomy:** An opening into the ileum (small bowel) for draining fecal material. Drainage is constant, it is liquid, and cannot be regulated. It contains digestive enzymes that may damage the skin. The appliance must be worn continuously. Odor is minimal because few bacteria are present.

## Potential Complications

- Skin breakdown around the stoma site
- Infection
- Constipation
- Fluid and electrolyte imbalances
- Stomal prolapse or retraction

## Associated NANDA Nursing Diagnoses

- Altered Sexual Patterns/Sexual Dysfunction
- Body Image Disturbance
- Bowel Incontinence
- Constipation
- Diarrhea
- Impaired Skin Integrity
- Ineffective Individual Management of Therapeutic Regimen
- Risk for Fluid Volume Imbalance
- Risk for Infection

## Assessing a Stoma

- Assess the stoma and the peristomal skin for irritation each time the appliance is changed and report unusual or abnormal findings to the primary health care provider.

- Assess the stoma for the following:
  - **Color**: Should be red; pale or dark-colored hues indicate impaired circulation
  - **Size and shape**: Most stomas protrude slightly from the abdomen. New stomas appear swollen for 2-3 weeks.
  - **Stomal bleeding**: Except for the immediate postoperative period, when the stoma may bleed slightly when touched, bleeding should be reported.
  - **Peristomal skin**: Check for redness and irritation.
  - **Amount and type of feces**: Assess amount, color, odor, consistency; inspect for pus and blood.
  - **Signs and symptoms**: Burning sensation under the faceplate may indicate skin breakdown; also assess for abdominal discomfort or distention.

## Procedure for Replacing an Ostomy Appliance

- Check the provider's orders for the type and size of the appliance and any barrier substance to use.
- Explain the procedure to the client.
- Review the institution's protocols for this procedure as well as the manufacturer's instructions for using the supplies.
- Change the appliance in the bathroom if the client is ambulatory to reduce embarrassment, and to prepare the client to manage the ostomy at home.
- Assemble the equipment.
  - The appropriate appliance or pouch
  - The prescribed skin barrier - paste, powder, water, or liquid skin
  - A stoma measuring guide
  - A pouch clip or other closure device
  - The appropriate adhesive if required
  - Clean gloves
  - Soap and washcloths
  - Warm water
  - Safety razor
  - A bedpan if working at the bedside
  - A bag for disposal of the used pouch
  - Any prescribed deodorizer
- Position the client comfortably.
- Wash your hands and put on the gloves.
- Empty the soiled pouch, then gently remove it by pressing on the skin while pulling on the pouch. Remove the pouch clip, then place the soiled pouch in the disposal bag.
- Clean the peristomal skin with soap and warm water or according to the provider's orders or the institution's protocols.
- Use the razor to remove any excess hair in the peristomal area, then dry the peristomal skin gently. Assess the peristomal skin and the stoma carefully.

- Check the pouch opening for a proper fit around the stoma.
- Apply the prescribed barrier or paste, then gently press the new pouch in place, fitting the opening carefully over the stoma.
- Add deodorant to the pouch as prescribed, and then apply the appliance closure device.
- Remove your gloves and dispose of them and any other materials in the proper receptacle.
- Make sure the client is comfortable.
- Wash your hands.
- Document in the client's record:
  - The condition of the stoma and the peristomal skin
  - The skin and stoma care provided
  - The appliance change
  - How the client tolerated the procedure
  - Any other concerns or observations
- If the client will go home with an ostomy appliance, demonstrate and explain each step of the procedure.
- Ask the client and a family member to perform a return demonstration.
- Arrange for a home health care nurse or an ostomy nurse to make a home visit and check on the client's progress.

## Therapeutic Nursing Interventions for Intestinal Ostomies

- Empty ostomy frequently to keep free of odors.
- Keep stoma site clean and dry. For ileostomies, special care must be taken to prevent skin breakdown.
- Apply a barrier such as karaya gum over the skin around the stoma to prevent contact with any excretions.
- Assess the pouch for correct fit if there is any skin irritation or leakage around the stoma.
- Measure client's intake and output while hospitalized.
- Disposable appliances can be kept on for up to seven days. They must be changed when they begin to leak.
- If feces leak onto the peristomal skin, the appliance should be removed and good skin care given to the peristomal area before applying a new appliance.
- If irritation persists at the stoma or on the surrounding skin, the appliance should be replaced every 24 to 48 hours. Good skin care and any prescribed treatments are priorities until the irritation subsides.
- Control odors because odor control is essential to client's self-esteem.
- Teach clients to include dark green vegetables in the diet (their chlorophyll content helps to deodorize the feces). Bismuth subgallates also help lessen fecal odor. A deodorizer can be placed in the pouch, and some appliances have a charcoal filter disk.
- Teach self-care and dietary considerations. Initially, clients should avoid high-

fiber foods and gas-producing foods.

- Teach client to avoid heavy lifting and contact sports.
- Address self-esteem and sexuality issues.
- Instruct client to avoid laxatives and enemas because they may cause severe fluid and electrolyte imbalance.

## Age-Related Changes—Gerontological Considerations

- Gastrointestinal motility, muscle tone and digestive enzymes decrease with age, predisposing the older adult to changes in bowel patterns, especially constipation.
- Older adults may not consume adequate amounts of fiber necessary to promote normal bowel elimination.
- The incidence of colorectal cancer increases with age and is characterized by changes in bowel patterns.
- Impaired vision, confusion, and arthritis of the hands may interfere with the older adult's ability to care for an ostomy appliance or irrigate a colostomy.

# Enemas

## Overview

An enema is the introduction of a solution into the rectum and large intestine. Enemas act by distending the intestine, and sometimes by irritating the intestinal mucosa, thereby increasing peristalsis and the expulsion of feces and flatus. They are administered to relieve constipation or fecal impaction, prevent contamination of the sterile field during surgery, promote visualization of the intestinal tract during x-ray, and treat parasite and worm infestations.

## Types of Enemas

**Cleansing enemas** are given to remove feces. They are used to (1) treat constipation and impaction, (2) prevent contamination of the sterile field during surgery, (3) promote visualization of the intestine for certain diagnostic tests (e.g., colonoscopy, x-ray), and (4) as a part of a bowel training program to establish regular bowel function. They use a variety of solutions.

**Hypertonic solutions** (e.g., saline) stimulate peristalsis by exerting osmotic pressure and drawing fluid from the interstitial space into the colon. A "Fleet Enema" is hypertonic.

- **Hypotonic solutions** (e.g., tap water) stimulate peristalsis by exerting a lower osmotic pressure than the interstitial fluid, causing water to move from the colon to the interstitial space. A tap water enema should not be repeated because of the risk for circulatory overload.

- **Isotonic solutions** (e.g., normal saline) do not create movement of fluid in or out of the colon. The volume of the solution stimulates peristalsis. A soapsuds enema increases the volume in the colon and irritates the mucosa. Pure soap (e.g., castile soap) must be used in order to minimize irritation of the mucosa.

- **Large volume** enemas are 500-1000 mL (for an adult). Others, primarily hypertonic solutions, are small volume. The volume depends on the age of the individual.

- **High (or low) enemas:** Cleansing enemas are also classified as high or low. A low enema is given to cleanse the rectum and sigmoid colon only; the client is in a left lateral position during administration. A high enema is given to cleanse as far up in the colon as possible. The client must change positions during administration—from left lateral to dorsal recumbent to right lateral–so the solution can follow the large intestine.

**Carminative enema** solutions release gas, distending the rectum and colon, and stimulating peristalsis. They are given to expel flatus. About 60-80 mL of fluid is used.

**Oil Retention enemas** are given to soften feces and lubricate the rectum and anal canal, facilitating passage of feces. About 90-120 mL of oil (e.g., mineral, olive, or cottonseed oil) is introduced into the rectum and sigmoid colon and retained for 1-3 hours.

**Return-flow enemas** are sometimes used to expel flatus. 100 to 200 mL of fluid is instilled and allowed to flow in and out of the rectum and colon to stimulate peristalsis. The process is repeated several times, until flatus is expelled and abdominal distention is relieved.

Other Types of Enemas:

- **Antibiotic enemas** are used to treat local infections.
- **Anthelmintic enemas** are used to kill helminths (worms and intestinal parasites).
- **Nutritive enemas** are used to administer fluids and nutrients to the rectum.

## Potential Complications of Enema Administration

- Perforated bowel
- Circulatory overload (hypotonic solutions)
- Fluid and electrolyte imbalance if given repeated hypotonic enemas
- Rectal mucosa irritation or damage

## Procedure and Techniques for Enemas

- Instruct the client about the purpose of the enema and where defecation should occur (e.g., on the bedpan, in the toilet).
- Protect against contact with body fluids (e.g., wear gloves).
- Minimize trauma to the sphincters (e.g., lubricate insertion tube).
- Use gravity: For example, (a) place the client in the left lateral (Sims) position so the fluid flows down into the sigmoid on the left side; (b) raise the solution container to increase the force of flow.
- The force of the solution is controlled by (a) the height of the solution container, (b) the size of the tubing, (c) the viscosity of the fluid, and (d) the resistance of the rectum. As a general rule, the solution should be about 12 inches above the rectum, the higher the container, the faster the flow and the greater the force.
- Minimize discomfort to the client (e.g., administer fluid slowly; if client complains of fullness or pain, clamp the tube for 30 seconds and restart at a slower rate).
- Measure the volume instilled and document the results of the enema.
- Enemas are contraindicated for clients immediately after rectal or anal surgery and for those with inflammatory bowel disorders.
- Procedure for **cleansing enema**:
  - Review the physician's order for specific type, volume, and time of administration.
  - Fill the enema container with the prescribed amount or 750-1000 mL of lukewarm (105-110°) water or solution ordered.
  - Hang the enema container on an IV pole and prime the tubing to remove air, then clamp the tubing.
  - With a bedpan nearby and the patient's privacy needs met, have the client move to his/her left side in Sims position (left lateral, semi-prone).
  - Lubricate the tip of the enema tube and gently insert the tubing into the rectum 3-4 inches. Have the client take slow deep breaths as the tube is inserted.

- Adjust the height of the enema bag (no greater than 18 inches for a "high enema" and from 12-18 inches for a "low enema;" in this case high and low describe how far the enema goes into the colon).
- Open the clamp. The fluid should flow slowly, using gravity, to protect from cramping or damage to the rectum.
- Never release the enema tube while it is in the rectum.
- Once the enema is complete, clamp the tube and gently remove the tubing.
- The client should be instructed to hold the fluid for at least 10-15 minutes (or according to the primary care provider's order) or as long as the client possibly can.
- Provide for the client's privacy and give the client any necessary assistance (e.g., bedpan, assist to bathroom or commode) while the fluid is expelled.
- Document the intervention and the results of the enema.

- Procedure for a **disposable enema**:
  - Review the provider's order for specific type and time of administration.
  - Review the directions on the packaging.
  - With a bedpan nearby and the patient's privacy needs met, have the client move to his/her left side in Sims position (left lateral, semi-prone).
  - Lubricate the tip of the enema tube and gently insert the tubing into the rectum 3-4 inches. Have the client take slow deep breaths as the tube is inserted.
  - Slowly empty the container into the rectum.
  - It is suggested that the container be rolled (like a toothpaste tube) as it is emptied and kept tightly rolled while removing to prevent the vacuum of the bottle from removing the enema solution.
  - Follow the manufacturer's recommendation or the primary care provider's orders when instructing the client in retaining the enema solution.
  - Provide for the client's privacy and give him/her any necessary assistance (e.g., bedpan, assist to bathroom or commode) while the fluid is expelled.
  - Document the intervention and the results of the enema.

## Age-Related Changes—Gerontological Considerations

- Because of impaired mobility, older adults may have difficulty getting to the bathroom in time to expel the enema in the toilet. The client can assume the supine position, with knees flexed, and head of bed elevated slightly; the nurse administers the enema with a bedpan under the client.
- Older clients may not be able to retain even small volumes of enema solutions. The nurse should administer the enema with the client on a bedpan and press the client's buttocks together to help retain the solution.

# Critical Thinking Exercise: Bowel Elimination – Ostomies and Enemas

1. You are caring for a client whose physician has ordered an opioid analgesic in order to manage severe pain associated with prostate cancer with extensive bone metastases. How might this regimen affect normal bowel elimination? What interventions might you recommend to promote normal bowel function?

2. As you perform a nursing assessment on a healthy 50-year-old female at a hospital based clinic, the client reports that she normally has a bowel movement every other day. "Is this normal?," she asks. How do you respond?

3. With the help of a clinical instructor, arrange an opportunity to examine the various appliances used with clients requiring fecal diversions. If possible, interview a wound and ostomy care nurse to learn about the various educational needs of clients with stomas.

# Normal Urinary Elimination

## Key Points

- Urinary elimination is the natural process in which the kidneys and bladder eliminate waste products and other materials.
- Normal urine volume is 500-2,500 mL/day, with an average of 1,200 mL.
- Normal urine is clear, light yellow in color, with a pH range of 4.6-8.0 and specific gravity of 1.010-1.025.
- Nursing measures to support normal voiding include:
  - Scheduling to support client's normal voiding pattern
  - Providing privacy
  - Positioning (to assume a normal voiding position, e.g., standing, for males)
  - Providing perineal hygiene as needed
  - Perineal care for women should proceed from front to back; teach clients to wipe from front to back after voiding or defecating.
- Acidic urine is irritating to the skin.
- When measuring urine, wear disposable gloves.
- Place calibrated container on flat surface at eye level to read measurements.
- Bladder capacity decreases by up to 50% in older adults; nocturia, urinary urgency, urinary frequency, and residual urine are common.
- Medications that can affect voiding (as either a therapeutic or side-effect) are: diuretics, anticholergics, antidepressants, antipsychotics, antiparkinsonism drugs, antihistamines, and antihypertensives
- **Key Concepts/Terms**: Kidneys, nephron, ureters, urinary bladder, detrusor muscle, urethra, urinary meatus, micturition (voiding, urination), diuretics, urinary retention, urinary tract infection (UTI), urinary incontinence, enuresis, polyuria, oliguria, anuria, reflux, basic urinary tract anatomy review, multiple factors that affect voiding, common urinary problems, nursing assessments, diagnoses, and therapeutic interventions

**Overview**

Urinary elimination is a natural process in which the kidneys and bladder eliminate waste products and other materials. Urinary elimination is usually taken for granted until a problem arises. Urinary habits are affected by culture (e.g., the need for or lack of privacy) and personal preferences and situations (e.g., the availability of a clean, private facility). Urinary elimination depends on effective functioning of the kidneys, ureters, bladder, and urethra. The normal volume of urine is about 500-2500 mL/day with the average being 1,200 mL. The color normally is clear, light yellow. Normal pH is about 6.0, with a range of 4.6 to 8. Normal specific gravity is 1.010 to 1.025.

**The kidneys** (one on either side of the spinal column, behind the peritoneal cavity) maintain the composition and volume of body fluids by filtering the blood, retaining those constituents that are needed and excreting those that are not. They are the primary regulators of fluid and acid-base balance. About 23% of the cardiac output (or 1200 mL of blood) passes through the kidneys each minute. The basic unit of kidney structure is the nephron. Each kidney is made up of about 1 million nephrons. Nephrons filter blood and remove waste. Every nephron has a glomerulus with porous capillaries that remove waste but will not allow plasma protein(s) or blood cells to be removed from the blood. This is the initial step in urine formation. The waste product that the nephrons form is called urine. Hormones such as antidiuretic hormone (ADH) and aldosterone affect the reabsorption of sodium and water, thereby influencing the amount of urine formed.

**The ureters** (one for each kidney) transport the urine from the renal pelvis to the bladder. Where they enter the bladder, a fold of mucous membrane functions as a valve to prevent reflux, which is the backflow of urine up the ureters.

**The urinary bladder** is a hollow organ made of smooth muscle. It acts as a reservoir for urine and as the organ of excretion. The bladder has three layers of muscle tissue; together, the three layers are called the detrusor muscle.

**The urethra** extends from the bladder to the urinary meatus (opening). The male urethra serves as a passageway for both semen and urine; it is about 5-7 inches long. The female urethra functions only in the excretory system and is 1-1/2 to 2-1/2 inches long. The internal sphincter muscle, at the base of the bladder, is under involuntary control. The external sphincter is under voluntary control, allowing the person to choose when to void.

**Micturition (voiding, urination)** is the process of emptying the bladder. Micturition is stimulated when enough urine (approximately 250-450 mL) collects in the bladder to stimulate stretch receptors. Impulses from the stretch receptors are transmitted through the spinal cord to the brain, causing the internal sphincter to relax (unconscious control) and allow urine to enter the posterior urethra. If the time and place are appropriate, the conscious portion of the brain relaxes the external sphincter and urination takes place. If not, the micturition reflex usually subsides until the bladder fills even more and again stimulates the reflex.

## Factors Affecting Voiding

**Developmental factors**: Voluntary control of micturition does not develop until between the ages of two and five years.

**Food and fluids**: Normal kidneys preserve a balance of fluid intake and output. When the body is dehydrated, they reabsorb fluid; the urine becomes more concentrated and decreases in amount. When there is fluid overload, the kidneys excrete a large amount of dilute urine. Caffeine (e.g., in coffee, tea, and colas) and alcohol produce a diuretic effect. High-sodium foods and beverages cause sodium and water reabsorption and retention, decreasing urine formation. Foods such as asparagus and onions affect the odor of the urine; beets cause urine to appear red.

**Psychologic variables:** People under stress void smaller amounts at frequent intervals. Stress may also make it difficult to relax the external urethral sphincter and perineal muscles, making it difficult to empty the bladder completely. People may voluntarily suppress urination because of time pressures. Others may be embarrassed about needing assistance with a bedpan or urinal.

**Activity and muscle tone:** Regular activity is needed to maintain bladder and sphincter tone and urinary control. People with indwelling catheters lose bladder muscle tone because the bladder is not allowed to fill with urine and stretch the muscles. Childbearing, menopause, and trauma can also contribute to loss of muscle tone.

**Pathologic conditions:** Renal and urologic problems (e.g., congenital abnormalities, polycystic kidney disease, urinary calculi, urinary tract infection, hypertension, diabetes, and gout) can affect the quantity and composition of urine produced. Diseases, such as arthritis and Parkinson's disease, reduce physical activity and lead to generalized weakness. Cognitive deficits and some psychiatric problems can interfere with a person's ability to control urination. Fever and diaphoresis cause the kidney to conserve body fluids, resulting in decreased urine production and concentrated urine. Congestive heart failure can lead to fluid retention and decreased urine output. Hypertrophy of the prostate gland, common in older men, may obstruct the urethra, interfering with urination.

**Medications:** Many prescription and over-the-counter drugs can cause kidney damage with long-term use or abuse (e.g., aspirin, antibiotics). Diuretics (e.g., furosemide and ethacrynic acid) are often prescribed for cardiac disease and hypertension. They increase urine production by preventing the reabsorption of water and electrolytes from the kidney tubules into the bloodstream.

**Urinary retention:** May be caused by (1) **anticholinergic** medications (e.g., atropine, belladonna, and papaverine), (2) **antidepressants** and **antipsychotics** (e.g., phenothiazines and MAO inhibitors), (3) **antiparkinsonism** drugs (e.g., levodopa, Artane, Cogentin), (4) **antihistamines**, and (5) **antihypertensives** (e.g., Apresoline and Aldomet).

## Urinary Problems

**Urinary retention** is the inability to empty the bladder completely. Urine collects in the bladder, stretching the walls and causing pressure. Symptoms include a lack of urinary output for several hours, a distended bladder, discomfort, restlessness, and diaphoresis. Urinary retention can lead to urinary stasis and urinary tract infections. Eventually, retention with outflow may develop. This is where pressure in the bladder becomes so high that the external sphincter is unable to hold back urine; the sphincter opens to allow less than 100 mL of urine to escape or leak. The client may void small amounts two or three times per hour with no relief of distention. To assess for urinary retention, the primary care provider may order a bladder scan or a post-voiding catheterization; both will allow for measurement of the amount of urine left in the bladder. If either of these tests are ordered, they are to be done immediately after the client voids. Some causes of retention are prostate gland enlargement, urethral edema after surgery, and spinal cord trauma.

**Lower urinary tract infections (UTI):** Microorganisms enter the urinary tract through the urethral meatus or the bloodstream, but most commonly through the urethral meatus. Women are especially susceptible to infection because of the proximity of the urethra to the anus and because the urethra is short. E. coli from the rectum is a common causative organism. Some risk factors for lower UTI are: poor perineal hygiene, frequent sexual intercourse, kinked urethral catheter, urinary retention, improperly handled instruments and urine receptacles. Symptoms include burning on urination and frequent voiding of small amounts.

**Urinary incontinence**: Loss of control over micturition. Leakage may be continuous or intermittent. See "Urinary Elimination: Urinary Incontinence" for further information.

**Enuresis**: Repeated involuntary urination in children who are old enough for voluntary control to be possible (usually around age 5). Enuresis occurs more commonly at night, during deep sleep. It may have an organic cause (e.g., UTI), or it may be of psychological origin.

**Polyuria**: Production of abnormally large amounts of urine by the kidneys. May be caused by excessive fluid intake or diseases such as diabetes mellitus or chronic nephritis. Can cause excessive fluid loss, intense thirst, and dehydration. Diuresis is another term for excretion of abnormally large amounts of urine. This term is used primarily when medications (diuretics) are given for the purpose of promoting urine output, or when the client has ingested caffeine or alcohol.

**Oliguria**: Low urine output (<500 mL a day or <30 mL an hour). May be a result of abnormal fluid losses, lack of intake, or impending renal failure. It should be reported promptly.

**Anuria**: Lack of urine production, with essentially no output. May be a result of abnormal fluid losses or lack of intake, or impending renal failure. It should be reported promptly.

## Related NANDA Nursing Diagnoses

- Altered urinary elimination
- Functional incontinence
- Reflex incontinence
- Stress incontinence
- Total incontinence
- Urge incontinence
- Urinary retention
- Risk for infection related to urinary retention and/ or invasive procedures (e.g., catheterization)
- Risk for impaired skin integrity related to prolonged skin dampness secondary to incontinence
- Self-care deficit: Toileting
- Self-esteem disturbance related to incontinence
- Social isolation related to incontinence

## Assessments

**The nursing history** should include questions about:

- Normal voiding pattern
- Appearance of the urine and any recent changes (e.g., color, odor)
- Past or current problems with urination (e.g., painful voiding, difficulty getting to the bathroom, leakage, past infections)
- Presence of urinary diversion ostomy and how the client manages it
- Factors that influence urinary elimination (see *"Urinary Problems"*)

**Physical assessment** should include: (1) percussion of the kidneys for tenderness, (2) palpation and percussion of the bladder, (3) inspection of the urethral meatus, if indicated by the history, (4) assessment of the skin for turgor, edema, and irritation.

## Procedure and Techniques for Urinary Elimination

**Supporting and maintaining normal voiding habits** requires attention to the following:

- **Schedule**: Support the client's normal voiding pattern as much as possible. Some clients become anxious if their voiding pattern is interrupted; others may experience urgency and need help with voiding.
- **Privacy**: Many people find it difficult to void in the presence of another person. Provide privacy unless you must stay for safety.
- **Position**: Help clients to assume a normal voiding position. Some males cannot use a urinal unless they are standing. Some females cannot void on a bedpan, but can void using a bedside commode.
- **Hygiene**: It is difficult for a client to perform perineal hygiene when confined to bed. The nurse may need to provide cleansing to promote comfort and prevent infection. Place the client on a bedpan and pour warm soapy water over the perineum, followed by clear water.

**Preventing infection**: Teach women to wipe from front to back after voiding or defecating, to avoid transferring microorganisms into the urethra; and to wash from front to back when performing perineal care.

**Protecting skin integrity**: Incontinent clients require meticulous skin care because acidic urine is irritating to the skin; the incontinent client is at risk for pressure ulcers.

**Measuring urinary output**

- Protect against contact with microorganisms or blood in the urine (e.g., wear gloves).
- Have the client void in a clean receptacle (e.g., bedpan, or toilet collection "hat" if client is ambulatory).
- Instruct client not to put toilet paper in the urine container.
- Instruct client to keep urine separate from feces.
- Pour urine into a calibrated container and place on a flat surface at eye level to read the amount.
- When the client has an indwelling catheter, take the calibrated container to the bedside, place it under the catheter bag so that the spout of the bag is above,

but not touching the container. The inside of the bag is sterile; the calibrated container is not; be careful not to contaminate the spout.

## Age-Related Changes—Gerontological Considerations

- Older people with impaired cognition may not be aware of the sensation of the need to urinate, leading to urinary incontinence.
- Urinary urgency and frequency are common (as a result of an enlarged prostate gland in men, and as a result of muscle weakness in women).
- Bladder capacity decreases by up to 50%; it also loses ability to empty completely, leading to nocturia and residual urine.
- Older adults may have difficulty reaching toilet facilities in time because of mobility and cognition problems, leading to incontinence.
- Excretory function of the kidneys decreases (about 30% of the nephrons are lost by age 80; renal blood flow decreases; ability to concentrate urine declines), but usually not significantly unless a disease process is present.

# Critical Thinking Exercise: Normal Urinary Elimination

1. You are caring for a 43-year-old female client with quadriplegia. Describe appropriate interventions that may be used to minimize the risk of urinary tract infections.

2. The son of an 82-year-old client who is taking atropine asks you about the potential adverse effects of this medication. With respect to urinary elimination, what risks are associated with this medication?

3. A 35-year-old female client has experienced recurrent lower urinary tract infections over the past year. After performing a thorough assessment and formulating an appropriate nursing diagnosis, develop a teaching plan to assist this client. What particular educational needs might you address?

# Diagnostic Testing/Collection of Urine Specimens

| Key Points |
| --- |

- Some diagnostic tests require sterile technique for specimen collection; others do not.
- Nurses should wear gloves when handling urine specimens and take care to not contaminate the outside of containers.
- Gravity, urinary pH, glucose, ketones, protein, occult blood
- Laboratory tests to evaluate urinary and renal function: urinalysis, blood urea nitrogen, creatinine clearance
- Visualization procedures: X-ray of the kidneys, ureters, and bladder (KUB); intravenous pyelography (IVP), retrograde pyelography, computerized tomography (CT), ultrasonography, and cystoscopy
- Some tests require the withholding of fluids for the 12 to 24 hours preceding the test
- Urine specimens must be free of fecal contamination.
- Laboratory analysis is difficult or impossible if there is paper in the specimen.
- Do not contaminate self or objects (e.g., the outside of the specimen container) with urine (e.g., wear gloves).
- Properly identify the specimen; never take an unlabeled specimen from the patient's room.
- Urine decomposes when left at room temperature; transport specimens to lab promptly.
- **Key Concepts/Terms:** Spectrometer, refractometer, ketoacidosis, urinometer, hydrometer, clean voided specimen, clean-catch (midstream) specimen, sterile specimen from indwelling catheter, 24-hour (timed) specimens

**Overview**

Nurses routinely collect sterile and unsterile urine specimens to determine abnormal constituents of the urine; determine the specific gravity; or obtain specimens for routine urinalysis, bacterial culture, and other laboratory studies. Nurses are also responsible for preparing clients before and for follow-up care after visualization procedures such as intravenous pyelography (IVP) and computerized tomography (CT scan). Nurses should wear gloves when handling urine specimens and take care not to contaminate the outside of containers (see section on *"Collecting Urine Specimens"*).

## Tests for Abnormal Constituents of Urine

The following tests are frequently performed by nurses on the unit, but they may be done in the laboratory, as well:

**Specific gravity**: Measures urine concentration. Normal range is 1.01-1.025. A high specific gravity may indicate fluid deficit, dehydration, or excess solutes (e.g., glucose) in the urine. If a urinometer or hydrometer is used, it is placed in a cylinder containing at least 20 mL of fresh urine. The reading should be done at eye level. If a spectrometer or refractometer is used, only one or two drops of urine are needed; the instrument gives a digital readout of the specific gravity.

**Urinary pH**: Measures the relative alkalinity or acidity of urine. Provides information about the client's acid-base status. Normally, urine is slightly acidic, with an average pH of 6 (7 is neutral, >7 is alkaline). Urine may be sent to the laboratory for a quantitative measurement of pH, or dipsticks or litmus paper may be used by the nurse. Many commercially prepared kits are available. Read the instructions and check the testing materials to be sure they are not outdated. Instruct clients to wash their hands with warm, soapy water before and after collecting and testing urine samples.

**Glucose**: Screens clients for diabetes mellitus and pregnancy-induced diabetes. A dipstick is used; normally, there is no amount of glucose in the urine.

**Ketones**: Reagent tablets or a dipstick are used to test for ketoacidosis in clients who are alcoholic, fasting, starving, or on a high-protein diet. Ketone bodies are produced by the breakdown of fatty acids; normally they are not present in the urine.

**Protein** should not be present in the urine. Protein molecules are too large to pass through glomerular capillaries into the urine unless the glomerular membrane has been damaged (e.g., because of infection or inflammation). Testing is usually done with a dipstick.

**Occult Blood**: Normal urine does not contain blood. Blood may be visible or invisible (occult). Commercial test strips are used to test for blood in the urine.

### Laboratory tests that are used to evaluate urinary and renal function

**Routine urinalysis**: A routine urinalysis usually includes all of the tests in the preceding section, as well as the appearance, color, and odor of the urine. In addition, it usually reports leukocyte esterase, and the presence of red and white blood cells, casts, crystals, and bacteria. A sterile specimen is not required.

- **Blood urea nitrogen (BUN)** tests for urea, the end product of protein metabolism. Urea is normally eliminated by the kidneys through filtration and tubular secretion.

- **Creatinine clearance**: Creatinine is produced in relatively constant quantities by the muscles and filtered by the kidneys. The creatinine clearance test uses 24-hour urine and serum creatinine levels to determine the glomerular filtration rate.

# Collecting Urine Specimens

## Overview

Nurses are responsible for collecting urine specimens for a variety of tests. Some tests require clean voided specimens, some require sterile specimens, and others require timed specimens. When handling body fluids, the nurse should wear gloves and take care not to contaminate other objects with the specimen. Care should also be exercised to see that the specimen is not contaminated with feces, menstrual blood, and so forth, during collection.

## Clean Voided Specimens

About 120 mL (4 oz.) of urine is required. Many clients do not need help to collect a clean voided specimen. Male clients can usually void directly into the specimen container (a clean, plastic cup with a lid). Female clients can sit or squat over the toilet, holding the container to catch the urine as they void. Clients who are physically incapacitated or disoriented may need to use a bedpan or urinal in bed; others may require assistance in the bathroom or on the commode. See "Therapeutic Nursing Interventions" for other instructions.

## Clean-Catch (Midstream) Specimen

A midstream-voided specimen is required for certain tests (e.g., to identify microorganisms causing a urinary tract infection). To obtain a midstream specimen, the external meatus is first cleansed with soap and water. Caution the client to hold the specimen container near, but not touching, the meatus. The client voids and discards a little urine (e.g., voids about 30 mL into the toilet); then the client voids into the specimen container; the last urine in the bladder is also discarded. If the client cannot void into the container, use a sterilized bedpan to collect the specimen.

Urine is collected in a sterile specimen container with a lid, although some contamination by skin bacteria may occur with this method. Disposable clean-catch kits are widely available. A clean-catch midstream specimen from a male is considered sterile; females must be catheterized to obtain a sterile specimen.

## Sterile Specimen from Indwelling Catheter

The specimen should be withdrawn from the catheter, not the collecting bag, using the special port for withdrawing urine. A urine culture requires about 3 mL of urine; a routine urinalysis requires at least 10 mL. Wearing disposable (not sterile) gloves, use a sterile 21-25-gauge needle and a sterile specimen container. If no urine is found in the tube below the collection port, clamp the tube briefly (for <30 minutes). Clean the port with an antiseptic swab, insert the sterile needle, aspirate urine, and transfer the specimen into the container.

## 24-Hour (Timed) Specimens

Some laboratory tests require collection of all urine voided over a specified period of time, usually 24 hours. The specimen container is usually either refrigerated, kept on ice, or contains a preservative. Depending on the type of test, the urine from each voiding may be kept in a separately marked container, on which the time of the voiding is recorded. All urine voided in the next 24 hours is collected. At the end of the 24 hours, have the client empty the bladder completely and keep this voiding as part of the specimen. The client and other members of the nursing team must understand the need for collecting all the urine voided in this period. It is a good idea to make notes in the chart and post a sign on the bathroom door to remind them not to discard any urine.

## Procedure and Techniques for Collecting Urine Specimens

**Urine specimens must be free of fecal contamination**. Advise clients to void in the urine receptacle and defecate in the toilet, or to collect the urine specimen at a time when they do not also need to defecate.

**Laboratory analysis is difficult if there is paper in the specimen**. Tell the client to discard toilet tissue in the toilet or a waste bag rather than in the bedpan.

**Take care not to contaminate other objects with urine** (e.g., wear gloves; place lids tightly on containers to prevent spilling and contamination of other objects). If the outside of a container has been contaminated with urine, clean it with a disinfectant.

**Take care that the specimen is properly identified**. Be sure that the specimen label and the laboratory requisition have the correct and identical information; attach them securely to the container before leaving the client's room, or at least place a piece of tape on the container and write the client's name; never take a specimen from the client's room without first labeling it.

**Urine decomposes when left at room temperature over time**. Transport specimens to the lab promptly, or refrigerate them. Find out from the laboratory whether a preservative is needed for 24-hour specimens.

## Visualization Procedures

"KUB": X-ray of the kidneys, ureters, and bladder

**Intravenous pyelography (IVP)**: Contrast medium is injected intravenously and x-rays are taken to evaluate urinary tract structures.

- **Preparation**: A laxative is given the evening before, and an enema the morning of the procedure. Food and fluids are withheld for the 12 hours before the examination.

- **Aftercare**: Observe for reaction to contrast material (e.g., rash); food and fluids may be given immediately.

**Retrograde pyelography**: Contrast medium is instilled, sometimes under anesthesia, directly into the kidney via the urethra, bladder, and ureters; x-rays are then taken to evaluate urinary tract structures.

- **Preparation**: Same as for IVP, except that the client should void before the examination.

- **Aftercare**: If anesthesia was used, check vital signs regularly and withhold food and fluids for several hours. Ureteral catheters may be in place and connected to drainage receptacles; observe the amount and character of drainage from each. Observe for reactions to contrast material.

**Computerized tomography (CT scan)**: An x-ray beam scans the body from different angles to provide a clear, computerized, cross-sectional image of the kidneys. It is effective in diagnosing diseases during their early stages. A contrast dye may or may not be used.

- **Preparation**: Usually NPO for eight hours before the test if a contrast dye is used. Remove metal objects (e.g., religious medals, rings). Check for allergy to shellfish and iodine.

- **Aftercare**: Observe for reaction to the contrast dye. Client can resume usual diet and activity.

**Ultrasonography (ultrasound)**: Uses high-frequency sound waves to produce an image or photograph of the kidneys (or other organs). This is a noninvasive procedure.

- **Preparation**: Food and fluids are restricted for eight to 12 hours prior to the test. The client should not smoke or chew gum, to prevent swallowing air.

- **Aftercare**: No special care required.

**Cystoscopy**: Direct visual examination of the urethra, bladder, and ureteral orifices with a cystoscope. Sedatives and analgesics may be prescribed prior to the procedure.

## Age-Related Changes—Gerontological Considerations

- Older adults, as well as debilitated and very young clients, may not tolerate the withholding of fluids for 12 to 24 hours before diagnostic tests.

## Sterile Specimen from Indwelling Catheter

The specimen should be withdrawn from the catheter, not the collecting bag, using the special port for withdrawing urine. A urine culture requires about 3 mL of urine; a routine urinalysis requires at least 10 mL. Wearing disposable (not sterile) gloves, use a sterile 21-25-gauge needle and a sterile specimen container. If no urine is found in the tube below the collection port, clamp the tube briefly (for <30 minutes). Clean the port with an antiseptic swab, insert the sterile needle, aspirate urine, and transfer the specimen into the container.

# Critical Thinking Exercise: Diagnostic Testing/Collection of Urine Specimens

**Situation:** A client with Type 1 diabetes has experienced an unexplained elevation in blood glucose levels (above 250 mg/dl) and will have her urine tested for the presence of ketones. Before providing a urine sample, the client asks you to explain the purpose of the testing. As you formulate your response, consider these questions:

1. What are ketones?

2. For a diabetic, what does the presence of ketones in a urine sample indicate?

3. When should the urine be tested for ketones?

# Urinary Devices and Their Care

| Key Points |
| --- |

- Urinary tract infections (UTIs) account for about half of all nosocomial infections; urinary catheters cause most nosocomial UTIs. **Sterile technique is essential when inserting, caring for, or removing a catheter!**
- Catheter size: the smaller the number, the smaller the lumen (usually #14–#16 for adults).
- Prevent contamination of the system:
  - Do not allow a catheter bag to lie on the floor; the floor is considered grossly contaminated.
  - When emptying the drainage bag, do not allow the drainage spout to touch the collection receptacle.
  - The less frequently a catheter is changed, the less the likelihood of infection.
  - Provide for gravity drainage (e.g., tubing not kinked, bag below level of the bladder).
- Client should void within four to six hours after an indwelling catheter is removed.
- Dorsal recumbent position is optimal for catheter insertion; however, Sim's position can be used as an alternative, when necessary.
- Irrigation is a washing out or flushing with a specified solution.
- Irrigation should not cause discomfort or pain.
- To perform open irrigation:
  - Obtain supplies
  - Establish a sterile field
  - Open system and irrigate
- Re-establish closed system
- Important principles:
  - Minimize client anxiety and embarrassment
  - Protect self from exposure to body fluids
  - Prevent transfer of microorganisms to the client
  - Maintain accurate record of intake and output
  - Assess status of urinary tract
- Urine drains continuously into the pouch so voluntary control is not possible.
- Nursing interventions are based on the following principles:
  - The pouch should be emptied and changed as needed (e.g., to keep the seal from loosening); usually this is several times a day.

- Change pouch when urine production is at its lowest.
- Protect self from contact with body fluids.
- Maintain skin integrity (e.g., clean and dry skin around stoma; apply skin protectant).
- Assure adhesion of the faceplate to the skin. Reusable pouches are washed with lukewarm soap and water and allowed to air dry.

- About 50% of those living in extended care facilities or receiving home care are incontinent.

- 80% of incontinence can be cured or significantly improved with treatment.

- Condom catheters have a comparatively low risk for infection and make self-care easier. On the other hand, they increase the risk for excoriation and impaired skin integrity.

- Independent nursing interventions for incontinence focus on:
  - Behavior-oriented continence training
  - Skin care
  - Application of an external drainage device (for males)

- **Key Concepts/Terms**: Indwelling urethral catheter (retention catheter, Foley catheter), straight catheter, suprapubic catherter, external urinary drainage device (condom catheter), closed drainage system, urinary tract infection (UTI), antiseptic handwash, closed irrigation, continuous closed irrigation, intermittent closed irrigation, ileal conduit, continent urinary diversion, urinary incontinence, total incontinence, functional incontinence, stress incontinence, urge incontinence, reflex incontinence, post-void residual, provacative stress test, voiding, diary, bladder training, habit training, prompted voiding, biofeedback, Kegel exercises, vaginal cones, electrical stimulation, condom catheter, maceration

## Overview

Urinary catheterization is the introduction of a catheter through the urethra into the bladder for the purpose of emptying urine. Catheters are used to keep incontinent clients dry, to relieve bladder distention, to assess accurate fluid balance, to keep the bladder from becoming distended during diagnostic tests and surgery, to collect a urine specimen, and to measure the residual urine left in the bladder after voiding. **Urinary tract infections (UTIs) account for about half of all nosocomial infections, and urinary catheters are responsible for most of those UTIs.**

## Types of Catheters

**Indwelling urethral catheter (retention catheter, Foley catheter):** Indwelling catheters are used when a catheter is needed for continuous drainage of urine, for gradual decompression of an over-distended bladder, and for intermittent bladder drainage and irrigation. The indwelling catheter is designed so that it does not slip out of the bladder. It has a balloon, which is inflated with sterile water after the catheter is inserted. Most have two lumens: a small lumen is connected to the balloon; the other, large lumen is the one through which the urine drains. A triple-lumen catheter has an additional lumen for instilling irrigating solutions. Retention catheters are usually connected to a **closed gravity drainage system**, consisting of the catheter, tubing, and a collection bag. Closed systems reduce the risk of microorganisms entering the system and infecting the urinary tract.

**Straight catheter**: Straight catheters are used to drain the bladder for short periods (e.g., 5-10 minutes). They are inserted and removed immediately after the urine is drained. A straight catheter is a single-lumen tube with a small opening about 1/2 inch from the insertion tip.

**Suprapubic catheter**: This type of catheter is inserted into the bladder through a small incision just above the pubic area. Care of clients with a suprapubic catheter is covered in medical-surgical texts.

**External urinary drainage devices (e.g., condom catheter)**: Condom catheters are commonly used for incontinent males because the risk of infection is minimal. Refer to the section on *"Urinary Incontinence."*

## Catheter Materials and Sizes

**Materials** are selected according to the length of time the catheter is expected to be in place. **Plastic** catheters are used for one week or less because they are inflexible. **Latex** or **rubber** catheters are used for periods of two to three weeks. **PVC** catheters are used for 4- to 6-week periods; they soften at body temperature. **Silicone** catheters are designed for long-term use (2-3 months) because there is less encrustation at the meatus; they are expensive.

**Length**: For females, use a 22-cm catheter; for males, a 40-cm catheter.

**Size (circumference)** of the catheter is determined by the size of the urethral canal. The smaller the number, the smaller the lumen. Use #8– #10 for children, #14– #16 for adults. Men frequently require an #18.

**Balloon size**: For adults, the usual balloon size is 10 mL. A 30-mL balloon or larger is used to achieve hemostasis of the prostatic area following prostate surgery.

## Potential Complications of Catheterization

**Urinary tract infection (UTI)**: This is the most common complication of catheter insertion. Signs and symptoms of UTI include burning on urination, urgency, frequency, cloudy urine, pelvic or back pain, nocturia, incontinence, hematuria, and sometimes confusion in older clients.

**Sepsis**: Most clients who have a catheter in place for more than two weeks will develop bacteriuria, which can lead to sepsis.

**Trauma** can result when the mucous membrane lining the urethra is damaged by the friction from the catheter. The male urethra, because of its length, is especially vulnerable.

## Basic Principles for Inserting a Urinary Catheter

- **Major requirements** for catheterization are proper aseptic technique, sterile equipment and antiseptic handwashing compliance.

- **All equipment is usually prepackaged** in a sterile, disposable kit.

- **Prevent anxiety and embarrassment.** Explain the procedure and drape the client.

- **Visualization of the meatus is essential.** Use the dorsal recumbent position, on a firm surface; if the client is in bed, supporting the buttocks on a firm cushion is helpful. The Sims' (lateral) position is sometimes used for clients with limited hip and knee mobility.

- **Insertion of the catheter**: In a female, initial insertion of the catheter should be 2-3 inches or until urine begins to flow. In a male, initial insertion of the catheter should be about 8 inches or until urine begins to flow.

- **Risk of urinary tract infection is great**. Meticulous sterile procedure must be used. In addition, the drainage bag must be kept off the floor (which is grossly contaminated) at all times.

- **Ensure that the indwelling catheter will not slip out.** Check for balloon patency before inserting the catheter: inflate balloon with the prefilled syringe and aspirate the fluid back into the syringe.

- **Prevent injury to the urethra.** Insert the catheter 1-2 inches beyond the point where urine flow occurs before inflating the balloon. If the client complains of pain on inflation, the balloon may be in the urethra. Remove fluid from the balloon, remove the catheter and replace it with another one.

- **Make use of gravity to assure adequate drainage.** Do not place tubing under the leg or above the level of the bladder. Be sure the tubing is not kinked.

- **Prevent pooling of urine in the drainage tubing.** Clip the tubing to the bottom bed sheet to keep the tube in place while the client is in bed.

## Procedure and Techniques for Inserting a Urinary Catheter

### Male Catheterization

- Explain the procedure to the client, and promote privacy by closing the door or drawing the curtain.
- Wash your hands for 15 to 30 seconds with an antiseptic.
- Obtain a sterile disposable prepackaged catheterization kit.
- Unfold the sterile wrap, creating a sterile field.
- Remove the pad and place it beneath the client.
- Put on your gloves, maintaining sterile technique.
- Remove the fenestrated drape from the tray and place it over the external genitalia so that only the penis is exposed.
- Saturate the three prep balls with the povidone-iodine solution.
- Put the lubricant in the compartment used for lubricating the catheter.
- Place the top of the tray to the side.
- Attach the syringe containing sterile water to the balloon inflation port and twist gently to insert.
- To begin prepping, elevate the penis with one hand. This hand is now contaminated and is committed to non-sterile activity.
- With the uncircumcised patient, the foreskin should be gently retracted.
- Using the forceps, cleanse the penis with the povidone-iodine solution in a circular motion starting at the meatus and working down.
- Use all three saturated cotton balls.

- After cleansing, wipe off the antiseptic with the two dry balls that remain.
- Remove the catheter from the wrap and apply lubricant. Insert the catheter into the urethral meatus using gentle downward pressure.
- Instruct the patient to take deep breaths and exhale slowly to make insertion easier.
- If during insertion there is an obstruction or difficulty. STOP. Do not force passage. Call for appropriate assistance or follow proper hospital protocol.
- When the catheter tip has entered the bladder, urine will appear in the drainage tube.
- Advance the catheter about two more inches.
- Inflate the catheter balloon with the entire 10 cc's of the sterile water provided in the syringe. Five cc's will remain in the inflation lumen, and five cc's will fully inflate the balloon. Use of less than 10 cc's will result in an asymmetrically inflated balloon. This can lead to incomplete retention of the catheter, accidental loss of the catheter, bladder spasms, leaking, and incomplete drainage of urine.
- Observe the client when inflating the balloon. If the patient reports pressure. STOP.
- Deflate the balloon, and gently advance the catheter and inflate it again.
- Once the balloon is inflated, ease the catheter back until slight tension is detected, indicating that the balloon is in place at the neck of the bladder.
- If proper catheterization is not accomplished, use a new catheter. Further attempts with a contaminated catheter may contribute to infection.
- After inserting the catheter, discard all materials in accordance with hospital protocol and remove contaminated gloves.
- Secure the catheter to the patient's thigh, using tape or a catheter leg strap. Securing the catheter is important for proper drainage and to help prevent trauma to the bladder or urethra.
- Hang the drainage bag below the bladder to promote the downhill flow of urine.
- Never allow the bag to touch the floor which could contaminate the bag.
- Arrange the tubing so it also promotes downhill flow. Make sure it is not kinked.
- Use the sheeting clip to secure the tubing to the bed. In addition to a plastic hook, the drainage bag is also equipped with a string that can be used when transporting a client. It will help you situate the bag below the bladder to prevent the backward flow of urine.
- Discard your gloves, wash your hands with an antiseptic hand wash for 15 to 30 seconds.
- When the patient no longer requires catheterization, remove the catheter in accordance with hospital protocol.

### Female Catheterization

- Explain the procedure to the client and promote privacy by closing the door or drawing the curtain.

- Wash your hands for 15 to 30 seconds with an antiseptic.
- Obtain a sterile disposable prepackaged catheterization kit.
- Position the female client in a dorsal recumbent position. The knees must be adequately separated to obtain a clear view of the perineum.
- Follow the same sterile technique as described for male catheterization.
- Using the forceps, cleanse the patient with the saturated prep balls.
- With one hand, separate the labia minora. This hand is now contaminated and committed to non-sterile aspects of the procedure.
- With a downward stroke, cleanse the right labia minora and discard the prep ball.
- Do the same for the left labia minora.
- With your last prep ball, cleanse the middle area between the labia minora.
- Use the dry prep balls to wipe away any excess antiseptic from the meatus.
- Remove the catheter from the wrap and apply lubricant. Insert the catheter into the urethral meatus, using gentle pressure until urine is seen in the drainage tube. Compared with the male urethra, the female urethra is much shorter and requires less maneuvering.
- After urine has begun to drain, inflate the catheter balloon using the 10 cc syringe. The inflated balloon will hold the catheter in place inside the bladder.
- Discard all materials in accordance with hospital protocol and remove contaminated gloves.
- Secure the catheter to the patient's thigh using tape or a catheter leg strap.
- Hang the drainage bag near the foot of the bed. The bag should always be positioned below the bladder and should never touch the floor.
- Arrange the tubing so it insures proper flow of urine.
- Use the sheeting clip to help secure the tubing to the bed.
- When the patient no longer requires catheterization, remove the catheter in accordance with hospital protocol.

## Caring for the Client with an Indwelling Catheter

- **Prevent contact with body fluids** (e.g., blood and urine). Wear disposable gloves when giving catheter care and handling the catheter, tubing, and collection bag.
- **Prevent transfer of microorganisms to clients**.
- Wash hands before and after catheter care.
- Never open the drainage system to obtain specimens or to measure urine.
- If tubing becomes disconnected, wipe both ends with antiseptic solution before reconnecting.
- When emptying the drainage bag, be sure the drainage spout does not touch the receptacle into which urine is being emptied.
- The less often a catheter is changed, the less the likelihood that infection will occur.

- **Keep perineal area clean**. Clean the perineal area with soap and water and rinse well, twice a day and after bowel movements.

- **Prevent urinary stasis**. Make sure the client has a fluid intake of about 3,000 mL/day.

- **Keep accurate intake and output record**. The calibrations on the collection bag are only approximate. Empty urine into a graduated container for accurate determination of output.

- **Gravity promotes drainage**. If the catheter is indwelling, the collection bag should always remain below the level of the bladder to prevent stasis and backflow of urine into the bladder (e.g., when client is ambulating). Secure the collection bag to the bed frame and not the side rails.

- **Avoid kinks in tubing**. Check that the tubing hangs straight down and does not fall below the top of the collection bag. Make sure that there are no loops or kinks in the tubing.

- **Acidify the urine**. Offer foods such as eggs, cheese, meat, poultry, whole grains, tomatoes, cranberries, plums, and prunes. Avoid milk and milk products.

- **Keep drainage system patent**. Blood clots and mucous plugs can obstruct urine flow. If a client is at risk for blood clots, the provider will usually prescribe bladder irrigations.

- **Observe for urinary tract infection**. Cloudy, dark foul-smelling urine typically signifies infection. Collect a urine specimen for laboratory analysis, notify the provider, and document your findings on the client's record.

## Techniques for Removing an Indwelling Catheter

**Prevent contact with body fluids** (e.g., blood and urine). Wear disposable gloves. Wrap the catheter in a towel or disposable drape after removing it.

**Prevent injury and pain to the client**. Be sure the balloon is deflated before removing the **catheter**. Check the size of the balloon so you will know how much fluid to remove; insert a syringe into the balloon valve and aspirate all the fluid in the balloon. **DO NOT cut the tubing with scissors**.

**Prevent urinary stasis and infection**. Clean the perineal area after removing the catheter. Assure a fluid intake of about 3,000 mL per day to keep the bladder flushed out. Observe the urine for any abnormalities.

**Observe for infection**. Report signs such as inability to void, a burning sensation when voiding, bleeding, and changes in vital signs.

**Be sure that voiding is satisfactorily re-established**. Record the time the catheter was removed. Record the client's intake and the time and amount of output for 24 hours (or according to agency policy). If the client does not void in four to six hours, palpate for bladder distention and assess for feelings of fullness.

## Discharge Planning for the Client with an Indwelling Catheter

- Prepare clients who are sent home with indwelling catheters in place.

- Teach clients and their families how to care for the catheter and drainage system and give them written instructions to refer to at home.

- Let clients know how to obtain the supplies they will use at home.

- Contact a home health care agency and arrange for an early home visit from a community health care nurse.
- Before discharge, make sure the client knows that he or she must:
  - Wash hands with soap and water before and after handling equipment.
  - Never pull on the catheter.
  - Check that the tubing is not twisted or kinked.
  - Keep the urine drainage bag below the level of the bladder at all times.
  - Take a shower instead of a bath to prevent contamination by bacteria in the bath water.
  - Drink sufficient fluids, especially water.
  - Report signs and symptoms of infection to the provider.

## Age-Related Changes—Gerontological Considerations

- Elderly women may by uncomfortable in the dorsal recumbent position used for catheterization; Sims position can be used as an alternative.
- Older adults with indwelling catheters lose bladder tone because the bladder muscle is not being stretched as the bladder fills with urine.
- Urinary tract infections are the main cause of nosocomial infections in the elderly.
- Blood flow to the kidneys decreases with cardiac output.
- Bladder capacity decreases by 50%.

# Bladder Irrigation

## Overview

Irrigation is a washing out or flushing with a specified solution, usually to wash out the bladder, but sometimes to apply a medication to the bladder mucosa. It may also be done to maintain or restore catheter patency (e.g., to remove blood clots blocking a catheter). Irrigations require a primary care provider's order. Irrigation should not cause discomfort or pain.

## Closed Irrigation Method

This is the preferred technique because it has less risk of urinary tract infection. Irrigations may be either continuous or intermittent. A 3-way (triple-lumen) catheter is usually used for closed irrigations. The irrigating solution flows into the bladder through the irrigation port of the catheter and out through the urinary drainage lumen. The urine drainage bag should be emptied before irrigation to allow accurate measurement of output after the irrigation is in place. Sterile technique is essential throughout the procedure.

**For continuous irrigation**, after the tubing is flushed and connected, open the flow and regulating clamp and adjust the flow rate to 40-60 drops per minute (if not specified by the primary care provider). Assess the drainage for amount, color, and clarity. Keep a record of the amount of irrigant infused.

**For intermittent irrigation**:

- If the solution is to remain in the bladder, close the clamp on the urinary drainage tubing.
- If the solution is not to remain in the bladder, open the flow clamp on the urinary drainage tubing.
- Next, open the flow clamp on the irrigation tubing, allow the specified amount to infuse, and clamp the tubing.
- If solution is to be retained, wait until the specified amount of time and then open the urine drainage tubing clamp and allow the bladder to empty. Record the amount of irrigant instilled.

## Open Irrigation Method

Open irrigation is sometimes necessary for restoring the patency of a double-lumen catheter. Open irrigation increases the risk of introducing microorganisms into the urinary tract. Strict sterile technique must be maintained.

**Obtain the necessary equipment**: A sterile Asepto or piston syringe, sterile basin, sterile irrigating solution at room temperature, sterile collection container, sterile waterproof drape, sterile protective cap for drainage tubing, sterile gloves, and antiseptic swabs. Disposable irrigation kits may be available.

**Establish a sterile field before beginning**: (a) Place sterile drape under client's hips; (b) apply sterile gloves; (c) clean the area between the catheter and the drainage tubing with antiseptic swabs; (d) disconnect the catheter from the drainage tubing; hold each about 1 inch from their ends and place them on a sterile surface (e.g., sterile pan or drape); (e) cover the open end of the drainage tube with a sterile protection cap.

**Irrigate**: Draw sterile irrigation fluid into the syringe and instill it slowly into the catheter. Remove the syringe and allow the solution to drain by gravity from the bladder into the collection basin. Repeat until the urine runs freely and drainage is clear.

**Reestablish closed system**: Reconnect the catheter and drainage tubing, maintaining sterility of the ends of the tubing and of the inside of the catheter.

## Therapeutic Nursing Interventions

**Minimize client anxiety and embarrassment**. Provide clear explanations; provide for privacy; drape the client.

**Protect from exposure to body fluids**. Wear clean, disposable gloves for irrigations.

**Prevent transmission of microorganisms to the client**. Wash hands before and after procedure.

**Prevent trauma to the urinary tract**. Connect the irrigation solution to the irrigation tubing and flush the tubing before connecting the tubing to the input port of the three-way catheter; this keeps air from being instilled into the bladder.

**Maintain accurate intake and output record**. Amount of drainage should equal the amount of irrigant entering the bladder plus the expected urine output. To calculate this, empty the drainage bag and measure the contents. Then subtract the amount of irrigant that has been instilled; the difference is the urine volume.

**Assess the status of the urinary tract**. Observe the amount, color, and clarity of the drainage. Note any abnormal constituents (e.g., pus, mucus, or blood clots).

**Assure that the urine drainage tubing is patent at all times.**

# Urinary Diversion Devices

## Overview

An **ileal conduit** is the surgical diversion of the ureters into a section of the ileum rather than into the bladder. A section of the ileum is removed and closed with sutures to create a pouch. The other end is brought out through the abdominal wall to create a stoma. The stoma is then fitted with an appliance to collect the urine. Other types of urinary diversion surgery are also performed; however, except for the location and type of the stoma, nursing care is similar.

Urinary diversions are usually permanent; urine drains continuously into the pouch; therefore, voluntary control is not possible. However, some clients may be candidates for a **continent urinary diversion** (also known as a Koch pouch), in which nipple valves are formed to prevent leakage and reflux of urine back from the pouch into the kidneys; the client inserts a clean catheter about every 4 hours to empty the pouch and wears a small dressing between catheterizations to protect the stoma and clothing.

## Types of Collection Appliances

The external appliance is a soft rubber or plastic pouch. Some are reusable, others are disposable. There is a firm faceplate with an opening the size of the stoma. The plate surface is secured around the stoma opening with a moisture-proof adhesive so that no urine leakage occurs. The lower end of the appliance may have a drainage valve for use in emptying the pouch.

## Associated NANDA Nursing Diagnoses

- Body Image Disturbance
- Risk for Impaired Skin Integrity
- Risk for Infection

## Therapeutic Nursing Interventions Key Procedural Points

**Empty and change the pouch and device as needed.** (1) The pouch must be emptied before it becomes heavy and causes the seal to loosen. This usually means several times a day; (2) Frequency for changing the appliance varies with the type being used.

**Change the appliance when urine production is at its lowest.** For example, after a time of low fluid intake, such as early morning.

**Protect the nurse from contact with body fluids.** Wear disposable gloves to change a stoma appliance on an ileal conduit.

**Procedural steps:**

- If the pouch is reusable, empty the pouch into a graduated container before removing it.
- Discard pouch if disposable; if reusable, wash pouch with soap and lukewarm water and allow to air dry.

**Maintain skin integrity.** Before applying the new pouch: (1) Clean skin around stoma with soap and water, removing all old adhesive from the skin; (2) gently pat dry;

(3) assess skin condition around the stoma; (4) place a gauze square over the stoma opening; (5) apply a skin protectant to a 5 cm radius around the stoma; allow to dry completely (about 30 seconds).

**Assure adhesion of the faceplate to the skin**. Apply adhesive to the faceplate (remove protective covering from a disposable faceplate), position carefully, and press it in place; remove the gauze squares from the stoma before applying the pouch.

# Urinary Incontinence

**Overview**

Urinary incontinence is loss of control over micturition. Leakage may be continuous or intermittent. Incontinence is widely underreported and underdiagnosed, but it is a treatable problem. About 50% of those who live in extended care facilities or who receive home care are incontinent. Twice as many women as men are incontinent. It is estimated that 80% of incontinence can be cured or significantly improved with treatment. Five types of incontinence are as follows:

**Total incontinence**: Total uncontrollable and continuous loss of urine at unpredictable times. The client is not aware of the bladder filling or of micturition.

**Functional incontinence**: Involuntary, unpredictable passage of urine in client with intact urinary and nervous systems. The client feels a strong urge to void, and voids before reaching an appropriate receptacle.

**Stress incontinence**: Increased intra-abdominal pressure (e.g., coughing, laughing, lifting) causes leaking of small amounts of urine.

**Urge incontinence**: Involuntary passage of urine after a strong urge to void. Bladder spasms result from alcohol or caffeine ingestion or increased fluid intake.

**Reflex incontinence**: Involuntary loss of urine that occurs at somewhat predictable intervals when a certain bladder volume is reached. The client lacks awareness of filling and has no urge to void. The bladder contracts or spasms at regular intervals. This type of incontinence is usually the result of spinal cord injury or disease.

**Related NANDA Nursing Diagnoses**

- Functional Incontinence
- Reflex Incontinence
- Stress Incontinence
- Total Incontinence
- Urge Incontinence
- Urinary Retention
- Risk for Infection related to urinary retention and/or invasive procedures (e.g., catheterization)
- Risk for Impaired Skin Integrity related to prolonged skin dampness secondary to incontinence
- Self-Care Deficit: Toileting
- Self-Esteem Disturbance related to odor, discomfort, and the embarrassment of wet clothing
- Social Isolation related to incontinence

## Diagnostic Evaluation

**Measurement of post-void residual (PVR)**: Immediately after the client voids, catheterization is performed to measure the amount of urine remaining in the bladder. A PVR of < 50 mL is acceptable. More than 200 mL is considered inadequate bladder emptying.

**Pelvic and rectal examinations** to determine muscle and sphincter tone

A **provocative stress test**: The client is instructed to cough while holding a paper towel or toilet tissue over the urinary meatus. Any leakage of urine indicates stress incontinence.

A **voiding diary** provides information about frequency, timing, and amounts of voiding.

## Treatment Options

**Surgery**: Type of surgery depends on the cause of the incontinence.

**Pharmacologic treatment**: Depends on the type of incontinence. Drugs may be given to suppress overactivity of the detrusor muscle (urge incontinence), or to improve urethral sphincter insufficiency (stress incontinence).

**Behavioral Techniques**

- **Bladder training** consists of education, scheduled voiding, and positive reinforcement. It requires the client to resist the sensation to void, postpone voiding, and urinate according to the schedule.

- **Habit training** focuses on voiding at regular intervals, but attempts to match the schedule to the person's natural voiding pattern.

- **Prompted voiding** is a strategy in which the nurse monitors continence status, reminds the client to try to void, and gives praise for remaining dry and/or attempting to void.

- **Biofeedback** uses electronic or mechanical devices to relay information to the client about physiologic activity occurring in the body.

- **Pelvic muscle exercises (Kegel exercises)** strengthen the periurethral and pelvic muscles, enhancing voluntary control.

- **Vaginal cones** of identical size and shape, but of increasing weight, are inserted and retained for up to 15 minutes twice a day.

- **Electrical stimulation** involves stimulation of the pelvic viscera and muscles or the nerve supply to these structures.

## External (Condom) Catheters

A **condom catheter** is commonly prescribed for incontinent males. It is a soft, pliable condom- like device made of plastic or rubberized material. It is applied externally to the penis and connected to tubing and a leg bag during the day. At night it is connected to a drainage bag so that the client does not need awaken to empty the bag.

- **Advantages**: Compared to an indwelling catheter, the risk for infection is very low, and the device makes self-care easier.

- **Disadvantages**: The risk for impaired skin integrity is high. Meticulous skin care is required to prevent excoriation

## Therapeutic Nursing Intervention

**Independent nursing interventions** focus primarily on: (a) behavior-oriented continence training, (b) skin care, and (c) application of an external drainage device (for males).

**Continence Training** (See "*Treatment Options: Behavioral Techniques,*" preceding):

- The goal is to gradually lengthen the intervals between urination.
- To help the client inhibit the urge to void, instruct the client to breathe slowly and deeply until the urge lessens or disappears.
- Regulate fluid intake, especially during evening hours; encourage fluids about 30 minutes before a scheduled voiding time.
- Use protective pads to keep the bed dry; use waterproof underwear; avoid using diapers, which are embarrassing and may suggest that incontinence is permissible.
- **To perform Kegel exercises**, have the client contract her perineal muscles and hold for a few seconds, and then gradually relax the area. When properly done, the muscles of the buttocks and thighs are not used. To identify the perineal muscles, have the client stop urination in midstream or to tighten her anal sphincter as if holding a bowel movement.

**Skin Care**

- *Both maceration and skin irritation predispose the area to skin breakdown and ulceration.* Incontinence is a risk factor for impaired skin integrity because (1) moist skin easily becomes macerated and (2) urine on the skin converts to ammonia, which is irritating to the skin.
- Wash the perineal area with soap and water after urination, dry thoroughly, and provide clean clothing and bed linen.
- Apply barrier creams (e.g., zinc oxide) to irritated skin.
- If it is necessary to pad clothing for protection, use products that absorb wetness and leave a dry layer next to the skin.
- Use incontinence drawsheets if available. These are double layered, with a quilted upper surface and an absorbent viscose rayon layer in the middle, and a waterproof backing on the underside.

**Care of an External (Condom) Catheter**

- Remove irritants, keep skin dry
  - Remove the condom daily, wash the penis with soap and water, dry carefully, and inspect for irritation.
  - Keep the tip of the tubing 1-2 inches beyond the tip of the penis to prevent irritation to the glans.
  - Position the tubing to draw urine away from the penis.
- Prevent leaking; promote circulation
  - Follow the manufacturer's instructions for applying the condom (they vary).
  - Fasten the condom securely enough to prevent leakage, but not so tightly as to constrict blood supply to the area.
  - Self-adhesive condom catheters are available; some, however, still require adhesive or wrapping a strip of elastic tape around the base of the penis over the condom.

- Maintain free urinary drainage. Monitor for kinked tubing.

## Age-Related Changes—Gerontological Considerations

Incontinence is a special problem for older adults because many experience decreased control over micturition; in addition, mobility and cognition problems make it difficult to reach a toilet in time to void.

# Critical Thinking Exercise: Urinary Devices and Their Care

Situation: You are caring for a hospice client who is incontinent of urine and is confined to bed. The client's caregiver asks you about the advisability of inserting a urinary catheter. Before responding, consider how you might weigh the risks and benefits of urinary catheterization for incontinent clients.

- Potential benefits of urinary catheterization
- Long-term effects of urinary catheterization

1. If a urinary catheter is inserted, what instructions will you provide the caregiver for managing catheter care?

# Normal Oxygenation

## Key Points

- The cerebral cortex can tolerate hypoxia for only three to five minutes before permanent damage occurs.
- Oxygenation has two components: ventilation and respiration.
- Normal oxygenation depends on proper functioning of the:
  - Airways transporting air to and from the lungs
  - Alveoli to exchange $CO_2$ and $O_2$
  - Cardiovascular system to circulate blood through the lungs
- When obtaining sputum specimens and throat cultures, it is important to protect self against contact with body fluids, and not contaminate the specimen.
- Be sure sputum is obtained from lower airways; be sure throat culture is obtained from far back in the throat, not from the mouth.
- Nursing interventions include:
  - Facilitating ventilation (e.g., by placing in Fowler's position)
  - Ensuring adequate hydration
  - Promoting a patent airway (e.g., by postural drainage or encouraging deep breathing)
  - Administering oxygen, per order
- Chemoreceptors that sense changes in oxygen, carbon dioxide, and hydrogen, which respond to maintain normal arterial blood gas levels, are located in the medulla, carotid bodies, and aortic bodies.
- **Key Concepts/Terms:** Ventilation, respiration, diffusion, perfusion, pulse oximetry, sputum, saliva, pulmonary function tests, chest x-ray, lung scan, pulmonary angiography, laryngoscopy, bronchoscopy, hypoxia, dyspnea, hypoventilation, hypercapnia, hyperventiliation, airway obstruction, cyanosis, hemoptysis, wheezing (rhonchus), orthopnea, bronchodilators, expectorants, cough suppressants, incentive spirometer, humidifier

## Overview

Oxygen ($O_2$) is a basic need; it is required for life. Adequate oxygenation is essential for cerebral functioning. The cerebral cortex can tolerate hypoxia for only three to five minutes before permanent damage occurs. Nurses frequently assist clients in meeting oxygen needs. The respiratory system moves $O_2$ from the atmosphere into the bloodstream and moves carbon dioxide ($CO_2$) from the bloodstream into the atmosphere. Oxygenation depends on two components: **ventilation**, which is the movement of air in and out of the lungs, and **respiration**, the process of gas exchange between the bloodstream and the air in the alveoli. Normal oxygenation depends on:

- **Properly functioning airways** to transport air to and from the lungs
- **Properly functioning alveoli** in the lungs to remove $CO_2$ from the blood and transport $O_2$ into the blood
- **Properly functioning cardiovascular system** to circulate blood through the lungs and carry nutrients to and wastes from the body
- **Ventilation** is accomplished through the act of breathing: **inhalation** (inspiration) and **exhalation** (expiration). Adequate ventilation depends on: (1) clear airways, (2) intact central nervous system, (3) intact thoracic cavity that can expand and contract, and (3) adequate lung compliance (the stretchability/expansibility of lung tissue).
- **Alveolar gas exchange (respiration):** The dense network of capillaries in the lungs and the thin alveolar walls facilitate gas exchange. **Diffusion** is the passive movement of $O_2$ and $CO_2$ between the air in the alveoli and the blood in the capillaries. Diffusion is affected by:
  - Any change in the surface area available for perfusion (e.g., a disease that destroys lung tissue)
  - Incomplete lung expansion or lung collapse (e.g., obstructed airways, tumors, pneumothorax)
  - Any condition that causes thickening of the alveolar-capillary membranes (e.g., pulmonary edema)
  - The solubility and molecular weight of the gas (e.g., $CO_2$ diffuses more rapidly than $O_2$.)
- **Perfusion** is the process in which the oxygenated capillary blood moves through the lung tissues. Perfusion depends on:
  - Whether the person is standing, sitting, or lying down
  - The amount of activity (more activity results in increased perfusion)
  - Adequate blood supply and cardiovascular functioning to carry $O_2$ and $CO_2$ to and from the lung tissues.

## Factors Affecting Oxygenation

**Health status**: Chronic illnesses can cause muscle wasting, including the muscles of the respiratory system. Renal and cardiac disorders create fluid overload, which affects respiratory functioning. Chest wall trauma impairs the ability to expand and contract the chest.

**Environment**: Altitude, heat, cold, and air pollution affect oxygenation. In high altitudes the respiratory rate and depth increase in response to a lowered partial pressure of the $O_2$ being breathed. In response to polluted air, healthy people experience headache, dizziness, coughing, choking, and stinging of the eyes. People with impaired respiratory function may not be able to perform self-care in a polluted environment.

**Lifestyle**: People who exercise routinely are better able to respond to stressors in respiratory health. Physical activity increases the rate and depth of respirations and the heart rate, and therefore oxygenation. Cigarette smoking is a major contributor to lung disease and respiratory problems. Certain occupations (e.g., asbestos workers, coal miners) predispose the person to lung disease.

Opioids: (Narcotic Analgesics) depress the medullary respiratory center, decreasing the rate and depth of respirations (especially true of morphine and meperidine [Demerol]).

Physical growth and development: Conditions such as scoliosis affect breathing patterns and cause air trapping. Obese people are often short of breath with activity and may exercise very little; so the alveoli at the base of the lungs are not stimulated to expand fully.

## Pulse Oximetry

A **pulse oximeter** is a noninvasive device for measuring the oxygen saturation ($SaO_2$), the percent of oxygenated hemoglobin in arterial blood. A small sensor is attached to the client's finger (or toe, nose, earlobe, or forehead). A preset alarm signals too-high or too-low $SaO_2$ measurements. Normal $SaO_2$ is 95-100%; <70% is life threatening.

## Diagnostic Studies

### Sputum Specimens

**Sputum** is the mucous secretion from the lungs, bronchi, and trachea. It is not the same as saliva, which is secreted in the mouth ("spit"). Healthy people do not secrete sputum. The client must cough in order to bring sputum up from the lower respiratory tract and expectorate it into a collecting container.

**Procedural key points for sputum collection**

- **Do not contaminate the specimen.** Provide mouth care before collecting the specimen. Ask the client to spit out the sputum; making sure it does not contact the outside of the container.

- **Protect against contact with body fluids.** Wear gloves. If sputum contacts the outside of the container, wash it with a disinfectant.

- **Obtain specimen from lower airways.** Have client breathe deeply and then cough; one to two tablespoons (15-30 mL) of sputum are needed.

- **Promote comfort.** Offer mouthwash afterward to remove any unpleasant taste.

### Throat Culture

These cultures are collected from the mucosa of the oropharynx and tonsils using a culture swab, to detect the presence of disease-producing microorganisms.

**Procedural key points for throat culture**

- **Do not contaminate the sputum.**

- **Protect against contact with body fluids.** Wear clean gloves.

- **Obtain specimen from far back in the throat**, not the mouth. If the posterior pharynx cannot be seen, use a flashlight/penlight and depress the tongue with a tongue blade.

- **Promote comfort**: decrease gag reflex. Have client sit upright, open the mouth, extend the tongue, and say, "ah". Take the specimen quickly.

### Pulmonary Function Tests (PFT)

Measure lung volume and capacity. PFTs are usually performed by a respiratory therapist. The tests are painless, but may be very tiring.

**Visualization procedures**: Various procedures are used to examine the respiratory tract:

- **X-ray of the chest**: Jewelry and clothing must be removed from the waist up.
- **Lung scan**: Radioisotope-tagged albumin is injected intravenously and recorded as it circulates through the lung. Both perfusion scans and ventilation scans are done. No radiation precautions are necessary because the amount of radioactivity is minimal.
- **Pulmonary angiography** assesses the pulmonary vascular system, particularly if pulmonary emboli are suspected.
- **Laryngoscopy and bronchoscopy** allow for direct visualization of the larynx and bronchi. Tissue samples may also be taken. Before the scope is inserted, a local anesthetic is sprayed on the pharynx to prevent gagging.

## Altered Respiratory Functioning

**Hypoxia** is a state in which inadequate oxygen is available to cells. Common signs of hypoxia are dyspnea, elevated blood pressure with a narrow pulse pressure, increased pulse and respiratory rates, pallor, and cyanosis. The hypoxic person usually appears tired, anxious, and may be confused. Hypoxia is treated by administering oxygen or by treating the underlying cause (e.g., shock).

**Hypoventilation** is a decreased rate or depth of air movement into the lungs. Hypoventilation results in hypercapnia.

**Hypercapnia** (hypercarbia) is elevation of the carbon dioxide levels in the blood.

**Hyperventilation** is an increased rate and depth of ventilation above the body's normal metabolic requirements. Because $CO_2$ is expired in larger than normal amounts, the $CO_2$ level in the arterial blood falls and respiratory halitosis can occur. Hyperventilation can be caused by anxiety, injuries to the respiratory center in the medulla, fever, and the body's attempt to compensate for metabolic acidosis. Signs and symptoms include: tachycardia, shortness of breath, dizziness, light-headedness, paresthesia, circumoral and extremity numbness, blurred vision.

**Airway obstruction** is a completely or partially blocked portion in the upper respiratory tract (e.g., in the nose, pharynx, or larynx) or in the lower respiratory tract (e.g., bronchi and lungs). Obstruction can arise because of aspirating a foreign object (e.g., food), because the tongue falls back into the oropharynx when the person is unconscious, or when secretions collect in the airways.

**Cyanosis** is a bluish discoloration of the skin, nailbeds, and mucous membranes due to reduced oxygen saturation of hemoglobin (In dark-skinned clients the best site to assess for cyanosis is the nail beds.).

**Hemoptysis** is the coughing up of blood from the respiratory tract.

**Wheezing (rhonchus)** is an abnormal breath sound characterized by a high-pitched musical quality. It is caused by air moving through a narrowed airway, as in asthma and acute bronchitis.

**Orthopnea** is the inability to breathe except in an upright sitting or standing position.

### Related NANDA Nursing Diagnoses

- Activity Intolerance
- Ineffective Airway Clearance

- Ineffective Breathing Pattern
- Impaired Gas Exchange
- NOTE: The preceding nursing diagnoses may be the etiology of other nursing diagnoses, for example: Anxiety, Fear, and Sleep Pattern Disturbance

## Medications

- **Bronchodilators** (sympathomimetics and xanthines) reduce bronchospasm.
- **Expectorants** (guaifenesin) liquefy mucus, making it easier to expectorate.
- **Cough suppressants** (e.g., codeine) are used when constant coughing interrupts sleep.

## Therapeutic Nursing Interventions for Oxygenation

**Facilitate ventilation:**

- Position the client in semi- or high-Fowler's position to allow for maximum chest expansion.
- Encourage or provide frequent position changes; encourage ambulation.
- **Incentive spirometers** measure and show the flow of air inhaled through the mouthpiece, thus providing an "incentive" to breathe deeply. Assist the client to a sitting position. The client takes a slow, deep breath to elevate balls in the device, and then holds the breath for two to six seconds to keep the balls elevated if possible. Clean the mouthpiece with water after use; change disposable mouthpieces every 24 hours.
- Discourage the use of over-the-counter medications, especially sleeping pills and narcotics.

**Ensure adequate hydration** (to maintain the moisture of the respiratory tract). **Humidifiers** add water vapor to the air in the form of a cool mist, to prevent drying and irritation of mucous membranes and loosen secretions.

**Promote a patent airway:**

- Provide percussion, vibration, and postural drainage (see section on "Postural Drainage" for more information).
- Provide care for artificial airways (see section on "Care of Artificial Airways").
- Encourage deep breathing exercises and coughing.
- Teach how to avoid infection and preventive measures (e.g., flu and pneumonia vaccinations).

**Provide oxygen.** See the section on *"Administering Oxygen"* for more information. Teach about use of cough medications.

**Clients with diabetes mellitus should avoid cough syrups that contain sugar or alcohol.**

**Be aware that even over-the-counter cough medications can have side effects (e.g., drowsiness).**

# Critical Thinking Exercise: Normal Oxygenation

**Situation**: You are performing a nursing history and physical assessment of a 52-year-old male client diagnosed with chronic bronchitis and emphysema. The client currently smokes two packs of cigarettes a day and is regularly exposed to environmental pollutants as a result of his work in housing construction. Consider the following questions as you develop a plan for the care of this client.

1. How will you focus your assessment to determine this client's health status and educational needs?

2. Use your assessment findings to formulate an applicable nursing diagnosis. Based on this diagnosis, what goals might be appropriate for this client?

# Administering Oxygen

> ## Key Points
>
> - The physician prescribes oxygen therapy, the method of delivery, and the liter flow per minute.
> - Oxygen is a dry gas, so humidifying devices are essential.
> - Oxygen delivery devices: Cannula, simple face mask, partial rebreather mask, non-rebreather mask, Venturi mask, face tent, transtracheal oxygen delivery, oxygen analyzer
> - Nursing interventions focus on:
>   - Promoting chest expansion (e.g., use Fowler's position)
>   - Promoting skin integrity (e.g., keep skin dry around masks, pad tubing)
>   - Promoting airway clearance (e.g., teach effective coughing, use humidifier)
>   - Ensuring adequate flow of $O_2$ (e.g., check for kinks in tubing, airtight connections, bubbles in the humidifier)
>   - Evaluating activity tolerance and anxiety
>   - Promoting safety. **Oxygen is flammable!**
> - Oxygen is combustible. Instruct the client to avoid wool and synthetic fabrics (which may cause static), use of electrical devices that are not in good working condition or are not grounded, use of volatile fluids (e.g., nail polish remover) and, of course, there should be no smoking in the area of the client receiving oxygen therapy.
> - Place warning signs on door and in room.
> - **Key Concepts/Terms**: Oxygen delivery, hypoxia, oxygen toxicity, hypercarbia

**Overview**

When a client has inadequate ventilation or impaired pulmonary gas exchange, oxygen ($O_2$) therapy may be needed to prevent hypoxia. The primary care provider prescribes $O_2$ therapy, the method of delivery, and the liter flow per minute. In hospitals and long-term care facilities, $O_2$ is usually piped into wall outlets at the client's bedside. In other facilities, pressurized tanks or cylinders of $O_2$ are used. Small, portable cylinders of $O_2$ are available for clients who require oxygen therapy at home. $O_2$ is a dry gas, so humidifying devices are essential to add water vapor to the inspired air, especially if the liter flow is >2 L/min.

## Oxygen Delivery Devices

**Cannula:** The cannula is a disposable plastic tube with two prongs for insertion into the nostrils. It fits around the head or loops over the ears to hold it in place and is connected by tubing to the $O_2$ source. It is easy to apply, relatively comfortable, and allows the client to eat and talk. The traditional nasal cannula should be used with oxygen flow rates of 1-6 L/minute. Unless the institution has the new larger-bore cannulas (labeled to support a greater flow) and the primary care provider's order supports it, the flow rate should not be greater than 6 L/min.

**Face masks** cover the client's nose and mouth. They have exhalation ports on the sides to allow exhaled carbon dioxide to escape. It is important that the mask be of appropriate size for the client. The nurse should inspect the skin under the oxygen mask and the elastic band at least once per shift.

- **Simple facemask** delivers $O_2$ concentrations of 40 to 60% at flows of 5-8 L/min, respectively.

- **Partial rebreather mask** delivers $O_2$ concentrations of 60 to 90 percent at flows of 6-10 L/min, respectively. The mask has an $O_2$ reservoir bag attached, which allows the client to rebreathe about the first third of the exhaled air together with the $O_2$. If the bag deflates totally during inspiration, carbon dioxide ($CO_2$) will build up. If this happens, the nurse increases the liter flow of the $O_2$.

- **Nonrebreather masks** deliver the highest possible $O_2$ concentration (95-100%), except for intubation or mechanical ventilation, at flows of 10-15 L/min. One-way valves on the mask and the reservoir bag keep the room air and the client's exhaled air from entering the bag, so only the $O_2$ in the bag is inhaled. If the bag deflates totally during inspiration, carbon dioxide ($CO_2$) will build up. If this happens, the nurse increases the liter flow of the $O_2$.

- **The Venturi mask** delivers $O_2$ concentrations of 24 to 50% at flows of 4-10 L/min. The mask has large tubing and color-coded jet adapters that correspond to a precise oxygen concentration and liter flow (e.g., a green adapter delivers a 35% concentration at 8 L/ min).

**Face tent**: Some clients do not tolerate masks well; they may respond with anxiety or even panic. A face tent is similar to a mask, but larger and open at the top. It fits snugly around the client's jaw line, but is open at the top over the nose. It delivers a concentration of 30- 50% at 4-8 L/min.

**Transtracheal oxygen delivery**: Oxygen is delivered through a small plastic cannula surgically inserted through the skin and into the trachea. It is held in place by a chain around the neck. This system is used for oxygen-dependent clients. All of the oxygen enters the lungs, rather than some of it being lost into the room as occurs with masks and cannulas. Therefore, a lower flow (0.5 to 2 L/min) is needed. The catheter must be kept patent by injecting saline and using a cleaning rod two or three times a day.

**Oxygen analyzer**: The analyzer is first used to measure the concentration of $O_2$ in the room. The sample tube is then placed next to the client's nose to measure the concentration of $O_2$ being received by the client. The oxygen flow rate is adjusted to obtain the desired percent of inspired oxygen ($FiO_2$).

## Safety Precautions

**Oxygen is flammable**. Place "No Smoking" and/or "Oxygen in Use" signs on the door, at the head of the bed, and on the oxygen equipment.

- Instruct the client and visitors about the fire hazard of smoking with $O_2$ in use.
- Be sure that electric devices (e.g., heating pads) are in good working order.

- Ground monitors and other electric machines (e.g., suction machines).
- Do not use blankets that generate static electricity (e.g., woolens and synthetics); use cotton. Caregivers should wear cotton fabrics as well.
- Do not use volatile, flammable materials (e.g., alcohol, acetone, greases) near clients receiving $O_2$.

## Assessment

**Signs of hypoxia**: Tachycardia, tachypnea, dyspnea, pallor, and cyanosis

**Signs of hypercarbia**: Restlessness, hypertension, and headache

**Signs of oxygen toxicity**: Tracheal irritation, cough, decreased pulmonary ventilation

## Therapeutic Nursing Management for Administering Oxygen

**Promote chest expansion.** (1) Place client in semi- or high-Fowler's position, if possible. (2) Instruct to use pillows as necessary for sleep. (3) Encourage smaller, more frequent meals.

**Promote skin integrity.** (1) For face masks and face tents, keep the client's facial skin dry and inspect for chafing. (2) Pad cannula and mask tubing and band over the ears and under chin as needed. (3) If client has a cannula, assess the nares for irritation, apply a water-soluble lubricant as needed.

**Promote safety.** (See *"Safety Precautions"*).

**Promote airway clearance.** (1) Teach how to splint thorax and cough effectively. (2) Teach to use a cool-mist humidifier. (3) Teach to avoid milk products if they cause excess mucus production.

**Ensure adequate flow of $O_2$:** (1) Check for kinks in tubing. (2) Be sure that connections are airtight. (3) Check for bubbles in the humidifier. (4) Feel to be sure there is oxygen at the outlets of the cannula, mask, or tent. (5) Check the liter flow and the level of water in the humidifier whenever giving care.

**Evaluate activity tolerance and anxiety.**

## Age-Related Changes—Gerontological Considerations

- Older adults are prone to dehydration that causes dry mucous membranes.
- Ciliary action decreases with age, causing decreased clearing of the airways.
- Muscular structures of the pharynx and larynx atrophy with age.
- Less ventilation in the lower lobes of the older adult causes secretions to pool and predisposes them to pneumonia.

# Critical Thinking Exercise: Administering Oxygen

**Situation:** A hospitalized client with chronic obstructive pulmonary disease (COPD) will require continuous oxygen therapy in the home setting. Devise a teaching plan to assure compliance with oxygen therapy on discharge.

1. Here are a few questions to help you develop your plan:

   • Why is oxygen prescribed?

   • Will I be restricted to home with this therapy?

   • What kind of monitoring is needed with oxygen therapy?

   • Is it okay to increase the flow rate?

   • Are there any special precautions to follow?

   Keep your explanation simple, but try to address all potential questions your client might have.

2. Enlist the help of your clinical instructor or a respiratory therapist and examine the various devices used to administer oxygen, such as simple masks, the Venturi mask, partial rebreather, and non-rebreather masks. Make sure you understand how each of these masks is used.

# Percussing, Vibrating and Postural Drainage

## Key Points

- Chest physiotherapy includes the techniques of percussion (clapping/cupping), vibrating, and postural drainage.
- For all three techniques it is important to:
  - Position the client correctly for comfort and to make use of gravity.
  - Perform the treatments for a specified length of time, and a specified number of times per day.
  - Promote client safety and comfort.
  - Evaluate the effects of treatment.
  - Do NOT perform the technique on clients who are pregnant, have rib or chest injuries, dizziness, fainting, head or neck injuries, pulmonary embolism, or abdominal surgery.
  - Increase oral fluid intake to liquefy secretions.
  - Teach the client how to cough effectively
- **Key Concepts/Terms**: Atelectasis, percussion, drainage and cough, chest physiotherapy

**Overview**

Nurses independently facilitate respiratory functioning by helping clients to breathe deeply and cough. Percussion, vibration, and postural drainage (together, referred to as **chest physiotherapy**), however, require a medical order. They are performed to loosen thick secretions and facilitate drainage from the lungs. Secretions that remain in the lungs and airways provide a medium for bacterial growth; they can also obstruct the smaller airways and cause **atelectasis** (incomplete expansion or collapse of a section of the lung; gas exchange cannot occur in atelectatic areas).

## Percussion ("Clapping," "Cupping")

Percussion is the manual striking of the chest with alternating cupped hands, mechanical percussion cups, or a vibrator. The vibrations through the chest wall help to loosen and dislodge pulmonary secretions.

### Key Procedural Points

- **Use correct technique.**
  - Cup the hands by holding the fingers and thumb together and flexing them slightly to form a dome, as one would to scoop up water.
  - Keep shoulders, elbows, and wrists relaxed; alternately flex and extend the wrists to rapidly strike the chest.

- Listen for a hollow, popping sound. If the sound is not hollow or the client is uncomfortable, the care provider probably needs to cup the hands more.
- **Use the correct amount of time.** Percuss each affected lung segment for 1-2 minutes (3-5 min. for those with tenacious secretions) several times a day.
- **Promote client comfort, safety, and relaxation.**
  - Have the client take slow, deep breaths as you percuss.
  - Do not percuss on bare skin.
  - Do not percuss over the spine, breasts, or below the ribs.
- **Make use of gravity.** Move from the lower ribs to the shoulders in back and from the lower ribs to the top of the chest in front.

## Vibration

Used after percussion, vibrating is done to increase the turbulence of the exhaled air, to aid in the expectoration of secretions. Vibration is a series of vigorous quiverings (a fine, shaking pressure) produced by contraction and relaxation of the arm and shoulder muscles while holding the hands flat against the client's chest wall. Vibrating increases the exhalation of trapped air and may shake mucus loose and induce coughing.

### Key Procedural Points

- **Use correct technique.**
  - Place hands on the client's chest wall. Either place them side-by-side with the fingers extended or place one hand on top of the other.
  - Ask the client to inhale deeply and then exhale slowly.
  - As the client exhales, tense all the hand and arm muscles and rapidly vibrate the hands, using primarily the heel of the hands. Strive for a frequency of about 200 vibrations per minute. Stop vibrating when the client inhales.
- **Use the correct amount of time.** Vibrate during five exhalations over each affected lung segment. Use vibration several times a day.
- **Promote client comfort, safety, and relaxation.** Do not vibrate over the spine, breasts, sternum, or below the ribs.

## Postural Drainage

In postural drainage, gravity is used to facilitate drainage of secretions from various lung segments. The client is positioned to promote drainage from smaller pulmonary branches into larger ones, where they can be removed by drainage or coughing. Often, percussion and vibration precede postural drainage. The client may also be given a bronchodilator medication or nebulization therapy 30 minutes prior to postural drainage. Several different positions are necessary to drain all segments of the lungs; however, not all positions will be required for every client—use only those positions needed to drain specific areas.

### Key Procedural Points

- **Position correctly (make use of gravity).** Position to promote drainage from specific areas of the lungs:
  - **Apical sections of upper lobes:** high-Fowler's position
  - **Anterior upper lobes:** supine position with pillow under the knees

- **Posterior sections of upper lobes**: side-lying, half on the abdomen and half on the side (right and left)
  - **Right lobe**: left side with a pillow under the chest wall
  - **Lower lobes**: Trendelenburg's position
- **Use the correct amount of time**. Have client assume each position for 10 to 15 minutes, if tolerated. The treatments are usually performed three or four times a day.
- **Promote client safety and comfort**.
  - Have tissues and emesis basin nearby for client to use when expectorating mucus.
  - Evaluate client's tolerance of the positions (assess stability of pulse and respirations; note signs of intolerance such as pallor, diaphoresis, dyspnea, and fatigue).
  - Clients may tolerate some positions but not others. If this occurs, use that position for a shorter time, or modify it (e.g., use minimal elevation in Trendelenburg's position).
  - Discontinue drainage if client feels weak or faint.
  - Try to schedule treatments before meals or at least 90 minutes after meals (after meals, postural drainage can be tiring for the client, and may induce vomiting).
- **Evaluate the effects of the treatment**. After postural drainage, auscultate the lungs, compare to baseline data, and document the amount, color, and character of expectorated secretions.

## Therapeutic Nursing Interventions for Percussing. Vibrating and Postural Drainage

- These three treatments should NOT be performed on clients who are pregnant, have rib or chest injuries, dizziness, fainting, head or neck injuries, pulmonary embolism or abdominal surgery.
- Increase oral fluid intake to liquefy secretions.
- Teach client how to cough, if necessary.

## Age-Related Changes—Gerontological Considerations

- Dry, brittle, porous bones predispose to fractures.
- Less ventilation at the bases of the lungs contributing to fluid accumulation and compromised ventilation.
- Cilia are less effective at clearing respiratory passages.
- Decreased strength of accessory muscles used to cough.
- Decreased cough reflex.
- Due to dehydration, pulmonary secretions may be thick and tenacious.
- Atrophy of muscular structures of the pharynx and larynx make airway clearance more difficult.
- Chronic pulmonary illnesses increase with age.
- Higher risk for developing cardiac dysrhythmias related to hypoxemia.
- Older adults, especially, may not tolerate the positions used for postural drainage.

# Critical Thinking Exercise: Percussing, Vibrating and Postural Drainage

**Situation:** In caring for a 9-year-old child diagnosed with cystic fibrosis, you are teaching the parents about the role of chest physiotherapy in managing pulmonary problems. The following question will help you address the educational needs of parents whose children require chest physiotherapy.

1. Why is the chest physiotherapy important for a child with cystic fibrosis? Might the pulmonary problems associated with this disease be adequately managed with medications alone? Why or why not?

2. What is the value of chest physiotherapy for a child diagnosed with asthma? And what are the limitations of this therapy for the child with asthma?

# Suctioning

| **Key Points** |

- Suction is performed to clear the airways; frequency is determined by the client's condition.
- Suctioning irritates the mucosa and removes oxygen from the respiratory tract.
- Suctioning should be painless and relieve respiratory distress.
- It is normal for suctioning to cause coughing, sneezing, or gagging.
- Key procedural points:
    - Use sterile technique.
    - Determine proper length of tube to insert.
    - Hyperventilate or oxygenate client before suctioning.
    - Do not apply suction while inserting the tube.
    - Restrict suction time to 5 to 15 seconds.
    - Encourage client to cough and deep breathe between suctions.
    - Allow to rest between suctions.
    - Protect self against exposure to body fluid.
    - Document amount, consistency, color, and odor of sputum; and respiratory status.
  - **Key Concepts/Terms:** Whistle-tipped catheter, open-tipped catheter, oropharyngeal suctioning, nasopharyngeal suctioning, orotracheal suctioning, nasotracheal suctioning, endotracheal tube

**Overview**

When a client cannot clear respiratory tract secretions by coughing, the nurse uses suctioning to clear the airways. Frequency of suctioning is determined by the client's condition. If inspection or auscultation indicates that secretions are present, suctioning is required. Suctioning irritates the mucosa and removes oxygen from the respiratory tract; to prevent hypoxemia, the client must be hyperoxygenated before suctioning. Performed correctly, suctioning should be painless and should relieve respiratory distress. Incorrectly performed, it can create anxiety, pain, and even cardiac arrest. Suctioning may precipitate coughing, sneezing, or gagging; this is normal.

## Signs/Symptoms of the Need for Suctioning

- Bubbling or rattling breath sounds
- Decreased breath sounds
- Dyspnea

- Pallor, cyanosis
- Decreased $O_2$ saturation ($SaO_2$ level)
- Drooling, vomitus in mouth

### Potential Complications (of Suctioning)

- Infection
- Cardiac arrhythmias
- Hypoxia
- Mucosa trauma
- Death

### Suction Catheters and Equipment

- Suction catheters may be open-tipped or whistle-tipped (the opening on the end of the catheter is cut on the diagonal). Whistle-tipped catheters are less irritating to mucosa, but the open-tipped catheter may be more effective for removing thick mucus. Most catheters have a port on the side, which is covered with the thumb or finger to control the suction.
- The catheter is connected via suction tubing to a collection chamber and suction control gauge.

## Therapeutic Nursing Interventions for Oropharyngeal and Nasopharyngeal Suctioning

**Oropharyngeal and nasopharyngeal suctioning** are done to remove secretions from the upper respiratory tract when the client can cough effectively, but is unable to expectorate or swallow them. The oropharynx extends behind the mouth from the soft palate and includes the tonsils. The nasopharynx is located behind the nose and extends to the level of the soft palate.

**Assess the need for suctioning.** (See *"Signs/Symptoms of the Need for Suctioning"*)

**Provide privacy.** Close door and pull bed curtain before suctioning.

**Promote client safety and comfort.**

- Explain what the procedure will do and what sensations to expect.
- **Oral:** Position in semi-Fowler's position with head turned to one side, if conscious with a functional gag reflex.
- **Nasal:** Position in semi-Fowler's position with neck hyperextended, if conscious with a functional gag reflex.
- If unconscious, place in side-lying position, facing the nurse.
- Place towel on pillow or under chin.
- Select proper suction pressure (for wall suction units, usually 110-150 mm/Hg for adults).
- Measure with thumb and forefinger the distance between client's ear lobe and tip of nose (about 5 in.); grasp catheter with dominant thumb and forefinger at that point (to assure that catheter does not go down into the trachea).
- Test the pressure of the suction and the patency of the catheter by applying a sterile gloved finger to the port to create suction.
- **Oral:** Moisten catheter tip with water. Gently insert catheter into one side of

mouth (reduces gag reflex) and glide into oropharynx, without applying suction (prevents damage to pharyngeal mucosa).

- **Nasal**: Lubricate the catheter tip with water-soluble lubricant. Gently insert catheter into one nostril and guide it medially along floor of the nasal cavity. Do not force; if one nostril is not patent, try the other.

- **After catheter is positioned**, use nondominant hand to occlude the suction port and apply suction. Gently rotate and withdraw catheter. Suction intermittently as catheter is withdrawn. This should take no longer than 15 seconds.

- Clear catheter by suctioning sterile water through it before reinserting it.

- Ask client to deep breathe and cough between catheter insertions.

- After suctioning is complete, suction secretions in mouth and under tongue.

**Prevent fatigue and hypoxia.** Allow client to rest 20 to 30 seconds before reinserting catheter. Replace oxygen cannula, if applicable, during that time.

**Prevent infection.** Protect against contact with body fluids.

- Use sterile gloves, catheter, and water (or normal saline). A clean glove, instead of a sterile glove, may be used on the nondominant hand. After each withdrawal of the suction catheter, flush the catheter by suctioning sterile water through it.

- Discard catheter by wrapping it around gloved hand and pulling glove off around catheter.

- Wash hands.

**Evaluate effectiveness of suctioning.** Assess and document pre- and post-suctioning respiratory status (e.g., skin color, lung sounds, dyspnea, anxiety); document amount, consistency, color, and odor of sputum.

## Therapeutic Nursing Interventions for Orotracheal or Nasotracheal Suctioning

**Orotracheal or nasotracheal suctioning** is necessary when the client is unable to cough and does not have an artificial airway. The catheter is passed through the nose or the mouth into the trachea. The procedure must be limited to 15 seconds from catheter passage to removal because adequate oxygen does not reach the lungs during this procedure. In order to reach the trachea, a longer length of catheter is inserted than in oro- or nasopharyngeal suctioning. Otherwise, the procedure is the same.

## Therapeutic Nursing Interventions for Suctioning a Tracheostomy or Endotracheal Suctioning

**Assess the need for suctioning.** (See "*Signs/Symptoms of the Need for Suctioning*")

**Provide privacy.** Close door, pull bed curtain.

**Promote client safety and comfort.**

- Explain that the procedure will make breathing easier and that it will cause coughing.

- Administer analgesia if necessary (e.g., coughing causes pain for clients who have had thoracic surgery).

- Place in semi-Fowler's position unless contraindicated.

- Select proper suction pressure (for wall suction units, usually 110-120 mm/Hg for adults).

- Place catheter tips in sterile saline solution, occlude the thumb control and suction a small amount of the sterile solution through the catheter.
- Quickly and gently insert the catheter without applying any suction.
- Insert catheter about 5 in. (12.5 cm) for adults, or until the client coughs or resistance is felt.
- Clear catheter by suctioning sterile water through it before reinserting it.
- Ask client to deep breathe and cough between catheter insertions.
- After each withdrawal of the suction catheter, flush the catheter by suctioning sterile water through it.
- After suctioning is complete, suction secretions in mouth and under tongue.

**Prevent fatigue, hypoxia and cardiac problems**.

- If client **does not** have copious secretions, hyperventilate the lungs with a resuscitation bag before suctioning. This is best done by a second person, who can use both hands to compress the Ambu bag.
- If the client **does** have copious secretions, increase the liter flow of the oxygen or adjust the $FO_2$ to 100% for several breaths before suctioning.
- After the catheter is inserted, apply intermittent suction for 5-10 seconds while rolling the catheter between your thumb and forefinger and slowly withdrawing it. Withdraw the catheter completely and release the suction.
- Allow client to rest two to three minutes between suctions. Hyperventilate the client before reinserting the catheter.
- Discontinue endotracheal suctioning if the client develops a cardiac dysrhythmia, and notify the primary care provider.

**Prevent infection**.

- Use sterile gloves, catheter, and water (or normal saline). A clean glove, instead of a sterile glove, may be used on the nondominant hand.

**Protect against contact with body fluids**.

- Wear goggles, mask, and gown if necessary.
- Discard catheter by wrapping it around gloved hand and pulling glove off around catheter.
- Wash hands.

**Evaluate effectiveness of suctioning**. Assess and document pre- and postsuctioning respiratory status (e.g., skin color, color of nail beds, vital signs, lung sounds, dyspnea, anxiety); document amount, consistency, color, and odor of sputum; document the amount of sterile solution instilled.

## Key Procedural Points for All Suctioning

- Use sterile technique.
- Determine proper length of tube to insert.
- Hyperventilate or oxygenate the client before suctioning.
- Do not apply suction while inserting the tube.

- Apply suction while rotating and withdrawing the catheter.
- Restrict suction time to 5-15 seconds to minimize oxygen loss.
- Encourage client to cough and deep breathe between suctions.
- Hyperventilate or oxygenate between suctions.
- Let client rest between suctions.
- Evaluate respiratory status before and after suctioning.
- Protect against exposure to body fluids (e.g., sputum).

# Critical Thinking Exercise: Suctioning

**Situation:** A fellow nursing student is caring for a client who requires regular suctioning of an endotraceal tube. As you cared for this client the previous day, your classmate poses some questions related to his care. Consider your response to the following questions:

1. How do you know when suctioning is necessary? Can unnecessary suctioning cause harm to this client?

2. What measures are necessary to prevent infection to a client while performing tracheal suctioning?

# Artifical Airway Care

> ## Key Points
>
> - A tracheostomy tube consists of an outer cannula, an inner cannula, and an obturator.
> - For oropharyngeal and nasopharyngeal airway care:
>   - Wear clean, disposable gloves.
>   - Do not tape airway in place; remove it when client begins to gag or cough.
>   - Provide frequent oral or nares care.
>   - Keep suction equipment at the bedside.
> - Key procedural points for tracheostomy care:
>   - Suction and care are sterile procedures.
>   - Be sure outer cannula is securely tied.
>   - Be sure the ties are not too tight.
>   - Be sure that client does not aspirate water or foreign materials (e.g., cotton balls, gauze threads, small needles).
>   - Protect the skin around the stoma and flange.
>   - Keep a suction set-up and obturator at the bedside for emergency use.
>   - Provide a means of communication (e.g., writing board); keep call light close at hand
> - **Key Concepts/Terms**: Orpharyngeal airway, nasopharyngeal airway, endotracheal tube, tracheostomy tube

**Overview**   Artificial airways may be inserted through the mouth, the nose, or through a tracheostomy.

## Oropharyngeal and Nasopharyngeal Airways

Oropharyngeal and nasopharyngeal airways are inserted in a spontaneously breathing client to keep the tongue clear of the airway and permit suctioning of secretions (e.g., for postoperative clients who have not yet regained consciousness). The oral airway extends from the teeth over the tongue to the oropharynx. Because they stimulate the gag reflex, oropharyngeal airways are used only for clients with an altered level of consciousness (e.g., drug overdose, head injury). Conscious clients tolerate a nasopharyngeal airway better than an oropharyngeal airway.

### Key Procedural Points

- Wear clean, not sterile, gloves.
- Do not tape the airway in place; remove it when the client begins to cough or gag.

- Provide mouth care at least every two to four hours.
- Keep suction equipment available at the bedside.
- For nasopharyngeal airways:
  - Provide frequent oral and nares care.
  - Reposition the airway in the other naris at least every eight hours.

## Endotracheal Tubes

An endotracheal tube is a polyvinyl chloride airway that is inserted through the nose or mouth into the trachea. It ends just above the bifurcation of the trachea into the bronchi. Commonly, the tube has an air-filled cuff to prevent air leakage and bronchial aspiration of foreign material. It is inserted by a primary care provider or advanced practice nurse, using a laryngoscope as a guide. It is used to administer oxygen by mechanical ventilator, to suction secretions, or to bypass upper airway obstructions. Because an endotracheal tube passes through the epiglottis and glottis, the client cannot speak while it is in place.

## Tracheostomy Tubes

A **tracheostomy** is an artificial opening made through the neck into the trachea of clients needing long-term airway support. A curved plastic or metal tube is inserted through the opening and into the trachea to maintain the airway. This is a surgical procedure. A tracheostomy can be temporary or permanent. Tracheostomy tubes can be cuffed or cuffless. The cuff, when inflated, seals the opening around the tube, preventing air leakage and aspiration, and permitting mechanical ventilation.

A **tracheostomy tube** consists of an outer cannula (or main shaft), an inner cannula, and an obturator.

- The **obturator** is inserted into the tube to guide the outer cannula during initial placement. It is kept at the bedside in case the outer cannula (or a new one) needs to be reinserted.
- The **outer cannula** is the "main shaft" that is inserted into the trachea. It has a flange at the external opening of the tracheostomy stoma. Tape or cloth is tied through the flange and around the client's neck to hold the outer cannula in place.
- The **inner cannula** fits into the outer cannula. The inner cannula is removed for cleaning, while the outer cannula remains in place.

**Procedure** for care and cleaning of tracheostomy tubes:

- Place client in Fowler's position (to promote lung expansion).
- Provide for communication (e.g., tell the client to raise a finger to indicate distress).
- Establish a sterile field; maintain sterile technique throughout.
- Don sterile gloves (or clean glove on nondominant and sterile glove on dominant hand).
- Suction the tracheostomy tube (see *"Oxygenation: Suctioning"*).
- Remove inner cannula and place in sterile hydrogen peroxide solution.
- Remove soiled tracheostomy dressing; then discard that glove and dressing.
- Re-glove with sterile gloves.

- Clean incision site and tube flange with sterile gauze or applicators. Remove encrustations; dry thoroughly.

- Clean inner cannula with brush or pipe cleaners; rinse in sterile saline.

- Dry the inside (to prevent aspiration), but not the outside (to provide lubrication), of the cannula.

- Replace and secure inner cannula.

- Apply a commercially prepared sterile dressing (of non-raveling material) under the flange of the tracheostomy tube. Do not use cotton-filled gauze; do not cut a 4x4 gauze (fibers can be aspirated, causing a tracheal abscess).

- **Changing ties**: Cut new tapes. Leave the old ties in place while threading the new tapes through the flange. Tie the new tapes, using square knots, and then remove the soiled tapes. Place a finger under the tape before tying it so that it is not too tight. Place a 4x4 gauze square under the tie now and apply tape over the knot (reduces skin irritation).

### Key Points of Tracheostomy Care

- Tracheostomy suction and care are sterile procedures.

- Ensure that the outer cannula is securely tied (use square knots) and does not become dislodged.

- Avoid movement of the outer cannula; this irritates the trachea.

- Ensure that the client does not aspirate water or foreign materials.

- Be sure that the ties are not too tight (interferes with coughing; places pressure on jugular vein).

- Protect skin integrity around the stoma and flange.

## Therapeutic Nursing Interventions for Artificial Airway Care

- Check cuff pressure regularly (on some types of tubes).

- Assure that oxygen is being delivered correctly and that it is warmed and humidified (to prevent drying of secretions).

- Keep foreign objects (e.g., cotton balls, loose threads from gauze, needles, and other small objects) and unsterile objects away from the opening.

- Suction to remove secretions (see "*Oxygenation: Suctioning*").

- Always have a suction set-up and obturator at the bedside for emergency use.

- Provide means of communication (e.g., writing board, vocabulary cards).

- Keep call light/bell close at hand.

- Provide frequent reassurance, explanations, and anticipation of needs to prevent anxiety.

- Clean the inner cannula once or twice a day (every one to two hours immediately after the surgery).

- Change the stoma dressing as often as needed.

- Deflate cuffed tubes before oral feeding unless the client is at high risk for aspiration.

# Critical Thinking Exercise: Artificial Airway Care

**Situation**: A nurse on a surgical unit is assigned to the care of a client who has undergone a tracheostomy to facilitate mechanical ventilation. Describe the postoperative nursing interventions that apply to the care of this client.

1. What major nursing objectives apply in the care of a client with a tracheostomy?

2. What potential complications are associated with tracheostomy tube management? Describe appropriate nursing interventions to minimize the risk of complications.

# Fluid Balance & Intravenous Therapy

## Key Points

- In a well person, fluid intake and fluid loss are balanced.
- As a "rule of thumb," urinary output usually is equal to the amount of fluids ingested (40-80 mL/hour or 1,500-2,000 mL/24 hr).
- The volumes and chemical composition of the body fluid compartments are regulated within narrow limits by: fluid intake, fluid output, and hormonal controls.
- The average adult drinks about 1,500 mL of fluids/day; an additional 1,000 mL is obtained from foods and oxidation of foods.
- Thirst is the primary regulator of fluid intake.
- The kidneys produce about 1500 mL of water output/day; remaining output is excreted from the skin, lungs, and GI tract.
- Assessments include: daily weights, vital signs, fluid intake and output, urine amount and appearance, skin turgor or edema, mucous membranes, breath sounds, and mental status.
- Foods (e.g., ice cream) that turn to liquid at room temperature are considered fluids.
- Do not use caffeinated drinks for fluid replacement because they cause diuresis.
- The physician prescribes the amount and kind of solution; the nurse initiates, monitors, and discontinues the therapy.
- Sterile technique is essential.
- Venipuncture:
  - Veins in the hand and arm are usually the most accessible.
  - Avoid antecubital and leg veins, and veins in surgical areas.
  - Avoid thin-walled and scarred veins.
  - Use a large vein for hypertonic, viscous, or irritating solutions, and for rapid infusion.
  - Choose a vein large enough to accommodate the catheter to be used.
  - Most agencies change the site about every 72 hours; for the first site, use the most distal site possible.
  - Label containers, tubing, and dressing.
  - Cleanse insertion site with antiseptic followed by an antimicrobial solution
  - Factors that influence flow rates: positon and patency of the IV tubing, position of the forearm, height of the infusion bag, infiltration or fluid leakage.

- As a rule, do not irrigate an IV catheter.
- Solution containers should be changed every 24 hours, even if they are not empty, to minimize risk of infection.
- Change IV dressings according to agency protocol (usually every 48 to 72 hours) and when damp or soiled.
- Potential problems of IV therapy: infiltration, phlebitis, thrombus, embolus, infection, speed shock, fluid overload
- Some agencies require the nurse to wear a mask when accessing or changing dressing of central lines.
- Veins of older adults are fragile, and risk of infiltration is higher.

- **Key Concepts/Terms:** intracellular fluid (ICF), extracellular fluid (ECF), intravascular fluid (plasma), interstitial fluid, transcellular fluid, sensible water loss, insensible water loss, antidiuretic hormone (ADH), aldosterone, renin, angiotensin II, fluid volume deficit (FVD), hypovolemia, third space syndrome, fluid volume excess (FVE), hypervolemia, edema, pitting edema, dehydration (hyperosmolar imbalance), overhydration (hypo- osmolar imbalance, water intoxication), isotonic, hypotonic, hypertonic, tubing, spike, drip chamber, roller clamp, slide clamp, needle adapter, inline filter, particulate matter, infusion pump, Angiocath, "butterfly" needle, central venous catheter (CVC), saline lock, heparin lock

**Overview**

Water is the main constituent of the body. By weight, about 50% of the human body is composed of about 50% water (for infants, it is 70-80%). Life can be sustained for only a few days without water. Water makes up a greater proportion of a lean person's body weight than obese persons because fat tissue is essentially free of water. The functions of water in the body are to:

- Act as a solvent for electrolytes and other substances
- Help maintain normal body temperature
- Provide a medium for transporting nutrients to and wastes from cells
- Provide a medium for transporting hormones, enzymes, red and white cells, and platelets
- Facilitate digestion and promote elimination
- Act as a tissue lubricant

Body fluids are divided into two compartments: intracellular and extracellular. **Intracellular fluid (ICF)** is found within the body cells; it makes up about 2/3 of the total body fluid. **Extracellular fluid (ECF)** is found outside the cells; it makes up about 1/3 of the total body fluid. Extracellular fluid is the transport system that carries nutrients to and wastes from the cells (e.g., plasma carries oxygen from the lungs to the capillaries; from there, the oxygen moves across the capillary membranes into the interstitial spaces and then across cellular membranes into the cells). The extracellular fluid is further subdivided into three compartments:

**Intravascular fluid (plasma)**, found within the vascular system (in the blood)

**Interstitial fluid**, surrounding the cells (it includes lymphatic fluid)

**Transcellular fluid**, such as cerebrospinal, pleural, peritoneal, and synovial fluids

# Fluid Balance

## Regulation of Body Fluids

In a well person, fluid intake and fluid loss are balanced, and the volumes and chemical composition of the fluid compartments are maintained within narrow limits. Illness can upset this balance. Body fluids are regulated by fluid intake, fluid output, and hormonal controls.

**Fluid Intake:** The average adult drinks about 1,500 mL per day; an additional 1,000 mL must be obtained from foods and from the oxidation of foods during metabolic processes. Thirst is the primary regulator of fluid intake. The thirst control center in the brain is stimulated when there is increased plasma osmolarity and decreased blood volume. Receptor cells monitor the osmotic pressure. When excess fluid is lost, the osmoreceptors activate the thirst center. Psychological factors and the sensation of a "dry throat" also cause thirst. Fluid intake requires an alert state; infants, some elderly persons, and those with altered levels of consciousness are at risk for dehydration.

**Fluid Output:** Sensible water loss is that which is perceived; insensible water loss is continuous and is not perceived by the person. Fluid output occurs through the kidneys, the skin, the lungs, and the gastrointestinal tract. The kidneys are the primary regulators of fluid balance (see "Normal Urinary Elimination"). They produce about 1,500 mL of water output per day. Water loss from the skin (perspiration) is mostly insensible, but excess sweating creates some sensible loss as well. The skin excretes about 400 mL of water per day. The lungs expire about 400 mL of water per day, an insensible loss. The average fluid loss from the gastrointestinal tract is about 100 mL/day—more if vomiting or diarrhea occurs.

**Hormonal Regulation**

- **Antidiuretic hormone (ADH)**, released by the posterior pituitary gland, reduces the production of urine by causing the kidney tubules to reabsorb water. When fluid volume deficit occurs (as in vomiting, diarrhea, or hemorrhage) the amount of ADH in the blood increases and the water reabsorbed by the kidney tubules is returned to the circulating blood volume.

- In response to decreased blood flow or pressure in the kidney, specialized receptors in the nephrons release **renin**. Renin causes the conversion of angiotensin I to **angiotensin II**, which acts directly on the nephrons to cause Na+ and water retention. Angiotensin II also stimulates the adrenal cortex to release **aldosterone**. This, too, causes the kidney tubules to excrete K+ and retain Na+; as a result, the kidneys reabsorb water and return it to the blood volume.

## Fluid Imbalances

**Fluid volume deficit (FVD)** occurs when the body loses both water and electrolytes (isotonic imbalance) from the ECF. Fluid is first lost from the intravascular compartment (as in hemorrhage), so it is also called **hypovolemia**. Risk factors for FVD include:

- Abnormal losses through the skin
- Anorexia

- "Third spacing"
- Nausea
- Excessive sweating
- Inability to access fluids
- Polyuria
- Impaired swallowing
- Fever
- Confusion
- Nasogastric suction
- Depression
- Bleeding
- Prolonged rapid respiration
- Losses from the gastrointestinal tract (vomiting, diarrhea)

Symptoms of FVD include: decreased blood pressure, decreased pulse volume and pressure, decreased urine output (<30 mL/hr), decreased skin turgor, dry mucous membranes, sunken eyeballs, elevated hematocrit, elevated temperature, increased pulse rate, concentrated urine (specific gravity >1.030), weakness, confusion.

**Third space syndrome** occurs when fluid shifts from the vascular space into a space where it is not accessible as extracellular fluid; it remains in the body, but is not available for use, causing an isotonic fluid volume deficit. For example, fluid may be trapped in the interstitial space as edema, or in the peritoneal or pleural cavities.

**Fluid volume excess (FVE) (hypervolemia)** occurs when both water and sodium are retained in normal proportions in the ECF. FVE always occurs because of an increase in total body sodium content, for example: (a) excessive intake of Na+Cl-, (b) rapid administration of sodium-containing infusions, and (c) diseases that impair regulatory mechanisms, such as cirrhosis of the liver, renal failure, and congestive heart failure.

- **Edema** is excess interstitial fluid. In FVE, both intravascular and interstitial fluids have an increased water and sodium content. Edema is most apparent around the eyes and in dependent tissues (e.g., the feet and legs). **Pitting edema** leaves a small depression (pit) after finger pressure is applied.

- **Symptoms of FVE** include weight gain (>2 lbs in 24 hrs), elevated blood pressure; full, bounding pulses; moist, labored respirations; large volume of pale urine; pitting edema; crackles auscultated in lungs; dyspnea; distended jugular veins; anxiety.

**Dehydration (hyperosmolar imbalance)** occurs when water is lost from the body, but there is no loss of electrolytes. This means that serum sodium concentration increases in the vascular fluid. In turn, water is drawn into the blood from the interstitial space and from the cells, resulting in cellular dehydration. Dehydration may be caused by hyperventilation, prolonged fever, diabetic ketoacidosis, and administering enteral feedings with insufficient water intake.

**Overhydration (hypo-osmolar imbalance, water intoxication)** occurs when more water is gained than electrolytes. This results in low serum osmolality and low serum Na+ levels. Water is drawn into the cells, causing them to swell. When cerebral edema occurs in the brain cells, neurologic function is impaired. Water intoxication is often caused by conditions in which both fluid and electrolytes are lost (e.g., through exercise and heavy sweating) but only water is replaced. It can also be caused by AIDS, some malignant tumors, head injury, or administration of certain drugs (e.g., oxytocin, barbiturates, anesthetics).

## Assessments

- Daily weights (>2 lbs in 24 hours is significant)
- Vital signs
  - Tachycardia is an early sign of hypovolemia.
  - Pulse volume will increase in FVE and decrease in FVD.
  - Blood pressure may fall significantly with FVD and hypovolemia.
- Fluid intake and output
  - Be sure to measure ice chips, tube feedings, foods that become liquid at room temperature (e.g., Jello, ice cream), tube irrigants, and intravenous medications as input.
  - Remember to measure urinary output, vomitus, liquid feces, tube drainage, wound drainage, draining fistulas and insensible losses that occur with diaphoresis and hyperventilation.
  - + As a "rule of thumb," urinary output is usually equivalent to the amount of fluids ingested (40-80 mL/hour or 1,500-2,000 mL in 24 hours).
- Urine amount and appearance
- Skin turgor (dehydration) or edema (FVE)
- Mucous membranes (dry, sticky, dehydration)
- Breath sounds (FVE)
- Mental status, confusion

## Related NANDA Nursing Diagnoses

- Fluid Volume Deficit
- Fluid Volume Excess
- Risk for Fluid Volume Deficit
- Risk for Fluid Volume Imbalance

## Therapeutic Nursing Interventions for Fluid Imbalance

- **Facilitate fluid intake (when fluids are needed):**
  - Establish a 24-hour plan.
  - Identify fluids the client likes.
  - Avoid caffeine (e.g., coffee and tea) because of the diuretic effect.
  - Help client select foods that become liquid at room temperature (e.g., gelatin, sherbet).
- **Help clients restrict fluid intake, when needed:**
  - Many clients need to be told that ice chips, gelatin, and ice cream are considered fluids.

- About half the total volume should be scheduled on the day shift.
- Place fluids in small containers so the client will see the container as full.
- Offer ice chips instead of water.
- Instruct client to not eat or chew salty or sweet foods (such as hard candy or gum) because they will cause thirst.
- Provide parenteral fluid replacement as prescribed.
- Assess and monitor (see "Assessments")
- Teach family members to offer fluids regularly to clients who cannot meet their own needs because of, for example, age, impaired mobility, impaired cognition, or impaired swallowing.
- Be alert for postural hypotension, teach client to change positions slowly, especially when moving from supine to sitting or standing.
- Place in Fowler's position (FVE) to facilitate breathing.
- Restrict dietary intake of sodium, if ordered (FVE).

## Age-Related Changes—Gerontological Considerations

- Older adults are at risk for dehydration because of decreased thirst sensation.
- Older adults may have mobility and cognition problems that prevent them from getting enough to drink.
- Many older adults have decreased renal function.
- Decreased sensible and insensible water loss related to a decrease in the number of functioning sweat glands.

# Intravenous Fluid Therapy

**Overview**

Intravenous (IV) therapy is used to manage fluid imbalances, as well as to administer medications (see *"Administering Medications: IV Medications"*). Blood and blood products are also administered intravenously; however, separate tubing is used. Eighty percent of hospitalized clients receive IV therapy, and many more receive it at home. The primary care provider prescribes the amount and kind of solution to be used; the nurse initiates, monitors and discontinues the therapy. Sterile technique is essential because any contaminants are introduced directly into the bloodstream. Most IV equipment (e.g., needles, infusion tubing) is disposable. Selection of equipment for IV administration is based on the type of fluid, the viscosity of the drug or blood product, and the client's condition and body size.

## Intravenous Solutions

| Isotonic Solutions | Uses | Comments |
|---|---|---|
| **D5W:** 5% dextrose in water | Supplies free water to aid in renal excretion of solutes and expands intracellular and extracellular fluid volumes<br>Supplies about 170 cal/L | • Contraindicated for clients with increased intracranial pressure<br>• Not used for fluid replacement in hypovolemia<br>• Limit volume for clients with increased ADH activity |
| **0.9% NaCl** normal saline, Lactated Ringer's | Expands vascular volume<br>Contains multiple electrolytes in physiological concentrations (except for Mg++ and PO4)<br>Used to treat hypovolemia, burns, and diarrhea<br>Used to treat mild metabolic acidosis | • Assess for hypervolemia (e.g., bounding pulse, shortness of breath).<br>• Does not provide calories<br>• Not used for routine maintenance because Na+ and Cl- are provided in excessive amounts<br>• Assess for signs of hypovolemia (e.g., bounding pulse, shortness of breath). |
| **Hypotonic Solutions** | Uses | Comments |
| **0.45% NaCl:** half normal saline | Provides Na+, Cl-, and free water<br>Treats cellular dehydration<br>Promotes waste elimination by the kidneys | • Contraindicated for clients at risk for increased intracranial pressure or third-space fluid shift |

| Isotonic Solutions | Uses | Comments |
|---|---|---|
| **0.33% NaCl**: 1/3 strength normal saline | See 0.45% NaCl, preceding. Used to treat hypernatremia because it dilutes plasma Na+, but not too rapidly | • Contraindicated for clients at risk for increased intracranial pressure or third-space fluid shift |
| **Hypertonic Solutions** | **Uses** | **Comments** |
| D51/2NS: 5% dextrose in 0.45% NaCl | Used to treat hypovolemia Used to maintain fluid intake Draws fluid out of the intracellular and interstitial spaces into the vascular space, expands vascular volume | • Do not use for clients with kidney or heart disease.<br>• Do not use for dehydration.<br>• Monitor for hypervolemia. |
| D5NS: 5% dextrose in normal saline (0.9% NaCl) | Replaces calories and electrolytes; draws fluid out of the intracellular and interstitial spaces into the vascular space, expands vascular volume Temporary treatment of hypovolemia if plasma expander not available. | • See D51/2NS, preceding. |
| D5LR: 5% dextrose in Lactated Ringer's | Replaces calories and electrolytes Draws fluid out of the intracellular and interstitial spaces into the vascular space, expands vascular volume. | • See D51/2NS, preceding. |
| D10W: 10% dextrose in water | Supplies 340 cal/L Used for peripheral parenteral nutrition | • See D51/2NS, preceding. |

## Equipment

**Equipment varies according to the manufacturer.**

**Solution containers**: Most solutions come in 500 mL or 1 liter flexible or rigid containers (most commonly, flexible plastic).

**Tubing**: A basic set has a **spike** on one end that inserts into the container (bag, bottle) with a twisting motion. Beneath the spike is a **drip chamber**, which permits the nurse to visualize the drops per minute as the solution flows from the container into the tubing. The number of drops per cc (or the size of the drops) varies among the different kinds of tubing. A roller clamp controls flow rate. A **slide clamp** is used to constrict the tubing and cut off flow completely. Most tubing also has a "Y" port for administering IV medications through the tubing. At the opposite end of the tubing from the spike, is a **needle adapter with a protective** cap, which is inserted into the IV catheter in the client's vein. Many infusion sets have an **inline filter** to trap air, particulate matter, and microorganisms (this is especially important for long-term IV therapy). Special infusion sets are usually necessary if the IV flow rate is to be regulated by an infusion pump.

### Needles/catheters

- **Angiocaths** are over-the-needle plastic catheters that fit over the needle used to pierce the skin and vein wall. After insertion, the metal needle is withdrawn and the catheter is left in place. Because they are flexible, catheters allow the client more mobility and are less likely to infiltrate than metal.

- **"Butterfly" needles** are short-beveled, thin-walled needles with plastic flaps attached to the shaft. The shafts are held together to aid with the angle of insertion; after insertion, the flaps are taped flat to the skin. Because they are inflexible, they are more likely to infiltrate.

### Other venous access devices

- **Central venous catheters (CVCs)** are frequently used in both hospital and home care. A CVC is inserted into the subclavian or internal jugular veins and terminate in the superior vena cava just above the right atrium. They have a single, double, or triple lumen entry port. A CVC can be either "short-term" or "long-term." Long-term CVCs are made of silicone rather than polyurethane to help prevent thrombus formation. Long-term CVCs are surgically inserted; however, a short-term CVC may be inserted at the bedside. The **Groshong long-term CVC** has pressure-sensitive valves to seal off each valve, simplifying routine maintenance. The insertion site of a CVC is protected by an occlusive (often transparent) dressing. Agency policy determines type of dressing and frequency of dressing change (usually 48 to 72 hrs). **Nursing responsibilities** with CVCs include: (1) meticulous sterile technique, (2) sterile dressing changes, (3) monitoring for infection, (4) changing injection caps on the lumens, and (5) flushing with a prescribed solution (e.g., saline and heparin) to maintain patency.

- **Peripherally inserted central catheters (PICCs)** are widely used for IV therapy in the home, as well as in the hospital. A PICC is introduced in a peripheral vein (ideally above or below the antecubital space) and advanced into the superior vena cava. Placement is verified by x-ray. They may be single- or multiple-lumen catheters. PICCs are used for long-term (two to six weeks) antibiotic therapy, parenteral nutrition, chemotherapy, continuous narcotic infusions, blood components, and long-term rehydration. There is less risk of pneumothorax than with CVCs, they are cost effective, and they cause less pain because multiple venipunctures are not needed. **Nursing responsibilities** include: (1) sterile dressing changes, (2) routine heparin or saline flushes to maintain patency, (3) observation for complications (e.g., phlebitis, infiltration, and infection).

## Venipuncture

Venipuncture is the insertion of a needle into a vein. Here, we will review venipuncture for the purpose of collecting a blood sample into an evacuated blood-collection tube, although it can also be done with a needle and a syringe. The median cubital vein at the antecubital fossa (the inside surface of the forearm at the level of the elbow) is most often accessed for blood sampling.

Other good choices are the other veins of the forearm, such as the basilic and the cephalic, and those accessed from the dorsum of the hand, such as the metacarpal veins.

Selection of the site for venipuncture varies with the age of the client and the viability of the client's vasculature. Choose a vein that is soft and resilient, that is bouncy when you release your palpation pressure. Avoid collecting blood from a vein that is very firm or "knotty", and avoid sites that are edematous, painful, or scarred, or have hematomas. Also, don't choose an arm that has an intravenous line in place, has an arteriovenous fistula or a vascular graft, or is on the same side as a mastectomy.

To collect a blood sample from an adult client, check the provider's order, confirm the client's identity, and explain the procedure to the client.

Gather the materials needed for the venipuncture. Prepare:

- A venipuncture kit and blood specimen tubes or
- Antiseptic pads
- An evacuated collection tube system
- Sterile cotton or gauze
- A touniquet
- Gloves
- An adhesive bandage
- Specimen tubes

Wash his/her hands, then help the client into a point in which the venipuncture site can be easily accessed. For a client who is in bed, it is helpful to raise the bed if the nurse will stand for the venipuncture. Open the packages, attaching the needle to the evacuated tube holder. Then apply the tourniquet above the planned insertion site and check for a palpable pulse distal to the tourniquet. Select the vein, and palpate it to be certain it doesn't "roll" or move away too readily. Then release the tourniquet and prepare the venipuncture site with antiseptic according to the institution's protocols. Reapply the tourniquet within three minutes of preparing the site. Next, apply gloves and remove the cap from the needle. Stabilize the vessel with the nondominant hand, and pull the skin at the site taut with a thumb. With the dominant hand, hold the needle at a 30-degree angle or less, and with the bevel up. Puncture the skin at the straightest portion of the vein with a steady, smooth movement of the needle and then observe for a "flashback" or blood at the hub of the needle. Then firmly push the evacuated collection tube through the tube holder so that its stopper is punctured by the distal end of the needle. As soon as blood flows into the collection tube, release the tourniquet. After the required amount of blood is collected and the evacuated tube is removed, hold a sterile gauze sponge or cotton ball over the site and withdraw the needle gently at the same angle it was inserted. He/she applies pressure to the puncture site for two to three minutes. When you are sure the bleeding has stopped, apply an adhesive bandage to the venipuncture site, then prepare the specimens for the laboratory according to the institution's protocols.

After you complete the blood collection procedure, make sure that the client is comfortable and lower the bed. Then wash your hands, and document the procedure and that the blood samples were sent to the laboratory.

Venipuncture presents unique challenges when the client is an older adult. The skin and veins become fragile with advancing age, so use the smallest needle possible to minimize injury.

You might not even need a tourniquet, because rising venous pressure can overstretch a fragile vein easily, and even a thin needle can rupture the vein's wall. If you need to use a tourniquet, apply it over clothing to reduce the risk of shearing the skin. Stabilize the vein by applying traction below the insertion site and use a small angle of insertion -5 to 15 degrees only. Even with these precautions, however, hematomas are common with older clients.

Venipuncture takes practice and can be difficult when the client is very ill. In some institutions, phlebotomists collect most blood specimens, while in others it is the nurse's responsibility. In teaching hospitals, it might be the medical students and interns who do a good deal of the specimen collection.

Nevertheless, this is a skill that nurses should become comfortable with and should take every opportunity to develop.

## Procedure and Techniques for Starting an IV

**Selecting a Site.** The site is selected on the basis of the type of solution to be used, the duration of the infusion, and the condition of the veins. Selection varies with the age of the client. Older clients have more fragile veins than younger clients. Choose a vein that is soft and resilient, that is bouncy when you release your palpation pressure. Select an insertion site that is naturally splinted by bone. Use the client's nondominant arm whenever possible.

- **Accessibility**: For adults, veins in the hand and arm (metacarpal, cephalic, accessory cephalic, and basilic veins) are usually used. For infants, the scalp and dorsal foot veins are used. When possible, select the client's nondominant hand/arm.

- **Avoid when possible**: (1) **antecubital veins** for long-term infusion, because flexion of the arm can displace the catheter; (2) leg veins because of the possibility of stagnant peripheral circulation and thrombus formation; (3) veins in surgical areas (e.g., do not place an IV in the left arm after a left mastectomy); (4) veins that are very firm or "knotty", and (5) sites that are edematous, painful, scarred, or have hematomas.

- **Condition of the veins**: Palpate to determine the condition of the vein. Thin-walled and scarred veins may make insertion and continued infusion a problem.

- **Type of solution**: A large vein should be used for hypertonic solutions, irritating medications, rapid infusion, and those with a high viscosity.

- **Duration of infusion**: The longer the IV will be in place, the more important it is to choose a site that restricts movement as little as possible. Choose a vein that is splinted by bone (e.g., the forearm).

- **Size of needle/catheter**: Choose a vein large enough to accommodate the needle to be used.

- **Sites are changed** about every 72 hours in most agencies. For the first site, start in the most distal site possible; move in a proximal direction, alternating arms, with each site change.

### Venipuncture and Initiating Infusion

- Be sure the solution is clear and contains no particles; check the expiration date.
- Wash hands well before starting the infusion.
- Spike the solution container.
- Place a timing label on the solution container, marked in hourly increments.
- Hang the solution on an IV pole.
- Squeeze the drip chamber to fill partially.
- Allow fluid to move through the tubing until all air is cleared from the line.
- Select site; shave around the site if the client is hairy.
- Dilate the vein: place arm in a dependent position; apply a tourniquet.
- Ask the client to open and close the fist while you palpate for a vein.
- Wear protective gloves.
- Cleanse the site with an antiseptic followed by an antimicrobial solution (e.g., alcohol and povidone-iodine).
- Insert the catheter, attach tubing, and start flow of the solution.
- Support catheter with a small piece of gauze under the hub; secure with narrow tape.
- Apply dressing and mark date, time, site, and type and size of catheter on the tape anchoring the tubing.
- Adjust flow rate.
- Label IV tubing with date, time of attachment, and nurse's initials.
- Return in 30 minutes to check flow rate and observe for infiltration.

## Regulating and Monitoring IVs

**Regulating flow**: Unless an infusion pump is used, the flow rate is regulated manually using the roller clamp. IV orders usually prescribe the total amount to be infused, the type of solution, and the rate in hours, for example:

- Give 1000 mL Lactated Ringer's over 24 hours
- Give 1000 mL normal saline at 125 mL/hr continuously
- Give 1000 mL $D_5$1/2NS every 8 hours X 3 bags

**To calculate the flow rate** (or drops per minute), the nurse must know: (1) the volume of fluid to be infused and (2) the specific time over which it is to be infused. The number of drops per milliliter of solution varies with different brands of infusion sets. Commonly, they have drop factors of 10, 12, 15, or 20 drops/mL; for a micro drip set, it is always 60 drops/mL. Refer to a fundamentals text or a drug calculation text for detailed instructions on calculating flow rates.

**Monitor the IV infusion** at least every hour: (a) count the drip rate, (b) check tubing (e.g., for kinking), (c) observe settings on pump if one is being used, (d) inspect site for swelling, pain, coolness, or pallor (which may indicate infiltration), (e) inspect site for redness, swelling, heat, and pain (which may indicate phlebitis).

Factors that influence flow rates include:

- **Position and patency of the IV tubing**: Flow rate will be inhibited if the tubing is lying under the client's body, is kinked, is dangling below the puncture site, or if a clamp is closed too tightly.
- **Position of the forearm**: Sometimes even a slight change in the position of the arm (e.g., slight supination or elevation) can change the position of the catheter in the vein and increase or decrease flow.
- **Height of the infusion bag**: Because of gravity, raising the height of the bag a few inches can create more pressure and increase the flow rate.
- **Infiltration or fluid leakage**
- **Because of the danger of dislodging a clot**, do not irrigate an IV catheter unless agency policy recommends it.

## Procedure and Technique for Changing Solutions and Tubing

### Changing Solutions

- IV solution containers should be changed when there is a small amount of fluid remaining in the neck of the container and drip chamber.
- IV solution containers should be changed every 24 hours, even if they are not empty, to minimize risk of infection.
- Verify solution type and flow rate.
- Label the new container.
- Wash hands.
- Clamp the old tubing and disconnect it from bag.
- Spike new bag and re-establish the infusion.

### Changing Tubing

- IV tubing is changed every 48 to 72 hours, depending on agency policy and procedures.
- Change tubing when you are hanging a new solution container.
- Verify solution type and flow rate.
- Label new container.
- Wash hands.
- Spike new container with new tubing, hang bag, and prime tubing.
- Don disposable gloves.
- Remove site dressing and all tape except the tape holding the IV catheter in place.
- Assess the IV site.
- Place sterile gauze square under catheter hub.
- Clamp and disconnect used tubing.
- Connect the new tubing and re-establish the infusion.
- Clean and redress the site.

## Caring for the Infusion Site

- Peripheral IV site dressings should be changed when damp or soiled, or every 48 to 72 hours, depending on agency protocol.
- Dressings are changed when tubing is replaced.
- Don clean gloves.
- Remove soiled dressing and all tape except that holding the catheter in place.
- Assess the site for infiltration or inflammation.
- Change the tubing and re-establish infusion (see preceding section).
- Remove the old tape securing the catheter.
- Clean the insertion site (usually alcohol and povidone-iodine swabs).
- Retape catheter.
- Apply antiseptic ointment per agency policy.
- Apply new dressing.
- Label the dressing and secure the IV tubing.

## Caring for Central Lines

- **For PICC lines**, change dressing 24 hours after insertion and about once a week thereafter (according to agency policy).
- **For CVC lines**, change dressing 24 hours after insertion and then every 48 to 72 hours thereafter (according to agency policy).
- Some agencies require the nurse and client to wear a mask when accessing a port or changing a dressing.
- Keep the external portion of the catheter coiled under the dressing.
- Change catheter caps according to agency policy (e.g., every three to seven days).
- Before accessing the port, clean an area 2 inches in diameter around the site with an alcohol-acetone solution; follow with povidone-iodine solution; allow to air dry.
- **Flush an open-ended PICC line** with heparin solution, using a 10 mL syringe, after intermittent use, or every 12 hours if not in use.
- **Flush a close-ended PICC line** with saline, using a 10 mL syringe, after each intermittent use, or once a week if not in use.
- Flush all lumens of a multiple lumen catheter.
- Do not take blood pressure in the involved arm.
- Monitor for signs of a dislodged catheter: pain in neck or ear on affected side, swishing sounds, palpitations. Notify primary care provider.
- Monitor for signs of a dislodged port: free movement of the port, swelling, and difficulty accessing the port. Notify primary care provider.
- Document: Site appearance, length of external catheter, dates of dressing and cap change, flushing frequency and routine.

## Converting a Primary Line to a Saline/Heparin Lock

When fluid replacement is no longer necessary, the primary IV line is sometimes converted to a heparin or saline lock, through which small volumes of intravenous medications can be given. A "lock" is a plastic tube with a sealed injection port on the end. The nurse dons clean gloves, and maintaining sterility, removes the IV tubing, connects the lock to the indwelling catheter, and flushes it with either saline or heparin. The nurse periodically injects it with heparin or saline to keep the catheter patent.

## Discontinuing the IV

- Wear protective gloves.
- Clamp infusion tubing (to prevent fluid from flowing out onto client or bed).
- Loosen tape at the site while stabilizing the catheter with the other hand.
- Hold a sterile gauze square above the puncture site.
- Withdraw the catheter, pulling in line with the vein and parallel to the skin.
- Immediately after withdrawing (not before) apply firm pressure to the site with the sterile gauze; this may take up to 3 minutes.
- If bleeding persists, elevate the limb above the level of the heart.
- Examine the catheter to make sure it is intact.
- Cover the venipuncture site with a sterile dressing.
- Document amount of fluid infused from the bag, the time of discontinuing the infusion, the appearance of the site, and the client's response.

## Potential Problems Associated with Intravenous Therapy

| Problem | Signs/Symptoms | Causes | Nursing Interventions |
|---|---|---|---|
| **Infiltration** (fluid escapes into the subcutaneous tissues) | Swelling, pallor, coldness, pain around infusion site; decrease in flow rate | Dislodged catheter; vessel wall penetrated by catheter | Discontinue infusion and restart at a different site. Use armboard to limit the movement of the extremity with the IV. |
| **Phlebitis** (inflammation of a vein) | Localized tenderness, redness, warmth, and mild edema of the vein above insertion site | Trauma from the catheter; chemical trauma from the solution; contamination by microorganisms | Discontinue infusion. Apply warm moist compresses. Restart infusion in a different vein. |
| **Thrombus** (blood clot) | Same as phlebitis. Also, flow may stop if clot obstructs catheter. | Trauma from catheter or needle | Same as for phlebitis. In addition, **do not rub or massage the affected area.** |

| Problem | Signs/Symptoms | Causes | Nursing Interventions |
|---------|----------------|--------|------------------------|
| **Embolus** (thrombus enters circulation) | Depends upon the location of any infarction caused by the thrombus (e.g., cerebral, pulmonary) perhaps dyspnea or sudden pain | Air enters vein through infusion line; thrombus dislodges from insertion site | Report any sudden pain or breathing difficulty. **Prevention:** Monitor site regularly; do not allow air to enter infusion line; treat phlebitis promptly and with caution. |
| **Infection** | Fever, chills, malaise; pain, swelling, inflammation, or pus at IV site | Poor sterile technique in inserting and caring for insertion site; contaminated solution | Discontinue infusion; notify primary care provider. **Prevention:** Always maintain sterile technique; change dressing and tubing every 24 hours if agency policy permits. |
| **Speed shock** | Pounding headache, rapid pulse rate, anxiety, chills, back pain, dyspnea | Infusing fluid too rapidly into the circulation | Discontinue infusion immediately. Report symptoms to primary care provider at once. Monitor vital signs. **Prevention:** Use micro drip tubing for pediatric and fluid intolerant clients; check flow rate often. |
| **Fluid overload** | Dyspnea, engorged neck veins, increased blood pressure | Too large a volume of fluid infused into circulation | Slow the infusion rate; notify primary care provider immediately; monitor vital signs. **Prevention:** Monitor flow rate and total volume; monitor intake and output. |

## Key Points for IV Therapy

- Wear clean, disposable gloves to protect against exposing the nurse to blood.
- Use sterile technique when giving medications, starting the IV, and caring for the IV.

- Scrupulous care of the infusion site, dressing changes, and tubing replacements are essential for preventing infection.
- Many agencies require the nurse to wear a mask when accessing or changing dressing of central lines.
- Monitor for infiltration, infection, and fluid imbalance (e.g., monitor intake and output).
- Never disconnect IV tubing when dressing and undressing a client—or for any other reason.
- Instruct client to call for help if any discomfort occurs at the site, container is almost empty, or the flow changes.
- Replace IV fluids every 24 hours, whether or not they are completely infused.

## Age-Related Changes—Gerontological Considerations

- Some older clients may be discharged from the hospital with IVs and may require teaching about flushing the ports or catheters.
- The veins of older adults are fragile, and the risk of infiltration is higher.
- Minimize damage to the veins by using the smallest needle possible.
- Older adults are more prone to fluid volume overload related to their cardiovascular status and decreased cardiac output.
- Older clients are also more prone to hematoma formation.

# Critical Thinking Exercise: Fluid Balance & Intravenous Therapy

**Situation**: A home care nurse is instructing a 72-year-old client diagnosed with congestive heart failure in self-care measures. The client questions the importance of taking his weight each day.

1. How might the nurse best explain the importance of daily weight to this client?

2. The client is also on a sodium-restricted diet and is taking 40 mg of Lasix PO daily for the treatment of heart failure. How might you explain the rationale for this regimen to encourage compliance?

3. The home care nurse next visits an 83-year-old client with limited mobility related to arthritis. With respect to maintaining fluid balance, what particular needs might this client have?

# Electrolyte & Acid-Base Imbalance

## Key Points

- Electrolytes regulate fluid balance, contribute to acid-base balance, facilitate enzyme reactions, and regulate neuromuscular excitability.
- The principal electrolytes in extracellular fluids are sodium, chloride, and bicarbonate.
- The principal cations in intracellular fluids are magnesium and potassium; the major anions are phosphate and sulfate.
- Most electrolytes are obtained through dietary intake and are excreted in the urine.
- Potassium and calcium are stored in the cells and in the bone. Sodium and chloride are not stored and must be taken in daily to maintain normal levels.
- Normal extracellular concentration of sodium is 135 to145 mEq/L.
- Normal serum potassium is 3.5 to 5.0 mEq/L.
- Acid-base balance depends on a balance between carbonic acid and bicarbonate in the body.
- Carbonic acid levels are regulated by the lungs, which retain or excrete carbon dioxide.
- Bicarbonate and hydrogen ion levels are regulated by the kidneys.
- Respiratory acidosis: High $PaCO_2$ due to alveolar hypoventilation (etiology examples: acute lung conditions, cardiac arrest).
- Respiratory alkalosis: Low $PaCO_2$ due to alveolar hyperventilation (etiology examples: hyperventilation, anxiety, high fe ver).
- Metabolic acidosis: Low bicarbonate, because either nonvolatile acid is present to use up $HCO_2$ too rapidly, or the $HCO_2$ is being lost in disproportionate amounts (etiology examples: impaired renal function, prolonged diarrhea).
- Metabolic alkalosis: High bicarbonate, when nonvolatile acid is lost and not using up $HCO_2$, or $HCO_2$ is gained in disproportionate amounts (etiology examples: vomiting, gastric suction, thiazide diuretics).
- **Key Concepts/Terms**: Cations, anions, hyponatremia, hypernatremia, hypokalemia, hyperkalemia, hypocalcemia, hypercalcemia, carbonic acid, bicarbonate, hydrogen ions, respiratory acidosis, respiratory alkalosis, metabolic acidosis, metabolic alkalosis, blood gas analysis, pH, $PaO_2$, $PaCO_2$, $HCO_2$, $SaO_2$

## Overview

Electrolytes are electrically charged ions dissolved in a solution. Many salts, when placed in water, break up into electrolytes. For example, sodium chloride breaks up into one ion of sodium (Na+) and one ion of chloride (Cl-). Electrolytes that carry a positive charge are cations; those that carry a negative charge are anions. Examples of **cations** and **anions** are:

**Cations**

Calcium (Ca++)

Magnesium (Mg++)

Potassium (K+)

Sodium (Na+)

**Anions**

Bicarbonate ($HCO_3$-)

Chloride (Cl-)

Phosphate ($HPO_4$-)

Sulfate ($SO_4$-)

Electrolytes regulate fluid balance, contribute to acid-base balance, facilitate enzyme reactions, and maintain balance in neuromuscular excitability. They are usually measured in milliequivalents per liter of water (mEq/L) or milligrams per 100 milliliters (mg/100 mL). Laboratory tests reflect the electrolyte concentration in the plasma, but it is not possible to directly measure electrolyte concentrations within the cells.

## Composition of Body Fluids

**Extracellular fluids (ECFs):** The principal electrolytes are Na+, Cl-, and $HCO_3$-. Other electrolytes are present but in much smaller quantities. Plasma and the interstitial fluid (both ECFs) are similar except that plasma contains large amounts of protein (albumin) and interstitial fluid contains little or no protein.

**Intracellular fluids (ICFs):** Are significantly different from ECF. The primary cations in ICF are Mg++ and K+. The major anions are phosphate and sulfate.

**Other body fluids:** (e.g., gastric and intestinal secretions) also contain electrolytes. Fluid and electrolyte imbalances can result from excessive losses (e.g., in severe vomiting or diarrhea, or in removal of gastric secretions by suctioning).

## Regulation of Electrolytes

Most electrolytes enter the body through dietary intake and are excreted in the urine. Some (e.g., K+ and Ca++) are stored in the cells and in bone. Others (e.g., Na+ and Cl-) are not stored and must be taken in daily to maintain normal levels.

**Sodium (Na+):** Sodium is the main electrolyte of the ECF. Diets in the U.S. average about six to 15 g of sodium per day; the average daily requirement is only about 500 mg. Excess sodium is eliminated primarily by the kidneys, but small amounts are lost in feces and perspiration. The normal extracellular concentration of sodium is 135–145 mEq/L; deviations rapidly cause a serious health problem. Sodium concentrations are affected by salt intake and the renin-angiotensin-aldosterone system (see *"Fluid Balance: Regulation of Body Fluids"*).

**Potassium (K+):** Potassium is the major cation of the ICF. Only a small amount is found in plasma and interstitial fluid. It is vital for transmission of nerve impulses and heart, skeletal, intestinal, and lung tissue. It also helps to regulate acid-base balance. An intake of 50 to 100 mEq per day is sufficient to maintain K+ balance. The normal range for serum K+ is 3.5 to 5 mEq/L. Potassium is excreted primarily by the kidneys; however, they have no effective way of conserving potassium, so deficits develop quickly if K+ is excreted in excess (as with diarrhea). Gastrointestinal (GI) secretions contain large quantities of potassium; some is also found in saliva and perspiration. K+ and Na+ have a reciprocal action—that is, an excess intake of one results in an excretion of the other.

**Calcium (Ca++):** About 99% of the total calcium in the body is found in the bones and teeth. It is the most abundant electrolyte in the body. The small amount of calcium found in the ECF is vital in regulating muscle, neuromuscular, and cardiac function. The average daily requirement for calcium is about 1 g for adults. Calcium is excreted in urine, feces, bile, digestive secretions, and perspiration.

- When calcium levels in the ECF fall, the parathyroid glands increase secretion of parathyroid hormone, which causes the bones to release Ca++ into the blood; it also causes the kidney tubules and the intestinal mucosa to increase reabsorption of calcium.
- Calcitonin, secreted by the thyroid gland, has an opposite effect, conserving calcium by opposing osteoclast bone resorption.
- A high serum phosphate concentration decreases serum calcium.
- Use of calcium is stimulated by vitamin D.

**Magnesium (Mg++):** Most magnesium is found in the skeleton and in the ICF. It is also present in heart, bone, nerve, and muscle tissues. It is necessary for intracellular metabolism (e.g., the production of ATP, the synthesis of DNA). Plasma concentrations of Mg++ range from 1.3 to 2.1 mEq/L. Magnesium is absorbed by the intestines and excreted by the kidneys.

**Chloride (Cl-):** Chloride is the main ECF anion; it is found in blood, interstitial fluid, gastric juice, and lymph (and in very small amounts in the ICF). It plays an important role in acid-base balance. The daily requirements are not known; normal serum chloride levels range from 95 to 105 mEq/L. It normally "follows" sodium—which means it is excreted or conserved with sodium by the kidneys.

## Related NANDA Nursing Diagnoses

- Acute Confusion
- Fluid Volume Deficit
- Risk for Impaired Skin Integrity
- Fluid Volume Excess

## Hyponatremia

| Risk Factors | Defining Characteristics | Therapeutic Nursing Interventions |
|---|---|---|
| • Loss of sodium: loss of gastrointestinal fluids, adrenal insufficiency, sweating, diuretics<br>• Gain of water: hypotonic tube feedings, oral ingestion of water, excess administration of D5W<br>• Disease states (e.g., head injury, AIDS, malignant tumors) | • Serum Na+ <135 mEq/ L<br>• Serum osmolality <280 mOsm/kg<br>• Headache<br>• Muscle twitching<br>• Seizures, coma<br>• Abdominal cramps<br>• Nausea, vomiting, anorexia<br>• Lethargy, confusion | • Monitor intake and output.<br>• Check urine specific gravity.<br>• Limit supplemental water given to clients receiving isotonic tube feedings.<br>• Closely assess clients receiving hypertonic saline.<br>• Encourage diet high in sodium (e.g., bacon, ham, cheese, table salt) unless contraindicated. |

## Hypokalemia

| Risk Factors | Defining Characteristics | Therapeutic Nursing Interventions |
|---|---|---|
| • Potassium loss: vomiting, gastric suction, diarrhea, heavy diaphoresis<br>• Use of K+ wasting drugs, (e.g., furosemide and thiazide diuretics)<br>• Steroid administration<br>• Low intake of K+ (e.g., alcoholism, anorexia, debilitation)<br>• Hyperaldosteronism<br>• Osmotic diuresis (as in uncontrolled diabetes mellitus | • Serum K+ <3.5 mEq/L<br>• ABGs may show alkalosis<br>• ECG shows flattened T-wave and depressed ST segment<br>• Muscle weakness<br>• Paresthesias, muscle soreness<br>• Leg cramps<br>• Fatigue, lethargy<br>• Anorexia<br>• Nausea, vomiting<br>• Decreased bowel sounds/ motility<br>• Cardiac arrhythmias | • Assess digitalized clients for symptoms of digitalis toxicity (hypokalemia potentiates the action of digitalis).<br>• Administer prescribed K+ supplements.<br>• **Prevention**: Encourage extra K+ intake for at-risk clients. |

## Hyperkalemia

| Risk Factors | Defining Characteristics | Therapeutic Nursing Interventions |
|---|---|---|
| • Decreased K$^+$ excretion: renal failure, K$^+$ conserving diuretics, hypoaldosteronism.<br>• High K$^+$ intake, especially in renal insufficiency: improper use of supplements, too rapid IV K$^+$ infusion, high dose K$^+$ penicillin, high oral intake of K$^+$<br>• Shift of K$^+$ out of cells: acidosis, burns, tissue trauma | • Serum K$^+$ >5.0 mEq/L<br>• ECG shows peaked T-wave and widened QRS, absent P wave<br>• First sign is usually vague muscle weakness<br>• Diarrhea, gastrointestinal hyperactivity, nausea<br>• Irritability, apathy, confusion<br>• Cardiac arrhythmias, arrest<br>• Absence of reflexes, flaccid muscle paralysis<br>• Numbness and paresthesias in extremities, face, and tongue | • Monitor cardiac status.<br>• Administer diuretics, glucose, insulin as ordered.<br>• Hold K$^+$ supplements and K$^+$-conserving diuretics.<br>• Monitor serum K$^+$ levels closely because a rapid drop may occur as K$^+$ shifts into the cells.<br>• Hyperkalemia is life threatening; it must be detected early.<br>• Teach at-risk clients to avoid foods high in K$^+$ (e.g., coffee, tea, cocoa, dried fruits, dried beans, and whole grain breads). |

# Hypocalcemia

| Risk Factors | Defining Characteristics | Therapeutic Nursing Interventions |
|---|---|---|
| • Hypoparathyroidism following thyroidectomy or radical neck surgery for cancer<br>• Malabsorption<br>• Vitamin D deficiency<br>• Acute pancreatitis<br>• Excessive administration of citrated blood<br>• Hyperphosphatemia<br>• Thyroid cancer<br>• Hypomagnesemia<br>• Hypoalbuminemia (e.g., as in nephrotic syndrome and starvation)<br>• Alkalosis<br>• Sepsis<br>• Alcohol abuse | • Total serum Ca++ <8.5 mg/dL or ionized level below 50%<br>• Numbness, tingling of circumoral region, fingers, and toes<br>• Hyperactive deep tendon reflexes<br>• Muscle tremors and cramps, progressing to tetany and convulsions if severe<br>• Cardiac arrhythmias<br>• Mental changes (e.g., confusion, mood and memory changes)<br>• Laryngeal spasm<br>• ECG shows prolonged QT interval<br>• **Chvostek's sign** (spasm of the facial muscles elicited by tapping the facial nerve in the region of the parotid gland)<br>• **Trousseau's sign** (carpal spasm can be elicited by compressing the upper arm and causing ischemia to the nerves distally) | • Take seizure precautions when hypocalcemia is severe.<br>• Monitor airway closely.<br>• Take safety precautions for confusion.<br>• Educate postmenopausal women about the need for dietary calcium; supplements may be needed. |

# Hypernatemia

| Risk Factors | Defining Characteristics | Therapeutic Nursing Interventions |
|---|---|---|
| • Water deprivation<br>• Hypertonic tube feedings with inadequate water supplement<br>• High insensible water loss (e.g., hyperventilation, fever)<br>• Unusually high salt intake<br>• Excess administration of hypertonic saline, sodium bicarbonate, or isotonic saline<br>• Profuse sweating, heatstroke<br>• Diabetes insipidus | • Serum Na+ >145 mEq/L<br>• Serum osmolality > 300 mOsm/kg<br>• Urinary Specific Gravity >1.015 (unless water loss is from kidneys)<br>• Thirst<br>• Dry, sticky mucous membranes<br>• Red, dry, swollen tongue<br>• Dyspnea<br>• Postural hypotension<br>• Fatigue<br>• Decreased consciousness<br>• Disorientation<br>• Convulsions | • Monitor for abnormal water loss or low water intake and for large gains of sodium (e.g., ingestion of Alka-Seltzer).<br>• Restrict dietary sodium<br>• Prevention: Offer fluids to debilitated clients at regular intervals. If fluid intake remains adequate obtain medical orders for alternate route (e.g., tube feedings).<br>• Prevention: If using tube feedings, give enough water to keep the serum Na+ and BUN levels in normal limits. |

## Hypercalcemia

| Risk Factors | Defining Characteristics | Therapeutic Nursing Interventions |
|---|---|---|
| • Hyperparathyroidism<br>• Prolonged immobility<br>• Megadoses of vitamin D<br>• Bone malignancy<br>• Paget's disease<br>• Thiazide diuretics<br>• Overuse of Ca++-containing antacids of Ca++ supplements | • Serum Ca++ >10.5 mg/dL or 5.5 mEq/L (total)<br>• ECG shows shortened QT interval<br>• Lethargy, weakness<br>• Depressed deep-tendon reflexes<br>• Constipation<br>• Anorexia, nausea, vomiting<br>• Polyuria and polydipsia<br>• Decreased memory and attention span; confusion<br>• Renal stones, flank pain<br>• Neuroses, psychoses (reversible)<br>• Cardiac arrest (in hypercalcemic crisis) | • Keep client well-hydrated.<br>• Discourage excessive dietary milk products.<br>• Institute safety precautions for confusion.<br>• Encourage adequate fiber in the diet to offset constipation.<br>• Be prepared to deal with cardiac arrest.<br>• In chronic hypercalcemia, transfer clients carefully because of weakened bones.<br>• Be alert for digitalis toxicity in digitalized clients.<br>• Encourage intake of acid-ash fluids (e.g., cranberry or prune juice).<br>• Prevention: Facilitate and encourage mobility to the extent possible. |

# Acid-Base Imbalance

## Overview

Acid-base balance depends on a proportionate balance of **carbonic acid** and **bicarbonate** levels in the body. **Carbonic acid** levels are normally regulated by the lungs, which retain or excrete carbon dioxide ($CO_2$). **Bicarbonate** and **hydrogen ion** levels are regulated by the kidneys. When there is a single primary cause, disturbances in carbonic acid-bicarbonate (acid-base) levels are referred to as respiratory acidosis or alkalosis and metabolic acidosis or alkalosis. The kidneys and the lungs work together to maintain acid-base balance. Arterial blood gases (ABGs) are most commonly used to evaluate acid-base balance.

**Respiratory acidosis and alkalosis** occur when a respiratory disturbance alters the carbonic acid level. Then the kidneys attempt to compensate (restore the balance) by either conserving or excreting bicarbonate.

**Metabolic acidosis and alkalosis** occur when a metabolic disturbance alters the bicarbonate levels. The lungs then try to compensate by conserving or excreting more $CO_2$, which is available in weakly ionized carbonic acid.

**Keep in mind** that combined clinical situations may occur when both respiratory and metabolic imbalances occur.

## Arterial Blood Gas Analysis

Arterial blood is used to evaluate the client's oxygenation and acid-base balance. Blood gases are drawn by laboratory technicians or nurses with specialized training. It is important to apply pressure to the puncture site for 1-2 minutes after the procedure to reduce bleeding and bruising. The six measurements commonly used to evaluate acid-base balance are:

- pH (norm 7.35 – 7.45) measures the relative acidity or alkalinity of the blood.
- $PaO_2$ (norm 80 -100 mm Hg) is an indirect measure of blood oxygen content. It is the pressure exerted by oxygen dissolved in the plasma of the arterial blood.
- $PaCO_2$ (norm 35 – 45 mm Hg) reflects the respiratory component of acid-base status. It is the partial pressure of carbon dioxide in arterial blood.
- HCO3- (bicarbonate; norm 22 – 26 mEq/L) measures the metabolic component of acid- base status.
- BE (base excess; norm -2 to +2 mEq/L)—also reflects the metabolic component of acid- base status. It is calculated from the bicarbonate level.
- $SaO_2$ (norm 95 – 98%) is the percentage of hemoglobin that is saturated with oxygen.

## Respiratory Acidosis

**Respiratory acidosis** is high $PaCO_2$ due to alveolar hypoventilation. Respiratory acidosis is caused by alveolar hypoventilation, which causes $CO_2$ to be retained. Carbonic acid levels increase and the pH falls below 7.35. The kidneys attempt to compensate by retaining bicarbonate; however, they are slow to respond, so compensation may require several days time.

Related Factors

- Acute lung conditions that impair alveolar gas exchange (e.g., acute pulmonary edema, aspiration of a foreign object, pneumonia)
- Chronic lung disease (e.g., emphysema, asthma, cystic fibrosis)
- Narcotic or sedative overdose that depresses the respiratory center in the brain
- Pneumothorax, hemothorax
- Cardiac arrest
- Improper regulation of mechanical ventilation
- Obesity
- Tight abdominal binders; postoperative pain
- Abdominal distention (e.g., from cirrhosis or bowel obstruction)

Defining Characteristics

- **Acute condition**: Increased pulse and respiratory rate, headache, dizziness, confusion, decreased level of consciousness, palpitations, warm flushed skin, ventricular fibrillation. **Arterial blood gases** (ABGs): pH < 7.35; $PaCO_2$ > 45 mmHg (primary), $HCO_3$- normal or slightly elevated.
- **Chronic condition**: Weakness, headache. **Arterial blood gases** (ABGs): pH < 7.35, $PaCO_2$ > 45 mm Hg (primary), $HCO_3$- > 26 mEq/L (compensatory).

Therapeutic Nursing Interventions

- Treatment focuses on improving ventilation.
- Medications include bronchodilators (to reduce bronchial spasm) and antibiotics (for respiratory infections).
- Assess respiratory status and lung sounds frequently.
- Perform pulmonary therapy measures such as inhalation therapy, percussion, and postural drainage.
- Monitor fluid intake and output, vital signs and arterial blood gases.
- Maintain adequate hydration (2-3 L of fluid per day).
- Administer supplemental oxygen as needed.
- A mechanical respirator may improve pulmonary ventilation; however caution must be used so that the elevated $PaCO_2$ is not decreased too rapidly.

# Respiratory Alkalosis

**Respiratory alkalosis** is low $PaCO_2$ due to hyperventilation. An increase in respiratory rate or depth causes the $CO_2$ to be excreted faster than normal. When the $CO_2$ level is high, the brain's respiratory center is stimulated; however, when the $CO_2$ level is low (as in respiratory alkalosis), the respiratory center depresses or even ceases respirations. The kidneys try to compensate by excreting more bicarbonate and retaining more hydrogen.

Related Factors

Hyperventilation can be caused by: anxiety (most common cause), hypoxemia, high fever, early salicylate overdose, central nervous system lesions, thyrotoxicosis, excessive ventilation by mechanical ventilators; pulmonary emboli, and gram-negative bacteremia.

### Defining Characteristics

- Light-headedness (low $PaCO_2$ causes cerebral vasoconstriction)
- Loss of concentration
- Hyperventilation syndrome: Tinnitus, palpitations, sweating, dry mouth, tremulousness, shortness of breath, chest tightness, nausea and vomiting, epigastric pain, blurred vision, convulsions
- ABGs: pH >7.45; $PaCO_2$ <35 mm Hg, $HCO_3$- under 22 mEq/L (compensatory)

### Therapeutic Nursing Interventions

- If the cause of the imbalance is anxiety, interventions should focus on relieving the anxiety. If anxiety is severe, a sedative will be required.
- Instruct the client to breathe more slowly or to breathe into a paper bag or apply a rebreather mask to inhale $CO_2$.
- If the cause is not anxiety, treatment will focus on correcting the underlying problem.
- Monitor vital signs and ABGs.

## Metabolic Acidosis

**Metabolic acidosis** is a low bicarbonate level (either nonvolatile acid is present to use up $HCO_3$- too rapidly, or the $HCO_3$- is being lost in disproportionate amounts). Metabolic acidosis occurs when bicarbonate levels are low in relation to the amount of carbonic acid in the body. This may happen because of renal failure and the inability of the kidneys to excrete hydrogen ions and produce bicarbonate. The respiratory center attempts to compensate by increasing the rate and depth of respirations, excreting $CO_2$ and causing carbonic acid levels to fall. Respiratory compensation occurs rapidly, within minutes of the pH imbalance. Metabolic acidosis may also occur when too much acid is being produced, as in diabetic ketoacidosis. In addition to respiratory compensation, the kidneys also try to compensate by retaining more bicarbonate and excreting more hydrogen.

### Related Factors

- Conditions that increase nonvolatile acids in the blood (e.g., diabetes mellitus, starvation, impaired renal function
- Conditions that cause loss of bicarbonate (prolonged diarrhea, intestinal fistulas)
- Excessive infusion of chloride-containing IV fluids (e.g., hyperalimentation)
- Acidifying drugs (e.g., ammonium chloride)
- Ingestion of toxins (e.g., salicylates, ethylene glycol, methanol)

### Defining Characteristics

- Headache
- Confusion, drowsiness
- Weakness
- Nausea and vomiting
- Peripheral vasodilation, causing warm, flushed skin
- Bradycardia, decreased cardiac output

- Hyperkalemia (frequently)
- Increased respiratory rate and depth (when $HCO_3^-$ is very low)
- ABGs: pH <7.35; $HCO_3^-$ <22 mEq/L, $PaCO_2$ <35 mm Hg (compensation by the lungs), base excess is always negative

### Therapeutic Nursing Interventions

- Treatment focuses on correcting the metabolic defect.
- IV sodium bicarbonate, if ordered
- Monitor ABGs.
- Monitor level of consciousness.
- Monitor intake and output.

## Metabolic Alkalosis

Metabolic alkalosis is a high bicarbonate level (nonvolatile acid is lost and not using up $HCO_3^-$ or $HCO_3^-$ is gained in disproportionate amounts). Metabolic alkalosis occurs because of excessive acid losses, or due to increased ingestion or retention of bases; for example, when someone ingests bicarbonate of soda as an antacid. When the metabolic alkalosis depresses the respiratory center, the lungs attempt to compensate—respirations become slow and shallow, $CO_2$ is retained, and $HCO_3^-$ levels increase. The kidneys also try to compensate by excreting potassium and sodium with the excess bicarbonate and by retaining hydrogen.

### Related Factors

- Excessive acid losses due to vomiting or gastric suction
- Excessive use of potassium-losing diuretics (e.g., thiazides, furosemide, ethacrynic acid)
- Ingestion of alkalis (bicarbonate-containing antacids)
- Abrupt relief of chronic respiratory acidosis
- Hypokalemia
- Hyperaldosteronism
- Cushing's syndrome

### Defining Characteristics

- Decreased respiratory rate and depth (compensatory)
- Dizziness
- Circumoral paresthesias, numbness and tingling of fingers and toes
- Carpopedal spasm
- Muscle hypertonicity
- Serum $Cl^-$ relatively lower than $Na^+$
- ABGs: pH >7.45; Bicarbonate over 26 mEq/L; $PaCO_2$ >45 mm Hg (compensatory), base excess always positive

# Critical Thinking Exercise: Electrolyte & Acid-Base Imbalance

**Situation:** A client who is taking digitalis in the treatment of congestive heart failure begins to demonstrate signs of hypokalemia, including fatigue, nausea, muscle weakness, and leg cramps. Diagnostic evaluation includes an abnormal ECG and serum potassium levels below normal levels.

1. Based on this, what do you think might be causing the client's potassium deficit, and what can be done to correct it?

2. Describe the function of potassium in the body. How might an imbalance of potassium manifest itself? Which clients are at particular risk for such imbalances?

3. In counseling a client taking diuretics, what foods might you recommend as good sources of potassium?

# Perioperative Nursing Care

## Key Points

- Types of surgery can be classified according to:
  - Degree of urgency
  - Seriousness
  - Purpose
- There are aspects of care that are common to all patients having surgery, regardless of type.
- Factors that increase surgical risk are: age, general health (e.g., obesity, cardiac conditions), and mental status (e.g., anxiety, dementia).
- Perioperative nursing stresses providing continuity of care for the surgical client.
- **Key Concepts/Terms:** Perioperative nursing, peroperative phase, intraoperative phase, postoperative phase, elective surgery, urgent surgery, emergency surgery, major surgery, minor surgery, diagnostic surgery, ablative surgery, palliative surgery, reconstructive surgery, transplant surgery

**Overview**

**Perioperative nursing**, a clinical specialty, refers to the role of the nurse during the **preoperative** (before surgery), **intraoperative** (during surgery), and **postoperative** (after surgery) phases of the client's surgical experience. Nursing care depends upon the type of surgery and such in-depth coverage is available in medical-surgical texts. However, there are aspects of care common to all clients having surgery. They are presented in some fundamentals courses and in this review. Surgeries have traditionally been performed in hospitals; however, it is common now for surgeries to be performed in outpatient units (day surgeries).

## Types of Surgery

**Degree of Urgency:**

- **Elective surgery** is performed when surgery is the preferred treatment and may improve the client's life, but is not essential for the person's health (e.g., facial plastic surgery, hernia repair, hip replacement surgery).
- **Urgent surgery** is necessary for the client's health and may prevent complications from developing (e.g., excision of a cancerous tumor, vascular repair for an obstructed artery).
- **Emergency surgery** must be done immediately to save the client's life or preserve function of a body part (e.g., control of internal hemorrhaging, repair of a fracture).

**Seriousness**: Surgery is classified according to the degree of risk to the client, which is determined by the client's age, nutritional status, general health, use of medications, and mental status.

- **Major surgery** involves a high degree of risk and/or involves extensive reconstruction or alteration of body parts (e.g., coronary bypass surgery, organ transplant, colon resection).
- **Minor surgery** involves minimal risk and minimal alteration of body parts. It is often done in day surgery (e.g., breast biopsy, tonsil removal, hernia repair).

**Purpose**:

- **Diagnostic surgery** is done to confirm/establish a diagnosis (e.g., breast biopsy).
- **Ablative surgery** is done to excise/remove a diseased body part (e.g., amputation, appendectomy).
- **Palliative surgery** is done to reduce pain or intensity of symptoms; it does not cure (e.g., colostomy, debridement of necrotic tissue).
- **Reconstructive surgery** restores appearance of or function to traumatized or malfunctioning tissues (e.g., breast implants, revision of facial scars).
- **Transplant surgery** replaces malfunctioning structures or organs (e.g., kidney transplant, hip replacement).

## Factors that Increase Surgical Risk

**Age**: Very young and elderly clients are at greater risk.

**General health**: The presence of infections or other pathophysiology increases risks. The following are examples:

- Malnutrition or dehydration (complications: delayed wound healing)
- Infection of any kind
- Obesity (complications: hypertension, cardiac problems, respiratory problems, delayed wound healing, wound infection)
- Cardiac conditions (complication: decreased cardiac output, poor tolerance of anesthesia)
- Respiratory disorders such as asthma (complications: postoperative lung infections, poor tolerance of general anesthesia)
- Renal disease (complications: fluid and electrolyte balance, inadequate excretion of drugs)
- Bleeding disorders (complications: hemorrhage, shock)
- Diabetes mellitus (complications: delayed healing, infection)
- Liver disease (complications: inability to detoxify medications, hemorrhage, delayed healing)
- Neurologic disease (complications: seizures)

**Mental status**: Mental illness, mental retardation, anxiety, and dementia affect the client's ability to cope with surgery. Dementia may cause an unpredictable response to anesthesia, as well.

## Nurse's Role for the Surgical Client

- Perioperative nursing stresses providing continuity of care for the surgical client. When possible, an operating room nurse should perform preoperative assessments, do preoperative teaching, and follow the client through the surgery and in recovery.

- The nurse is responsible for providing safe, consistent, effective nursing care before, during, and after surgery.

# Critical Thinking Exercise: Perioperative Nursing Care

**Situation:** As you are preparing a 50-year-old client for a diagnostic cardiac catheterization, he admits to you that he has never undergone a surgical procedure before and is quite anxious.

1. In a situation like this, what nursing diagnosis might apply? And what specific nursing interventions might be useful with this client?

2. In a client who is fearful about an impending surgery, how might this fear manifest itself? Do you suppose such a client will always appear openly anxious? Why or why not?

3. In what ways might a client's spirituality help him or her cope with the surgical experience? What other resources might help an anxious client at this time?

# Preoperative Nursing Care

## Key Points

- The preoperative period begins with the decision to have surgery and ends when the client is on the operating table.
- The nurse's role encompasses preoperative assessments, physical preparations, teaching, emotional support, and confirmation of informed consent.
- Routine preoperative screening tests include: complete blood count, serum electrolyte analysis, coagulation studies, serum creatinine and blood urea nitrogen, urinalysis, chest x-ray, electrocardiogram, blood typing and cross matching, and blood glucose.
- Most agencies have a preoperative checklist to use prior to surgery to ensure that all necessary preparation has been done.
- Preparation for surgery depends on the type of surgery being performed, and commonly has implications for the following:
  - Nutrition and hydration (e.g., usually clients are "NPO;" intake and output are measured)
  - Elimination (e.g., client may need a urinary catheter or an enema)
  - Rest and sleep (e.g., a sedative may be ordered for the night before surgery)
  - Hygiene (e.g., remove hair pins, have the client bathe)
  - Medications (e.g., routine meds may be discontinued; anticholinergics or sedatives may be   administered)
  - Antiembolism stockings
  - Personal "valuables" sent for safekeeping (e.g., jewelry, money)
  - Prostheses removed (e.g., dentures, contact lenses)
  - Special skin preparation (e.g., scrubbing with antibacterial soap, shaving the site)
- The client must sign a consent form before surgery. The surgeon is legally responsible for obtaining informed consent; the nurse may witness the client's signature, and the nurse checks to be sure that a signed consent form is included in the chart.
- Consent is "informed" only when:
  - The client understands the information
  - The client is not a minor
- **Key Concepts/Terms**: Transcutaneous electrical nerve stimulation (TENS), patient controlled analgesics (PCA), sedatives, tranquilizers, narcotic analgesics, anticholinergics, histamine-receptor antihistamines, neuroleptanalgesics

## Overview

The **preoperative period** begins with the decision to have surgery and ends when the client is on the operating table. Preoperative preparation may occur several days before the surgery. For example, laboratory tests and X-rays may be performed in a primary care provider's office or an outpatient laboratory. Depending on the type of procedure expected, the client may enter the hospital the day before the surgery, or early on the morning of the surgery for preoperative tests and preparation. The nurse's role in the preoperative period encompasses preoperative assessments, physical preparations, teaching, emotional support, and confirmation of informed consent.

## Preoperative Assessments

**Nursing history** should include current health status (e.g., presence of any chronic diseases, physical limitations); allergies; current medications; medical history; previous surgeries; understanding and expectations of the surgery; use of alcohol, caffeine, or other drugs; family and social support; occupation; and emotional health (coping resources, anxiety).

**Physical examination** is brief but complete, and focuses on systems that could affect the client's response to the surgery or to anesthesia. Data from the physical exam provide an important baseline for comparison during and after surgery.

- **Vital signs**: Help to rule out any fluid and electrolyte abnormalities or infections.
- **Head and neck**: Condition of the oral **mucous membranes** reveals the level of hydration. Inspection of the **oro- and nasopharynx** can reveal sinus drainage and respiratory infection. Distended **jugular** veins may indicate fluid excess of heart disease.
- **Skin turgor**: Reflects level of hydration. During surgery, the client must lie in one position, sometimes for several hours. Skin that has poor turgor or is thin and dry is a risk factor for pressure sores over bony prominences.
- **Thorax and lungs**: Auscultation of the lungs and heart may reveal a respiratory infection or fluid excess. Observation of chest excursion may alert the nurse to a possible decline in ventilatory capacity.
- **Heart and vascular system**: Surgical stress can cause cardiac dysrhythmias. Auscultation can alert the nurse to pre-existing problems. Adequacy of peripheral pulses is important for clients having vascular surgery, casts, or constrictive bandages.
- **Abdomen**: Especially important if the client is having abdominal surgery, but also important as a baseline for assessing the return of bowel function after general anesthesia.
- **Neurological status**: General anesthesia and analgesics alter the client's level of consciousness and spinal anesthesia alters motor function, so it is important to know the preoperative status for comparison after surgery.

## Preoperative Screening Tests

The surgeon orders diagnostic tests to screen for pre-existing abnormalities; if tests reveal problems, the surgery may be canceled until the condition can be treated. The nurse must be sure that all tests have been completed and, in some instances, to be sure the client is prepared properly for each test. As results become available, the nurse also alerts the surgeon of his/her findings. The following are routine preoperative screening tests that are common to most surgeries:

**Complete blood count (CBC):** A CBC uses venous blood to measure red and white blood cell count, hemoglobin concentration, and hematocrit. An abnormal CBC may reflect many conditions (e.g., infection, malnutrition, anemia, impaired immune function) that increase the risk of cardiovascular and pulmonary complications. If the hemoglobin is <10 g/dL and hematocrit <33%, blood may be given preoperatively to maintain volume and increase tissue oxygenation during surgery.

**Serum electrolyte analysis:** Na+, K+, Ca++, Mg++, Cl-, and $HCO_3$- in peripheral venous blood are checked to evaluate fluid and electrolyte status. Surgery creates the potential for fluid and electrolyte imbalance, so electrolyte replacement is sometimes necessary before surgery.

**Coagulation study (prothrombin time, partial thromboplastin time, platelet count):** Results identify clients who are at risk for hemorrhage or thrombus formation.

**Serum creatinine test and blood urea nitrogen (BUN):** Results are used to assess renal function, which is important in regulating fluid and electrolyte levels.

**Urinalysis:** A urinalysis measures color, pH, and specific gravity; it also detects the presence of protein, glucose, ketones, and blood. It screens for urinary tract infection, renal disease, and diabetes mellitus.

**Chest x-ray** reflects the condition of the heart and lungs and, therefore, the ability to tolerate sedatives and inhalation anesthetics.

**Electrocardiogram (ECG):** An ECG is done if the client is over age 40 or has heart disease.

**Blood typing and cross matching** are done if the client is likely to lose a large amount of blood.

**Fasting blood glucose** is done to detect presence of diabetes mellitus.

## Associated NANDA Nursing Diagnoses

- Anxiety/Fear related to threat of loss of body part or unknown effects of surgery on usual functions and roles
- Fear related to anticipation of postoperative pain
- Knowledge Deficit (preoperative and postoperative routines/care) related to no prior experience
- Sleep Pattern Disturbance related to hospital routines, stress, and anxiety
- Anticipatory Grieving related to anticipated surgical loss of body part

## Physical Preparation for Surgery

Most agencies have a preoperative checklist for use on the day of surgery to ensure that all necessary records (e.g., laboratory reports, consent form) are in the chart and that all physical preparation has been done to assure client safety. Preparation depends on the type of surgery being performed.

### Nutrition and Hydration

- Diet order depends on the type of surgery and the anesthesia to be used.
- Measure intake and output, including intravenous fluids.

- Usually clients are to have "nothing by mouth" (NPO) for six to 12 hours prior to surgery because anesthetics decrease gastrointestinal functioning.
- Remove food and fluids from the bedside and place an "NPO" sign at the bedside. If the client does ingest anything, notify the surgeon.

### Elimination

- Insert a retention catheter, if ordered, to keep the bladder empty and prevent injury during surgery.
- If the client does not have a catheter, have him/her empty the bladder before administering the preoperative medications.
- An enema may be necessary if bowel surgery is planned.

**Rest and Sleep:** Promote rest and sleep the night before surgery. A sedative is sometimes ordered for this purpose.

### Hygiene

- The client may need to bathe or shower and shampoo the night or morning before surgery (or both) to reduce the risk of infection.
- Remove cosmetics so that nail beds, skin, and lips can be used to assess circulation during and after surgery.
- Have the client don a surgical cap (depending on agency policy) to cover the hair.
- Remove all hair pins and clips that might cause injury while the client is unconscious.
- Have the client remove personal clothing and don an operating room gown.

### Medications

- The client's "routine" medications may be temporarily discontinued.
- Preoperative medications are either given on the hospital unit or in the operating room. They may include:
  - **Sedatives and tranquilizers** to reduce anxiety and ease anesthetic induction (e.g., secobarbital, diazepam)
  - **Narcotic analgesics** (e.g., morphine, meperidine) to sedate the client and reduce the amount of anesthetic needed
  - **Anticholinergics** (e.g., atropine, scopolamine, Robinul) to reduce oral secretions and prevent laryngospasm
  - **Histamine-receptor antihistamines** (e.g., Tagamet, Zantac) to reduce gastric acidity and volume
  - **Neuroleptanalgesics** (e.g., Innovar) to create calm and sleepiness

**Antiembolism stockings** are firm elastic stockings that compress the veins of the legs and facilitate the return of venous blood to the heart. They are sometimes applied before surgery, as well as postoperatively, to help prevent edema of the feet and legs and to facilitate return of venous blood to the heart. (See "*Postoperative Care*")

**Personal "valuables"** such as jewelry and money should be labeled and sent home with the family or placed in safekeeping. Wedding rings can usually be taped in place if there is no danger of the fingers swelling after surgery. They must be removed, for example, if there is to be a mastectomy or a cast applied to that arm.

Prostheses (partial or complete dentures, contact lenses, artificial eyes, artificial limbs, eyeglasses, wigs, and false eyelashes) must be removed. Check also for loose teeth or presence of chewing gum in the mouth, which might be aspirated during anesthesia. Hearing aids are sometimes left in place.

Special orders: Depending on the surgery, there may be special orders to implement (e.g., inserting a nasogastric tube, administering insulin).

Special skin preparation (e.g., scrubbing the operative site with antibacterial soap or solution) is frequently done in surgery, but may also be done with the bath or shower. The incisional area is usually shaved before surgery, either on the unit or in the preop holding area.

# Preoperative Teaching

Preoperative teaching has been shown to decrease postoperative complications. Timing of the teaching is important. When done too far in advance or when the client is anxious, it is not as effective. Teaching should include information about:

**Surgical events and sensations**

- **Surgical events** include, for example, the scheduled time for surgery; how long it will last; what procedures and medications will be given before, during, and afterward; what equipment will be used (e.g., ECG machine); who will be present; where visitors can wait; presence of surgical dressings, casts, and so on.

- **Sensations** include how the preoperative medication will make the client feel; anticipating a sore throat from the endotracheal tube; incisional pain; a dry mouth, and so on.

**Pain management**: The client should be taught about PRN orders (the need to request pain medications), how to splint the incision, and how to use the planned method of pain control. For example:

- Pain medications may be given by injection while the client is NPO, and then changed to PO meds.

- Alternative methods of pain control include:

  - **Transcutaneous electrical nerve stimulation (TENS)**, in which electrodes are placed on the skin along each side of the surgical incision. The client pushes a button to control the electric current when he/she feels pain.

  - **Client controlled analgesia (PCA)** delivers analgesia by pump; the client pushes the button to administer a preset dose of analgesic.

- When narcotics are used to manage pain, there is little danger of addiction.

- The use of relaxation techniques (e.g., deep breathing) increases the effectiveness of pain medications.

- Pain medications are more effective if given before the pain becomes too severe, so PRN medications should be requested before pain becomes intense.

- Pain medications will help the client to ambulate, breathe deeply, and perform other activities necessary for recovery.

**Physical activities**: Physical activities to prevent the most common postoperative cardiovascular and pulmonary complications (e.g., atelectasis, pneumonia, thrombophlebitis, and emboli) are taught in the preoperative period. They include:

- Deep (diaphragmatic) breathing (see "*Postoperative Breathing Exercises*")

- Coughing (see *"Postoperative Breathing Exercises"*)
- Incentive spirometry (see *"Oxygenation: Normal Oxygenation"*)
- **Leg exercises** increase venous return through flexion and contraction of the quadriceps and gastrocnemius muscles, helping to prevent thrombophlebitis and emboli. Teach the client to:
    - Alternately dorsiflex and plantar flex the toes, and then make a circle with the toes.
    - Flex and extend the knees, pressing them down toward the mattress when extending.
    - Raise and lower each leg while keeping the knee straight.
    - Repeat the exercises every hour or two.
- **Turning in bed** promotes venous return, gastrointestinal peristalsis, and respiratory function. Explain that the client will be asked to turn from side to side every two hours. Incisional pain makes this difficult, so the procedure should be practiced before surgery. First show the client how to splint the incision with a small pillow or folded blanket. Have the client raise one knee and reach across to grasp the side rail on the side to which she/he is turning; then roll over while pushing with the bent leg and pulling on the side rail.

## Emotional Support

- Therapeutic communication and client/family teaching are important methods of providing emotional support.
- Encourage the client to verbalize fears and concerns.
- Use active listening skills to identify anxiety and fear.
- Use touch, as appropriate, to show caring.
- Be informed about the client's surgery and be prepared to answer client and family questions (e.g., "When can I go back to work?" "How long before I can eat?" What will the scar look like?")
- Avoid false reassurance; do not say, "I'm sure you will be fine." This denies the client's emotional needs and blocks therapeutic communications—and it may not be true!

## Obtaining Informed Consent

- **Informed consent** is when, after receiving—and understanding—the following information, the client voluntarily agrees to undergo a particular treatment:
    - A description of the treatment/procedure
    - Purpose of the treatment/procedure
    - The name and qualifications of the person who will perform the treatment/procedure
    - Explanation of the risks involved (e.g., the potential for damage, disfigurement, and death)
    - The chances for success
    - Possible alternative treatments/approaches
    - An explanation that the client has the right to refuse the treatment and the right to withdraw consent.

- The client must sign a consent form before surgery or any invasive treatment. The surgeon has the legal responsibility for ensuring that the client is giving informed consent; however, the nurse may witness the client's signature on the form. The nurse should ensure that the client understands the procedure and risks. If it is not clear that this is so, the nurse must contact the surgeon before the surgery proceeds.
- The nurse checks to be sure that a consent form is signed and included in the chart.
- Consent is "informed" only when: (1) the client understands the information (e.g., speaks the language, is conscious, not sedated, and mentally competent) and (2) is not a minor.

## Therapeutic Nursing Interventions

See *"Physical Preparation for Surgery," "Preoperative Teaching," "Emotional Support,"* and *"Obtaining Informed Consent."*

# Critical Thinking Exercise: Preoperative Nursing Care

**Situation**: You are preparing a 45-year-old client for major abdominal surgery. Devise an appropriate teaching plan that will address each of the following factors:

- Deep breathing and coughing
- Postoperative mobility
- Pain control

1. With respect to the educational needs of this client, what is your major goal?

# Postoperative Breathing Exercises

## Key Points

- The following contribute to poor alveolar inflation, possible alveolar collapse, and retention of secretions—creating the potential for pulmonary infection and atelectasis:
  - During surgery, the cough reflex is suppressed, the lungs are not fully ventilated, and mucous accumulates in the air passages.
  - Postoperative pain interferes with deep breathing, leading to shallow respirations.
  - Analgesia and anesthesia decrease the respiratory center and level of consciousness.
  - Coughing and deep breathing should be done with the client in semi-Fowler's position.
  - Breathing exercises should be done every one to two hours, while awake, for the first 24 to 48 hours postop.
- Teach and help the client to "splint" the incision during coughing and breathing exercises.
- Provide pain medications prior to breathing exercises.
- Be sure the client coughs deeply and fully, and does not merely clear the throat.
- **Key Concepts/Terms**: Cough and deep breathing, alveoli, splinting

## Overview

Breathing exercises are designed to assist the client in achieving more effective and controlled breathing, to decrease the workload of breathing and to correct respiratory and circulatory problems. During surgery, the cough reflex is suppressed, the lungs are not fully ventilated, and mucous accumulates in the air passages. Postoperative pain interferes with deep breathing, leading to shallow respirations. Analgesia and anesthesia decrease the respiratory center and the level of consciousness. As a result, alveoli do not inflate well and may collapse; and secretions are retained, creating the potential for pulmonary infection and atelectasis. Deep-breathing exercises hyperventilate the alveoli, improve lung expansion and volume, facilitate oxygenation of tissues, and help to get rid of anesthetic gases and mucous. Breathing exercises may be used for purposes other than surgery. Postoperative breathing exercises are best taught preoperatively, not too far in advance of the surgery.

## Procedure and Technique for Deep Breathing Exercises

- Place the client in semi-Fowler's position, with support for the neck and shoulders.

- Have the client place the hands over the rib cage in order to feel the chest rise as the lungs expand.
- Have the client:
  - Exhale gently and completely.
  - Inhale through the nose gently and completely, using the abdominal muscles.
  - Hold the breath for several seconds (count to three).
  - Exhale the breath slowly through pursed lips, as completely as possible
  - Repeat three times.
- The exercise should be done every one to two hours while the client is awake for the first 24 to 48 hours post surgery, and as necessary thereafter.
- Support incision with pillows, blanket, or hands during breathing exercises.
- Provide pain medications prior to performing breathing exercises.
- Monitor effectiveness of breathing treatments by using pulse oximetry and auscultation of breath sounds.

## Procedure and Technique for Coughing

Coughing is usually taught with deep breathing. It helps to remove mucous from the respiratory tract. Because coughing is painful after surgery, teach the client how to splint the incision with a pillow or folded bath blanket.

- Place client in semi-Fowler's position, leaning forward.
- Ask the client to:
  - Inhale slowly and deeply and exhale slowly through the nose three times.
  - Take a deep breath and hold it while mentally counting to three.
  - Cough fully for two to three consecutive coughs without inhaling between coughs.
  - Take another deep breath.
- Repeat every two hours while awake.
- Remind client not to just clear the throat, but to cough deeply and fully.

# Critical Thinking Exercise: Postoperative Breathing Exercises

**Situation**: You are caring for a client who has undergone thoracic surgery. With respect to postoperative breathing exercises, how will you address each of the following important topics?

- Breathing techniques

- Positioning for optimal breathing

- Improving airway clearance

# Intraoperative Nursing Care

> ### Key Points
>
> - The intraoperative period begins with admission of the client to the surgical area and ends with transfer to the recovery area.
> - Nursing roles include maintaining surgical asepsis in the room, assessing, positioning, assisting with monitoring devices, obtaining additional supplies, recording, and participating in counts of instruments, needles, and sponges.
> - Types of anesthesia include: general anesthesia and regional anesthesia (topical, local, infiltration, nerve block, infiltration block, spinal, epidural)
> - Proper positioning should:
>   - Provide visualization and access to the operative site
>   - Provide correct position for the specific procedure
>   - Protect bony prominences
>   - Avoid strain or injury to muscles, bones, and joints
>   - Protect the skin
>   - Prevent complications of various positions (e.g., lithotomy predisposes to thrombophlebitis)
> - **Key Concepts/Terms**: Scrub nurse, circulating nurse, surgical skin preparation

**Overview**

The intraoperative period begins with the admission of the client to the surgical area and ends when the client is transferred to the recovery area. The type of surgery (e.g., ambulatory, major, emergency) determines the assessments and interventions needed; however, the nursing role emphasizes a number of specialized procedures designed to create a safe, therapeutic environment for the client. During the intraoperative period, nurses function as either scrub nurses or circulating nurses. A **scrub nurse** assists the surgeon, handles ("sets up") instruments and supplies, and is responsible for maintaining surgical asepsis in the room. A **circulating nurse** assesses the client in the operating room, assists with positioning the client on the operating table, helps with monitoring devices, obtains additional supplies, records events in the operating room, and participates in the count of instruments, needles, and sponges used in the surgery.

## Types of Anesthesia

**General anesthesia** acts by blocking awareness centers in the brain to induce amnesia (loss of memory), analgesia (insensibility to pain), hypnosis (artificial sleep), and relaxation. It produces total loss of sensation and consciousness, along with protective reflexes such as the cough and gag reflexes. General anesthetics are usually given by intravenous infusion or by inhalation through an endotracheal tube. The main disadvantage is that general anesthesia depresses the respiratory and circulatory systems.

**Regional anesthesia** blocks the transmission of nerve impulses to and from a specific region of the body. The client loses sensation and motor ability in that area of the body, but remains conscious. Regional anesthesia commonly uses medications of the "-caine" family (e.g., Marcaine, lidocaine, benzocaine).

- **Topical anesthesia** is applied directly to the surface of the skin and mucous membranes, wounds, and burns. Lidocaine (Xylocaine) and benzocaine are most commonly used.

- **Local anesthesia (infiltration)** is injected into a specific area. Lidocaine or tetracaine 0.1% is used for minor procedures such as suturing a small cut or performing a biopsy.

- **A nerve block** is a technique in which the anesthetic is injected into and around a nerve or small group of nerves that supply sensation to a small area (e.g., a facial nerve).

- **An infiltration block** is usually used for procedures involving the arm, wrist, and hand. The anesthesia is injected intravenously and a tourniquet is applied to the extremity to prevent the infiltration and absorption of the medication beyond the involved limb.

- **Spinal anesthesia** requires a lumbar puncture through one of the interspaces between L2 and S1. The anesthetic agent is injected into the subarachnoid space around the spinal cord. Spinal anesthesia blocks the motor and sensory nerve impulses in the perineal and rectal areas (**low spinals**), below the level of the umbilicus (**mid spinals**: T10), or below the nipple line (**high spinals**: T4).

- **Epidural anesthesia** is an injection of the anesthetic agent into the epidural space, the area inside the spinal column but outside the dura mater. The effect is similar to that of spinal anesthesia.

## Assessments

- Check the client's identity on admission to the surgical suite.
- Assess physical and emotional status.
- Verify the information on the preoperative checklist.
- Assess the client's knowledge about the events to follow.
- Assess client response to the preoperative medication.
- Check the placement and patency of all tubes and lines (e.g., urinary catheters, IVs).
- Continuous assessment is needed throughout surgery, to rapidly identify any adverse responses to surgery or anesthesia. The nurse and anesthetist continuously monitor:
  - Vital signs
  - ECG
  - Oxygen saturation
  - Fluid intake and urinary output
  - Blood loss (estimated)

- Arterial and venous pressures
- Pulmonary artery pressure

## Related NANDA Nursing Diagnoses

- Hyperthermia
- Latex Allergy Response
- Risk for Latex Allergy Response
- Risk for Perioperative Positioning Injury

## Procedure and Technique for Surgical Skin Preparation

This involves cleaning the surgical site, shaving if necessary, and applying an antimicrobial agent. The purpose is to **reduce the risk of wound infection by removing soil and transient microbes from the skin, reducing the resident microbial count, and inhibit rapid regrowth of microbes.**

- Clean the surgical site and surrounding areas (either by having the client shower beforehand or by washing the surgical site in the surgical area).
- Assess the site for moles, warts, rashes, pustules, abrasions, and exudate before skin preparation. Document findings.
- Remove hair from the site only when necessary. Electric clippers or depilatory cream are preferred to a razor to reduce the risk of traumatizing the skin.
- Prepare the site and surrounding area with an antimicrobial agent. The agent should be nontoxic and have a broad range of germicidal action.
- Document surgical skin preparation.

## Procedure and Technique for Positioning the Surgery Client

Responsibility for positioning is shared by the nurse, surgeon, and anesthetist. It is performed after anesthesia is given and before surgical draping.

**Provide visualization and access to the site.** Positioning should provide:

- The best possible visualization of and access to the surgical site
- The best possible access for assessing and maintaining anesthesia and vital functions (e.g., vital signs, respirations)
- Protection of the client from injury

**Provide correct position for the specific procedure.** The exact position depends upon the type of operation and surgical approach (e.g., a supine position is used for abdominal surgery).

**Protect bony prominences.** Pad bony prominences.

**Avoid strain or injury to muscles, bones, and joints.** Maintain good body alignment.

**Protect the skin.** Lift rather than pull or roll the client into position (to avoid shearing force on the skin).

**Be aware of the potential complications of various positions.** For example:

- **Lithotomy position** (legs elevated in stirrups) causes the blood to pool in the legs. Potential complications are thrombophlebitis and foot drop (from damage to the peroneal nerve).

- **Trendelenburg's position** (head lower than feet) displaces the abdominal viscera, decreasing diaphragm movement and respiratory exchange. Blood pooling in the upper torso increases the blood pressure. On return to supine position, the client may experience hypotension.

## Nurse's Role During Surgery

- Drape the client to establish a sterile field around the operative site. The only area left exposed is the incision site.
- Assist in preparing and maintaining the sterile field.
- Open and distribute sterile packages and items during surgery.
- Provide medications and solutions for the sterile field.
- Monitor and maintain sterile environment.
- Manage catheters, tubes, drains, and specimens.
- Perform sponge, instrument, and "sharps" (e.g., needles) counts to avoid leaving any supplies inside the client.
- Document care provided and client responses (e.g., item counts, monitor data, positioning, medications given).
- After surgery, move the client carefully from the operating table to the stretcher for transfer to the recovery room. Sudden, rough handling at this time can cause hypotension or even cardiac or respiratory arrest.
- Communicate relevant pre- and postoperative assessments and interventions to the recovery room nurses.

# Critical Thinking Exercise: Intraoperative Nursing Care

**Situation**: An operating room nurse is caring for a client who is about to undergo general anesthesia. How would you describe the role of a nurse as client advocate within the setting of the operating room?

1. Interview an operating room nurse to develop a better sense of this nursing speciality. Here are a few questions to consider when planning your interview:

   - What made you choose to be an operating room nurse?

   - What are the rewards associated with your job?

   - What are some of the challenges?

   - What does it take to be a successful operating room nurse?

2. Based on your interview, write a brief advertisement seeking the ideal candidate for an operating room nurse.

# Immediate Postoperative Period

## Key Points

**Potential complications of the immediate postop period include:**

- Respiratory obstruction
- Hypertension or hypotension
- Central nervous system depression, confusion
- Fluid volume excess or deficit
- Hemorrhage from operative site
- Seizures, injury from involuntary movements while unconscious

**Nursing interventions, in general, are to:**

- Provide continual assessment of body systems, wound, and pain.
- Keep all lines (e.g., IV) patent.
- Assure that monitors and equipment (e.g., oxygen) are functioning.
- Position to:
  - Keep airway open.
  - Enhance chest expansion.
- Arouse and orient the client.
- Facilitate oxygenation (e.g., suction, oxygen).
- Treat hypotension.
- Provide for safety (e.g., bedrails up).
- Provide for comfort (e.g., keep warm).
- **Key Concepts/Terms:** Central nervous system (CNS), depression, respiratory depression, airway patency, fluid status

## Overview

The postoperative period begins with admission of the client to the postanesthesia recovery area and ends when healing is complete. The postoperative period consists of two stages: **immediate care** (usually given in the recovery area) and **ongoing postoperative care**, which lasts from return to the unit until healing is complete. Goals of nursing care are to: prevent complications, promote healing, facilitate coping with altered structure/function, perform teaching, and plan for home care.

## Potential Complications of Surgery and Anesthesia

**Respiratory** obstruction is the most common recovery room emergency (e.g., accumulation of secretions, laryngospasm, obstruction by the tongue, or laryngeal edema).

**Cardiovascular**

- Transient hypertension, caused by anesthetic
- Hypotension can result from anesthesia, preoperative medications, position changes, blood loss, respiratory alterations, and peripheral blood pooling
- Shock (tachycardia is an early symptom)

**Central nervous system**: Some clients may be difficult to arouse or may experience confusion after anesthesia.

**Fluid status**

- Fluid volume excess (e.g., stress causes sodium and water retention)
- Fluid volume deficit (e.g., blood loss, inadequate hydration)

**Wound status**: In the immediate postoperative period, the primary danger is from hemorrhage.

**Physical safety**: Seizures, injury from moving about before return of consciousness

## Immediate Postoperative Care

Nursing care in the immediate postanesthetic phases focuses on preventing complications from anesthesia or the surgery. Assessments are initially made every 10 to 15 minutes, and usually continue for about two hours, depending on the surgery and length of anesthesia. The client is transferred from the recovery area when physical status and level of consciousness are stable.

**Provide continual assessment of body systems.**

- Respiratory status
  - Airway patency (e.g., observe for wheezing or crowing sounds in addition to signs of ineffective ventilation)
  - Oxygen saturation
  - Effectiveness of ventilation (e.g., observe for restlessness, cyanosis, shallow respirations, unequal chest expansion, use of accessory muscles)
- Cardiovascular status
  - Blood pressure (anesthesia can cause transient hypertension)
  - All pulses (for bilateral equality, rhythm, rate, and character). Irregular rhythm, absence of pulses, and tachycardia are especially significant.
  - Color (e.g., cyanosis, pale, blotchy, jaundiced)
  - Cool skin temperature
- Edema
  - Decreased urine output
- Central nervous system status
  - Level of consciousness
  - Orientation to time, place, person
  - Reflexes (gag, cough)
  - Ability to move extremities
- Fluid status
  - Intravenous fluid intake

- Urine output
- Wound drainage
- Drainage from catheters, tubes, and drains (amount and character)
- Skin turgor or edema
- Vital signs
  - Status of Wound
    - Condition of dressing (dry, intact)
    - Drainage (amount, type, color)
  - Pain (type, location, intensity)
  - Nausea and vomiting

**Keep all lines patent** (e.g., intravenous, urinary catheter, chest tubes, oxygen tubing).

**Assure that monitors and equipment are functioning** (e.g., ventilators, electrocardiogram, suction equipment, oxygen).

**Positioning**:

- **Keep airway open.**
  - Place an unconscious client on his side; face slightly down, with no pillow under the head. This keeps the tongue forward, preventing occlusion of the pharynx and allowing for drainage from the mouth.
  - Maintain artificial airway in place; suction until cough and swallowing reflexes return.
  - After removing the airway, help the client to turn, cough and deep breathe.
  - After spinal anesthesia, the client may have to remain flat for a specified time.
- **Enhance chest expansion.** Place upper arm on a pillow.

**Help arouse and orient the client.** After reflexes return (e.g., swallowing, gagging), arouse clients by calling by name in a normal tone of voice and telling them repeatedly that surgery is over and that they are in the recovery room.

**Facilitate oxygenation** by positioning, administering oxygen, and suctioning.

**Treat hypotension** by administering oxygen, having the client breathe deeply and do leg exercises, and maintaining intravenous flow rates.

**Provide for safety.** Keep bedrails up; use side rails; use restraints as necessary.

**Provide for comfort**:

- Most clients are cold postoperatively; cover with blankets.
- Provide psychologic comfort by reassuring the client that the surgery is over and by constantly reorientating.

# Critical Thinking Exercise: Immediate Postoperative Period

1. You are caring for a postoperative client who is becoming increasingly agitated and has made several attempts to dislodge her intravenous catheter. What nursing interventions are appropriate in this situation?

2. You are caring for a client who is recovering from an uncomplicated hernia repair. When this client refuses to participate in her postoperative care, how might you respond?

3. In your work on a medical-surgical unit of an acute care hospital, you frequently care for clients who have undergone abdominal surgery. What are some of the risk factors for postoperative pulmonary complications following abdominal surgery?

# Ongoing Postoperative Care

| Key Points |
| --- |

- The goals of nursing care are to: assess for and prevent complications, promote healing, facilitate coping with altered structure/function, teach self-care, and plan for home-care.
- Potential complications of surgery include: atelectasis, pneumonia, pulmonary embolism, hemorrhage, shock, thrombophlebitis, postoperative ileus, urinary retention or infection, dehydration, wound infection, dehiscence, evisceration, and depression. The nurse should assess for signs/symptoms of these complications.
- General interventions to prevent a variety of complications are to: evaluate vital signs, turn every two hours, ambulate as early as possible, maintain fluid volume and hydration, and monitor responses to narcotic analgesics.
- Interventions for preventing specific complications include:
  - **Respiratory**: Turn, cough, deep breathe.
  - **Cardiovascular**: Leg exercise, turn every two hours, antiembolic stockings, early ambulation, no pillows under knees
  - **Elimination**: Early ambulation, intake/output, catheterize if unable to void.
  - **Fluids & nutrition**: Intake/output, maintain IV flow rate.
  - **Comfort & rest**: Keep environment clean and quiet; offer ice chips; teach to splint incision.
  - **Psychological**: Orient the client to time, person, and place frequently.
- **Key Concepts/Terms**: Atelectasis, pulmonary embolism, thrombophlebitis, antiembolic stockings, postoperative ileus, dehiscence, evisceration, shock

**Overview**

The postoperative period begins with admission of the client to the postanesthesia recovery area and ends when healing is complete. The postoperative period consists of two stages: **immediate care** (usually given in the recovery area) and **ongoing postoperative care**, which lasts from return to the unit until healing is complete. Nursing interventions are aimed at preventing complications, promoting healing, facilitating coping with altered structure or function, teaching, and planning for home care.

## Potential Complications of Surgery

### Respiratory

- **Atelectasis** is the incomplete expansion and collapse of the alveoli, resulting in poor gas exchange. Etiologies: inadequate lung expansion (e.g., from analgesics, immobility, and positioning).

- **Pneumonia** is inflammation or infection of the alveoli. Etiologies include: aspiration, infection, depressed cough reflex, dehydration, and immobility.

- **Pulmonary embolism** occurs when a blood clot dislodges and travels through the bloodstream until it lodges in a small pulmonary vessel. Pulmonary embolus is a life-threatening condition. Etiologies include: immobility and resulting stasis of venous blood, venous injury, and pre-existing circulatory disorders.

### Cardiovascular

- **Hemorrhage** is excessive internal or external bleeding, which may lead to hypovolemic shock. Etiologies include: pre-existing conditions (e.g., clotting disorders), a disrupted suture, or insecure ligation of blood vessels.

- **Shock** occurs when inadequate tissue perfusion results from (a) an alteration in circulatory control or (b) a loss of circulating fluid. **Hypovolemic** shock is the result of a decrease in blood volume; it is the most common type of shock in the postoperative client.

- **Thrombophlebitis** is inflammation of a vein associated with a blood clot, usually in the legs. Etiologies include: slowed venous blood flow due to immobility or injury to a vein.

### Elimination

- **Constipation** is the inability to defecate within a normal length of time (e.g., 48 hours after resuming a solid diet). Etiologies include: anesthetic and narcotic analgesic side effects (decreased intestinal motility), immobility, manipulation of organs during surgery, and altered food and fluid intake.

- **Postoperative ileus** is intestinal obstruction characterized by absence of peristalsis. Etiologies include: handling of the bowel during surgery, anesthesia, wound infection, and electrolyte imbalance.

- **Urinary retention** is the accumulation of urine in the bladder and the inability to empty the bladder. Etiologies include: side effects of narcotics and anesthetics (decreased bladder muscle tone), and handling of tissues during surgery.

- **Urinary tract infection** is inflammation of the bladder, ureters, or urethra. Etiologies include: limited fluid intake, instrumentation of the urinary tract (e.g., diagnostic tests and surgery), immobility (e.g., horizontal position), and catheterization.

### Fluids

- **Dehydration** is inadequate volume of water in the body. Etiologies include: blood loss in surgery, electrolyte imbalance, fever, and restricted oral intake of water.

### Wound

- **Infection** (inflammation and infection of the incision or drain site). Etiologies include: poor aseptic technique and poor nutritional status preoperatively.

- **Dehiscence** is the separation of the suture line before healing occurs. Etiologies include: malnutrition, obesity, poor circulation, and strain on the suture line.

- **Evisceration** is the separation of the suture line and the extrusion of internal organs and tissues through the incision. Etiologies include: malnutrition, obesity, poor circulation, infection, and strain on the suture line.

#### Psychological

- **Depression** is not unusual postoperatively. It includes feelings of sadness and lethargy, often accompanied by decreased activity and other functional changes. Etiologies include: pain, illness, poor prognosis, having emergency surgery, physiological response to the particular surgery, and reactions to medications.
- **Confusion** in the postoperative period is always a potential complication. In addition to confusion the client may have memory loss and/or anxiety. Etiologies include: effects of medications, anesthesia, pain, hypoxia, metabolic or fluid imbalances, over-or under- stimulation environmentally (in the PACU, ICU, or CCU overstimulation is common; the client has interrupted sleep and may develop confusion sometimes referred to as ICU psychosis).

## Related NANDA Nursing Diagnoses

- Altered Urinary Elimination
- Body Image Disturbance
- Impaired Gas Exchange
- Impaired Verbal Communication
- Impaired Skin Integrity
- Impaired Physical Mobility
- Ineffective Airway Clearance
- Ineffective Breathing Pattern
- Nausea
- Pain
- Risk for Fluid Volume Deficit
- Risk for Infection
- Risk for Injury
- Urinary Retention

## Assessments

Most hospitals have protocols for postoperative assessments. For example, when a client is returned to the unit after surgery, assessments may be made every 15 minutes for the first hour, every 30 minutes for the second hour, and then every four hours for the next 48 hours. Assessments should, of course, be made more frequently if the client's condition requires. Assessments include: pain level, level of consciousness; vital signs; skin color and temperature; comfort; fluid balance; dressings and bedclothes; appearance of wound; and tubes and drains.

**Level of consciousness**: Orientation to time, place, and person, reaction to verbal stimuli, ability to move extremities

#### Respiratory

- Assessments include respiratory rate, depth, and character; skin color; auscultation of the lungs; and monitoring for responses to narcotic analgesics.
- Assess for signs of **pneumonia**: cough, blood-tinged or purulent sputum, dyspnea, chest pain, fever, chills, crackles and wheezes.

- Assess for signs of **atelectasis**: decreased lung sounds, dyspnea, cyanosis, crackles, restlessness, and apprehension.
- Assess for signs of **pulmonary embolus**: dyspnea, chest pain, cough, cyanosis, rapid respirations, tachycardia, and anxiety.

### Cardiovascular

- Assessments include vital signs, intake and output, skin color and turgor, hydration of mucous membranes, amount and type of wound drainage, daily calf and thigh measurements, and mental status.
- Assess for signs of **hemorrhage**: frank bleeding (e.g., dressing saturated with bright red blood, blood in drains or tubes), restlessness, anxiety, increasing abdominal girth, increasing pain, swelling or bruising around the incision, and signs of shock.
- Assess for signs of **shock**: rapid weak pulse; tachypnea; dyspnea; restlessness and apprehension; urine output <30 mL/hr; hypotension; cool, clammy skin; thirst; pallor.
- Assess for signs of **thrombophlebitis**: calf or thigh pain; redness and swelling in the affected area; hardness of the vein; Homan's sign (discomfort in calf when foot is dorsiflexed); elevated temperature; increase in the diameter of the involved extremity.

### Elimination

- Assessments include: intake and output record; palpation of urinary bladder; assessing client's ability to pass flatus or stool; and inspection, percussion, and auscultation of abdomen.
- Assess for **constipation**. Auscultate bowel sounds every four hours until peristalsis returns (peristalsis should return in 24 to 48 hours). Ask client whether he/she has been passing flatus or stool. Observe for abdominal distention, tympany on percussion, and pain.
- Assess for **postoperative ileus**: abdominal pain/distention, absence of bowel sounds, constipation, and vomiting.
- Assess for **urinary retention**: Inability to pass urine within eight hours after surgery, distended bladder, suprapubic discomfort, restlessness, voiding small amounts frequently.
- Assess for **urinary tract infection**: burning sensation when voiding, urgency and frequency, cloudy urine.

### Fluids

- Monitor intake and output.
- Monitor tolerance of diet (e.g., clear liquids).
- Monitor skin turgor and condition of oral mucous membranes.
- Monitor for electrolyte imbalances, if drains are present or gastric suction is being used.
- Monitor temperature (fever can cause dehydration).

### Wound

- Monitor for signs of **infection**: fever; redness, tenderness, purulent exudate from wound; and wound odor.

- Monitor for wound **dehiscence**: increased serosanguineous drainage from wound (especially between postoperative days five and 12); underlying tissue becoming visible along the incision.

- Monitor for **evisceration**: increased serosanguineous drainage from wound (especially between postoperative days five and 12), incision dehiscence and visible protruding organs (e.g., viscera).

### Psychological

- Monitor for signs of depression: anorexia, tearfulness, withdrawal, sleep disturbances, lack of participation in therapies and healthcare decisions.

- Monitor for signs of confusion by assessing orientation to person, place, and time.

## Therapeutic Nursing Interventions

**Nursing interventions are mostly preventive in nature**. For monitoring interventions, see "*Assessments*," preceding.

**General interventions for a variety of complications include**:

- Compare vital signs to preoperative baseline.

- Turn q 2hrs.

- Ambulate as ordered (as early as possible; ambulation usually begins the evening of surgery and increases as tolerated).

- Maintain fluid volume and hydration.

- Monitor responses to narcotic analgesics (e.g., respiratory rate and depth).

**Respiratory** (in addition to "*General interventions*," preceding)

- Position to increase ventilation (e.g., semi-Fowler's, pillow under upper arm when in lateral position).

- Assist client in deep breathing, coughing, and incentive spirometry q 2hrs.

**If pulmonary embolus occurs**, the goals are to stabilize respiratory and cardiovascular function and prevent further emboli:

- Notify the primary care provider immediately.

- Keep client on bedrest in semi-Fowler's position.

- Do not overhydrate with intravenous fluids.

- Administer oxygen.

- Administer anticoagulant medications as ordered.

- Administer analgesic medications as ordered (use narcotics with caution).

- Assess vital signs frequently.

- Provide emotional support.

- Prevent intrathoracic pressure and increased emboli (e.g., have client avoid Valsalva's maneuver, as in straining to have a bowel movement).

**If pneumonia occurs**, the goals are to treat the infection, support respiratory function, and prevent spread of the microorganisms:

- Place in Fowler's position to increase full aeration of lungs.

- Maintain nutritional status.

- Keep well-hydrated.
- Assist with deep breathing and coughing every two hours.
- Provide frequent oral hygiene.
- Teach proper disposal of tissues and sputum.
- Promote rest and comfort.
- Provide emotional support.
- Administer medications as ordered (antibiotics, expectorants, analgesics).

**If atelectasis occurs**, the goals are to prevent further atelectasis, expand lung tissue, and ensure tissue oxygenation:

- Place in semi-Fowler's position.
- Administer oxygen as needed.
- Assist with deep breathing, coughing, and incentive spirometry every two hours.
- Perform leg exercises every two hours, or ambulate as ordered.
- Keep well hydrated.
- Provide emotional support.
- Administer prescribed medications (analgesics).

**Cardiovascular** (in addition to *"General interventions,"* preceding):

- Provide extra blankets as needed to prevent chilling.
- Assist with leg exercises and turn in bed every two hours.
- Monitor tolerance of ambulation (e.g., blood pressure, pulse, and respiratory rates).
- Apply antiembolic stockings or compression devices if ordered.
- Position to promote venous return (e.g., do not raise the knee gatch on the bed or place pillows under the knees).

**If shock occurs**, the goals are to improve and maintain tissue perfusion by eliminating the cause of the shock:

- Keep airway patent.
- Place client supine with legs elevated to 45 degrees (not in Trendelenburg position).
- Administer oxygen as indicated.
- Administer fluids and whole blood or blood components, as ordered.
- Keep the client warm.
- Continuously monitor vital signs and general condition.
- Provide emotional support for client and family.
- Administer prescribed medications.

**If hemorrhage occurs**, the goals are to stop the bleeding and replace blood volume:

- Apply a pressure dressing to the site of the bleeding.
- Prepare to return the client to the operating room if bleeding is massive or cannot be stopped.
- Provide care as in *"Shock,"* preceding.

**If thrombophlebitis occurs,** the goals are: (a) to prevent the clot from breaking loose, and (b) to prevent further clot formation:

- Maintain bed rest.
- Apply antiembolic stockings (see "*Postoperative Care: Antiembolism Stockings*") or sequential pneumatic compression devices.
- Elevate the affected leg to the level of the heart.
- Do not rub or massage the legs.
- Give medications as ordered: anticoagulants, analgesics.
- Apply warm moist heat to the leg as ordered.
- Measure bilateral calf or thigh circumference every eight hours.
- Provide emotional support.

**Elimination** (in addition to "*General Interventions,*" preceding):

- Help client move in bed and ambulate to help relieve gas pains.
- Encourage intake of fluids and fiber foods as soon as ordered.
- Provide privacy for use of bedpan or commode.
- Administer flatus tubes, suppositories, or enemas as ordered.
- Help client assume a normal position for voiding (e.g., place client in upright sitting position when using bedpan; assist to commode or bathroom when able; help male clients stand upright to void).
- Maintain correct intravenous rate.
- Catheterize, as ordered, if unable to void.

**Fluids and nutrition** (in addition to "*General Interventions,*" preceding):

- Maintain intravenous flow rate as ordered.
- Provide oral hygiene before meals.
- Gradually increase diet from surgical (clear) liquids to solid food as tolerated. Assess tolerance by asking the client about nausea and auscultating for the return of peristalsis (without peristalsis to move intake from the stomach, nausea and vomiting may occur).
- Keep bedside articles clean and neat; eliminate odors.
- Encourage client to sit up in bed or in a chair to have meals.
- Encourage family to participate in meals.
- Be aware that clients with drains, gastric suction, fever, and IV therapy are at risk for fluid and electrolyte imbalances.

**Comfort and rest:**

- Provide personal hygiene.
- Keep bed linens and environment clean.
- Keep the environment quiet during rest periods and at night.
- Allow family members to remain with the client if possible.

**For nausea and vomiting** that occur as a result of decreased peristalsis (e.g., from anesthesia, analgesics, or handling of viscera during surgery) and from side effects of certain medications:

- Maintain a clean, odor-free environment.
- Provide oral hygiene.
- Offer food and fluids frequently in small amounts after being NPO.
- Monitor side effects of antibiotics and analgesics.
- Administer antiemetic medications if ordered.

**For thirst**:

- Provide oral hygiene.
- Offer sips of water or ice chips (if allowed).

**For hiccups**: Have client breathe into a paper bag, take several swallows of water while holding his/her breath, or eat a teaspoon of sugar.

**For incisional pain**:

- Assess for pain every two to four hours during the first 24 to 48 hours postoperative.
- Offer analgesics every two to four hours for the first 48 hours postoperative.
- Offer position changes, back rubs, and relaxation techniques (e.g., guided imagery).
- Teach client how to splint the incision when moving.

**Wound** (in addition to "*General Interventions*," preceding):

- Monitor, especially for elevated temperature.
- Encourage diet high in carbohydrates, proteins, calories, and vitamins.
- Use and teach meticulous hand washing.
- Maintain aseptic technique for dressing changes and tube/drain care.
- Keep wound clean and dry.
- **If dehiscence or evisceration occurs**: Cover the wound area with sterile towels soaked in saline solution; notify the primary care provider immediately; expect prompt surgical repair.

**Psychological**:

- Provide adequate rest.
- Provide opportunity to express negative feelings (e.g., anger).
- Orient the client to person, place and time frequently by:
  - Using a calendar and a clock in the room and assuring the client of the time, date, and place (where they are) frequently
  - Wearing a name tag large enough for the client to read and telling them your name regularly
  - Addressing the client by name
  - Keeping familiar items in the client's line of vision (e.g., family pictures or personal items)

# Critical Thinking Exercise: Ongoing Postoperative Care

1. A client who has undergone a permanent colostomy expresses concern about the appearance of the new stoma. What nursing diagnosis might apply to this client? How might you support the client in adjusting to the stoma?

2. From your own clinical experiences, think of a challenging situation you may have encountered in caring for a postoperative client.? What particular skills, knowledge, or interventions did you employ in meeting this challenge?

3. Now write a diary entry describing your experiences with this client.

# Antiembolism Stockings

| Key Points |
| --- |

- Antiembolism stockings help to prevent thrombophlebitis by promoting venous return from the legs.
- A physician's order is required.
- The client must be measured to assure proper fit of the stockings.
- Apply stockings initially in the morning before the client gets out of bed, and after elevating the legs for at least 15 minutes.
- Remove stockings for 20 to 30 minutes every eight hours.
- Inspect for skin breakdown.
- **Key Concepts/Terms**: Venous return, thrombus, Homan's sign

**Overview**

Antiembolism stockings are knee- or thigh-high elastic hose used for clients with limited activity to prevent phlebitis and thrombus formation. These stockings help force blood from the superficial veins into the deeper veins and prevent stagnation of blood in the veins. They also promote venous return to the heart. A primary care provider's order is required for the use of antiembolism stockings.

## Procedure and Technique for Antiembolism Stockings

- Measure the calf and/or thigh to determine appropriate size.
- Assess skin integrity and circulation of the legs and feet prior to applying the hose and during every shift.
- Apply stockings initially in the morning before the client is out of bed, while the client is supine.
  - Have client lie down with legs elevated for at least 15 minutes before applying.
  - Be sure the heel is well-positioned in the stocking.
  - After applying, pull on the toes of the stockings to relieve pressure on the toes.
- Check Homan's sign at least every shift, assessing for signs and symptoms related to thrombus formation.
- Remove stockings for 20 to 30 minutes every shift; inspect the legs, and reapply before the client gets out of bed.
- Inspect legs regularly for redness, blistering, swelling, and pain.
- Launder stockings as necessary, at least every three days. Dry on flat surface to prevent stretching.

## Complications

- Improper fit and ineffectiveness if measured incorrectly
- Skin breakdown under the stockings

# Critical Thinking Exercise: Antiembolism Stockings

**Situation**: An elderly postoperative client is participating willingly in all aspects of her care. She is using an incentive spirometer, practicing breathing exercises and coughing, and ambulating with assistance. However, this client refuses to wear antiembolism stockings ordered by her physician. How do you respond?

1. In the postoperative period, what factors represent the greatest risk for deep vein thrombosis (DVT)?

2. You are caring for a 40-year-old obstetrical client with numerous varicose veins in both legs. As you assess this client, what findings might make you suspect a potential thrombosis?

# Critical Thinking Exercise Answer Keys

---

**Situation**: A hospital-based clinic serves a range of clients (e.g., children scheduled for routine immunizations, women seeking prenatal care, and individuals requiring treatment for a variety of chronic conditions). The following questions relate to health care delivered in this and other settings.

---

1. What factors distinguish primary care from secondary care? What levels of care would you expect to see provided in a clinic?

   **Levels of preventive care include primary, secondary, and tertiary prevention. Primary prevention is aimed at health promotion and the early detection of illness; secondary prevention is concerned with diagnosis and appropriate intervention. In a clinic setting, you would ordinarily expect to see the clients receive both primary care (e.g. routine immunizations) and secondary care (e.g. diagnosis and management of chronic conditions such as diabetes mellitus).**

2. Define tertiary care. In what setting would you expect to see tertiary care provided? Give at least two examples.

   **Tertiary care is concerned mainly with rehabilitation. Although this may occur in a variety of settings, it would typically take place in a community or hospital setting.**

3. Who is eligible for Medicare? Who is eligible for Medicaid? What are these programs intended to accomplish?

   **Medicare is a federally funded program of health care coverage for individuals 65 years of age or older. Medicaid is a federally funded, state-administered health care assistance program providing benefits to the poor. Eligibility for coverage under Medicaid varies from state to state.**

---

**Situation:** A 45-year-old male with a family history of cardiovascular disease seeks advice about quitting smoking. The client, who smokes approximately one pack a day, admits that several previous attempts to stop smoking have been unsuccessful. Keep in mind the goals of Healthy People 2010, when answering the following questions?

---

1. In counseling this client, explain the risks associated with tobacco use. For a long-time smoker, what is the benefit of quitting?

   Cigarette smoking is strongly linked to a number of cancers, including cancers of the lung, head and neck, esophagus, and pancreas. In addition, it is a significant risk factor for a number of other diseases, including coronary artery disease, peptic ulcer disease, and COPD. Women who smoke during pregnancy can increase the risk of a low-birth weight infant or a premature delivery.

   The benefits of quitting are both immediate and long-term. Shortly after quitting, the former smoker will experience a more efficient breathing, better circulation, and an improved sense of taste and smell. The long-term benefits (those that accrue between 5 and 15 years after quitting) are most important. These include a significantly reduced risk of cancer, heart disease, and stroke.

2. What are the two-fold goals of Healthy People 2010? How might they apply to the client described?

   The two-fold goals of this initiative are to increase the length and quality of life for Americans and, second, to eliminate health disparities within the population. By quitting smoking, the client described in the scenario will have taken an important step toward reducing his risk of many diseases, including lung cancer and cardiovascular disease, and increased his chances for a long life.

3. Describe the three levels of prevention--primary, secondary, and tertiary. Provide an example of each.

Primary care precedes illness, injury, or dysfunction and refers to persons considered physically and mentally healthy. Primary prevention is aimed at health promotion and disease prevention efforts to maintain or improve well-being of individuals, families or communities. Wellness activities typically include early detection measures, health education programs, immunization, and nutritional and fitness activities.

Secondary care focuses on persons who are experiencing physical or mental illness or are at risk for health problems. Secondary prevention activities are focused on diagnosis and prompt treatment to reduce the severity of the illness and promote recovery. Typically, the care delivered within acute health care institutions and outpatient surgical service facilities are examples of secondary care.

Tertiary care occurs when a long-term or permanent disability or dysfunction is present and restricts the individual from performing activities of daily living. Tertiary prevention involves preventing complications or deterioration in status. Interventions are directed towards rehabilitation rather than diagnosis and treatment. The goal of tertiary care is to implement measures to improve the quality of life and increase the level of functioning. Rehabilitation activities include occupational therapy, physical therapy, hospice care, and some types of skilled nursing care.

# WORD SEARCH

Here are 12 terms associated with *Healthy People 2010*. Can you find them below?

| | | |
|---|---|---|
| ARTHRITIS | FOOD SAFETY | NUTRITION |
| CANCER | HEARING | OSTEOPOROSIS |
| DIABETES | HIV | STROKE |
| DISABILITY | IMMUNIZATION | VISION |

| D | I | S | A | B | I | L | I | T | Y | L | R |
|---|---|---|---|---|---|---|---|---|---|---|---|
| I | M | L | R | P | M | L | H | I | V | N | N |
| A | C | T | T | F | M | P | S | I | H | U | Y |
| B | A | E | H | E | U | I | S | N | H | T | T |
| E | N | N | R | H | N | I | S | T | E | R | E |
| T | C | F | I | P | I | C | I | K | A | I | F |
| E | C | F | T | N | Z | V | O | I | R | T | A |
| S | R | J | I | T | A | R | I | L | I | I | S |
| C | T | O | S | R | T | I | O | O | N | O | D |
| L | E | R | U | S | I | P | U | R | G | N | O |
| S | I | S | O | R | O | P | O | E | T | S | O |
| V | I | S | I | O | N | S | T | L | M | P | F |

---

**Situation**: As a nursing student, you have been asked to address fellow students on the multifaceted roles related to contemporary nursing practice. In preparing your talk, consider the various ways nursing has been defined and the forces that have shaped the profession.

---

1. In what ways is nursing both an art and a science? Describe a situation from your own clinical experience to support your answer.

   **Competency in nursing requires a solid foundation in the biological and physical sciences and a broad knowledge of the physiological and pathophysiological functions that govern human health and illness. Nurses must also be adept in the art of interpersonal communication. Their interactions with others must demonstrate both sensitivity and compassion. Nursing is therefore both an art and a science.**

2. Nurses assume a variety of roles: from teacher, caregiver, communicator, counselor. Choose one of the many roles and describe how it might be carried out in contemporary practice?

   **Depending on the role described, answers will vary. As a caregiver, for example, a nurse promotes the process of healing by addressing specific needs of the client. These needs not only encompass the client's physical care, but include emotional, social, and spiritual aspects of care as well.**

**Situation**: A classmate expresses difficulty in understanding the relationship between theoretical concepts and actual nursing practice. Before responding, consider the purpose of nursing theories.

1. Nursing theory is intended to guide practice. Choose one of the theories presented in this chapter and describe how it is reflected in your own practice or in a clinical setting you have observed.

   **Responses will vary depending on the theory selected. A response based on the theory of Hidegard Peplau, for example, would focus on the client-nurse relationship, which is both interpersonal and therapeutic. In fostering such a relationship, a nurse collaborates with a patient to meet his or her individual needs. Peplau's theory, first published in 1952, helped guide the practice of psychiatric nursing.**

2. Write a diary or journal entry based on a particularly memorable incident from your own clinical experience. Be sure to include elements that reflect how theory affects actual nursing practice.

   **Diary entries will vary but should reflect personal experience and refer to a relevant nursing theory.**

**Situation:** On an initial visit to a new client, a hospice nurse is greeted by family members with a barrage of questions ranging from advance directives to end-of-life care. The following questions relate to this and similar situations.

1. What is an advance directive? When is it implemented?

   **An advance directive is a legal document in which an individual states his or her intentions regarding the provision of medical care at some time in the future. It is implemented when an individual can no longer communicate his or her desires regarding medical treatment. The durable power of attorney, by which one person assigns to another the responsibility for health care decisions in his or her behalf, is one example of an advance directive.**

2. In many situations, a nurse cares for both client and family. How might it be possible to resolve conflicts that arise?

   **In caring for both client and family, a nurse must recognize that situations may arise in which the desires of a client conflict with those of the family. Although each situation is unique, a nurse's first obligation is to his or her client.**

3. What ethical principles apply to nurses in caring for clients and their families? For example, what does the principle of veracity require of a nurse?

   **A number of ethical issues apply as nurses provide care for clients and their families. Among these are autonomy, beneficence, confidentiality, and veracity. Veracity concerns a nurse's obligation to truthfulness.**

**Situation:** A nurse who is present at the scene of an automobile accident provides assistance until emergency aid arrives. The following questions relate to nurses' responsibilities under the law.

1. What law governs a nurse's actions in a situation like this? With respect to the obligation to offer assistance, is it a legal one or a moral one? Explain by citing the law that applies to this situation.

   **The Good Samaritan Laws apply in situations in which a nurse provides assistance at the scene of an accident. With respect to a nurse's obligation to offer assistance, it is primarily a moral one.**

2. What is the source of law that governs nursing practice?

   **Laws governing nursing practice are derived from a number of sources--statutory law, regulatory law, and common law. Statutory law is created by state and federal legislators. Regulatory law is created by bodies such as state boards of nursing. Common law is based on judicial decisions; an example of common law is the right to refuse treatment.**

3. Legal responsibility is intended to promote a high standard of nursing practice. Explain how it does this. Consider one area of focus, such as the right to privacy, professional conduct, or documentation, when giving your response.

   **By holding nurses accountable for the care they provide, laws governing nursing practice intend to promote the highest level of client care. For example, nurses understand the importance of accurate and timely documentation and are aware that a client's chart is an essential means of communication among members of the health care team.**

**Situation:** As part of his cardiac rehabilitation, a 58-year-old male recovering from an acute myocardial infarction quits smoking and modifies his diet but is reluctant to adopt a program of moderate exercise. Overall, does this client exhibit an internal or an external locus of control?

1. Describe an individual with an internal locus of control regarding health beliefs. How would this individual behave differently from someone with an external locus of control? How might circumstances like a hospitalization affect one's locus of control?

    **Individuals who take positive steps to promote health, such as adopting good nutritional habits and a program of regular exercise tailored to their condition, are exhibiting an internal locus of control. They recognize the potential of their actions for maintaining and restoring good health. An event such as a hospitalization may be associated with vulnerability and a certain lack of control over one's daily routine. Potentially, it could cause an individual to experience an external locus of control.**

    **Overall, the man described in the scenario is exhibiting an internal locus of control; he recognizes that modifying his lifestyle is an essential element in cardiac rehabilitation.**

2. Compare the definition of health offered by the World Health Organization with that of the American Nurses Association. Which best reflects your own definition of health? Draw on your own observations and experiences to support your response.

    **The WHO definition focuses on "complete" health, while the ANA describes health as a dynamic state of well-being. Both definitions emphasize the many aspects of health, from the physical to the social to the spiritual. Student's definitions of health will vary.**

3. Distinguish disease from illness. In what ways might it be possible to have a disease without having an illness?

A disease is an objective change in bodily function which often results in diminished capacities or a shortened life span. Illness, on the other hand, is a subjective state in which an individual's perception of well-being is altered. Illness may or may not accompany a disease process.

# Caring for Individuals, Families and Communities Answer Key

---

**Situation:** A community health nurse coordinates the care of a 75-year-old woman following hip replacement surgery. The client lives with her daughter and son-in-law, both of whom work full time and who admit to difficulty in coping with the care of an elderly parent. The following questions apply to the nursing care of individuals, families, and communities.

---

1. In caring for individuals, why is it important to include the family and the community in your assessment?

   **Every individual lives within a social context, of which the family and the community are basic units. As an individual's physical, emotional, and spiritual health is affected by the family and the community, it is vital to consider these elements in a nursing assessment.**

2. Select a nursing diagnosis that applies to the family described above. Based on this diagnosis, how might a nurse assist individuals and families in a similar situation?

   **Possible nursing diagnoses include "Caregiver Role Strain" and "Family Coping: Compromised." Nurses might guide individuals and families to resources in the community that might be of assistance (e.g. adult day care, volunteer organizations providing companionship, food delivery services) and help them to make optimal use of their own resources (e.g. support groups, reaching out to extended family for assistance).**

3. Define homeostasis. Give an example of an individual's homeostatic response to the environment.

   **Homeostasis is a dynamic state of equilibrium or balance. Its purpose is to maintain stability within an organism. For example, when an individual is in a cold environment, blood vessels in the skin constrict, minimizing heat loss through the skin and conserving heat within the body. In this way, relatively stable internal conditions are maintained.**

**Situation**: You have been assigned to care for a client whose language and culture are different from your own. Consider the principles that will guide a cultural assessment of your client.

1. What awareness must a nurse first develop in order to provide transcultural care?

   **In order to provide transcultural care, nurses must first develop an awareness of their own culture and how it affects their view of the world. From this understanding, they can better understand individuals from other cultures.**

2. Work together with a fellow student to refine your understanding of transcultural nursing. Below is a list of topics and questions to guide your discussion.

   - Choose a cultural concept, such as attitudes toward nonverbal communication, touch, or pain response. Reflect on variations among cultural groups. For example, how does your own culture view these concepts?

   - Do you believe it is possible for a nurse to remain culturally neutral? Why or why not?

   - Define diversity. In what ways might you expect to find evidence of diversity within a cultural group?

   - Although not all stereotyping is negative, what dangers are inherent in stereotyping individuals?

   **Although responses will vary, students should recognize the validity of cultural differences and how they affect an individual's functioning and interactions with others. They should recognize that complete cultural neutrality may be difficult to achieve, but respect for differences is essential. They should also recognize that diversity refers to differences in a variety of areas, including language, food, and religious beliefs. Although stereotyping is not always negative, there is a danger in not looking at the individual beyond the group. Indeed, an individual is always distinct from the cultural group to which he or she belongs.**

**Situation**: During a home care visit, you are conducting an admission interview with an elderly client. In response to a question about spirituality, the client states, "Oh, I don't really get to church too often." How might you proceed in your assessment of this client?

1. What kinds of questions might you ask in obtaining a spiritual assessment?

   During a spiritual assessment, you might ask about the influence of religion or religious practices in a person's life. In what ways are these influences and practices important? If the person is not religious, what values or beliefs guide his or her life? In the scenario described, the client is not a churchgoer but most likely has a variety of spiritual concerns, such as maintaining harmonious relationships with others and finding purpose in life. In such a case, you might direct your assessment to the principles, hopes, and ideas by which the individual lives.

2. As a nurse, what is your role in promoting a client's spiritual health? Once you have completed a spiritual assessment, how might this information direct your care?

   Spirituality is an important component of a client's health. The information obtained in a spiritual assessment can therefore help guide the client's care. For example, if the client is experiencing a strained relationship with a friend or family member, resolving this relationship may be an important part of promoting spiritual health.

3. What resources might you utilize in planning a client's spiritual care?

   A chaplain or pastoral care coordinator within a hospital, church, or other organization may be an excellent resource.

4. Explain the signs of spiritual distress. Under what circumstances might an individual experience spiritual distress? How might you respond?

> The signs of spiritual distress vary, but may include hopelessness, actions inconsistent with one's character or beliefs, and questioning of formerly held beliefs. Any stressful event, including chronic or terminal illness, may cause or contribute to spiritual distress. The proper response begins with identifying the cause of distress through a spiritual assessment and helping the individual or family to find the help needed to resolve distress and achieve peace.

**Situation:** At a health fair, a middle-aged woman expresses interest in holistic healing. "But there are so many different therapies" she states. "I don't know where to start." How might you offer guidance?

1. What is the underlying principle or belief that governs all holistic practices? Explain this briefly in your own words.

> Although a variety of holistic practices exist, there is an underlying assumption about the interdependence of mind, body, and spirit. Most holistic practices are based on this premise.
>
> In the scenario described, it might be helpful to encourage the woman to follow her own interests. For example, if she enjoys scent, she might benefit from aromatherapy. If she is attracted to slow, meditative types of exercise, yoga might be a good choice. The key is to base the intervention on the interests of the individual. Those expressing interest in alternative therapies should be reminded that these are not substitutes for needed medical care.

2. Select a holistic healing therapy that interests you and design a brochure or poster to illustrate its benefits. In developing your advertisement, consider your target audience, describe the therapy, and outline its benefits.

> Use your own interests to guide this project. For example, if you had the opportunity to engage in one holistic therapy, which would it be? Suppose you choose yoga. In order to target an older population, you might describe the benefits of chair yoga or some other modification of a basic yoga practice. Whatever therapy you choose, you might want to interview a teacher or practitioner to gain insight into its value.

**Situation**: A classmate is about to begin a clinical assignment in a long-term care facility. Before reporting, she wonders whether the nursing process will apply in this facility as it did in an acute care setting. How might you respond?

1. Explain how the steps of the nursing process can be utilized in a variety of clinical situations and settings.

   **The nursing process forms the framework by which nurses systematically identify and address client's health care and nursing needs. The components of the nursing process are assessment, diagnosis, planning, implementation, and evaluation. The nursing process can be used in a variety of settings, from the home to acute care facilities, and is effective both with individuals and with families.**

2. To reinforce your own appreciation for the nursing process, explain to a classmate how utilizing the process requires creativity and flexibility. Draw on your own experiences in formulating your response.

   **Students will draw on their own experiences in responding to this question. They should recognize that the nursing process provides a framework to which they must apply ingenuity, skill, and caring in addressing clients' needs.**

3. Identify two important benefits derived from using the nursing process?

   **By using the nursing process, nurses are able to provide individualized care and to promote both coordination and continuity of care.**

**Situation:** A fellow student enlists your help in understanding the concepts of growth and development. Use the questions below to guide your discussion.

1. How is the process of growth different from that of development? Respond by defining and providing an example of each term.

   Growth refers to a physical change and increase in size. An example is the increase in height and weight in a child between 12 and 18 months. Development refers to a behavioral change and the progressive acquisition of skill and function. An example is a child's learning to walk and develop language skills.

2. Identify the five dimensions of growth and development. What factors influence growth and development?

   The five dimensions of growth and development are physiological, psychosocial, cognitive, moral, and spiritual. A number of factors, including cultural influences and interactions with others and the environment, influence growth and development.

3. Select one of the various psychosocial, cognitive, spiritual, and moral development theories. What does this theory seek to explain? In what ways might this theory either complement or conflict with other developmental theories?

   Responses will vary with students' choice of theory. If one of the moral developmental theories of Kohlberg or Gilligan is chosen, students should be aware that Gilligan found Kohlberg's theory inadequate to describe the spiritual development of women, for whom caring and responsibility must be emphasized over justice and reason.

4. Why is it important for a nurse to understand the theoretical basis for growth and development?

Growth and development occur in a sequential manner throughout the lifespan. Nurses must recognize the normal, expected patterns of growth and development from infancy to mature adulthood in order to assist clients at each of these stages.

1. In working with parents to promote the health of children at all stages of development, nurses address a variety of issues and concerns. Develop a chart highlighting the major teaching concerns unique to each stage of childhood development.

   - Neonates and infants
       - **Focus on understanding developmental norms, instituting appropriate safety measures, providing appropriate sensory stimulation.**

   - Toddlers
       - **Continue focus on safety, choosing appropriate toys, setting limits.**

   - Preschoolers
       - **Focus on accident prevention, dental health, adequate sleep.**

   - School-age children
       - **Focus on adequate nutrition, adjustment to school, accident prevention, dental health.**

2. Select one of the theories described in this chapter, and briefly outline the developmental tasks accomplished at each stage from birth to adolescence.

   **Responses will vary with students' choices of theory. If the cognitive developmental theory of Piaget is outlined, five phases of development will be outlined:**

   **Sensorimotor Phase (Infancy through 24 Months): Infants progress from reflex behavior to activity requiring repetition and imitation.**

   **Preconceptual Phase (Ages 2 through 4): Children make the transition from typically self- satisfying behaviors of infancy to acquiring rudimentary social skills. This period is characterized by an increasing use of language.**

   **Preoperational Phase (Ages 2 through 7): Subdivided into the preconceptual phase (ages 2 through 4) and the phase of intuitive thought (ages 4 through 7), this phase is characterized by the transition from egocentric thought to a growing social awareness.**

   **Concrete-operational Period (Ages 7 through 11): Children develop increasingly more flexible thought processes by which they come to a greater understanding of the world and an acceptance of a variety of points of view.**

   **Formal Operational Thinking (Ages 11 through 14): Adolescents develop the ability to think abstractly and to make use of a future-time reference.**

---

**Situation:** As part of a routine pre-employment screening, a nurse interviews a healthy 25-year-old male. How might the interview be tailored to the specific health concerns of this age group?

---

1. What are the major health problems of young adults? What nursing interventions are appropriate to address in these areas? With these concerns and interventions in mind, design a one-page handout to promote healthful behaviors among young adults.

   **Major health problems of young adults include accidents, sexually-transmitted diseases, and such malignancies as testicular cancer and breast and cervical cancer. Appropriate nursing interventions include recommending a schedule of physical examinations and immunizations, teaching testicular and breast self-examination, and promoting safe driving measures.**

2. One of the defining characteristics of young adulthood is maturity. What guidelines signal that an individual has reached maturity?

   **Maturity is characterized by a knowledge of self and a sense of purpose, acceptance of others, and taking responsibility for one's actions.**

3. Identify and explain one or two of the major developmental tasks of young adulthood.

   **A young adult must assume new roles both in the work place and at home. Additionally, a young adult must develop a set of values to live by.**

**Situation**: A 48-year-old male inquires about recommended health screening for his age group. Base your response on your knowledge of the specific health concerns of middle adulthood.

1. As individuals move into middle adulthood, what health problems are they likely to experience? Identify the most appropriate nursing interventions to address these concerns. Use this information to develop a chart of recommended health screenings of middle adults

    In middle adulthood, individuals are at increasing risk for cancer and cardiovascular disease. It is also a time when chronic illnesses like diabetes mellitus begin to emerge. Appropriate interventions include recommending regular physical and dental examinations, screening for cancer and cardiovascular disease, and a tetanus booster every ten years.

2. What is meant by a "midlife crisis"? What do individuals begin to realize at this stage of life?

    A "midlife crisis" may be precipitated by the realization of the passage of time and the loss of youthfulness, and the abilities and opportunities once taken for granted.

**Situation:** In planning the discharge of an 82-year-old client from an acute care setting to home, a nurse takes into account the special needs of older adults.

1. Based on your knowledge of older adults, how might you adapt your discharge instructions to their specific needs?

   **Because older adults are more likely to take multiple medications, it is essential that you ensure that they know the purpose of their medications and how and when to take them. Impaired vision may add to the challenge of safe medication use. By patiently reinforcing medication teaching with clients, you can promote safe medication use. Additionally, you must ensure that the client is being discharged to a safe environment and knows how and when to follow up with health care providers.**

2. Devise a safety plan for older adults living at home. Take into consideration limitations they are likely to experience and specific safety challenges encountered in the home.

   **Older adults may experience limited mobility and vision impairment and are therefore susceptible to falls and other accidents. Safety in the home includes having adequate lighting, placing grab bars in bathrooms when appropriate, eliminating scatter rugs and clutter, and encouraging individuals to take advantage of available community support services.**

**Situation**: In an effort to assist an anxious client, a nurse utilizes the principles of therapeutic communication. Keep these principles in mind as you respond to the following exercises.

1. What is the purpose of the therapeutic relationship?

   **The purpose of the therapeutic relationship is to resolve specific problems and to promote the client's health. It is a collaborative, helping relationship between a nurse and client.**

2. What distinguishes therapeutic communication from other forms of communication?

   **Therapeutic communication is purposeful. It focuses on addressing a client's needs. Because it is a professional relationship, a nurse must set appropriate boundaries from the beginning.**

3. Describe the phases of the therapeutic relationship. Do you believe it is possible to skip any of the phases without compromising the relationship?

   **The phases of the therapeutic relationship include a preinteractive phase, an orientation phase, a working phase, and a termination phase. Although some phases may be shortened, each is essential to the therapeutic relationship.**

**Situation**: An elderly woman experiencing shortness of breath is accompanied by her adult children to an emergency room. The client, who speaks limited English, relies on her children as interpreters. In order to meet the needs of the client and her family, a nurse utilizes techniques of effective communication.

1. When a client speaks limited English, what strategies might a nurse employ to facilitate communication?

   **In a situation like this, you might seek an interpreter through your facility. Additionally, it is appropriate to speak slowly in a normal volume, use simple vocabulary and sentence structure, and make use of gestures to communicate your ideas.**

2. Working with a fellow student, identify and discuss the major factors influencing communication. Use the following questions to guide your discussion.
   - In addition to the spoken or written word, by what means does communication occur?
   - Identify attitudes or actions that serve as barriers to effective communication.
   - Describe techniques and nursing interventions that can be used to foster effective communication.

   **Although responses will vary, students should recognize that communication occurs in a number of ways, not only through words but also through body language, touch, tone of voice, gesture, and eye contact. Barriers to communication include not listening actively, dismissing a client's concerns, and hurrying a client. An attitude of trust and respect is essential to fostering effective communication.**

**Situation**: You are responsible for teaching a 60-year-old client about a new medication. Before formulating your teaching plan, consider the various factors that influence learning.

1. Before you begin client teaching, what must you first do?

> Before you begin teaching, you must assess the client's readiness to learn. With that knowledge, you can tailor your teaching to the client's individual needs.

2. What strategies promote effective learning? How might you overcome obstacles to learning?

> As a strategy to promote effective learning, you can assess the client's knowledge base and then build on this. In order to overcome obstacles to learning, target the client's strengths and encourage active participation in the learning process.

3. Describe the role of the nurse in client education.

> Client education is a principal and essential element in the nurse-client relationship. Almost every interaction between nurse and client can be used as a teaching opportunity.

**Situation:** Imagine that you are about to interview for your first position as a staff nurse. In preparing for this meeting, consider the qualities you will look for in a prospective manager. Enlist the help of a classmate in formulating questions that will guide you in assessing the leadership potential of this individual.

1. With your classmate assuming the role of the manager and you that of the prospective employee, role play the interview. Focus on questions that will help you discern the manager's leadership style and its effect on the workplace environment.

   **Possible questions include:**

   - What are the main challenges of working on this unit?
   - How would you describe the morale of your staff?
   - What do you do to promote staff morale?
   - What are your principal expectations of your staff?
   - How do you help them to meet these expectations?
   - How would you describe your leadership style?

2. What distinguishes a leader from a manager? What qualities must both possess? What is unique to each?

   A manager directs the day-to-day work of staff. A leader inspires others to work collaboratively to achieve important goals. Both must exercise authority and responsibility. Although not every manager is a leader, the roles are not mutually exclusive.

3. Consider the advantages and the disadvantages of various styles of leadership. When, for example, might an autocratic style be necessary and appropriate? Under what circumstances might it be detrimental to the good of individuals or of the organization? Do you consider any one style of management superior to another? Be prepared to support your answer.

> **An autocratic style might be necessary in an emergency situation, such as during a code when direct orders must be given and carried out. In most situations, however, individuals benefit from participating in decisions that affect their work and that of coworkers. An autocratic style might potentially limit valuable input from staff. Responses to the last question will vary.**

4. You have been hired by an advertising agency to write a 25-to-50 word job description for a nurse manager on a medical-surgical unit. Describe the ideal candidate.

> **Responses will vary, but students should emphasize the importance of clinical knowledge, the ability to motivate a staff, and excellent communication skills in the job description of a nurse manager.**

---

**Situation:** In an effort to improve staffing, a community hospital has hired several new nurses. How might current staff and management work together with new nurses to promote effective change?

---

1. Imagine that, as a new member of a nursing staff, you encounter a lack of support and recognition from senior staff. What steps might you take to overcome resistance and implement positive change?

   **Begin by recognizing that most efforts to implement change are met by some degree of resistance. Seek to communicate openly with those staff members resistant to change. Find out their reasons for resistance. Help them to understand both the negative consequences of resisting change and the potential benefits of the proposed change. Foster an atmosphere of trust and good will in which change can be implemented.**

2. In your own experience, have you ever worked to promote a needed change in your family, school, or community? If so, what qualities enabled you to act as an agent of change? What resistance did you encounter? How were you able to respond to or perhaps even overcome this resistance?

   **Responses will vary, but you should recognize the importance of hard work and persistence in bringing about positive change.**

3. Interview an experienced nurse on the topic of change. In preparing for your interview, you may either adapt the questions below or develop your own.

   • How long have you been a nurse, and what patterns of change have you seen during your career?

   • Have most of the changes you've witnessed been more positive than negative? Explain why you think so.

   • As a nurse, how have you participated in the process of change? What advice would you give a new nurse about implementing positive change?

   **Responses to interviews will vary.**

---

**Situation:** In caring for clients across the life span, a family nurse practitioner in a small clinic routinely performs health assessments. Use the following exercises to guide you in understanding the purpose and components of a general health assessment.

---

1. Enlist the help of a family member or close friend in refining your health assessment skills. With this individual's permission, obtain a comprehensive nursing health history. Structure your interview to include all essential elements related to your subject's health status.

   **The comprehensive nursing history should focus on any problems the client is currently experiencing along with a detailed health history and a review of systems. After you have completed the interview, ask your friend or family member for appropriate feedback.**

2. What kind of information are you attempting to ascertain in a general health assessment? How will you integrate this information with the findings of a physical assessment?

   **In a general health assessment, you are attempting to ascertain a client's general health status along with past or current health problems. A head-to-toe physical assessment should confirm your general health assessment and uncover any potential problems.**

---

**Situation:** An 82-year-old male arrives at an emergency room with a body temperature of 101° F in combination with an elevated heart and respiratory rate, restlessness, and skin that is warm to touch. Describe specific nursing interventions you will use in caring for this and other febrile clients.

---

1. Discuss with a classmate how an understanding of aging and its effects on body temperature can guide your assessments. For example, how does the baseline body temperature of an elderly individual compare with that of a young or middle-aged adult?

   **In an older person, baseline body temperature is generally lower by approximately one degree Fahrenheit than in young or middle-aged adults. Because a fever may not be as readily apparent in older individuals, nurses must be alert to symptoms that may indicate an infection.**

   **Nursing interventions for febrile clients, including careful monitoring of vital signs and administration of antipyretics as ordered, are intended to support normal physiological processes, reduce fever, and prevent complications.**

2. The selection of a site for assessing body temperature may be guided by agency policy or left to the discretion of individual nurses. In the absence of a specific policy, identify the site you would select to assess temperature in these individuals and describe the rationale for your choice.

   * A newborn

   * A confused 82-year-old client

   * A healthy 60-year-old female

   **Avoid use of oral site for confused clients or young children. For these clients, an axillary or tympanic measurement may be a better choice. Rectal temperatures also provide accurate measurements but may be unpleasant for clients. Oral temperatures are convenient for alert and oriented clients; however, the readings will not be accurate if they have smoked or recently eaten hot or cold foods.**

**Situation**: Before administering digoxin, a nurse assesses a client's heart rate and rhythm. "Why are you doing that?" the client asks. How would you explain the purpose of your assessment? Under what circumstances is it appropriate to withhold digoxin? Describe your teaching plan for a client who will continue taking digoxin after discharge from an acute care setting.

**Before administering digoxin, assess heart rate and rhythm. If the rate is below 60 beats per minute or a change in rhythm is detected, withhold the medication and notify the client's physician. Clients who are taking digoxin at home should be taught to monitor heart rate and rhythm and to notify their physician of any significant changes.**

1. Using a simple drawing or diagram of the adult human body, indicate the location of each of the nine sites where pulse is commonly assessed. Using a friend or classmate as a model, demonstrate how you would locate and assess each of these pulse sites.

   **From head to toe, your diagram should indicate the following pulse sites.**

   - **Temporal**
   - **Carotid**
   - **Apical**
   - **Brachial**
   - **Radial**
   - **Femoral**
   - **Popliteal**
   - **Posterior tibial**
   - **Dorsalis pedis**

   **When palpating the carotid artery, do not palpate both right and left arteries at the same time as this may result in diminished blood supply to the brain and cause fainting.**

2. In addition to pulse rate, rhythm, and volume, what other factors are most useful in evaluating a client's cardiovascular status?

**A number of factors affect cardiovascular status. It is, therefore, important to perform a routine assessment of such indicators of cardiovascular status as general appearance, level of consciousness, and blood pressure.**

3. In an acute care setting, as you auscultate a client's apical pulse while simultaneously palpating the radial pulse, you discover a disparity between these two rates. What are some possible causes of such a disparity, or pulse deficit?

**A disparity between an apical pulse and a peripheral may accompany dysrythmias such as atrial fibrillation, atrial flutter, and premature ventricular contractions. Such findings must be reported to the client's physician.**

**Situation**: You are caring for an elderly postoperative client who is receiving IV Demerol for pain management by means of patient-controlled analgesia. Within an hour of starting the PCA, the client's respiratory rate declines from 14 breaths/minute to 8 breaths/minute. Consider the action you will take.

1. Identify the parameters of normal respiration. When assessing respiration, what special considerations apply to clients receiving narcotic analgesia? Specifically, what adverse effects might these medications have on a client's respiration? What medications can be used to reverse postoperative opioid effects?

   **Normal respirations are approximately 12-18 beats per minute for adults and 40-60 beats per minute for infants. Narcotic analgesia may cause respiratory depression.**

2. Imagine you are caring for a client with chronic obstructive pulmonary disease (COPD). What physical manifestations of COPD might you observe? When assessing respiration, what deviations from normal would you expect to see? Describe the way in which individuals with COPD typically breathe. How might you promote a more efficient pattern of breathing?

   **Common manifestations of COPD include a "barrel chest," abnormal breathing patterns which may be characterized by shallow, rapid respirations from the upper chest, use of accessory muscles during inspiration, fatigue, and shortness of breath. The client with COPD may benefit from practicing diaphragmatic breathing, learning to pace their activities, and performing routine self-care measures.**

3. One of the best ways to learn a skill is by teaching it to someone else. To refine your own assessment skills, teach a classmate how to perform a respiratory assessment. Before you begin, consider the components of the assessment. What techniques will you use? How can you distinguish normal from abnormal patterns of respiration? Encourage your classmate to ask any questions that come to mind. When you have finished, reverse roles and allow your classmate to teach you how to assess respiration.

   **A respiratory assessment should proceed in an orderly fashion and include inspection, palpation, percussion, and auscultation. Note the rate, rhythm, and depth of respirations. Assess skin color for cyanosis. Note the presence of a cough, respiratory problems, or lifestyle factors such as smoking that may affect respiration.**

**Situation:** You are an adult nurse practitioner working in a collaborative practice with a physician. A healthy 40-year-old female arrives for a yearly physical examination. Although all other findings are normal, the client's blood pressure is measured as 145/90. How do you interpret this measurement?

1. Describe several factors affecting blood pressure. Is a single elevated blood pressure measurement sufficient to support a diagnosis of hypertension? Why or why not?

> **A number of factors affect blood pressure. Among these include age, exercise and physical condition, anxiety level, obesity, and disease process. With the array of physical and emotional factors that affect blood pressure, a single elevated reading, such as described in the scenario, is usually not sufficient for a diagnosis of hypertension. An elevated reading requires follow-up in order to determine if a pattern of elevated reading is present, which might require medical treatment.**

2. Design an information packet for clients called "Tips for Maintaining Normal Blood Pressure." In your packet, include the following information:

- Why maintaining normal blood pressure is so important.
- Foods to choose and foods to avoid.
- Adopting an exercise program that works.
- Proven methods for reducing stress.
- When and why medical management might be needed.

Can you think of any other information to include in your packet? Consult the American Heart Association for additional ideas.

> **Information packets will vary in appearance and content but should be easy to read, informative, and accurate.**

**Situation**: A nurse working in a hospital-based clinic is about to perform a physical assessment on an 80-year-old female. System by system, what changes might the nurse expect to find in this client compared to a younger client?

1. Describe a logical sequence for performing a head-to-toe assessment. How might this sequence be adapted to a client's needs? What measures might you take to assure a client's comfort during the assessment?

   **Physical assessment should proceed in an orderly, systematic fashion, normally from head to toe, and include an examination of all systems and a measurement of vital signs. Patients should be carefully draped during the examination and drafts should be avoided. All instruments should be gathered before the assessment.**

   **In assessing elderly clients, it is important to distinguish normal patterns of aging from pathological findings. In the elderly, it is normal to find a decrease in the elasticity and turgor of the skin, changes in visual and auditory acuity, some alteration in reflex responses, and decreased endurance. Although some memory impairments may accompany aging, be alert for changes in mental status which may require further investigation.**

2. Identify the four basic techniques used in physical assessment. When you are assessing a client's cardiac status, for instance, what is the usual order in which these techniques are performed? Can you think of a specific situation in which it would be necessary to vary this order?

   **The four basic techniques used in physical assessment are inspection, palpation, percussion, and auscultation, normally performed in this order. However, when assessing the abdomen, auscultate before palpating or percussing. In this way, you will not alter the normal bowel sounds before auscultating.**

**Situation:** You have been invited by the infection control specialist of an acute care facility to participate in developing an in-service for new employees on the prevention of nosocomial infections. What essential information do you want to convey about this important topic?

1. As a nurse, what specific measures can you take to prevent the spread of infection? Why are these measures so important?

   **Thorough handwashing is vital to preventing the spread of infection, as is the use of gloves and the careful observance of medical and surgical asepsis. Failure to follow these measures contribute to poor outcomes, client deaths, and rising health care costs.**

2. With respect to a specific pathogen, such as Mycobacterium tuberculosis, identify the six links in the chain of infection. What interventions might be effective in breaking the chain of infection?

   **The six links of infection for any pathogen are:**
   - **An infectious agent (e.g. bacteria, viruses, fungi)**
   - **A reservoir (e.g. people, medical equipment)**
   - **A portal of exit (e.g. broken skin)**
   - **A method of transmission (e.g. airborne)**
   - **A portal of entry to a susceptible host (e.g. respiratory tract)**
   - **A susceptible host (e.g. an immunocompromised individual)**

   **Appropriate interventions on the part of the health care worker can break the chain of infection and reduce the infection rate. For example, proper handling and sterilization of equipment can break the chain of transmission at the reservoir, thus reducing the risk of a spread of infection.**

3. What are the goals of the Centers for Disease Control and Prevention (CDC) and the Occupational Safety and Health Administration (OSHA) regarding protecting the public from the spread of infectious disease?

> The Centers of Disease Control (CDC) acts as a resource for nurse and other health care professionals and public health officials with the goal of reducing the spread of disease. OSHA, on the other hand, is concerned with the reduction of risk exposure. This agency's objective is to protect health care workers in particular industries from hazards.

**Situation**: As a nurse in a skilled nursing facility, you are responsible for supervising patient care assistants. In response to a recent increase in the incidence of infection among clients, you have designed a 20-minute presentation on the importance of medical asepsis. Consider the key points you wish to communicate and the visual aids that will enhance your talk. The following are a few topics to consider including in your presentation:

Once you have developed a presentation that is clear and comprehensive, practice delivering it to a friend, classmate, or family member.

- The best way to prevent the spread of infection.
- The importance of personal hygiene.
- The handling of bed linen and other equipment.
- Cleaning soiled objects.

**Presentations will vary but should emphasize proper hand washing as the single most important element in preventing the spread of microorganisms. CDC guidelines regarding hand hygiene and the use of gloves should be covered along with basic principles and practices of medical asepsis.**

**Situation**: A nursing student accompanies a home care nurse on a series of visits. The student is surprised to learn how often procedures requiring sterile technique are performed in the home. Outside the acute care setting, in what situations might it be necessary to maintain surgical asepsis?

1. With respect to the importance of establishing and maintaining a sterile field, what specific challenges might nurses encounter in serving clients in the community? Explain resourceful strategies a nurse working outside acute care might use to meet these challenges.

> **Nurses working in the community are likely to encounter some challenging situations, from homes without soap to unsanitary conditions of varying degree. Planning ahead for these situations means carefully assessing the environment in advance and being prepared-carry liquid antimicrobial soap and other antimocrobial hand-washing agents, provide adequate supplies for establishing a sterile field, and observe meticulous bag technique.**
>
> **In the home, as the nursing student in our scenario discovered, the principles of surgical asepsis apply in a number of situations. A few examples include inserting a urinary catheter, administering intravenous infusions, and changing a PICC line dressing.**

2. Imagine that, as a nurse educator, you are preparing an inservice for nurses on the principles of surgical asepsis. Ask a group of your classmates to assume the role of participants and plan an interactive session. If possible, obtain permission from your school to obtain various supplies that may be needed, such as packages of sterile gloves, a sterile field, and perhaps a urinary catheter insertion kit. Once you have described the rationale for using sterile technique, practice the skills involved: open a sterile package, don sterile gloves, and set up a sterile field. Critique each other's technique throughout the practice session until all participants are confident of their skills.

> **Use this "inservice" as an opportunity to practice, practice, practice until you are confident of your technique.**

**Situation:** You are caring for a 40-year-old male diagnosed with pulmonary tuberculosis. From your interactions with the client, you assess that both he and his family have limited knowledge about how TB is spread. Consider how you might best respond to questions about the reason for isolation in the hospital and the kind of follow-up that will be needed when he returns home. Develop a teaching plan for the client and his family with the goal of preventing the transmission of the TB microorganism.

**Your primary goal in a situation like this is to safeguard both your client and his family. Emphasize this when explaining the necessity of isolation precautions until there is sufficient clinical evidence of reduced infectiousness. Be sure your client and his family understand that the TB microorganism is spread by airborne droplets. When an infected individual sneezes, coughs, laughs, or even speaks, these droplets are released and may be inhaled by susceptible individuals. Those at greatest risk for acquiring the infection include anyone who is immunocompromised, is elderly, is an IV drug user or alcoholic, or has a chronic illness. Emphasize the importance of total compliance with the treatment regimen to its completion along with careful observance of hygienic measures, such as covering the mouth and nose while sneezing, disposing properly of tissues, and frequent and thorough hand washing.**

1. Soon after a client with Vancomycin-Resistant Enterococcus (VRE) is admitted to a medical unit, a patient care assistant asks about specific measures to reduce risk of transmission. Explain what VRE is and how it is transmitted, and describe the special precautions that apply in this situation.

   **Explain that enterococcus is a bacterium that exists in the gastrointestinal tract and has a significant potential to cause infection. This infectious microorganism is of particular concern (along with methicillin-resistant Staphylococcus Aureus (MRSA) and other multi-drug resistant microorganisms) because of its resistance to antimicrobial drugs. To reduce the risk of transmitting VRE, health care workers must observe meticulous attention to appropriate hand washing and glove use. Additionally, equipment used with patients suspected of VRE must not be used with other patients until it is appropriately disinfected. To learn more about preventing the transmission of VRE, investigate the policy of your facility and CDC guidelines for risk reduction.**

2. Investigate the policies and procedures regarding isolation precautions in one of the clinical settings from your own experience. Who developed these policies, and where can they be found? How are nurses and other facility personnel made aware of these policies? In what specific situations have you observed nurses applying these principles?

> **A nursing administrator or nurse manager would be a good resource in helping you locate this information.**

**Situation**: A nurse in a rehabilitation facility is planning the discharge of a 75-year-old client who had a left total hip replacement several weeks earlier. The client, who lives with his son and daughter-in-law, can walk approximately 50 feet with the use of a walker and will continue to receive physical therapy in the home. On the day of discharge, the client's son asks about specific measures to promote a safe environment in the home. Before you respond, consider the specific needs and abilities of this client.

1. Working with a classmate, prepare a basic checklist to be used by home care or community health nurses to promote safety in the home. Or, if possible, prepare an audio or video presentation called "Hidden Dangers in the Home." Develop a script for your presentation. You might even want to ask your classmate to videotape you walking through the home, pointing out common dangers and suggesting ways to promote safety.

   **In a situation like the one described above, it is essential that the client demonstrate competence in using the walker or other assistive device. With careful planning, the client's family can arrange the environment to minimize the potential for falls. Risk factors in the home of particular concern to the elderly or those with impaired mobility include improperly-fitting shoes or clothing, scatter rugs or hazardous flooring, pets in the individual's path or environment, and poor lighting.**

   **If possible, enlist the help of a home care nurse, physical therapist, or occupational therapist in planning your presentation. In addition to the suggestions presented here, they may have some excellent tips for promoting safety in the home.**

2. In both acute and subacute settings, restraints may at certain times be used. Describe the basic rationale for applying restraints and discuss the following topics related to restraint use:

   - Alternatives to restraints

   - When restraints may be used

   - Guidelines governing their use

   - Basic measures that must be taken to ensure the safety and well-being of clients in restraints
     **Remember that restraints are a "last resort". They require a physician's order and are intended to assure the safety of the client and others when all other measures have failed.**

When using restraints, it is essential to use the least restrictive device required. A nurse is obliged to explore alternatives to restraints, such as having a family member remain with the client to reduce anxiety. Remember also that the client in restraints is vulnerable physically and psychologically and must be carefully monitored, with particular attention to range-of-motion and skin care. Familiarize yourself with the policies regarding restraints in your facility.

---

**Situation**: A nursing colleague confides that she can no longer tolerate "all this heavy lifting and pulling". In observing this nurse with clients, what principles do you keep in mind as you assess the proper use of body mechanics? Consider how you might use this information to assist your colleague.

---

1. You are participating in a health fair. As a result of a number of questions about back pain, you decide to create a handout called "The Key to Preventing Back Injuries". What essential information will you include in this teaching tool? Here are a few topics to get you started.

   • Maintaining good posture

   • The importance of exercise

   • Choosing clothing and shoes

   • The right mattress

   • How to move heavy objects

   **Emphasize the importance of good posture in avoiding musculoskeletal injury. Suggest that individuals check their posture in a mirror to ensure that abdominal and gluteal muscles are contracted to produce proper alignment. Advise clients to avoid high-heeled shoes as these may cause or exacerbate injury and to perform regular, moderate exercise that will strengthen back muscles, with emphasis on stretching, flexibility, and strengthening. (Clients should consult their physicians before beginning any exercise program.)**

   **With the nurse described in the scenario and with others at risk for back injury, reinforce the importance of proper body mechanics when moving heavy objects, such as using a wide base of support with bent knees and contracted abdominal muscles and avoiding twisting and overreaching.**

2. Each day your best friend carries a schoolbag filled with heavy books over her left shoulder. You notice that this practice is beginning to affect her posture. What advice might you offer your friend?

   **Carrying heavy objects that compromise posture should be avoided. Perhaps a lightweight luggage carrier or similar device might be helpful in a situation like this.**

# Hygiene Care: Bathing, Oral Hygiene, and Linen Change Answer Key

1. You are caring for an 82-year-old postoperative client who is now able to ambulate with the assistance of a nurse. At home, the client is independent with all activities of daily living. The client is reluctant to accept your help and wants to take a shower "on my own". What would you do?

   **In a situation like this, your first responsibility is to assure the client's safety. A patient, understanding attitude will go a long way in helping you and your client reach a compromise that is safe for everyone.**

2. Nurses in all practice settings encounter a variety of factors, ranging from culture to religious practice to personal preference, influencing clients' hygienic practices. For example. how might an individual's culture affect his or her attitude to hygiene? Begin by considering your own culture. What is the standard of hygiene accepted in your culture? In what ways might this differ from the standard of other cultures? How might you adapt this knowledge to promote the goal of optimal hygiene?

   **Keep in mind that a variety of cultural, religious, and personal factors may influence a client's hygienic practices. For example, in some cultures it is essential to maintain privacy while bathing, while in others it is not. Some cultures, such as our own, place a high priority on frequent bathing, while others do not. A nurse must be sensitive to cultural differences and yet maintain standards that will promote the well-being of all clients.**

3. Consider how a nurse might integrate ongoing assessment into the process of bathing a client. What kinds of information can a nurse gather about a client's health status during this process?

   **The process of bathing a client is a perfect opportunity to gather important information about an individual's health status. Here are a few areas to target in your assessment:**
   - **Neurological function**
   - **Affect**
   - **Circulation (assess peripheral pulses, note any edema)**
   - **Skin condition, mobility, turgor, color, temperature, texture, presence of lesions**
   - **Mobility**
   - **Self-care abilities**

1. You are caring for a febrile 75-year-old client with peripheral neuropathy and a healing Stage II pressure ulcer on her sacrum. A cooling bath has been ordered. What special precautions apply in this situation?

   **Because applications of cold result in vasoconstriction, blood flow to an area is impaired, which may inhibit wound healing. Therefore, special care must be taken in the application of cold to any client with impaired skin integrity. Additionally, a client with peripheral neuropathy is at increased risk for tissue damage due to cold applications. (Of course, a cooling bath is normally not extreme or prolonged.)**

2. A 16-year-old soccer player sprains his ankle as a result of a twisting movement on a fall. With respect to the application of heat or cold, what is the best way to manage the swelling of the ankle and foot? Explain the rationale for your action.

   **In the event of a sprain, moist or dry cold is applied to the affected area for periods of 20 to 30 minutes intermittently for the initial 24 to 48 hours following the injury. This is intended to produce vasoconstriction, which will minimize bleeding, swelling, and discomfort.**

3. Conduct a literature search on the therapeutic uses of heat and cold applications. Focusing on the findings of one study, share and discuss its implications with a classmate.

   **Responses to this activity will vary based upon the study.**

4. Develop a one-page "Safety Sheet" to guide you in the appropriate application of heat and cold. Share these guidelines with classmates.

**Caution in the application of heat and cold is especially important for clients with:**

- **Neurosensory impairments. (They may be unable to detect the damaging effects of the application.)**
- **An altered level of consciousness. (These individuals require more vigilant monitoring to assure safety.)**
- **Impaired circulation. (These individuals have a limited ability to dissipate heat.)**
- **Open wounds. (Because cold can reduce blood supply to an area, it may impede healing.)**

1. As you are preparing to administer an oral medication to a 78-year-old nursing home resident who has dementia, the client states "I don't remember taking that before." What action is appropriate in this situation.

   **Never assume that a client is wrong when they question a medication. In a case like this, it is important to re-check the client's medication record and, if further clarification is needed, contact the prescriber.**

2. Discuss with a classmate the nature and extent of a nurse's responsibility with regard to medication administration. What part do the "five rights" play in this responsibility? For example, does the responsibility begin and end with the actual administration of the medication? Explain.

   **Safe medication administration begins with, but is not limited to, observing the "five rights." Nurses are obligated to be knowledgeable about pharmacology, to understand the intended effects and adverse reactions of various medications, to monitor clients closely after a medication is given, to teach clients about safe medication use, and to work with other members of the health care team, including pharmacists and physicians, to assure the safety and well-being of their clients.**

3. You are reviewing the medications of a home care client, who is currently taking an oral hypoglycemic, two cardiac medications and a nonsteroidal anti-inflammatory drug for arthritic pain. The client points to several bottles of herbal preparations and states "My wife has been taking these for months, so I decided to give them a try." How might you best respond?

   **You might consider this an opportunity to teach your client about the safe use of medications. Explain that herbal preparations are not inert substances and can interact with a client's medications. Emphasize the importance of checking with one's physician regarding any changes in medications. Additionally, stress the danger of "swapping" medications or other substances with other individuals.**

# Oral Medication Preparations Answer Key

**Situation**: A 55-year-old client refuses to take his antihypertensive medications. "I feel just fine," the client states. "Anyway, my blood pressure has been okay for at least a month. I just don't want to take any medication I don't need." Consider your response to this client.

1. What do you tell your client about hypertension to address his statement "I feel just fine?"

   **Help clients understand that the way they feel is not a reliable guide to their blood pressure and that untreated hypertension can lead to heart disease, kidney disease, blindness, and stroke. Urge clients to work closely with their health care providers in adhering to a regimen to treat hypertension.**

2. An elderly client is being discharged from your facility with several new medications. The client's family is concerned that the client, who lives alone, may have difficulty remembering to take so many medications. How might you assist this client and family?

   **Advise the client's family to obtain a box for organizing medications from their pharmacy. This convenient box is arranged by days of the week and can be filled in advance. In this way, caregivers can monitor that the client is taking the right medications at the right time. Make sure the client's physician is aware of the situation as well; perhaps the medication regimen can be streamlined.**

**Situation**: You are caring for a hospice client who is experiencing severe chronic pain associated with metastatic cancer. A transdermal fentalyl (Duragesic) patch has been ordered for pain management. On your visit to this client, you apply the patch and instruct the caregiver and other family members on the use of this medication. Consider how you might organize your teaching in order to anticipate any questions the family may have. For example, how would you respond to a caregiver who poses the following questions?

1. How does the patch "work"?

2. What is the advantage of using a patch? How is it different from just taking oral medication?

3. What do I need to know to apply the patch correctly?

4. Will the patch work immediately, or will other pain medication be needed until the effects of the patch are felt?

5. How long should the patch be left in place?

6. What are the best sites for applying the patch?

7. When I am ready to change the patch, can I apply it to the same place?

8. What adverse effects are associated with this medication?

9. What safety considerations should I be aware of?

10. How is the medication discarded?

> The fentanyl (Duragesic) and other transdermal patches are applied to the skin of the upper torso. This medication is indicated for severe, chronic pain, such as cancer pain. The advantage of the transdermal route is that it allows for a consistent opioid level in the serum. As the medication is slowly released from the patch and absorbed across the skin, it may take up to 48 hours to reach therapeutic levels. Because of this, other pain medications may be needed until it takes effect. For most patients, patches are replaced every 72 hours, but some clients will require a new patch in 48 hours. Care must be taken in applying the patch as it contains a powerful narcotic agent. (Carefully read all directions supplied with the medication and ask your pharmacist or health care provider for any needed clarification about administering this medication.) When changing the patch, rotate sites for maximum absorption. As with the other opioids, adverse effects of transdermal fentanyl include respiratory depression, sedation, nausea, and constipation. This and all medications should be kept out of the reach of children. Normally, used patches can be flushed down the toilet.

**Situation:** An oncology nurse is preparing a client for discharge from an inpatient unit. At home, the client will self-administer epoetin alfa (Epogen) subcutaneously in an initial dose of 150 U/kg three times a week to treat chemotherapy-induced anemia. The client, a 56-year-old woman, is motivated to participate in her own care and has several questions related to self-administration of this drug. Work with a classmate to develop a teaching plan that responds to a variety of potential concerns about self-administration of subcutaneous medications and about this medication in particular. You might even want to role play the scenario with your classmate assuming the role of the client and you the role of the nurse. Here are a sample of client questions to guide your teaching.

1. Why am I taking Epogen?

2. What does the medication do?

3. Why can't I take this medication orally?

4. How do I draw up and inject this drug?

5. Should I use the same site each time?

6. What should I know about handling the vials?

7. How do I store the drug?

8. What about the used syringes? Can I throw them out with the trash, or is there a particular way to dispose of them?

9. What kind of monitoring will I need while I'm taking the medication?

> **Explain that epoetin alfa (Epogen) has been prescribed to treat an anemia that developed as a result of chemotherapy for a nonmyeloid malignancy. It is used to boost hematocrit levels and thereby alleviate the symptoms of anemia. This medication cannot be taken orally as it is a protein and would be degraded in the gastrointestinal tract. Carefully instruct the client on drawing up and administering the medication and on proper storage. Explain that vials should not be shaken as this could cause the protein to become denatured (to lose its shape). Make sure the client is aware of the need for routine monitoring with this medication.**

1. You are caring for a 62-year-old client who is receiving intravenous antibiotics following orthopedic surgery. As you assess the client's left forearm, you note that the area is cool to touch, there is swelling around the insertion site, and the flow rate has decreased significantly. Based on these findings, what do you think the problem might be? What action is appropriate?

> These findings suggest that the IV cannula may have become dislodged from the vein and that intravenous solution has infiltrated subcutaneous tissue around the insertion site. In this situation, it is appropriate to stop the infusion and discontinue the IV. After a careful assessment of the site, it may be helpful to apply a warm compress and elevate the affected arm. The IV infusion should be started at another site. Selection of an appropriate cannula is important to prevent infiltration. Careful monitoring of the insertion site for redness, swelling, coolness, and blood return allows for early detection and treatment of potential complications.

2. A nurse working for an infusion company is instructing the parents of a 12-year-old who will be receiving short-term intravenous therapy at home. The parents are concerned about their ability to manage the therapy successfully. Develop a teaching plan to support the caregivers and address any potential concerns they might have.

> Intravenous therapy is increasingly common in the home. If family and caregivers are to feel confident in the day-to-day management of infusions, they will require both education and support. Educational efforts should teach caregivers not only how to manage the infusions but also to recognize signs of potential complication. In addition to direct teaching of the family and caregivers, written instructions must be supplied and the infusion and home care agencies must be available as resources at all times.

3. A client complains of tenderness around his IV insertion site. You assess the site and detect redness, swelling and warmth in the area. What does your assessment suggest? What will you do next?

> These findings suggest an inflammation of the vein, or phlebitis. Like infiltration, it is one of several potential complications of IV therapy. In a situation like this, you would discontinue the IV and apply a warm, moist compress to the affected area. The IV can be restarted at another site.

**Situation**: You are preparing a 70-year-old client for cataract surgery. The client asks, "Will I be able to drive home after the surgery, or should I arrange for a ride?" Based on his current knowledge, formulate an appropriate teaching plan for your client following surgery. Be sure to include the following topics in your discussion. You might even choose to develop a one-page handout of self- care activities following cataract surgery.

- The need for eye medication

- Instilling eye drops

- Activities to avoid

- Signs and symptoms to report

- Scheduling follow-up visits

After reviewing these topics, the client's wife remarks "I have eye drops left over from the cataract surgery I had last year. Maybe we can use these." What additional teaching might be needed?

**Important areas to review with clients include self-care measures, activity limitations, and use of prescribed eye medications. Make sure client has arranged for transportation home following surgery as driving will be restricted for approximately one week. Bending and lifting heavy objects is also usually restricted at this time. Instruct the client on how to instill eye drops and to wash hands thoroughly before instilling the drops. Stress the importance of using all medications as prescribed, following up with one's physician, and reporting unusual signs or symptoms, such as pain not managed by prescribed pain medication, persistent headaches, redness, inflammation, drainage, and changes in visual acuity. Finally, emphasize to clients the importance of taking only those medications prescribed for them.**

**Situation**: As you inspect the ear of a four-year-old child, you note the absence of cerumen. The child's parent mentions that she uses a cotton swab to clean deep within the ear. How would you respond to this information?

1. You are instructing a parent on how to instill ear drops. Describe how the technique is adapted to a child under 3 years of age and an older child. What other information would you convey to the parent about the correct way to instill ear drops?

> **Instruct the parents to place the child in a supine position and turn the head to the appropriate side. For children under 3 years of age, gently pull the pinna downward and straight back to instill drops. Have the child remain in this position for a few minutes to insure proper installation. The parent may gently massage the area anterior to the ear to facilitate movement of the drops into the ear canal.**

3. Unscramble the following words related to the ear.

- TOCI - **OTIC**

- INNPA - **PINNA**

- PATYNICM - **TYMPANIC**

- TOPCOSOE - **OTOSCOPE**

- RUCEMEN - **CERUMEN**

**Situation**: A diabetes educator is working with a 12-year-old boy recently diagnosed with Type 1 diabetes mellitus and his family to promote adherence to an insulin regimen and maintain normal blood glucose levels. Before meeting with his clients, the educator will gather written and visual materials to enhance learning. With this in mind, design a brochure introducing basic concepts related to diabetes management. Here are topics to include in your brochure:

- The goals of insulin therapy

- Glucose monitoring with ease

- Changing insulin requirements

- Injecting insulin

- Rotating sites for optimal absorption (develop a simple chart for illustration)

Edit your brochure so that it is clear and concise, focusing on essential concepts of diabetes management.

> In your brochure, emphasize that the goal of insulin therapy is prevention of the complications of diabetes through maintaining normal blood glucose levels. Stress the value of home blood glucose monitoring in managing diabetes. To minimize the pain of finger sticks, use the ring finger or thumb, puncturing the finger just to the side of the finger pad as this area has fewer nerve endings. Be sure to address signs and symptoms of hypoglycemia, including nervousness, weakness, sweating, dizziness, and headaches. Emphasize the important role of the diabetes educator as a partner in the management of diabetes.

1. After checking medication orders, you are preparing to administer subcutaneous heparin to a 68-year-old postoperative client. The client asks, "I used to take warfarin (Coumadin). Isn't this the same thing?" How do you respond?

   **Explain to the client that heparin is used in situations where a rapid onset of anticoagulant effects is required. The size and polarity of heparin prevent its absorption across the GI tract, so it must be given by injection. Like warfarin (Coumadin), heparin is used to prevent clot formation. Warfarin, however, is most often used for its long-term prophylactic effects. Although there are other differences between the two medications, these are the important ones for the client to know.**

2. As part of a preadministration assessment, what baseline data should you obtain for a client on heparin therapy?

   **Heparin must be used with extreme caution in clients at risk for bleeding or those with severe disease of the liver or kidney.**

3. With a classmate, identify those clients who are at greatest risk for deep vein thrombosis (DVT). Describe your assessment of these clients. What measures can be taken to prevent thrombus formation?

   **Among risk factors for deep vein thrombosis are leg trauma, surgery, immobilization, obesity, advanced age, and coagulopathy. Preventative measures include elastic stockings or intermittent pheumatic compression devices, anticoagulant therapy, active and passive leg exercises, and early ambulation after surgery.**

1. While changing a surgical dressing, you use the opportunity to assess your client's readiness to manage dressing changes and incision care after discharge. What signs might indicate a client's readiness to manage his or her own wound care?

> **Clues of a client's readiness to learn include looking at the incision site, asking questions related to wound care, and helping with the dressing change. By encouraging questions and participation, you are helping prepare clients to manage their own care after discharge.**

2. As a teaching tool to provide postoperative clients on discharge, develop a set of instructions on wound care. Keep your instructions concise but be sure to cover areas of potential concern to the client, such as:

- Signs of infection
- Managing soreness or discomfort
- Reducing swelling
- Signs to report to health care provider

> **Classic signs of infection are redness, significant swelling, tenderness, and warmth around the incision. Emphasize the importance of reporting these signs to one's health care provider. A cool, dry compress and acetaminophen may be prescribed for soreness. As a certain amount of swelling may be expected, clients may be advised to elevate the affected area appropriately while sitting or reclining.**

3. On a visit to a home care client, a caregiver expresses concern about a sacral pressure ulcer that is slow to heal. The client, an 86-year-old female diagnosed with Alzheimer's disease, ambulates with assistance but spends the majority of the day in a wheelchair. The client is continent of urine and feces. The client is repositioned frequently and pressure-relieving devices are used regularly. From what the caregiver tells you, you suspect that the delay in healing may be related to the client's nutritional status. How might you educate the caregiver on the importance of adequate nutrition for wound healing?

> **A client's nutritional status is of particular importance in treating pressure ulcers and promoting wound healing. Wound healing requires optimal nutrition and hydration, with an emphasis on adequate protein, vitamin A, B vitamins, ascorbic acid (vitamin C), zinc, and sulfur. Along with a nutritional assessment and appropriate dietary recommendations, carefully monitor the client's hemoglobin and albumin levels and weight on an ongoing**

**Situation**: You are caring for a 50-year-old woman who has just undergone a mastectomy. Immediately after surgery, and as you prepare your client for discharge, what special considerations will apply with respect to incision and drain care?

Special considerations following surgery include inspecting the dressing and drains for bleeding and the extent of drainage and maintaining the patency of surgical drains. During dressing changes, a nurse can discuss the incision with the client, teach signs and symptoms of infection, and promote important self-care measures.

1. Suppose the client described in this scenario is discharged with drains still in place. How will you assure continuity of care between the hospital and home?

   Clients are often discharged with drains still in place and will need guidance in the proper care of the incision site, dressings, and drains, including how to empty the reservoir and measure drainage. Instruction should be provided to both client and caregiver and include signs of potential complication to report. Follow-up phone calls and a visit from a homecare nurse may help assure continuity of care.

2. In your clinical setting, familiarize yourself with the drainage systems in use, such as the Hemovac or Jackson-Pratt. Ask your clinical instructor if you might be assigned to a client requiring incision and drain care.

   Following your assignment, discuss your experience with your clinical instructor and classmates.

**Situation:** In accompanying a homecare nurse on a series of visits, you encounter a high percentage of clients with one or more risk factors for pressure ulcers, including immobility or limited mobility, incontinence, vascular disease, diabetes mellitus, and advanced age. Providing education on appropriate preventative measures is a major aspect of nursing responsibility. What kind of instructions will you provide to assist these clients and their caregivers?

1. Develop a brochure to help clients and caregivers cope with the challenge of preventing pressure ulcers. Include written and visual materials as part of your instructions. In concise, direct language, address the following questions:

   • What causes pressure ulcers?

   • Who is at risk?

   • How to recognize early signs of pressure formation.

   • Where to assess the skin.

   • What can be done to prevent pressure ulcers?

   • What measures can be taken to promote healing once a pressure ulcer has formed?

Think of this brochure as a teaching tool to reinforce your instructions on pressure ulcer prevention and care.

> Pressure ulcers, which may range from nonblanchable reddened areas of intact skin to full-thickness skin losses with extensive damage to underlying bone and muscle, are caused by prolonged, unrelieved pressure resulting in damage to underlying tissue. Risk factors for pressure ulcer formation include immobility, extended confinement to bed, neurosensory deficits, impaired mental status, poor nutrition, incontinence, and advanced age. Among early signs of pressure ulcer formation, look for an erythematous, nonblanchable area of intact skin. Frequent assessment of skin condition is essential, with particular attention to bony prominences, where pressure ulcers are most likely to form. To prevent pressure ulcers and promote healing of existing ulcers, careful attention must be given to proper positioning, providing pressure-relieving devices, improving mobility, improving nutritional status, and providing meticulous hygiene for incontinent clients.
>
> In the situation described, make sure caregivers are aware of the risk factors for pressure ulcer formation and understand the importance of frequent turning and positioning, pressure-relieving devices, and other preventive measures discussed above.

2. Consider the clients you have cared for during your most recent clinical experiences. Choose one particular client who either had a pressure ulcer or was at greatest risk for developing pressure ulcers. What risk factors were present in this individual? What interventions were used either to prevent pressure ulcer formation or to promote healing? Were the interventions successful? What nursing diagnosis applied to this client?

> **Responses will vary depending on students' clinical experiences. A nursing diagnosis that might apply is: impaired skin integrity (related to immobility, decreased nutritional status, etc.).**

1. During your clinical experience on a medical-surgical unit, a classmate who is caring for a client with a fresh surgical incision asks you the most important measures for preventing wound infection? How do you respond?

   **Emphasize to your classmate the necessity for using aseptic technique when performing dressing changes along with meticulous attention to proper handwashing, cleanliness, and appropriate environmental disinfection.**

2. A number of hospital-acquired infections are related to surgical wounds. What is the function of the inflammatory response that occurs with wound infection? As you assess postoperative clients, what signs of infection would typically accompany the inflammatory response? How long after surgery would you expect to see these signs?

   **Inflammation is a normal physiological response to injury which initiates a process of tissue repair. Inflammation is not the same as infection, but an infectious agent may cause an inflammatory response. The cardinal signs of inflammation are redness, heat, pain, swelling, and compromised function. Most often, you would expect to see these signs two to seven days following surgery.**

3. Not all surgical clients experience the same risk of wound infection. What particular factors increase a client's risk of wound infection? From your own clinical experience, describe a client who is at increased risk of developing a wound infection. How do you respond?

   **Clients at greater risk include those who have experienced a traumatic wound with delayed repair, gross spillage from the GI tract, or any major break in aseptic technique.**

**Situation:** You are caring for a 38-year-old male client who underwent an emergency appendectomy five days earlier. As you enter the client's room, he holds his hands across the incision site and tells you "I think something has given way here." What has most likely occurred?

1. What warning signs suggest impending dehiscence? What immediate action is necessary if dehiscence or evisceration occurs?

   **In the event of impending dehiscence, with or without evisceration, there might be a gush of serosanguinous peritoneal fluid from the wound. In a situation like the one described above, a disruption of the surgical incision has most likely occurred. Cover the area with a sterile dressing and notify the surgeon immediately.**

2. What factors predispose a surgical client to dehiscence or evisceration? When are these events most likely to occur?

   **A number of factors, including strenuous coughing and the presence of infection or cardiovascular or pulmonary disease, may dispose a surgical client to dehiscence or evisceration.**

# Sensory Functioning and Sensoriperceptual Alterations Answer Key

1. You are caring for a 65-year-old client with peripheral neuropathy secondary to diabetes mellitus. As you assess the client, what changes in the sensory functioning would you expect to find? What particular risks does the client face related to the disease process? How might you focus your assessment and teaching to promote the client's safety and well being?

   **Sensorimotor neuropathy (peripheral neuropathy) is one type of neuropathy associated with diabetes. It most often affects the lower extremities and may cause reduced sensation in the feet. These individuals are therefore at increased risk for injury and infections of the foot. Stress the importance of routine preventive foot care and regular examination by a physician or podiatrist.**

2. In a skilled nursing facility, you are assigned to care for a 78-year-old woman with a significant hearing impairment. Describe the specific measure you will take in your interaction with this client to promote optimal communication.

   **Give this individual your full attention as you try to communicate with her. Make sure you are directly facing her, and speak slowly and clearly. Focus on the essential context of the message. When possible, use written communication to clarify any lapses. Exercise both patience and creativity in your attempts to communicate with hearing-impaired individuals.**

3. A neighbor who is caring for an elderly parent tells you that "Dad just doesn't seem as sharp these days as he used to be." She wonders whether the changes she has observed are a normal part of aging or indicate an actual disease process. How do you respond to her concern?

   **Sensory losses that occur with advanced age can affect an individual's interaction with others and the environment in many ways. Although a certain degree of sensory changes normally accompany aging, it is important to distinguish normal changes due to aging from pathology. Ask for specific examples of the changes your neighbor's father is experiencing and encourage a thorough evaluation by his physician.**

Situation: You are caring for a 71-year-old client in an ICU who demonstrates an increasing level of anxiety and disorientation to time and place. Based on a nursing diagnosis of risk for sensoriperceptual alterations related to sensory overload, you develop a plan of care.

4. What nursing interventions might be most effective to reduce the potential for sensory overload in this client?

> **Because the environment of an intensive care unit can increase the potential for sensory overload, it is particularly important to implement appropriate interventions in this setting Such nursing interventions include providing adequate pain relief, planning care to promote adequate rest and sleep, minimizing unnecessary environmental stimuli, and providing continuity of care.**

5. You accompany a home care nurse on a visit to a 78-year-old woman who has recently undergone orthopedic surgery. The client lives alone following the death of her husband four months ago. Although the client is progressing well with physical therapy and is now able to walk short distances without an assistive device, she expresses a lack of interest in activities she once found pleasurable and is becoming increasingly isolated. Based on these findings, formulate a plan to reduce the client's risk of social isolation.

> **Social isolation places an individual at risk for sensory deprivation, which can lead to boredom, confusion, disorientation, and anxiety. With clients at risk for social isolation, explore both family and community resources that may offer assistance. Encourage clients to maintain existing relationships and to remain open to forming new ones. Although grief following the recent death of a spouse is normal, clients may need additional support. You can assist by identifying appropriate resources, such as support groups and the services of bereavement counselors.**

6. A colleague working in a nursing home facility expresses concern for clients who become increasingly agitated in the evening. Based on her description, you recognize the signs of sundowner's syndrome. What are the possible causes of sundowning? What interventions might be useful to assist individuals who may wander or become disoriented around sunset?

> **Sundowners syndrome involves disorientation or wandering that occurs in early evening hours, after sunset. If an individual demonstrates behavior that arouses suspicion of sundowner's syndrome, try to identify and alleviate possible causes, such as untreated pain. Speak to the client in a calm, reassuring manner and take measures to promote a safe environment.**

7. A neighbor tells you that his father has recently been diagnosed with Alzheimer's disease. Although the family has met with their father's physician, your neighbor asks you to clarify some questions about the disease process. What causes Alzheimer's? Why does his father have difficulty concentrating? How might family members respond when their father seems confused? What can he and other family members do to assist their father and promote a safe environment?

> **Explain that Alzheimer's is a degenerative neurological disease characterized by a gradual loss of cognitive function and changes in behavior. As the disease gradually progresses, cognitive and other changes, including memory loss, will become increasingly evident. For individuals with Alzheimer's disease, it is important to establish a regular routine and to promote physical safety by eliminating environmental hazards as much as possible. Because of the great demands involved in caring for an individual with Alzheimer's disease, caregivers will need particular support. Assist caregivers in locating appropriate resources, such as support groups, respite care, and adult day care, which may be available through the Alzheimer's Association, a national advocacy organization with many local chapters.**

**Situation**: A school nurse shares her concerns with the parents of a 13-year-old boy who is experiencing extreme anxiety over school performance. The boy is an honor roll student who works hard at his studies but lacks close friendships or outside interests. In an interview with the nurse, the boy states "I have no time for friends or sports because I have to get straight A's so that I can get into a good school and make my Mom and Dad happy.".

1. Based on the child's presentation, what nursing diagnosis might apply? How might the nurse assist this family to help the child develop a more positive self-concept?

   **"Self-esteem disturbance" is an applicable nursing diagnosis. Assisting this child and his family involves working together to help the child develop more balanced, realistic goals along with a more realistic self-concept.**

2. Body image is an important component in an individual's self-concept. Think of a client you have cared for who has experienced an altered self-concept related to a change in body image or function. How were you able to assist this individual maintain or recover a more positive self- image?

   **Responses will vary according to students' experiences.**

3. Discuss with a classmate how your own self-concept as a nurse affects your relationship with clients. As a caregiver, what measures can you take to promote a positive personal self image?

   **Responses will vary, but students should recognize that one's self-concept as a nurse will affect all client-nurse relationships.**

4. You are caring for a 54-year-old client who has undergone a modified radical mastectomy. Your client is reluctant to look at the incision and expresses concern about how her husband will view her altered appearance and the effect it will have on their relationship. How might you respond to her concerns about her health status and any factors that may result in altered sexual functioning?

**Recognize that it is normal for this client to experience concern about her altered appearance and how it will affect her relationship with her spouse. A nurse can initiate a discussion about how the client sees herself and the possible effects of breast surgery, including reduced libido related to fatigue, anxiety, or other factors, and offer specific suggestions where applicable. Encourage open discussion to help reduce your client's and her spouse's anxieties.**

5. A nurse working in a clinic performs a routine physical examination on an 18-year-old male. In response to the nurse's question about testicular self-exam, the client asks: What has that got to do with me? I'm too young to have cancer." How might a nurse educate this client about the importance of testicular self-examination?

**With this client a nurse's primary responsibility is to dispel misinformation and encourage appropriate self-care measures. Inform your client that testicular cancer is the most common cancer in young men between the ages of 15 and 35 and instruct him in how to perform a monthly testicular self-examination.**

6. Obtaining information about a client's sexual health history is an important component of a nursing assessment. How might a nurse use this information to identify factors that can place an individual at risk for sexual dysfunction?

**As with other information that emerges from the client's health history, a nurse can use this to identify potential problems and informational deficits and to promote the client's sexual health. In many situations, a nurse must initiate discussion about sexuality, at all times respecting an individual's cultural, religious, and personal beliefs. Additionally, a nurse must recognize situations that require the intervention of a clinical nurse specialist or sex therapist and make the appropriate referrals.**

**Situation:** During a routine physical examination, a 50-year-old client reports chronic stress related to the many demands of his business. He also reports skipping breakfast on a regular basis, a lack of exercise, an increasing irritability with his family, and an inability to relax at the end of the work day. Physical examination reveals a blood pressure of 144/90 and a 10-pound client weight gain over the past year. The client acknowledges his unhealthful lifestyle, but does not know how to implement change. What might you suggest to help this individual cope with stress and promote good health?

**Your client will be most successful in implementing and maintaining positive lifestyle changes if he makes these changes gradually. Help the client to plan small changes, including eating breakfast each day and getting moderate exercise, such as walking, on a regular basis. Be sure your client is aware of the dangers of untreated hypertension and encourage him to have his blood pressure monitored on a regular basis. Encourage gradual, manageable changes to reduce stress and promote optimum health.**

1. Stress is an unavoidable part of everyday living. It cannot be entirely eliminated. However, there are many ways to minimize the effects of stress, such as adopting a program of regular exercise and a healthful diet. Develop a pamphlet that can be used as a simple teaching tool for clients who wish to incorporate stress-reducing activities into their daily lives.

   **Student's responses will vary, but should incorporate suggestions related to such areas as the need for adequate rest and relaxation, what constitutes a healthful diet, and ways to promote regular exercise and fitness.**

2. A client who has been experiencing a number of stressors asks you to explain the mechanism that mediates the stress response. In simple terms, identify the adaptive hormones that are involved in the stress response and the changes they produce in the body. What measures can the individual take to cope with unavoidable stressors?

   **Help the client understand the "fight-or-flight" mechanism that is triggered when an individual experiences a stressful situation. In response to a real or perceived stressor, the body releases a cascade of hormones (e.g. epinephrine, norepinephrine, cortisol) that, among other effects, increase heart rate, blood pressure, blood glucose levels, and tension in skeletal muscles. Although this response is appropriate and necessary, in certain situations, high levels of many of the hormones associated with this response can lead to**

depressed immune function and ultimately lead to illness. For this reason, it is important for individuals to learn to manage stress. Techniques for managing stress include relaxation training, meditation, and the appropriate use of humor.

**Situation**: A hospice home care nurse is making an initial visit to a client diagnosed with lung cancer. At the door, the client's family greets the nurse and states "Dad doesn't know his diagnosis. For his own good, we don't want him to know." The nurse finds the client alert, oriented, and experiencing good pain control. The client asks no questions and doesn't mention his prognosis in the presence of the family.

3. What is the nurse's responsibility in this case? How might the nurse best respond to this potential conflict?

Hospice agencies generally have "truth-telling" policies and will not withhold information from a client who has asked a direct question. However, hospice nurses also recognize the stages of grieving and will therefore respond to questions tactfully and with a view to a client's and a family's readiness to receive specific information. Hospice nurses and other team members respect family dynamics while promoting an acknowledgment of the truth of a client's situation.

4. Interview a staff member of a hospice, whether a chaplain or a nurse or a volunteer. Ask this person to explain the benefits derived from participating in hospice care. Your questions might include:

- What is hospice care?

- What distinguishes hospice from other forms of care offered at the end-of life?

- How is the role of "team" emphasized in the delivery of hospice care?

Hospice is a comprehensive program of care which is intended to improve the quality of life for an individual and his or her family at life's final stage. The goals of hospice are palliative and not curative. Hospice offers an interdisciplinary team approach, emphasizing the collaboration of nurse physician, therapists, pastoral counselor, social worker, and volunteers in the care of client and family.

5. Although death represents the ultimate loss, each of us experiences a variety of losses daily, both large and small. Think of a loss you have experienced-perhaps the loss of a pet, or a friend who moved away, or a failure to meet a goal. Consider your own grieving process as you came to terms with your loss. In a brief diary entry, identify the loss and explore its impact on your life.

**Students' diary entries will vary.**

6. A 65-year-old hospice client starts taking inventory of a number of keepsakes collected over the years and begins giving them to friends and family. How might a nurse encourage life review in such a process? At the same time, the client expresses a desire to buy a new winter coat. Based on this information, what can a nurse infer about the client's grieving process?

**In the many interactions with a client, a hospice nurse encourages life review. Looking over keepsakes is a perfect opportunity for such review. The client described here appears to have achieved acceptance in the grieving process but is perhaps still doing a little bargaining. Since the grieving process is seldom linear, this is perfectly normal.**

7. Read "Tuesdays with Morrie," Mitch Albom's best-selling chronicle of a young man's weekly visits to his former college professor and mentor who has been diagnosed with amyotrophic lateral sclerosis and is coming to terms with impending death. If possible, form a discussion group with classmates to share your reactions to this book. In particular, what lessons might nurses and future nurses learn from this book?

**Students' responses to the book will vary. Lessons learned from the book might have to do with discovering what is most important in life.**

1. A woman who is caring for her terminally ill mother at home asks what to expect as death approaches. In your attempt to prepare the family for an impending death, explain some of the common signs and symptoms that often accompany the dying process.

   As death approaches, changes can be observed in virtually all systems of the body. Such changes may include a progressive weakness of the muscles, with a resultant inability to speak or move, a decrease in blood pressure, mottling of the extremities, and changes in breathing patterns characterized by alternating periods of apnea and deep breathing (Cheyne-Stokes respirations). Most family members find it helpful to be prepared in advance for these and other changes.

2. You are at the bedside of an 85-year-old client whose death appears imminent. Although the client is comatose, his son is present and asks if his father might still be able to hear. You encourage the son to speak to his father. Explain the appropriateness of this action.

   In a situation like this, it is presumed that the client can still hear and family members are therefore encouraged to speak to and gently touch the client as appropriate.

3. The daughter of a critically-ill client has earlier met with her father's physician to discuss implementing a DNR. Later in the day, the daughter approaches you and expresses confusion about how having a DNR will affect the nursing care her father receives. How would you respond to her concern?

   Explain that a DNR order means that hospital staff will not attempt to resuscitate the client in the event of cardiac or respiratory arrest. However, the client will continue to receive the care and attention of the nursing staff without any changes.

1. You are preparing a client with a total hip replacement for discharge from an acute care facility. With respect to mobility, what home care considerations apply to this client?

> Your goal is to promote both continuity of care and compliance with the therapeutic regimen so the client will achieve full rehabilitation. The importance of daily exercise at home is emphasized, and physical therapy may be continued once the client has returned home. Specific instructions include the safe use of assistive devices and appropriate body mechanics, such as avoiding crossing the legs, lifting heavy objects, and unnecessary bending and twisting movement. A gradual increase in activities is encouraged as the client's condition improves.

2. A healthy 45-year-old woman asks your advice about starting an exercise program. Because of the demands of caring for family and work, she admits to being "a real couch potato" and wonders if a membership at a local gym might help her conquer her sedentary habits. How might you respond?

> If this woman has not exercised regularly for some time, encourage her to consult her physician before adopting an exercise program. Encourage her also to choose a form of exercise she likes and to start gradually, exercise regularly, and build up her fitness and endurance levels incrementally.

3. A client whose physician has prescribed antilipidemics has also been advised on the importance of regular exercise. The client asks you: "I really don't have time to exercise. Isn't it enough that I take the medication?" What might you say to this client about the importance of exercise.

> Although medications are often the first-line therapy in the treatment of LDL-cholesterol, the importance of regular exercise must be emphasized. A sedentary lifestyle is associated with an increased risk of CAD, while regular exercise can lower this risk. Encourage a program of regular exercise within the client's limits.

**Situation:** Following the recent death of her spouse, a healthy 70-year-old woman expresses concern about her inability to sleep restfully at night. After several hours of lying awake, she eventually falls asleep. By the next morning, she is so exhausted that she remains in bed the next morning until approximately 10 or 11 a.m. What advice might you offer?

1. What nursing diagnosis applies in this situation?

   **An appropriate nursing diagnosis might be "Sleeping pattern disturbance related to normal grieving."**

2. What interventions might be most helpful with this woman?

   **Although disruptions in sleep patterns often accompany the normal grieving process, a number of interventions may be helpful to this client. Encourage the client to maintain an adequate activity level during the day and to engage in some moderate exercise, such as walking. Participating in a support group, along with maintaining existing relationships may assist with the grieving process. Even if the client has not slept well during the night, she should be encouraged to get up at a regular time earlier in the morning.**

1. You are caring for a 75-year-old male who is experiencing severe pain related to a cancer diagnosis. Along with other medications, a narcotic analgesic has been prescribed to treat the pain. Your client expresses reluctance to take the medication and states "I don't want to become addicted." How might your respond to your client's concern?

   **Reassure the client that addiction is unlikely to result from taking needed pain medications that have been prescribed by a physician.**

2. You are preparing to administer a scheduled dose of narcotic analgesic to a client on an inpatient hospice unit. As you enter the room, you find the client sleeping. His daughter suggests that you skip the scheduled medication by explaining "My father can't be in any pain; he's sleeping." How do you respond?

   **If narcotic analgesics are to be effective, they must be taken as scheduled. Advise caregivers not to skip a dose because an individual is sleeping. The fact that an individual is sleeping is not a reliable guide to his or her comfort level.**

3. In 1968, McCaffery advanced a definition of pain that emphasized its subjective nature. "Pain" she told us, "is whatever the patient says it is." What implications does this definition have for nursing practice? Discuss this issue with classmates.

   **Foremost, it obligates the nurse to believe the client's report of pain and to intervene appropriately. The use of this definition has a great potential for encouraging all health care providers to institute appropriate pain control measures.**

# Metabolic Function and Basic Nutrition: Glucose Monitoring Answer Key

Situation: A homecare nurse is instructing a client in the use of a glucometer. The client, a 59-year-old woman recently diagnosed with Type II Diabetes Mellitus, poses this question: "Isn't it enough that I follow up with my doctor? Do I really need to do this every day?" Based on this question, how might you convince your client of the importance of regular glucose monitoring? Keep these factors in mind when formulating your response:

- Self-management strategies in the treatment of diabetes

- Regular glucose monitoring

- Preventing diabetes-related complications

**Emphasize to this client the importance of self-management of diabetes for encouraging optimum quality of life and preventing the many potential complications associated with this disease process. Regular glucose monitoring is the cornerstone of self- management.**

1. During a routine physical assessment, a healthy 28-year-old female tells you that she has recently become a "total vegetarian." Based on your knowledge of basic nutrition, what potential deficits might be associated with this diet?

   **A principal danger with a vegetarian diet is that it may not supply complete proteins, such as are found in fish, poultry, milk, and eggs. Emphasize the need for complementary proteins, such as are found in legumes and grains, in this diet.**

2. Take stock of your own nutritional status. Using the food pyramid and your understanding of basic nutritional needs, make a careful assessment of your own diet. What are your nutritional strengths? What deficits might you work to overcome?

   **Student's responses to this activity will vary. Almost everyone, though, will find some room for improvement.**

3. After you've completed your nutritional self-assessment, create a weekly sample menu based on foods from the food pyramid. Make sure your menu represents foods that are both balanced and nutritious.

**Once again, responses to this activity will vary.**

4. A homecare nurse visits an 85-year-old client who is independent in all activities of daily living (ADLs) and until recently, loved to cook for herself. Her children, however, have asked her to stop cooking because she occasionally forgets to turn off the oven or burners. Your client is now concerned that her nutritional status will decline. What suggestions might you offer?

**You might suggest available community resources to this client, such as a "Meals on Wheels" program or similar service. In addition, family members might prepare meals in advance and deliver these to your client.**

**Situation**: You are caring for a postoperative client who has a nasogastric tube in place. Describe your ongoing assessment of this client along with appropriate interventions to manage such potential complications as:

- Fluid volume deficit

- Pulmonary complications

- Irritations related to the presence of the tube

1. As you remove the tube, what measures are appropriate to assure the client's comfort?

At your assessment, pay particular attention to signs of fluid volume deficit, including dryness of the skin and mucous membranes, reduced urinary output, and lethargy. Auscultate all lung fields to assess for pulmonary complications, and encourage coughing and deep breathing. Provide good oral hygiene and inspect mucous membranes for signs of irritation. When the tube is to be removed, you might use a towel to conceal it so that your client is not disturbed. Do not use force if the tube is not removed easily, and report the problem.

1. You are caring for a client whose physician has ordered an opioid analgesic in order to manage severe pain associated with prostate cancer with extensive bone metastases. How might this regimen affect normal bowel elimination? What interventions might you recommend to promote normal bowel function?

   **Since constipation may result from the use of opioid analgesics, an appropriate bowel regimen should be instituted, which may include a preparation such as Senakot and a stool softener. Clients should also be encouraged to make sure they are getting adequate fluids and fiber. Physical mobility is encouraged as tolerated.**

2. As you perform a nursing assessment on a healthy 50-year-old female at a hospital based clinic, the client reports that she normally has a bowel movement every other day. "Is this normal?," she asks. How do you respond?

   **Explain to the client that not every individual has a bowel movement each day. In the absence of other physical problems, this is not normally a cause for concern.**

3. With the help of a clinical instructor, arrange an opportunity to examine the various appliances used with clients requiring fecal diversions. If possible, interview a wound and ostomy care nurse to learn about the various educational needs of clients with stomas.

   **Arrange this with the approval of nursing management at a clinical facility or through your school.**

1. You are caring for a 43-year-old female client with quadriplegia. Describe appropriate interventions that may be used to minimize the risk of urinary tract infections.

   **Encourage adequate fluid intake to promote an adequate flow of urine. Emphasize the importance of emptying the bladder regularly and of careful attention to personal hygiene and wearing cotton undergarments to minimize the risk of infection ascending to the bladder or kidneys. The client should recognize the signs of urinary tract infection and report these promptly.**

2. The son of an 82-year-old client who is taking atropine asks you about the potential adverse effects of this medication. With respect to urinary elimination, what risks are associated with this medication?

   **An adverse effect of this medication and other muscarinic antagonists is urinary hesitancy or retention. This effect can be minimized by encouraging clients to void before taking the medication. The client should be carefully monitored for urinary retention.**

3. A 35-year-old female client has experienced recurrent lower urinary tract infections over the past year. After performing a thorough assessment and formulating an appropriate nursing diagnosis, develop a teaching plan to assist this client. What particular educational needs might you address?

   **Emphasize the importance of prompt treatment with an appropriate antimicrobial agent and of careful attention to personal hygiene to minimize the risk of recurrent infection.**

Situation: A client with Type 1 Diabetes has experienced an unexplained elevation in blood glucose levels (above 250 mg/dl) and will have her urine tested for the presence of ketones. Before providing a urine sample, the client asks you to explain the purpose of the testing. As you formulate your response, consider these questions:

- What are ketones?

- For a diabetic, what does the presence of ketones in a urine sample indicate?

- When should the urine be tested for ketones?

    Explain to the client that ketone bodies are compounds formed from the rapid breakdown of stored fat. This occurs when there is insufficient insulin for the body to absorb glucose. The accumulation of ketone bodies in the blood or urine is a sign of deteriorating control of Type 1 diabetes. For individuals with Type 1 diabetes, it is important to test the urine for ketones when there is an unexplained elevation of blood glucose levels, with glucosuria, and during illness or pregnancy.

Situation: You are caring for a hospice client who is incontinent of urine and is confined to bed. The client's caregiver asks you about the advisability of inserting a urinary catheter. Before responding, consider how you might weigh the risks and benefits of urinary catheterization for incontinent clients.

- Potential benefits of urinary catheterization

- Long-term effects of urinary catheterization

1. If a urinary catheter is inserted, what instructions will you provide the caregiver for managing catheter care?

   With urinary catheterization, there is always the potential for infection. However, urinary incontinence is also associated with the risk of skin breakdown. If maintaining skin integrity is an ongoing problem, it may be appropriate to insert a urinary catheter. The decision, however, must be made on an individual basis with careful attention to the client's needs. If a urinary catheter is inserted, strict asepsis must be maintained during insertion. Emphasize hand washing before handling the catheter, tubing, or drainage bag. The bag is emptied at least every eight hours, and caregivers should never disconnect the tubing as this may cause contamination of a closed system. Caregivers should also be instructed to recognize and promptly report signs of infection.

Situation: You are performing a nursing history and physical assessment of a 52-year-old male client diagnosed with chronic bronchitis and emphysema. The client currently smokes two packs of cigarettes a day and is regularly exposed to environmental pollutants as a result of his work in housing construction. Consider the following questions as you develop a plan for the care of this client.

1. How will you focus your assessment to determine this client's health status and educational needs?

   Assess for factors associated with COPD, such as cigarette smoking and environmental pollutants. Observe for dyspnea, which may occur either on exertion or at rest. Observe the client's posture. For example, does he routinely lean forward in such a way that will compress the abdomen and facilitate increased air expiration? Auscultate the lung fields for wheezing, and evaluate the client's cough and sputum production. Additional assessment focus areas include the physical appearance of the chest and the character and rate of the client's respirations.

2. Use your assessment findings to formulate an applicable nursing diagnosis. Based on this diagnosis, what goals might be appropriate for this client?

   An appropriate nursing diagnosis might be "Ineffective airway clearance related to thick secretions." Appropriate goals include maintaining adequate gas exchange and having the client remain free of infections.

Fundamentals for Nursing

---

**Situation**: A hospitalized client with chronic obstructive pulmonary disease (COPD) will require continuous oxygen therapy in the home setting. Devise a teaching plan to assure compliance with oxygen therapy on discharge.

---

1. Here are a few questions to help you develop your plan:

   • Why is oxygen prescribed?

   • Will I be restricted to home with this therapy?

   • What kind of monitoring is needed with oxygen therapy?

   • Is it okay to increase the flow rate?

   • Are there any special precautions to follow?

Keep your explanation simple, but try to address all potential questions your client might have.

> Oxygen is prescribe in the treatment of hypoxemia. It is a nursing responsibility to monitor the effectiveness of the oxygen therapy and to promote client compliance. The most important home care consideration is safety. Since there is a danger of fire when oxygen is used, patients must be advised to avoid smoking while using oxygen or around a compressor. Portable devices for delivering oxygen are available, so clients are not restricted to the home. While using oxygen, the client's status should be carefully monitored by health care professionals. The flow rate should never be increased without obtaining a physician's order or excessive oxygen may have damaging effects on the lungs and central nervous system.

2. Enlist the help of your clinical instructor or a respiratory therapist and examine the various devices used to administer oxygen, such as simple masks, the Venturi mask, partial rebreather, and non-rebreather masks. Make sure you understand how each of these masks is used.

> Be sure you have the appropriate permission before disassembling and examining any of this equipment.

**Situation:** In caring for a 9-year-old child diagnosed with cystic fibrosis, you are teaching the parents about the role of chest physiotherapy in managing pulmonary problems. The following question will help you address the educational needs of parents whose children require chest physiotherapy.

1. Why is the chest physiotherapy important for a child with cystic fibrosis? Might the pulmonary problems associated with this disease be adequately managed with medications alone? Why or why not?

    **Chest physiotherapy is a cornerstone in the treatment of cystic fibrosis. Although medication is important, a daily routine of chest physiotherapy is needed to maintain pulmonary hygiene. Chest physiotherapy is usually performed on rising and in the evening; bronchodilators may be used before therapy in a child who is wheezing or exhibits evidence of reactive airway disease.**

2. What is the value of chest physiotherapy for a child diagnosed with asthma? And what are the limitations of this therapy for the child with asthma?

    **In the child with asthma, chest physiotherapy-along with breathing exercises and physical training-has a number of potential positive effects, including promoting relaxation, improving posture, and developing more efficient patterns of breathing.**

---

**Situation:** A fellow nursing student is caring for a client who requires regular suctioning of an endotraceal tube. As you cared for this client the previous day, your classmate poses some questions related to his care. Consider your response to the following questions.

---

1. How do you know when suctioning is necessary? Can unnecessary suctioning cause harm to this client?

    **Suctioning is necessary when an obvious accumulation of secretions impair respirations and when adventitious breath sounds are detected. Unnecessary suctioning can cause bronchospasms or damage of tracheal mucosa.**

2. What measures are necessary to prevent infection to a client while performing tracheal suctioning?

    **Suctioning is necessary when an obvious accumulation of secretions impair respirations and when adventitious breath sounds are detected. Unnecessary suctioning can cause bronchospasms or damage of tracheal mucosa.**

**Situation:** A nurse on a surgical unit is assigned to the care of a client who has undergone a tracheostomy to facilitate mechanical ventilation. Describe the postoperative nursing interventions that apply to the care of this client.

1. What major nursing objectives apply in the care of a client with a tracheostomy?

   **A major objective for this client is to alleviate apprehension and to facilitate effective communication. Because clients with a tracheostomy are both physically and psychologically vulnerable, ongoing reassurance and a calm presence are needed. Providing a pad of paper and a pencil can facilitate communication.**

2. What potential complications are associated with tracheostomy tube management? Describe appropriate nursing interventions to minimize the risk of complications.

   **Several potential complications are associated with a tracheostomy, including such early complications as bleeding, air embolism, and aspiration, and long-term complications, including airway obstruction related to accumulation of secretions. Postoperatively, clients will require careful monitoring and assessment to assure a patent airway and proper ventilation.**

**Situation:** A home care nurse is instructing a 72-year-old client diagnosed with congestive heart failure in self-care measures. The client questions the importance of taking his weight each day.

1. How might the nurse best explain the importance of daily weight to this client?

   **Instruct clients to weigh themselves at the same time each day and to take diuretics as prescribed. A weight gain of two pounds or more in this period may indicate recurring cardiac failure and should be reported promptly to one's physician.**

2. The client is also on a sodium-restricted diet and is taking 40 mg of Lasix PO daily for the treatment of heart failure. How might you explain the rationale for this regimen to encourage compliance?

   **Sodium restriction is important for the prevention and management of edema, hypertension, and cardiac failure. Encourage clients and caregivers to read labels carefully and be alert for the presence of sodium in many pre-packaged foods. Diuretics may be needed to promote excretion of sodium and water by the kidneys.**

3. The home care nurse next visits an 83-year-old client with limited mobility related to arthritis. With respect to maintaining fluid balance, what particular needs might this client have?

   **Since an elderly client may have a diminished thirst sensation, adequate fluids should be encouraged and made available, particularly among clients with limited mobility or cognitive impairments.**

**Situation:** A client who is taking digitalis in the treatment of congestive heart failure begins to demonstrate signs of hypokalemia, including fatigue, nausea, muscle weakness, and leg cramps. Diagnostic evaluation includes an abnormal ECG and serum potassium levels below normal levels.

1. Based on this, what do you think might be causing the client's potassium deficit, and what can be done to correct it?

   **Clients taking digitalis who are at risk for hypokalemia should be carefully monitored, because hypokalemia potentiates the action of digitalis. Supplemental potassium may be needed. A potassium-sparing diuretic may also be indicated.**

2. Describe the function of potassium in the body. How might an imbalance of potassium manifest itself? Which clients are at particular risk for such imbalances?

   **Potassium affects both neuromuscular function and cardiac muscle activity. A deficiency in this electrolyte has widespread effects and may ultimately lead to respiratory or cardiac arrest. Signs of hypokalemia include fatigue, anorexia, muscle weakness, nausea and vomiting. Clients at risk include the debilitated elderly, those with poor nutrition, alcoholics, and those taking diuretics that promote loss of potassium.**

3. In counseling a client taking diuretics, what foods might you recommend as good sources of potassium?

   **Dietary sources of potassium include oranges, bananas, avocados, beans, and potatoes.**

Situation: As you are preparing a 50-year-old client for a diagnostic cardiac catheterization, he admits to you that he has never undergone a surgical procedure before and is quite anxious.

1. In a situation like this, what nursing diagnosis might apply? And what specific nursing interventions might be useful with this client?

   Appropriate nursing diagnoses include "Fear related to the surgical experience" and "Knowledge deficit regarding preoperative procedures." Appropriate interventions include allowing clients to express their fears and providing preoperative education. Persistent fears may require the intervention of other health professionals.

2. In a client who is fearful about an impending surgery, how might this fear manifest itself? Do you suppose such a client will always appear openly anxious? Why or why not?

   Fear may express itself in different ways in different individuals. Some individuals may not appear openly anxious yet might hold unexpressed fears about the upcoming surgery. For this reason, a careful nursing assessment is important, with particular attention to answering clients' questions and alleviating fear.

3. In what ways might a client's spirituality help him or her cope with the surgical experience? What other resources might help an anxious client at this time?

   Spiritual beliefs can help an individual cope with anxiety. In a situation like this, it might be useful to enlist the support of a pastor or member of the clergy.

**Situation**: You are preparing a 45-year-old client for major abdominal surgery. Devise an appropriate teaching plan that will address each of the following factors:

- Deep breathing and coughing

- Postoperative mobility

- Pain control

1. With respect to the educational needs of this client, what is your major goal?

Your major goal with this client is to provide appropriate preoperative teaching to help your client achieve an uncomplicated recovery from surgery. With this in mind, you will teach the client the importance of deep breathing and coughing in the preoperative period. The importance of mobility will also be emphasized as a means to promote circulation, prevent venous stasis, and support optimal respiratory function. Although the client will be expected to ambulate as soon as possible, appropriate leg exercises will be taught; maintaining muscle tone facilitates early ambulation. Clients should be reminded that pain medication will be available and are encouraged to ask for medication as needed.

**Situation**: You are caring for a client who has undergone thoracic surgery. With respect to postoperative breathing exercises, how will you address each of the following important topics?

- Breathing techniques

- Positioning for optimal breathing

- Improving airway clearance

  A major goal for this client is to promote effective gas exchange and breathing. Clients are expected and encouraged to practice such techniques as diaphragmatic and pursed lip breathing to expand alveoli and prevent atelectasis. Incentive spirometry will also be used to promote lung inflation. Proper positioning improves breathing, and the head of the bed may be elevated appropriately when the client is stabilized. (The surgeon is consulted about client positioning in the postoperative period.) Other measures are used to insure a patent airway, including the coughing technique and chest physiotherapy.

**Situation**: An operating room nurse is caring for a client who is about to undergo general anesthesia. How would you describe the role of a nurse as client advocate within the setting of the operating room?

1. Interview an operating room nurse to develop a better sense of this nursing speciality. Here are a few questions to consider when planning your interview:

   - What made you choose to be an operating room nurse?

   - What are the rewards associated with your job?

   - What are some of the challenges?

   - What does it take to be a successful operating room nurse?

   **In acting as a client advocate, an operating room nurse provides physical privacy, maintains confidentiality, provides for client safety and comfort, and uses communication skills to inform clients about the various phases of the intraoperative experience. In this and in many other ways, a nurse utilizes the nursing process throughout the intraoperative period.**

2. Based on your interview, write a brief advertisement seeking the ideal candidate for an operating room nurse.

   **Students' advertisements will vary but should emphasize the many skills required to be an operating room nurse.**

1. You are caring for a postoperative client who is becoming increasingly agitated and has made several attempts to dislodge her intravenous catheter. What nursing interventions are appropriate in this situation?

   **In a situation like this, your priority is the safety of your client. Try to provide a calm and reassuring presence. If possible, cover the intravenous site so its appearance does not disturb your client. Try to arrange for ancillary nursing staff to remain with the client. If all attempts to calm your client and provide a safe environment fail, you may then request a physician's order to apply restraints.**

2. You are caring for a client who is recovering from an uncomplicated hernia repair. When this client refuses to participate in her postoperative care, how might you respond?

   **Make sure your client's pain is well managed so that she is better able to participate in her own care. Provide encouragement for a client to participate in activities intended to promote an uneventful recovery.**

3. In your work on a medical-surgical unit of an acute care hospital, you frequently care for clients who have undergone abdominal surgery. What are some of the risk factors for postoperative pulmonary complications following abdominal surgery?

   **Among risk factors for pulmonary complications are prolonged surgery, extended bed rest, obesity, dehydration, malnutrition, and immunosuppression.**

1. A client who has undergone a permanent colostomy expresses concern about the appearance of the new stoma. What nursing diagnosis might apply to this client? How might you support the client in adjusting to the stoma?

   **An appropriate nursing diagnosis in this situation is "Body image disturbance related to colostomy." Encourage this client to verbalize concerns and anxieties. A supportive attitude on the part of the nurse is vital in helping the client adapt to the changes resulting from the surgery. It may also be helpful for this client to speak with others who have also undergone a colostomy.**

2. From your own clinical experiences, think of a challenging situation you may have encountered in caring for a postoperative client.? What particular skills, knowledge, or interventions did you employ in meeting this challenge?

   **Students' responses to this exercise will vary.**

3. Now write a diary entry describing your experiences with this client.

   **Diary entries will vary.**

---

**Situation**: An elderly postoperative client is participating willingly in all aspects of her care. She is using an incentive spirometer, practicing breathing exercises and coughing, and ambulating with assistance. However, this client refuses to wear antiembolism stockings ordered by her physician. How do you respond?

---

**Make sure your client understands the importance of wearing antiembolism stockings in her postoperative care. Explain that her physician has ordered the stockings to promote adequate venous return and thereby to prevent thrombophlebitis.**

1. In the postoperative period, what factors represent the greatest risk for deep vein thrombosis (DVT)?

   **Risk is greatest for clients following many urological, neurosurgical or orthopedic procedures, for obese or elderly clients, for obstetrical clients with varicose veins or other predisposing factors, and for clients with a stroke.**

2. You are caring for a 40-year-old obstetrical client with numerous varicose veins in both legs. As you assess this client, what findings might make you suspect a potential thrombosis?

   **Assessment findings that bear watching include a positive Homan's sign, increased warmth, redness, tenderness or swelling of the legs.**

Abrams, A. C. (2001). *Clinical drug therapy*. (6th ed.). Philadelphia: Lippincott Williams & Wilkins.

Agency for Healthcare Research and Quality. (1992, February). *Clinical practice guideline, acute pain management: Operative or medical procedures and trauma*. AHCPR Publication No. 92-0032. Rockville, MD: USDHHS.

Agency for Healthcare Research and Quality. (1994, March). *Clinical practice guideline, management of cancer pain*. AHCPR Publication No. 94-0592. Rockville, MD: USDHHS.

Agency for Healthcare Research and Quality. (1994, December). *Clinical practice guideline, pressure ulcer treatment*. AHCPR Publication No. 95-0653. Retrieved April 28, 2003 from http://www.ahcpr.gov

AHCPR, Acute Pain Management Guideline Panel. (1992, February). *Clinical practice guideline, acute pain management: Operative or medical procedures and trauma*. AHCPR Pub. No. 92-0032. Rockville, MD: USDHHS.

AHCPR, Management of Cancer Guideline Panel. (1994, March). *Clinical practice guideline, management of cancer pain*. AHCPR Pub. No. 94-0592. Rockville, MD: USDHHS.

American Association of Colleges of Nursing. (1986). *Essentials of college and university education for professional nursing*. Washington D.C.: Author.

American Nurses Association. (1980). *Nursing: A social policy statement*. Kansas City, MO: Author.

American Nurses Association. (1985). *Code for nurses with interpretive statements*. Kansas City, MO: ANA.

American Nurses Association. (1991). *Standards of clinical nursing practice*. Washington, DC: ANA.

American Nurses Association. (1995). *Nursing policy statement*. Washington DC: ANA.

Annon, J. (1974). *The behavioral treatment of sexual problems: Vol. I. Brief therapy*. New York: Harper & Row.

Becker, M. H. (Ed.). (1974). *The health belief model and personal health behavior*. Thorofare, NJ: Charles B. Slack.

Benton, R. (1978). *Death and Dying*. New York: Van Nostraand Reinhold Co.

Bloom, B. S. (Ed.) (1956). *Taxonomy of education objectives*. Book 1, Cognitive domain. New York: Longman.

Carson, V. B. (1989). *Spiritual dimensions of nursing practice*. Philadelphia: Saunders.

Centers for Disease Control. (2003). *Guideline for hand-hygiene in health care settings*. Retrieved April 28, 2003, from http://www.cdc.gov/handhygiene

Christensen, B.L. & Kockrow, E.O. (1999). *Foundations of nursing*. (3rd ed.). St. Louis, MO: Mosby.

DeWit, Susan C., (2001). *Fundamental concepts and skills for nursing*. (1st ed.). Philadelphia: W.B Saunders Co.

Ehrlich, A. & Schroeder, C. (2001). *Medical terminology for health professions* (4th ed.). Albany, NY: Delmar Publishers.

Erickson, E.H. (1963). *Childhood and society* (2nd ed.). New York: Norton.

Erickson, E.H. (1982). *The life cycle completed: A review*. New York: Norton.

*Fluids and electrolytes made incredibly easy!* (1997). Springhouse, PA: Springhouse Corporation.

Fowler, J. (1981). *Stages of faith: The psychology of human development and the quest for meaning*. New York: Harper & Row.

Freud, S. (1923). *The ego and the id*. London: Hogarth Press.

Gawel, S. H., Hertz, J. E., & Yocom, C. J. (2000). *Linking the NCLEX-RN national licensure examination to practice: 1999 Practice analysis of newly licensed registered nurses in the U. S.* Chicago: National Council of State Boards of Nursing.

Gilligan, C. (1982). *In a different voice: Psychological theory and women's development*. Cambridge, MA: Harvard University Press.

Havinghurst, R. J. (1972). *Developmental tasks and education* (3rd ed.). New York: Longman Publishers.

Henderson, V. (1966). *The nature of nursing: A definition and its implications for practice, research, and education*. New York: Macmillan.

International Council of Nurses (1999). ICNP Update. Available Internet http://icn.ch/icnpupdate.htm

Johnson, M., Moorhead, S. & Maas, M. (Eds.). (2000). *Nursing outcomes classification* (NOC) (2nd ed.). St. Louis, MO: Mosby.

Kalish, R. A. (1983). *The psychology of human behavior* (5th ed.). Monterey, CA: Brooks/Cole.

Kee, J. L., & Hayes, E. R. (1997). *Pharmacology: A nursing process approach* (2nd ed.). Philadelphia: W. B. Saunders Co.

Keegan, L. (1994). *The nurse as healer*. Albany, NY: Delmar.

Kohlberg, L. (1981). *The psychology of moral development: Moral stages and the idea of justice*. San Francisco: Harper & Row.

Kozier, B., Erb, G., Berman, A. J., & Burke, K. (2000). *Fundamentals of nursing* (6th ed.). Upper Saddle River, NJ: Prentice-Hall, Inc.

Kozier, B., Erb, G., & Blais, K. 1995). *Fundamentals of Nursing: Concepts, Process, and Practice* (5th ed.). Redwood City, California: Addison-Wesley.

Kubler-Ross, E. (1969). *On death and dying*. New York: Macmillan.

Leavell, H. R. & Clark, E. G. (1965). *Preventive medicine for the doctor in the community* (3rd ed.). New York: McGraw-Hill.

Lehne, R.A. (2004) *Pharmacology for Nursing Care* (5th ed.). Philadelphia: Mosby.

Leininger, M. M. (1988, November). Leininger's theory of nursing: Cultural care diversity and universality. *Nursing Science Quarter*, 14, 152-160.

LeMone, P., Lillis, C., & Taylor, C. (1997). *Fundamentals of nursing: The art & science of nursing care* (3rd ed.). Philadelphia: Lippincott.

Lewis, S.M., Heitkemper, M.M., & Dirksen, S.R. (2004). Older Adults. In M.M. Heitkemper (Ed.) *Medical-Surgical Nursing: Assessment and Management of Clinical Problems* (6th ed.). St. Louis, MO: Mosby.

Maslow, A. (1968). *Toward a psychology of being* (2nd ed.). New York: Van Nostrand Reinhold.

McCaffery, M., & Beebe, A. (1989). *Pain: Clinical manual for nursing practice*. St. Louis, MO: Mosby.

McCloskey, J. & Bulechek, G. (1999). *Nursing interventions classification* (3rd ed.). St. Louis, MO: Mosby.

National Pressure Ulcer Advisory Panel. (1993). *Pressure ulcer prevention points*. Retrieved April 29, 2003, from http://www.npuap.org

National Pressure Ulcer Advisory Panel. (1998). *Stage I assessment in darkly pigmented skin*. Retrieved April 29, 2003, from http://www.npuap.org

Neuman, B. (1974). The Betty Neuman health care system model: A total person approach to patient problems. Riehl, J. P. & Roy, C.(Eds.) *Conceptual models for nursing practice*. New York: Appleton-Century-Crofts.

Nightingale, F. (1860). *Notes on nursing. What it is, and what it is not*. Commemorative Edition. Philadelphia: Lippincott.

Orem, D. E. (1971). *Nursing: Concepts of practice*. Hightstown, NJ: McGraw-Hill.

Parsons, T. (1972). Definitions of health and illness in the light of American values and social structure. In Jaco, E. G. (Ed.). *Patients, primary care providers, and illness* (2nd ed.). New York: Free Press

Pender, N. (1996). *Health Promotion in Nursing Practice*. (3rd ed.). Stamford, CN: Appleton & Lange.

Peplau, H. E. (1952). *Interpersonal relations in nursing*. New York: Putnam.

Peplau, H. E. (1963). Interpersonal relationships and the process of adaptations. *Nursing Science*, 1(4):272-279.

Peplau, H. E. (1980). The Peplau development model for nursing practice. In Riehl J. P. & Roy, C. (Eds.). *Conceptual models for nursing practice* (2nd ed.). New York: Appleton-Century-Crofts, pp. 53-75.

Phipps, W., Sands, J., & Marek, J. (1999). *Medical surgical nursing concepts & clinical practice* (6th ed.). St. Louis, MO: Mosby-Year Book, Inc.

Piaget, J. (1966). *Origins of intelligence in children*. New York: Norton.

Potter, P.A. & Perry, A.G. (2001). *Fundamentals of Nursing* (5th ed.). St. Louis, MO: Mosby

Potter, P.A. & Perry, A.G. (2003). *Basic nursing essentials for practice*, (5th ed.). St. Louis: Mosby.

President's Commission on Health Needs of the Nation. (1953). *Building Americans' Health*. Vol. 2. Washington, DC: U. S. Government Printing Office.

Rogers, M. E. (1970). *An introduction to the theoretical basis of nursing*. Philadelphia: F. A. Davis

Rosdahl, C.B., & Kowalski, M.T. (2003). *Textbook of basic nursing* (4th ed.). Philadelphia: Lippincott Williams & Wilkins.

Rosdahl, C.B. & Kowalski, M. T., (2003). *Textbook of basic nursing*. (8th ed.). Philadelphia: Lippincott Williams & Wilkins.

Rosenstock, I. M. (1974). Historical origin of the health belief model. In Becker M. H. (Ed.). *The health belief model and personal health behavior*. Thorofare, NJ: Charles B. Slack.

Roy, C. (1976). *Introduction to nursing: An adaptation model*. Englewood Cliffs, NJ: Prentice-Hall.

Scanlon, C. (1995, October). *Ethical issues on the national level*. Presentation at the Fourth Annual Clinical Ethics Institute for Nurses, conducted by Midwest Bioethics Center, Kansas City, MO.

Scherer, J. C., Smith, N. E., & Timby, B. K. (1999). *Introductory medical-surgical nursing* (7th ed.). Philadelphia: Lippincott Williams & Wilkins.

Selye, H. (1956). *The stress of life*. New York: McGraw-Hill.

Selye, H. (1976). *The stress of life* (revised. ed.). New York: McGraw-Hill.

Smeltzer, S.C. & Bare, B.G. (1996). *Brunner & Suddarth's Textbook of Medical- Surgical Nursing* (8th ed.). Philadelphia: Lippincott.

Smith, S., Duell, D., & Martin, B. (2002). *Photo guide of nursing skills*. Upper Saddle River, NJ: Pearson Education, Prentice Hall.

Stevens, B. (1975, February). Effecting change. *Journal of Nursing Administration*, 5, 25.

Stotts, N. A. (1990). Seeing red, yellow, and black. The three-color concept of wound care. *Nursing 90*, 20, 59-61.

Suchman, E. A. (1972). States of illness and medical care. In Jaco E. G. (Ed.), *Patients, physicians, and illness* (2nd ed.). New York: Free Press.

Taylor, C., Lillis, C., and LeMone, P. (1997). *Fundamentals of nursing* (3rd ed.). Philadelphia: Lippincott Williams & Wilkins.

U. S. Department of Health and Human Services. (1997). Revised Guideline for isolation precautions in hospitals. Part II. Recommendations for isolation precautions in hospitals. *American Journal of Infection Control*, 24, 32-52.

U. S. Department of Health and Human Services, PHS. (2000). *Healthy People 2010*. Washington. DC: U.S. Government Printing Office.

U. S. Department of Health and Human Services. (2002). *Healthy people 2010*. Retrieved May 1, 2003 from http://www.healthypeople.gov

U. S. Department of Labor, Occupational Safety and Health Administration. (1991, December 6). Occupational exposure to blood-borne pathogens: Final rule. 29 CFR, Part 1910.1030. *Federal Register*, 56(235), 64175-82.

Watson, J. (1979). *Nursing: The philosophy and science of caring*. Boston: Little, Brown.

Westerhoff, J. (1976). *Will our children have faith?* New York: Seabury Press.

Wilkinson, J. M., (2001). *Nursing process and critical thinking*. Upper Saddle River, NJ: Prentice Hall Health.

Wold, G. (1993). *Basic geriatric nursing*. St. Louis, MO: Mosby-Year Book.

Wong, D.L. (1999). *Whaley & Wong's Nursing Care of Infants and Children* (6th ed.). St. Louis, MO: Mosby.

World Health Organization. (1947). *Constitution of the World Health Organization: Chronicle of the World Health* Organization 1Health. Geneva, Switzerland: World Health Organization.

World Health Organization. (1948). *Preamble to the Constitution of the World Health Organization*. Retrieved December 31, 2003 from http://www.who.int/en

World Health Organization. (1975). *Education and treatment in human sexuality: The training of health professionals*. Geneva, Switzerland: World Health Organization.

Wound, Ostomy and Continence Nurses Society. (1996). *Staging pressure ulcers*. Retrieved April 29, 2003 from http://www.wocn.org

# Look for these additional products by ATI

## PN Review Modules and ATI-PLAN™ DVD Series

**Medical-Surgical**

**Mental Health**

**Maternal-Newborn**

**Nursing Care
of Children**

**Nutrition**

**Pharmacology**

**Leadership and
Management**

**Fundamentals**

## ATI NurseNotes Series

**Pediatrics**

**Maternal-Newborn**

**Medical-Surgical**

**Psychiatric
Mental Health**

# Fundamentals for Nursing